Nonlinear Systems of Partial Differential Equations

Applications to Life and Physical Sciences

Anthony W Leung
University of Cincinnati, USA

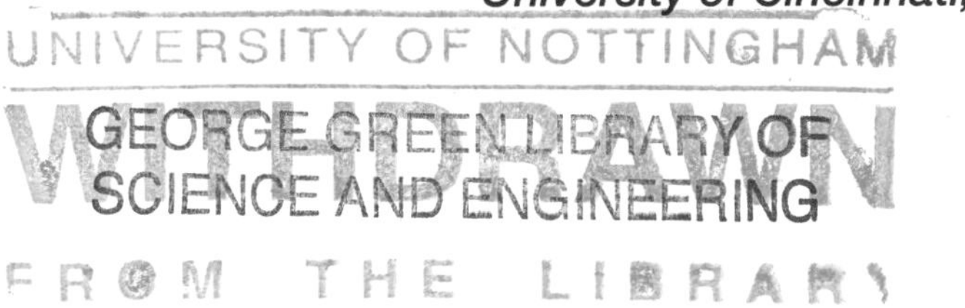

World Scientific

NEW JERSEY • LONDON • SINGAPORE • BEIJING • SHANGHAI • HONG KONG • TAIPEI • CHENNAI

Published by

World Scientific Publishing Co. Pte. Ltd.
5 Toh Tuck Link, Singapore 596224
USA office: 27 Warren Street, Suite 401-402, Hackensack, NJ 07601
UK office: 57 Shelton Street, Covent Garden, London WC2H 9HE

British Library Cataloguing-in-Publication Data
A catalogue record for this book is available from the British Library.

NONLINEAR SYSTEMS OF PARTIAL DIFFERENTIAL EQUATIONS
Applications to Life and Physical Sciences

ISBN-13 978-981-4277-69-3
ISBN-10 981-4277-69-X

Printed in Singapore by Mainland Press Pte Ltd

Nonlinear Systems of Partial Differential Equations

Applications to Life and Physical Sciences

TO MY

FAMILY

Preface

Substantial progress had been made in the last two decades in the theory of nonlinear systems of partial differential equations. Much of the developments are motivated by applications to the natural sciences of biology, physics and chemistry. There is a considerable amount of results concerning positive solutions for the study of ecological and medical sciences. Other applications involve reactor dynamics, fluid, plasma, display technology etc. There are several excellent books published in such topics in the last decade; however, due to numerous recent developments of new methods and results there is a need for a book to collect them for convenient reference and study. The gathering of many existing theorems enhances the understanding of the subject and leads to directions for further research or applications. In the mean time, the demand for reliable applications encourages deeper understanding of the underlying mathematical methods of nonlinear partial differential equations. Many of the problems were introduced in my first book in 1989. In the last twenty years, there is tremendous progress in the mathematical formulation for studies in cancer, cardiology, epidemiology and cell development etc., leading to larger systems of nonlinear partial differential equations. More thorough understanding of the interaction between a few components is crucial for building to large systems. Serious efforts are made to make this book self-contained. Although many theorems used in the book are presented in other books or papers, we include their explanations in the Appendices so that this book is more readable to many graduate students and researchers who are not specialists in these topics.

For the study of positive solutions, several methods are used extensively in this book. Topological degree theory method is extremely fruitful in proving the existence of positive equilibrium for several coupled elliptic systems. One of the most important tools in nonlinear analysis is the Leray-Schauder degree. Due to the fact that the positive cone is a retract of a Banach space, it is possible to define a fixed point index for compact maps as introduced by H. Amann. The fixed point index is equivalent to Leray-Schauder degree. Many existence theorems follow from the property of homotopic invariance of degree. Another powerful method in nonlinear analysis is the use of bifurcation theory as developed by Crandall and Rabinowitz. Bifurcation of solutions may occur at points where the implicit function theorem does not apply. Estimation of solutions by means of maximum principle combined with global bifurcation theory provides convenient analysis of the behavior of positive equilibria as various parameters changes. Many diagrams are included describing the range of parameters so that coexistence can occur. Through the use of maximum principle for $W^{2,p}$ solutions, the book considers the theory of non-classical solutions for interacting species. By means of weak upper-lower solutions, it studies solutions with discontinuous and highly spatially varying growth rates.

For the study of parabolic time-dependent problems, we use both semigroup methods and the classical Schauder's theory. The semigroup method provides existence of solution of initial value problems in various function spaces. Combined with the spectral analysis of the related linearized parabolic system, we obtain many time-stability results for positive equilibria. Comparison theorems for parabolic systems under various boundary conditions also provide estimates of solutions by means of upper and lower solutions.

A significant part of the book is devoted to the study of optimal control of systems of nonlinear partial differential equations as developed by J. L. Lions. They are systems motivated by applications involving equilibrium or time dependent problems. The object is to control the coefficients of the systems so that certain properties of the subsequent solutions are maximized. Both the theories of weak and classical solutions are used. A larger optimality system of equation is deduced for the optimal control. Combined with the method of upper-lower solutions, we construct monotone sequences converging to estimates of the optimality system.

The book also describes results concerning systems of nonlinear wave equations and traveling wave solutions for parabolic systems. The system of wave equations is analyzed by semigroup method. In contrast to the popular method of finding traveling wave solutions by means of dynamical system theory, we carefully explain the method of finding traveling solutions for parabolic systems by using upper-lower solution in an unbounded domain. Other topics studied include invariant manifolds for coupled parabolic-hyperbolic systems, cross-diffusion for elliptic systems, persistence, blow-up due to boundary inflow, coupled elliptic-parabolic system related to display technology, degenerate diffusive systems and other related topics. Although the systems are motivated by applications, the techniques of analyzing such types of problems are carefully explained. They involve extensions of methods described in the above paragraphs.

Chapter 1 considers systems of two coupled nonlinear elliptic or parabolic equations. The nonlinear terms incorporate the interactions between two life species occupying a common domain. The cases of competition, cooperation or prey-predator relationship are covered in detail in separate sections. The boundary values are given in the Dirichlet type. The trivial vector function is always a solution of the systems. The major concern is the additional possibility of coexistence solutions when both species survive together. We use the methods described in the last few paragraphs to find coexistence states under various configuration of interaction parameters, diffusion rates and size of the environment. Many results are related to the principal eigenvalues of various scalar problems induced by the original larger system. The time-stability of the coexistence states for all the three cases are discussed in the last section of the chapter. Some other long-time behavior of the corresponding reaction-diffusion systems are also studied. Most of the results are published by many researchers in the nineties or afterwards, and cannot be found in other books. Chapter 2 extends our discussion of problems in the first chapter to larger systems of equations. The species components may now be classified into groups inside which they interact in competition, cooperative or food-chain manner, while the different groups interact in various ways. The conditions for coexistence becomes more complex. However, we see the methods developed in the first chapter can be extended to cover many different cases. For practical applications, we consider analysis of epidemics, fission reactor engineering and other problems.

Chapter 3 studies the optimal control of nonlinear systems analyzed in the first two chapters. We control the interaction parameters or boundary conditions in order to optimize an expression involving the solutions of the systems. Using the understanding of the uncontrolled systems in the last chapters, we deduce conditions when optimal control is possible. We consider the control of elliptic, parabolic and time-periodic systems. For biological systems, we maximize the economic return of species-harvesting; and for reactor problems, we optimize the target temperature profile. From the original systems together with the optimization criteria we deduce larger optimality systems which describe the optimal controls. We further analyze the solution of the optimality systems by means of monotone convergence schemes. So far, results for such systems have not been gathered coherently in a book form.

Chapter 4 emphasizes on other aspects of the solutions of the reaction-diffusion systems. We consider conditions on the equations when certain components can persist indefinitely in time. We study the effect of diffusion rates which may depend on the concentrations of other species. Such self and cross-diffusion property can have significant effect on the coexistence problem. Questions concerning blow-up, extinction, degenerate diffusion rates and others are also investigated by various methods described above. Chapter 5 first considers traveling wave solutions for competitive parabolic systems. There had been numerous results on such topics over two decades ago found by means of dynamical systems technique. Here, we present some very new recent results found by means of upper-lower solution method in an unbounded domain. We also study a system of hyperbolic equations and the stability of their equilibrium. We further discuss the problem of invariant manifold for solutions of coupled Navier-Stokes and wave equations. Roughly speaking, we find a relationship between the fluid velocity field and the magnetic field so that it is invariant as time changes. Finally, we consider a coupled elliptic-parabolic problem motivated by research on plasma display technology. We estimate whether the sizes of the ion concentrations can reach a high enough level for light emission.

We painstakingly itemize in the Appendices those theories and theorems used in deducing the results in Chapters 1 to 5. They include many standard theorems in scalar and systems of partial differential equations, methods in linear and nonlinear functional analysis and topology. In real world applications, models are very complicated and they have to be continuously improved with deeper investigation. It is therefore important to understand how the applicable results in Chapters 1 to 5 are deduced from the more fundamental theories in the Appendices. With the standard tools conveniently displayed, one can then readily modify the theorems in the first five chapters to forms more suitable for proper utilization. The stress of this book is consequently different from others with similar titles existing in the literature. On the other hand, the presentation of the topics in the first five chapters are motivated by practical applications. Consequently, the results can be applied to real world problems by non-specialists, even if the rigorous proofs presented are not completely understood. Finally, this book can only cover those topics which have interest me, my friends and colleagues. Many other subjects concerning systems of nonlinear partial differential equations are beyond our present scope. I hope that this book is helpful for researchers who will continue to explore on the subject.

I am grateful to many colleagues, students and friends who had discussed various topics with me. They include in alphabetical order: G. Chen, R. Cantrell, C. Cosner, E. Dancer, Q. Fan, F. He, X. Hou, P. Korman, A. Lazer, S. Lenhart, L. Li, W. Ni, L. Ortega, C. Pao, S. Stojanovic, B. Villa, Q. Zhang, B. Zhang and many others. Their inspirations and encouragements are valuable in the development of the subject matter of this book. I would also like to thank my wife, Soleda, for the design of the book cover and her joint preparation of some figures in the book with Z. Kang. I also appreciate the help of R. Chalkley, D. Mueller and L. F. Kwong for the efficient production of this manuscript.

Anthony W. Leung
Cincinnati, 2009

Contents

Chapter 1

Positive Solutions for Systems of Two Equations

1.1 Introduction

In this chapter, we consider a system of two partial differential equations describing two interacting population species. Each species diffuse from location of higher to lower concentration, and they interact with each other in a prey-predator, competing or cooperating relationship. We emphasize the situation when the species must have zero concentration at the boundary of the environment. These are known as reaction-diffusion equations with homogeneous Dirichlet boundary condition. The boundary condition is known as "hostile" in some ecological studies. We first consider the possibility of positive coexistence equilibrium for the case of prey-predator in Section 1.2, competing species in Section 1.3, and cooperating species in Section 1.4. They are thus systems of elliptic partial differential equations of the form:

$$(1.1)\qquad \begin{cases} \sigma_1 \Delta u + u(a_1 + f_1(u,v)) = 0 & \\ & \text{in } \Omega, \\ \sigma_2 \Delta v + v(a_2 + f_2(u,v)) = 0 & \\ u = v = 0 & \text{on } \partial\Omega. \end{cases}$$

In this chapter, we always assume that Ω is a bounded domain in $R^N, N \geq 2$, unless otherwise stated. If $N > 1$, we assume that the boundary $\partial\Omega \in C^{2+\alpha}, 0 < \alpha < 1$; that is, the boundary has local representation whose second order partial derivatives are Hölder continuous with exponent α. The symbol Δ denotes the Laplacian operator:

$$\sum_{i,j=1}^{N} \frac{\partial^2}{\partial x_i \partial x_j}.$$

The constants a_1, a_2 are respectively the intrinsic growth rates of the species whose population concentrations at the position x are denoted by $u = u(x)$ and $v = v(x)$. The parameters σ_1, σ_2 are positive diffusion coefficient constants. The reaction functions $f_1(u, v), f_2(u, v)$ involve many other parameters reflecting interaction rates and self-crowding effects of the species. We shall investigate the ranges of these parameters and their sizes relative to that of the size of the environment domain Ω so that coexistence states are possible. We shall use various methods of nonlinear analysis to study these problems, including upper-lower solutions, monotone schemes, bifurcation, degree theory and their generalizations. We usually begin with the simplest cases in order to illustrate how the various methods are used in obtaining the results.

In Section 1.5, we consider the time dependent parabolic system associated with system (1.1):

$$(1.2) \qquad \begin{cases} u_t = \sigma_1 \Delta u + u(a_1 + f_1(u, v)) & \\ & \text{in } \Omega \times (0, \infty), \\ v_t = \sigma_2 \Delta v + v(a_2 + f_2(u, v)) & \\ & \\ u = v = 0 & \text{on } \partial\Omega. \end{cases}$$

The main emphasis is to analyze the long time behavior of the system, and to find whether the solutions are tending to the equilibria described in the previous sections.

We now proceed to introduce some symbols which will be used repeatedly in this and later chapters. For any real $q(x)$ in $C^\alpha(\bar{\Omega})$ and $\sigma > 0$, the linear eigenvalue problem:

$$(1.3) \qquad -\sigma \Delta u + q(x) u = \rho u \text{ in } \Omega, \quad u = 0 \text{ on } \partial\Omega$$

has an infinite sequence of eigenvalues, $\rho_1 < \rho_2 < \rho_3 < \ldots$, which are bounded below. It is also known that the first eigenvalue:

$$(1.4) \qquad \rho = \rho_1 = \rho_1(-\sigma\Delta + q(x))$$

is simple, and all solutions of (1.3) with $\rho = \rho_1(-\sigma\Delta + q(x))$ are multiples of a particular eigenfunction, which does not change sign on Ω and has its normal derivatives never vanish on the boundary $\partial\Omega$.

For convenience, we define

$$(1.5) \qquad \lambda_1 := \rho_1(-\Delta),$$

and denote by $\omega(x)$, a positive eigenfunction of the operator $-\Delta$ on Ω with boundary condition $u = 0$ on $\partial\Omega$. Similarly, for any real $\hat{q}(x)$ in $C^\alpha(\bar{\Omega})$ and $\sigma > 0$, the linear eigenvalue problem:

$$(1.6) \qquad \sigma \Delta u + \hat{q}(x) u = \hat{\rho} u \text{ in } \Omega, \quad u = 0 \text{ on } \partial\Omega$$

has an infinite sequence of eigenvalues, $\hat{\rho}_1 > \hat{\rho}_2 > \hat{\rho}_3 > \dots$, which are bounded above. We denote the largest eigenvalue by:

$$\hat{\rho} = \hat{\rho}_1 = \hat{\rho}_1(\sigma\Delta + \hat{q}(x)), \tag{1.7}$$

which is simple. As shown in Sections 1.4 and 1.5 below, most of the results in this chapter are valid if the Laplacian operator is replaced with the uniform elliptic operator:

$$L \equiv \sum_{i,j=1}^{N} a_{ij}(x)\frac{\partial^2}{\partial x_i \partial x_j} + \sum_{i,j=1}^{N} b_i(x)\frac{\partial}{\partial x_i} + c(x),$$

where $a_{ij}(x)$, $b_i(x), c(x)$ are in $C^\alpha(\bar{\Omega}), 0 < \alpha < 1$, $c(x) \leq 0$ in $\bar{\Omega}$, and

$$\sum_{i,j=1}^{N} a_{ij}(x)\xi_i\xi_j \geq \mu_0 \sum_{i=1}^{N} \xi_i^2, \ \mu_0 > 0,$$

for all $x \in \bar{\Omega}$, all $(\xi_1, ..., \xi_N) \in R^N$. For simplicity, we present most of the results using the Laplacian.

For convenience, we state a simple direct consequence of the maximum principle, which will be used repeatedly to assert that in many instances non-negative non-constant solutions in $\bar{\Omega}$ are actually strictly positive in Ω.

Lemma 1.1. *Let $u \in C^2(\bar{\Omega})$ be a non-negative non-constant solution of:*

$$Lu + h(x)u = 0 \ \text{in } \Omega, \quad u = 0 \ \text{on } \partial\Omega,$$

where L is the operator described above and $h(x)$ is bounded, then u must satisfy $u(x) > 0$ for all $x \in \Omega$.

Proof. Let P be a positive constant such that $h(x) - P \leq 0$ for all $x \in \Omega$, and define $v = -u$. Then we have

$$Lv + [h(x) - P]v = -Pv \geq 0 \ \text{ in } \Omega, \ v = 0 \ \text{ on } \partial\Omega.$$

From the maximum principle, we obtain $v(x) < 0$ in Ω, since $v(x)$ is not a constant function. This means $u(x) > 0$ in Ω.

In this chapter, we avoid the consideration of zero outward normal derivative:

$$\frac{\partial u}{\partial \nu} = \frac{\partial v}{\partial \nu} = 0 \quad \text{on } \partial\Omega.$$

Such homogeneous Neumann boundary condition, which represents no flux of species across the boundary, has been studied more extensively in other books in the literature, e.g. Smoller [209] and Leung [125].

1.2 Strictly Positive Coexistence for Diffusive Prey-Predator Systems

Let $u(x)$ and $v(x)$ be respectively the density of prey and predator at the point x in a bounded domain Ω. We first consider an earliest result concerning coexistence equilibrium when both species are restricted to vanish on the boundary. We consider the following homogeneous Dirichlet boundary value problem for the coupled Volterra-Lotka type reaction-diffusive system.

$$(2.1)\qquad \begin{cases} \sigma_1\Delta u + u(a - bu - cv) = 0 & \\ & \text{in } \Omega, \\ \sigma_2\Delta v + v(e + fu - gv) = 0 & \\ u = v = 0 & \text{on } \partial\Omega. \end{cases}$$

Here, $\sigma_1, \sigma_2, a, b, c, e, f, g$ are positive constants. The parameters a, e are the intrinsic growth rates and b, c, f, g are interaction rates. Note that the prey-predator relation is reflected by the signs of $-c$ and $+f$.

Part A: Early Results via Upper-Lower Solutions and Bifurcation.

The following theorem concerning the coexistence of both species can be readily deduced by means of upper-lower solutions method for a system of elliptic equations.

Theorem 2.1. *The boundary value problem (2.1) under hypotheses:*

$$(2.2)\qquad \begin{cases} a > \sigma_1\lambda_1, \; e > \sigma_2\lambda_1, & \\ cf < gb, & \text{and} \\ a > \frac{\sigma_1 gb}{gb - cf}[\lambda_1 + \frac{ce}{g\sigma_1}] & \end{cases}$$

has a solution with each component strictly positive in Ω. Here λ_1 is defined in (1.5).

Proof. The last two inequalities of hypotheses (2.2) imply that $a(1 - \frac{cf}{gb}) > \sigma_1(\lambda_1 + \frac{ce}{g\sigma_1})$; hence $a > \sigma_1\lambda_1 + \frac{c}{g}(e + f\frac{a}{b})$. It follows that for each fixed v, $0 \le v \le \frac{1}{g}(e + f\frac{a}{b})$, the function $u_1 := \delta\omega(x) > 0$ is a lower solution of the first equation in (2.1), for $\delta > 0$ sufficiently small. (Here $\omega(x)$ is described in Section 1.1.) That is, we have for each such v:

$$(2.3)\qquad \begin{aligned} &\sigma_1\Delta u_1 + u_1(a - bu_1 - cv) \ge 0 && \text{in } \Omega, \text{ and} \\ &u_1 \le 0 && \text{on } \partial\Omega. \end{aligned}$$

On the other hand, the function $u_2(x) :\equiv a/b$ is an upper solution for the first equation in (2.1). That is,

$$\sigma_1 \Delta u_2 + u_2(a - bu_2 - cv) \le 0 \qquad \text{in } \Omega, \text{ and}$$

$$u_1 \ge 0 \qquad \text{on } \partial\Omega.$$

Similarly, for each fixed u, $0 \le u \le a/b$, the functions $v_1 := \delta\omega(x)$, for sufficiently small positive δ and $v_2(x) :\equiv \frac{1}{g}(e + f\frac{a}{b})$ are respectively lower and upper solutions for the second equation in (2.1). By means of an intermediate-value type theorem (see Tsai [221] or Theorem 1.4-2 in Leung [125]), we assert that there exists a solution $(u^*(x), v^*(x))$ of (2.1) satisfying $u_1(x) \le u^*(x) \le u_2, v_1(x) \le v^*(x) \le v_2$ for all $x \in \bar{\Omega}$. Note that since $v_1 > 0$, we have (2.3) valid for all v satisfying $v_1 < v < v_2$.

Remark 2.1. The proof of Theorem 2.1 is simple. However, it uses an intermediate-value type theorem, whose proof requires Leray-Schauder degree theory. Observe also that the inequalities in (2.2) are more readily satisfied for large domains, because λ_1 will then be small.

We next use a more sophisticated procedure to see how the various sizes of the parameters a and e lead to different results of existence and non-existence of positive solutions. We first consider the boundary value problem:

$$-\sigma\Delta u + q(x)u = u(a - bu) \text{ in } \Omega, \quad u = 0 \text{ on } \partial\Omega, \tag{2.4}$$

where $\sigma > 0, q(x), a$ and b are as described above. Suppose that $a \le \rho_1(-\sigma\Delta + q(x))$. Let $\phi(x) > 0$ be an eigenfunction for (2.4) with $\rho = \rho_1(-\sigma\Delta + q(x))$. Using the family of upper solutions $\epsilon\phi(x), \epsilon > 0$, for (2.4) and the sweeping principle described in Theorem 1.4-3 [125] one readily deduces that $u = 0$ is the only non-negative solution of (2.4) if $a \le \rho_1(-\sigma\Delta + q(x))$. On the other hand, suppose $a > \rho_1(-\sigma\Delta + q(x))$. We use large constant as upper solution and small multiple of $\phi(x)$ as lower solution for (2.4) to deduce the existence of a solution which is positive in Ω. Furthermore, such positive solution is unique, when $a > \rho_1(-\sigma\Delta + q(x))$. (See Lemma 5.2-2 in [125].) We will state the above observation in a slightly more general situation, which will be used repeatedly in many chapters.

Lemma 2.1. *Let $q(x)$ be in $C^\alpha(\bar{\Omega})$, $0 < \alpha < 1$; $G \in C^1([0, \infty)), G' < 0$ in $(0, \infty)$ and there exists some $c_0 > 0$ such that $G(c_0) < 0$. Consider the boundary value problem*

$$-\sigma\Delta u + q(x)u = uG(u) \text{ in } \Omega, \quad u = 0 \text{ on } \partial\Omega. \tag{2.5}$$

(i) If $G(0) \le \rho_1(-\sigma\Delta + q(x))$, then $u \equiv 0$ is the only non-negative solution of the problem.

(ii) If $G(0) > \rho_1(-\sigma\Delta + q(x))$, then the problem has a unique strictly positive solution in Ω.

We consider the problem (2.1) under the simplest situation when the growth rate a of the prey is small. In such situation, no prey population can survive as described below.

Theorem 2.2. *Suppose $a \leq \rho_1(-\sigma\Delta)$ and (u, v) is a non-negative solution of (2.1). Then the following are true:*
(i) $u \equiv 0$ in $\bar{\Omega}$.
(ii) If $e \leq \rho_1(-\sigma_2\Delta)$, then we also have $v \equiv 0$ in $\bar{\Omega}$; if $e > \rho_1(-\sigma_2\Delta)$, then either $v \equiv 0$ in $\bar{\Omega}$ or v is the unique positive solution of

$$\sigma_2\Delta v + v(e - gv) = 0 \text{ in } \Omega, \quad v = 0 \text{ on } \partial\Omega. \tag{2.6}$$

Proof. Multiplying the first equation of (2.1) by u, and integrating over Ω, we obtain

$$-\sigma_1 \int_\Omega u\Delta u dx \leq a \int_\Omega u^2 dx - b \int_\Omega u^3 dx. \tag{2.7}$$

On the other hand, the characterization of the first eigenvalue gives

$$\rho_1(-\sigma_1\Delta) \int_\Omega u^2 dx \leq \int_\Omega \sigma_1 |\nabla u|^2 dx = -\sigma_1 \int_\Omega u\Delta u dx. \tag{2.8}$$

Inequalities (2.7) and (2.8) imply that $\int_\Omega u^2 dx < \frac{a}{\rho_1(-\sigma_1\Delta)} \int_\Omega u^2 dx$ if $u \not\equiv 0$. Thus we must have $u \equiv 0$ in $\bar{\Omega}$. Consequently, assertion (ii) follows from the discussion for single equation above analogous to (2.4), with $q \equiv 0$, and σ, a, b respectively replaced by σ_2, e, f.

We next use bifurcation technique to analyze problem (2.1) as the parameters e or a varies. This will eventually lead to Theorem 2.3 and Theorem 2.4. The approach involves decoupling the two equations in (2.1). We write the first equation in (2.1) in the form;

$$-\sigma_1\Delta u + cvu = u(a - bu) \text{ in } \Omega, \quad u = 0 \text{ on } \partial\Omega, \tag{2.9}$$

which can be regarded as a special case of (2.4) with $\sigma = \sigma_1, q(x) = cv(x)$. Thus, if $a \leq \rho_1(-\sigma_1\Delta + cv(x))$, then (2.9) has no positive solution; while if $a > \rho_1(-\sigma_1\Delta + cv(x))$, then (2.9) has a unique positive solution in Ω. Let v be an arbitrary function in $C^1(\bar{\Omega})$, we define $u(v)$ as a function on $\bar{\Omega}$ by:

$$u(v) = \begin{cases} 0 \text{ if } a \leq \rho_1(-\sigma_1\Delta + cv), \\ \\ \text{unique solution of problem (2.9) if } a > \rho_1(-\sigma_1\Delta + cv). \end{cases} \tag{2.10}$$

Clearly, if v satisfies the single equation:

$$-\sigma_2 \Delta v = v(e - gv + fu(v)) \text{ in } \Omega, \quad v = 0 \text{ on } \partial\Omega, \tag{2.11}$$

then the pair $(u(v), v)$ will be a solution of (2.1). To analyze (2.11), we first obtain the following properties of the mapping $v \to u(v)$.

Lemma 2.2. *(i) The mapping: $v \to u(v)$ defined by (2.10) considered as a function from $C^1(\bar{\Omega})$ to $C^1(\bar{\Omega})$ is continuous;*
(ii) if $v_1 \geq v_2$ in $\bar{\Omega}$, then $u(v_1) \leq u(v_2)$ in $\bar{\Omega}$.

The proof of this lemma can be found in Brown [15] or p. 360 in Leung [125].

To study more interesting situations, we now suppose that

$$a > \rho_1(-\sigma_1 \Delta). \tag{2.12}$$

In the following Theorem 2.3, we let the parameter e varies, while all other parameters are held fixed. Problem (2.1) has two non-negative solutions $(0, 0)$ and $(u(0), 0)$ for all values of e. We consider the global bifurcations as e varies in the decoupled equation (2.11). This leads to bifurcation from the line of solution $(u(0), 0)$ to solution of (2.1) with both components positive in Ω. Let L be the operator defined by

$$Lv = -\sigma_2 \Delta v - fu(0)v.$$

Without loss of generality, we may assume that $\rho_1(-\sigma_2\Delta - fu(0)) \neq 0$. Otherwise, we replace L by $L+k$ for an appropriate constant k. For each h in $C^1(\bar{\Omega})$, let Kh denote the unique solution of the problem: $Lu = h$ in Ω, $u = 0$ on $\partial\Omega$. The map $K : C^1(\bar{\Omega}) \to C^1(\bar{\Omega})$ is a compact linear operator. Let $F : C^1(\bar{\Omega}) \to C^1(\bar{\Omega})$ be defined by

$$F(v) = -gv^2 + f[u(v) - u(0)]v.$$

By Lemma 2.2, F is continuous and $||F(v)|| = o(||v||)$ as $v \to 0$ in $C^1(\bar{\Omega})$, where $|| \cdot ||$ denotes the norm in $C^1(\bar{\Omega})$. We now write (2.11) in the form:

$$v - eKv - KF(v) = 0. \tag{2.13}$$

Since $||KF(v)|| = o(||v||)$ as $v \to 0$ in $C^1(\bar{\Omega})$, we can apply the global bifurcation results of Rabinowitz [190] as the parameter e varies. We can also apply results concerning bifurcation from simple eigenvalues described in Crandall and Rabinowitz [33], Blat and Brown [11] or [125] to obtain properties concerning the local behavior of the bifurcation solutions. It is shown that in a neighborhood of the bifurcation point $(\rho_1(-\sigma_2\Delta - fu(0)), 0)$, all non-trivial solutions (e, v) of (2.13) lie on a curve of the form $\{(\bar{e}(\alpha), \phi(\alpha)) : -\delta \leq \alpha \leq \delta\}$ in $R \times C^1(\bar{\Omega})$, where $\bar{e}(0) = \rho_1(-\sigma_2\Delta - fu(0))$ and $\phi(\alpha) = \alpha\phi_1 +$ terms of higher "order" in α. Here, ϕ_1 is a positive principal eigenfunction for the eigenvalue $\rho_1(-\sigma_2\Delta - fu(0))$.

From the fact that $\frac{\partial \phi_1}{\partial \nu} < 0$ on $\partial\Omega$ where ν is the outward unit normal at the boundary, we thus conclude that for α sufficiently small and positive, the corresponding non-trivial solution v lies in the cone

$$P = \{v \in C^1(\bar{\Omega}) : v(x) > 0 \ for \ x \in \Omega, \ \frac{\partial v}{\partial \nu}(x) < 0 \ for \ x \in \partial\Omega\}.$$

Moreover, the closure of the set of non-trivial solutions (e, v) of (2.13) contains a component S (i.e. a maximal connected subset) such that either S joins $(\rho_1(-\sigma_2\Delta - fu(0)), 0)$ to ∞ in $R \times C^1(\bar{\Omega})$ or S joins $(\rho_1(-\sigma_2\Delta - fu(0)), 0)$ to $(\bar{\rho}, 0)$, where $\bar{\rho}$ is some other eigenvalue of L. More precisely, we can further deduce (see [125] or [11]) the following properties for the set S.

Lemma 2.3. *The component S contains a connected subset $S^+ \subset S - \{(\bar{e}(\alpha), \phi(\alpha)) : -\delta \le \alpha \le 0\}$ with the following properties:*
(i) S^+ is contained in $R \times P$;
(ii) $\{\rho \in R : (\rho, v) \in S^+\} = (\rho_1(-\sigma_2\Delta - fu(0)), +\infty)$.

Let (u, v) be any solution of (2.1) with each component non-negative in $\bar{\Omega}$. Suppose that v is not the trivial function, then v is the unique positive solution of the equation

$$-\sigma_2\Delta v - fuv = v(e - gv) \text{ in } \Omega, \ v = 0 \text{ on } \partial\Omega. \tag{2.14}$$

Let λ_1 and ω_1 be the principal eigenvalue and the corresponding eigenfunction with $\max\{\omega_1(x) : x \in \Omega\} = 1$. It is readily checked that if $e > \sigma_2\lambda_1$, then the function $g^{-1}(e - \sigma_2\lambda_1)\omega_1$ is a lower solution of the problem (2.14), and that any sufficiently large positive constant is an upper solution. Since v must be between the upper and lower solutions, we conclude that if $e > \sigma_2\lambda_1$ we have $v \ge g^{-1}(e - \sigma_2\lambda_1)\omega_1 := k(e)\omega_1$ in $\bar{\Omega}$, where $k(e) := g^{-1}(e - \sigma_2\lambda_1) \to \infty$ as $e \to \infty$. Now, consider the eigenvalue problem

$$-\sigma_1\Delta u + ck(e)\omega_1 u = \lambda u \text{ in } \Omega, \ u = 0 \text{ on } \partial\Omega. \tag{2.15}$$

The least eigenvalue $\lambda = \hat{\lambda}_1(e)$ has the characterization:

$$\hat{\lambda}_1(e) = inf. \{\int_\Omega \sigma_1|\nabla u|^2 + ck(e)\omega_1 u^2 dx : u \in W_0^{1,2}(\Omega), \int_\Omega u^2 dx = 1\}.$$

Thus from the limit of $k(e)$, we can deduce that $\hat{\lambda}_1(e) \to \infty$ as $e \to \infty$. Consequently, we have $\hat{\lambda}_1(e) > a$, if e is large enough.

Next, from the characterization of first eigenvalue and comparing with (2.15), we find that the first eigenvalue of

$$-\sigma_1\Delta w + cvw = \lambda w \text{ in } \Omega, \ w = 0 \text{ on } \partial\Omega$$

is greater that a. Hence the only non-negative solution of

$$-\sigma_1 \Delta u + cvu = u(a - bu) \text{ in } \Omega, \quad u = 0, \quad \text{on } \partial\Omega$$

is the zero function. We have proved that if e is large enough and v is not the trivial function, then $u \equiv 0$. From Lemma 2.3, we see that the only way the continuum of solutions S^+ can join the bifurcation point $(\rho_1(-\sigma_2\Delta - fu(0)), 0)$ on the (e, v) plane to ∞ is by $u(v)$ becoming equal to zero for e sufficiently large. However, when $u(v) \equiv 0$, then clearly v satisfies (2.6). If we consider the bifurcation diagram on the $e - (u, v)$ plane, the continuum of solutions $\{(e, u(v), v) : (e, v) \in S^+\}$ for (2.1) must join up with the continuum of solutions $\{(e, 0, v) : (e, v)$ is a solution of (2.6)$\}$, Solutions of (2.6) are discussed in Theorem 2.2(ii). From the above arguments, we obtain the following theorem.

Theorem 2.3. *(i) Suppose:*

$$a > \sigma_1 \lambda_1. \tag{2.16}$$

Then there exists $\lambda^* > \sigma_2 \lambda_1 = \rho_1(-\sigma_2\Delta)$ such that if e satisfies: $\rho_1(-\sigma_2\Delta - fu(0)) < e < \lambda^*$, that is:

$$\hat{\rho}_1(\sigma_2\Delta + e + fu(0)) > 0 \quad \text{and} \quad e < \lambda^*, \tag{2.17}$$

the boundary value problem (2.1) has a solution with each component strictly positive in Ω. Moreover, there exists $\tilde{\lambda} \geq \lambda^*$ such that if $e > \tilde{\lambda}$, then any non-negative solution (u,v) of problem (2.1) with $v \not\equiv 0$ must have $u \equiv 0$. (Recall the definition of $\hat{\rho}_1$ in (1.7).)
(ii) Suppose:

$$a < \sigma_1 \lambda_1. \tag{2.18}$$

Then any non-negative solution (u,v) of problem (2.1) must have $u \equiv 0$.

In the following theorem, we let the parameter a varies, while all other parameters are held fixed. We write the second equation in (2.1) in the form of (2.14). Analogous to Lemma 2.2, we define a map from $C^1(\bar{\Omega})$ to $C^1(\bar{\Omega})$ by:

$$v(u) = \begin{cases} 0 \text{ if } e \leq \rho_1(-\sigma_2\Delta - fu), \\ \text{unique solution of problem (2.14) if } e > \rho_1(-\sigma_2\Delta - fu). \end{cases} \tag{2.19}$$

We can show as in Lemma 2.2 that $u \to v(u)$ is a continuous function from $C^1(\bar{\Omega})$ to $C^1(\bar{\Omega})$ and that $u \to v(u)$ is an increasing function.

Theorem 2.4. *(i) Suppose:*

$$\begin{cases} e > \sigma_2 \lambda_1 \qquad \text{and} \\ \hat{\rho}_1(\sigma_1\Delta + a - cv(0)) > 0, \end{cases} \tag{2.20}$$

then the boundary value problem (2.1) has a solution with each component strictly positive in Ω.
(ii) Suppose that $e \le \sigma_2\lambda_1$. Then, provided that a is sufficiently large, the problem (2.1) has a solution with each component strictly positive in Ω.

Proof. Let $e > \sigma_2\lambda_1$, then problem (2.1) has a solution $(u, v) = (0, v(0))$ with $v(0)$ non-trivial. We write the first equation of (2.1) as:

$$(2.21) \qquad -\sigma_1\Delta u + cv(0)u = au - bu^2 - c[v(u) - v(0)]u \text{ in } \Omega, \;\; u = 0 \text{ on } \partial\Omega,$$

and bifurcate with the parameter a at $a = \rho_1(-\sigma_1\Delta + cv(0))$ when $(u, v) = (0, v(0))$. As in Lemma 2.3, we can show that there exists a continuum of solutions S^+ of (2.21) contained in $R \times P$, i.e. $u \ge 0$ whenever $(a, u) \in S^+$, and that $\{a : (a, u) \in S^+\} = (\rho_1(-\sigma_1\Delta + cv(0)), \infty)$. If $(a, u) \in S^+$, then $u \ge 0$ and so $v(u) \ge v(0)$, i.e. v(u) is not the trivial function. Consequently, the continuum of solutions $\{(a, u, v(u)) : (a, u) \in S^+\}$ for the system (2.1) cannot connect with the continuum of solutions $\{(a, u(0), 0) : a > \sigma_1\lambda_1\}$. This leads to the assertion of part (i).

For part (ii), suppose that $e \le \sigma_1\lambda_1$. We have $u(0)$ satisfies

$$-\sigma_1\Delta u = au - bu^2 \text{ in } \Omega, \;\; u = 0 \text{ on } \partial\Omega.$$

Let λ_1 and ω_1 be the principal eigenvalue and the corresponding eigenfunction with $\max\{\omega_1(x) : x \in \Omega\} = 1$. Using a/b and $b^{-1}(a - \sigma_1\lambda_1)\omega_1$ for a large enough as upper and lower solutions respectively, we find that $b^{-1}(a - \sigma_1\lambda_1)\omega_1 \le u(0) \le ab^{-1}$. Comparing the least eigenvalue of

$$(2.22) \qquad -\sigma_2\Delta w - fu(0)w = \lambda w \text{ in } \Omega, \;\; w = 0 \text{ on } \partial\Omega,$$

with that of

$$-\sigma_2\Delta v - fb^{-1}(a - \sigma_1\lambda_1)\omega_1 v = \lambda v \text{ in } \Omega, \;\; v = 0 \text{ on } \partial\Omega,$$

by means of Rayleigh's quotient, we conclude that the first eigenvalue $\rho_1(-\sigma_2\Delta - fu(0))$ of (2.22) tends to $-\infty$ as $a \to +\infty$. That is we have $\rho_1(-\sigma_2\Delta - fu(0)) \le e \le \sigma_2\lambda_1$ for a sufficiently large. Thus by Theorem 2.3(i), we assert that problem (2.1) has a solution which is positive in both components.

Part B: General Results via Degree Theory.

We next consider a prey-predator system with more general type of interaction than quadratic (or Lotka-Volterra type). Moreover, we will obtain somewhat sharper results, and find necessary and sufficient conditions for the existence of positive solutions. We shall use degree theory method of cone index to prove that the conditions in parts (ii) and (iii) of the following Theorem 2.5 is sufficient for the existence of positive solution. More precisely, for a constant $d > 0$, we will consider the boundary value problem:

$$(2.23) \qquad \begin{cases} \Delta u + uM(u,v) = 0 & \\ & \text{in } \Omega, \\ d\Delta v + v(h(u) - m(v)) = 0 & \\ u = v = 0 & \text{on } \partial\Omega, \end{cases}$$

where $M(u,v)$ and its first partial derivatives are continuous in the first closed quadrant. Moreover, it satisfies

$$(2.24) \qquad M_v(u,v) < 0 \text{ for } u, v \geq 0; \;\; M_u(u,0) < 0 \text{ for } u \geq 0,$$

(2.25)
$M(0,0) > 0$; there exists a constant $C_0 > 0$ such that $M(u,0) < 0$ for $u > C_0$.

(2.26) The functions h and m belong to $C^1([0,\infty))$, with each function strictly increasing in $[0,\infty)$.

A solution (u,v) of problem (2.23) is called a positive solution if both components are ≥ 0 and $\not\equiv 0$ on Ω. A common assumption in ecological studies is to set the rate $M(u,v) = \sigma_1^{-1}(a - bu - \frac{vc}{k+u})$ where σ_1, a, b, c, k are positive constants or other rates involving ratios of u and v. Such type of growth rate is called Holling's type. From the smoothness of M, h and m, the positive solutions of (2.23) are classical solutions with components in $C^2(\bar{\Omega})$, if they exist.

Theorem 2.5. *Assume hypotheses (2.24) to (2.26) and there exists a positive number B_0 such that*

$$(2.27) \qquad m(B_0) > h(C_0).$$

Then all positive solutions (u,v) of (2.23) must satisfy $0 \leq u \leq C_0, 0 \leq v \leq B_0$. Moreover:
(i) Suppose $M(0,0) \leq \lambda_1$, and $h(0) \leq \lambda_1 d + m(0)$, then (0,0) is the only non-negative solution of (2.23).
(ii) Suppose $h(0) < \lambda_1 d + m(0)$, then problem (2.23) has a positive solution (u,v) iff:

$$(2.28) \qquad M(0,0) > \lambda_1; \;\; \textit{and} \;\; \hat{\rho}_1(d\Delta + (h(u_0) - m(0))) > 0.$$

(iii) If $h(0) > \lambda_1 d + m(0)$, and $M(0,v) \geq M(u,v)$ for $u, v \geq 0$. Then (2.23) has a positive solution (u,v) iff

$$(2.29) \qquad M(0,0) > \lambda_1; \;\; \textit{and} \;\; \hat{\rho}_1(\Delta + M(0,v_0)) > 0.$$

(Note that the assumption on $h(0)$ in case (iii) already implies that the second inequality in (2.28) is true.) Furthermore, from the boundedness of the positive solution of (2.23), we have each component of the positive solution strictly positive in Ω, by Lemma 1.1.

Remark 2.2. In (ii) above, the function u_0 is the unique positive solution of $\Delta u + uM(u,0) = 0$ in Ω, $u = 0$ on $\partial\Omega$. Such solution exists provided $M(0,0) > \lambda_1$, by Lemma 2.1. In (iii) above, the function v_0 is the unique positive solution of $d\Delta v + v(h(0) - m(v)) = 0$ in Ω, $v = 0$ on $\partial\Omega$. Such solution exists provided $h(0) > \lambda_1 d + m(0)$, by Lemma 2.1.

Example 2.1. In the usual Volterra-Lotka model, we let

$$M(u,v) = \sigma_1^{-1}(a - bu - cv),\ d = \sigma_2,\ h(u) = e + fu,\ m(v) = gv, \tag{2.30}$$

where $\sigma_1, \sigma_2, a, b, c, f, g$ are positive constants and e is any constant, then Theorem 2.5 readily leads to the following corollary.

Corollary 2.6. *Consider problem (2.23) with $M(u,v), d, h(u), m(v)$ as given in (2.30).*
(i) If $a \le \sigma_1\lambda_1$, $e \le \lambda_1\sigma_2$, then (0,0) is the only non-negative solution of (2.23).
(ii) If $e < \sigma_2\lambda_1$, then problem (2,23) has a positive solution iff

$$a > \sigma_1\lambda_1 \ \textit{and}\ \hat{\rho}_1(\sigma_2\Delta + e + fu_0) > 0. \tag{2.31}$$

(iii) If $e > \sigma_2\lambda_1$, then (2.23) has a positive solution iff

$$\hat{\rho}_1(\sigma_1\Delta + a - cv_0) > 0. \tag{2.32}$$

(Note that part (ii) is closely related to Theorem 2.4(ii). When a is sufficiently large, the second inequality in (2.31) will be satisfied. Note also part (iii) above is closely related to Theorem 2.4(i) and Theorem 2.3(i); when e is large enough, (2.32) cannot hold.)

The following lemma is very useful for proving Theorem 2.5.

Lemma 2.4. *Assume $a(x) \in L^\infty(\Omega)$. Let u be an arbitrary function satisfying $u \ge 0, \not\equiv 0$ in Ω with $u = 0$ on $\partial\Omega$.*
(i) If $0 \not\equiv (\Delta + a(x))u \ge 0$ in Ω, then $\hat{\rho}_1(\Delta + a(x)) > 0$.
(ii) If $0 \not\equiv (\Delta + a(x))u \le 0$ in Ω, then $\hat{\rho}_1(\Delta + a(x)) < 0$.
(iii) If $(\Delta + a(x))u \equiv 0$ in Ω, then $\hat{\rho}_1(\Delta + a(x)) = 0$.

Proof. (i) Let $\theta(x) > 0$ in Ω be an eigenfunction corresponding to the principal eigenvalue $\bar{\rho} = \hat{\rho}_1(\Delta + a(x))$. Then $0 < \int_\Omega (\Delta + a(x))u\theta\, dx = \bar{\rho}\int_\Omega u\theta\, dx$, hence we must have $\bar{\rho} > 0$.
(ii) We simply reverse the sign in the argument in part (i) involving the integral.

(iii) We have $0 = \bar{\rho} \int_\Omega u\theta\, dx$ with $u \geq 0, u \not\equiv 0$, $\theta > 0$ in Ω. This implies that $\bar{\rho} = \rho_1(\Delta + a(x)) = 0$.

Proof (of Part (i) and necessity part of (ii), (iii) of Theorem 2.5 for the existence of positive solution). We first prove the existence of an a-priori bound for all non-negative solutions of (2.23). For any given $v \geq 0$ in $\bar{\Omega}$, we have by (2.24) and (2.25) a family of upper solutions $w \equiv \bar{C}$ in Ω, $\bar{C} \geq C_0$ for the first equation in (2.23), i.e.

$$\Delta w + wM(w, v) < 0 \text{ in } \Omega, \quad w \geq 0 \text{ on } \partial\Omega.$$

By the sweeping principle,any positive solution of problem (2.23) must have $0 \leq u \leq C_0$. Let $(\tilde{u}, \tilde{v})$ be a positive solution of (2.23). Suppose $x_0 \in \Omega$ such that $\tilde{v}(x_0) = max_{x \in \bar{\Omega}} \tilde{v}(x) > 0$. Then from the second equation in (2.23), $\tilde{v}(x_0)[h(\tilde{u}(x_0)) - m(\tilde{v}(x_0))] = -d\Delta\tilde{v}(x_0) \geq 0$ at the interior maximum point. Thus we must have

$$m(\tilde{v}(x_0)) \leq h(\tilde{u}(x_0)) \leq h(C_0). \tag{2.33}$$

By the increasing property of the function m, we must have $\tilde{v}(x_0) < B_0$.

We now prove the necessity assertion for part (ii) and (iii) of Theorem 2.5. Suppose $h(0) < \lambda_1 d + m(0)$, and $(\tilde{u}, \tilde{v})$ is a positive solution of (2.23). Since $\tilde{v} \geq 0$ in Ω, and $d\Delta\tilde{v} + \tilde{v}(h(\tilde{u}) - m(\tilde{v})) = 0$ in Ω, $\tilde{v} = 0$ on $\partial\Omega$, we can obtain by maximum principle that $\tilde{v}$ cannot have a nonpositive minimum in Ω. Consequently, we must have $\tilde{v} > 0$ in Ω (cf. Lemma 1.1). Next, consider the scalar problem: $\Delta w + wM(w, 0) = 0$ in Ω, $w = 0$ on $\partial\Omega$. The constant function $C_0 + \epsilon$ is a upper solution. On the other hand, the fact that $\Delta\tilde{u} + \tilde{u}M(\tilde{u}, 0) \geq \Delta\tilde{u} + \tilde{u}M(\tilde{u}, \tilde{v}) = 0$ in Ω, implies the function $\tilde{u}$ is a lower solution. We conclude that $\tilde{u} \leq u_0 \leq C_0 + \epsilon$. By Lemma 1.1, we have $u_0 > 0$ in Ω, and Lemma 2.1(i) implies $M(0, 0) > 0$. The fact that $u_0 \geq \tilde{u}$ implies that

$$d\Delta\tilde{v} + \tilde{v}(h(u_0) - m(0)) \geq d\Delta\tilde{v} + \tilde{v}(h(\tilde{u}) - m(\tilde{v})) = 0$$

in Ω. By Lemma 2.4(i), the above inequalities implies the second inequality in (2.28).

Next, suppose $h(0) > \lambda_1 d + m(0)$. Consider the scalar problem: $d\Delta w + w(h(0) - m(w)) = 0$ in Ω, $w = 0$ on $\partial\Omega$. We can verify readily as above that $\tilde{v}$ and $\delta\omega$ are respectively upper and lower solutions, where $\delta > 0$ is sufficiently small and moreover $\tilde{v} > \delta\omega$ in Ω. (Recall the definition of ω in (1.5).) The uniqueness of positive solution of this scalar problem leads to $\delta\omega \leq v_0 \leq \tilde{v}$ in Ω. Consequently, we have $\Delta\tilde{u} + \tilde{u}M(0, v_0) \geq \Delta\tilde{u} + \tilde{u}M(\tilde{u}, v_0) \geq \Delta\tilde{u} + \tilde{u}M(\tilde{u}, \tilde{v}) = 0$. By Lemma 2.4(i) again, we conclude that the second inequality of (2.29) is valid.

Before we begin to prove the sufficiency part of Theorem 2.5, we need to introduce some concepts in cone index method. Roughly speaking, we will apply the theory to compact operators on the cone of positive vector functions. Let E be a Banach space, $W \subset E$ is called a wedge in E if W is a closed convex set and $\alpha W \subseteq W$ for every real $\alpha \geq 0$. A wedge is called a cone if $W \cap \{-W\} = \{0\}$. For $y \in W$, define

$$W_y = \{x \in E : y + \theta x \in W, \text{ for } 0 \leq \theta \leq \gamma, \text{ for some } \gamma > 0\}.$$

One readily verifies that W_y is convex, and $W_y \supseteq \{y\} \cup \{-y\} \cup \{W\}$. Moreover, the set $\bar{W}_y$,(the closure of W_y), is also a wedge. Let $S_y = \{x \in \bar{W}_y : -x \in \bar{W}_y\}$; we easily see that S_y is a linear subspace of E.

A nonempty subset A of a metric space X is called a retract of X if there exists a continuous map $r : X \to A$ (called a retraction), such that $r|A = id_A$. By a theorem of Dugundji [53], [54], every nonempty closed convex subset of a Banach space E is a retract of E. Let X be a retract of a Banach space E. For every open subset U of X and every compact map $f : \bar{U} \to X$ which has no fixed points on ∂U, there exists an integer $i_X(f, U)$ defined by

$$i_X(f, U) = i_E(f \circ r, r^{-1}(U)) = deg(id - f \circ r, r^{-1}(U), 0),$$

where $i_E(f \circ r, r^{-1}(U))$ is the well-known Leray-Schauder degree. This definition is independent of the choice of the retraction. The integer $i_X(f, U)$ is called the fixed point index of f (over U with respect to X). This index satisfies the normalization, additivity, homotopic invariance and permanence properties as the Leray-Schauder degree (cf. Theorem A2-1 in Chapter 6). If $x_0 \in U$ is an isolated fixed point of f, and x_0 is the only fixed point of f in $x_0 + \rho B, \rho > 0$, where B is the open unit ball of E. We define the fixed point index by

$$index_X(f, x_0) := i_X(f, x_0 + \rho B).$$

Definition 2.1. *Let $L : E \to E$ be a compact linear operator such that $L(\bar{W}_y) \subseteq \bar{W}_y$. L is said to have property (α) on $\bar{W}_y$ if the following holds:*
(α) There exists $t \in (0, 1)$ and a $w \in \bar{W}_y \backslash S_y$ such that $w - tLw \in S_y$.

Remark 2.3. If $I - L$ is invertible in $\bar{W}_y$, an important consequence of property (α) on $\bar{W}_y$ is given below in Lemma 2.5, asserting that there exists $z \in \bar{W}_y$ such that the equation $x - Lx = z$ has no solution for x in $\bar{W}_y$. Under appropriate circumstances, this in turn leads to $index_W(A, y_0) = 0$, by Lemma 2.6(i), where L is the Fréchet derivative of A at y_0 in W.

Lemma 2.5. *Let $L : E \to E$ be a compact linear operator such that $L(\bar{W}_y) \subseteq \bar{W}_y$. Assume $I - L$ is invertible in $\bar{W}_y$ (in the sense that $h \neq Lh$ if $h \in \bar{W}_y \backslash \{0\}$). If L has property ($\alpha$) on $\bar{W}_y$, then there exists $z \in \bar{W}_y$ such that the equation $x - Lx = z$ has no solution for x in $\bar{W}_y$.*

Proof. Since L has property (α), there exists $v \in \bar{W}_y \backslash S_y$ and $t \in (0,1)$ such that $v - tLv = h \in S_y$. Thus $-v \notin \bar{W}_y, -h \in S_y$ and $-v - L(-v) = -v + Lv = -v + tLv + (1-t)Lv = -h + (1-t)Lv \in \bar{W}_y$. Let $z := -h + (1-t)Lv = -(v - Lv)$. If there exists $q \in \bar{W}_y$ such that $q - Lq = z$. Then we have $v + q - L(v+q) = v - Lv + q - Lq = -z + z = 0$. This implies $v + q = 0$, and $-v = q \in \bar{W}_y$. Thus we have $v \in S_y$, which is a contradiction.

Lemma 2.6. *Let W be a wedge in Banach space E, $E_W := W - W$ is dense in E, and D is an open set in W. Suppose that $A : \bar{D} \to W$ is a compact map with fixed point $y_0 = Ay_0 \in D$, and the Fréchet derivative of A at y_0 in W, denoted by $L = A'_+(y_0)$, is compact on E. Then L maps $\bar{W}_{y_0}$ into itself. Moreover:*
(i) Assume that $I - L$ is invertible in $\bar{W}_{y_0}$(in the sense that $h \neq Lh$ if $h \in \bar{W}_{y_0} \backslash \{0\}$). If there exists an element $z \in \bar{W}_{y_0}$ such that the equation $x - Lx = z$ has no solution for x in $\bar{W}_{y_0}$, then $index_W(A, y_0) = 0$.
(ii) Assume $I - L$ is invertible in $\bar{W}_{y_0}$. If L does not have property (α) on $\bar{W}_{y_0}$, then $index_W(A, y_0) = index_{S_{y_0}}(L, 0) = (-1)^{\sigma(y_0)} = \pm 1$. Here, $\sigma(y_0)$ is the sum of multiplicities of the eigenvalues of L in S_{y_0} which are greater than one.

Proof. The proof can be found in Dancer [37] and Li [148]. More explanations are found in Remark 2.1(i) and (ii) in Ruan and Feng [194]. Remark 2.1(ii) in [194] is same as Theorem A2-3 in Chapter 6 (Appendices).

From Lemmas 2.5 and 2.6, we obtain the following lemma.

Lemma 2.7. *Under the hypotheses of Lemma 2.6, let $I - L$ be invertible on $\bar{W}_{y_0}$ as described in Lemma 2.6.*
(i) If L has property (α) on $\bar{W}_{y_0}$, then $index_W(A, y_0) = 0$.
(ii) If L does not have property (α) on $\bar{W}_{y_0}$, then $index_W(A, y_0) = index_{S_{y_0}}(L, 0)$ $= \pm 1$.

We are now ready to prove the sufficiency part (ii) and (iii) of Theorem 2.5. Let $[C_0(\bar{\Omega})]^2 := \{(u_1, u_2) : u_i \in C(\bar{\Omega}),\ and\ u_i = 0\ on\ \partial\Omega,\ \ for\ i = 1, 2\}$. Let $[C_0^+(\bar{\Omega})]^2 := \{(u_1, u_2) : u_i \in C_0(\bar{\Omega}), u_i \geq 0\ in\ \Omega,\ for\ i = 1, 2\}$, $B_1 = max.\{C_0, B_0\}$, and $[E(B_1)]^2 := \{(u_1, u_2) : u_i \in C(\bar{\Omega}),\ |u_i| < B_1\ in\ \Omega,\ for\ i = 1, 2\}$, with closure $[\bar{E}(B_1)]^2$. For each $(u_1, u_2) \in [C(\bar{\Omega})]^2, \theta \in [0, 1]$, define the operator $A_\theta : [C_0(\bar{\Omega})]^2 \cap [\bar{E}(B_1)]^2 \to [C_0(\bar{\Omega})]^2$ by $A_\theta(u_1, u_2) = (v_1, v_2)$ where

$$\begin{cases} v_1 = (-\Delta + P)^{-1}[\theta u_1 M(u_1, u_2) + Pu_1] \\ v_2 = (-\Delta + P)^{-1}[\theta d^{-1} u_2(h(u_1) - m(u_2)) + Pu_2]. \end{cases} \tag{2.34}$$

Here, the inverse operator is taken with homogeneous Dirichlet boundary condition on $\partial\Omega$, and $P > 0$ is a large enough constant such that the operator A_θ is positive, compact and Fréchet differentiable on $[C_0^+(\bar{\Omega})]^2 \cap [\bar{E}(B_1)]^2$. For convenience, let K denotes the cone $K := [C_0^+(\bar{\Omega})]^2$, as described above, and

$D := [C_0^+(\bar{\Omega})]^2 \cap [E(B_1)]^2$. The bound on the solution implies that the operators A_θ has no fixed point on the boundary ∂D in the relative topology, i.e. on the intersection of boundary of $[E(B_1)]^2$ with K. We can further use a familiar cut-off procedure (see Li [148]) to extend A_θ to be defined outside D as a compact positive mapping from the cone K into itself. For convenience, we will denote $A := A_1$. We will denote the fixed point index of A_θ over D with respect to the cone K by $i_K(A_\theta, D)$. By homotopy invariance principle, we obtain $i_K(A, D) = i_K(A_1, D) = i_K(A_0, D)$. From definition, the i-th component of $A_0(u_1, u_2)$ is $(-\Delta + P)^{-1}(Pu_i)$. One readily verifies by maximum principle that $A_0(u) \neq \lambda u$ for every $u = (u_1, u_2) \in \partial D$ and $\lambda \geq 1$. Hence, by Theorem A2-4 in Chapter 6 (Appendices), we conclude by contraction argument that $i_K(A, D) = i_K(A_0, D) = 1$.

Let y be an isolated fixed point of the map A_θ in K, we denote the local index of A_θ at y with respect to K by $index_K(A_\theta, y)$.We now show that $index_K(A, (0,0)) = 0$ for both cases (ii) and (iii). For $y \in K$, define

$$K_y := \{p \in [C(\bar{\Omega})]^2 : y + sp \in K \text{ for some } s > 0\}, \text{ and}$$

$$S_y := \{p \in \bar{K}_y : -p \in \bar{K}_y\}.$$

Here $\bar{K}_y$ denotes the closure of K_y. We have $\bar{K}_{(0,0)} = K, S_{(0,0)} = \{(0,0)\}$. Let $A'_+((0,0))$ be the Fréchet derivative of A at $(0,0)$ in K. The first component of $A'_+((0,0))(u_1, u_2)$ is $(-\Delta + P)^{-1}(M(0,0) + P)u_1$. Hence $[I - A'_+((0,0))]u = 0$ for $u = (u_1, u_2) \in K$ implies that $[\Delta + M(0,0)]u_1 = 0$, $u_1 \in C_0^+(\bar{\Omega})$. Thus the assumption $M(0,0) > \lambda_1$ in (2.28) or (2.29) implies that $u_1 = 0$. Similarly, we have for the second component $[d_1\Delta + h(0) - m(0)]u_2 = 0$, $u_2 \in C_0^+(\bar{\Omega})$. Thus the assumption $h(0) \neq \lambda_1 d + m(0)$ in (2.28) or (2.29) implies that $u_2 = 0$. We thus conclude $I - A'_+((0,0))$ is invertible in $\bar{K}_{(0,0)}$. Further, the assumption $M(0,0) > \lambda_1$ implies that $\hat{\rho}_1(\Delta + tM(0,0) + (t-1)P)$ is positive when $t = 1$ and negative when $t = 0$. From the continuity in $t \in [0,1]$ for the eigenvalue $\hat{\rho}_1(\Delta + tM(0,0) + (t-1)P)$, there must exist some $t \in (0,1)$ and a nontrivial function $\bar{u} \in C_0^+(\bar{\Omega})$ such that $(-\Delta + P)\bar{u} = t(M(0,0) + P)\bar{u}$ or $\bar{u} - t(-\Delta + P)^{-1}(M(0,0) + P)\bar{u} = 0$ in Ω. We thus have $[I - tA'_+((0,0))](\bar{u}, 0) = (0,0) \in S_{(0,0)}$, with $(\bar{u}, 0) \in \bar{K}_{(0,0)} \backslash S_{(0,0)}$. We thus conclude by Lemma 2.7(i), with $\bar{W}_{y_0}$ replaced by $\bar{K}_{(0,0)}$, that $index_K(A, (0,0)) = 0$.

We will show that for both cases (ii) and (iii), we have $index_K(A, (u_0, 0)) = index_K(A, (0, v_0)) = 0$.

Consider case (ii) when $h(0) < \lambda_1 d + m(0)$, and assume (2.28). Let $L = A'_+((u_0, 0))$ be the Fréchet derivative of A at $(u_0, 0)$ in K. Suppose that

$(I-L)(u_1,u_2)=0$, for some $u_1 \geq 0, u_2 \geq 0$ in $\bar{\Omega}$, and $u_1=u_2=0$ on $\partial\Omega$. Then

$$(2.35) \quad \begin{cases} \Delta u_1 + [M(u_0,0)+u_0M_u(u_0,0)]u_1 + u_0M_v(u_0,0)u_2 = 0 & \\ & \text{in } \Omega, \\ d\Delta u_2 + [h(u_0)-m(0)]u_2 = 0 & \\ u_1 = u_2 = 0 & \text{on } \partial\Omega. \end{cases}$$

Thus the second assumption in (2.28) and the second equation above imply that $u_2 \equiv 0$. We then consider the first equation above again. Since $\hat{\rho}_1(\Delta + M(u_0,0)) = 0$, and $u_0M_u(u_0,0) < 0$ by (2.24), we have $\hat{\rho}_1(\Delta + M(u_0,0) + u_0M_u(u_0,0)) < 0$. Hence, all the eigenvalues ρ of the problem:

$$\Delta u + [M(u_0,0)+u_0M_u(u_0,0)]u = \rho u \text{ in } \Omega, \ u=0 \text{ on } \partial\Omega,$$

satisfy $\rho < 0$. However, u_1 satisfies this problem with $\rho = 0$. Thus $u_1 \equiv 0$. That is the operator $(I-L)$ is invertible on $\bar{K}_{(u_0,0)}$.

We next show that the operator L has property (α) on $\bar{K}_{(u_0,0)}$. Let $P>0$; observe that the eigenvalue $\hat{\rho}_1(d\Delta - dP + t[h(u_0)-m(0)+dP])$ is negative when $t=0$, and is positive when $t=1$. By continuity, there exists some $t^* \in (0,1)$, such that $\hat{\rho}_1(d\Delta - dP - t^*[h(u_0)-m(0)+dP]) = 0$. There exists $u_2^* > 0$ in Ω, vanishing on $\partial\Omega$ such that $(-d\Delta + dP)u_2^* - t^*[h(u_0)-m(0)+dP]u_2^* = 0$ in Ω. Since $S_{(u_0,0)} = C_0(\bar{\Omega}) \times \{0\}$, we can readily verify that if we let $w=(0,u_2^*)$, then we have $w - t^*Lw \in S_{(u_0,0)}$ with $w \in \bar{K}_{(u_0,0)} \backslash S_{(u_0,0)}$. Consequently, by Lemma 2.7(i), with $\bar{W}_{y_0}$ replaced by $\bar{K}_{(u_0,0)}$, we conclude that $index_K(A,(u_0,0)) = 0$.

Next, consider the case (iii) when $h(0) > \lambda_1 d + m(0)$, and assume (2.29). Let $L = A'_+((u_0,0))$ and (u_1,u_2) be as described above for case (ii), and thus obtain (2.35) again. The assumption $h(0) > \lambda_1 d + m(0)$ and increasing property of the function h imply the validity of the second assumption in (2.28). Thus we obtain $u_2 \equiv 0$ as before. We then follow the same argument as before to conclude that $I-L$ is invertible in $\bar{K}_{(u_0,0)}$. We then prove the operator L has property (α) on $\bar{K}_{(u_0,0)}$ exactly as in case (ii) above. Thus we conclude that $index_K(A,(u_0,0)) = 0$.

We now consider the point $(0,v_0)$. For case (ii), that is, $h(0) < \lambda_1 d + m(0)$, we must have $v_0 \equiv 0$. Thus $(0,v_0) = (0,0)$, and the index of A at this fixed point has been shown to be 0.

For case (iii) when $h(0) > \lambda_1 d + m(0)$, let $\tilde{L} = A'_+((0,v_0))$ be the Fréchet derivative of A at $(0,v_0)$.Suppose that $(I-\tilde{L})(u_1,u_2) = 0$, for some $u_1 \geq 0, u_2 \geq 0$ in $\bar{\Omega}$, and $u_1 = u_2 = 0$ on $\partial\Omega$. Then

$$\begin{cases} \Delta u_1 + M(0,v_0)u_1 = 0 & \\ & \text{in } \Omega, \\ d\Delta u_2 + v_0h'(0)u_1 + [h(0) - v_0m'(v_0) - m(v_0)]u_2 = 0 & \\ u_1 = u_2 = 0 & \text{on } \partial\Omega. \end{cases}$$

The second assumption in (2.29) implies that $\hat{\rho}_1(\Delta + M(0, v_0)) \neq 0$; thus we conclude from the first equation above that $u_1 \equiv 0$. Since $\hat{\rho}_1(d\Delta + h(0) - m(v_0)) = 0$ and $m'(v_0) \geq 0$ is not the trivial function, we have $\hat{\rho}_1(d\Delta + h(0) - v_0 m'(v_0) - m(v_0)) < 0$. We thus deduce from the second equation above that $u_2 \equiv 0$. We thus conclude that the operator $I - \tilde{L}$ is invertible in $\bar{K}_{(0,v_0)}$.

From the second assumption in (2.29), we deduce that for $P > 0$, the eigenvalue $\hat{\rho}_1(\Delta - P + t[M(0, v_0) + P])$ is negative if $t = 0$ and is positive if $t = 1$. Hence, there exist $t^* \in (0, 1)$ and a non-trivial, non-negative function u_1^* vanishing on $\partial\Omega$, such that

$$-\Delta u_1^* + P u_1^* - t^*(M(0, v_0) + P)u_1^* = 0 \text{ in } \Omega.$$

Since $S_{(0,v_0)} = \{0\} \times C_0(\bar{\Omega})$, we can readily verify that if we let $\tilde{w} = (u_1^*, 0)$, then we have $\tilde{w} - t^* \tilde{L}\tilde{w} \in S_{(0,v_0)}$ with $\tilde{w} \in \bar{K}_{(0,v_0)} \backslash S_{(0,v_0)}$. Consequently, by Lemma 2.7(i), we conclude that $index_K(A, (0, v_0)) = 0$.

From the above paragraphs, we thus have

$$i_K(A, D) = 1, \ index_K(A, (0, 0)) = index_K(A, (u_0, 0)) = index_K(A, (0, v_0)) = 0.$$

If one component of a solution of (2.23) in K is identically zero, there are at most three solutions $(0, 0)$, $(u_0, 0)$ and $(0, v_0)$ in K. In order to avoid contradicting the additive property of the indices of the map on disjoint open subsets, there must be at least another fixed point of A in D. (See Theorem A2-1(ii) in Chapter 6.) This complete the proof of Theorem 2.5(ii) and (iii).

In some interesting applications, the predator v may have no crowding effect on its own growth rates. This lead to the following theorem.

Theorem 2.7. *Let $N = 1, 2$ or 3. Assume hypotheses (2.24) to (2.26) except that here $m \equiv 0$. Moreover*

$$h(0) < 0; \textit{ and there exists } \theta > 0, \textit{ such that } M_v(u, v) < -\theta \textit{ for } 0 \leq u \leq C_0, v > 0.$$

Then all positive solutions (u, v) of (2.23) must satisfy $0 \leq u \leq B_1, 0 \leq v \leq B_2$, for some positive constants B_1, B_2. Moreover

(i) If $M(0, 0) \leq \lambda_1$, then (0,0) is the only non-negative solution of (2.23).

(ii) Problem (2.23) has a positive solution iff

$$M(0, 0) > \lambda_1; \textit{ and } \hat{\rho}_1(d\Delta + h(u_0)) > 0. \tag{2.36}$$

Proof. We first prove the existence of an a-priori bound for all non-negative solutions of (2.23). For any given $v \geq 0$ in $\bar{\Omega}$, we have by (2.24) and (2.25) a family of upper solutions $w \equiv \bar{C}$ in Ω, $\bar{C} \geq C_0$ for the first equation in (2.23), i.e.

$$\Delta w + wM(w, v) < 0 \text{ in } \Omega, \;\; w \geq 0 \text{ on } \partial\Omega.$$

By the sweeping principle, any positive solution of problem (2.23) must have $0 \leq u \leq C_0$. Suppose there is no a-priori bound for v. Then there exists a sequence of positive solutions (u_n, v_n) for (2.23) satisfying:

$$||v_n||_{L^\infty} \to \infty, \;\; \text{as } n \to \infty, \;\; 0 \leq u_n \leq C_0 \text{ in } \bar{\Omega}.$$

Let $\bar{v}_n = v_n/||v_n||_{L^\infty}$; it satisfies $0 \leq \bar{v}_n \leq 1$ and:

$$d\Delta \bar{v}_n + h(0)\bar{v}_n = -[h(u_n) - h(0)]\bar{v}_n \text{ in } \Omega. \tag{2.37}$$

By $W^{2,p}(\Omega)$ estimates and appropriate embedding, we can find subsequence, again denoted as $\{\bar{v}_n\}$ such that $\bar{v}_n \to \bar{v}_0 \in C^{1,\alpha}(\bar{\Omega})$ uniformly for some $\alpha \in (0,1)$, and $v_0(x) \geq 0, \not\equiv 0$ in $\bar{\Omega}$.

Next, let $\tilde{u}_n = u_n/||u_n||_{L^2} \geq 0$ in Ω. Divide the first equation satisfied by (u_n, v_n), multiply by $\tilde{u_n}$, and integrate over Ω, we obtain:

$$-M(0,0) < \int_\Omega |\nabla \tilde{u}_n|^2 dx - \int_\Omega M(0,0)\tilde{u}_n^2 dx = \int_\Omega [M(u_n, v_n) - M(0,0)]\tilde{u}_n^2 dx \leq 0. \tag{2.38}$$

The last inequality above is due to assumption (2.24) on M_v, M_u. From (2.38), we obtain a uniform bound on the $W^{1,2}(\Omega)$ norm on $\tilde{u_n}$. We can select subsequence, denoted the same way, such that $\tilde{u_n}$ converge weakly in $W^{1,2}(\Omega)$ and strongly in $L^q(\Omega)$, $q < 2N(N-2)^{-1}$ to a non-negative function $\tilde{u}_0 \in W^{1,2}(\Omega)$, if $N > 2$ (by Rellich-Kondrachov Compactness Theorem, see e.g. p. 272 in Evans [57]). If the space dimension $N = 3$, the inequality $q < 2N(N-2)^{-1}$ is satisfied if we choose $q = N = 3$. If the space dimension $N = 2$, using $pN(N-p)^{-1} \to \infty$ as $p \to N^-$, we can also assume $\tilde{u}_n$ converge to $\tilde{u}_0$ in $L^q(\Omega), q = 2$. We also have $||\tilde{u}_0||_{L^2} = lim_{n\to\infty}||\tilde{u}_n||_{L^2} = 1$, and we may assume $||u_n||_{L^2} \to k \geq 0$, as $n \to \infty$. Taking limit in (2.37), and using $W^{2,q}$ theory, we find $\bar{v}_0 \in W_0^{2,q} = W_0^{2,N}$ is a strong solution of

$$d\Delta \bar{v}_0 + h(0)\bar{v}_0 = -[h(k\tilde{u}_0) - h(0)]\bar{v}_0 \tag{2.39}$$

for $N = 3$ or 2. Moreover the increasing property of $h(u)$ and equation (2.37) implies that

$$d\Delta \bar{v}_0 + h(0)\bar{v}_0 \leq 0 \;\; \text{in } \Omega. \tag{2.40}$$

Since $h(0) < 0$, we obtain by maximum principle for the strong solution that $\bar{v}_0(x) > 0$ for all $x \in \Omega$. (See e.g. Gilbarg and Trudinger [71] or Theorem A3-1 in Chapter 6.) For the case $N = 1$, the convergence on the right of (2.37) to

(2.39) is valid in $C^{0,\gamma}(\bar{\Omega}), \gamma = 1 - \frac{N}{2}$, by Morrey's inequality (see e.g. p. 266 in [57]). Thus the solution $\bar{v}_0$ of (2.39) is in $C^2(\bar{\Omega})$ by Schauder's theory. Therefore we can also conclude that $\bar{v}_0(x) > 0$ for all $x \in \Omega$ by means of (2.40).

Consider the integral on the right side of (2.38). We use (2.24) and the assumption concerning M_v in the statement of Theorem 2.7 to obtain

$$\begin{aligned} \int_\Omega [M(u_n, v_n) - M(0,0)]\tilde{u}_n^2\,dx &\leq \int_\Omega -\theta v_n \tilde{u}_n^2\,dx \\ &= -\theta ||v_n||_{L^\infty} \int_\Omega \bar{v}_n \tilde{u}_n^2\,dx. \end{aligned} \tag{2.41}$$

However, we have

$$\int_\Omega \bar{v}_n \tilde{u}_n^2\,dx \to \int_\Omega \bar{v}_0 \tilde{u}_0^2\,dx \ > 0, \text{ as } n \to \infty. \tag{2.42}$$

Taking limit as $n \to \infty$ in (2.38) and using (2.41) and (2.42), we obtain the contradiction $-M(0,0) \leq -\infty$ if $||v_n||_{L^\infty} \to \infty$. Consequently, we must have an a-priori bound for all positive solutions of problem (2.23).

Parts (i) and (ii) of this Theorem follow readily from parts (i) and (ii) of Theorem 2.5 respectively, with the role of C_0 and B_0 respectively replaced by B_1 and B_2.

Example 2.2. Let $M(u,v) = \sigma_1^{-1}(a - bu - cv)$, $d = \sigma_2$, $h(u) = u - \gamma. m \equiv 0$, where $\sigma_1, \sigma_2, a, b, c, \gamma$ are positive constants. Note that $h(0) = -\gamma < 0$. This is a very common model, when the predator has negative intrinsic growth rate, and there is no crowding effect of the population of predator on itself. Here, we can apply Theorem 2.7. The conditions in (2.36) becomes

$$a > \sigma_1 \lambda_1; \ \text{ and } \ \hat{\rho}_1(\sigma_2 \Delta + u_0 - \gamma) > 0. \tag{2.43}$$

Example 2.3. We can also apply Theorem 2.7 to models not of Volterra-Lotka type reaction. The popular Holling's type of growth rate assumption may be assumed. Let $M(u,v) = \sigma_1^{-1}(a - bu - \frac{vc}{\delta+u})$, $d = \sigma_2$, $h(u) = ku\frac{c}{\delta+u} - \gamma$, $m \equiv 0$, where $\sigma_1, \sigma_2, a, b, c, d,\ k, \gamma, \delta$ are positive constants.

Part C: Coexistence Regions in Parameter Space.

We now return to the diffusive Volterra-Lotka model, and describe a region on the (a, e) intrinsic growth rate parameter plane when positive solutions always exist while the other parameters are fixed. This leads to Theorem 2.8, Fig. 1.2.1 and Fig. 1.2.2. More precisely, consider problem (2.1) with $\sigma_1 = \sigma_2 = 1$ for simplicity. Let the parameters b, c, f and g be fixed positive constants. We can readily use Theorem 2.5(i) and (ii) to obtain a region in the (a, e) plane between certain lines or curves so that positive solutions will always exist. Here, we have

(2.23) with $\sigma_1 = \sigma_2 = d = 1$. The conditions $h(0) < \lambda_1 d + m(0)$ and (2.28) in Theorem 2.5(ii) are the same as

$$\lambda_1 > e > \rho_1(-\Delta - fu_0), \quad a > \rho_1(-\Delta + cv_0), \quad v_0 \equiv 0. \tag{2.44}$$

The conditions $h(0) > \lambda_1 d + m(0)$ and (2.29) in Theorem 2.5(iii) are the same as

$$e > \lambda_1 := \rho_1(-\Delta) > \rho_1(-\Delta - fu_0), \quad a > \rho_1(-\Delta + cv_0). \tag{2.45}$$

Consequently, if $\sigma_1 = \sigma_2 = 1$, $a \neq \lambda_1$, $e \neq \lambda_1$, and

$$a > \rho_1(-\Delta + cv_0), \quad e > \rho_1(-\Delta - fu_0), \tag{2.46}$$

then Theorem 2.5(ii) and (iii) imply that problem (2.1) has a solution with each component strictly positive in Ω. We next use (2.44) to (2.46) and the characterization of principal eigenvalue to obtain very simple description, in terms of the interaction parameters and the size of Ω, of a region on the (a, e) plane where positive solutions always exist. By avoiding the reference to u_0 and v_0, the description is easier to use.

Theorem 2.8. *Consider the boundary value problem (2.1), under the assumptions $\sigma_1 = \sigma_2 = 1$, $a \neq \lambda_1$, $e \neq \lambda_1$,*
(i) Suppose $c \geq g$ and

$$a > \lambda_1 + (e - \lambda_1)\frac{c}{g}, \quad e > \lambda_1 - (a - \lambda_1)\frac{f}{b}. \tag{2.47}$$

Then (2.1) has a solution with each component strictly positive in Ω.
(ii) Suppose $c < g$ and

$$\begin{cases} e > \lambda_1, \quad a > min\{\lambda_1 + \frac{ec}{g}, e[1 - (1 - \frac{c}{g})(1 - \frac{\lambda_1}{e})^3 K]\}; \quad or \\ \lambda_1 > e > \lambda_1 - (a - \lambda_1)\frac{f}{b}, \quad a > \lambda_1, \end{cases} \tag{2.48}$$

where $K := |\Omega|^{-1} \int_\Omega \phi^3 dx$, $|\Omega| =$ measure of Ω and ϕ is the positive principal eigenfunction of $-\Delta$ with $max_\Omega \phi = 1$. Then problem (2.1) has a positive solution with each component strictly positive in Ω.

Proof. (i) Suppose $c \geq g$. First, assume $e > \lambda_1$. We now show that in this case

$$\lambda_1 + (e - \lambda_1)\frac{c}{g} \geq \rho_1(-\Delta + cv_0). \tag{2.49}$$

By the characterization of principal eigenvalue, we have

$$\rho_1(-\Delta + cv_0) = inf_{u \in S}\{\int_\Omega |\nabla u|^2 dx + cv_0 u^2 dx\},$$

where $S := \{u \in H_0^1(\Omega),\ \int_\Omega u^2 dx = 1\}$. Consequently,

$$\rho_1(-\Delta + cv_0) \le ||v_0||^{-2}_{L^2(\Omega)}\{\int_\Omega |\nabla v_0|^2 dx + c\int_\Omega v_0^3 dx\}. \tag{2.50}$$

From the equation satisfied by v_0, we have

$$g\int_\Omega v_0^3 dx = e\int_\Omega v_0^2 dx - \int_\Omega |\nabla v_0|^2 dx. \tag{2.51}$$

From (2.50) and (2.51), we obtain

$$\rho_1(-\Delta + cv_0) \le ||v_0||^{-2}_{L^2(\Omega)}\{(1-\frac{c}{g})\int_\Omega |\nabla v_0|^2 dx + e\frac{c}{g}\int_\Omega v_0^2 dx\}. \tag{2.52}$$

Since $1 - \frac{c}{g} \le 0$, we obtain (2.49) as follows:

$$\rho_1(-\Delta + cv_0) \le (1-\frac{c}{g})\lambda_1 + e\frac{c}{g} = \lambda_1 + (e-\lambda_1)\frac{c}{g}.$$

Thus by (2.45) and (2.49), we find that strictly positive solutions must exist for (a, e) satisfying:

$$e > \lambda_1 \text{ and } a > \lambda_1 + (e-\lambda_1)\frac{c}{g}. \tag{2.53}$$

Next, let $e < \lambda_1$. We obtain from Theorem 2.5(ii) above that (2.44) is sufficient for existence of positive solutions. By the characterization of principal eigenvalue, we find

(2.54)

$$\begin{aligned}\rho_1(-\Delta - fu_0) &= inf_{u\in S}\{(1+\tfrac{f}{b})\textstyle\int_\Omega |\nabla u|^2 dx - \tfrac{f}{b}[\int_\Omega |\nabla u|^2 dx + b\int_\Omega u_0 u^2 dx]\} \\ &\le inf_{u\in S}\{(1+\tfrac{f}{b})\textstyle\int_\Omega |\nabla u|^2 dx\} - \tfrac{f}{b}\, inf_{u\in S}\{\int_\Omega |\nabla u|^2 dx + b\int_\Omega u_0 u^2 dx\} \\ &= (1+\tfrac{f}{b})\lambda_1 - \tfrac{f}{b}\rho_1(-\Delta + bu_0) \\ &= (1+\tfrac{f}{b})\lambda_1 - \tfrac{f}{b}a = \lambda_1 - \tfrac{f}{b}(a-\lambda_1).\end{aligned}$$

Thus by (2.44) and (2.54), if $e < \lambda_1$, we find that strictly positive solutions must exist for (a, e) satisfying

$$\lambda_1 > e > \lambda_1 - \frac{f}{b}(a-\lambda_1),\ \ a > \lambda_1. \tag{2.55}$$

The assertion of part (i) follows from (2.53) and (2.55). (See Fig. 1.2.1.)

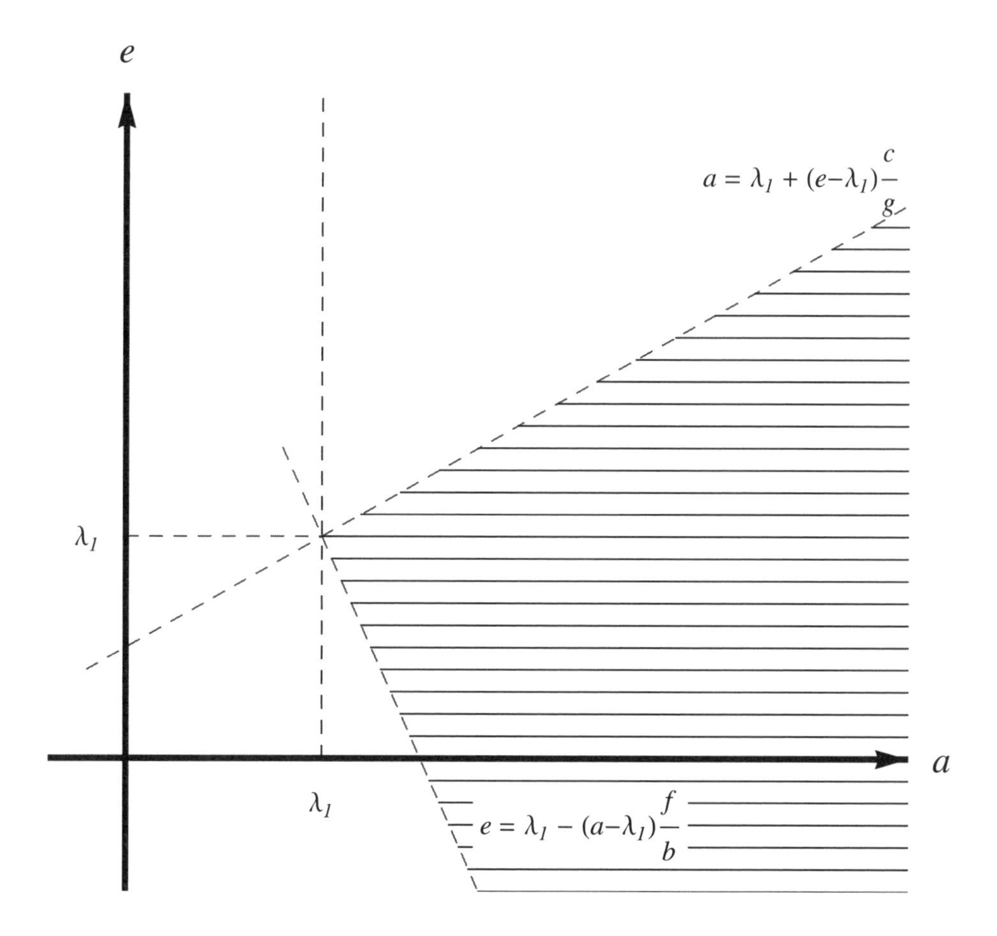

Figure 1.2.1: Coexistence Region in (a, e) Parameter Space, for case $c \geq g$.

We next consider part (ii) and assume $c < g$. Suppose $e > \lambda_1$, then Theorem 2.5(iii) asserts that if a satisfies (2.45) then problem (2.1) has positive solutions. It suffices to show that

$$\rho_1(-\Delta + cv_0) \leq min\{\lambda_1 + \frac{ec}{g},\ e[1 - (1 - \frac{c}{g})(1 - \frac{\lambda_1}{e})^3 K]\}. \tag{2.56}$$

By the characterization of the principal eigenvalue and the fact that $v_0 \leq e/g$ we find

$$\begin{aligned} \rho_1(-\Delta + cv_0) &= inf_{u \in S}\{\int_\Omega |\nabla u|^2 dx + \int_\Omega cv_0 u^2 dx\} \\ &\leq inf_{u \in S}\{\int_\Omega |\nabla u|^2 dx + \tfrac{ec}{g} \int_\Omega u^2 dx\} \\ &= \lambda_1 + \tfrac{ec}{g}. \end{aligned} \tag{2.57}$$

Next, observe that (2.52) remains valid in the present case $c < g$, i.e. $1 - cg^{-1} > 0$. Thus using (2.51) and (2.52) we obtain

$$\rho_1(-\Delta + cv_0) \leq [1 - \frac{c}{g}][e - (g\int_\Omega v_0^3 dx)(\int_\Omega v_0^2 dx)^{-1}] + \frac{ec}{g}. \tag{2.58}$$

Using the fact that $g^{-1}(e - \lambda_1)\phi$ is a lower solution for the problem satisfied by v_0, we have

$$(\frac{e - \lambda_1}{g})^3 \int_\Omega \phi^3 dx \leq \int_\Omega v_0^3 dx. \tag{2.59}$$

From $v_0 \leq eg^{-1}$, we also have

$$\int_\Omega v_0^2\, dx \leq (\frac{e}{g})^2 |\Omega|. \tag{2.60}$$

Letting $K = |\Omega|^{-1} \int_\Omega \phi^3 dx$, we deduce readily from (2.58) to (2.60) that

$$\begin{aligned} \rho_1(-\Delta + cv_0) &\leq [1 - \tfrac{c}{g}][e - (e - \lambda_1)^3 e^{-2} K] + \tfrac{ec}{g} \\ &= e[1 - (1 - \tfrac{c}{g})(1 - \tfrac{\lambda_1}{e})^3 K]. \end{aligned} \tag{2.61}$$

Thus, from (2.57) and (2.61), we conclude that if (a, e) satisfy the first line of the inequalities in (2.48), there must exist positive solutions to problem (2.1).

Next, assume $e < \lambda_1$, we obtain the inequalities in the second line of (2.48) as sufficient condition for the existence for positive solution of (2.1) in exactly the same way as obtaining the second inequality of (2.47) as sufficient condition in part (i).

Remark 2.4. If we define

$$\hat{a}(e) = e[1 - (1 - \frac{c}{g})(1 - \frac{\lambda_1}{e})^3 K],$$

where K is defined in Theorem 2.8. It can be shown by calculus that
(1) The graph of $(\hat{a}(e), e)$ and $(\lambda_1 + (e - \lambda_1)\frac{c}{g}, e)$ do not intersect when $e > \lambda_1$.
(2) The graphs of $(\hat{a}(e), e)$ and $(\lambda_1 + \frac{ec}{g}, e)$ intersect at one point $(a_0(\frac{c}{g}), e_0(\frac{c}{g}))$, when $e > \lambda_1$, and

$$\lim_{c/g \to 1} a_0(\frac{c}{g}) = \infty, \quad \lim_{c/g \to 1} e_0(\frac{c}{g}) = \infty;$$

$$\lim_{c/g \to 0} a_0(\frac{c}{g}) = \lambda_1, \quad \lim_{c/g \to 0} e_0(\frac{c}{g}) = \lambda_1.$$

(3) $\lim_{c/g \to 1} [\hat{a}(e) - (\lambda_1 + (e - \lambda_1)\frac{c}{g})] = 0$ uniformly on compact subsets of $e \in [\lambda_1, \infty)$.

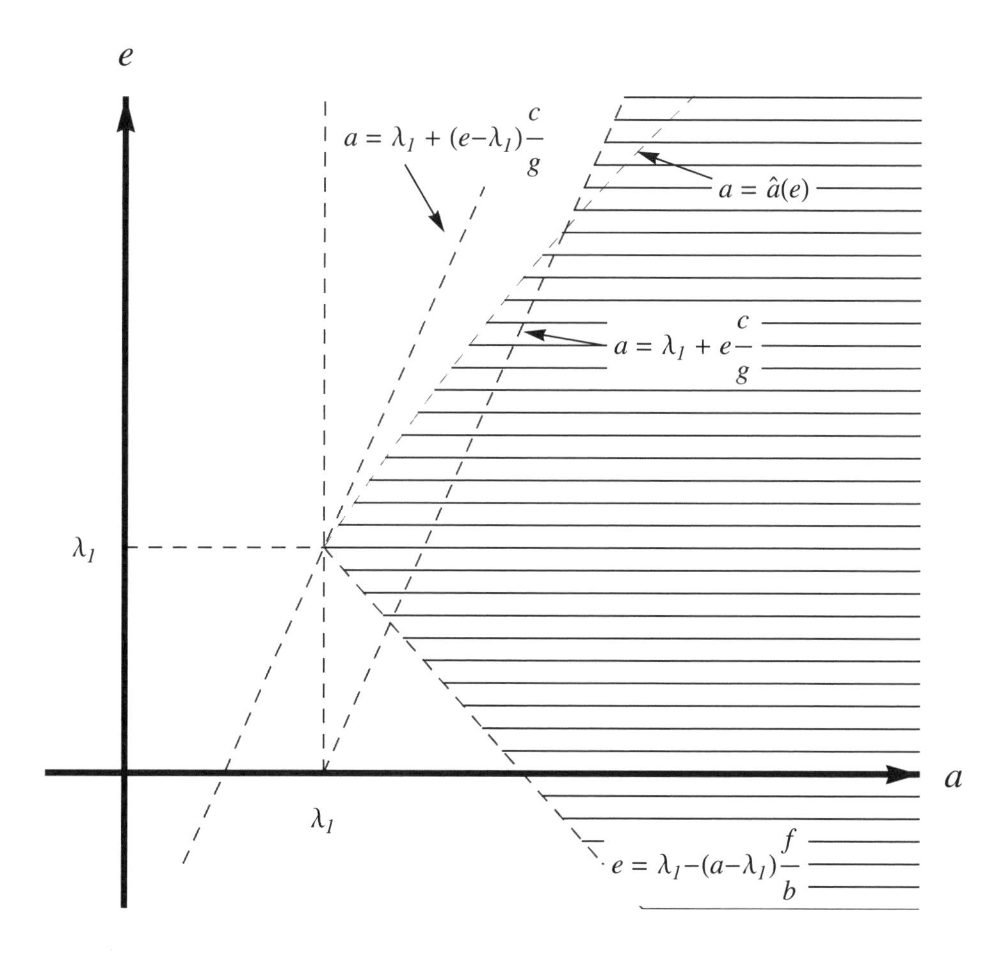

Figure 1.2.2: Coexistence Region in (a, e) Parameter Space, for case $c < g$.

The situation is illustrated in Fig 1.2.2 above. The details can be found in J. López-Gómez and R. Pardo San Gil [162].

Notes.

Theorem 2.1 was first proved in Leung [123], and Theorem 2.2 to Theorem 2.4 are due to Blat and Brown [11]. Theorem 2.5 and Corollary 2.6 are found in Li [148]. Lemmas 2.5 to 2.7 concerning indices of fixed point are obtained from Dancer [37] and Li [148]. Theorem 2.7 is a modification of a theorem in [148]. Theorem 2.8 is obtained from López-Gómez and Pardo San Gil [162]. More recent extension of the theory in this section for the case of ratio-dependent interaction rates can be found in Ryu and Ahn [197].

1.3 Strictly Positive Coexistence for Diffusive Competing Systems

In this section we study problem (1.1) when the functions $f_1(u,v)$ and $f_2(u,v)$ simulate competition between the two species populations $u(x)$ and $v(x)$ in a bounded domain Ω, with conditions as described in Section 1.1. More precisely, we first assume:

$$(3.1) \qquad \begin{cases} f_i(0,0) = 0, \;\; i = 1,2, \\ \frac{\partial f_i}{\partial u}, \frac{\partial f_i}{\partial v} < 0, \;\; \text{for } u \geq 0, v \geq 0, \;\; i = 1,2, \;\; \text{and} \\ a_i > \sigma_i \lambda_1, \;\; i = 1,2. \end{cases}$$

Part A: General Results.

The following theorem can be readily obtained by the method of upper and lower solution for a system of semilinear elliptic equations.

Theorem 3.1. *Assume that f_i, $i = 1,2$, are in $C^1([0,\infty) \times [0,\infty))$ and hypotheses (3.1) are valid. Suppose there exist positive constants k_1, k_2 such that the following inequalities:*

$$(3.2) \qquad \begin{cases} a_1 - \sigma_1\lambda_1 + f_1(0,k_2) > 0, \\ a_2 + f_2(0,k_2) < 0, \\ a_2 - \sigma_2\lambda_1 + f_2(k_1,0) > 0, \\ a_1 + f_1(k_1,0) < 0 \end{cases}$$

are satisfied. Then problem (1.1) has a positive solution $(\bar{u}(x), \bar{v}(x))$ with $\bar{u}(x) > 0, \bar{v}(x) > 0$ in Ω.

Proof. Let $\omega(x)$ be the positive principal eigenfunction in Ω for the eigenvalue problem $\Delta u + \lambda u = 0$ in Ω, $u = 0$ on $\partial\Omega$. For $r_1 > 0$ small enough, we have $\sigma_1\Delta(r_1\omega) + r_1\omega[a_1 + f_1(r_1\omega, v)] = r_1\omega[a_1 - \sigma_1\lambda_1 + f_1(r_1\omega, v)] > 0$, and $\sigma_1\Delta k_1 + k_1[a_1 + f_1(k_1,v)] < 0$ for $0 \leq v \leq k_2$. Also, for $r_2 > 0$ small enough, we have $\sigma_2\Delta(r_2\omega) + r_2\omega[a_2 + f_2(u, r_2\omega)] = r_2\omega[a_2 - \sigma_2\lambda_1 + f_2(u, r_2\omega)] > 0$, and $\sigma_2\Delta k_2 + k_2[a_2 + f_2(u,k_2)] < 0$ for $0 \leq u \leq k_1$. The pair of functions $(r_1\omega(x), k_1), (r_2\omega(x), k_2)$ form a coupled upper-lower solution for the system (1.1). Thus by e.g. Theorem 1.4-2 in Leung [125], there exists a solution $(\bar{u}(x), \bar{v}(x))$ to (1.1) with $r_1\omega(x) \leq \bar{u}(x) \leq k_1, r_2\omega(x) \leq \bar{v}(x) \leq k_2$ in $\bar{\Omega}$.

The above theorem applies immediately to the following Volterra-Lotka competition system.

$$(3.3)\qquad \begin{cases} \sigma_1 \Delta u + u(a - bu - cv) = 0 & \\ & \text{in } \Omega, \\ \sigma_2 \Delta v + v(e - fu - gv) = 0 & \\ & \\ u = v = 0 & \text{on } \partial\Omega, \end{cases}$$

where $\sigma_1, \sigma_2, a, b, c, e, f$ and g are positive constants.

Corollary 3.2. *Suppose that*

$$(3.4)\qquad a > \sigma_1 \lambda_1 + c\frac{e}{g}, \quad \textit{and} \quad e > \sigma_2 \lambda_1 + f\frac{a}{b}.$$

Then the boundary value problem (3.3) has a positive solution $(\bar{u}(x), \bar{v}(x))$ with $\bar{u}(x) > 0, \bar{v}(x) > 0$ in Ω.

Proof. Identify a, e respectively with a_1, a_2 and let $f_1(u, v) = -bu - cv$, $f_2(u, v) = -fu - gv$. Consider (3.3) as a special case of (1.1) under hypotheses (3.1) and (3.2). Choose $k_1 = \frac{a}{b} - \epsilon$ and $k_2 = \frac{e}{g} - \epsilon$, for $\epsilon > 0$ sufficiently small, then we can verify that (3.2) is satisfied. The results follows from Theorem 3.1.

Remark 3.1. If $a > \sigma_1 \lambda_1, e > \sigma_2 \lambda_1$, then the inequalities in (3.4) are readily satisfied when competition between the two species are relatively weak, in the sense of small c and f. The conditions in (3.4) are very easy to verify.

By using bifurcation method, we can follow the procedures as in Theorem 2.3(i) to obtain positive solutions as the growth rate of the second species e varies.

Theorem 3.3. *Suppose*

$$(3.5)\qquad a > \sigma_1 \lambda_1.$$

Then there exist μ_1, μ_2, μ_3 satisfying $\sigma_1 \lambda_1 < \mu_1 \leq \mu_2 < \mu_3$ with the following properties:

(i) If $\mu_1 \leq e \leq \mu_2$, then the boundary value problem (3.3) has at least one solution with each component strictly positive in Ω.

(ii) If $e > \mu_3$, then every non-negative solution of the boundary value problem (3.3) has at least one component identically equal to zero.

Proof. The proof is analogous to that of Theorem 2.3. Let (3.5) be satisfied. For any $v \in C^1(\Omega)$, define $u(v)$ as in (2.9) and (2.10). Then, for all values e, problem (3.3) has the solution $(u, v) = (u(0), 0)$. We consider the bifurcation of solution as the parameter e varies in the problem:

$$-\sigma_2 \Delta v = ev - gv^2 - fu(v)v \ \text{ in } \Omega, \quad v = 0 \text{ on } \partial\Omega. \tag{3.6}$$

Bifurcation occurs when $e = \rho_1(-\sigma_2\Delta + fu(0))$. As in the proof of Theorem 2.3, we can show that there exists a continuum S^+ of solutions (e, v) of (3.6) such that $v > 0$ for all $v \in S^+$ and S^+ intersects with the curve corresponding to the zero solution only when $e = \rho_1(-\sigma_2\Delta + fu(0))$. Multiplying (3.6) by v and integrating by parts, we find $e > \rho_1(-\sigma_2\Delta)$ for all e such that $(e, v) \in S^+$. By means of sweeping principle argument, we find $v \le e/f$ for all v with $(e, v) \in S^+$. Thus for all v such that $(e, v) \in S^+$, there exists a bound in $C^1(\bar{\Omega})$ which is dependent on e. Since S^+ connects $(\rho_1(-\sigma_2\Delta + fu(0)), 0)$ with ∞ in $C^1(\bar{\Omega})$, it follows that $\{e : (e, v) \in S^+\} \supseteq (\rho_1(-\sigma_2\Delta + fu(0)), \infty)$. (More details can be found in Blat and Brown [11].)

As in Theorem 2.3, we can show that there exists a constant $\mu_3 > 0$ such that if $e > \mu_3$, then all solutions of (3.3) have at least one component identically equal to zero. As in the proof of Theorem 2.3, we will need to obtain a lower bound for v in terms of e. For this purpose, we need a slight change in the definition of $k(e)$ for analyzing (2.15). The eigenvalue λ_1 and eigenfunction ω_1 will be respectively replaced with the least eigenvalue and principal eigenfunction of

$$-\sigma_2\Delta\phi + fu(0)\phi = \lambda\phi \text{ in } \Omega, \quad \phi = 0 \text{ on } \partial\Omega.$$

The remaining part of the proof follows the argument in the last part for the proof of Theorem 2.3(i).

One can obtain similar bifurcation results as above, when the growth rate parameter a varies. For competitive system more general than Volterra-Lotka type, we can obtain the following existence theorem by cone-index method. The conditions are in terms of the signs of principal eigenvalues of appropriate related scalar equations. The results are analogous to the sufficiency part of Theorem 2.5 in the last section.

Theorem 3.4. *Assume that $f_i, i = 1, 2$, are in $C^1([0, \infty) \times [0, \infty))$. Consider the boundary value problem (1.1), under assumptions:*

$$\begin{cases} (i)\ f_i(0,0) = 0,\ lim_{u\to\infty} f_1(u, 0) = -\infty,\ lim_{v\to\infty} f_2(0, v) = -\infty, \\ (ii)\ \partial f_i/\partial u < 0, \partial f_i/\partial v < 0 \ for\ \ u \ge 0, v \ge 0, i = 1, 2. \end{cases} \tag{3.7}$$

Suppose that

$$a_1 > \rho_1(-\sigma_1\Delta), \ a_2 > \rho_1(-\sigma_2\Delta), \tag{3.8}$$

and one of the following two situations hold:

(i) $\hat{\rho}_1(\sigma_1\Delta + a_1 + f_1(0, v_0)) > 0$, *and* $\hat{\rho}_1(\sigma_2\Delta + a_2 + f_2(u_0, 0)) > 0$*;*

(ii) $\hat{\rho}_1(\sigma_1\Delta + a_1 + f_1(0, v_0)) < 0$, *and* $\hat{\rho}_1(\sigma_2\Delta + a_2 + f_2(u_0, 0)) < 0$*;*

Then the boundary value problem (1.1), has a positive solution with each component strictly positive in Ω*.*

Remark 3.2. In Theorem 3.4, u_0 is the unique positive solution of $\sigma_1 u + u[a_1 + f_1(u, 0)] = 0$ in Ω, $u = 0$ on $\partial\Omega$; v_0 is the unique positive solution of $\sigma_2 v + v[a_2 + f_2(0, v)] = 0$ in Ω, $v = 0$ on $\partial\Omega$.

Proof. Let B_1, B_2 be positive numbers such that $f_1(B_1, 0) = -a_1, f_2(0, B_2) = -a_2$. Using assumption (ii) in (3.7), we can show by applying sweeping principle to each scalar equation of (1.1) for a fixed positive function assigned to the other component to conclude that all positive solutions of (1.1) satisfy $0 \le u \le B_1$ and $0 \le v \le B_2$ in $\bar{\Omega}$. Let $[C_0^+(\bar{\Omega})]^2 := \{(u_1, u_2) : u_i \in C(\bar{\Omega}), u_i \ge 0 \ in \ \Omega, \ and = 0 \ on \ \partial\Omega, \ for \ i = 1, 2\}$, $B = max.\{B_1, B_2\}$, and $[E(B)]^2 := \{(u_1, u_2) : u_i \in C(\bar{\Omega}), |u_i| < B \ in \ \Omega, \ for \ i = 1, 2\}$, with closure $[\bar{E}(B)]^2$. For each $(u_1, u_2) \in [C(\bar{\Omega})]^2, \theta \in [0, 1]$, define the operator $A_\theta : [C_0(\bar{\Omega})]^2 \cap [\bar{E}(B)]^2 \to [C_0(\bar{\Omega})]^2$ by $A_\theta(u_1, u_2) = (v_1, v_2)$ where

$$\begin{cases} v_1 = (-\sigma_1\Delta + P)^{-1}[\theta u_1(a_1 + f_1(u_1, u_2)) + Pu_1] \\ v_2 = (-\sigma_2\Delta + P)^{-1}[\theta u_2(a_2 + f_2(u_1, u_2)) + Pu_2]. \end{cases} \tag{3.9}$$

Here, the inverse operator is taken with homogeneous Dirichlet boundary condition on $\partial\Omega$, and $P > 0$ is a large enough constant such that the operator A_θ is positive, compact and Fréchet differentiable on $[C_0^+(\bar{\Omega})]^2 \cap [\bar{E}(B)]^2$. Let K be the cone $K := [C_0^+(\bar{\Omega})]^2 := \{(u_1, u_2) : u_i \in C(\bar{\Omega}), u_i \ge 0 \ in \ \Omega, \ and = 0 \ on \ \partial\Omega, \ for \ i = 1, 2\}$, and $D := [C_0^+(\bar{\Omega})]^2 \cap [E(B)]^2$. The bound on the solution implies that the operators A_θ has no fixed point on the boundary ∂D in the relative topology, i.e. on the intersection of boundary of $[E(B)]^2$ with K. We can further use a familiar cut-off procedure to extend A_θ to be defined outside D as a compact positive mapping from the cone K into itself. For convenience, we will denote $A := A_1$. We will denote the fixed point index of A_θ over D with respect to the cone K by $i_K(A_\theta, D)$. As in the proof of Theorem 2.5(ii), (iii), we obtain $i_K(A, D) = i_K(A_0, D) = 1$.

Let y be an isolated fixed point of the map A_θ in K, we denote the local index of A_θ at y with respect to K by $index_K(A_\theta, y)$. We now show that $index_K(A, (0,0)) = 0$ for each case (i) and (ii). For $y \in K$, define

$$K_y := \{p \in [C(\bar{\Omega})]^2 : y + sp \in K \ \text{ for some } s > 0\}, \text{ and}$$

$$S_y := \{p \in \bar{K}_y : -p \in \bar{K}_y\}.$$

Here $\bar{K}_y$ denotes the closure of K_y. We have $\bar{K}_{(0,0)} = K, S_{(0,0)} = \{(0,0)\}$. Let $A'_+((0,0))$ be the Fréchet derivative of A at $(0,0)$ in K. The first component of $A'_+((0,0))(u_1, u_2)$ is $(-\sigma_1\Delta + P)^{-1}(a_1 + P)u_1$. Hence $[I - A'_+((0,0))]u = 0$ for $u = (u_1, u_2) \in K$ implies that $[\sigma_1\Delta + a_1]u_1 = 0$, $u_1 \in C_0^+(\bar{\Omega})$. Thus the assumption $a_1 > \rho_1(-\sigma_1\Delta)$ implies that $u_1 = 0$. Similarly, we have for the second component $[\sigma_2\Delta + a_2]u_2 = 0$, $u_2 \in C_0^+(\bar{\Omega})$. Thus the assumption $a_2 > \rho_1(-\sigma_2\Delta)$ implies that $u_2 = 0$. We thus conclude $I - A'_+((0,0))$ is invertible in $W_{(0,0)}$. Further, the assumption $a_1 > \rho_1(-\sigma_1\Delta)$ and the continuity in $t \in [0,1]$ for the eigenvalue $\rho_1(\sigma_1\Delta + ta_1 + (t-1)P)$ imply that there exists some $t \in (0,1)$ and a nontrivial function $\bar{u} \in C_0^+(\bar{\Omega})$ such that $(-\sigma_1\Delta + P)\bar{u} = t(a_1 + P)\bar{u}$ or $\bar{u} - t(-\sigma_1\Delta + P)^{-1}(a_1 + P)\bar{u} = 0$ in Ω. We thus have $[I - tA'_+((0,0))](\bar{u}, 0) = (0,0) \in S_{(0,0)}$, with $(\bar{u}, 0) \in \bar{K}_{(0,0)} \backslash S_{(0,0)}$. We thus conclude by Lemma 2.7(i) that $index_K(A, (0,0)) = 0$.

We next show that for case (i), we have

$$index_K(A, (u_0, 0)) = index_K(A, (0, v_0)) = 0.$$

Let $L = A'_+((u_0, 0))$ be the Fréchet derivative of A at $(u_0, 0)$. Suppose that $(I - L)(u_1, u_2) = 0$, for some $u_1 \geq 0, u_2 \geq 0$ in $\bar{\Omega}$, and $u_1 = u_2 = 0$ on $\partial\Omega$. Then

(3.10)
$$\begin{cases} \sigma_1\Delta u_1 + [a_1 + f_1(u_0,0) + u_0\frac{\partial f_1}{\partial u}(u_0,0)]u_1 + u_0\frac{\partial f_1}{\partial v}(u_0,0)u_2 = 0 & \\ & \text{in } \Omega, \\ \sigma_2\Delta u_2 + [a_2 + f_2(u_0,0)]u_2 = 0 & \\ & \\ u_1 = u_2 = 0 & \text{on } \partial\Omega. \end{cases}$$

Thus the second equation above and the second assumption in situation (i) implies that $u_2 \equiv 0$. We then consider the first equation above again. Since $\hat{\rho}_1(\sigma_1\Delta + a_1 + f_1(u_0, 0)) = 0$, and $u_0\frac{\partial f_1}{\partial u}(u_0, 0) < 0$ by (3.7), we have $\hat{\rho}_1(\sigma_1\Delta + a_1 + f_1(u_0, 0) + u_0\frac{\partial f_1}{\partial u}(u_0, 0)) < 0$. Hence, all the eigenvalues ρ of the problem:

$$\sigma_1\Delta u + [a_1 + f_1(u_0,0) + u_0\frac{\partial f_1}{\partial u}(u_0,0)]u = \rho u \text{ in } \Omega, \ u = 0 \text{ on } \partial\Omega,$$

satisfy $\rho < 0$. However, u_1 satisfies this problem with $\rho = 0$. Thus $u_1 \equiv 0$. That is the operator $(I - L)$ is invertible on $\bar{K}_{(u_0,0)}$.

We next show that the operator L has property (α) on $\bar{W}_{(u_0,0)}$. Let $P > 0$; observe that the eigenvalue $\hat{\rho}_1(\sigma_2\Delta - P + t[a_2 + f_2(u_0, 0) + P])$ is negative when $t = 0$, and is positive when $t = 1$. By continuity, there exists some $t^* \in (0, 1)$, such that $\hat{\rho}_1(\sigma_2\Delta - P + t^*[a_2 + f_2(u_0, 0) + P]) = 0$. There exists $u_2^* > 0$ in Ω, vanishing on $\partial\Omega$ such that $(-\sigma_2\Delta + P)u_2^* - t^*[a_2 + f_2(u_0, 0) + P]u_2^* = 0$ in Ω. Since $S_{(u_0,0)} = C_0(\bar{\Omega}) \times \{0\}$, we can readily verify that if we let $w = (0, u_2^*)$, then we have $w - t^*Lw \in S_{(u_0,0)}$ with $w \in \bar{K}_{(u_0,0)} \backslash S_{(u_0,0)}$. Consequently, by Lemma 2.7(i), we conclude that $index_K(A, (u_0, 0)) = 0$.

We now consider the point $(0, v_0)$ and let $\tilde{L} = A'_+((0, v_0))$ be the Fréchet derivative of A at $(0, v_0)$ in K. Suppose that $(I - \tilde{L})(u_1, u_2) = 0$, for some $u_1 \geq 0, u_2 \geq 0$ in $\bar{\Omega}$, and $u_1 = u_2 = 0$ on $\partial\Omega$. Then

$$\begin{cases} \sigma_1\Delta u_1 + [a_1 + f_1(0, v_0)]u_1 = 0 & \\ & \text{in } \Omega, \\ \sigma_2\Delta u_2 + v_0\frac{\partial f_2}{\partial u}(0, v_0)u_1 + [a_2 + f_2(0, v_0) + v_0\frac{\partial f_2}{\partial v}(0, v_0)]u_2 = 0 & \\ u_1 = u_2 = 0 & \text{on } \partial\Omega. \end{cases}$$

From the first equation above and the first assumption in situation (i), we obtain $u_1 \equiv 0$. Since $\hat{\rho}_1(\sigma_2\Delta + a_2 + f_2(0.v_0)) = 0$ and $v_0\frac{\partial f_2}{\partial u_2}(0, v_0) \leq 0$ is not the trivial function, we have $\hat{\rho}_1(\sigma_2\Delta + a_2 + f_2(0, v_0) + v_0\frac{\partial f_2}{\partial v}(0, v_0)) < 0$. We thus deduce from the second equation above that $u_2 \equiv 0$. We thus conclude that the operator $I - \tilde{L}$ is invertible in $\bar{K}_{(0,v_0)}$.

From the first assumption in (i), we deduce that for $P > 0$, the eigenvalue $\hat{\rho}_1(\sigma_1\Delta - P + t[a_1 + f_1(0, v_0) + P])$ is negative if $t = 0$ and is positive if $t = 1$. Hence, there exist $t^* \in (0, 1)$ and a nontrivial, non-negative function u_1^* vanishing on $\partial\Omega$, such that

$$-\sigma_1\Delta u_1^* + Pu_1^* - t^*(a_1 + f_1(0, v_0) + P)u_1^* = 0 \quad \text{in } \Omega.$$

Since $S_{(0,v_0)} = \{0\} \times C_0(\bar{\Omega})$, we can readily verify that if we let $\tilde{w} = (u_1^*, 0)$, then we have $\tilde{w} - t^*\tilde{L}\tilde{w} \in S_{(0,v_0)}$ with $\tilde{w} \in \bar{K}_{(0,v_0)} \backslash S_{(0,v_0)}$. Consequently, by Lemma 2.7(i), we conclude that $index_K(A, (0, v_0)) = 0$.

We thus have

$$i_K(A, D) = 1, \ index_K(A, (0, 0)) = index_K(A, (u_0, 0)) = index_K(A, (0, v_0)) = 0.$$

In order to avoid contradicting the additive property of the indices of the map on disjoint open subsets, there must be at least another fixed point of A in D. Hence for case (i), there must be more positive solution in D other than $(0, 0), (u_0, 0)$ or $(0, v_0)$.

We next consider the proof of case (ii). Let $L = A'_+((u_0, 0))$ be the Fréchet derivative of A at $(u_0, 0)$. Suppose that $(I - L)(u_1, u_2) = 0$, for some $u_1 \geq 0, u_2 \geq 0$ in $\bar{\Omega}$, and $u_1 = u_2 = 0$ on $\partial\Omega$. Then (u_1, u_2) satisfies (3.10) again. The second equation in (3.10) and the second assumption in (ii) implies that $u_2 \equiv 0$. We then obtain from the second equation in (3.10) that $u_1 \equiv 0$ in the same way as in case (i) above. Hence the operator $(I - L)$ is invertible on $\bar{K}_{(u_0,0)}$.

We next show that the operator L does not have property (α) on $\bar{K}_{(u_0,0)}$. We have $S_{(u_0,0)} = C_0(\bar{\Omega}) \times \{0\}$, and $\bar{W}_{(u_0,0)} \backslash S_{(u_0,0)} = C_0(\bar{\Omega}) \times \{C_0^+(\bar{\Omega}) \backslash \{0\}\}$. Suppose the operator L has property (α) on $\bar{K}_{(u_0,0)}$. Then there exists some $t^* \in (0, 1)$ and $(u_1^*, u_2^*) \in \bar{K}_{(u_0,0)} \backslash S_{(u_0,0)}$, such that

$$u_1^* - t^*(-\sigma_1\Delta + P)^{-1}([a_1 + f_1(u_0, 0) + u_0 \tfrac{\partial f_1}{\partial u}(u_0, 0) + P]u_1^*$$

$$+ u_0 \tfrac{\partial f_1}{\partial v}(u_0, 0)u_2^*) \in C_0(\bar{\Omega}),$$

$$u_2^* - t^*(-\sigma_2\Delta + P)^{-1}(\, [a_2 + f_2(u_0, 0) + P]u_2^* \,) = 0.$$

The first equation above is always satisfied. The second equation above implies that

$$Tu_2^* = \frac{1}{t^*}u_2^* > u_2^* \text{ in } C_0(\bar{\Omega}),$$

where $T := (-\sigma_2\Delta + P)^{-1}[a_2 + f_2(u_0, 0) + P]$. By Theorem A2-6(i) in Chapter 6, we obtain $r(T) > 1$ for the spectral radius $r(T)$. On the other hand the second assumption in (ii) implies that there exists a positive eigenfunction ϕ for the negative eigenvalue $\beta := \hat{\rho}_1(\sigma_2\Delta + a_2 + f_2(u_0, 0))$ such that

$$(-\sigma_2\Delta + P)\phi - (a_2 + f_2(u_0, 0) + P)\phi = -\beta\phi > 0 \text{ in } \Omega.$$

We thus have $\phi - T\phi > 0$ in $C_0(\bar{\Omega})$. Thus by Theorem A2-6(ii) in Chapter 6, we obtain $r(T) < 1$. From this contradiction we conclude that L cannot have property α. Consequently, using Lemma 2.7, we have

$$index_K(A, (u_0, 0)) = index_{C_0(\bar{\Omega}) \times \{0\}}(L, (0, 0)) = \pm 1.$$

In order to calculate $index_{C_0(\bar{\Omega}) \times \{0\}}(L, (0, 0))$, we use Theorem A2-3 in Chapter 6 to find $index_{C_0(\bar{\Omega}) \times \{0\}}(L, (0, 0)) = (-1)^m$, where m is the sum of multiplicities of eigenvalues of L greater than 1. Suppose $(\phi, \psi) \in C_0(\bar{\Omega}) \times \{0\}$ is an eigenvector of L with λ as eigenvalue. Then

$$(-\sigma_1\Delta + P)^{-1}[a_1 + f_1(u_0, 0) + u_0 \frac{\partial f_1}{\partial u}(u_0, 0) + P]\phi = \lambda\phi.$$

Let $\tilde{T} := (\sigma_1\Delta + P)^{-1}[a_1 + f_1(u_0, 0) + u_0\frac{\partial f_1}{\partial u}(u_0, 0) + P]$. We have $\tilde{\beta} := \hat{\rho}_1(\sigma_1\Delta + a_1 + f_1(u_0, 0) + u_0\frac{\partial f_1}{\partial u}(u_0, 0)) < 0$, and

$$(-\sigma_1\Delta + P)\tilde{\phi} - (a_1 + f_1(u_0, 0) + u_0\frac{\partial f_1}{\partial u}(u_0, 0) + P)\tilde{\phi} = -\tilde{\beta}\tilde{\phi} > 0 \text{ in } \Omega$$

for some positive eigenfunction $\tilde{\phi}$ for the eigenvalue $\tilde{\beta}$. We thus have $\tilde{\phi} - \tilde{T}\tilde{\phi} > 0$, and by Theorem A2-6(ii) in Chapter 6 again we obtain $r(\tilde{T}) < 1$. Consequently $\lambda < 1$ and $m = 0$. Thus we have $index_K(A, (u_0, 0)) = (-1)^0 = 1$. Similarly, we can show that $index_K(A, (0, v_0)) = 1$.

We thus have

$$i_K(A, D) = 1, \ index_K(A, (0, 0)) = 0,$$
$$index_K(A, (u_0, 0)) = index_K(A, (0, v_0)) = 1.$$

In order to avoid contradicting the additive property of the indices (Theorem A2-1(ii) in Chapter 6), there must exist a positive solution of problem (1.1) in D other than $(0, 0), (u_0, 0)$ or $(0, v_0)$.

For each case (i) or (ii), the positive solution $(u(x), v(x))$ in D, other than $(0, 0), (u_0, 0)$ or $(0, v_0)$, has each component $\not\equiv 0, \geq 0$ in $\bar{\Omega}$. From the boundedness of the coefficients $[a_i + f_i(u(x), v(x))]$, we obtain from each equation in (1.1) and Lemma 1.1 that each component of $(u(x), v(x))$ is strictly positive in Ω.

For necessary conditions, we consider the special case when the intrinsic growth rates a_1 and a_2 of each species are the same

$$\sigma_1 = \sigma_2 = \sigma, \ \ a_1 + f_1(u, v) = p(u) - q(v), \ \ a_2 + f_2(u, v) = r(v) - s(u), \tag{3.11}$$

where $p(0) = a_1 = a_2 = r(0)$, $q(0) = s(0) = 0$, $p, r, -q, -s$ are $C^1([0, \infty))$ non-increasing functions, and $p' < 0, r' < 0$. Moreover, there exist constants c_1, c_2 such that

$$p(u) < 0 \text{ for } u > c_1; \ \ r(v) < 0 \text{ for } v > c_2. \tag{3.12}$$

Theorem 3.5. *Consider the boundary value problem (1.1), under the special conditions (3.11) and (3.12). Suppose further that both $p' + s', r' + q'$ have the same constant sign on $(0, c)$ where $c = max.\{c_1, c_2\}$. If the boundary value problem has a positive solution, then $p(0) > \lambda_1$, $r(0) > \lambda_1$, and one of the following three situations must hold:*

(i) $\hat{\rho}_1(\sigma\Delta + p(0) - q(v_0)) > 0$, *and* $\hat{\rho}_1(\sigma\Delta + r(0) - s(u_0)) > 0$;
(ii) $\hat{\rho}_1(\sigma\Delta + p(0) - q(v_0)) < 0$, *and* $\hat{\rho}_1(\sigma\Delta + r(0) - s(u_0)) < 0$;
(iii) $\hat{\rho}_1(\sigma\Delta + p(0) - q(v_0)) = 0$, *and* $\hat{\rho}_1(\sigma\Delta + r(0) - s(u_0)) = 0$.

Proof. First assume that

$$p' + s' > 0, \;\; r' + q' > 0. \tag{3.13}$$

Let $(\tilde{u}, \tilde{v})$ be a positive solution of the boundary value problem (1.1), (1.3) under the hypotheses of this theorem. The function $\tilde{u}$ is a positive lower solution to the scalar problem:

$$\sigma \Delta u + up(u) = 0 \;\text{ in } \Omega, \;\; u = 0 \text{ on } \partial\Omega; \tag{3.14}$$

and the constant function $u \equiv c_1$ is an upper solution for (3.14). Thus problem (3.14) has a positive solution, and Lemma 2.1 implies that $p(0) > \sigma\lambda_1$. Moreover, we have $r(0) = a_2 = a_1 = p(0) > \sigma\lambda_1$, and we have a positive solution u_0 for (3.14) and a positive solution v_0 for $\sigma\Delta v + vr(v) = 0$ in Ω, $v = 0$ on $\partial\Omega$. We have

$$p(0) - q(v_0) = r(0) - q(v_0) = r(0) + q(0) - q(v_0) < r(v_0),$$

where the last inequality is due to the second part of (3.13). Similarly, from the first inequality in (3.13), we obtain

$$r(0) - s(u_0) < p(u_0).$$

From the two inequalities above, we find

$$\hat{\rho}_1(\sigma\Delta + p(0) - q(v_0)) < \hat{\rho}_1(\sigma\Delta + r(v_0)) = 0,$$

$$\hat{\rho}_1(\sigma\Delta + r(0) - s(u_0)) < \hat{\rho}_1(\sigma\Delta + p(u_0)) = 0.$$

That is, we obtain the second conclusion (ii) of the statement of the theorem. If we reverse both inequalities in (3.13), the arguments above lead to conclusion (i) of the theorem. If both inequalities in (3.13) are changed to equality, then we obtain conclusion (iii).

Remark 3.3. When (3.13) holds, there are strong competitions between the two species, we obtain case (ii) in Theorem 3.5 when both eigenvalues involved are negative. If both inequalities in (3.13) are reversed, there are weaker competitions, and we obtain case (i) above when both eigenvalues involved are positive. The assumptions in Theorem 3.5 are very restrictive. In Theorem 3.11 to Theorem 3.13 below, we will consider cases when one species is much stronger than the other.

For fixed $f_i(u, v), i = 1, 2$, satisfying (3.7) and suppose a_1, a_2 satisfy (3.8), we now utilize Theorem 3.4 and comparison method to deduce a more detailed description of the set:

(3.15)
$$\Lambda := \{(a_1, a_2) |\; a_i > \rho_1(-\sigma_i\Delta), i = 1, 2; \text{ the boundary value problem (1.1) has a strictly positive solution in } \Omega.\}$$

The analysis here is more general than that for the Volterra-Lotka prey-predator case given in the last section. By assumption (3.8), there exist positive solutions $u = u_0(a_1), v = v_0(a_2)$ respectively satisfying the following:

$$\sigma_1 \Delta u + u[a_1 + f_1(u, 0)] = 0 \ \text{ in } \Omega, \ \ u = 0 \ \text{ on } \partial\Omega,$$

$$\sigma_2 \Delta v + v[a_2 + f_2(0, v)] = 0 \ \text{ in } \Omega, \ \ v = 0 \ \text{ on } \partial\Omega.$$

Define $\underline{v}, \bar{u}, \underline{u}, \bar{v}$ to be the maximal non-negative solutions of the following scalar boundary value problems:

$$(3.16) \qquad \begin{cases} \sigma_2 \Delta \underline{v} + \underline{v}(a_2 + f_2(u_0(a_1), \underline{v})) = 0 \text{ in } \Omega, \ \underline{v} = 0 \ \text{ on } \partial\Omega; \\ \sigma_1 \Delta \bar{u} + \bar{u}(a_1 + f_1(\bar{u}, \underline{v})) = 0 \text{ in } \Omega, \ \bar{u} = 0 \ \text{ on } \partial\Omega; \\ \sigma_1 \Delta \underline{u} + \underline{u}(a_1 + f_1(\underline{u}, v_0(a_2)) = 0 \text{ in } \Omega, \ \underline{u} = 0 \ \text{ on } \partial\Omega; \\ \sigma_2 \Delta \bar{v} + \bar{v}(a_2 + f_2(\underline{u}, \bar{v})) = 0 \text{ in } \Omega, \ \bar{v} = 0 \ \text{ on } \partial\Omega. \end{cases}$$

The four functions are not always nontrivial, and are completely determined by the two constants a_1 and a_2. For each fixed v, $\underline{v} \le v \le v_0(a_2)$, the functions $\bar{u}$ and 0 are respectively upper and lower solutions of

$$(3.17) \qquad \sigma_1 \Delta u + u(a_1 + f_1(u, v)) = 0 \text{ in } \Omega, \ u = 0 \ \text{ on } \partial\Omega.$$

For each fixed u, $0 \le u \le \bar{u}$, the functions $v_0(a_2)$ and $\underline{v}$ are respectively upper and lower solutions of

$$(3.18) \qquad \sigma_2 \Delta v + v(a_2 + f_2(u, v)) = 0, \ \text{ in } \Omega, \ v = 0 \ \text{ on } \partial\Omega.$$

Similarly, for each fixed v, $0 \le v \le \bar{v}$, the functions $u_0(a_1)$ and $\underline{u}$ are respectively upper and lower solutions of problem (3.17). For each fixed u, $\underline{u} \le u \le u_0(a_1)$, the functions $\bar{v}$ and 0 are respectively upper and lower solutions of problem (3.18). Let $(u_1(x,t), v_1(x,t))$ and $(u_2(x,t), v_2(x,t))$ be respectively solutions of the initial boundary value problem (1.2) with initial conditions:

$$(3.19) \qquad (u_1(x, 0), v_1(x, 0)) = (\bar{u}, \underline{v}), \ \text{ and } (u_2(x, 0), v_2(x, 0)) = (\underline{u}, \bar{v}).$$

One can show by comparison that as $t \to +\infty$, $u_1(x,t)$ and $v_1(x,t)$ respectively tend from above and below to some $\bar{u}^s(x)$ and $\underline{v}^s(x)$ in $\bar{\Omega}$, since the initial conditions are upper and lower solutions for the steady state problem. Similarly as $t \to +\infty$, $u_2(x,t)$ and $v_2(x,t)$ respectively tend from below and above to some $\underline{u}^s(x)$ and $\bar{v}^s(x)$ in $\bar{\Omega}$, since the initial conditions are lower and upper solutions for the steady state problem. Note that $\bar{u}^s, \underline{v}^s, \underline{u}^s, \bar{v}^s$ are completely determined by the parameters a_1, a_2. Next, we define

$$
\begin{aligned}
G_1(a_1, a_2) &:= \rho_1(-\sigma_2\Delta - f_2(\underline{u}^s, 0)) > \rho_1(-\sigma_2\Delta), \\
G_2(a_1, a_2) &:= \rho_1(-\sigma_1\Delta - f_1(0, \underline{v}^s)) > \rho_1(-\sigma_1\Delta).
\end{aligned} \tag{3.20}
$$

For given (a_1, a_2) satisfying (3.8) and let (u, v) be a corresponding solution of the steady state problem (1.1), we can readily deduce by comparison that:

$$
\underline{u}^s \le u \le \bar{u}^s, \ \ \underline{v}^s \le v \le \bar{v}^s. \tag{3.21}
$$

If $(a_1, a_2) \in \Lambda$, then we have $\bar{u}^s \ge u > 0$ in Ω. Taking limit as $t \to +\infty$, we also find that $\bar{u}^s$ is a positive solution of:

$$
\sigma_1 \Delta \bar{u}^s + \bar{u}^s(a_1 + f_1(\bar{u}^s, \underline{v}^s)) = 0 \text{ in } \Omega, \ \bar{u}^s = 0 \text{ on } \partial\Omega.
$$

Comparing with (3.20), we must have:

$$
a_1 > G_2(a_1, a_2).
$$

Similarly $\bar{v}^s$ is a positive solution of:

$$
\sigma_2 \Delta \bar{v}^s + \bar{v}^s(a_2 + f_2(\underline{u}^s, \bar{v}^s)) = 0 \text{ in } \Omega, \ \bar{v}^s = 0 \text{ on } \partial\Omega.
$$

We conclude that

$$
a_2 > G_1(a_1, a_2).
$$

Define

$$
\begin{aligned}
H_1(a_1) &:= inf.\{\beta > \rho_1(-\sigma_2\Delta) : \beta > G_1(a_1, \beta)\}, \\
H_2(a_2) &:= inf.\{\alpha > \rho_1(-\sigma_1\Delta) : \alpha > G_2(\alpha, a_2)\}.
\end{aligned}
$$

Using comparison arguments as given in the last paragraph and more careful analysis by means of Theorem 3.4, one can obtain a more precise description of the set Λ as follows.

Theorem 3.6. *Consider problem (1.1) under assumptions (3.7) and (3.8). The set Λ defined in (3.15) is a connected region bounded by the two curves:*

$$
\Gamma_1 : a_1 = H_2(a_2), \ \ \Gamma_2 : a_2 = H_1(a_1)
$$

in the following sense: for each $a_2 > \rho_1(-\sigma_2\Delta)$, the horizontal slice $\{a_1 : (a_1, a_2) \in \Lambda\}$ is a nonempty interval whose left endpoint is on Γ_1; and for each $a_1 > \rho_1(-\sigma_1\Delta)$, the vertical slice $\{a_2 : (a_1, a_2) \in \Lambda\}$ is a nonempty interval whose lower endpoint is Γ_2.

Details of the proof can be found in Ruan and Pao [195]. Moreover, the theorem is actually true for more general boundary conditions for the functions u, v respectively of the form $B_i = \alpha_i(x)\frac{\partial}{\partial \nu} + \beta_i(x), i = 1, 2$. Here, α_i and β_i are non-negative functions in $C^{1+\alpha}(\partial\Omega), 0 < \alpha < 1$, with either $\alpha_i = 0, \beta_i > 0$ or $\alpha_i > 0, \beta_i \geq 0$. The set Λ is illustrated in Fig. 1.3.1 below.

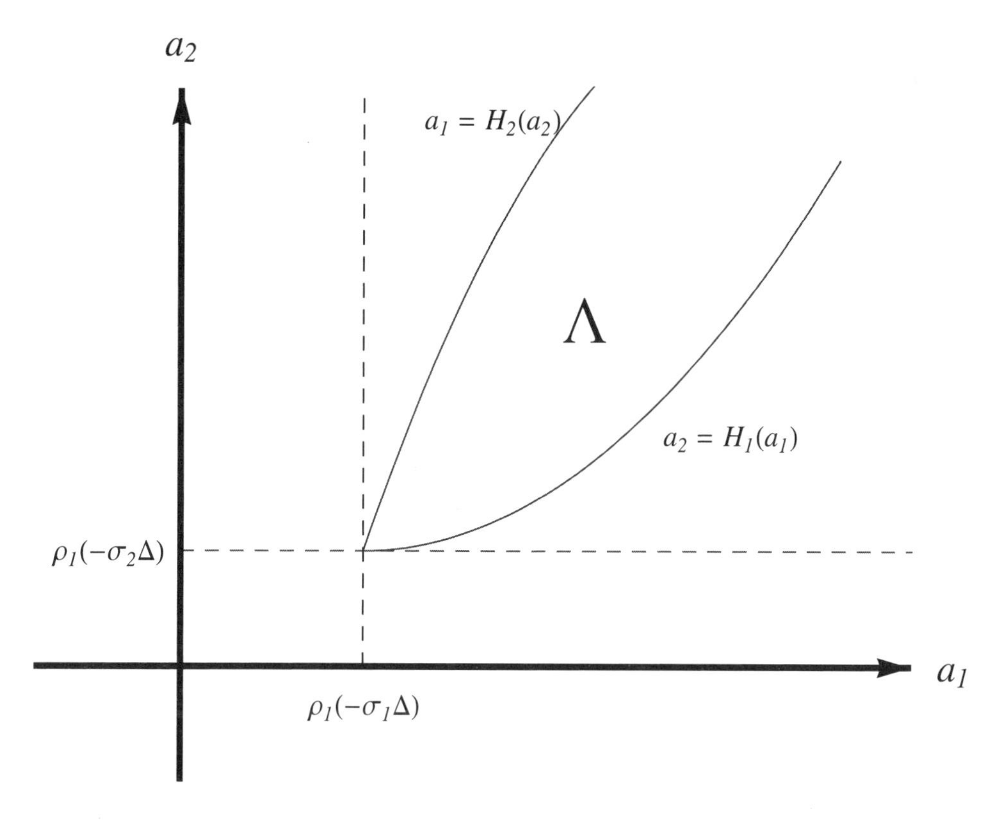

Figure 1.3.1: Coexistence Region in (a_1, a_2) Parameter Space.

Applying the results of Theorem 3.4 to system the Volterra-Lotka system (3.3), we see that when the interaction rates c and f are small, we will have case (i) when both of the related principal eigenvalues are positive. Thus Theorem 3.4 asserts the existence of positive coexistence solution. Actually, we can also obtain the existence of such solution when both c and f are small from the result in Corollary 3.2. On the other hand, when c and f are both large, we have strong competition between the species. In this case, we will have case (ii) in Theorem 3.4 when both related principal eigenvalues are negative. Thus we have coexistence positive solution again.

Part B: Extreme Strong Competition.

We now make a more careful study of the situation when both competition interaction parameters c and f are large. We shall see that the species may tend to segregate from each other as they coexist. For simplicity we restrict to $\sigma_1 = \sigma_2 = 1$, $a > \lambda_1, e > \lambda_1, b = g = 1$ and Dirichlet boundary condition. That is, we consider:

$$(3.22) \qquad \begin{cases} \Delta u + u(a - u - cv) = 0 & \\ & \text{in } \Omega, \\ \Delta v + v(e - v - fu) = 0 & \\ & \\ u = v = 0 & \text{on } \partial\Omega. \end{cases}$$

Recall that by Lemma 2.1, there cannot be any non-negative solution of the above problem with $u \not\equiv 0$ if $a < \lambda_1$ (or $v \not\equiv 0$ if $e < \lambda_1$). When both c and f are large, there is one type of positive solution to the problem (3.22) closely related to the positive solutions of the reduced problem:

$$(3.23) \qquad \begin{cases} \Delta u + u(a - v) = 0 & \\ & \text{in } \Omega, \\ \Delta v + v(e - u) = 0 & \\ & \\ u = v = 0 & \text{on } \partial\Omega, \end{cases}$$

with $a > \lambda_1, e > \lambda_1$. Moreover, the existence of positive solution of the reduced problem (3.23) is related to the trivial solution of the problem:

$$(3.24) \qquad \Delta w + aw^+ + ew^- = 0 \quad \text{in } \Omega, \;\; w = 0 \quad \text{on } \partial\Omega.$$

We will find that if (3.24) has only the trivial solution which has nonzero index, then (3.23) has a positive solution. Moreover, for each isolated positive solution $(\hat{u}, \hat{v})$ of (3.23), there is a positive solution (u, v) of (3.22) with c, f large, fu close to $\hat{u}$ and cv close to $\hat{v}$.

There is another type of positive solution of (3.22) when $c, f \to \infty, c^{-1}f \to \alpha \in (0, \infty)$. Here, the corresponding positive solution (u, v) of (3.22) has the property that $c||u||_\infty, f||v||_\infty$ both tending to infinity. More precisely, if there is an isolated solution w_0 of the problem:

$$(3.25) \qquad \Delta w + w^+(a - \alpha^{-1}w^+) + w^-(e + w^-) = 0 \;\; \text{in } \Omega, \quad w = 0 \;\; \text{on } \partial\Omega$$

which changes sign in Ω and has non-zero index, then (3.22) has a positive solution near $(\alpha^{-1}w_0^+, -w_0^-)$ for c, f large. In this situation, one species is segregated to near where $w_0^+ \neq 0$ and the other species is segregated to near where $-w_0^- \neq 0$ in Ω. We now describe more carefully the first type of positive solution of (3.22).

Theorem 3.7 (Extreme Strong Competition for Both Species, I). *Suppose $(\hat{u}, \hat{v})$ is an isolated positive solution of (3.23) with non-zero index. Then for any $\epsilon > 0$, there exists a large $M > 0$, such that for any $c, f \geq M$, problem (3.22) has at least one positive solution (u, v) satisfying:*

$$||fu - \hat{u}||_\infty < \epsilon, \ \ ||cv - \hat{v}||_\infty < \epsilon. \tag{3.26}$$

By the index of $(\hat{u}, \hat{v})$, we mean the fixed point index denoted by $index_P(B, (\hat{u}, \hat{v}))$, where P represents the natural positive cone in $C(\bar{\Omega}) \times C(\bar{\Omega})$, and $B : C(\bar{\Omega}) \times C(\bar{\Omega}) \to C(\bar{\Omega}) \times C(\bar{\Omega})$ is defined by

$$B(u, v) = (-\Delta + k)^{-1}(ku + au - uv, kv + ev - uv),$$

with homogeneous Dirichlet boundary condition. Here, k is a large positive constant so that B maps some neighborhood N of $(\hat{u}, \hat{v})$ in P into P. (Recall the definition of fixed point index in Part B of Section 1.2.)

Proof. Let $\bar{u} = fu, \bar{v} = cv$. We readily verify that (u, v) is a positive solution of (3.22) if and only if $(\bar{u}, \bar{v})$ solves:

$$\begin{cases} \Delta u + u(a - v) - f^{-1}u^2 = 0 & \\ & \text{in } \Omega, \\ \Delta v + v(e - u) - c^{-1}v^2 = 0 & \\ u = v = 0 & \text{on } \partial\Omega. \end{cases} \tag{3.27}$$

Consequently, in order to prove this theorem, it suffices to prove (3.27) has a positive solution near $(\hat{u}, \hat{v})$ when f and c are large. Comparing (3.27) with (3.23), we see that this can be achieved readily by homotopy invariance of degree argument.

Theorem 3.8. *Suppose problem (3.24) has only the trivial solution $w \equiv 0$ with $index_{C_0^1(\bar{D})}(\tilde{B}_1, 0) \neq 0$, where $\tilde{B}_1 w = (-\Delta)^{-1}(aw^+ + ew^-)$. Then the problem (3.23) has at least one positive solution. Moreover, there exists a constant $M > 0$ such that any solution (u, v) of problem (3.23) satisfies:*

$$||u||_\infty + ||v||_\infty \leq M;$$

and the sum of the indices of all the positive solutions of (3.23) is equal to $index_{C_0^1(\bar{\Omega})}(\hat{B}_1, 0)$. Here, $C_0^1(\bar{\Omega})$ denotes the functions in $C^1(\bar{\Omega})$ with zero boundary value on $\partial\Omega$.

Proof. The proof of this theorem can be divided into three steps.
Step 1. We first show that there exists $M > 0$ such that any positive solution (u, v) of the problem:

$$\begin{cases} -\Delta u = tau + (1-t)a(u-v)^{+} - uv & \\ & \text{in } \Omega, \\ -\Delta v = tev + (1-t)e(v-u)^{+} - uv & \\ u = v = 0 & \text{on } \partial\Omega, \\ 0 \leq t \leq 1, & \end{cases} \tag{3.28}$$

must satisfy

$$||u||_{\infty} + ||v||_{\infty} < M. \tag{3.29}$$

Observe that neither component of a non-negative solution of (3.28) can vanish identically unless both components vanish identically.

Suppose, by contradiction, there exist $t_n \in [0, 1]$ and positive solutions (u_n, v_n) of (3.28) with $t = t_n$ such that

$$||u_n||_{\infty} + ||v_n||_{\infty} \to \infty.$$

Then from the equation we find

$$-\Delta\tilde{u}_n \leq a\tilde{u}_n, \ -\Delta\tilde{v}_n \leq e\tilde{v}_n, \ \tilde{u}_n|_{\partial\Omega} = \tilde{v}_n|_{\partial\Omega} = 0, \tag{3.30}$$

where

$$\tilde{u}_n = (||u_n||_{\infty})^{-1}u_n, \ \tilde{v}_n = (||v_n||_{\infty})^{-1}v_n.$$

From (3.30), we obtain

$$\int_{\Omega} |\nabla\tilde{u}_n|^2 dx \leq a \int_{\Omega} \tilde{u}_n^2 dx \leq a \ mes.(\Omega),$$

$$\int_{\Omega} |\nabla\tilde{v}_n|^2 dx \leq e \int_{\Omega} \tilde{v}_n^2 dx \leq e \ mes.(\Omega).$$

Here $mes.(\Omega)$ is the measure of the domain Ω. Thus $\{\tilde{u}_n\}, \{\tilde{v}_n\}$ are bounded in $W_0^{1,2}(\Omega)$, which is a Hilbert space. By compact embedding in $L^2(\Omega)$, we can choose a subsequence such that

$$\tilde{u}_n \to \tilde{u}, \ \tilde{v}_n \to \tilde{v} \text{ weakly in } W_0^{1,2}(\Omega) \text{ and strongly in } L^2(\Omega).$$

Moreover, we can deduce by taking $(-\Delta)^{-1}$ with Dirichlet boundary conditions on both sides of (3.30) that if $\tilde{u} = 0$, then $\tilde{u}_n \to 0$ in $C(\bar{\Omega})$. This contradicts $||\tilde{u}_n||_{\infty} = 1$. Therefore we have $\tilde{u} \neq 0$. Similarly, $\tilde{v} \neq 0$.

Since $||u_n||_\infty + ||v_n||_\infty \to \infty$, without loss of generality we suppose $||v_n||_\infty \to \infty$. From (3.28), we find

$$-\Delta \tilde{u}_n = t_n a \tilde{u}_n + (1-t_n)a(\tilde{u}_n - \frac{v_n}{||u_n||_\infty})^+ - ||v_n||_\infty \tilde{u}_n \tilde{v}_n.$$

Multiplying both sides by $\phi \in C_0^\infty(\Omega)$ and integrating over Ω, we obtain

$$\int_\Omega \tilde{u}_n(\Delta\phi)dx + t_n \int_\Omega a\tilde{u}_n \phi dx + (1-t_n)\int_\Omega a\left(\tilde{u}_n - \frac{v_n}{||u_n||_\infty}\right)^+ \phi dx$$
$$= ||v_n||_\infty \int_\Omega \tilde{u}_n \tilde{v}_n \phi dx.$$

Since $\int_\Omega \tilde{u}_n \tilde{v}_n \phi dx \to \int_\Omega \tilde{u}\tilde{v}\phi dx, ||v_n||_\infty \to \infty$ and the left side of the above equation is uniformly bounded, we obtain $\int_\Omega \tilde{u}\tilde{v}\phi dx = 0$. Since ϕ is arbitrary, we conclude $\tilde{u}\tilde{v} = 0$ a.e. in Ω.

Let $\alpha_n = ||u_n||_\infty/||v_n||_\infty$. Without loss of generality we may assume $\alpha_n \to \alpha \in [0,\infty)$ (otherwise, we consider $||v_n||_\infty/||u_n||_\infty$). We also assume $t_n \to \bar{t} \in [0,1]$. From the equations in (3.28), we obtain

$$-\Delta(\alpha_n \tilde{u}_n - \tilde{v}_n) = t_n[a(\alpha_n \tilde{u}_n) - e\tilde{v}_n] + (1-t_n)[a(\alpha_n \tilde{u}_n - \tilde{v}_n)^+ - e(\tilde{v}_n - \alpha_n \tilde{u}_n)^+].$$

Multiplying the above equation by $\phi \in C_0^\infty(\Omega)$, integrating over Ω and passing to the limit, we obtain

$$\int_\Omega \nabla(\alpha\tilde{u} - \tilde{v})\nabla\phi dx = \bar{t}\int_\Omega [a(\alpha\tilde{u}) - e\tilde{v}]\phi dx + (1-\bar{t})\int_\Omega [a(\alpha\tilde{u} - \tilde{v})^+ - e(\tilde{v} - \alpha\tilde{u})^+]\phi dx.$$

Since $\tilde{u}, \tilde{v} \geq 0$ and $\tilde{u}\tilde{v} = 0$, we have

$$(\alpha\tilde{u} - \tilde{v})^+ = \alpha\tilde{u}, \ \ (\tilde{v} - \alpha\tilde{u})^+ = \tilde{v};$$

and hence

$$\int_\Omega \nabla(\alpha\tilde{u} - \tilde{v})\nabla\phi \, dx = \int_\Omega [a(\alpha\tilde{u}) - e\tilde{v}]\phi \, dx.$$

Let $w_0 = \alpha\tilde{u} - \tilde{v}$. We have $w_0^+ = \alpha\tilde{u}, w_0^- = -\tilde{v}$ and so

$$\int_\Omega \nabla w_0 \nabla\phi \, dx = \int_\Omega (aw_0^+ + ew_0^-)\phi \, dx, \quad \text{for all } \phi \in C_0^\infty(\Omega).$$

This means $w_0 = \alpha\tilde{u} - \tilde{v}$ is a bounded weak solution of (3.24) and hence a classical solution. Since $w_0 \not\equiv 0$, we arrive at a contradiction. This completes the proof of step 1.

Step 2. Let B_M denotes the ball in $C(\bar{\Omega}) \times C(\bar{\Omega})$ centered at 0, with radius M as described in (3.29). Let P be the natural positive cone in $C(\bar{\Omega}) \times C(\bar{\Omega})$, and the operator B be defined as in Theorem 3.7. Here, we assume the positive constant k has been chosen sufficiently large for the definition of B so that B maps $P \cap B_M$ to P. We will show

$$deg_P(I - B, P \cap B_M, 0) = index_{C_0^1(\bar{\Omega})}(\tilde{B}_1, 0), \tag{3.31}$$

where $\tilde{B}_1$ is the mapping defined in the statement of this theorem.

First, by means of (3.29), the homotopy

$$\begin{aligned} H_t(u,v) = (-\Delta + k)^{-1}(tau + (1-t)a(u-v)^+ - uv + ku, \\ tev + (1-t)e(v-u)^+ - uv + kv) \end{aligned}$$

with $k > 0$ leads to

$$deg_P(I - B, P \cap B_M, 0) = deg_P(I - H_0, P \cap B_M, 0). \tag{3.32}$$

Next, we consider another homotopy

$$\begin{cases} -\Delta u = a(u-v)^+ - tuv + (1-t)\epsilon_0 & \\ & \text{in } \Omega, \\ -\Delta v = e(v-u)^+ - tuv + (1-t)\epsilon_0 & \\ u = v = 0 & \text{on } \partial\Omega, \end{cases} \tag{3.33}$$

where $t \in [0,1]$. Here, ϵ_0 is a fixed positive number. If (u,v) is a non-negative solution of (3.33), then $u - v$ satisfies

$$-\Delta(u-v) = a(u-v)^+ + e(u-v)^- \; in \; \Omega, \; (u-v)|_{\partial\Omega} = 0.$$

Thus, by the assumption of the theorem we obtain $u = v$; and hence u satisfies

$$-\Delta u = (1-t)\epsilon_0 - tu^2 \; in \; \Omega, \; u|_{\partial\Omega} = 0. \tag{3.34}$$

Using an upper and lower solution argument and noting that the right hand side of (3.34) is concave with respect to u, we see that (3.34) has a unique non-negative solution ψ_t for $t \in [0,1], 0 \le \psi_t \le (1-t)\epsilon_0(-\Delta)^{-1}(1)$ and $\psi_t > 0$ in Ω for $0 \le t < 1$. Here, $(-\Delta)^{-1}$ is taken with zero Dirichlet boundary condition. Thus (3.33) has a unique non-negative solution $(u,v) = (\psi_t, \psi_t)$. Define

$$\tilde{H}_t(u,v) = (-\Delta+k)^{-1}(a(u-v)^+ - tuv + (1-t)\epsilon_0 + ku, e(v-u)^+ - tuv + (1-t)\epsilon_0 + kv)$$

with $k > 0$ large and ϵ_0 chosen sufficiently small so that $||\psi_0||_\infty < M/2$. Then we obtain by homotopy invariance of degree that

$$deg_P(I - \tilde{H}_1, P \cap B_M, 0) = deg_P(I - \tilde{H}_0, P \cap B_M, 0) = index_P(\tilde{H}_0, (\psi_0, \psi_0)).$$

Since $\tilde{H}_1 = H_0$, we combine with (3.32) to find

$$deg_P(I - B, P \cap B_M, 0) = index_P(\tilde{H}_0, (\psi_0, \psi_0)). \tag{3.35}$$

Let $\tilde{P}$ denote the natural positive cone in $C_0^1(\bar{\Omega}) \times C_0^1(\bar{\Omega})$ and $j : \tilde{P} \to P$ denote the inclusion. Since $\tilde{H}_0$ maps a neighborhood of (ψ_0, ψ_0) in P continuously into $\tilde{P}$, the commutativity property of the fixed point index (see Nussbaum [178]) leads to

$$index_P(\tilde{H}_0, (\psi_0, \psi_0)) = index_P(j\tilde{H}_0, (\psi_0, \psi_0)) = index_{\tilde{P}}(\tilde{H}_0 j, (\psi_0, \psi_0)). \tag{3.36}$$

Since $(\psi_0, \psi_0) \in int(\tilde{P})$, due to maximum principle, we further find that

$$index_{\tilde{P}}(\tilde{H}_0 j, (\psi_0, \psi_0)) = index_{\tilde{E}}(\tilde{H}_0 j, (\psi_0, \psi_0)), \tag{3.37}$$

where $\tilde{E}$ denotes $C_0^1(\bar{\Omega}) \times C_0^1(\bar{\Omega})$.

Now consider the homeomorphism $h : \tilde{E} \to \tilde{E}$ defined by $h(u, v) = (u, u - v)$. Clearly $h^{-1} = h$ and h maps a neighborhood of (ψ_0, ψ_0) into a neighborhood of $(\psi_0, 0)$. Since (ψ_0, ψ_0) is an isolated fixed point of $\tilde{H}_0 j$, $(\psi_0, 0)$ is an isolated fixed point of $h^{-1}\tilde{H}_0 jh$. By the commutativity of the fixed point index, we have

$$index_{\tilde{E}}(\tilde{H}_0 j, (\psi_0, \psi_0)) = index_{\tilde{E}}(h^{-1}\tilde{H}_0 jh, (\psi_0, 0)), \tag{3.38}$$

where one readily verifies that

$$h^{-1}\tilde{H}_0 jh(u, w) = (-\Delta + k)^{-1}(aw^+ + \epsilon_0 + ku,\ aw^+ + ew^- + kw).$$

Since by assumption the problem $-\Delta w = aw^+ + ew^-$ *in* Ω, $w|_{\partial\Omega} = 0$ has only the trivial solution, we can use the homotopy $\hat{H}_t(u, w) = (-\Delta + tk)^{-1}(taw^+ + \epsilon_0 + tku,\ aw^+ + ew^- + tkw), 0 \le t \le 1$ to find

$$index_{\tilde{E}}(h^{-1}\tilde{H}_0 jh, (\psi_0, 0)) = index_{\tilde{E}}(\hat{H}_0, (\psi_0, 0)). \tag{3.39}$$

Now $\hat{H}_0(u, w) = ((-\Delta)^{-1}(\epsilon_0), (-\Delta)^{-1}(aw^+ + ew^-))$, and by the product theorem of degree, (cf. [75] or [178]), we find

(3.40)

$$\begin{aligned} & index_{\tilde{E}}(\hat{H}_0, (\psi_0, 0)) \\ &= index_{C_0^1(\bar{\Omega})}((-\Delta)^{-1}(\epsilon_0), \psi_0)) \cdot index_{C_0^1(\bar{\Omega})}((-\Delta)^{-1}(aw^+ + ew^-), 0) \\ &= index_{C_0^1(\bar{\Omega})}(\tilde{B}_1, 0). \end{aligned}$$

From (3.35) to (3.40), we obtain

$$deg_P(I - B, P \cap B_M, 0) = index_{C_0^1(\bar{\Omega})}(\tilde{B}_1, 0).$$

This proves the validity of (3.31).

Step 3. This last step will complete the proof of the theorem. By taking $t = 1$ in (3.28), the argument in step 1 above shows that any positive solution (u, v) of (3.23) satisfies $||u||_\infty + ||v||_\infty \le M$. Since $a, e > \lambda_1$, one can show as in Theorem 2.5 or 3.4 that (0,0) is a solution of (3.23) with

$$index_P(B, (0,0)) = 0.$$

Choose a small ball B_r such that

$$deg_P(I - B, B_r \cap P, 0) = index_P(B, (0,0)).$$

Then by the additivity of degree (cf. Theorem A2-1 in Chapter 6) and (3.31) we obtain

$$deg_P(I - B, (B_M \backslash \bar{B}_r) \cap P, 0) = index_{C_0^1(\bar{\Omega})}(\tilde{B}_1, 0) \neq 0.$$

Hence (3.23) has at least one non-negative solution in $(B_M \backslash \bar{B}_r) \cap P$, and the sum of the indices of all such solutions is equal to $index_{C_0^1(\bar{\Omega})}(\tilde{B}_1, 0)$. Since $a, e > \lambda_1$, (3.23) has no non-negative solution (u, v) with only one component identically zero. Since any non-negative solution of (3.23) must be in $B_M \cap P$ by step 1, and B_r is chosen so small that (3.23) has only the trivial solution in B_r, we see that all solutions with each component strictly positive in Ω of (3.23) are contained in $(B_M \backslash \bar{B}_r) \cap P$. Consequently, the sum of indices of such positive solutions of (3.23) is equal to $index_{C_0^1(\bar{\Omega})}(\tilde{B}_1, 0)$. This completes the proof of the theorem.

Corollary 3.9. *Suppose problem (3.24) has only the trivial solution $w \equiv 0$ with $index_{C_0^1(\bar{D})}(\tilde{B}_1, 0) \neq 0$, where $\tilde{B}_1 w = (-\Delta)^{-1}(aw^+ + ew^-)$. Then there exist large positive constants M and N such that for any $f, e \ge N$, problem (3.22) has at least one positive solution (u, v) satisfying:*

$$f||u||_\infty + c||v||_\infty \le M.$$

The next Theorem describes the second type of positive solution of (3.22) mentioned above.

Theorem 3.10 (Extreme Strong Competition for Both Species, II). *Let $\alpha \in (0, \infty)$. Suppose problem (3.25) has an isolated solution w_0 in $L^2(\Omega)$, which changes sign and has non-zero index. Then there exist respectively very large and small positive constants N and ϵ such that for any c, f satisfying*

$$c \ge N, \quad |c^{-1}f - \alpha| \le \epsilon,$$

the problem (3.22) has a positive solution (u, v) near $(\alpha^{-1}w_0^+, -w_0^-)$ in $L^2(\Omega) \times L^2(\Omega)$.

(The question of uniqueness and stability of the steady-state will be considered in Theorems 5.10 and 5.11 in Section 1.5 below).

Outline of Proof. The proof is similar to that of Theorem 3.8, thus we will only outline the main ideas. First, we consider the homotopy:

(3.41)

$$\begin{cases} -\Delta u = tau + (1-t)a(u-\alpha^{-1}v)^+ - tu^2 - (1-t)((u-\alpha^{-1}v)^+)^2 - cuv \\ \qquad\qquad\qquad\qquad\qquad\qquad\qquad\qquad\qquad\qquad\qquad\qquad \text{in } \Omega, \\ -\Delta v = tev + (1-t)e(v-\alpha u)^+ - tv^2 - (1-t)((v-\alpha u)^+)^2 \\ \qquad -(t\beta + (1-t)\alpha)cuv \\ u = v = 0 \qquad\qquad\qquad\qquad\qquad\qquad\qquad\qquad\qquad \text{on } \partial\Omega, \end{cases}$$

where $0 \le t \le 1$. Here $\beta > 0$ is fixed. If (u, v) is any non-negative solution of (3.41), we can readily show that

$$-\Delta u \le \frac{a^2}{4}, \; -\Delta v \le \frac{d^2}{4} \text{ in } \Omega, \; u = v = 0 \text{ on } \partial\Omega,$$

since the function $g(s) := \lambda s - s^2$ is bounded by $\lambda^2/4$ for $s > 0$. We thus obtain that if (u, v) is any non-negative solution of (3.41), then

$$0 \le u \le M, \; 0 \le v \le M \tag{3.42}$$

for some $M > 0$. For simplicity, we next denote the right hand side of the first equation in (3.41) by $f_1(u, v, t)$, and that for the second equation by $f_2(u, v, t)$. Let

$$u_M = min.\{u, M\}, \; v_M = min.\{v, M\}.$$

Define

$$\tilde{f_1}(u, v, t) = f_1(u_M, v_M, t), \; \tilde{f_2}(u, v, t) = f_2(u_M, v_M, t).$$

Choose $\delta > 0$ so small such that in the neighborhood $N_\delta(w_0)$ in $L^2(\Omega)$, w_0 is the only solution of (3.25). Then choose $\delta_1 > 0$ so that $(u, v) \in \partial N_{\delta_1}(\alpha^{-1}w_0^+, -w_0^-)$ implies that $u \ne 0, v \ne 0$ and $\alpha u - v \in N_\delta(w_0)$. Here $\partial N_{\delta_1}(\alpha^{-1}w_0^+, -w_0^-)$ denotes the boundary of the δ_1-neighborhood $N_{\delta_1}(\alpha^{-1}w_0^+, -w_0^-)$ of $(\alpha^{-1}w_0^+, -w_0^-)$ in $L^2(\Omega) \times L^2(\Omega)$. We then show there exist positive N_1 large and ϵ small such that problem (3.41), with the first and second line on right hand side respectively replaced by $\tilde{f_1}(u, v, t)$ and $\tilde{f_2}(u, v, t)$, has no non-negative solution (u, v) with $(u, v) \in \partial N_{\delta_1}(\alpha^{-1}w_0^+, -w_0^-)$ whenever $c \ge N_1$, $|\beta - \alpha| \le \epsilon$ and $0 \le t \le 1$.

For any $c \ge N_1$, let $M_c > 0$ be large enough such that

$$\tilde{f_1}(u, v, t) + M_c u \ge 0, \; \tilde{f_2}(u, v, t) + M_c v \ge 0,$$

for any $u, v \geq 0$ and $t \in [0, 1]$. Define the mapping

$$A_t(u, v) = (-\Delta + M_c)^{-1}(\tilde{f}_1(u, v, t) + M_c u, \tilde{f}_2(u, v, t) + M_c v)$$

which is completely continuous and maps the natural positive cone P in $L^2(\Omega) \times L^2(\Omega)$ into itself. We show that for large c, the problem with $t = 0$:

$$\begin{cases} -\Delta u = a(u - \alpha^{-1}v)^+ - ((u - \alpha^{-1}v)^+)^2 - cuv & \text{in } \Omega, \\ -\Delta v = e(v - \alpha u)^+ - ((v - \alpha u)^+)^2 - \alpha cuv & \\ u = v = 0 & \text{on } \partial\Omega, \end{cases} \tag{3.43}$$

has a unique non-negative solution in $N_{\delta_1}(\alpha^{-1}w_0^+, -w_0^-)$, and the solution denoted by $(u_c, v_c) = (u_c, \alpha u_c - w_0)$ tends to $(\alpha^{-1}w_0^+, -w_0^-)$ in $L^2(\Omega) \times L^2(\Omega)$ as $c \to \infty$.

As in the proof of Theorem 3.8, we use the regularity of A_0, the homeomorphism $h(u, v) = (u, \alpha u - v)$ in $\tilde{E} = C_0^1(\bar{\Omega}) \times C_0^1(\bar{\Omega})$, the commutativity of the fixed point index and the product formula (cf. [75] or [178]) to obtain:

$$\begin{aligned} °_P(I - A_1, P \cap N_{\delta_1}(\alpha^{-1}w_0^+, -w_0^-), 0) \\ &\qquad = deg_P(I - A_0, P \cap N_{\delta_1}(\alpha^{-1}w_0^+, -w_0^-), 0) \\ &\qquad = index_P(A_0, (u_c, v_c)) \\ &\qquad = index_{\tilde{E}}(A_0, (u_c, v_c)) \\ &\qquad = index_{C_0^1(\bar{\Omega})}(\tilde{B}_2, w_0) \cdot index_{C_0^1(\bar{\Omega})}(B, u_c) \\ &\qquad = index_{C_0^1(\bar{\Omega})}(\tilde{B}_2, w_0) \neq 0. \end{aligned}$$

Here, $\tilde{B}_2 w = (-\Delta)^{-1}[w^+(a - \alpha^{-1}w^+) + w^-(e + w^-)]$, $Bu = (-\Delta)^{-1}[a\alpha^{-1}w_0^+ - (\alpha^{-1}w_0^+)^2 - cu(\alpha u - w_0)]$, and we can obtain from uniqueness property that $index_{C_0^1(\bar{\Omega})}(B, u_c) = 1$.

This shows that $(u, v) = A_1(u, v)$ has at least one solution in the set $P \cap N_{\delta_1}(\alpha^{-1}w_0^+, -w_0^-)$ for $c \geq N > N_1$ and $|\beta - \alpha| \leq 1$. The solution of (3.22) will satisfy (3.41) for $t = 1$. For more details, see Dancer and Du [41].

Part C: One Much Stronger Competitor.

In the remaining part of this section, we finally consider the case of existence of positive solution for problem (3.22) when none of the conditions (i) or (ii) of

Theorem 3.4 is satisfied. (Here, in the notation of Theorem 3.4, $f_1(u,v) := -u - cv$, $f_2(u,v) := -fu - v, a_1 = a > \rho_1(-\Delta), a_2 = e > \rho_1(-\Delta), \sigma_1 = \sigma_2 = 1$. Also recall the definition of u_0 and v_0 in Remark 3.2.)

Define $\bar{c}, \bar{f}$ to be positive constants when

$$\hat{\rho}_1(\Delta + a - \bar{c}v_0) = 0, \quad \hat{\rho}_1(\Delta + e - \bar{f}u_0) = 0. \tag{3.44}$$

For convenience, we define coexistence parameter sets as follows:

$$T^+ := \{(c,f) : c > \bar{c},\ 0 \le f < \bar{f}, \text{ and problem (3.22) has a strictly positive solution}\},$$

$$T^- := \{(c,f) : f > \bar{f},\ 0 \le c < \bar{c}, \text{ and problem (3.22) has a strictly positive solution}\}.$$

Let:

$$g_1(c) = \int_\Omega h^3\,dx - \bar{f}c\int_\Omega h^2(-\Delta - (a - 2u_0))^{-1}(u_0 h)\,dx, \tag{3.45}$$

where h is the positive eigenfunction which spans the kernel of $-\Delta - (e - \bar{f}u_0)$ and normalized so that $||h||_{L^2(\Omega)} = 1$. Similarly, define

$$g_2(f) = \int_\Omega k^3\,dx - \bar{c}f\int_\Omega k^2(-\Delta - (e - 2v_0))^{-1}(v_0 k)\,dx, \tag{3.46}$$

where k is the positive eigenfunction which spans the kernel of $-\Delta - (a - \bar{c}v_0)$ and normalized so that $||k||_{L^2(\Omega)} = 1$.

The coefficients c and f in (3.22) can be interpreted as coefficients of competition. The following theorem describes situations of coexistence when the competition coefficient of one species is relatively large compared with the other.

Theorem 3.11 (Positive Solution with One Competitor much Stronger). *Consider problem (3.22) with $a > \rho_1(-\Delta), e > \rho_1(-\Delta)$ and $\bar{c}, \bar{f}$ as defined in (3.44). The coexistence parameter set described above has the following properties.*
(i) The set T^+ is nonempty if either $g_1(\bar{c}) > 0$ or $g_2(\bar{f}) < 0$.
(ii) For almost all (a,e) in $(\lambda_1,\infty) \times (\lambda_1,\infty)$, either T^+ is nonempty or T^- is nonempty.
(Here, g_1, g_2 are defined in (3.45) and (3.46).)

Proof. Linearizing equations (3.22) at $(u_0, 0)$ leads to the system:

$$\begin{cases} -\Delta y + (2u_0 - a)y = -cu_0 z & \\ & \text{in } \Omega, \\ -\Delta z + fu_0 z = ez & \\ y = z = 0 & \text{on } \partial\Omega. \end{cases} \tag{3.47}$$

Since $u_0 > 0$ in Ω, by comparison we have $\rho_1(-\Delta + (2u_0 - a)) > \rho_1(-\Delta + (u_0 - a)) = 0$. Thus by [3], the operator $[-\Delta + (2u_0 - a)]^{-1}$ exists and is a compact positive operator on $C_0^{1,\alpha}(\bar{\Omega}), 0 < \alpha < 1$. Equation (3.47) is thus equivalent to the system:

$$(3.48) \qquad \begin{aligned} y &= [-\Delta + (2u_0 - a)]^{-1}(-cu_0 z), \\ z &= f[\Delta + e]^{-1}(u_0 z). \end{aligned}$$

For convenience, let $A_1 = [\Delta + e]^{-1}, M_a z = u_0 z, A_2 = [-\Delta + (2u_0 - a)]^{-1}$, $M_{ca} z = -cu_0 z$, and $q = (y, z)^T$ then (3.48) can be written as

$$q = fB_1 q + B_2 q,$$

where

$$B_1 = \begin{bmatrix} 0 & 0 \\ 0 & A_1 M_a \end{bmatrix}, \qquad B_2 = \begin{bmatrix} 0 & A_2 M_{ca} \\ 0 & 0 \end{bmatrix}$$

are compact operators on the Banach space $[C_0^{1,\alpha}(\Omega)]^2$. Thus $I - fB_1 - B_2$ is a Fredholm operator on $[C_0^{1,\alpha}(\Omega)]^2$, with index 0. Furthermore, since $ker(I - \bar{f}B_1 - B_2)$ has dimension 1, $\bar{f}$ will be a simple eigenvalue of the pair $(I - B_2, B_1)$ provided

$$(3.49) \qquad B_1\phi \notin Range(I - B_2 - \bar{f}B_1),$$

where ϕ is any element in the $ker(I - \bar{f}B_1 - B_2)$ (cf. Chow and Hale [28]). To verify (3.49), let $\phi = (y, z)^T$ and

$$(3.50) \qquad \begin{aligned} y &= [-\Delta + (2u_0 - a)]^{-1}(-cu_0 z), \\ z &= \bar{f}[\Delta + e]^{-1}(u_0 z), \end{aligned}$$

with $z \neq 0$. Then the second component of $B_1\phi$ is $[\Delta + e]^{-1}u_0 z = \bar{f}^{-1}z$. For any $\phi^* = (y^*, z^*) \in [C_0^{1,\alpha}(\Omega)]^2$, the second component of $(I - B_2 - \bar{f}B_1)\phi^*$ is $I - \bar{f}[\Delta + e]^{-1}(u_0 z^*)$. Hence if $B_1\phi \in Range(I - B_2 - \bar{f}B_1)$, then $z \in Range(I - \bar{f}[\Delta + e]^{-1}M_a)$. However, by (3.50), the kernel $ker(I - \bar{f}[\Delta + e]^{-1}M_a)$ is spanned by z, leading to a contradiction.

The analysis above justifies the application of the bifurcation theorem of Crandall and Rabinowitz [33], with fixed $a > \lambda_1, e > \lambda_1, c > 0$, while the parameter f varies across $\bar{f}$. For all f near $\bar{f}$, $(u_0, 0)$ is a solution of problem (3.22). There exists $\delta_0 > 0$ and smooth functions $f : (-\delta_0, \delta_0) \to R$, $u : (-\delta_0, \delta_0) \to C_0^{1,\alpha}(\bar{\Omega})$, $v : (-\delta_0, \delta_0) \to C_0^{1,\alpha}(\bar{\Omega})$ such that:

$$f(0) = \bar{f}, \quad u(s) = u_0 + sy_0 + \tilde{y}(s), \quad v(s) = sz_0 + \tilde{z}(s),$$

where

$$z_0 \text{ spans the } ker(I - \bar{f}[\Delta + e]^{-1}M_a),$$

$$z_0(x) > 0, \text{ for } x \in \Omega, \quad \int_\Omega z_0^2\, dx = 1,$$

$$y_0 = [-\Delta + (2u_0 - a)]^{-1}(-cu_0 z_0);$$

$$||\tilde{y}(s)||_{C_0^{1,\alpha}(\Omega)} = o(|s|), \quad ||\tilde{z}(s)||_{C_0^{1,\alpha}(\Omega)} = o(|s|), \text{ as } s \to 0.$$

Moreover, in a sufficiently small neighborhood of $(\bar{f}, u_0, 0)$ in $R \times C_0^{1,\alpha}(\bar{\Omega}) \times C_0^{1,\alpha}(\bar{\Omega})$, the triples $(f(s), u(s), v(s)), |s| \leq \delta_0$ are the only solutions to (3.22) other than $(f, u_0, 0)$. In particular, we have positive solutions to (3.22) when $s > 0$.

We now let $\lambda(s) = f(s) - \bar{f}$ and calculate $\lambda'(0)$. For this purpose, we consider the equation

$$-\Delta v(s) = ev(s) - v^2(s) - f(s)u(s)v(s),$$

which is equivalent to:

(3.51)

$$\begin{aligned}(-\Delta + \bar{f}u_0 - e)(sz_0 + \tilde{z}(s)) &= -\lambda(s)(u_0 + sy_0 + \tilde{y}(s))(sz_0 + \tilde{z}(s)) \\ &\quad - (sz_0 + \tilde{z}(s))^2 - \bar{f}(sy_0 + \tilde{y}(s))(sz_0 + \tilde{z}(s)).\end{aligned}$$

Differentiating (3.51) with respect to s yields

(3.52)

$$\begin{aligned}&(-\Delta + \bar{f}u_0 - e)(\tilde{z}'(s)) \\ &\quad = \lambda'(s)(u_0 + sy_0 + \tilde{y}(s))(sz_0 + \tilde{z}(s)) - \lambda(s)(y_0 + \tilde{y}'(s))(sz_0 + \tilde{z}(s)) \\ &\quad\quad - \lambda(s)(u_0 + sy_0 + \tilde{y}(s))(z_0 + \tilde{z}'(s)) - 2(sz_0 + \tilde{z}(s))(z_0 + \tilde{z}'(s)) \\ &\quad\quad - \bar{f}(y_0 + \tilde{y}'(s))(sz_0 + \tilde{z}(s)) - \bar{f}(sy_0 + \tilde{y}(s))(z_0 + \tilde{z}'(s)).\end{aligned}$$

If we differentiate with respect to s once more and evaluate at $s = 0$, we obtain

$$\begin{aligned}(-\Delta + \bar{f}u_0 - e)(\tilde{z}''(s)) &= -2\lambda'(0)u_0 z_0 - 2(z_0 + \tilde{z}'(0))^2 \\ &\quad - 2\bar{f}(y_0 + \tilde{y}'(0))(z_0 + \tilde{z}'(0)).\end{aligned} \tag{3.53}$$

We deduce $\tilde{z}'(0) = \tilde{y}'(0) = 0$ and obtain from (3.53)

$$\int_\Omega z_0(-\Delta + \bar{f}u_0 - e)(\tilde{z}''(s))\, dx = \int_\Omega [-2\lambda'(0)u_0 z_0^2 - 2(z_0)^3 - 2\bar{f}y_0 z_0^2]\, dx. \tag{3.54}$$

Integrating by parts, we obtain

$$\lambda'(0) = -[\int_\Omega u_0 z_0^2\, dx]^{-1}[\int_\Omega (z_0^3 + \bar{f}y_0 z_0^2)\, dx]. \tag{3.55}$$

Identifying z_0 with h in (3.45), and noting that $y_0 = [-\Delta + (2u_0 - a)]^{-1}(-cu_0 z_0)$, we find

$$\lambda'(0) = -[\int_\Omega u_0 z_0^2\, dx]^{-1} g_1(c), \tag{3.56}$$

where $g_1(c)$ is given in (3.45). Suppose $g_1(\bar{c}) > 0$, then $g_1(c)$ changes sign at some $\tilde{c} > \bar{c}$, and for $c = c_1 \in (\bar{c}, \tilde{c})$, we have $g_1(c_1) > 0$, and $\lambda'(0) < 0$. Consequently for $\delta > 0$ sufficiently small, (3.22) has a positive solution if $(c, f) = (c_1, \bar{f} - \delta)$. That is T^+ is non-empty.

If $g_1(\bar{c}) > 0$ and c_0 is slightly less than $\bar{c}$, positive solution bifurcates to the left of $\bar{f}$ at $(c_0, \bar{f})$. If $g_1(\bar{c}) < 0$, then positive solution bifurcates to the right of $\bar{f}$ at $(c_0, \bar{f})$. By symmetry, if $g_2(\bar{f}) < 0$ and f_0 is slightly less than $\bar{f}$, positive solution bifurcates to the right of $\bar{c}$ at $(\bar{c}, f_0)$. Thus T^+ is also nonempty. This proves part (i).

We shall not show the proof of part (ii), which can be found in Dancer [40].

The following lemma can be proved readily, and can be used for applying part (i) of Theorem 3.11 to find positive coexistence states.

Lemma 3.1. *Consider problem (3.22) with $a > \lambda_1, e > \lambda_1$ and $\bar{c}, \bar{f}, g_1(\bar{c}), g_2(\bar{f})$ be as described in Theorem 3.11. If e is sufficiently large, then $g_1(\bar{c}) > 0$ and $g_2(\bar{f}) < 0$.*

Proof. We first deduce a more convenient expression for $g_1(\bar{c})$ and $g_2(\bar{f})$. By definition in (3.45)

$$-\Delta h = eh - \bar{f} u_0 h \text{ in } \Omega, \quad h = 0 \text{ on } \partial\Omega.$$

We have

$$-\Delta h - (a - 2u_0)h = (e - a)h + (2 - \bar{f})u_0 h \quad \text{in } \Omega.$$

That is

$$h = (e - a)Lh + (2 - \bar{f})L(u_0 h) \tag{3.57}$$

where L is the operator $[-\Delta - (a - 2u_0)]^{-1}$, under zero Dirichlet boundary conditions. Using (3.57), we can rewrite $g_1(\bar{c})$ by means of (3.45) as:

$$g_1(\bar{c}) = (2 - \bar{f})^{-1}(2 - \bar{f} - \bar{f}\bar{c}) \int_\Omega h^3\, dx - (2 - \bar{f})^{-1}(a - e)\bar{f}\bar{c} \int_\Omega h^2 Lh\, dx, \tag{3.58}$$

if $\bar{f} \neq 2$. Note that h is positive and L is a positive operator. Similarly, we obtain:

$$k = (a - e)L_2 k + (2 - \bar{c})L_2(v_0 k),$$

where L_2 is the operator $[-\Delta - (e - 2v_0)]^{-1}$, under zero Dirichlet boundary conditions. Moreover from (3.46), we have

$$g_2(\bar{f}) = (2 - \bar{c})^{-1}(2 - \bar{c} - \bar{c}\bar{f}) \int_\Omega k^3\, dx - (2 - \bar{c})^{-1}(e - a)\bar{c}\bar{f} \int_\Omega k^2 Lk\, dx. \tag{3.59}$$

It is easy to see that $g_2(\bar{f}) < 0$ if $\bar{c} < 2, 2 - \bar{c} - \bar{f}\bar{c} < 0$ and $e > a$. For fixed $a > \lambda_1$, equations (3.44) indicate that if e is large, then both $\bar{f}$ and v_0 are increased. This in turn leads to smaller $\bar{c}$. We thus have to carefully estimate $\bar{f}\bar{c}$ as e becomes large.

Let $\tilde{v} = e^{-1}v_0$. Then $\tilde{v}$ is a solution of

$$-e^{-1}\Delta\tilde{v} = \tilde{v}(1-\tilde{v}) \text{ in } \Omega, \quad \tilde{v} = 0 \text{ on } \partial\Omega.$$

It can be shown that as $e \to \infty$, we have $\tilde{v} \to 1$ in $L^p(\Omega)$ for each $p \in (1,\infty)$ (see e.g. Dancer [38] or similar proof in Theorem 5.2 in Section 1.5). Let $\hat{c} = e\bar{c}$. Note that $\bar{c}$ depends on e, and the spectral radius satisfies $r((-\Delta)^{-1}(a-\hat{c}\tilde{v})I) = 1$. Now, if $c > 0$, $r((-\Delta)^{-1}(a-c\tilde{v})I) \to r((-\Delta)^{-1}(a-c)I) = \lambda_1^{-1}(a-c)$ as $e \to \infty$, since $\tilde{v} \to 1$ in $L^p(\Omega)$. Suppose $\hat{c} = e\bar{c}$ is unbounded as $e \to \infty$. Let $e_i \to \infty$ and $\hat{c}(e_i) := e_i\bar{c} > c^*$ for some $c^* > a - \lambda_1$. Then

$$1 = r((-\Delta)^{-1}(a-\hat{c}\tilde{v})I) < r((-\Delta)^{-1}(a-c^*\tilde{v})I) \to \lambda_1^{-1}(a-c^*)$$

as $e_i \to \infty$. This implies $c^* < a - \lambda_1$, contradicting the assumption on c^*. We may thus assume $\hat{c} = e\bar{c} \to \alpha$ for some α as $e \to \infty$. Since $1 = r((-\Delta)^{-1}(a - \hat{c}\tilde{v})I) \to \lambda_1^{-1}(a-\alpha)$ as $e \to \infty$. We must have $e\bar{c} \to a - \lambda_1$ as $e \to \infty$.

We next consider the change of $\bar{f}$ for large e. Let $\tilde{r} > 0$ be an arbitrary number. We will show that if e is large, then

$$r((-\Delta)^{-1}(e-\tilde{r}eu_0)I) > 1, \text{ that is } \hat{\rho}_1(\Delta + (e-\tilde{r}eu_0)I) > 0. \tag{3.60}$$

Thus, from the definition of $\bar{f}$ in (3.44) and comparison, we must have $\bar{f} > \tilde{r}e$. Since $\tilde{r}$ is arbitrary, it follows that $e^{-1}\bar{f} \to \infty$ as $e \to \infty$. To prove (3.60), it suffices to find a $\mu_e > 1$ and a non-negative nontrivial function $w = s_e \in W_0^{1,2}(\Omega)$ which is a weak lower solution (as described in Section 9.3 in Chapter 9 of Evans [57]) for the problem:

$$-\Delta w = \mu_e^{-1}(e - reu_0)w \text{ in } \Omega, \quad w = 0 \text{ on } \partial\Omega. \tag{3.61}$$

This follows because a simple calculation shows that this implies that $(-\Delta + KI)^{-1}(e + K - \tilde{r}eu_0)s_e \geq \beta s_e$ for some $\beta > 1$. Here, $K > \tilde{r}eu_0$ is chosen to ensure the operator acting on s_e is positive. Thus by p. 265, in Schaefer [205] or a theorem similar to Theorem A2-6 in Chapter 6, we find the spectral radius satisfies

$$r((-\Delta + KI)^{-1}(e + K - \tilde{r}eu_0)I) \geq \beta > 1.$$

In order to construct the weak lower solution as described above, we choose a neighborhood N of $\partial\Omega$ in Ω such that $u_0(x) \leq (2\tilde{r})^{-1}$ in N. Let z denote the principal non-negative eigenfunction of $(-\Delta)$ on N, with zero Dirichlet boundary conditions on ∂N. Define $s_e(x)$ to be $z(x)$ on N and to be zero otherwise. Then $s_e \in W_0^{1.2}(\Omega)$. Since $e - \tilde{r}eu_0(x) \geq (1/2)e$ on N, we have

$$-\Delta s_e \leq \mu_e^{-1}(e - \tilde{r}eu_0)s_e \tag{3.62}$$

pointwise in N if e is large. Moreover (3.62) is trivially valid in $\Omega\backslash\bar{N}$ pointwise. We can then deduce as in Section 4.4 of Chapter 4 or Lemma 1.1 in Berestyki and Lions [7] that s_e is a weak lower solution for (3.61). This completes the proof that $e^{-1}\bar{f} \to \infty$ as $e \to \infty$. It follows that $\bar{f}\bar{c} \to \infty$ as $e \to \infty$. Since $\bar{c} \to 0$ as $e \to \infty$, the comments after (3.59) imply that $g_2(\bar{f}) < 0$ for large e.

By means of further analysis of the asymptotics for $\bar{f}, \bar{c}$ as $e \to \infty$ using formula (3.58), we can deduce as above that $g_1(\bar{c}) > 0$ as $e \to \infty$. For more details see Dancer [40].

The following theorem provides more information concerning the coexistence states as the parameters (c, f) changes near $(\bar{c}, \bar{f})$ as described in (3.44).

Theorem 3.12. *Consider problem (3.22) with hypotheses as described in Theorem 3.11. Assume that $(c_1, f_1) \in T^+$. Suppose further that $0 \le c \le c_1, f_1 \le f < \bar{f}$ and either $c < c_1$ or $f_1 < f$, then problem (3.22) has a strictly positive solution which is an "asymptotically stable" solution of the corresponding parabolic problem.*

Remark 3.4. Here, by an "asymptotically stable" solution, we mean a solution (u, v) such that the spectral radius satisfies $r(A'(u, v)) \le 1$, (u, v) is an isolated solution and has index 1 in

$$D = \{(u, v) : C_0(\bar{\Omega}) \times C_0(\bar{\Omega}), 0 \le u \le a, 0 \le v \le e \ in \ \Omega\},$$

where A is the map whose fixed points are solutions of (3.22), described in (3.65) below. $A'(u, v)$ denotes the Fréchet derivative of A at (u, v).

Proof. We first prove the existence of a strictly positive solution. Since $(c_1, f_1) \in T^+$, there exists a strictly positive solution (u_1, v_1) satisfying

$$\begin{cases} -\Delta u_1 = u_1(a - u_1 - c_1 v_1) & \\ & \text{in } \Omega, \\ -\Delta v_1 = v_1(e - v_1 - f_1 u_1) & \\ u_1 = v_1 = 0 & \text{on } \partial\Omega. \end{cases}$$

By comparison, we have $0 \le u_1 \le u_0 \le a, 0 \le v_1 \le v_0 \le e$ in $\bar{\Omega}$. Since $c \le c_1$, we have

$$-\Delta u_1 \le u_1(a - u_1 - cv_1) \text{ in } \Omega.$$

Moreover, strict inequality holds in Ω if $c < c_1$, since $u_1 > 0$ and $v_1 > 0$ in Ω. Thus

$$u_1 \le (-\Delta + \hat{k}I)^{-1}(u_1(a + \hat{k} - u_1 - cv_1)), \tag{3.63}$$

and equality does not hold if $c < c_1$. Similarly

$$v_1 \ge (-\Delta + \hat{k}I)^{-1}(v_1(e + \hat{k} - v_1 - fu_1)), \tag{3.64}$$

and equality does not hold if $f > f_1$. Here $\hat{k}$ is a positive constant satisfying $\hat{k} \geq max.\{a + ce, e + fa\}$ such that the mapping:

$$A(u, v) = (-\Delta + \hat{k})^{-1}(u(a + \hat{k} - u - cv), v(e + \hat{k} - v - fu)) \tag{3.65}$$

is monotone on the set:

$$D = \{(u, v) \in C_0(\bar{\Omega}) \times C_0(\bar{\Omega}) : 0 \leq u \leq a, 0 \leq v \leq e \ in \ \bar{\Omega}\}.$$

If we define $(u_2, v_2) = A(u_1, v_1)$, we see from (3.63) and (3.64) that

$$u_0 \geq u_2 \geq u_1, \ \ 0 \leq v_2 \leq v_1.$$

Defining $(u_{n+1}, v_{n+1}) = A(u_n, v_n), n = 2, 3, 4 \ldots,$ we have

$$u_0 \geq u_{n+1} \geq u_n, \ \ 0 \leq v_{n+1} \leq v_n, \ n = 2, 3, 4 \ldots$$

By theory explained in Leung [125], the function $(\tilde{u}, \tilde{v}) = lim_{n\to\infty}(u_n, v_n)$ is a strictly positive solution of (3.22), unless $\tilde{v} = 0$. In this case $(\tilde{u}, \tilde{v}) = (u_0, 0)$. It is also shown in Dancer [40] that

$$\left.\begin{array}{l} \tilde{u} \leq \hat{u} \\ \tilde{v} \geq \hat{v} \end{array}\right\} \text{ in } \bar{\Omega},$$

if $(\hat{u}, \hat{v})$ is any solution of (3.22) satisfying:

$$\left.\begin{array}{l} u_1 \leq \hat{u} \leq u_0 \\ 0 \leq \hat{v} \leq v_1 \end{array}\right\} \text{ in } \bar{\Omega}.$$

Let $C = \{(u, v) \in D : u_1 \leq u \leq u_0, 0 \leq v \leq v_1\}$. If the limit is such that $(\tilde{u}, \tilde{v}) = (u_0, 0)$, then $(u_0, 0)$ is the only fixed point of the mapping A in C. The set C is closed and convex (thus contractible); and $AC \subseteq C$ by monotonicity. Hence by basic properties of fixed point index, the sum of the indices of fixed points of A in C (counted relative to C) is 1, see Amann [3]. If $(u_0, 0)$ is the only fixed point in C, then we must have $index_C(A, (u_0, 0)) = 1$. On the other hand, by using Theorem 1 and Lemma 2 in Dancer [37] and part of Proposition 1 in Dancer [39], we can show that $index_C(A, (u_0, 0)) = 0$, provided we have

$$r(A'(u_0, 0)) > 1, \text{and } A'(u_0, 0)(h, k) \neq (h, k) \text{ if } (h, k) \in (C_0^+(\bar{\Omega}) \times C_0^+(\bar{\Omega})) \backslash \{0, 0\}. \tag{3.66}$$

Note that the first property above is related to property (α) of Definition 2.1, as indicated in the proof of case (ii) in Theorem 3.4. In order to analyze the spectral radius indicated in (3.66), we note that

$$A'(u_0, 0)(h, k) = (-\Delta + \hat{k}I)^{-1}(a + \hat{k} - 2u_0)h - cu_0k, (e + \hat{k} - fu_0)k).$$

We thus have the following relationship for the various spectrum

$$\sigma(A'(u_0,0)) = \sigma((-\Delta + \hat{k}I)^{-1}(a+\hat{k}-2u_0)I) \cup \sigma((-\Delta+\hat{k}I)^{-1}(e+\hat{k}-fu_0)I).$$

It follows that

$$r(A'(u_0,0)) \geq r((-\Delta+\hat{k}I)^{-1}(e+\hat{k}-fu_0)I) > 1.$$

The last inequality above is due to the fact that $f < \bar{f}$. The second property of (3.66) can be proved by procedures as in the proof of Theorem 3.4. We can thus conclude that we also have $index_C(A,(u_0,0)) = 0$, contradicting the fact that $index_C(A,(u_0,0)) = 1$ deduced above. Consequently, we must have $(\tilde{u},\tilde{v})$ is a strictly positive solution of (3.22).

The details of the proof of the "asymptotic stability" of the solution as described in the above remark will be given later in Section 1.5 of this chapter. By applying Remark 4 on p. 58 of Dancer [39], with $E = L^p(\Omega) \times L^p(\Omega)$ for large p, we can deduce that the solution is actually asymptotically stable with respect to the corresponding parabolic problem in the space $X^\alpha \times X^\alpha$, where X^α is a fractional power space in the sense of p. 29 in Henry [84] (cf. Section 6.4 in Chapter 6). For more explanations, see Dancer [40].

By means of Theorem 3.12, we can obtain the following more detailed information.

Theorem 3.13. *Consider problem (3.22) with hypotheses described in Theorem 3.11. Assume T^+ is nonempty. Then there exist $\mu > \bar{c}$, $\nu \in (0,\bar{f})$ and a continuous strictly increasing function $g^+ : [\bar{c},\mu] \to (0,\bar{f}]$ such that $g^+(\bar{c}) = \nu, g^+(\mu) = \bar{f}$, and $T^+ = \{(c,f) : c > \bar{c}, 0 < f < \bar{f}, f \geq g^+(c)\}$. Moreover, if $(c,f) \in int\, T^+$, then problem (3.22) has at least two solutions, at least one of which is "asymptotically stable". Furthermore, problem (3.22) has a strictly positive asymptotically stable solution if $c = \bar{c}$, $\nu < f < \bar{f}$, and a strictly positive solution if $f = \bar{f}$, $\bar{c} < c < \mu$.*

Notes.

Theorem 3.1 and Corollary 3.2 are found in Leung [121] and Pao [182]. Theorem 3.3 is due to Blat and Brown [11]. Theorems 3.4 and 3.5 are obtained from Li and Logan [151]. Theorem 3.6 is due to Ruan and Pao [195]. Theorems 3.7, 3.8, Corollary 3.9 and Theorem 3.10 are results in Dancer and Du [41]. Theorem 3.11 is given in Dancer [40] and the proof of part (i) follows an argument in Cantrell and Cosner [18]. Lemma 3.1, Theorem 3.12 and Theorem 3.13 are obtained from Dancer [40].

1.4 Strictly Positive Coexistence for Diffusive Cooperating Systems

In this section we study problem (1.1) when the functions $f_1(u,v)$ and $f_2(u,v)$ simulate cooperation or mutualism between the two species populations $u(x)$ and $v(x)$ in a bounded domain Ω, with conditions as described in Section 1.1. For simplicity, we first consider the Volterra-Lotka type of interaction when f_1 and f_2 are linear. More precisely, we consider

$$(4.1)\qquad \begin{cases} \sigma_1 \Delta u + u(a - bu + cv) = 0 & \\ & \text{in } \Omega, \\ \sigma_2 \Delta v + v(e + fu - gv) = 0 & \\ u = v = 0 & \text{on } \partial\Omega, \end{cases}$$

where a, b, c, e, f and g are all positive constants. The signs of the interaction coefficients $+c$ and $+f$ indicate mutualism. The first theorem shows that when each species can survive by itself (i.e. they satisfy (4.2)), and the cooperation coefficients are not very large (i.e. they satisfy (4.3)), then there will be coexistence equilibrium state. The main idea in the proof is that condition (4.3) will impose a bound on both populations.

Theorem 4.1. *Suppose*

$$(4.2)\qquad a > \sigma_1 \lambda_1 \quad \textit{and} \quad e > \sigma_2 \lambda_1,$$

then the boundary value problem (4.1) has a solution with each component strictly positive in Ω if and only if

$$(4.3)\qquad cf < bg.$$

Proof. By hypothesis (4.3), there exists (x_0, y_0) in the first open quadrant where

$$a - bx_0 + cy_0 \leq 0, \quad e + fx_0 - gy_0 \leq 0.$$

Define $(\bar{u}(x), \bar{v}(x)) \equiv (x_0, y_0)$ for $x \in \Omega$. Let $\omega(x) > 0$ be the principal eigenfunction for the operator $(-\Delta)$ on Ω with principal eigenvalue $\lambda_1 > 0$ and zero

Dirichlet boundary conditions. We readily verify that for $\delta > 0$ sufficiently small,

(4.4)
$$\begin{cases} \sigma_1\Delta\bar{u} + \bar{u}(a - b\bar{u} + cv) \le 0 \text{ in } \Omega, \text{ for } \delta\omega(x) \le v \le \bar{v}, \\ \sigma_2\Delta\bar{v} + \bar{v}(e + fu - g\bar{v}) \le 0 \text{ in } \Omega, \text{ for } \delta\omega(x) \le u \le \bar{u}, \\ \sigma_1\Delta(\delta\omega) + \delta\omega(a - b\delta\omega + cv) \\ \qquad = \delta\omega(-\sigma_1\lambda_1 + a - b\delta\omega + cv) \ge 0 \text{ in } \Omega, \text{ for } \delta\omega(x) \le v \le \bar{v}, \\ \sigma_2\Delta(\delta\omega) + \delta\omega(e + fu - g\delta\omega) \\ \qquad = \delta\omega(-\sigma_1\lambda_1 + e + fu - g\delta\omega) \ge 0 \text{ in } \Omega, \text{ for } \delta\omega(x) \le u \le \bar{u}. \end{cases}$$

Thus the functions $(\bar{u}(x), \bar{v}(x))$ and $(\delta\omega(x), \delta\omega(x))$ form a pair of coupled ordered upper-lower solutions for the boundary value problem (4.1). By Theorem 1.4-2 in Leung [125], the problem (4.1) has a solution $(u(x), v(x))$ satisfying

$$\delta\omega(x) \le u(x) \le x_0, \quad \delta\omega(x) \le v(x) \le y_0 \text{ for } x \in \bar{\Omega}.$$

To prove the converse, suppose (4.1) has a nontrivial positive solution $(\tilde{u}, \tilde{v})$ and $cf \ge bg$. Choose k such that

$$\frac{b}{c} \le k \le \frac{f}{g}.$$

Define $(u_\alpha(x), v_\alpha(x)) = (\alpha\omega(x), \alpha k\omega(x))$ for $x \in \bar{\Omega}$. Considering the equations (4.1) in a neighborhood of the boundary, we readily obtain by maximum principle that the outward normal derivative of $\tilde{u}$ and $\tilde{v}$ are strictly negative at the boundary. Thus we have

$$\tilde{u}(x) \ge u_{\alpha_0}(x), \ \tilde{v}(x) \ge v_{\alpha_0}(x), \quad x \in \bar{\Omega} \text{ for some } \alpha_0 > 0. \tag{4.5}$$

By the choice of k, we readily verify that

$$\begin{cases} \sigma_1\Delta u_\alpha + u_\alpha(a - bu_\alpha + cv_\alpha) \ge 0 \text{ in } \Omega, \\ \sigma_2\Delta v_\alpha + v_\alpha(e + fu_\alpha - gv_\alpha) \ge 0 \text{ in } \Omega \end{cases} \tag{4.6}$$

for all $\alpha \ge \alpha_0$. Using (4.5), (4.6) and the sweeping principle for quasimonotone nondecreasing system by means of a family of lower solutions, we assert that

$$\tilde{u}(x) \ge u_\alpha(x), \ \tilde{v}(x) \ge v_\alpha(x), \quad x \in \bar{\Omega} \text{ for all } \alpha > \alpha_0.$$

(The sweeping principle is an extension of Theorem 1.4-2 for the scalar case described in Leung [125]. The extension to quasimonotone nondecreasing system is described in Theorem A3-9 in Chapter 6.) We thus obtain a contradiction by letting $\alpha \to \infty$. Consequently, we must have (4.3).

We next consider the more general cooperating system:

$$\begin{cases} \Delta u + uM(u,v) = 0 & \\ & \text{in } \Omega, \\ \Delta v + vN(u,v) = 0 & \\ u = v = 0 & \text{on } \partial\Omega, \end{cases} \tag{4.7}$$

where $M, N \in C^1(R \times R)$,

$$M_v(u,v) > 0,\ N_u(u,v) > 0 \text{ for } u, v \geq 0. \tag{4.8}$$

(4.9) For $u, v \geq 0, -D \leq M_u \leq 0,\ -D \leq N_v \leq 0,$ for some $D > 0$; moreover, either M_u or N_v is not identically zero.

Let Γ_M and Γ_N be points on the open uv-plane defined respectively by the equations $M(u,v) = 0$ and $N(u,v) = 0$. For convenience, define the functions $M_1(u,v) = M(u,v) - M(0,0), N_1(u,v) = N(u,v) - N(0,0)$; and let Γ_{M_1} and Γ_{N_1} be points on the open uv-plane defined respectively by the equations $M_1(u,v) = 0$ and $N_1(u,v) = 0$. We will assume that

(4.10)
Γ_M and Γ_N are two distinct curves; and the set Γ_{M_1} and Γ_{N_1} are represented by two distinct positive functions $u = \phi_1(v), u = \psi_1(v)$ respectively for $v \geq 0$.

Theorem 4.2. *Under hypotheses (4.8) to (4.10), suppose $M(0,0) > \rho_1(-\Delta) = \lambda_1$, and $N(0,0) > \rho_1(-\Delta) = \lambda_1$.*

(i) If Γ_M and Γ_N intersect at a point (x_0, y_0) in the first open quadrant, then problem (4.7) has a solution with each component strictly positive in Ω.

(ii) If the problem (4.7) has a positive solution (with each component strictly positive in Ω), then

$$sup_{x>0} \frac{\psi_1(x)}{x} > inf_{x>0} \frac{\phi_1(x)}{x}. \tag{4.11}$$

Proof. We have $M(x_0, y_0) = N(x_0, y_0) = 0$. Define $(\bar{u}(x), \bar{v}(x)) \equiv (x_0, y_0)$ for $x \in \bar{\Omega}$. Let $\omega(x), \lambda_1$ be as defined in the proof of Theorem 4.1. For $\delta > 0$ sufficiently small, we have:

(4.12)

$$\begin{cases} \Delta\bar{u} + \bar{u}M(\bar{u}, v) = x_0 M(x_0, v) \le x_0 M(x_0, y_0) = 0 \text{ in } \Omega, \text{ for } \delta\omega(x) \le v \le \bar{v}, \\ \\ \Delta\bar{v} + \bar{v}N(u, \bar{v}) = y_0 N(u, y_0) \le y_0 N(x_0, y_0) = 0 \text{ in } \Omega, \text{ for } \delta\omega(x) \le u \le \bar{u}, \\ \\ \Delta(\delta\omega) + \delta\omega M(\delta\omega, v) \ge \Delta(\delta\omega) + \delta\omega M(\delta\omega, 0) \text{ in } \Omega, \text{ for } \delta\omega(x) \le v \le \bar{v}, \\ \qquad = -\lambda_1\delta\omega + \delta\omega M(\delta\omega, 0) \\ \qquad\quad > -\lambda_1\delta\omega + \delta\omega\lambda_1 = 0 \quad \text{in } \Omega, \text{ for } \delta > 0 \text{ sufficiently small}, \\ \\ \Delta(\delta\omega) + \delta\omega N(u, \delta\omega) \ge \Delta(\delta\omega) + \delta\omega N(0, \delta\omega) \text{ in } \Omega, \text{ for } \delta\omega(x) \le u \le \bar{u}, \\ \qquad = -\lambda_1\delta\omega + \delta\omega N(0, \delta\omega) \\ \qquad\quad > -\lambda_1\delta\omega + \delta\omega\lambda_1 = 0 \quad \text{in } \Omega, \text{ for } \delta > 0 \text{ sufficiently small}. \end{cases}$$

Thus the functions $(\bar{u}(x), \bar{v}(x))$ and $(\delta\omega(x), \delta\omega(x))$ form a pair of coupled ordered upper-lower solutions for the boundary value problem (4.7). By Theorem 1.4-2 in [125], the problem (4.7) has a solution $(u(x), v(x))$ satisfying

$$\delta\omega(x) \le u(x) \le x_0, \quad \delta\omega(x) \le v(x) \le y_0 \quad \text{for } x \in \bar{\Omega}.$$

This proves (i). For part (ii), suppose (4.7) has a positive solution $(\tilde{u}(x), \tilde{v}(x))$ in $\bar{\Omega}$ and (4.11) is false. That is, assume

$$inf_{x>0}\frac{\phi_1(x)}{x} \ge sup_{x>0}\frac{\psi_1(x)}{x}.$$

Then there exists a constant $\tau > 0$ such that $\phi_1(\sigma\omega(x)) \ge \tau\sigma\omega(x) \ge \psi_1(\sigma\omega(x))$ for all $x \in \Omega$, and all $\sigma > 0$. Note that by definition and (4.9), we have $0 = M(\phi_1(\sigma\omega(x)), \sigma\omega(x)) - M(0,0) \le M(\tau\sigma\omega(x), \sigma\omega(x)) - M(0,0)$. Thus, $M(0,0) \le M(\tau\sigma\omega(x), \sigma\omega(x))$ for $x \in \Omega$. Similarly, we obtain $N(0,0) \le N(\tau\sigma\omega(x), \sigma\omega(x))$ for $x \in \Omega$. We thus arrive at the following inequalities for $x \in \Omega$, all $\sigma > 0$:

$$\begin{cases} -\Delta(\tau\sigma\omega) = \tau\lambda_1\sigma\omega < \tau\sigma\omega M(0,0) \le \tau\sigma\omega M(\tau\sigma\omega, \sigma\omega), \\ \\ -\Delta(\sigma\omega) = \lambda_1\sigma\omega < \sigma\omega N(0,0) \le N(\tau\sigma\omega, \sigma\omega). \end{cases} \tag{4.13}$$

Moreover, by means of maximum principle at the boundary, we can verify that $\tilde{u} \ge \sigma_0\tau\omega, \tilde{v} \ge \sigma_0\omega$, for $x \in \bar{\Omega}$, $\sigma_0 > 0$ sufficiently small. Thus using the family of lower solutions $(\sigma\tau\omega, \sigma\omega), \sigma \ge \sigma_0$ for the quasimonotone nondecreasing system (4.7), we obtain a contradiction, $\tilde{u} \ge \sigma\tau\omega, \tilde{v} \ge \sigma\omega$, as $\sigma \to \infty$.

For a simple special case for (4.7), we consider

$$(4.14)\qquad \begin{cases} \Delta u + u(m_1(v) - m_2(u)) = 0 & \\ & \text{in } \Omega, \\ \Delta v + v(n_1(u) - n_2(v)) = 0 & \\ u = v = 0 & \text{on } \partial\Omega. \end{cases}$$

Here $m_1, m_2, n_1, n_2 \in C^2(R); m_2(0) = n_2(0) = 0$. As in (4.8) and (4.9), we assume

$$(4.15)\qquad m_i' > 0,\ n_i' > 0,\ \ for\ i = 1, 2.$$

$$(4.16)\qquad |m_2'| \le D, |n_2'| \le D,\ \ for\, some\, constant\, D > 0.$$

Theorem 4.3. *Assume the hypotheses on m_i, n_i above and (4.15) and (4.16). Suppose $m_1(0) > \rho_1(-\Delta)$, and $n_1(0) > \rho_1(-\Delta)$, and further*

$$(4.17)\qquad m_1'', n_1'' \le 0,\ \ m_2'', n_2'' \ge 0.$$

Then the problem (4.14) has a solution with each component strictly positive in Ω if and only if the two simultaneous equations: $m_1(v) = m_2(u), n_1(u) = n_2(v)$ has a solution in the first open quadrant in the uv-plane.

Proof. Hypotheses (4.15) to (4.17) ensures that conditions (4.8) to (4.10) are satisfied for problem (4.7) with $M(u, v) = m_1(v) - m_2(u), N(u, v) = n_1(u) - n_2(v)$,(except that the function $\psi_1(v)$ may possibly be only defined in a bounded subinterval of $v \ge 0$). If the simultaneous equations: $m_1(v) = m_2(u), n_1(u) = n_2(v)$ have a solution in the first open quadrant, then we can apply the same proof as the first part of Theorem 4.2 to assert that problem (4.14) has a positive solution.

Next, assume that (4.14) has a positive solution. Recall that $M_1(u, v) = m_1(v) - m_1(0) - m_2(u), N_1(u, v) = n_1(u) - n_1(0) - n_2(v)$. We verify $\phi_1(v) = m_2^{-1}(m_1(v) - m_1(0)), \psi_1(v) = n_1^{-1}(n_2(v) + n_1(0))$. The functions ϕ_1 is concave down and ψ_1 is concave up. Therefore, $\frac{\phi_1(x)}{x}$ is nonincreasing and $\frac{\psi_1(x)}{x}$ is non-decreasing. Note also that both ϕ_1 and ψ_1 are increasing functions. For $v \ge 0$, let $\phi(v) = m_2^{-1}(m_1(v)), \psi(v) = n_1^{-1}(n_2(v))$, we have $\phi(0) = m_2^{-1}(m_1(0)) > m_2^{-1}(0) = 0 > n_1^{-1}(0) = \psi(0)$. If the $lim_{u\to\infty} n_1(u)$ is finite, then $\psi(v)$ tends to ∞ as v tends to a finite number. Then there must be a number x^* where $\psi(x^*) > \phi(x^*)$. Hence ψ and ϕ must be equal at some positive number, and the simultaneous equations must have a solution in the first open quadrant.

If $lim_{u\to\infty} n_1(u) = \infty$, then $\psi_1(x)$ is defined for all $x > 0$, and by Theorem 4.2, we have (4.11):

$$sup_{x>0} \frac{\psi_1(x)}{x} > inf_{x>0} \frac{\phi_1(x)}{x}.$$

The last inequality implies that there exist $x_0 > 0$ such that $\psi_1(x_0) > \phi_1(x_0)$. Since $\psi_1(0) = 0 = \phi_1(0)$, there must be a $x_1 \in (0, x_0)$ at which $\psi_1'(x_1) > \phi_1'(x_1)$. This can be written as

$$\frac{n_2'(x_1)}{n_1'(n_1^{-1}(n_2(x_1) + n_1(0)))} > \frac{m_1'(x_1)}{m_2'(m_2^{-1}(m_1(x_1) - m_1(0)))}. \tag{4.18}$$

Since $n_2'' \geq 0, n_2' \geq n_2'(0) > 0$, we must have $n_2(x) \to \infty$ as $x \to \infty$. Consequently, there must exists $x_2 \in (x_1, \infty)$ such that $n_2(x) > n_2(x_1) + n_1(0)$ and $m_1(x) > m_1(x_1) - m_1(0)$ for $x > x_2$. It then follows readily from (4.18) that

$$\begin{aligned} \frac{n_2'(x)}{n_1'(n_1^{-1}(n_2(x)))} &\geq \frac{n_2'(x_1)}{n_1'(n_1^{-1}(n_2(x_1)+n_1(0)))} \\ &> \frac{m_1'(x_1)}{m_2'(m_2^{-1}(m_1(x_1)-m_1(0)))} \geq \frac{m_1'(x)}{(m_2'(m_2^{-1}(m_1(x)))}. \end{aligned} \tag{4.19}$$

for $x \geq x_2$. This means

$$\frac{d}{dx} n_1^{-1}(n_2(x)) > \frac{d}{dx} m_2^{-1}(m_1(x))$$

for $x \geq x_2$. Note that $\frac{d}{dx} n_1^{-1}(n_2(x))$ is nondecreasing in x while $\frac{d}{dx} m_2^{-1}(m_1(x))$ is nonincreasing. Consequently, we must have $n_1^{-1}(n_2(\tilde{x})) \geq m_2^{-1}(m_1(\tilde{x}))$ for some $\tilde{x} > x_2$; i.e. $\psi(\tilde{x}) \geq \phi(\tilde{x})$. We can then conclude the proof as in the last paragraph.

We next consider a generalization of Theorem 4.1, when the interaction coefficients b, c, f and g may change with position, and the Laplacian is replaced by two second order uniformly elliptic operators as follow:

$$L_k = \sum_{i,j=1}^{N} a_{ijk}(x)\partial_i\partial_j + \sum_{j=1}^{N} b_{jk}(x)\partial_j - c_k(x), \quad k = 1, 2 \tag{4.20}$$

with

$$a_{ijk} \in C(\bar{\Omega}), \quad b_{jk}, c_k \in L^\infty(\Omega), \quad i, j = 1, ..., N, \quad k = 1, 2, \tag{4.21}$$

We will consider the problem:

$$\begin{cases} L_1 u + u[a - b(x)u + c(x)v] = 0 & \\ & \text{in } \Omega, \\ L_2 v + v[e + f(x)u - g(x)v] = 0 & \\ u = v = 0 & \text{on } \partial\Omega, \end{cases} \tag{4.22}$$

where $b, c, f, g \in C(\bar{\Omega})$ satisfy $b(x) > 0, g(x) > 0$ for each $x \in \bar{\Omega}$, and $c \geq 0, f \geq 0$ in Ω, $c \not\equiv 0, f \not\equiv 0$; the parameters $a, e \in R$ are constants. For any function $h \in L^\infty(\Omega)$, we denote

$$h_L := ess\, inf_\Omega\, h, \quad h_M := ess\, sup_\Omega\, h.$$

We will consider solutions of (4.22) with u, v in $W^{2,p}(\Omega), p > N$, and the equations are satisfied almost everywhere. By the Sobolev embedding, we have $W^{2,p}(\Omega) \subset C^{2-N/p-\epsilon}$ for any small $\epsilon > 0$. Moreover, the functions $u, v \in W^{2,p}(\Omega)$ is twice classically differentiable almost everywhere in Ω. Actually, many of the theorems in the last two sections can be extended in analogous fashions as below. For convenience, we will let $w = \theta_{[-L_k, p(x), q(x)]}$ denote the positive solution of

$$-L_k w = w[p(x) - q(x)w] \text{ in } \Omega, \quad w = 0 \text{ on } \partial\Omega,$$

if $\hat{\rho}_1(L_k + p(x)) > 0$. Otherwise, let $\theta_{[-L_k, p(x), q(x)]} \equiv 0$ in $\bar{\Omega}$. (Here, we assume $q(x) > 0$ in $\bar{\Omega}$.) Recall the definition of the principal eigenvalues, $\rho_1(-\sigma\Delta + p(x))$ and $\hat{\rho}_1(\sigma\Delta + \tilde{p}(x))$, given in (1.4) and (1.7) in Section 1.1. We now extend the definitions naturally when $\sigma\Delta$ is replaced by a second order uniformly elliptic operator. Moreover, for the corresponding Dirichlet problem in a different domain G, the principal eigenvalues are denoted by $\rho_1^G(-L_k + p(x))$ and $\hat{\rho}_1^G(L_k + \tilde{p}(x))$. For more detailed description of the properties of such solutions, the maximum and comparison theorems in $W^{2,p}(\Omega)$ theory, the reader is referred to Theorems A3.1 to A3.5 in Section 6.3 in Chapter 6. The following theorem is an extension of Theorem 4.1 of this section.

Theorem 4.4 (Positive Solution under Weak Cooperation). *Suppose*

$$(4.23) \quad \begin{cases} c_M f_M < b_L g_L; \text{ and} \\ \hat{\rho}_1(L_1 + a + c(x)\theta_{[-L_2, e, g(x)]}) > 0, \ \hat{\rho}_1(L_2 + e + f(x)\theta_{[-L_1, a, b(x)]}) > 0, \end{cases}$$

then the boundary value problem (4.22) has a solution with each component strictly positive in Ω.

Proof. From (4.22), we see that if (u, v) is a positive solution of problem (4.22), then

$$u = \theta_{[-L_1, a+cv, b(x)]}, \quad v = \theta_{[-L_2, e+fu, g(x)]}.$$

By comparison, we readily deduce

$$\theta_{[-L_1, a+cv, b(x)]} \leq \theta_{[-L_1, a+c_M v_M, b_L]} \leq \frac{a + c_M v_M - (c_1)_L}{b_L}.$$

Thus,

$$(4.24) \qquad u_M \leq \frac{a + c_M v_M - (c_1)_L}{b_L}.$$

Similarly, we deduce

$$v_M \leq \frac{e + f_M u_M - (c_2)_L}{g_L}. \tag{4.25}$$

From (4.24) and (4.25) we obtain a bound of any positive solution of problem (4.22) in terms of a and e as follows:

$$\begin{cases} u_M \leq \frac{(a-(c_1)_L)g_L+(e-(c_2)_L)c_M}{b_L g_L - c_M f_M}, \\ v_M \leq \frac{(e-(c_2)_L)b_L+(a-(c_1)_L)f_M}{b_L g_L - c_M f_M}. \end{cases} \tag{4.26}$$

As in (2.14) we consider the problem:

$$-L_2 v - f(x)uv = v(e - g(x)v) \text{ in } \Omega, \quad v = 0 \text{ on } \partial\Omega. \tag{4.27}$$

Define the map $v(u)$ from $C^1(\bar{\Omega})$ to $C^1(\bar{\Omega})$ as in (2.19) with $\rho_1(-\sigma_2\Delta - fu)$ replaced by $\rho_1(-L_2 - f(x)u)$.

Note that if both $a \leq \rho_1(-L_1)$ and $e \leq \rho_1(-L_2)$, then the second and third inequality of assumptions (4.23) cannot be satisfied. Suppose $e > \rho_1(-L_2)$, we write the first equation of (4.22) as

$$-L_1 u - c(x)v(0)u = au - b(x)u^2 + c(x)[v(u) - v(0)]u \text{ in } \Omega, \quad u = 0 \text{ on } \partial\Omega \tag{4.28}$$

and bifurcate with increasing parameter a at $a = \rho_1(-L_1 - c(x)\theta_{[-L_2,e,g(x)]})$ when $(u, v) = (0, \theta_{[-L_2,e,g(x)]})$. From the bound (4.26) of positive solutions in terms of a, we can show as in Lemma 2.3 that there exists a continuum of solutions S^+ of (4.28) contained in $R \times P$, i.e. $u \geq 0$ whenever $(a, u) \in S^+$ and $\{a \in R : (a, u) \in S^+\} = (\rho_1(-L_1 - c(x)\theta_{[-L_2,e,g(x)]}, +\infty)$.

If $(a, u) \in S^+$, then $u \geq 0$ and so $v(u) \geq v(0)$, i.e. $v(u)$ is not the trivial solution. Consequently, the continuum of solutions $\{(a, u, v(u)) : (a, u) \in S^+\}$ for system (4.22) cannot be connected with the continuum of solutions $\{(a, u(0), 0) : a > \rho_1(-L_1)\}$. Thus both components of the solutions of (4.22) on the continuum $\{(a, u, v(u)) : (a, u) \in S^+\}$ are positive in Ω; and by comparison, both the second and third inequality in (4.23) are satisfied.

Similarly, suppose $a > \rho_1(-L_1)$, we bifurcate with e to obtain a solution of problem (4.22) with both components positive in Ω for each $e > \rho_1(-L_2 - g(x)\theta_{[-L_1,a,b(x)]})$. This completes the proof

Remark 4.1. From the methods in the previous two sections, we can deduce that the last two inequalities in (4.23) imply that the related indices of both solutions $(0, \theta_{[-L_2,e,g(x)]})$ and $(\theta_{[-L_1,a,b(x)]}, 0)$ of (4.22) are zero.

The previous theorems in this section essentially concentrate on finding steady states when cooperative interaction coefficients between the different species are relatively small. These are reflected for instance in assumptions (4.3) and the first part of (4.23). We next concentrate on situations when the cooperative coefficients are relatively large. For simplicity, we will assume

$$L_1 = L_2 = L. \tag{4.29}$$

The conditions on large cooperative coefficients will be imposed in the form:

$$c_L f_L - b_M g_M > b_M c_M - b_L c_L, \tag{4.30}$$

or

$$c_L f_L - b_M g_M > g_M f_M - g_L f_L. \tag{4.31}$$

In the next theorem, we will see for instance that under condition (4.30), for any given fixed $a < \rho_1(-L)$, there exists a constant $e(a)$ such that for $e > e(a)$, the problem (4.22), (4.29) cannot have positive equilibrium. Roughly speaking, the cooperation rates and growth rate of one species is too large for any possible coexistence equilibrium.

Theorem 4.5 (Nonexistence under Strong Cooperation). *Assume (4.29) for problem (4.22).*
(i) Suppose (4.30) holds, then for any fixed $a < \rho_1(-L)$, there exists a number $e = e(a)$ such that $a > \rho_1(-L - c(x)\theta_{[-L,e(a),g(x)]})$, and problem (4.22) does not have any coexistence positive solution if $e > e(a)$.
(ii) Suppose (4.31) holds, then for any fixed $e < \rho_1(-L)$, there exists a number $a = a(e)$ such that $e > \rho_1(-L - f(x)\theta_{[-L,a(e),b(x)]})$, and problem (4.22) does not have any coexistence positive solution if $a > a(e)$.

Before proving this theorem, we first consider the following two lemmas which estimate the sizes of the solutions, and will be used to prove the theorem.

Lemma 4.1. *Assume (4.29), and let (u, v) be any positive coexistence solution of (4.22). Then,*
(i) If $e \geq a$, then

$$u \leq \frac{c_M + g_M}{f_L + b_L} v. \tag{4.32}$$

(ii) If $a \geq e$, then

$$v \leq \frac{f_M + b_M}{c_L + g_L} u. \tag{4.33}$$

Proof. Assume (4.29), $e \geq a$, and let (u, v) be any coexistence positive solution of (4.22). Define

$$w = (c_M + g_M)v - (f_L + b_L)u. \tag{4.34}$$

We can deduce from (4.22) that we have in Ω,

$$(-L - a + b_L u + g_M v)w \geq 0. \tag{4.35}$$

Moreover, from the second equation of (4.22), we find

$$e = \rho_1(-L - fu + gv). \tag{4.36}$$

Thus the monotonic dependence of the principal eigenvalue on the potential implies that

$$a \leq e \leq \rho_1(-L - f_L u + g_M v).$$

This gives $\rho_1(-L - a - f_L u + g_M v) \geq 0$, and

$$\rho_1(-L - a + b_L u + g_M v) > \rho_1(-L - a - f_L u + g_M v) \geq 0. \tag{4.37}$$

Consequently, (4.35), (4.37), the strong maximum principle (cf. Theorem A3-1 in Chapter 6), and the argument for strong maximum principle for Theorem 3.5 in p. 35 of [71] imply that $w \geq 0$. This completes the proof of part (i) of this Lemma. Part (ii) is proved similarly.

Lemma 4.2. *(i) For a fixed $a < \rho_1(-L_1)$, let $e_0(a) > \rho_1(-L_2)$ be such that*

$$a > \rho_1(-L_1 - c(x)\theta_{[-L_2,e,g]}) \quad \textit{for each} \quad e > e_0(a). \tag{4.38}$$

Assume that there exists a sequence of positive coexistence solutions (e_n, u_n, v_n) of (4.22), $n \geq 1$, such that $e_n > max\{e_0(a), 0\}$ for each $n \geq 1$ and $lim_{n\to\infty} e_n = \infty$. Then, for any compact subset $K \subset \Omega$ there exists a positive constant $\alpha = \alpha(K) > 0$ such that for each $n \geq 1$

$$\frac{v_n}{e_n} \geq \alpha \quad \textit{in} \quad K. \tag{4.39}$$

(ii) Similarly, for a fixed $e < \rho_1(-L_2)$, let $a_0(e) > \rho_1(-L_1)$ be such that

$$e > \rho_1(-L_2 - f(x)\theta_{[-L_1,a,b]}) \quad \textit{for each} \quad a > a_0(e). \tag{4.40}$$

Assume that there exists a sequence of positive coexistence solutions (a_n, u_n, v_n) of (4.22), $n \geq 1$, such that $a_n > max\{a_0(e), 0\}$ for each $n \geq 1$ and $lim_{n\to\infty} a_n = \infty$. Then, for any compact subset $K \subset \Omega$ there exists a positive constant $\beta = \beta(K) > 0$ such that for each $n \geq 1$

$$\frac{u_n}{a_n} \geq \beta \quad \text{in} \quad K. \tag{4.41}$$

Proof. We first prove the existence of $e_0(a)$ with property as stated in inequality (4.38). Since $c \in C(\bar{\Omega}), c \geq 0, c \not\equiv 0$, there exists a ball B with $\bar{B} \subset \Omega$ such that

$$\tilde{c}_L := min_{\bar{B}} c > 0.$$

On the other hand, by Theorem 3.4 in [45] or Theorem A3-4 in Chapter 6

$$lim_{e\to\infty}\frac{\theta_{[-L_2,e,g]}}{e} = g^{-1} \text{ uniformly in } \bar{B};$$

and hence, there exists $\hat{e}$ such that for $e > \hat{e}$, we have

$$\theta_{[-L_2,e,g]} > \frac{e}{2\,max_{\bar{B}}g} \quad \text{in } \bar{B}.$$

Consequently, by comparison of principal eigenvalues (Theorem 2.3 in [45] or Theorem A3-5 in Chapter 6), we obtain

$$\rho_1(-L_1 - c\theta_{[-L_2,e,g]}) < \rho_1^B(-L_1 - c\theta_{[-L_1,e,g]}) < \rho_1^B(-L_1) - \frac{c_L}{2\,max_{\bar{B}}g}e$$

for each $e > \hat{e}$. Thus, for a fixed $a < \rho_1(-L_1)$, there must exists $e_0(a) > \rho_1(-L_2)$ such that inequality (4.38) is satisfied.

Let $(e_n, u_n, v_n), n \geq 1$, be a sequence of positive solutions of (4.22) with $e_n > max\{e_0(a), 0\}$ and $lim_{n\to\infty}e_n = \infty$. Then, from the second equation of (4.22), we find

$$-L_2 v_n = e_n v_n - g v_n^2 + f u_n v_n \geq e_n v_n - g v_n^2 \ \text{ in } \Omega,$$

with $f \not\equiv 0$; thus v_n is a strict positive upper solution of

$$-L_2 w = e_n w - g w^2 \ \text{ in } \Omega, \ \ w = 0 \ \text{ on } \partial\Omega.$$

By Lemma 3.2 in Delgado, López-Gómez and Suarez [45] (cf. Theorem A3-3 in Chapter 6), we find

$$v_n \geq \theta_{[-L_2,e_n,g]}. \tag{4.42}$$

Substituting (4.42) into the first equation of (4.22) and repeating the previous arguments, we obtain

$$u_n \geq \theta_{[-L_1-c(x)\theta_{[-L_2,e_n,g]},a,b(x)]}. \tag{4.43}$$

Note that the function on the right of the above inequality is well defined and positive because of (4.38). From (4.43) we find

$$lim\,inf_{n\to\infty}\frac{u_n}{e_n} \geq lim\,inf_{n\to\infty}\frac{\theta_{[-L_1-c(x)\theta_{[-L_2,e_n,g]},a,b(x)]}}{e_n}. \tag{4.44}$$

We now show that

$$lim\,inf_{n\to\infty}\frac{\theta_{[-L_1-c(x)\theta_{[-L_2,e_n,g]},a,b(x)]}}{e_n} \geq \frac{c_L}{b_M g_M} \tag{4.45}$$

uniformly in compact subsets of Ω. Let Ω_1, Ω_2 be two arbitrary subdomains of Ω such that

$$\bar{\Omega}_1 \subset \Omega_2, \quad \bar{\Omega}_2 \subset \Omega.$$

Define

$$\Theta_n := \frac{\theta_{[-L_1 - c(x)\theta_{[-L_2, e_n, g]}, a, b(x)]}}{e_n},$$

which is the unique positive solution of

$$(4.46) \qquad -\frac{1}{e_n} L_1 w = \left(\frac{a}{e_n} + c\frac{\theta_{[-L_2, e_n, g]}}{e_n}\right) w - bw^2 \quad \text{in } \Omega, \quad w = 0 \text{ on } \partial\Omega.$$

By Theorem 3.4 in [45] or Theorem A3-4 in Chapter 6,

$$lim_{n\to\infty} \frac{\theta_{[-L_2, e_n, g]}}{e_n} = g^{-1} \quad \text{uniformly in } \bar{\Omega}_2.$$

Thus, for any $\epsilon > 0$, there exists $n_0 = n_0(\epsilon)$ such that for each $n \geq n_0$ we have

$$(4.47) \qquad \frac{a}{e_n} + c\frac{\theta_{[-L_2, e_n, g]}}{e_n} \geq \frac{c_L}{g_M} - \epsilon \ \text{ in } \Omega_2.$$

Since Θ_n is the unique positive solution of (4.46), it follows from (4.47) that for each $n \geq n_0$, the function Θ_n is a strict positive upper solution of the problem:

$$(4.48) \qquad -\frac{1}{e_n} L_1 w = \left(\frac{c_L}{g_M} - \epsilon\right) w - bw^2 \quad \text{in } \Omega_2, \quad w = 0 \text{ on } \partial\Omega_2.$$

Suppose that $\epsilon > 0$ is sufficiently small such that $\frac{c_L}{g_M} - \epsilon > 0$, then for n sufficiently large, we have

$$\frac{c_L}{g_M} - \epsilon > \rho_1^{\Omega_2}\left(\frac{-1}{e_n} L_1\right) = \frac{\rho_1^{\Omega_2}(-L_1)}{e_n} \to 0 \quad \text{as} \quad n \to \infty.$$

Consequently (4.48) has a unique positive solution, say $\Theta_n^{\Omega_2}$; and by comparison we have

$$\Theta_n \geq \Theta_n^{\Omega_2} \quad \text{in} \ \Omega_2$$

for all n sufficiently large. Moreover, from (4.48) we obtain from Theorem 3.4 in [45] or Theorem A3-4 in Chapter 6 that

$$lim_{n\to\infty} \Theta_n^{\Omega_2} = \frac{c_L}{bg_M} - \frac{\epsilon}{b} \quad \text{uniformly in } \Omega_1.$$

Thus,

$$lim\,inf_{n\to\infty} \Theta_n \geq \frac{c_L}{b_M g_M} - \frac{\epsilon}{b_L} \quad \text{uniformly in } \ \Omega_1.$$

Since the above is valid for any $\epsilon > 0$, we obtain (4.45) uniformly in any compact subset of Ω. We then obtain from (4.44) that

$$liminf_{n\to\infty}\frac{u_n}{e_n} \geq \frac{c_L}{b_M g_M} \tag{4.49}$$

uniformly in any compact subset of Ω, and in particular in $\bar{\Omega}_1$. We next define

$$\hat{u}_n := \frac{u_n}{e_n}, \quad \hat{v}_n := \frac{v_n}{e_n},$$

and obtain from the second equation of (4.22) that

$$\frac{-1}{e_n}L_2\hat{v}_n = \hat{v}_n - g\hat{v}_n^2 + f\hat{u}_n\hat{v}_n.$$

Consequently, from (4.49) we see that for any $\epsilon > 0$, there exists $n_0 = n_0(\epsilon)$ such that $\hat{v}_n$ is a strict positive upper solution of the problem

$$\frac{-1}{e_n}L_2 w = (1 + \frac{f_L c_L}{b_M g_M} - \epsilon)w - gw^2 \text{ in } \Omega_1, \quad w = 0 \text{ on } \partial\Omega_1 \tag{4.50}$$

for each $n \geq n_0$. Choose $\epsilon > 0$ sufficiently small so that $1 + \frac{f_L c_L}{b_M g_M} - \epsilon > 0$. then we see that for n sufficiently large, problem (4.50) has a unique positive solution, which we denote by $\hat{\Theta}_n^{\Omega_1}$. Moreover, by Lemma 3.2 in [45] or Theorem A3-3 in Chapter 6, we find

$$\hat{v}_n = \frac{v_n}{e_n} \geq \hat{\Theta}_n^{\Omega_1}, \tag{4.51}$$

for sufficiently large n.

Let K be an arbitrary compact subset of Ω, we choose subdomains $\Omega_1 \subset \Omega_2$ as described above, and $K \subset \Omega_1$. Then by Theorem 3.4 in [45] or Theorem A3-4 in Chapter 6,

$$\lim_{n\to\infty} \hat{\Theta}_n^{\Omega_1} = (1 + \frac{f_L c_L}{b_M g_M} - \epsilon)g^{-1} \text{ uniformly in } K.$$

Since the limit above is bounded away from zero in K, we obtain (4.39) as described in part (i). Part (ii) is proved similarly.

Proof of Theorem 4.5.

Assume (4.29), (4.30) and fix $a < \rho_1(-L)$. Suppose there exists a sequence of positive coexistence solutions of (4.22), $(e_n, u_n, v_n), n \geq 1$, such that $e_n > max\{e_0(a), 0\}$ and $lim_{n\to\infty}e_n = \infty$. Let $\Omega_1 \subset \Omega$ be an arbitrary subdomain of Ω with $\bar{\Omega}_1 \subset \Omega$. By Lemma 4.2, there exists $\alpha = \alpha(\Omega_1) > 0$ such that for each $n \geq 1$

$$\frac{v_n}{e_n} \geq \alpha \text{ in } \Omega_1.$$

Moreover, by Lemma 4.1, we have for each $n \geq 1$

$$\frac{u_n}{e_n} \leq \frac{c_M + g_M}{f_L + b_L} \frac{v_n}{e_n}.$$

Thus by (4.30), there exists $\epsilon > 0$ such that for each $n \geq 1$

$$\frac{u_n}{e_n} \leq \frac{c_L v_n}{b_M e_n} - \epsilon \text{ in } \Omega_1.$$

That is, for all $n \geq 1$, we have

$$b_M u_n - c_L v_n \leq -\epsilon b_M e_n \text{ in } \Omega_1. \tag{4.52}$$

On the other hand, we find from the first equation in (4.22) that

$$a = \rho_1(-L + bu_n - cv_n) \leq \rho_1^{\Omega_1}(-L + b_M u_n - c_L v_n).$$

Consequently, we find from (4.52) that

$$a \leq \rho_1^{\Omega_1}(-L) - \epsilon b_M e_n \to -\infty \quad \text{as } n \to \infty.$$

This contradiction shows that problem (4.22) cannot have any positive coexistence state for e large enough. This completes the proof of Theorem 4.5.

The following theorem concerning a-priori uniform bound for positive solutions of (4.22) will lead to sufficient conditions for coexistence state in the case of large cooperative coefficients.

Theorem 4.6. *Assume (4.29) for problem (4.22) and $N \leq 5$. Suppose that*

$$c_L f_L > b_M g_M,$$

and for some $\alpha > 0$

$$max\{|a|, |e|\} \leq \alpha;$$

then there exists a constant $C = C(\alpha, \Omega, b, c, f, g)$ such that

$$||u||_{L^\infty(\Omega)} \leq C, \quad ||v||_{L^\infty(\Omega)} \leq C,$$

for any positive coexistence solution (u, v) of problem (4.22).

Proof. We shall prove this theorem under the condition $a \geq e$. By symmetry, the result can be proved similarly if $e \geq a$. Suppose that the conclusion of the theorem is false, and there exists a sequence of positive coexistence solutions $(a_k, e_k, u_k, v_k), k \geq 1$ with $-\alpha \leq e_k \leq a_k \leq \alpha$, such that

$$lim\, sup_{k \to \infty}(||u_k||_{L^\infty(\Omega)} + ||v_k||_{L^\infty(\Omega)}) = \infty. \tag{4.53}$$

We claim that

$$\text{(4.54)} \qquad lim\, sup_{k\to\infty}||u_k||_{L^\infty(\Omega)} = \lim\, sup_{k\to\infty}||v_k||_{L^\infty(\Omega)} = \infty.$$

Otherwise, if $\{||v_k||_{L^\infty(\Omega)}\}_{k\geq 1}$ is bounded by a positive constant β, then the first equation in (4.22) leads to

$$-Lu_k \leq (\alpha + c_M\beta)u_k - bu_k^2 \ \text{ in } \ \Omega;$$

and by comparison, we deduce that $\{||u_k||_{L^\infty(\Omega)}\}_{k\geq 1}$ is also bounded. However, this contradicts (4.53). Similarly, if $\{||u_k||_{L^\infty(\Omega)}\}_{k\geq 1}$ is bounded, then $\{||v_k||_{L^\infty(\Omega)}\}_{k\geq 1}$ is also bounded. Consequently, (4.54) must hold. By choosing a subsequence, if necessary, we may assume that

$$\text{(4.55)} \qquad lim_{k\to\infty}||u_k||_{L^\infty(\Omega)} = \infty, \ \ lim_{k\to\infty}(a_k, e_k) = (a_\infty, e_\infty),$$

for some $(a_\infty, e_\infty) \in R^2$ satisfying $-\alpha \leq e_\infty \leq a_\infty \leq \alpha$. From Lemma 4.1(ii), we obtain

$$\text{(4.56)} \qquad v_k \leq \frac{f_M + b_M}{c_L + g_L} u_k, \quad \text{for all } \ k \geq 1.$$

For each $k \geq 1$, let $x_k \in \Omega$ be such that

$$\text{(4.57)} \qquad u(x_k) = M_k := ||u_k||_{L^\infty(\Omega)}.$$

Since Ω is bounded, we may assume without loss of generality that

$$\text{(4.58)} \qquad lim_{k\to\infty}x_k = x_\infty \in \bar{\Omega}.$$

We now consider the two different cases where (i) $x_\infty \in \Omega$, or (ii) $x_\infty \in \partial\Omega$.

For case (i), denote

$$\delta := d(x_\infty, \partial\Omega)/2 > 0, \ \ \epsilon_k := M_k^{-1/2}, \ \ k \geq 1.$$

Since $lim_{k\to\infty}M_k = \infty$, we have $lim_{k\to\infty}\epsilon_k = 0$. The change of variables

$$\text{(4.59)} \qquad y := \frac{x - x_k}{\epsilon_k}, \ \ (z_k, w_k) := \epsilon_k^2(u_k, v_k), \ \ k \geq 1,$$

transforms the system (4.22) into

$$\text{(4.60)} \qquad \begin{cases} \mathcal{A}_k z_k = \epsilon_k^2 a_k z_k - b(x_k + \epsilon_k y)z_k^2 + c(x_k + \epsilon_k y)z_k w_k, \\ \\ \mathcal{A}_k w_k = \epsilon_k^2 e_k w_k - g(x_k + \epsilon_k y)z_k^2 + f(x_k + \epsilon_k y)z_k w_k, \end{cases}$$

where

$$\text{(4.61)} \ \ \mathcal{A}_k = -\Sigma_{i,j=1}^N a_{ij1}(x_k + \epsilon_k y)\partial_i\partial_j - \epsilon_k\Sigma_{j=1}^N b_{j1}(x_k + \epsilon_k y)\partial_j + \epsilon_k^2 c_1(x_k + \epsilon_k y),$$

provided $x_k + \epsilon_k y \in \Omega$. If $y \in R^N$ satisfies $|y| \leq \frac{\delta}{\epsilon_k}$, then $x = x_k + \epsilon_k y \in \Omega$, and thus (4.60) holds. For any $\rho > 0$, let B_ρ be the ball of radius ρ centered at the origin, we have $B_\rho \subset B_{\delta/\epsilon_k}$ for k sufficiently large, since $lim_{k\to\infty}\epsilon_k = 0$. From definition (4.59) we have $z_k = u_k/M_k$, thus

$$(4.62) \qquad ||z_k||_{L^\infty(B_\rho)} = 1, \;\; z_k(0) = 1, \;\; \text{for all} \;\; k \geq 1.$$

Moreover, from (4.56) and (4.62), we have

$$(4.63) \qquad ||w_k||_{L^\infty(B_\rho)} \leq \frac{f_M + b_M}{c_L + g_L}, \quad \text{for all} \;\; k \geq 1.$$

Using compactness argument as in Section 5.1 or Section A.3 in [125], we can choose subsequence, again labeled by k, such that there exists non-negative functions $z, w \in W^{2,p}(B_\rho) \cap C^{1,\nu}(B_\rho), 0 < \nu < 1$, $p > N$ sufficiently large, with

$$lim_{k\to}(z_k, w_k) = (z, w) \;\; \text{in} \;\; (W^{2,p}(B_\rho) \cap C^{1,\nu}(B_\rho))^2.$$

We thus have $z(0) = 1$, and passing to limit as $k \to \infty$ in (4.60), we find (z, w) satisfies:

$$(4.64) \qquad \begin{cases} -\Sigma_{i,j=1}^N a_{ij1}(x_\infty)\partial_i\partial_j z = -b(x_\infty)z^2 + c(x_\infty)zw, \\ \\ -\Sigma_{i,j=1}^N a_{ij1}(x_\infty)\partial_i\partial_j w = -g(x_\infty)z^2 + f(x_\infty)zw. \end{cases} \qquad \text{in} \;\; B_\rho.$$

Since ρ is arbitrary, by a standard diagonal sequence argument we can assert that $z, w \in W^{2,p}_{loc}(R^N)$ and (4.64) holds in the whole R^N. Moreover, standard elliptic regularity theory implies that $z, w \in C^2(R^N)$. Furthermore, by a linear change of coordinates, (4.64) can be reduced to

$$(4.65) \qquad \begin{cases} -\Delta z = -b(x_\infty)z^2 + c(x_\infty)zw \\ \\ -\Delta w = -g(x_\infty)w^2 + f(x_\infty)zw \end{cases} \qquad \text{in} \;\; R^N.$$

From (4.65), we obtain

$$(4.66) \qquad (-\Delta + b(x_\infty)z + g(x_\infty)w)(w - \frac{f(x_\infty) + b(x_\infty)}{c(x_\infty) + g(x_\infty)}z) = 0 \;\; \text{in} \;\; R^N.$$

Since the functions z, w are non-negative and $z(0) = 1$, the potential coefficient $V := b(x_\infty)z + g(x_\infty)w$ of the above equation has the property:

$$V \geq 0, \;\; V \not\equiv 0 \;\; \text{in} \;\; R^N.$$

By a Liouville type Theorem (see Lemma 7.5 in [45] or Theorem A3-6 in Chapter 6), the bounded solution in R^N of (4.66) must satisfy:

$$(4.67) \qquad w - \frac{f(x_\infty) + b(x_\infty)}{c(x_\infty) + g(x_\infty)}z = 0 \quad \text{in} \;\; R^N.$$

Using the relation (4.67), the first equation in (4.65) becomes

$$(4.68)\qquad -\Delta z = \frac{c(x_\infty)f(x_\infty) - b(x_\infty)g(x_\infty)}{c(x_\infty) + g(x_\infty)} z^2 \quad \text{in } R^N.$$

Since $c_L f_L > b_M g_M$, we have $c(x_\infty)f(x_\infty) > b(x_\infty)g(x_\infty)$. By Theorem 1.1 in Gidas and Spruck [70] or Theorem A3-7 in Chapter 6, the non-negative solution of the above equation must satisfy $z = 0$ in R^N, because $N \leq 5$. This contradicts the fact that $z(0) = 1$; therefore we must have case (ii) with $x_\infty \in \partial\Omega$.

For case (ii), we use the same argument as in the second part of the proof of Theorem 1.1 in [70] or Theorem A3-7 in Chapter 6 to show that the problem:

$$(4.69)\qquad \begin{cases} -\Delta z = -b(x_\infty)z^2 + c(x_\infty)zw \\ \\ -\Delta w = -g(x_\infty)w^2 + f(x_\infty)zw \end{cases} \quad \text{in } R^N_+,$$

where $R^N_+ = \{x \in R^N : x_N \geq 0\}$, has a non-negative solution (z, w) with $z(0) = 1$. Then, using the same argument as above with R^N replaced with R^N_+, we arrive again at a contradiction. We thus conclude that there must exist a-prior bound for the positive coexistence solution of (4.22) as described in the statement of this Theorem.

Theorem 4.7 (Positive Solution under Strong Cooperation). *Consider problem (4.22) with $L_1 = L_2$ and $N \leq 5$.*
(i) Suppose
(4.70)

$$\begin{cases} c_L f_L - b_M g_M > b_M c_M - b_L c_L, \text{ and} \\ \\ \hat{\rho}_1(L_1 + a + c(x)\theta_{[-L_2,e,g(x)]}) < 0, \text{ i.e. } a < \rho_1(-L_1 - c(x)\theta_{[-L_2,e,g(x)]}); \end{cases}$$

then the boundary value problem (4.22) has a solution with each component strictly positive in Ω.
(ii) Suppose
(4.71)

$$\begin{cases} c_L f_L - b_M g_M > g_M f_M - g_L f_L; \text{ and} \\ \\ \hat{\rho}_1(L_2 + e + f(x)\theta_{[-L_1,a,b(x)]}) < 0, \text{ i.e. } e < \rho_1(-L_2 - f(x)\theta_{[-L_1,a,b(x)]}); \end{cases}$$

then the boundary value problem (4.22) has a solution with each component strictly positive in Ω.

(Outline of Proof.) Let $G(e) := \rho_1(-L_1 - c\theta_{[-L,e,g]})$, $G(e)$ is a decreasing function of e. For a fixed $a < \rho_1(-L_1)$, we find from Theorem 4.5 above that there exists a number $e(a)$ such that if $a > \rho_1(-L_1 - c(x)\theta_{[-L_2,e(a),g(x)]}) = G(e(a))$, then there is no coexistence state, and if $e > e(a)$ there is no coexistence. Since

$\theta_{[-L_2,e,g(x)]} = 0$ if $e = 0$ and $a < \rho_1(-L_1 - c(x)\theta_{[-L_2,e,g(x)]})$ when $e = 0$, there is a number e_a such that $a = G(e_a)$. Theorem 4.6 above gives uniform bound for all solutions under first inequality in (4.70). Hence with e as the bifurcation parameter, the branch of unbounded curve of solutions has to connect e to minus infinity. That is bifurcating (e, u, v) at $(e_a, 0, \theta_{[-L_2,e_a,g(x)]})$, the branch of positive solutions connects e from e_a to minus infinity. However, if $e < e_a$, then $G(e) > G(e_a) = a$. This means the second inequality in (4.70) is satisfied. This proves part (i). The second part is proved in the same way by symmetry. (See Fig. 1.4.1 and Theorem 4.9 below for clarification, we allow both e and a to be $\leq \rho_1(-L_1)$ simultaneously.)

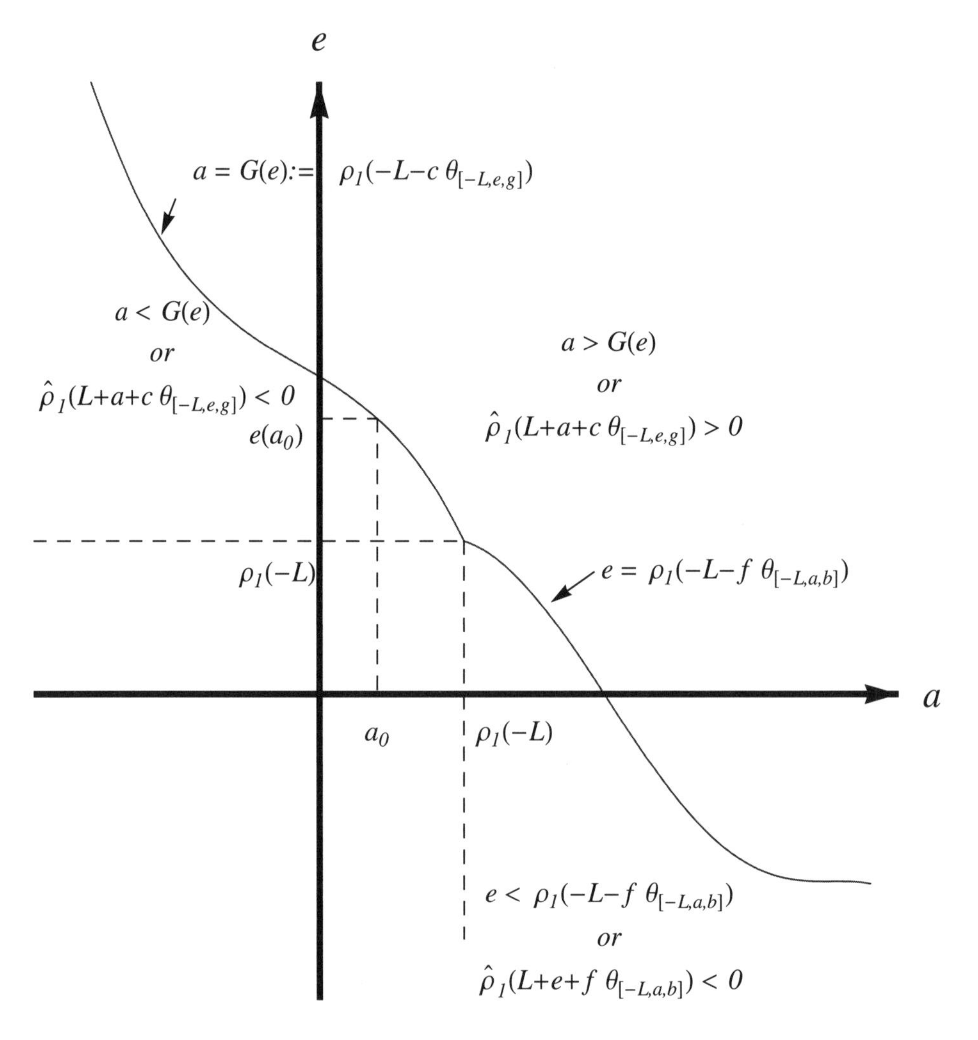

Figure 1.4.1: (For large c, f) Curves bounding regions of coexistence on (a, e) plane when b, c, f, g are fixed, and $L = L_1 = L_2$.

Remark 4.2. Roughly speaking, suppose

$$c_L f_L - b_M g_M > max.\{b_M c_M - b_L c_L, g_M f_M - g_L f_L\} \tag{4.72}$$

and some of the semitrivial positive solution is "linearly stable" (with index 1), then there exists positive coexistence state to (4.22).

The following theorems describe more carefully the set of parameters a, e when one or more coexistence state may occur. For fixed a, we define interval for the parameter e so that there exist coexistence solution(s) by I_2^a. For fixed e, define interval for parameter a so that there exist coexistence solution(s) by I_1^e. The following theorem first considers the case when the cooperative coefficients are relatively small.

Theorem 4.8. *Assume the first inequality in (4.23) for problem (4.22), and let I_1^e, I_2^a be defined as above.*

(i) Suppose $e > \rho_1(-L_2)$, then either $I_1^e = (\rho_1(-L_1 - c(x)\theta_{[-L_2,e,g(x)]}, \infty)$ or there exists $a_* \leq \rho_1(-L_1 - c(x)\theta_{[-L_2,e,g(x)]})$ such that $I_1^e = [a_*, \infty)$. If $a_* < \rho_1(-L_1 - c(x)\theta_{[-L_2,e,g(x)]})$, then there exists at least two positive coexistence states for $a \in (a_*, \rho_1(-L_1 - c(x)\theta_{[-L_2,e,g(x)]}))$.

(ii) Suppose $a > \rho_1(-L_1)$, then either $I_2^a = (\rho_1(-L_2 - f(x)\theta_{[-L_1,a,b(x)]}, \infty)$ or there exists $e_* \leq \rho_1(-L_2 - f(x)\theta_{[-L_1,a,b(x)]})$ such that $I_2^a = [e_*, \infty)$. If $e_* < \rho_1(-L_2 - f(x)\theta_{[-L_1,a,b(x)]})$, then there exists at least two positive coexistence states for $e \in (e_*, \rho_1(-L_2 - f(x)\theta_{[-L_1,a,b(x)]}))$.

Remark 4.3. The details of the proof of the above theorem can be found in Theorems 8.8 and 8.14 in Delgado, López-Gómez and Suarez [45]. The idea of the proof of part (i) is as follows. In case I_1^e is larger than $(\rho_1(-L_1 - c(x)\theta_{[-L_2,e,g(x)]}, \infty)$, then there exists a coexistence state (u_*, v_*) when $a = a_*$. Moreover for such e there will be a maximal coexistence state (u^e, v^e) satisfying $u_* \leq u^e \leq K_1, v_* \leq v^e \leq K_2$ for some large constants K_1, K_2. Using degree theory method as in the last chapter, it can be shown that the index of this maximal coexistence solution is 1. In order to satisfy the homotopy invariance of degree in an appropriate set of positive functions, there must be at least one more positive coexistence solution.

Similar multiplicity results can also be obtained in the case for large cooperative coefficients. In this case, we use homotopy invariance and show that the index of an appropriate minimal coexistence state is 1 to conclude that there must be another positive coexistence solution.

Theorem 4.9. *Consider problem (4.22) with $L_1 = L_2 = L$ and $N \leq 5$.*

(i) Assume the first inequality in (4.70) and $a < \rho_1(-L)$. Then either $I_2^a = (-\infty, e_a)$ or $I_2^a = (-\infty, e^*]$ for some $e^* \geq e_a$ where e_a is the unique value of

e satisfying $a = \rho_1(-L - c(x)\theta_{[-L,e_a,g(x)]})$. *If* $I_2^a = (-\infty, e^*]$ *with* $e^* > e_a$, *then problem (4.22) has at least two coexistence state for each* $e \in (e_a, e^*)$.
(ii) Assume the first inequality in (4.71) and $e < \rho_1(-L)$. *Then either* $I_1^e = (-\infty, a_e)$ *or* $I_1^e = (-\infty, a^*]$ *for some* $a^* \geq a_e$ *where* a_e *is the unique value of a satisfying* $e = \rho_1(-L - f(x)\theta_{[-L,a_e,b(x)]})$. *If* $I_1^e = (-\infty, a^*]$ *with* $a^* > a_e$, *then problem (4.22) has at least two coexistence state for each* $a \in (a_e, a^*)$.

The details of the proof of the theorem above can be found in Theorems 8.9 and 8.10 in [45].

Notes.

Theorem 4.1 is due to Korman and Leung [107]. Theorems 4.2 and 4.3 are found in Li and Ghoreishi [149]. Theorems 4.4 to 4.9 are obtained from Delgado, López-Gómez and Suarez [45].

1.5 Stability of Steady-States as Time Changes

In this section, we discuss the stabilities of the steady states found in the previous sections. Here, stability can be interpreted slightly differently in various cases. We might prove directly that certain smooth solutions of the corresponding parabolic problem stay close and tend to the steady state. Sometimes, only linearized stabilities are considered, and the steady states are stable or unstable with respect to solutions of the corresponding parabolic problem in appropriate functions spaces by means of applying standard stability theorems. In case that the linearized problem has zero as its eigenvalue, more sophisticated theorem will be applied. We will call nontrivial non-negative steady-state solutions with one component identically zero semi-trivial solutions.

Part A: Prey-Predator Case.

We first consider the prey-predator case discussed in Section 1.2. Before discussing the stability of the coexistence states, we note a very remarkble necessary and sufficient condition relating the existence of positive coexistence state and the linearized stability of the trivial and semi-trivial non-negative solutions.

Theorem 5.1. *Consider problem (2.23) under hypotheses (2.24) to (2.27) and additionally:*

$$h(0) - m(0) \neq \lambda_1 d, \text{ and } M_u(u, v) \leq 0 \text{ if } u, v \geq 0.$$

Then problem (2.23) has a positive solution iff the point spectrum of the linearized system at each of its trivial and semi-trivial non-negative solutions contains a positive number.

Proof. We first prove necessity, and assume (2.23) has a positive solution $(\bar{u}, \bar{v})$. The possible trivial or semi-trivial non-negative solutions are $(0,0), (u_0, 0), (0, v_0)$. We thus have to consider the linearization of the operator:

$$F : \begin{bmatrix} u \\ v \end{bmatrix} \rightarrow \begin{bmatrix} \Delta u + uM(u,v) \\ d\Delta v + v(h(u) - m(v)) \end{bmatrix} \tag{5.1}$$

at these three solutions. By comparison, we have $u_0 \geq \bar{u}$; thus by Lemma 2.1, we have $M(0,0) > \lambda_1$. That is, $\hat{\rho}_1(\Delta + M(0,0)I) > 0$. The Fréchet derivative $F'(0,0)$ is given by:

$$F'(0,0) = \begin{bmatrix} \Delta + M(0,0)I & 0 \\ 0 & d\Delta + (h(0) - m(0))I \end{bmatrix}. \tag{5.2}$$

Hence, the spectrum of $F'(0,0)$ contains a positive real number. We next consider the Fréchet derivative $F'(u_0, 0)$:

$$F'(u_0,0)\begin{pmatrix} w \\ z \end{pmatrix} = \begin{bmatrix} \Delta w + (M(u_0,0) + u_0 M_u(u_0,0))w + u_0 M_v(u_0,0)z \\ d\Delta z + (h(u_0) - m(0))z \end{bmatrix}. \tag{5.3}$$

We see that $F'(u_0,0)$ has only pure point spectrum σ_p given by $\sigma_p = \{\xi_1, \xi_2, \dots\} \cup \{\theta_1, \theta_2, \dots\}$ where $\{\xi_1, \xi_2, \dots\}$ is the point spectrum of the operator $\Delta + (M(u_0,0) + u_0 M_u(u_0,0))$, while $\{\theta_1, \theta_2, \dots\}$ is the point spectrum of the operator $d\Delta + (h(u_0) - m(0))$. By Theorem 2.5 (ii), (iii) we have $\theta_1 = \hat{\rho}_1(d\Delta + (h(u_0) - m(0)) > 0$. This means σ_p contains a positive number.

In case the solution $(0, v_0)$ of (2.23) exists with v_0 nontrivial, then Lemma 2.1 implies $h(0) > \lambda_1 d + m(0)$. We apply Theorem 2.5(iii) to assert $\hat{\rho}_1(\Delta + M(0, v_0)) > 0$. We then consider the Fréchet derivative: $F'(0, v_0)$

$$F'(0,v_0)\begin{pmatrix} w \\ z \end{pmatrix} = \begin{bmatrix} \Delta w + M(0,v_0)w \\ v_0 h'(0)w + d\Delta z + (h(0) - m(v_0) - v_0 m'(v_0))z \end{bmatrix}. \tag{5.4}$$

As in the above case, we deduce that the spectrum of $F'(0, v_0)$ contains a positive number.

We next prove the sufficiency part of this Theorem, and assume the point spectrum of the linearized system at each of the trivial and semi-trivial solutions contains a positive number. First, consider the point $(0,0)$. From the representation (5.2) for $F'(0,0)$, we must have either $\hat{\rho}_1(\Delta + M(0,0)I) > 0$ or $\hat{\rho}_1(d\Delta + (h(0) - m(0))I) > 0$. There are thus three possible cases (a) $M(0,0) > \lambda_1$, $h(0) < \lambda_1 d + m(0)$; (b) $M(0,0) > \lambda_1$, $h(0) > \lambda_1 d + m(0)$; or (c) $M(0,0) \leq \lambda_1$, $h(0) > \lambda_1 d + m(0)$.

We first consider case (a). Since $M(0,0) > 0$, we have a solution $(u_0, 0)$ with u_0 nontrivial. Consider the linearization of the operator $u :\rightarrow \Delta u + uM(u,0)$ at u_0. The principal eigenvalue for the operator $\Delta + M(u_0, 0)$ is zero. By comparison, the principal eigenvalue ξ_1 of the corresponding linear operator $\Delta+[M(u_0,0)+u_0M_u(u_0,0)]$ must have $\xi_1<0$. From (5.3), the spectrum σ_p of $F'(u_0,0)$ satisfies $\sigma_p=\{\xi_1,\xi_2,\dots\}\cup\{\theta_1,\theta_2,\dots\}$ where $\{\xi_1,\xi_2,\dots\}$ is the point spectrum of the operator $\Delta+[M(u_0,0)+u_0M_u(u_0,0)]$, while $\{\theta_1,\theta_2,\dots\}$ is the point spectrum of the operator $d\Delta+(h(u_0)-m(0))$. Thus $\xi_1<0$ implies that the principal eigenvalue θ_1 of the operator $d\Delta+(h(u_0)-m(0))$ must be >0. From Theorem 2.5(ii), we conclude that problem (2.23) has a positive solution.

We next consider case (b). Since $h(0)>\lambda_1 d+m(0)$, we have a solution $(0,v_0)$ with v_0 nontrivial. Consider the linearization of the operator $v:\to d\Delta v+v(h(0)-m(v))$ at v_0. By comparison, the principal eigenvalue $\tilde{\xi}_1$ of the corresponding linear operator $d\Delta+[h(0)-m(v_0)-v_0m'(v_0)]$ must have $\tilde{\theta}_1\le 0$. From (5.4), the spectrum σ_p of $F'(0,v_0)$ satisfies $\sigma_p=\{\tilde{\xi}_1,\tilde{\xi}_2,\dots\}\cup\{\tilde{\theta}_1,\tilde{\theta}_2,\dots\}$ where $\{\tilde{\xi}_1,\tilde{\xi}_2,\dots\}$ is the point spectrum of the operator $\Delta+M(0,v_0)I$, while $\{\tilde{\theta}_1,\tilde{\theta}_2,\dots\}$ is the point spectrum of the operator $d\Delta+[h(0)-m(v_0)-v_0m'(v_0)]$. Thus $\tilde{\theta}_1\le 0$ implies that the principal eigenvalue $\tilde{\xi}_1$ of the operator $\Delta+M(0,v_0)I$ must be >0. From Theorem 2.5(iii), we conclude that problem (2.23) has a positive solution.

Finally, we now show that case (c) cannot occur. Since $h(0)>\lambda_1 d+m(0)$, we have a solution $(0,v_0)$ with v_0 nontrivial. As in the last paragraph, the principal eigenvalue $\tilde{\xi}_1$ of the linear operator $\Delta+[h(0)-m(v_0)-v_0m'(v_0)]$ must have $\tilde{\theta}_1\le 0$, and the principal eigenvalue $\tilde{\xi}_1$ of the operator $\Delta+M(0,v_0)I$ must be >0. However, by assumption $M(0,0)\ge M(0,v_0)$; thus $\hat{\rho}_1(\Delta+M(0,0))>0$. This contradicts the condition $M(0,0)\le\lambda_1$ of case (c).

This completes the proof of Theorem 5.1

The problem of uniqueness and stability of positive solutions of (2.23) is usually quite difficult. We now consider the uniqueness and stability of positive solution for a special case of (2.23) when the diffusion parameters are small. This is a singular perturbation problem. The result can also be used to study the situation when the space domain is large. We will see that the effect of the boundary condition will become less significant. More precisely, consider

$$\begin{cases} \epsilon\Delta u+u(a-bu-cv)=0 & \\ & \text{in } \Omega, \\ \epsilon\Delta v+d^{-1}v(h(u)-m(v))=0 & \\ & \\ u=v=0 & \text{on } \partial\Omega. \end{cases} \tag{5.5}$$

(5.6) The functions h and m belong to $C^1(R)$, with $h'>0$ and $m'\ge 0$; a,b,c,d and ϵ are positive constants.

For convenience, we denote

$$F(u,v)=(u(a-bu-cv),d^{-1}v(h(u)-m(v)), \tag{5.7}$$

$X = L^p(\Omega) \times L^p(\Omega)$, $p > 1$, $A = diag(\Delta, \Delta)$ is an operator on X.

Theorem 5.2 (Uniqueness near Constant Equilibrium). *Assume that the equation $F(u,v) = 0$ has an isolated root $w_0 = (\hat{u}, \hat{v})$ in the first open quadrant in R^2, and there exists a constant B_0 such that $m(B_0) > h(a/b)$. Then the problem (5.5) has a unique solution w_ϵ in a neighborhood $N(w_0)$ of the constant function w_0 in X for sufficiently small $\epsilon > 0$. Moreover, $||w_\epsilon - w_0||_X \to 0$ as $\epsilon \to 0^+$.*

Proof. From the proof of Theorem 2.5. we have an a-priori bound on the values of all positive solutions of (5.5), independent of $\epsilon > 0$. We can modify the function $F(u,v)$ for large $|u| + |v|$, and for $u < 0$ or $v < 0$, without affecting the equilibrium positive solutions we are seeking. We may thus assume without loss of generality that $F(u,v)$ and all its first partial derivatives are bounded for all $(u,v) \in R^2$, and the first or second component of F is zero when $u \leq 0$ or $v \leq 0$ respectively. By comparison and sweeping principle argument for scalar equations, we can readily justify that the solutions we found will be positive solutions of the original problem.

Let p be a large positive number greater that $max\{2, dim\,\Omega\}$, and consider the operator A on $L^p(\Omega) \times L^p(\Omega)$:

$$A = \begin{bmatrix} \Delta & 0 \\ 0 & \Delta \end{bmatrix} \tag{5.8}$$

with domain $D(A) = (W^{2,p}(\Omega) \cap W_0^{1,p}(\Omega)) \times (W^{2,p}(\Omega) \cap W_0^{1,p}(\Omega))$. We may consider the functions F to be a mapping from $L^p(\Omega) \times L^p(\Omega)$ into $L^\infty(\Omega) \times L^\infty(\Omega)$, and thus into $L^r(\Omega) \times L^r(\Omega)$ for any $r \in [1, \infty)$. Due to the structure of F, we can assert that the operator is continuous from $L^p(\Omega) \times L^P(\Omega)$ into $L^r(\Omega) \times L^r(\Omega)$. (See Theorem 19.2 in Vainberg [222] or Theorem A4-1 in Chapter 6). Let F' be the Jacobian matrix of F, we can similarly obtain the mapping from $L^p(\Omega) \times L^p(\Omega)$ into the entries of F' in $L^r(\Omega)$ is continuous. Moreover, using Hölder's inequality *i.e.*$||fg||_q \leq ||f||_p ||g||_r$ for $1/q = 1/p + 1/r$, and the argument in Section 20 in [222], we can show that F maps $L^p \times L^p$ into $L^q \times L^q$ with continuous Gateaux derivatives expressible by means of $F' \in M_r^2$ where M_r^2 denotes 2×2 matrices with entries in $L^r(\Omega)$. Since its Gateaux derivative F' is continuous, the map F is Fréchet differentiable as a mapping from $L^p(\Omega) \times L^p(\Omega)$ into $L^q(\Omega) \times L^q(\Omega)$ (cf. Theorem 3.3 in Vainberg [222] or Theorem A4-2 in Chapter 6). More precisely, we obtain

$$\begin{aligned} &F(w) - F(w_1) = F'(w_1)(w - w_1) + \tilde{\theta}(w), \\ &\text{with } F'(w_1) \in M_r^2, \;\; ||\tilde{\theta}(w)||_q = o(||w - w_1||_p), \end{aligned} \tag{5.9}$$

where w, w_1 are elements of $L^p(\Omega) \times L^p(\Omega)$. Note that r can be chosen arbitrarily large so that q can be made large and satisfies $q > max\{2,\, dim\,\Omega\}$. Moreover we have $q < p$ and $L^q \supset L^p$. The operator A defined in (5.8) can be extended from $L^p(\Omega) \times L^p(\Omega)$ into $L^q(\Omega) \times L^q(\Omega)$, with domain $D(A) = (W^{2,q}(\Omega) \cap W_0^{1,q}(\Omega) \times (W^{2,q}(\Omega) \cap W_0^{1,q}(\Omega))$. We will denote by A as operator in both $L^p(\Omega) \times L^p(\Omega)$ and $L^q(\Omega) \times L^q(\Omega)$ without causing confusion.

The function w_0 is a constant function in Ω, and thus $F'(w_0)$ is a constant matrix which commutes with the operator A. We have:

$$F'(w_0) = \begin{bmatrix} -b\hat{u} & -c\hat{u} \\ d^{-1}\hat{v}h'(\hat{u}) & -d^{-1}\hat{v}m'(\hat{v}) \end{bmatrix}. \tag{5.10}$$

Let μ_1, μ_2 be eigenvalues of the matrix $F'(w_0)$. Then we have

$$\mu_1 + \mu_2 = -b\hat{u} - d^{-1}\hat{v}m'(\hat{v}) < 0,$$

$$\mu_1 \cdot \mu_2 = bd^{-1}\hat{u}\hat{v}m'(\hat{v}) + cd^{-1}\hat{u}\hat{v}h'(\hat{u}) > 0.$$

This implies that $Re\,\mu_1 < 0$ and $Re\,\mu_2 < 0$. Thus we have the spectrum $\sigma(F'(w_0)) \subset \{z : z \in C,\ Re\,z \le -\bar{c} < 0\}$ for some constant $\bar{c}$.

The C_0 semigroup $U(t)$ generated by the bounded operator $F'(w_0)$ satisfies $||U(t)|| \le Me^{-\bar{c}t}$ for some constant $M > 0$. For each $\epsilon > 0$, the operator ϵA generate a C_0 semigroup T_ϵ with $||T_\epsilon|| \le M_0$, where $M_0 \ge 1$. Since ϵA commutes with $F'(w_0)$, we have $T_\epsilon(t)U(t) = U(t)T_\epsilon(t)$, and $||(T_\epsilon(t)U(t))^n|| = ||(T_\epsilon(nt)U(nt))|| \le M_0 Me^{-\bar{c}nt}$. Let $S_\epsilon(t)$ be the C_0 semigroup generated by $\epsilon A + F'(w_0)$, then the Trotter product formula (Corollary 5.5, in Chapter 3 of Pazy [184] or Theorem A4-6 in Chapter 6) yields for all $x \in L^q(\Omega) \times L^q(\Omega)$:

$$S_\epsilon(t)x = lim_{n\to\infty}(T_\epsilon(t/n)U(t/n))^n x = lim_{n\to\infty}(T_\epsilon(t)U(t))x. \tag{5.11}$$

We thus have

$$||S_\epsilon(t)|| \le M_0 Me^{-\bar{c}t}, \tag{5.12}$$

which is independent of $\epsilon > 0$. We can thus assert that 0 is not an element of $\sigma(\epsilon A + F'(w_0))$, and the resolvent operator satisfies:

$$||(\epsilon A + F'(w_0))^{-1}|| \le \frac{MM_0}{\bar{c}} \tag{5.13}$$

by using the general version of Hille-Yosida Theorem or Theorem A4-3 in Chapter 6.

For a small $\delta > 0$, let $N_\delta(w_0) = \{w \in L^p(\Omega) \times L^p(\Omega) : ||w - w_0||_p < \delta\}$ be the δ-neighborhood of w_0 in $L^p(\Omega) \times L^p(\Omega)$. We can consider problem (5.5) as finding a solution of

$$-\epsilon Aw = F(w) \tag{5.14}$$

in a neighborhood $N_\delta(w_0)$ of $X = L^p(\Omega) \times L^p(\Omega)$. Since $F(w_0) = 0$, we can set w_1 to be w_0 in (5.9) to rewrite (5.14) as

$$(5.15) \qquad (\epsilon A + F'(w_0))w = F'(w_0)w_0 + \theta(w),$$

where $||\theta(w)||_q = o(||w - w_0||_p)$ (Here, the function θ is determined by w_0). For $w \in X$, let

$$(5.16) \qquad Q_\epsilon(w) := (\epsilon A + F'(w_0)^{-1}[F'(w_0)w_0 + \theta(w)].$$

We now show that for $\epsilon > 0$ sufficiently small, Q_ϵ maps $N_\delta(w_0)$ into itself. Note that $\theta(w) \in L^q(\Omega)\times L^q(\Omega)$, and thus $\theta_1(w) := (\epsilon A+F'(w_0))^{-1}\theta(w) \in W^{2,q}(\Omega)\times W^{2,q}(\Omega)$, by the regularity theory of elliptic equations. Since $q > dim\,\Omega$, we obtain by Sobolev embedding that $||\theta_1(w)||_p \leq c_1||\theta_1(w)||_{W^{2,q}} \leq c_2||\theta(w)||_q = o(||w - w_0||_p$, for some constants c_1, c_2. Hence, for $\delta > 0$ sufficiently small, we have

$$(5.17) \qquad ||\theta_1(w)||_p < \delta/2 \quad \text{for} \quad w \in N_\delta(w_0).$$

We next consider the term $(\epsilon A+F'(w_0)^{-1}F'(w_0)w_0$ in (5.16). The C_0 semigroup generated by $\epsilon A+F'(w_0)$ satisfies (5.12), and $(\epsilon A+F'(w_0))x \to F'(w_0)x$ as $\epsilon \to 0$ for any $x \in D(A)$, where $D(A)$ is the domain of A, which is dense in X. By the The Trotter-Neveu-Kato Semigroup Convergence Theorem (see Theorem 7.2, on p. 44 in Goldstein [74] or Theorem A4-7 in Chapter 6), we find $S_\epsilon(t)x \to U(t)x$ for all $t \geq 0$, $x \in X$. Moreover, the resolvent satisfies $R(\lambda, \epsilon A + F'(w_0))x \to R(\lambda, F'(w_0))x$ for any $x \in X$ as $\epsilon \to 0$ with $\lambda > -\bar{c}$, where $R(\lambda, A)$ denotes the operator $(\lambda - A)^{-1}$. Putting $\lambda = 0$, we find $(\epsilon A + F'(w_0))^{-1}x \to (F'(w_0))^{-1}x$. In particular

$$(5.18) \qquad (\epsilon A + F'(w_0))^{-1}F'(w_0)z \to z \quad \text{for any} \quad z \in X,$$

as $\epsilon \to 0$. Thus for sufficiently small ϵ, we have $||(\epsilon A + F'(w_0))^{-1}F'(w_0)w_0 - w_0||_p < \delta/2$. Since

$$||Q_\epsilon(w) - w_0||_p \leq ||(\epsilon A + F'(w_0))^{-1}F'(w_0)w_0 - w_0||_p + ||\theta_1(w)||_p,$$

we find from (5.17) and the last inequality that Q_ϵ maps $N_\delta(w_0)$ into itself. Note that for $w_1, w_2 \in N_\epsilon(w_0)$, we have

$$Q_\epsilon(w_1) - Q_\epsilon(w_2) = (\epsilon A + F'(w_0))^{-1}(\theta(w_1) - \theta(w_2)).$$

By means of Sobolev embedding and elliptic $W^{2,q}$ estimates, we find

$$||Q_\epsilon(w_1) - Q_\epsilon(w_2)||_p \leq K||\theta(w_1) - \theta(w_2)||_q,$$

for some constant K. Using the continuity of Gateaux derivative F', Lagrange' formula, and Hölder's inequality as mentioned above, we deduce:

$$||\theta(w_1) - \theta(w_2)||_q \leq \beta||w_1 - w_2||_p$$

where β is arbitrarily small if w_1, w_2 are close enough to w_0 in X (cf. [222]). From the last two inequalities we find that Q_ϵ is a contraction in $N_\delta(w_0)$ for sufficiently small δ. We thus obtain in the neighborhood a unique fixed point w_ϵ, which is the solution of the problem.

Remark 5.1. The solution in $[W^{2,p}(\Omega) \cap W_0^{1,p}(\Omega)]^2$ found in the above theorem is actually a classical solution. However, it converges to the constant solution as $\epsilon \to 0$ only in $L^p(\Omega) \times L^p(\Omega)$. The fact that the product of the eigenvalues of $F'(w_0)$ is positive is a consequence of the prey-predator interaction. It leads to the fact that all eigenvalues have negative real parts.

In order to study the stability of the steady-state found for (5.5), we now consider the parabolic problem:

$$(5.19) \qquad \begin{cases} \bar{u}_t(\bar{x}, t) = R^{-2}\Delta\bar{u} + \bar{u}(a - b\bar{u} - c\bar{v}) & \\ & (\bar{x}, t) \in \Omega \times (0, t), \\ \bar{v}_t(\bar{x}, t) = R^{-2}\Delta\bar{v} + d^{-1}\bar{v}(h(\bar{u}) - m(\bar{v})) & \\ \bar{u} = \bar{v} = 0 & (\bar{x}, t) \in \partial\Omega \times \{0\}, \end{cases}$$

$$(5.20) \qquad (\bar{u}(\bar{x}, 0), \bar{v}(\bar{x}, 0)) = (\bar{u}_0(\bar{x}), \bar{v}_0(\bar{x})) \qquad \bar{x} \in \Omega.$$

For $R > 0$ sufficiently large, we have an equilibrium solution $(\bar{u}_R(\bar{x}), \bar{v}_R(\bar{x}))$ of (5.19), which is in an arbitrary small neighborhood of w_0 in X. Here, we may define $(u_R(x), v_R(x)) := (\bar{u}_R(\bar{x}), \bar{v}_R(\bar{x}))$ for $x \in R\Omega$ where $\bar{x} = x/R \in \Omega$.

Let

$$(5.21) \qquad B := \{(\bar{u}, \bar{v}) : (\bar{u}, \bar{v}) \in C(\bar{\Omega}) \times C(\bar{\Omega}),\ \bar{u} = \bar{v} = 0 \ on \ \partial\Omega\}.$$

The operator $A_1 := diag.(R^{-2}\Delta, R^{-2}\Delta)$ is an infinitesimal generator of an analytic semigroup on B for $t \geq 0$, with domain $D(A_1) = \{(\bar{u}_1, \bar{u}_2) : \bar{u}_i \in W^{2,p}(\Omega)$ for all p, $\bar{u}_i = 0$ and $\Delta\bar{u}_i = 0$ on $\partial\Omega$, $i = 1, 2\}$. If $(\bar{u}_0, \bar{v}_0) \in B$, we can consider the solution of the initial boundary value problem (5.19), (5.20) as a function

$$(5.22) \qquad (\bar{u}(\cdot, t), \bar{v}(\cdot, t)) \in C([0, T], B) \cap C^1((0, T], B),$$

with $(\bar{u}(\cdot, 0), \bar{v}(\cdot, 0)) \in B$, $(\bar{u}(\cdot, t), \bar{v}(\cdot, t)) \in D(A_1)$ for all $t \in (0, T]$. We have the following stability theorem.

Theorem 5.3 (Asymptotic Stability under Small Diffusion). *Assume all the hypotheses concerning $F(u,v)$ in Theorem 5.2. For $R > 0$ sufficiently large, the equilibrium solution $(\bar{u}_R(\bar{x}), \bar{v}_R(\bar{x}))$ of (5.19) is asymptotically stable in the sense that any solution of the initial boundary value problem (5.19), (5.20), with initial condition in B, considered as a function described in (5.22) will satisfy:*

$$||\bar{u}(\cdot,t), \bar{v}(\cdot,t) - (\bar{u}_R(\bar{x}), \bar{v}_R(\bar{x}))||_B \to 0, \text{ as } t \to +\infty, \tag{5.23}$$

provided that $||(\bar{u}_0, \bar{v}_0) - (\bar{u}_R(\bar{x}), \bar{v}_R(\bar{x}))||_B$ is sufficiently small, where

$$(\bar{u}(\bar{x},0), \bar{v}(\bar{x},0) = (\bar{u}_0(\bar{x}), \bar{v}_0(\bar{x})) \qquad \bar{x} \in \Omega. \tag{5.24}$$

Proof. (Outline) To prove the asymptotic stability of $(\bar{u}_R, \bar{v}_R)$, we apply a stability result of Mora [176] or Theorem A4-9 in Chapter 6. We see that it suffices to show that the spectrum of the linearization $A_R + B_R$ of the elliptic system corresponding to (5.19) at $(\bar{u}_R, \bar{v}_R)$ is in a subset of $\{z : Re\, z \leq -c_1\}$ for some $c_1 > 0$ where

$$A_R = \begin{bmatrix} R^{-2}\Delta & 0 \\ 0 & R^{-2}\Delta \end{bmatrix},$$

$$B_R = \begin{bmatrix} a\bar{u}_R - 2b\bar{u}_R - c\bar{v}_R & -c\bar{u}_R \\ d^{-1}\bar{v}_R h'(\bar{u}_R) & d^{-1}(h(\bar{u}_R) - \bar{v}_R m'(\bar{v}_R) - m(\bar{v}_R)) \end{bmatrix}$$

are operators on B given in (5.21). Let $S_M := \{z : Re\, z \leq -M\}$. For any large $M > 0$, the spectrum of the operator A_R is contained in S_M for all large enough $R > 0$. Since the functions $\bar{u}_R$ and $\bar{v}_R$ are uniformly bounded for all R, the norms of the operators B_R are uniformly bounded as operators on the Banach space B. Moreover, the operators A_R and B_R commute. We thus obtain from the semicontinuity of the spectrum of closed operator (see Sections 3.1-3.2 of Chapter 4 in Kato [102] or Theorem A4-10 in Chapter 6.) that the spectrum of $(A_R + B_R)$ is contained in a closed subset of the left open complex plane of the form $\{z : Re\, z \leq -c_1\}$, for some $c_1 > 0$, provided R is sufficiently large.

The last theorem gives the stability of a steady-state for only a very special situation. More general theorem will be more elaborate. In view of Theorems 5.2 and 5.3, one is interested in conditions which insure the uniqueness of positive solutions for certain prey-predator systems. In the case of Volterra-Lotka type of interaction, there are some simple conditions. Without loss of generality, we consider problem (2.1) with $\sigma_1 = \sigma_2 = 1$ as follows:

$$\begin{cases} \Delta u + u(a - bu - cv) = 0 & \\ & \text{in } \Omega, \\ \Delta v + v(e + fu - gv) = 0 & \\ & \\ u = v = 0 & \text{on } \partial\Omega, \end{cases} \tag{5.25}$$

where a, b, c, e, f and g are positive constants.

Theorem 5.4 (Uniqueness under Weak Interaction). *Consider the boundary value problem (5.25) under hypotheses:*

$$\begin{cases} a > \lambda_1, \ e > \lambda_1, \\ cf < gb, \\ a > gb(gb - cf)^{-1}[\lambda_1 + ce/g]. \end{cases} \quad \text{and} \tag{5.26}$$

There exists a positive constant $k < 1$ such that if

$$cf < k(bg), \tag{5.27}$$

then (5.25) has a unique coexistence solution with each component strictly positive in Ω, and in $C^{2+\alpha}(\bar{\Omega})$.

Proof. Let C, F be positive constants such that $c \le C, f \le F$ and

$$CF < gb, \quad a > gb(gb - CF)^{-1}[\lambda_1 + Ce/g]. \tag{5.28}$$

Let $\hat{U}, \tilde{U}, \hat{V}, \tilde{V} \in C^{2+\alpha}(\Omega)$ be strictly positive functions in Ω satisfying the following scalar problems:

$$\begin{aligned} &\Delta\hat{U} + \hat{U}(a - b\hat{U}) = 0 \ \text{ in } \Omega, \ \hat{U} = 0 \ \text{ on } \partial\Omega, \\ &\Delta\hat{V} + \hat{V}(e + \tfrac{Fa}{b} - g\hat{V}) = 0 \ \text{ in } \Omega, \ \hat{V} = 0 \ \text{ on } \partial\Omega, \\ &\Delta\tilde{U} + \tilde{U}(a - b\tilde{U} - C\hat{V}) = 0 \ \text{ in } \Omega, \ \tilde{U} = 0 \ \text{ on } \partial\Omega, \\ &\Delta\tilde{V} + \tilde{V}(e - g\tilde{V}) = 0 \ \text{ in } \Omega, \ \tilde{V} = 0 \ \text{ on } \partial\Omega. \end{aligned} \tag{5.29}$$

Note that $\hat{U}, \hat{V}, \tilde{V}$ exist because $a, e, e + Fa/b$ are $> \lambda_1$; and $\hat{U}, \hat{V}, \tilde{V}$ are $\ge \delta\phi > 0$ in Ω for sufficiently small $\delta > 0$. One can readily deduce by upper lower solutions method that $\hat{V}(x) \le \frac{1}{g}(e + \frac{Fa}{b})$, hence $a - C\hat{V} \ge a - \frac{C}{g}(e + \frac{Fa}{b}) > \lambda_1$ for all $x \in \bar{\Omega}$. Consequently, we obtain

$$0 < \delta\phi \le \tilde{U} \le \hat{U}, \ 0 < \delta\phi \le \tilde{V} \le \hat{V}$$

for $x \in \Omega, \delta > 0$ sufficiently small. Since the outward normal derivatives of ϕ are negative on the boundary, there must exist a constant $\bar{K} > 0$ such that

$$\hat{U} \le \bar{K}\tilde{U}, \ \hat{V} \le \bar{K}\tilde{V}, \ \hat{U} \le \bar{K}\tilde{V}, \ \hat{V} \le \bar{K}\tilde{U} \tag{5.30}$$

for all $x \in \bar{\Omega}$.

Define $u_1 := \hat{U}$. Let v_1 be the positive solution of

$$\Delta v_1 + v_1(e + fu_1 - gv_1) = 0 \text{ in } \Omega, \quad v_1 = 0 \text{ on } \partial\Omega,$$

and $u_i, v_i, i = 2, 3, \dots$ be defined inductively by:

$$\begin{aligned} &\Delta u_i + u_i(a - bu_i - cv_{i-1}) = 0 \\ &\Delta v_i + v_i(e + fu_i - gv_i) = 0 \\ &u_i = v_i = 0 \quad \text{on } \partial\Omega. \end{aligned} \quad \text{in } \Omega, \tag{5.31}$$

As described in Leung [123] or Section 5.3 in [125], the sequence satisfies:

$$\begin{aligned} &\tilde{U} \le u_2 \le u_4 \le u_6 \le \cdots \le u_5 \le u_3 \le u_1 \le \hat{U}, \\ &\tilde{V} \le v_2 \le v_4 \le v_6 \le \cdots \le v_5 \le v_3 \le v_1 \le \hat{V} \end{aligned} \tag{5.32}$$

for all $x \in \bar{\Omega}$. From (5.31), we find for $i \ge 1$:

$$\begin{aligned} 0 &= \int_\Omega (u_{2i+2}\Delta u_{2i+1} - u_{2i+1}\Delta u_{2i+2})dx \\ &= -\int_\Omega u_{2i+1}u_{2i+2}[b(u_{2i+2} - u_{2i+1}) + c(v_{2i+1} - v_{2i})]dx, \end{aligned}$$

which implies

$$b\int_\Omega (u_{2i+1} - u_{2i+2})u_{2i+1}u_{2i+2}dx = c\int_\Omega (v_{2i+1} - v_{2i})u_{2i+1}u_{2i+2}dx. \tag{5.33}$$

Also for $i \ge 1$, we have

$$\begin{aligned} 0 &= \int_\Omega (v_{2i+1}\Delta v_{2i} - v_{2i}\Delta v_{2i+1})dx \\ &= -\int_\Omega v_{2i}v_{2i+1}[f(u_{2i} - u_{2i+1}) + g(v_{2i+1} - v_{2i})]dx, \end{aligned}$$

which implies

$$g\int_\Omega (v_{2i+1} - v_{2i})v_{2i}v_{2i+1}dx = f\int_\Omega (u_{2i+1} - u_{2i})v_{2i}v_{2i+1}dx. \tag{5.34}$$

Moreover, for $i \ge 1$

$$0 = \int_\Omega (u_{2i}\Delta u_{2i+1} - u_{2i+1}\Delta u_{2i})dx$$

$$= -\int_\Omega u_{2i}u_{2i+1}[b(u_{2i}-u_{2i+1})+c(v_{2i-1}-v_{2i})]dx,$$

$$0 = \int_\Omega (v_{2i-1}\Delta v_{2i} - v_{2i}\Delta v_{2i-1})dx$$

$$= -\int_\Omega v_{2i-1}v_{2i}[f(u_{2i}-u_{2i-1})+g(v_{2i-1}-v_{2i})]dx,$$

respectively gives

$$b\int_\Omega (u_{2i+1}-u_{2i})u_{2i}u_{2i+1}dx = c\int_\Omega (v_{2i-1}-v_{2i})u_{2i}u_{2i+1}dx, \tag{5.35}$$

$$g\int_\Omega (v_{2i-1}-v_{2i})v_{2i-1}v_{2i}dx = f\int_\Omega (u_{2i-1}-u_{2i})v_{2i-1}v_{2i}dx. \tag{5.36}$$

Using (5.33), (5.34) and (5.30), (5.32) we deduce that:

$$\begin{aligned}&\textstyle\int_\Omega (u_{2i+1}-u_{2i+2})u_{2i+1}u_{2i+2}dx = \frac{c}{b}\int_\Omega (v_{2i+1}-v_{2i})u_{2i+1}u_{2i+2}dx\\ &\textstyle\le \frac{c}{b}\int_\Omega \bar{K}^2(v_{2i+1}-v_{2i})v_{2i}v_{2i+1}dx = \bar{K}^2\frac{cf}{bg}\int_\Omega (u_{2i+1}-u_{2i})v_{2i}v_{2i+1}dx.\end{aligned} \tag{5.37}$$

Then, we use (5.35), (5.36) and (5.30), (5.32) again to obtain:

$$\begin{aligned}&\textstyle\int_\Omega (u_{2i+1}-u_{2i})u_{2i}u_{2i+1}dx = \frac{c}{b}\int_\Omega (v_{2i-1}-v_{2i})u_{2i}u_{2i+1}dx\\ &\textstyle\le \frac{c}{b}\int_\Omega \bar{K}^2(v_{2i-1}-v_{2i})v_{2i-1}v_{2i}dx = \bar{K}^2\frac{cf}{bg}\int_\Omega (u_{2i-1}-u_{2i})v_{2i-1}v_{2i}dx.\end{aligned} \tag{5.38}$$

Combining (5.37), (5.38) and (5.30), (5.32) once more, we obtain:

$$\int_\Omega (u_{2i+1}-u_{2i+2})u_{2i+1}u_{2i+2}dx = \bar{K}^8(\frac{cf}{bg})^2\int_\Omega (u_{2i-1}-u_{2i})u_{2i-1}u_{2i}dx \tag{5.39}$$

for each integer $i \ge 1$. From (5.39), we conclude that if

$$cf < (\bar{K})^{-4}(bg),$$

then $\lim_{i\to\infty}\int_\Omega (u_{2i+1}-u_{2i+2})u_{2i+1}u_{2i+2}dx = 0$. By dominated convergence, and $lim_{i\to\infty}u_{2i+1} = u^* > 0$ in Ω, $lim_{i\to\infty}u_{i+2} = u_* > 0$ in Ω, $lim_{i\to\infty}(u_{2i+1} - u_{2i+2}) = u^* - u_* \ge 0$ in Ω, we conclude that $u^* = u_*$ for all $x \in \Omega$. Similarly, from $lim_{i\to\infty}(v_{2i+1} - v_{2i+2}) = v^* - v_* \ge 0$ in Ω, we deduce $v^* = v_*$. By [125], the solution of (u, v) of (5.25) satisfies $u_* \le u \le u^*, v_* \le v \le v^*$ in Ω. We thus obtain (5.27) by choosing $k = \bar{K}^{-4}$.

Part B: Competing Species Case.

We next discuss some stability results for the competing species case, and consider system (1.2) with initial conditions:

$$u(x,0) = u^0(x),\ v(x,0) = v^0(x) \qquad \text{for } x \in \Omega. \tag{5.40}$$

We assume that

(5.41)
The functions f_i have Hölder continuous partial derivatives up to second order in compact sets, $i = 1,2$; a_1, a_2, σ_1 and σ_2 are positive constants.

Moreover

$$\begin{cases} f_i(0,0) = 0,\ i = 1,2; \\ \frac{\partial f_i}{\partial u}, \frac{\partial f_i}{\partial v} < 0,\ i = 1,2 \text{ for } (u,v) \text{ in the first open quadrant.} \end{cases} \tag{5.42}$$

Under appropriate conditions, we will prove the local asymptotic stability of steady states by the method of upper lower solutions for the corresponding parabolic problem. The main assumption essentially means the competitions between the species are relatively small. The method of proof here avoids the problem of locating the spectrum of the linearized equation. It may not be readily justified that the spectrum is on the right half plane as in proof of Theorem 5.3 above.

Theorem 5.5 (Asymptotic Stability under Weak Competition). *Consider the initial-boundary value problem (1.2), (5.40), under hypotheses (5.41), (5.42) and*

$$a_i > \sigma_i \lambda_1,\ i = 1,2. \tag{5.43}$$

Suppose $(u,v) = (\bar{u}_1(x), \bar{u}_2(x))$ is an equilibrium solution of (1.2) with each $\bar{u}_i$ in $C^{2+\alpha}(\bar{\Omega})$, $\bar{u}_i(x) > 0$ in Ω, $\partial \bar{u}_i/\partial \nu < 0$ on $\partial\Omega$, for $i = 1,2$, and

$$\begin{aligned} &sup_{x\in\Omega}\left|\frac{\bar{u}_i(x)\cdot(\partial f_j/\partial u_i)(\bar{u}_1(x),\bar{u}_2(x))}{\bar{u}_j(x)\cdot(\partial f_j/\partial u_j)(\bar{u}_1(x),\bar{u}_2(x))}\right| \\ &< inf_{x\in\Omega}\left|\frac{\bar{u}_i(x)\cdot(\partial f_i/\partial u_i)(\bar{u}_1(x),\bar{u}_2(x))}{\bar{u}_j(x)\cdot(\partial f_i/\partial u_j)(\bar{u}_1(x),\bar{u}_2(x))}\right| < \infty \end{aligned} \tag{5.44}$$

for each $1 \le i,j \le 2, i \ne j$, then $(\bar{u}_1(x), \bar{u}_2(x))$ is asymptotically stable. Here asymptotically stable means that if $(u,v) = (u_1(x,t), u_2(x,t))$ is a solution of (1.2), (5.40) with $u_i \in C^{2+\alpha,1+\alpha/2}(\bar{\Omega} \times [0,T])$, each $T > 0, i = 1,2$, then $u_i(x,t) \to \bar{u}_i(x)$ uniformly as $t \to +\infty, i = 1,2$ in $\bar{\Omega}$, provided that $(u_1(x,0)$,

$u_2(x,0)) = (u^0(x), v^0(x))$ and its first partial derivatives are close enough to that of $\bar{u}_i(x)$ respectively for all $x \in \bar{\Omega}, i = 1,2$.

Remark 5.2. Recall that in Section 1.3, there are many theorems giving sufficient conditions for the existence of positive equilibrium. For (3.3) with sufficiently small c and f, one can show that there exists equilibrium with property as described in (5.44).

Proof. Hypothesis (5.44) implies that there are constants ρ_1, ρ_2 close enough to 1, with $\rho_1 < 1 < \rho_2$ such that for each $x \in \Omega$,

$$(5.45)\quad \begin{aligned} 0 &< \frac{\bar{u}_i(x)\cdot max_{\rho_1 \le s,\tau \le \rho_2}|(\partial f_j/\partial u_i)(s\bar{u}_1(x), \tau\bar{u}_2(x))|}{\bar{u}_j(x)\cdot min_{\rho_1 \le s \le 1}|(\partial f_j/\partial u_j)(s\bar{u}_1(x), \bar{u}_2(x))|} \\ &< inf_{x\in\Omega}\frac{\bar{u}_i(x)}{\bar{u}_j(x)}\{\frac{min_{\rho_1 \le s,\tau \le \rho_2}|(\partial f_i/\partial u_i)(s\bar{u}_1(x), \tau\bar{u}_2(x))|}{max_{\rho_1 \le s \le 1}|(\partial f_i/\partial u_j)(s\bar{u}_1(x), \bar{u}_2(x))|}\} - \epsilon_1 < \infty \end{aligned}$$

for each $1 \le i,j \le 2, i \ne j$, where ϵ_1 is a small positive number. We will construct appropriate lower and upper solutions v_i, w_i, and apply a comparison theorem to obtain the results here. Let

$$G(x) = \frac{\bar{u}_2(x)\, min_{\rho_1 \le s \le \tau \le \rho_2}|\frac{\partial f_2}{\partial u_2}(s\bar{u}_1(x), \tau\bar{u}_2))|}{\bar{u}_1(x)\, max_{\rho_1 \le s \le 1}|\frac{\partial f_2}{\partial u_1}(s\bar{u}_1(x), \bar{u}_2(x))|},$$

for $x \in \Omega$; and let α be a number, $1 < \alpha < \rho_2$ such that $(1-\rho_1) > (\alpha - 1)inf_{x\in\Omega}G(x)$. Define $w_2(x,t) = p(x,t)\bar{u}_2(x), p(x,t) = 1+(\alpha-1-\epsilon_4\bar{u}_2(x))e^{-mt}$, where ϵ_4 and m are positive constants to be determined later(one condition on ϵ_4 is $\epsilon_4 max_{x\in\bar{\Omega}}\bar{u}_2(x) < \alpha - 1$). On the other hand, define $v_1(x,t) = q(x,t)\bar{u}_1(x)$, $q(x,t) = 1 - (1-\beta(x))e^{-mt}$, where $\beta(x) = 1 - (\alpha-1)inf_{x\in\Omega}G(x) + \epsilon_2(\alpha - 1) + \epsilon_3(\alpha-1)\bar{u}_1(x)$, ϵ_2 and ϵ_3 are small positive constants satisfying $\epsilon_2 + \epsilon_3 max_{x\in\bar{\Omega}}\bar{u}_1(x) < \epsilon_1 < inf_{x\in\Omega}G(x)$. (Observe that $\rho_1 < \beta(x) < 1$). We have

(5.46)

$$\begin{aligned} &\sigma_2\Delta w_2[a_2 + f_2(v_1, w_2)] - \frac{\partial w_2}{\partial t} \\ &= p(x,t)\bar{u}_2[f_2(v_1,w_2) - f_2(v_1,\bar{u}_2) + f_2(v_1,\bar{u}_2) - f_2(\bar{u}_1,\bar{u}_2)] \\ &\quad + e^{-mt}[m(\alpha - 1 - \epsilon_4\bar{u}_2(x))\bar{u}_2 - \bar{u}_2\sigma_2\epsilon_4\Delta\bar{u}_2 - 2\sigma_2\epsilon_4\textstyle\sum_{i=1}^n \bar{u}_{2x_i}^2] \\ &\le p(x,t)\bar{u}_2[max_{1\le\tau\le\rho_2}\{\frac{\partial f_2}{\partial u_2}(v_1,\tau\bar{u}_2)\}\{(\alpha-1)\bar{u}_2e^{-mt} - \epsilon_4\bar{u}_2^2e^{-mt}\} \\ &\quad - min_{\rho_1\le s\le 1}\{\frac{\partial f_2}{\partial u_1}(s\bar{u}_1,\bar{u}_2)\}\cdot\{(1-\hat{\beta})\bar{u}_1e^{-mt} - \epsilon_3(\alpha-1)\bar{u}_1^2e^{-mt}\}] + e^{-mt}[\cdots] \end{aligned}$$

where $[\cdots]$ represents the terms inside the brackets immediately before the inequality sign $\le$, and $\hat{\beta} = 1 - (\alpha-1)inf_{x\in\Omega}G(x) + \epsilon_2(\alpha-1)$. Set $\epsilon_4 = m = \epsilon_3$;

thus

$$|p(x,t)\bar{u}_2[max_{1\le\tau\le\rho_2}\{\tfrac{\partial f_2}{\partial u_2}(v_1,\tau\bar{u}_2)\}\{(-\epsilon_4\bar{u}_2^2e^{-mt})$$
$$+min_{\rho_1\le s\le 1}\{\tfrac{\partial f_2}{\partial u_1}(s\bar{u}_1,\bar{u}_2)\}\epsilon_3(\alpha-1)\bar{u}_1^2e^{-mt}]$$
$$+e^{-mt}[m(\alpha-1-\epsilon_4\bar{u}_2(x))\bar{u}_2-\bar{u}_2\sigma_2\epsilon_4\Delta\bar{u}_2]|\le\epsilon_4e^{-mt}\bar{u}_2(x)K_1$$

for all $x\in\Omega$, where K_1 is some positive constant. In a neighborhood $\emptyset$ of $\partial\Omega$ in Ω, we have $-2\sigma_2\epsilon_4\sum_{i=1}^n\bar{u}_{2x_i}^2e^{-mt}+\epsilon_4e^{-mt}\bar{u}_2(x)K_1<0$, for all $t\ge 0$, since $\bar{u}_2=0$ on $\partial\Omega$. Further,

$$max_{1\le\tau\le\rho_2}\{\tfrac{\partial f_2}{\partial u_2}(v_1(x,t),\tau\bar{u}_2(x))\}(\alpha-1)\bar{u}_2(x)$$
$$-min_{\rho_1\le s\le 1}\{\tfrac{\partial f_2}{\partial u_1}(s\bar{u}_1(x),\bar{u}_2(x))\}(1-\hat{\beta})\bar{u}_1(x)$$
$$\le max_{1\le s\le\tau\le\rho_2}\{\tfrac{\partial f_2}{\partial u_2}(s\bar{u}_1(x),\tau\bar{u}_2(x))\}(\alpha-1)\bar{u}_2(x)$$
$$+max_{\rho_1\le s\le 1}|\tfrac{\partial f_2}{\partial u_1}(s\bar{u}_1(x),\bar{u}_2(x))|((\alpha-1)G(x)-\epsilon_2(\alpha-1))\bar{u}_1(x)$$
$$=-min_{1\le s\le\tau\le\rho_2}|\tfrac{\partial f_2}{\partial u_2}(s\bar{u}_1(x),\tau\bar{u}_2(x))|(\alpha-1)\bar{u}_2(x)$$
$$+\bar{u}_2(x)min_{\rho_1\le s\le\tau\le\rho_2}|\tfrac{\partial f_2}{\partial u_2}(s\bar{u}_1(x),\tau\bar{u}_2(x))|(\alpha-1)$$
$$-\epsilon_2(\alpha-1)\bar{u}_1(x)max_{\rho_1\le s\le 1}|\tfrac{\partial f_2}{\partial u_1}(s\bar{u}_1(x),\bar{u}_2(x))|<0,$$

for all $x\in\Omega, t\ge 0$. Consequently, we have $\sigma_2\Delta w_2+w_2[a_2+f_2(v_1,w_2)]-\partial w_2/\partial t<0$, for $x\in\emptyset,,t\ge 0$. For $x\in\Omega\backslash\emptyset$, two terms in (5.46) satisfy the inequality:

$$p(x,t)\bar{u}_2[max_{1\le\tau\le\rho_2}\{\tfrac{\partial f_2}{\partial u_2}(v_1,\tau\bar{u}_2)\}(\alpha-1)\bar{u}_2e^{-mt}$$
$$-min_{\rho_1\le s\le 1}\{\tfrac{\partial f_2}{\partial\bar{u}_1}(s\bar{u}_1,\bar{u}_2)\}\cdot(1-\hat{\beta})\bar{u}_1e^{-mt}]<-\epsilon_2K_2e^{-mt},$$

for some $K_2>0$, all $t\ge 0$; and for such (x,t), the sum of all the other remaining terms after the inequality sign $\le$ in (5.46) can be reduced to less than $(1/2)\epsilon_2K_2e^{-mt}$ in absolute value, by choosing $\epsilon_4=m=\epsilon_3$ sufficiently small. We therefore have $\sigma_2\Delta w_2+w_2[a_2+f_2(v_1,w_2)]-\partial w_2/\partial t<0$, for $(x,t)\in\Omega\times[0,\infty)$, and $w_2(x,t)$ is an upper solution.

For v_1, we have the inequality:
(5.47)

$$\sigma_1 \Delta v_1[a_1 + f_1(v_1, w_2)] - \frac{\partial v_1}{\partial t}$$

$$= q(x,t)\bar{u}_1[f_1(v_1, w_2) - f_1(v_1, \bar{u}_2) + f_1(v_1, \bar{u}_2) - f_1(\bar{u}_1, \bar{u}_2)]$$

$$+ e^{-mt}[-m(1-\beta(x))\bar{u}_1 + \bar{u}_1\sigma_1\epsilon_3(\alpha-1)\Delta\bar{u}_1 + 2\sigma_1\epsilon_3(\alpha-1)\sum_{i=1}^{n}\bar{u}_{1x_i}^2]$$

$$\geq q(x,t)\bar{u}_1[min_{1\leq\tau\leq\rho_2}\{\frac{\partial f_1}{\partial u_2}(v_1, \tau\bar{u}_2)\}\{(\alpha-1)\bar{u}_2 e^{-mt} - \epsilon_4\bar{u}_2^2 e^{-mt}\}$$

$$- max_{\rho_1\leq s\leq 1}\{\frac{\partial f_1}{\partial u_1}(s\bar{u}_1, \bar{u}_2)\} \cdot \{(1-\hat{\beta})\bar{u}_1 e^{-mt} - \epsilon_3(\alpha-1)\bar{u}_1^2 e^{-mt}\}] + e^{-mt}[\cdots]$$

where $[\cdots]$ represents the terms inside the brackets immediately before the inequality sign $\geq$. Due to the choice of $\epsilon_4 = m = \epsilon_3$ made previously, one has the inequality

$$|\, q(x,t)\bar{u}_1[min_{1\leq\tau\leq\rho_2}\{\frac{\partial f_1}{\partial u_2}(v_1, \tau\bar{u}_2)\}(-\epsilon_4\bar{u}_2^2 e^{-mt})$$

$$- max_{\rho_1\leq s\leq 1}\{\frac{\partial f_1}{\partial u_1}(s\bar{u}_1, \bar{u}_2)\}(-\epsilon_3(\alpha-1)\bar{u}_1^2 e^{-mt})]$$

$$+ e^{-mt}[-m(1-\beta(x))\bar{u}_1 + \bar{u}_1\sigma_1\epsilon_3(\alpha-1)\Delta\bar{u}_1]| \leq \epsilon_4 e^{-mt}\bar{u}_1 K_3$$

for all $x \in \Omega$, where K_3 is some positive constant. In a neighborhood $\tilde{\emptyset}$ of $\partial\Omega$ in Ω, we have $2\sigma_1\epsilon_3(\alpha-1)\sum_{i=1}^{n}\bar{u}_{1x_i}^2 e^{-mt} - \epsilon_4 e^{-mt}\bar{u}_1(x)K_3 > 0$, for all $t \geq 0$, since $\bar{u}_1 = 0$ on $\partial\Omega$. Further,

$$min_{1\leq\tau\leq\rho_2}\{\frac{\partial f_1}{\partial u_2}(v_1, \tau\bar{u}_2)\}\{(\alpha-1)\bar{u}_2 - max_{\rho_1\leq s\leq 1}\{\frac{\partial f_1}{\partial u_1}(s\bar{u}_1, \bar{u}_2)\}(1-\hat{\beta})\bar{u}_1$$

$$\geq -max_{\rho_1\leq s,\tau\leq\rho_2}|\frac{\partial f_1}{\partial u_2}(s\bar{u}_1, \tau\bar{u}_2)|(\alpha-1)\bar{u}_2 + min_{\rho_1\leq s\leq 1}|\frac{\partial f_1}{\partial u_1}(s\bar{u}_1, \bar{u}_2)|(1-\hat{\beta})\bar{u}_1$$

$$\geq -(\alpha-1)\bar{u}_1\, min_{\rho_1\leq s\leq 1}|\frac{\partial f_1}{\partial u_1}(s\bar{u}_1, \bar{u}_2)|(inf_{x\in\Omega}G(x) - \epsilon_1)$$

$$+ min_{\rho_1\leq s\leq 1}|\frac{\partial f_1}{\partial u_1}(s\bar{u}_1, \bar{u}_2)|(1-\hat{\beta})\bar{u}_1$$

$$= -\bar{u}_1(x)min_{\rho_1\leq s\leq 1}|\frac{\partial f_1}{\partial u_1}(s\bar{u}_1, \bar{u}_2)| \cdot (\epsilon_2 - \epsilon_1)(\alpha-1) > 0,$$

for all $x \in \Omega, t \geq 0$. The second $\geq$ sign in the last sentence is due to hypothesis (5.45). Consequently, we have $\sigma_1\Delta v_1 + [a_1 + f_1(v_1, w_2)] - \partial v_1/\partial t > 0$, for $x \in \tilde{\emptyset}, t \geq 0$. For $x \in \Omega\backslash\tilde{\emptyset}$, two terms in (5.47) satisfies the inequality:

$$q(x,t)\bar{u}_1[min_{1\leq\tau\leq\rho_2}\{\frac{\partial f_1}{\partial u_2}(v_1, \tau\bar{u}_2)\} \cdot (\alpha-1)\bar{u}_2 e^{-mt}$$

$$- max_{\rho_1\leq s\leq 1}\{\frac{\partial f_1}{\partial u_1}(s\bar{u}_1, \bar{u}_2)\}(1-\hat{\beta})\bar{u}_1 e^{-mt}] > (\epsilon_1 - \epsilon_2)K_4 e^{-mt}$$

for some $K_4 > 0$, all $t \geq 0$; and for such (x,t), the sum of all the other remaining terms after the inequality sign $\geq$ in (5.47) can be reduced to less than $(1/2)(\epsilon_1 - \epsilon_2)K_4 e^{-mt}$ in absolute value, by reducing the size of $\epsilon_4 = m = \epsilon_3$. We therefore have $\sigma_1 \Delta v_1 + [a_1 + f_1(v_1, w_2)] - \partial v_1/\partial t > 0$, for $(x,t) \in \Omega \times [0,\infty)$, and $v_1(x,t)$ is a lower solution.

Since all the first partial derivatives of f_1 and f_2 have the same sign, we can interchange the role of $\bar{u}_1, f_1$ with $\bar{u}_2, f_2$ respectively and construct lower and upper solutions v_2, w_1 in exactly the same manner as before. Here v_2, w_1 are of the form $v_2 = \tilde{q}(x,t)\bar{u}_2, w_1 = \tilde{p}(x,t)\bar{u}_1(x)$ with $\tilde{p}(x,t), \tilde{q}(x,t)$ analogous to $p(x,t), q(x,t)$ respectively. ($\tilde{p}(x,t) \to 1^+, \tilde{q}(x,t) \to 1^-$, as $t \to \infty$, all $x \in \bar{\Omega}$).

Finally, we have $v_i(x,t) \to \bar{u}_i(x)$ from below, and $w_i(x,t) \to \bar{u}_i(x)$ from above, as $t \to \infty$, uniformly for $x \in \bar{\Omega}, i = 1,2$. When the initial conditions $u_i(x,0)$ and their partial derivatives are close to that of $\bar{u}_i(x)$ in the sense described in the theorem, we have $v_i(x,0) \leq u_i(x,0) \leq w_i(x,0), x \in \bar{\Omega}$. (Note that we have $\partial \bar{u}_i/\partial \nu < 0$ on $\partial\Omega$). Applying appropriate comparison or differential inequalities as in Section 1.2 in [125], we obtain

$$v_i(x,t) \leq u_i(x,t) \leq w_i(x,t) \qquad \text{for } (x,t) \in \bar{\Omega} \times [0,\infty),$$

and thus we have $(\bar{u}_1(x), \bar{u}_2(x))$ as an asymptotically stable equilibrium solution.

In the situation when the intrinsic growth rate of one species is small, we can prove the following theorem in a similar fashion.

Theorem 5.6. *Consider the initial-boundary value problem (1.2), (5.40), under hypotheses (5.41), (5.42) and*

$$a_1 < \sigma_1 \lambda_1, \quad a_2 > \sigma_2 \lambda_1; \tag{5.48}$$

and $u_2^*(x) \in C^{2+\alpha}(\bar{\Omega})$ is a solution of

$$\sigma_2 \Delta v + v[a_2 + f_2(0,v)] = 0 \text{ in } \Omega, \ v = 0 \text{ on } \partial\Omega, \tag{5.49}$$

with $u_2^*(x) > 0$ for $x \in \Omega$. Let $(u,v) = (u_1(x,t), u_2(x,t))$ be a solution of (1.2), (5.40) with $u_i \in C^{2+\alpha, 1+\alpha/2}(\bar{\Omega} \times [0,T])$, each $T > 0, i = 1,2$, where $u^0(x), v^0(x)$ are both non-negative functions in $C^{2+\alpha}(\bar{\Omega})$ satisfying compatibility conditions as described in Ladyzhenskaya, Solonnikov and Ural'ceva [113] or Section 1.3 in [125]. Then $(u_1(x,t), u_2(x,t)) \to (0, u_2^*(x))$ as $t \to \infty$, uniformly for $x \in \bar{\Omega}$, provided that u^0, v^0 and all their first partial derivatives are close enough to $0, u_2^*$ respectively and their corresponding first partial derivatives.

Under the stronger assumption of uniqueness of positive steady state, we can obtain a global stability result as follows. (We shall discuss the problem of such uniqueness in later theorems in this section.)

Theorem 5.7 (Global Asymptotic Stability in case of Uniqueness). *Consider problem (3.3) restricted to $\sigma_1 = \sigma_2$. Assume condition (3.4) is satisfied and that problem (3.3) only has a unique solution $(u^*(x), v^*(x))$ with both components strictly positive in Ω. Let $(u(x,t), v(x,t)$ be a solution of the initial boundary value problem:*

$$(5.50)\qquad \begin{cases} u_t = \Delta u + u(a - bu - cv) & \\ & \text{in } \Omega \times [0, \infty), \\ v_t = \Delta v + v(e - fu - gv) & \\ u = v = 0 & \text{on } \partial\Omega \times [0, \infty), \\ u(x,0) = u^0(x), v_(x,0) = v^0(x) & \text{in } \Omega \end{cases}$$

with both $u^0, v^0 \geq 0, \not\equiv 0$ in $C^\alpha(\bar{\Omega}), 0 < \alpha < 1$, and vanishing on $\partial\Omega$, then

$$(u(x,t), v(x,t)) \to (u^*(x), v^*(x)), \quad \text{as } t \to \infty$$

uniformly in $\bar{\Omega}$.

Proof. We first choose numbers a_1 and e_1 such that

$$(5.51)\qquad a_1 > a, \quad e_1 > e$$

and

$$(5.52)\qquad a > \lambda_1 + \frac{ce_1}{g}, \quad e > \lambda_1 + \frac{fa_1}{b}.$$

By hypothesis (3.4), such a_1, e_1 must exist.

For convenience, we introduce the following notation: If $w \in C^1(\bar{\Omega}), w(x) > 0$ for all $x \in \Omega$, and $\partial w/\partial\nu < 0$ everywhere on $\partial\Omega$, we write $w >> 0$. If $w, z \in C^1(\bar{\Omega})$,we write $w << z$ if $z - w >> 0$. We first prove the theorem under the additional conditions $u^0, v^0 \in C^1(\bar{\Omega})$,

$$(5.53)\qquad u^0 >> 0, \quad v^0 >> 0,$$

and for all $x \in \bar{\Omega}$,

$$(5.54)\qquad u^0(x) \leq \frac{\theta_{a_1}}{b}, \quad v^0(x) \leq \frac{\theta_{e_1}}{g}.$$

Here, for any $A > \lambda_1$, the symbol θ_A denotes the unique positive solution of

$$\Delta Z + Z[A - Z] = 0 \text{ in } \Omega, \quad Z|_{\partial\Omega} = 0.$$

Let ϕ_1 be the positive principal eigenfunction for the Laplacian on Ω with Dirichlet boundary condition. Choose $\epsilon > 0$ so small such that

$$\epsilon\phi_1(x) \leq u^0(x), \quad \epsilon\phi_1(x) \leq v^0(x) \tag{5.55}$$

and

$$\begin{aligned} a &> \lambda_1 + \tfrac{ce_1}{g} + b\epsilon\phi_1(x), \\ e &> \lambda_1 + \tfrac{fa_1}{b} + g\epsilon\phi_1(x), \end{aligned} \tag{5.56}$$

for all $x \in \bar{\Omega}$. If we set $\bar{u} = \theta_{a_1}/b, \bar{v} = \theta_{e_1}/g$ and $\underline{u} = \underline{v} = \epsilon\phi_1$, then

$$\Delta\bar{u} + \bar{u}[a - b\bar{u} - c\underline{v}] = (a - a_1)\bar{u} - c\bar{u}\underline{v} < 0$$

for $x \in \Omega$; and since $\bar{u} \leq a_1/b$,

$$\begin{aligned} \Delta\underline{v} + \underline{v}[e - f\bar{u} - g\underline{v}] &= \underline{v}[e - \lambda_1 - f\bar{u} - g\underline{v}] \\ &\geq \underline{v}[e - \lambda_1 - fa_1/b - g\underline{v}] > 0 \end{aligned}$$

on Ω. Similarly, we have

$$\begin{aligned} &\Delta\underline{u} + \underline{u}[a - b\underline{u} - c\bar{v}] > 0, \\ &\Delta\bar{v} + \bar{v}[e - f\underline{u} - g\bar{v}] < 0. \end{aligned}$$

The conclusion of the theorem follows from the uniqueness assumption, the inequalities $\underline{u}(x) \leq u^0(x) \leq \bar{u}(x), \underline{v}(x) \leq v^0(x) \leq \bar{v}(x), x \in \bar{\Omega}$, and comparison with solutions of the differential system (5.50) with initial conditions replaced at the steady-state upper lower solutions $(\bar{u}(x), \underline{v}(x))$. Solutions with such initial conditions converges monotonically to a maximum-minimum pair of steady - state. (See Pao [183] or Theorem 1.3 in [32].)

We next remove condition (5.54) on the initial functions $u^0(x), v^0(x)$. First, observe that there exists large $K > 1$, such that

$$u^0(x) < \frac{K\theta_a(x)}{b}, \quad v^0(x) < \frac{K\theta_e(x)}{g}$$

on Ω. Define $(\bar{U}(x,t), \underline{V}(x,t))$ to be the solution of problem (5.50) with initial conditions replaced with

$$(\bar{U}(x,0), \underline{V}(x,0)) = (\frac{K\theta_a(x)}{b}, 0).$$

It is clear that $\underline{V} \equiv 0$, $\bar{U}$ is non-negative in $\Omega \times [0,\infty)$ and

$$lim_{t\to\infty}\bar{U}(x,t) = \frac{\theta_a(x)}{b}. \tag{5.57}$$

Moreover, the convergence above is monotone, because $\bar{U}(x,0), \underline{V}(x,0)$ satisfies

$$\Delta\bar{U}(x,0) + \bar{U}(x,0)[a - b\bar{U}(x,0) - c\underline{V}(x,0)] = b(\bar{U}(x,0))^2[K^{-1} - 1] < 0,$$

$$\Delta\underline{V}(x,0) + \underline{V}[e - f\bar{U} - g\underline{V}] = 0.$$

The convergence in (5.57) is also in $C^1(\bar{\Omega})$ norm by using $W^{2,p}$ estimates, compact embedding and equations (5.50). (See e.g. pp. 87–89 in Fife [59]). Similarly, define $(\underline{U}(x,t), \bar{V}(x,t))$ to be the solution of problem (5.50) with initial conditions replaced with

$$(\underline{U}(x,0), \bar{V}(x,0)) = (0, \frac{K\theta_e(x)}{g}).$$

We have $\underline{U} \equiv 0$, $\bar{V}$ is non-negative in $\Omega \times [0,\infty)$ and the monotone $C^1(\bar{\Omega})$ convergence

$$lim_{t\to\infty}\bar{V}(x,t) = \frac{\theta_e(x)}{g}. \tag{5.58}$$

On the other hand, one readily verifies that the functions $\underline{U}(x,t), \bar{U}(x,t), \underline{V}(x,t), \bar{V}(x,t)$ satisfies:

$$\begin{aligned} &\Delta\bar{U} + \bar{U}[a - b\bar{U} - c\underline{V}] - \partial\bar{U}/\partial t < 0 \\ &\Delta\underline{V} + \underline{V}[e - f\bar{U} - g\underline{V}] - \partial\underline{V}/\partial t \geq 0 \\ &\Delta\bar{V} + \bar{V}[e - f\underline{U} - g\bar{V}] - \partial\bar{V}/\partial t < 0 \\ &\Delta\underline{U} + \underline{U}[a - b\underline{U} - c\bar{V}] - \partial\bar{U}/\partial t \geq 0 \end{aligned} \tag{5.59}$$

for $(x,t) \in \Omega \times (0,\infty)$, and

$$\begin{aligned} &0 = \underline{U}(x,0) \leq u^0(x) \leq \bar{U}(x,0) = \tfrac{K\theta_a(x)}{b} \\ &0 = \underline{V}(x,0) \leq v^0(x) \leq \bar{V}(x,0) = \tfrac{K\theta_e(x)}{g} \end{aligned} \tag{5.60}$$

for $x \in \bar{\Omega}$. From comparison theorems (cf. pp. 24–26 in [125]), we assert that

$$0 = \underline{U}(x,t) \leq u(x,t) \leq \bar{U}(x,t), \quad 0 = \underline{V}(x,t) \leq v(x,t) \leq \bar{V}(x,t) \tag{5.61}$$

for $(x,t) \in \Omega \times [0,\infty)$. We next observe that $\Delta\bar{u} + \bar{u}[a - b\bar{u}] = (a - a_1)\bar{u} < 0$ in Ω, $\bar{u}|_{\partial\Omega} = 0$, thus $\bar{u} = \theta_{a_1}/b$ is a strict upper solution of the problem

$$\Delta z + z[a - bz] = 0 \text{ in } \Omega, \quad z|_{\partial\Omega} = 0.$$

Similarly, θ_{e_1}/g is a strict upper solution of the problem

$$\Delta z + z[e - gz] = 0 \text{ in } \Omega, \quad z|_{\partial\Omega} = 0.$$

By monotone iteration and comparison, we obtain

$$\frac{\theta_a}{b} << \frac{\theta_{a_1}}{b}, \quad \frac{\theta_e}{g} << \frac{\theta_{e_1}}{g}. \tag{5.62}$$

For $s > 0$, let $u^s(x) = u(x,s), v^s(x) = v(x,s)$ for $x \in \bar{\Omega}$. We obtain from (5.57), (5.58), (5.61) and (5.62) that for $s > 0$ sufficiently large

$$u^s(x) \leq \frac{\theta_{a_1}(x)}{b}, \quad v^s(x) \leq \frac{\theta_{e_1(x)}}{g} \tag{5.63}$$

for $x \in \bar{\Omega}$. On the other hand for $s > 0$, we obtain from the theory of parabolic equations and strong maximum principle that u^s, v^s are in $C^1(\bar{\Omega})$ and

$$u^s >> 0, \quad v^s >> 0. \tag{5.64}$$

Comparing (5.63) and (5.64) respectively with (5.54) and (5.53), we obtain the conclusion of this theorem by using the beginning part of the proof.

When the intrinsic growth rates of both species are the same, the following theorem gives sufficient conditions for uniqueness of coexistence solution. It reflects the situation when the crowding effect of each species on itself is greater than its competing effecting on the growth rate on the other species.

Theorem 5.8 (Uniqueness under Weak Competition). *Consider problem (3.3) with $\sigma_1 = \sigma_2 = 1$. Suppose that*

$$a = e > \lambda_1, \ b > f, \text{ and } c < g, \tag{5.65}$$

then (3.3) has a unique coexistence solution with each component in $C^{2+\alpha}(\bar{\Omega})$ and strictly positive in Ω.

Proof. Suppose $\sigma_1 = \sigma_2 = 1$, (5.65) holds and (u, v) is a solution of (3.3) with each component in $C^{2+\alpha}(\bar{\Omega})$ and strictly positive in Ω. We claim that if z is a function in $C^1(\bar{\Omega})$ satisfying

$$\Delta z + z[a - bu - gv] = 0 \text{ in } \Omega, \quad z = 0 \text{ on } \partial\Omega, \tag{5.66}$$

then $z \equiv 0$ in Ω. Note that the eigenvalue problem

$$\Delta w + w[a - bu - cv] + \lambda w = 0 \text{ in } \Omega, \ w = 0 \text{ on } \partial\Omega, \tag{5.67}$$

had eigenvalue $\lambda = 0$ with eigenfunction $w = u$ which is strictly positive in Ω. It follows that $\lambda = 0$ is the smallest eigenvalue of the problem (5.67). Thus from

Rayleigh's quotient, we find that for any nontrivial function $\psi \in C^1(\bar{\Omega})$ which vanishes at $\partial\Omega$, we have

$$0 \leq \frac{\int_\Omega |\nabla\psi|^2 - [a - bu - cv]\psi^2)dx}{\int_\Omega \psi^2 ds}. \tag{5.68}$$

Suppose z satisfies (5.66), we integrate by parts and obtain from (5.68) that

$$\begin{aligned} 0 &= \int_\Omega(-z\Delta z - z^2[a - bu - gv])dx \\ &= \int_\Omega(|\nabla z|^2 - z^2[a - bu - cv])dx + \int_\Omega v(g - c)z^2 dx \geq \int_\Omega v(g - c)z^2 dx. \end{aligned}$$

Since $g > c$ and $v > 0$ in Ω, we justify the claim that $z \equiv 0$ in $\bar{\Omega}$.

The differential equations in (3.3) can be written as:

$$\begin{cases} \Delta u + u[a - bu - gv] + (g - c)uv = 0 \\ \Delta v + v[a - bu - gv] + (b - f)uv = 0 \end{cases} \quad \text{in } \Omega.$$

Multiplying the first and second equation above respectively by $(b - f)$ and $(g - c)$ and subtracting, we obtain $\Delta\psi + \psi[a - bu - gv] = 0$ in Ω, where $\psi = (b - f)u - (g - c)v$. We thus have $\psi \equiv 0$; that is $v \equiv ru$, where $r = (b - f)/(g - c)$. From the first equation in (3.3), we obtain

$$\Delta u + u[a - \frac{bg - cf}{g - c}u] = 0 \text{ in } \Omega, \;\; u = 0 \text{ on } \partial\Omega.$$

Hence the function $\theta = \frac{bg - cf}{g - c}u$ satisfies

$$\Delta\theta + \theta[a - \theta] = 0 \text{ in } \Omega, \;\; \theta = 0 \text{ on } \partial\Omega. \tag{5.69}$$

Consequently (u, v) must satisfy

$$(u, v) = (\frac{g - c}{bg - cf}\theta, \frac{b - f}{bg - cf}\theta) \text{ in } \bar{\Omega}.$$

where θ is uniquely defined as the solution of problem (5.69).

Other sufficient conditions for unique positive coexistence state even for $a \neq e$ can also be found.

Theorem 5.9 (Uniqueness under Weak Competition). *Consider problem (3.3) with $\sigma_1 = \sigma_2 = 1$ and assume (3.4) is satisfied. Suppose that*

$$4bg > \frac{gc^2\theta_a}{b\theta_{(e-af/b)}} + 2cf + \frac{bf^2\theta_e}{g\theta_{(a-ec/g)}}, \tag{5.70}$$

then (3.3) has a unique coexistence solution with each component in $C^{2+\alpha}(\bar{\Omega})$ and strictly positive in Ω.

Remark 5.3. Here, for any $A > \lambda_1$, θ_A denotes the unique positive solution of (5.69) where a is replaced by A. Thus, by (3.4), $\theta_{(e-af/b)}$ and $\theta_{(a-ec/g)}$ are positive functions in Ω. For fixed a, b, e and g, hypothesis (5.70) will be satisfied for c, f sufficiently small. This is true because $\theta_{(e-af/b)}$(or $\theta_{(a-ec/g)}$) increases as f (or c) decreases for $x \in \Omega$. Thus $\frac{gc^2\theta_a}{b\theta_{(e-af/b)}}$ (or $\frac{bf^2\theta_e}{g\theta_{(a-ec/g)}}$) decreases as f (or c) decreases.

Proof. Assume all the hypotheses of this theorem, and $(u_1, v_1), (u_2, v_2)$ are two strictly positive solutions of (3.3) in Ω. Let $p = u_1 - u_2, q = v_1 - v_2$, then

$$\text{(5.71)} \qquad \begin{cases} \Delta p + [a - bu_1 - cv_1]p - bu_2 p - cu_2 q = 0 & \\ & \text{in } \Omega, \\ \Delta q + [e - fu_2 - gv_2]q - fv_1 p - gv_1 q = 0 & \\ & \\ p = q = 0 & \text{on } \partial\Omega. \end{cases}$$

Since u_1 is a strictly positive solution of

$$\text{(5.72)} \qquad \begin{cases} \Delta\psi + [a - bu_1 - cv_1]\psi + \alpha\psi = 0 & \text{in } \Omega, \\ \psi = 0 & \text{on } \partial\Omega, \end{cases}$$

with $\alpha = 0$, the number $\alpha = 0$ must be the smallest eigenvalue of the above problem. Moreover, by variational properties, we have

$$\text{(5.73)} \qquad \int_\Omega z(-\Delta z - [a - bu_1 - cv_1]z)dx \geq 0,$$

for any $z \in C^2(\bar{\Omega})$ which vanishes on $\partial\Omega$. Similarly, since v_2 is a strictly positive solution of

$$\text{(5.74)} \qquad \begin{cases} \Delta\psi + [e - fu_2 - gv_2]\psi + \alpha\psi = 0 & \text{in } \Omega, \\ \psi = 0 & \text{on } \partial\Omega, \end{cases}$$

with $\alpha = 0$, the number $\alpha = 0$ must be the smallest eigenvalue of the above problem. Moreover,

$$\text{(5.75)} \qquad \int_\Omega z(-\Delta z - [e - fu_2 - fv_2]z)dx \geq 0,$$

for any $z \in C^2(\bar{\Omega})$ which vanishes on $\partial\Omega$. Multiplying the first equation of (5.71) by $-p$, the second by $-q$, integrating over Ω and adding, we deduce from (5.73) and (5.75) that

$$\text{(5.76)} \qquad \int_\Omega (bu_2 p^2 + (cu_2 + fv_1)pq + gv_1 q^2)dx \leq 0.$$

By comparison of scalar equations using upper and lower solutions we can readily obtain for $i = 1, 2$, $x \in \Omega$,

$$(5.77) \qquad \begin{aligned} &(\tfrac{1}{b})\theta_{(a-ec/g)} \leq u_i \leq (\tfrac{1}{b})\theta_a, \\ &(\tfrac{1}{g})\theta_{(e-af/b)} \leq v_i \leq (\tfrac{1}{g})\theta_e. \end{aligned}$$

From (5.77), we have

$$(5.78) \qquad \frac{c^2 g\theta_a}{b\theta_{(e-af/b)}} + 2cf + \frac{f^2 b\theta_e}{g\theta_{(a-ec/g)}} > c^2\frac{u_2}{v_1} + 2cf + f^2\frac{v_1}{u_2}$$

in Ω. Thus by hypothesis (5.70), the quadratic expression in the integrand of (5.76) is positive definite for each $x \in \Omega$. Consequently, we must have p and q identically equal to zero in Ω. That is $(u_1, v_1) \equiv (u_2, v_2)$ in $\bar{\Omega}$.

Theorem 5.9 is relevant for weak competition (i.e. small c, f). In case of strong competition, we consider problem (3.22) in Section 1.3 with large c, f. With a modification of the proof of Theorem 3.10 and more carefully analysis of the indices we can extend Theorem 3.10 to obtain the following "uniqueness" result.

Theorem 5.10 (Local Uniqueness of Segregated Coexistence under Strong Competition). *Suppose $w_0 \in C_0^1(\bar{\Omega})$ is a non-degenerate solution of (3.25) which changes sign. Let $max\{2, N/2\} < p < \infty$, then there exist respectively very large and small positive constants $\tilde{N}$ and ϵ such that for any c, f satisfying*

$$(5.79) \qquad c \geq \tilde{N}, \quad |cf^{-1} - \alpha| \leq \epsilon,$$

the problem (3.22) has a "unique" positive solution (u, v) near $(\alpha^{-1}w_0^+, -w_0^-)$ in $L^p(\Omega) \times L^p(\Omega)$. (Recall that α is any fixed number satisfying $\alpha \in (0, \infty)$.)

We now discuss the stability of the "unique" positive solution described in Theorem 5.10. Consider the problem:

$$(5.80) \qquad \begin{cases} u_t = \Delta u + u(a - u - cv) & \\ & \text{in } \Omega \times [0, \infty), \\ \tau u_t = \Delta v + v(e - v - fu) & \\ u = v = 0 & \text{on } \partial\Omega. \end{cases}$$

where $a > \lambda_1, e > \lambda_1$ and $\tau > 0$. The stability problem of the positive steady-state solution of the system is reduced by the following theorem to the study of the stability of the steady-state w_0 with non-zero index of the scalar problem (3.25) described in Theorem 3.10 of Section 1.3, Part B.

Theorem 5.11 (Stability of the Segregated Coexistence Solution). *Assume that the hypotheses of Theorem 5.10 are valid. Then the "unique" positive solution (u, v) for problem (3.22) found in Theorem 5.10 is a stable steady state solution for the parabolic problem (5.80) if w_0 described in Theorem 5.10 is a stable solution of (3.25) (for the corresponding parabolic problem); and it is unstable if w_0 is unstable. (Here, stable or unstable is interpreted as in Theorem A4-11 or Theorem A4-12 in Chapter 6 for solutions in the fractional power space $X^{\tilde{\alpha}} \times X^{\tilde{\alpha}}, 0 \le \tilde{\alpha} < 1, X = L^p(\Omega)$).*

Proof. Suppose w_0 is a stable solution, and there exist $i \to \infty$ such that the unique positive solution (u_i, v_i) of problem (3.22) with $(c, f) = (c_i, f_i)$ satisfying (5.79) and $u_i \to \alpha^{-1} w_0^+, v_i \to (-w_0^-)$ in $L^p(\Omega)$ for $p > max\{2, N/2\}$ is unstable. (Recall in Theorem 3.10, we consider the solution (u, v) for (3.22) near $(\alpha^{-1} w_0^+, -w_0^-)$ in L^2; however, by Lemma 1.4 in Dancer and Guo [42], convergence in L^2 together with $|| \cdot ||_\infty$ bound imply convergence in $L^p, p > 2$). Consider the eigenvalues for linearized problem of (3.22) at (u_i, v_i) with the second equation multiplied by τ^{-1}:

$$(5.81) \qquad \begin{cases} \Delta h_i + (a - 2u_i - c_i v_i)h_i - c_i u_i k_i = \lambda h_i & \\ & \text{in } \Omega, \\ \Delta k_i - f_i v_i h_i + (e - 2v_i - f_i u_i)k_i = \lambda \tau k_i & \\ h_i = k_i = 0 \quad \text{on } \partial\Omega. & \end{cases}$$

If we let $w_i = -k_i$, (5.81) becomes:

$$(5.82) \qquad \begin{cases} \Delta h_i + (a - 2u_i - c_i v_i)h_i + c_i u_i w_i = \lambda h_i & \\ & \text{in } \Omega, \\ \Delta w_i + f_i v_i h_i + (e - 2v_i - f_i u_i)w_i = \lambda \tau w_i & \\ h_i = k_i = 0 \quad \text{on } \partial\Omega. & \end{cases}$$

By the stability theorems in Henry [84] (cf. Theorem A4-11 in Chapter 6), and the assumption that (u_i, v_i) is unstable, we deduce that the principal eigenvalue $\tilde{\lambda}_i$ of (5.82) must satisfy

$$\tilde{\lambda}_i \ge 0.$$

Let the eigenfunctions corresponding to $\lambda = \tilde{\lambda}_i$ in (5.82) be $(\tilde{h}_i, \tilde{w}_i) \in K \backslash \{0, 0)\}$, where K is the cone of non-negative functions in $L^p(\Omega) \times L^p(\Omega)$, $||\tilde{h}_i||_p + ||\tilde{w}_i||_p = 1$. We first show that $\{\tilde{\lambda}_i\}$ is uniformly bounded. Suppose $\tilde{\lambda}_i \to \infty$ as $i \to \infty$, we obtain from (5.82)

$$(5.83) \quad -\Delta(\beta_i \tilde{h}_i + \tilde{w}_i) = (a - 2u_i)\beta_i \tilde{h}_i + (e - 2v_i)\tilde{w}_i - \tilde{\lambda}_i(\beta_i \tilde{h}_i + \tau \tilde{w}_i) \ \text{ in } \Omega,$$

where $\beta_i = f_i/c_i$. Hence,

$$\begin{aligned} \int_\Omega |\nabla(\beta_i \tilde{h}_i + \tilde{w}_i|^2 dx &= \int_\Omega [\beta_i(a - \tilde{\lambda}_i)\tilde{h}_i + (e - \tilde{\lambda}_i \tau)\tilde{w}_i](\beta_i \tilde{h}_i + \tilde{w}_i)dx \\ &\quad - 2\int_\Omega (\beta_i u_i \tilde{h}_i + v_i \tilde{w}_i)(\beta_i \tilde{h}_i + \tilde{w}_i)dx < 0. \end{aligned} \tag{5.84}$$

Here, we use $a - \tilde{\lambda}_i < 0$, $e - \tilde{\lambda}_i \tau < 0$ for large i and $\tilde{h}_i, \tilde{w}_i \geq 0, \not\equiv 0$. This is a contradiction, and thus $\{\tilde{\lambda}_i\}$ is uniformly bounded. Consequently, we may assume without loss of generality that $lim_{i\to\infty}\tilde{\lambda}_i = \tilde{\lambda}$ with $\tilde{\lambda} \geq 0$. Note that the $L^p(\Omega)$ norm of the expression on the right of (5.83) is uniformly bounded, we thus assert by regularity theory that $\{||\beta_i \tilde{h}_i + \tilde{w}_i||_{2,p}\}$ is uniformly bounded. By compact embedding, there exists a subsequence (still denoted as $\{\beta_i \tilde{h}_i + \tilde{w}_i\}$) such that $\beta_i \tilde{h}_i + \tilde{w}_i \to y$ in $L^p(\Omega)$ as $i \to \infty$, and $y \geq 0$. We must have $y \not\equiv 0$; otherwise it follows readily from $0 \leq \beta_i \tilde{h}_i \leq \beta_i \tilde{h}_i + \tilde{w}_i, 0 \leq \tilde{w}_i \leq \beta_i \tilde{h}_i + \tilde{w}_i$ that $||\tilde{h}_i||_p + ||\tilde{w}_i||_p \to 0$ as $i \to \infty$, contradicting $||\tilde{h}_i||_p + ||\tilde{w}_i||_p = 1$. We also know that there exist $\tilde{h}, \tilde{w} \in L^p(\Omega)$ such that $\tilde{h}_i \to \tilde{h}, \tilde{w}_i \to \tilde{w}$ weakly in $L^p(\Omega)$. Hence, we have $y = \alpha \tilde{h} + \tilde{w}$. Note that by Sobolev embedding, $||\tilde{h}_i||_\infty$ and $||\tilde{w}_i||_\infty$ are also uniformly bounded. Let ϕ be a C^2 function with compact support in Ω, and multiply the first equation in (5.81) by ϕ when $(h_i, k_i, \lambda) = (\tilde{h}_i, \tilde{k}_i, \tilde{\lambda}_i), \tilde{k}_i = -\tilde{w}_i$, and integrate by parts, we find

$$(\tilde{h}_i, -\Delta\phi) = (a - 2u_i - c_i v_i - \tilde{\lambda}_i, \tilde{h}_i \phi) - (c_i u_i, \tilde{k}_i \phi), \tag{5.85}$$

where $(\cdot, \cdot)$ denotes the integral of the product over Ω. Dividing both sides above by c_i, we find

$$\frac{1}{c_i}(\tilde{h}_i, -\Delta\phi) = (\frac{a}{c_i} - \frac{2}{c_i}u_i - v_i - \frac{\tilde{\lambda}_i}{c_i}, \tilde{h}_i \phi) - (u_i, \tilde{k}_i \phi). \tag{5.86}$$

Passing to the limit as $i \to \infty$ and noting that $c_i \to \infty$, $u_i \to \alpha^{-1} w_0^+, v_i \to -w_0^-$ in $L^r(\Omega)$ for any $r > 2$, we obtain

$$(w_0^- \tilde{h} - \alpha^{-1} w_0^+ \tilde{k}, \phi) = 0, \tag{5.87}$$

where $\tilde{k} = \tilde{w}$. Since the C^2 functions ϕ satisfying (5.87) are dense in $L^q(\Omega)$ for $1/q + 1/p = 1$, we obtain

$$w_0^- \tilde{h} = \alpha^{-1} w_0^+ \tilde{k} \quad \text{in } \Omega. \tag{5.88}$$

Let

$$D_1 = \{x : w_0(x) < 0\}, \;\; D_2 = \{x : w_0(x) > 0\}.$$

Since $w_0 \in C_0^1(\bar{\Omega})$ and w_0 changes sign on Ω, both D_1 and D_2 are not empty we must have the property

$$\tilde{h} \equiv 0 \;\text{ in }\; D_1, \;\; \tilde{w} \equiv 0 \;\text{ in }\; D_2. \tag{5.89}$$

Let ϕ be a C^2 function with compact support in Ω, if we multiply (5.83) by ϕ and integrate by parts, we obtain

$$
\begin{aligned}
(5.90)\qquad (\beta_i\tilde{h}_i + \tilde{w}_i, -\Delta\phi) &= ((a\beta_i\tilde{h}_i + e\tilde{w}_i), \phi) \\
&\quad -2((\beta_i u_i\tilde{h}_i + v_i\tilde{w}_i), \phi) - \tilde{\lambda}_i(\beta_i\tilde{h}_i + \tau\tilde{w}_i), \phi),
\end{aligned}
$$

where $(\cdot,\cdot)$ denotes the integral of the product over Ω. Since $u_i \to \alpha^{-1}w_0^+, v_i \to (-w_0^-)$ in $L^\gamma(\Omega)$ for any $\gamma \in (2,\infty)$, we can pass to the limit above as $i \to \infty$ to obtain

$$
\begin{aligned}
(5.91)\qquad (\beta_i\tilde{h} + \tilde{w}, -\Delta\phi) &= ((a\alpha\tilde{h} + e\tilde{w}), \phi) \\
&\quad -2((w_0^+\tilde{h} + (-w_0^-)\tilde{w}), \phi) - \tilde{\lambda}(\alpha\tilde{h} + \tau\tilde{w}), \phi).
\end{aligned}
$$

Note that $y = \alpha\tilde{h} + \tilde{w}$, and since C^2 functions ϕ in (5.91) above are dense in $L^q(\Omega)$, where $1/q + 1/p = 1$, we find by means of property (5.89) that

$$
\begin{aligned}
-\Delta y &= (a\alpha\tilde{h} + e\tilde{w}) - 2(w_0^+\tilde{h} + (-w_0^-)\tilde{w}) - \tilde{\lambda}(\alpha\tilde{h} + \tau\tilde{w}) \\
(5.92)\qquad &= [(a - 2\alpha^{-1}w_0^+)sgn^+w_0 + (e + 2w_0^-)sgn^-w_0 \\
&\quad - \tilde{\lambda}(sgn^+w_0 + \tau sgn^-w_0)]y \equiv B(\tilde{\lambda})y,
\end{aligned}
$$

and $y = 0$ on $\partial\Omega$. Here, sgn^+w_0 (or sgn^-w_0) is the function with value 1 (or 0) and 0 (or 1) respectively at points where w_0 is positive or negative. The expression $(B(\lambda)y, y)$ defined above is decreasing in λ for λ real. Hence by (5.91) and the fact that $\tilde{\lambda} \geq 0$, we deduce $(\Delta y + B(0)y, y) \geq 0$. It follows from the characterization of eigenvalues that $\Delta + B(0)I$ has a non-negative real eigenvalue. By our non-degeneracy assumption, this eigenvalue must be positive. Thus by Theorem A4-12 in Chapter 6, we find w_0 is not stable as a solution of the corresponding parabolic equation. This contradicts the assumption in the beginning that w_0 is stable, unless the unique positive solutions (u_i, v_i) for large i are all stable.

We next prove the converse part of the theorem, and assume w_0 is unstable. Suppose the conclusion is false; then there exists a sequence of stable solutions (u_i, v_i) with corresponding $c_i \to \infty, c_i^{-1}f_i \to \alpha$, and principal eigenvalues $\tilde{\lambda}_i$ for corresponding linearized problem (5.82) satisfying $\tilde{\lambda}_i \leq 0$. Hence, there exists $(\tilde{h}_i, \tilde{w}_i) \in K\backslash\{(0,0)\}$ with $||\tilde{h}_i||_p + ||\tilde{w}_i||_p = 1$ such that $(\tilde{h}_i, \tilde{w}_i, \tilde{\lambda}_i)$ satisfies (5.82). We first show $\{\tilde{\lambda}_i\}$ is uniformly bounded. Suppose, not, there exists a subsequence, denoted again by $\{\tilde{\lambda}_i\}$ such that $\tilde{\lambda}_i \to -\infty$ as $i \to \infty$. Let $\beta_i = f_i/c_i$, then

$$
(5.93)\qquad -\Delta(\beta_i\tilde{h}_i + \tilde{w}_i) = (a - 2u_i)\beta_i\tilde{h}_i + (e - 2v_i)\tilde{w}_i - \tilde{\lambda}_i(\beta_i\tilde{h}_i + \tau\tilde{w}_i) \text{ in } \Omega.
$$

Therefore, from the non-negativity of $\tilde{h}_i, \tilde{w}_i$, we find

$$-\Delta(\beta_i \tilde{h}_i + \tilde{w}_i) \geq (-\theta_i - \tilde{\lambda}_i \alpha_1)(\beta_i \tilde{h}_i + \tilde{w}_i) \text{ in } \Omega, \tag{5.94}$$

where $\theta_i = max\{||(a - 2u_i)||_\infty, ||(e - 2v_i)||_\infty\}$, $\alpha_1 = min\{1, \tau\}$. From the proof of Theorem 3.10, we have uniform bound for $\{||u_i||_\infty\}$, $\{||v_i||_\infty\}$, and thus $\{\theta_i\}$ is uniformly bounded. Thus $-\theta_i - \tilde{\lambda}_i \to +\infty$ as $i \to \infty$. Let ϕ_1 be a positive principal eigenfunction of the Δ, we obtain from (5.94)

$$\lambda_1 \int_\Omega (\beta_i \tilde{h}_i + \tilde{w}_i)\phi_1 dx \geq (-\theta_i - \tilde{\lambda}_i \alpha_1) \int_\Omega (\beta_i \tilde{h}_i + \tilde{w}_i)\phi_1 dx. \tag{5.95}$$

This is impossible as $i \to \infty$. Thus $\{\tilde{\lambda}_i\}$ is uniformly bounded; then, we use the same argument as in the proof of the stable case to obtain:

$$\beta_i \tilde{h}_i + \tilde{w}_i \to y \text{ in } L^p(\Omega), \; y \not\equiv 0 \text{ in } L^p(\Omega) \text{ and } y \geq 0,$$

since $\tilde{h}_i \geq 0, \tilde{w}_i \geq 0$. Moreover, we have $\tilde{\lambda}_i \to \tilde{\lambda}, \tilde{\lambda} \leq 0$, and $y = \alpha \tilde{h} + \tilde{w}$ satisfies

$$\begin{cases} -\Delta y = [(a - 2\alpha^{-1} w_0^+) sgn^+ w_0 + (e + 2w_0^-) sgn^- w_0 \\ \qquad -\tilde{\lambda}(sgn^+ w_0 + \tau sgn^- w_0)] y \quad \text{in } \Omega, \\ y = 0 \quad \text{on } \partial\Omega. \end{cases} \tag{5.96}$$

Since $y \geq 0, \not\equiv 0$, the characterization of eigenvalues implies that the eigenvalue problem for λ in

$$\begin{cases} -\Delta h = [(a - 2\alpha^{-1} w_0^+) sgn^+ w_0 + (e + 2w_0^-) sgn^- w_0 \\ \qquad -\tilde{\lambda}(sgn^+ w_0 + \tau sgn^- w_0) + \lambda] h \quad \text{in } \Omega, \\ h = 0 \quad \text{on } \partial\Omega. \end{cases} \tag{5.97}$$

has principal eigenvalue equal to zero. Moreover, since $-\tilde{\lambda}(sgn^+ w_0 + \tau sgn^- w_0) \geq 0$, we obtain by comparison that the principal eigenvalue $\lambda = \bar{\lambda}$ of the problem

$$\begin{cases} -\Delta h = [(a - 2\alpha^{-1} w_0^+) sgn^+ w_0 + (e + 2w_0^-) sgn^- w_0 + \lambda] h \quad \text{in } \Omega, \\ h = 0 \quad \text{on } \partial\Omega, \end{cases} \tag{5.98}$$

must have $\bar{\lambda} \geq 0$. However, from the fact that w_0 is non-degenerate, we have $\bar{\lambda} \neq 0$. Consequently we have $\bar{\lambda} > 0$. This implies that w_0 is stable, contradicting the assumption of the second half of the proof. This completes the proof of this theorem.

We now come to the discussion of the case when the competition coefficients c, f for system (3.22) are not both small or both large. Recall that in Theorem 3.12 in Section 1.3, we find the existence of positive solutions when c and f are such that $\hat{\rho}_1(\Delta + a - cv_0)$ and $\hat{\rho}_1(\Delta + e - fu_0)$ are of different signs, where $(u_0, 0)$ and $(0, v_0)$ are the semi-trivial solutions. We shall now prove the "asymptotic stability" of the positive solution asserted by Theorem 3.11.

Recall that for Theorem 3.12, we set

$$D = \{(u, v) : C_0(\bar{\Omega}) \times C_0(\bar{\Omega}), 0 \leq u \leq a, 0 \leq v \leq e \text{ in } \Omega\},$$

where A is the map given by (3.65) whose fixed points are solutions of (3.22). By an "asymptotically stable" solution (u, v) in Theorem 5.12 below, we mean the spectral radius satisfies

(5.99)

$$r(A'(u, v)) \leq 1, \ (u, v) \text{ is an isolated solution, and } index_D(A, (u, v)) = 1.$$

Note that the case when $r(A'(u, v)) = 1$ is usually undetermined. However, if we also find that the index is 1 and the solution is isolated, then we can use a relevant theorem involving stability on the "center manifold" to obtain the solution is asymptotically stable with respect to flows in an appropriate function subspace $X^\alpha \times X^\alpha$ of $X \times X, X = L^p(\Omega)$. This will be explained in the proof of the following theorem.

Theorem 5.12. *Under the hypotheses of Theorem 3.12, one of the positive solution for (3.22) found in Theorem 3.12 is asymptotically stable in the sense described in (5.99) if* $0 \leq c \leq c_1, f_1 \leq f < \bar{f}$ *and either* $c < c_1$ *or* $f_1 < f$. *Here* $(c_1, f_1) \in T^+$ *is defined for Theorem 3.11, so that positive solution exist.*

Proof. (Outline) For convenience, we define the cone $\tilde{K} = \{(u, v) \in C_0(\bar{\Omega}) \times C_0(\bar{\Omega}) : u \geq 0 \text{ in } \Omega, v \leq 0 \text{ in } \Omega\}$ and denote the corresponding induced order by $\geq_S$. Recall that in the proof of Theorem 3.12, we choose k to be a positive constant satisfying $k \geq max.\{a + ce, e + fa\}$, and define that the mapping:

$$A(u, v) := (-\Delta + k)^{-1}(u(a + k - u - cv), v(e + k - v - fu))$$

on the set:

$$D := \{(u, v) \in C_0(\bar{\Omega}) \times C_0(\bar{\Omega}) : 0 \leq u \leq a, 0 \leq v \leq e \ in \ \bar{\Omega}\}.$$

For $(c_1, f_1) \in T^+$, we have a strictly positive solution (u_1, v_1) satisfying

$$\begin{cases} -\Delta u_1 = u_1(a - u_1 - c_1 v_1) & \\ & \text{in } \Omega, \\ -\Delta v_1 = v_1(e - v_1 - f_1 u_1) & \\ & \\ u_1 = v_1 = 0 & \text{on } \partial\Omega. \end{cases}$$

We can write $C := \{(u,v) \in D : u_1 \leq u \leq u_0, 0 \leq v \leq v_1\}$ as an order interval $C = [(u_1, v_1), (u_0, 0)]$ in D with order $\geq_S$ induced by $\tilde{K}$. The mapping A is increasing on the order interval C.

We have shown in the proof of Theorem 3.12 that $A(u_1, v_1) >_S (u_1, v_1)$(where $>_S$ means $\geq_S$ and equality does not hold). Let $w = A(u_1, v_1) - (u_1, v_1) >_S 0$. Since A is increasing, the map A_t defined by $A_t(u,v) = A(u,v) - tw$, for $0 < t < 1$, is an increasing C^1 map of C into itself. Let x_t denote its minimal fixed point in C, which can be obtained by iterating from (u_1, v_1). Moreover, by iteration, x_t increases as t decreases. Since $\{x_t : t \in (0,1)\}$ lies in a compact set (by the boundedness of C and the compactness of A), we readily see that $x_0 = \lim_{t \to 0^+} x_t$ exists, is in C, and is a fixed point of A.

We will prove the solution x_0 is an "asymptotically stable" solution. Since $x_t \geq_S (u_1, v_1)$, the first component of x_t is positive in Ω. Since, $x_t = Ax_t - tw \leq_S A(u_0, 0) - tw = (u_0, 0) - tw$, and both components of w are positive in Ω, we find that the second component of x_t must be positive in Ω. By argument as in the proof of the last theorem $A'(x_t)$ is a demi-interior operator to $\tilde{K}$. That is, for any $y \in \tilde{K} \backslash \{(0,0)\}$, we have $f(A'(x_t)y) > 0$ for all $f \in \tilde{K}^* \backslash \{0\}$, where $\tilde{K}^* = \{g \in (C_0(\bar{\Omega}) \times C_0(\bar{\Omega}))^* : g(z) \geq 0 \text{ for all } z \in \tilde{K}\}$. (Note that this is true by using the Riesz representation of linear functional, and the fact that such $A'(x_t)y$ is positive in Ω by the maximum principle applied to the linearized system of the form (5.82)). However, as described in p. 50 of Dancer [39], if $A'(x_t)$ is a demi-interior operator, then $(\lambda I - A'(x_t))^{-1}$ is a demi-interior operator for some $\lambda > r(A'(x_t))$. Then, using the geometric expansion for $(\lambda I - A'(x_t))^{-1}$ as described in the appendix of Schaefer [205], we can obtain $f(y) > 0$ if $y \in \tilde{K} \backslash \{(0,0)\}$ for any $f \in \tilde{K}^* \backslash \{0\}$, which is an eigenfunction corresponding to the eigenvalue $r(A'(x_t))$. Moreover, we have $r(A'(x_t))$ is a simple eigenvalue of $A'(x_t)$ and is the only non-zero eigenvalue to which there corresponds a positive eigenfunction (cf. Lemma 2.4 in Dancer and Guo [42] or Theorem 3.2 on p.632 of Amann [3], or equivalently Theorem A3-8 in Chapter 6). Then, applying a variant of the remark on p. 143 of Dancer [37] to the increasing mapping $A_t : C \to C$ with minimal solution x_t in C and the fact that $f(x_t) > 0$ as deduced above, we find $r(A'(x_t)) \leq 1$. Further, from the continuity of spectral radius, we obtain $r(A'(x_0)) \leq 1$ as $t \to 0^+$.

From Theorem 3.12, we have $index_C(A, (u_0, 0)) = 0$ and $r(A'(u_0, 0)) > 1$. Thus, from the conclusion of the above paragraph, $x_0 \neq (u_0, 0)$. Hence, x_0 is a strictly positive solution. By the arguments given in the last paragraph, we find $A'(x_0)$ is a demi-interior operator. If $r(A'(x_0)) < 1$, then we can use Theorem A2-3 in Chapter 6 to obtain $index_C(A, x_0) = 1$. Moreover, this implies that the principal eigenvalue of the linearized equation at x_0 is negative. By the first theorem in Chapter 5 in Henry [84] or Theorem A4-11 in Chapter 6, we obtain asymptotic stability with respect to solutions of parabolic problems corresponding to (3.22) with initial conditions in the subspace $X^\alpha \times X^\alpha, 0 \leq$

$\alpha < 1$, with $X = L^p(\Omega)$, for large p.

On the other hand, suppose $r(A'(x_0)) = 1$. Then all fixed points x_t of A_t in a neighborhood of $(x,t) = (x_0, 0)$ are represented by $(x_t, t) = (x_0 + \alpha h + z(\alpha), \phi(\alpha))$, where z and ϕ are C^1 functions, with $\phi : (-\epsilon, \epsilon) \to R, \phi(0) = 0, z(0) = 0, h$ spans $N(I - A'(x_0)), f$ spans $N(I - A'(x_0)^*)$ and $f(w(\alpha)) = 0$, for all small α (cf Dancer [37]). Moreover, we have $\phi(\alpha) > 0$ when $\alpha \in (-\epsilon, 0)$. We choose a number $\alpha_0 < 0$, with $\tau := \phi(\alpha_0) > 0$ where $\phi'(\alpha_0) \neq 0, I - A'(x_\tau)$ is invertible. Since we also have $r(A'(x_\tau)) \leq 1$ and by Krein-Rutman theorem $r(A'(x_\tau))$ is in the spectrum of $A'(x_\tau)$, thus $r(A'(x_\tau)) < 1$ (cf. Theorem A2-5 in Chapter 6), and we obtain $index_C(A_\tau, x_\tau) = 1$. We next deduce that x_0 is isolated. Suppose not, then we obtain from the analyticity of A that $\phi(\alpha) = 0$ for all small α. Thus, any solution of $x = A(x) - tw$ near $(x_0, 0)$ must has $t = 0$, contradicting (x_t, t) are solutions. To calculate the $index_C(A, x_0)$, we can construct a neighborhood V containing $x_t, 0 \leq t \leq \tau$ so that by homotopy invariance:

$$deg_C(A_\tau, V) = deg_C(A, V) = index_C(A, x_0). \tag{5.100}$$

Then, by means of the functional f and the isolated property of x_0, we can construct appropriate neighborhood to show:

$$deg_C(A_\tau, V) = index_C(A_\tau, x_\tau) = 1. \tag{5.101}$$

By means of (5.100) and (5.101), we obtain $index_C(A, x_0) = 1$. For more details, see the arguments for proving Proposition 3 in p. 144 and Remark 4 in p. 146 of Dancer [37].

We next observe that the $index_C(A, x_0)$ is the same as $index_D(A, x_0)$, when we assume $\partial\Omega$ is smooth. To see this, we use the space $V \times V$ where V denotes the space of functions u in $C_0(\bar{\Omega})$ for which $\phi_1^{-1}u$ extends to a continuous function on $\bar{\Omega}$ with the norm $||u|| := sup_{x\in\Omega}|\phi_1^{-1}(x)u(x)|$, where ϕ_1 denotes the positive eigenfunction corresponding to the principal eigenvalue for $-\Delta$ on Ω with Dirichlet boundary condition. The set V is a Banach space under the norm $||\cdot||$. The functions u for which $inf_{x\in\Omega}\phi_1^{-1}u(x) > 0$ are interior elements of the cone $K \cap V$, where K is the usual cone in $C_0(\bar{\Omega})$. In particular, this holds for $u \in C^1(\bar{\Omega})$ with $u(x) > 0$ in Ω and $\partial u/\partial\nu < 0$ on $\partial\Omega$. The mapping A is completely continuous from $E := C_0(\bar{\Omega}) \times C_0(\bar{\Omega})$ into $V \times V$. Moreover, if x_0 is an isolated fixed point of A in E, the commutativity theorem for degree (see Granas [75] or Nussbaum [178]) ensures that $index_E(A, x_0) = index_{V\times V}(A, x_0)$. Similarly, if $x_0 \in C$, the index of x_0 in C is the same as that in $C \cap (V \times V)$. That is, we only need to prove our results for indices in the space $V \times V$. In this case, we simply have to prove that the fixed point is interior to $C \cap (V \times V)$ or $D \cap (V \times V)$, and the result then follows. This can be readily justified by using maximum principle for the corresponding system as explained before.

Next, suppose (i) $r(A'(x_0)) = 1$, (ii) the fixed point x_0 is isolated and (iii) the index $index_D(A, x_0) = 1$. From (i) we have the principal eigenvalue of the corresponding linearized system at x_0 is $= 0$; then we justify as in above that the eigenspace (center manifold) is one dimensional. From Theorem 6.2.1 in Chapter 6 of Henry [84], we assert that if x_0 is asymptotically stable in the center manifold, then it is asymptotically stable in $X^\alpha \times X^\alpha$. We can use the argument in Theorem 9.3.2 of Chow and Hale [28] to assert the stability of x_0 on the manifold is determined by Liapunov Schmidt reduction since the related function F there is C^2. In particular, x_0 is stable if 0 has index 1 for the Liapunov Schmidt reduction. The Liapunov Schmidt reduction for $I - A$ is equivalent to that for $L - F$. Moreover, by Theorem 24.2 in Krasnosel'skii and Zabrieko [109], we can relate the $index_{L^p \times L^p}(A, x_0)$ or $index_D(A, x_0)$ with the index of the 0 of the bifurcation equation, and find they are equal in this simple case. Thus by property (iii), the index of 0 of the bifurcation equation is 1, and x_0 is asymptotically stable on the manifold and $X^\alpha \times X^\alpha$ in $L^p \times L^p$. The details are too technical to be included here (cf. Dancer [40]).

Part C: Cooperative Species Case.

We now come to the discussion of the stability of some of the positive steady states (i.e. coexistence states) for cooperating species found in Section 1.4. We will use the operators L_1 and L_2 as defined in (4.20) and (4.21). Recall the general cooperative system (4.22), and the part concerning weak cooperation in assumptions (4.23). Also, recall the symbol $\theta_{[-L,a,b]}$ defined immediately before Theorem 4.4, denoting the solution for the related scalar problem. Under some further restrictive conditions, we have the following uniqueness and stability theorem.

Theorem 5.13 (Uniqueness under Weak Cooperation and Asymptotic Stability). *Assume hypotheses (4.23),*

$$c(x) > 0, f(x) > 0 \ \ for \ x \in \Omega,$$

and that any coexistence state (u^*, v^*) of problem (4.22) satisfies

$$sup_\Omega(u^*/v^*) \cdot sup_\Omega(v^*/u^*) < inf_\Omega(b/c) \cdot inf_\Omega(g/f). \tag{5.102}$$

Then the boundary value problem (4.22) possesses a unique coexistence solution. Moreover, it is asymptotically stable.

Proof. (Outline) Under the assumptions (4.23), we can obtain uniform a-priori bound for all non-negative solutions of (4.22). Thus we can use the fixed point cone index method to study non-negative solutions as in Sections 2 and 3. As in Theorem 2.5 and part (i) of Theorem 3.4, we can show that both semi-trivial positive solutions $(\theta_{[-L_1,a,b]}, 0)$ and $(0, \theta_{[-L_2,e,g]})$ have zero local index. Moreover, the solution $(0,0)$ has index zero and the global index of the related mapping equals one. Consequently, it suffices to show that under condition (5.102), all the eigenvalues of the corresponding linearized problem at (u^*, v^*) have real parts less than some negative constant. By Theorem A2-3 in Chapter 6, we can then infer that the solution (u^*, v^*) has local index one; and by the additivity of indices, such positive solution must be unique. The linearized problem at (u^*, v^*) is given by:

$$\begin{cases} L_1 h + (a - 2bu^* + cv^*)h + cu^* k = \lambda h & \\ & \text{in } \Omega, \\ L_2 k + fv^* h + (e + fu^* - 2gv^*)k = \lambda k & \\ h = k = 0 & \text{on } \partial\Omega. \end{cases} \tag{5.103}$$

By Theorem A3-8 in Chapter 6 (or Theorem 8.3 and 8.4 in [45]), the principal eigenvalue of problem (5.103) is negative if we can find functions $h^* > 0, k^* > 0$ in Ω satisfying

$$L_1 h^* + (a - 2bu^* + cv^*)h^* + cu^* k^* < 0, \quad L_2 k^* + fv^* h^* + (e + fu^* - 2gv^*)k^* < 0 \tag{5.104}$$

in Ω. Then, by Theorem A3-8 in Chapter 6 again, the real parts of all other eigenvalues of problem (5.103) are less than some negative number. In order to construct h^*, k^* satisfying (5.104), we first find positive constants α, β such that

$$\frac{inf_\Omega(b/c)}{sup_\Omega(v^*/u^*)} > \frac{\beta}{\alpha} > \frac{sup_\Omega(u^*/v^*)}{inf_\Omega(g/f)}. \tag{5.105}$$

Such constants exist due to assumption (5.102). Then, define

$$h^* := \alpha u^*, \quad k^* := \beta v^*.$$

We have

$$L_1 h^* + (a - 2bu^* + cv^*)h^* + cu^* k^* = u^*(\beta cv^* - \alpha bu^*). \tag{5.106}$$

Moreover, for each $x \in \Omega$, we obtain from (5.105)

$$\begin{aligned} (\beta cv^* - \alpha bu^*)\tfrac{1}{cu^*} &= \beta \tfrac{v^*}{u^*} - \alpha \tfrac{b}{c} \\ &\leq \beta sup_\Omega(\tfrac{v^*}{u^*}) - \alpha\, inf_\Omega(\tfrac{b}{c}) < 0. \end{aligned} \tag{5.107}$$

Thus we obtain the first inequality in (5.104) from (5.106) and (5.107). Similarly, we can verify the second inequality in (5.104). The assertion on asymptotic stability of the solution (u^*, v^*) can be deduced from Theorem A4-11 in Chapter 6. This completes the proof of Theorem 5.13.

The following corollary is more readily applicable, using only information from the coefficients of the system (4.22). Recall the definitions of c_M, g_M, b_L and g_L etc. for Theorem 4.4.

Corollary 5.14. *Assume $L_1 = L_2$, $c(x) > 0, f(x) > 0$ for each $x \in \Omega$,*

$$c_M f_M < b_L g_L,\ \rho_1(-L_1) > 0,\ a > \rho_1(-L_1),\ e > \rho_1(-L_2), \tag{5.108}$$

and

$$\frac{b_M g_M}{16 b_L g_L (b_L g_L - c_M f_M)^2} \cdot \frac{(g_L a^2 + c_M e^2)(b_L e^2 + f_M a^2)}{(a-\rho_1(-L_1))(e-\rho_1(-L_1))} \cdot (sup_\Omega \frac{\phi}{\psi})^2 < \frac{1}{c_M f_M}, \tag{5.109}$$

where $\phi > 0$ is the principal eigenfunction associated with $\rho_1(-L_1)$, normalized so that $||\phi||_\infty = 1$ and $\psi > 0$ is the unique solution of

$$-L_1 \psi = 1, \quad in\ \Omega, \quad \psi|_{\partial\Omega} = 0.$$

Then the boundary value problem (4.22) has exactly one coexistence solution. Furthermore, it is asymptotically stable.

Proof. We first show that the family of functions $(\bar{u}_t, \bar{v}_t), t \geq 1$, defined by

$$\bar{u}_t := \frac{t(g_L a^2 + c_M b^2)}{4(b_L g_L - c_M f_M)} \psi, \quad \bar{v}_t := \frac{t(b_L e^2 + f_M a^2)}{4(b_L g_L - c_M f_M)} \psi, \tag{5.110}$$

are upper solutions to problem (4.22). To verify this, it suffices to show for $x \in \Omega$,

$$\begin{aligned} 1 &\geq \psi \cdot [a - t(b(x)K_1 - c(x)K_2)\psi], \\ 1 &\geq \psi \cdot [e - t(g(x)K_2 - f(x)K_1)\psi], \end{aligned} \tag{5.111}$$

where

$$K_1 := \frac{(g_L a^2 + c_M e^2)}{4(b_L g_L - c_M f_M)}, \quad K_2 := \frac{(b_L e^2 + f_M a^2)}{4(b_L g_L - c_M f_M)}.$$

For positive A and B, we have $sup_{\xi \geq 0}(A - B\xi)\xi = A^2/(4B)$. Thus we find that for each $t \geq 1$,

$$\psi \cdot [a - t(b(x)K_1 - c(x)K_2)\psi] \leq \frac{a^2}{4t(b(x)K_1 - c(x)K_2)} \leq \frac{a^2}{4(b_L K_1 - c_M K_2)}. \tag{5.112}$$

Similarly, we find

$$\psi \cdot [e - t(g(x)K_2 - f(x)K_1)\psi] \leq \frac{e^2}{4(b_L K_2 - c_M K_1)}. \tag{5.113}$$

From the definition of K_1 and K_2, we have:

$$a^2 = 4(b_L K_1 - c_M K_2), \quad e^2 = 4(b_L K_2 - c_M K_1). \tag{5.114}$$

We thus obtain (5.111) from (5.112), (5.113) and (5.114). From the generalized sweeping principle or Theorem A3-9 in Chapter 6, substituting at $t = 1$ in (5.110) we obtain the estimates:

$$u^* \leq \frac{(g_L a^2 + c_M b^2)}{4(b_L g_L - c_M f_M)}\psi, \quad v^* \leq \frac{(b_L e^2 + f_M a^2)}{4(b_L g_L - c_M f_M)}\psi, \tag{5.115}$$

for any positive solution (u^*, v^*) of (4.22). By comparison with the scalar equations, we readily obtain

$$u^* \geq \theta_{[-L_1,a,b]} \geq \frac{a - \rho_1(-L_1)}{b_M}\psi, \quad v^* \geq \theta_{[-L_2,e,g]} \geq \frac{e - \rho_1(-L_1)}{g_M}\psi. \tag{5.116}$$

We can readily verify that (5.102) is satisfied by using (5.109), (5.115) and (5.116). Consequently, we can apply Theorem 5.13 to complete the proof of this corollary.

As in the above corollary, we can apply the generalized sweeping principle and Theorem 5.13 to deduce other uniqueness and asymptotic stability results as follows. (Since the technique is similar, we will omit the details which is given in Corollary 9.5 in [45].)

Corollary 5.15. *Assume* $L_1 = L_2$, $c(x) > 0$, $f(x) > 0$ *for each* $x \in \Omega$,

$$c_M f_M < b_L g_L, \quad a \geq e > \rho_1(-L_1),$$

and

$$\frac{N_1}{M_2} \cdot \frac{N_2}{M_1} \cdot (sup_{\bar{\Omega}} \frac{\theta_{[-L_1,a,b(x)]}}{\theta_{[-L_2,e,g(x)]}})^2 < \frac{b_L g_L}{c_M f_M}. \tag{5.117}$$

Then problem (4.22) has a unique coexistence state, which is asymptotically stable. Here

$$N_1 = \frac{b_M(g_L + c_M)}{b_L g_L - c_M f_M}, \quad N_2 = \frac{b_M(b_L + f_M)}{b_L g_L - c_M f_M},$$

$$M_1 = max\{\frac{g_L(c_L + g_M)}{b_M g_M - c_L f_L}, \frac{(c_L + g_L)[g_M(b_M + f_M) - f_L g_L]}{b_M[g_M(b_M + f_M) - f_L(c_L + g_L)]}\},$$

$$M_2 = max\{\frac{g_L(f_L + b_M)}{b_M g_M - c_L f_L}, \frac{g_M(b_M + f_M)}{g_M(b_M + f_M) - f_L(c_L + g_L)}\}.$$

In the Volterra-Lotka model (4.1) for cooperating species with constant coefficients, we may perform stretching of variables in u and v to attain without loss of generality $\sigma_1 = \sigma_2 = b = g = 1$. In this case, Corollary 5.15 simplifies into the following result as in Theorem 3.3 and 3.4 in Korman and Leung [107].

Corollary 5.16. *Consider problem (4.1) with*

$$\sigma_1 = \sigma_2 = b = g = 1. \tag{5.118}$$

Suppose

$$cf < 1, \ \ a \geq e > \rho_1(-\Delta),$$

and

$$(sup_{\bar{\Omega}} \frac{\theta_{[-\Delta,a,1]}}{\theta_{[-\Delta,e,1]}})^2 < \frac{1}{cf}. \tag{5.119}$$

Then problem (4.1) has a unique coexistence state, which is asymptotically stable.

Proof. Under the assumptions of this corollary, the constants M_1, M_2, N_1 and N_2 of Corollary 5.15 satisfies

$$N_1 = M_1 = \frac{1+c}{1-cf}, \quad N_2 = M_2 = \frac{1+f}{1-cf}.$$

Consequently hypothesis (5.117) becomes (5.119). The result follows from Corollary 5.15.

There are some results for global attractivity of positive solution. In the situation when we have uniform a-priori bound as in Theorem 5.13, we can apply the following topological result in Hirsch [86] to the parabolic system associated with (4.22).

Theorem 5.17. *Assume that T is a strongly positive monotone continuous dynamical system on X where the cone K has non-empty interior and X is separable. Moreover, assume that the closure of the positive semi-orbit $O(x)$ of x is compact for each $x \in X$. Then, there exists a dense subset A of X such that if $x \in A$, then $\omega(x)$ (the ω-limit of x), is contained in the set of stationary points.*

Due to excessive technicalities, we will omit the details of the above theorem. As a consequence of the theorem, we have the following result.

Theorem 5.18 (Global Attractivity). *Assume that $c_M f_M < b_L g_L$, $a > \rho_1(-L_1)$, $e > \rho_1(-L_2)$, $c(x) > 0$, $f(x) > 0$ for each $x \in \Omega$, and that problem (4.22) has a unique coexistence state, say (u^*, v^*). Consider the following* *corresponding parabolic problem:*

$$(5.120)\qquad \begin{cases} u_t = L_1 u + u[a - b(x)u + c(x)v] & \\ & in\ \Omega \times (0,\infty), \\ v_t = L_2 v + v[e + f(x)u - g(x)v] & \\ u = v = 0 & on\ \partial\Omega \times (0,\infty), \\ u(x,0) = \hat{u}_0(x),\ v(x,0) = \hat{v}_0(x) & x \in \Omega, \end{cases}$$

where $\hat{u}_0, \hat{v}_0 \in C(\bar{\Omega})$. *Then the solution of (5.120) is defined for all* $t > 0$ *and there exists a dense subset* A *of* $(C(\bar{\Omega}))^2$ *such that if* $\hat{u}_0 > 0, \hat{v}_0 > 0$ *and* $(\hat{u}_0, \hat{v}_0) \in A$, *then*

$$(5.121)\qquad lim_{t\to\infty}||u(x,t) - u^*||_\infty = lim_{t\to\infty}||v(x,t) - v^*||_\infty = 0.$$

The idea of the proof is as follows. If $(\hat{u}_0, \hat{v}_0) \in A$, we first show by means of Theorem 5.17 that the solution of the parabolic problem (5.120) converges to some steady state. Then we use comparison method to show that if $\hat{u}_0 > 0$ and $\hat{v}_0 > 0$, the solution converges to a positive coexistence state. Finally, (5.121) follows from the assumption on uniqueness of positive coexistence state. More details can be found in the proof of Theorem 9.8 in [45]. They are omitted in order to condense the length of this section.

Notes.

Theorem 5.1 and Theorem 5.2 can be found respectively in Li [148] and Li and Ghoreishi [149]. Theorem 5.3 is obtained from Li and Ramm [152], and Theorem 5.4 is adopted from methods in Leung [123]. Theorem 5.5 is found in Leung [122]; Theorem 5.7 to Theorem 5.9 are due to Cosner and Lazer [32]. Theorems 5.10 and 5.11 are found in Dancer and Guo [42]. Theorem 5.12 is obtained from Dancer [40]. Theorem 5.13, Corollaries 5.14 to 5.16 and Theorem 5.18 are due to Delgado, López-Gómez and Suarez [45]. Theorem 5.17 is obtained by Hirsch [86].

Chapter 2

Positive Solutions for Large Systems of Equations

2.1 Introduction

Most practical or real world reaction-diffusion problems involve more than two interacting species. It is therefore crucial to extend the results in the last chapter to various cases which arise in applicable problems. The challenge is to use sophisticated methods to obtain rigorous reliable results which are useful for fundamental understanding and application. An example for study is the migration of numerous North American mammals to interact with those of South American origin (see, e.g. Darlington [43] and Young [240]). Will all species coexist in the same environments? This leads to the study of the Section 2.2 where we study the coexistence of all species when various subgroups of species are mixed together. Within each subgroup, the species may have mutualistic, competing or food-chain relationship. It is shown that appropriate properties of the subsystems will insure that the full system has a steady-state solution which is strictly positive in each components. On the other hand, in a well-known experiment, Paine [181] found that the removal of predator starfish Pisaster orchraceus from an area resulted in the reduction of a fifteen-species community to an eight-species community. The question is how the presence of the starfish enables the coexistence of all fifteen-species. As another example of application for Section 2.4 concerning interaction of three species, we consider a problem in neurochemistry. The process of inducing and suppressing pain in humans is related to the interactions between certain biochemical substances distributing along neuro-path nets. It is known that two neurotransmitters acetylcholine and triethylcholine compete for chemical receptors. The first one induces pain while the second one serves as an suppressor. These two are both predated through deprivation of receptors by a third neuropeptide of a choline with a special structure (cf. Xu [237]). The positive coexistence of such three species will affect the physiological sensitivity of an individual in response to pain stimulation. In Section 2.3 we study the spread of several bacterial infections among many interacting species. Conditions are found concerning the growth, infection, and interaction rates such that it is possible to attain positive coexistence of the infected species and the various bacterias.

In Section 2.6, we consider applications of reaction-diffusion system to the study the dynamics of nuclear fission reactors. The neutrons are divided into m energy groups. The fission, removal, absorption and transfer rates of the neutrons in different groups are dependent on temperature, which is described by the $(m+1)$-st equation (see, e.g. Duderstadt and Hamilton [52] and Glasstone and Sesonke [72]). We use bifurcation method to find critical size of reactor core when positive steady-state can occur for the system which is not symmetric. The asymptotic stability of the steady-state is also analyzed by means of a stability theorem for sectorial operators.

Recently, various models are developed for the treatment of tumor growth by means of drug and chemotherapy. They involve large systems of reaction-diffusion equations with many interacting species (cf. Cui [35], Jackson [95], Cui and Friedman [36], Friedman and Reitich [64], Ward and King [228] etc.). Such new models in medical sciences necessitate systematic rigorous study of large systems of nonlinear partial differential equations.

In every section in this chapter, Ω is a bounded domain in R^N with boundary $\partial\Omega$ belonging to $C^{2+\mu}, 0 < \mu < 1$. The symbol Δ denotes the Laplacian operator. We have limit our discussion only on bounded domains, the extension to unbounded domains (see e.g. Furusho [66]) is too lengthy for our present purpose.

2.2 Synthesizing Large (Biological) Diffusive Systems from Smaller Subsystems

In this section, large systems of steady-state reaction-diffusion equations describing many interacting species are studied. In every case, two uncoupled related subsystems are first constructed and analyzed. We assume that all species of each subgroup coexist among themselves when the other group is absent. The problem is to determine whether all species can coexist when the two subgroups are mixed together in the same environment. This section is motivated by studies of many researchers on large ecosystem models (see e.g. Goh [73] and Simberloff [207]). Here, we follow the presentations in Leung and Ortega [139]. It is shown that appropriate properties of the subsystems will insure that the full system has a steady-state solution which is strictly positive in each component. The method of bifurcation and upper-lower solutions are used in the analysis. Bifurcation theory is also used in the construction of lower solutions. More precisely, we consider the system:

$$
(2.1)\qquad \begin{cases} \Delta u_i + u_i[p_i + \Sigma_{j=1}^m a_{ij}u_j + \Sigma_{j=1}^n r_{ij}u_{m+j}] = 0 \quad \text{in} \quad \Omega, \\ \qquad\qquad i = 1, ..., m, \\ \Delta u_{m+i} + u_{m+i}[q_i + \Sigma_{j=1}^m s_{ij}u_j + \Sigma_{j=1}^n b_{ij}u_{m+j}] = 0 \quad \text{in } \Omega, \\ \qquad\qquad i = 1, ..., n, \\ u_i = 0 \qquad\qquad \text{on } \partial\Omega, \ \ i = 1, ..., m+n. \end{cases}
$$

Here, Ω is a bounded domain in R^N with boundary $\partial\Omega$ belonging to $C^{2+\mu}, 0 < \mu < 1$. The functions $u_1(x), ..., u_m(x)$ represent the concentrations of the first group of m species, while the functions $u_{m+1}(x), ..., u_{m+n}(x)$ denote those of the second group of n species. The constants $p_i, q_i, a_{ij}, r_{ij}, b_{ij}, s_{ij}$ are intrinsic growth rates and interaction rates of the species. The system (2.1) describes the $m+n$ species in diffusive equilibrium while undergoing Volterra-Lotka type of interaction.

It is convenient to introduce constant matrices:

$$
A = \{a_{ij}\}, \ 1 \le i, j \le m, \quad B = \{b_{ij}\}, \ 1 \le i, j \le n,
$$

$$
R = \{r_{ij}\}, \ \ 1 \le i \le m, \ 1 \le j \le n,
$$

and

$$
S = \{s_{ij}\}, \ \ 1 \le i \le n, \ 1 \le j \le m.
$$

Letting $u_i = w_i, i = 1, ..., m$ and $u_{m+i} = z_i, i = 1, ..., n$, system (2.1) can be written in the form:

$$
(2.2)\qquad \begin{cases} \Delta w_i + w_i[p_i + (Aw)_i + (Rz)_i] = 0 \quad \text{in } \Omega, \ i = 1, ..., m, \\ \Delta z_i + z_i[q_i + (Sw)_i + (Bz)_i] = 0 \quad \text{in } \Omega, \ i = 1, ..., n; \\ w_i = 0 \ \text{ on } \partial\Omega, \ \ i = 1, ..., m, \ \text{ and} \\ z_i = 0 \ \text{ on } \partial\Omega, \ \ i = 1, ..., n. \end{cases}
$$

We will assume that the two groups of species, I and II, competes with each other, and each species inhibits its own growth rate. More precisely, we assume::

$$\begin{cases} r_{ij} < 0 \text{ for } 1 \leq i \leq m, \ 1 \leq j \leq n, \\ s_{ij} < 0 \text{ for } 1 \leq i \leq n, \ 1 \leq j \leq m, \\ a_{ii} < 0 \text{ for } 1 \leq i \leq m, \text{ and } b_{ii} < 0 \text{ for } 1 \leq i \leq n. \end{cases} \tag{2.3}$$

Assumption (2.3) will be made in Theorem 2.1 to Theorem 2.3 and all the examples in this section. The following two uncoupled systems will serve as starting points for analyzing the full system (2.1):

$$\begin{cases} \Delta w_i + w_i[p_i + (Aw)_i] = 0 \quad \text{in } \Omega, \\ w_i = 0 \quad \text{on } \partial\Omega, \qquad \text{for } i = 1, ..., m; \end{cases} \tag{2.4}$$

and

$$\begin{cases} \Delta z_i + z_i[q_i + (Bz)_i] = 0 \quad \text{in } \Omega, \\ z_i = 0 \quad \text{on } \partial\Omega, \qquad \text{for } i = 1, ..., n. \end{cases} \tag{2.5}$$

The methods of bifurcation and upper-lower solutions will be used in the analysis. The lower solutions are constructed by applying bifurcation theory at the first eigenvalue of the Laplacian operator.

In the first theorem, we assume that within each of the two groups of species the interactions among themselves are cooperative. That is, we assume:

$$a_{ij} > 0, \ \text{ for } i \neq j, \ 1 \leq i, j \leq m; \tag{2.6}$$

and

$$b_{ij} > 0, \ \text{ for } i \neq j, \ 1 \leq i, j \leq n. \tag{2.7}$$

For convenience, we define

$$\underline{p} = min_{1 \leq i \leq m}\, p_i \quad \text{and} \quad \underline{q} = min_{1 \leq i \leq n}\, q_i.$$

The principal eigenvalue for the operator $-\Delta$ on the domain Ω with zero Dirichlet boundary data is denoted by λ_0. We use λ_0 in this section rather than the usual λ_1 to avoid possible confusion.

Theorem 2.1 (Synthesizing Competing Groups of Cooperative Species). *Assume hypotheses (2.3), (2.6) and (2.7). Suppose that there exist* $\bar{w} = (\bar{w}_1(x), ...,$

$\bar{w}_m(x)), \bar{z} = (\bar{z}_1(x), ..., \bar{z}_n(x))$ whose components are in $C^{\alpha+\mu}(\bar{\Omega}), 0 < \mu < 1$, and are positive for all $x \in \Omega$ such that

$$\begin{cases} \Delta \bar{w}_i + \bar{w}_i[p_i + (A\bar{w})_i] \leq 0 \quad in\ \Omega, \\ \\ \bar{w}_i = 0 \quad on\ \partial\Omega, \qquad for\ i = 1, ..., m; \end{cases} \tag{2.8}$$

and

$$\begin{cases} \Delta \bar{z}_i + \bar{z}_i[q_i + (B\bar{z})_i] \leq 0 \quad in\ \Omega, \\ \\ \bar{z}_i = 0 \quad on\ \partial\Omega, \qquad for\ i = 1, ..., n. \end{cases} \tag{2.9}$$

Further, assume

$$c^* + \lambda_0 \leq \underline{p} \quad and \quad d^* + \lambda_0 \leq \underline{q}, \tag{2.10}$$

where

$$c^* = max_{1 \leq i \leq m}\{sup(-R\bar{z}(x))_i : x \in \bar{\Omega}\},$$

$$d^* = max_{1 \leq i \leq n}\{sup(-S\bar{w}(x))_i : x \in \bar{\Omega}\}.$$

Then there exist a solution $(u_1, u_2, ..., u_{m+n}) \in [C^{2+\mu}(\bar{\Omega}]^{m+n}$ of (2.1) such that $u_i > 0$ in $\Omega, i = 1, ..., m+n$.

Remark 2.1. Hypotheses (2.8) and (2.9) can be rephrased vaguely as assuming that the subsystems (2.4) and (2.5) each has a positive upper solution. Also note that, for hypotheses (2.10), c^* and d^* can be made arbitrarily small by assuming $-R$ and $-S$ are small.

Before proving this theorem, we first observe the following lemma, which is a direct result of the bifurcation Theorem A1-4 in Chapter 6.

Lemma 2.1. *(a) For each $i = 1, ..., m$, the point $(\lambda, v) = ([\lambda_0 + c^*]\underline{p}^{-1}, 0)$ is a bifurcation point for the scalar equation:*

$$-\Delta v + c^* v = v[\lambda \underline{p} + a_{ii}|v|] \ in\ \Omega, \ \ v = 0 \ on\ \partial\Omega. \tag{2.11}$$

(That is, in every neighborhood of $([\lambda_0 + c^*]\underline{p}^{-1}, 0)$ in $R \times C^{\mu}(\bar{\Omega})$, there is a solution to the equation with $v \not\equiv 0$).

(b) For each $i = 1, ..., n$, the point $(\lambda, v) = ([\lambda_0 + d^*]\underline{q}^{-1}, 0)$ is a bifurcation point for the scalar equation:

$$-\Delta v + d^* v = v[\lambda \underline{q} + b_{ii}|v|] \ \text{in}\ \Omega, \ \ v = 0 \ \text{on}\ \partial\Omega. \tag{2.12}$$

For convenience we will denote $C^{\mu}(\bar{\Omega})$ by $\Im$ in the rest of this section.

Proof of Theorem 2.1. By the above lemma, there exist sequences $\{\lambda_k^i, w_k^i)\}_{k=1}^{\infty}$ and $\{\hat{\lambda}_k^i, z_k^i)\}_{k=1}^{\infty}$ in $R \times \Im\backslash\{0\}$, with $(\lambda_k^i, w_k^i) \to ([\lambda_0 + c^*]\underline{p}^{-1}, 0)$ and $(\hat{\lambda}_k^i, z_k^i) \to ([\lambda_0 + d^*]\underline{q}^{-1}, 0)$ as $k \to \infty$, where

$$(2.13) \qquad \begin{cases} -\Delta w_k^i + c^* w_k^i = w_k^i[\lambda_k^i \underline{p} + a_{ii}|w_k^i|] & \text{in } \Omega, \\ w_i^k = 0 \text{ on } \partial\Omega, \quad \text{for } i = 1, ..., m, & \text{and} \end{cases}$$

$$(2.14) \qquad \begin{cases} -\Delta z_k^i + d^* z_k^i = z_k^i[\hat{\lambda}_k^i \underline{q} + b_{ii}|z_k^i|] & \text{in } \Omega, \\ z_i^k = 0 \text{ on } \partial\Omega, \quad \text{for } i = 1, ..., n. \end{cases}$$

Let $\hat{w}_k^i = w_k^i/||w_k^i||_\Im, \hat{z}_k^i = z_k^i/||z_k^i||_\Im$. Since $(\Delta + c^*)^{-1}$ and $(\Delta + d^*)^{-1}$ are compact operators, there are subsequences of $\{\hat{w}_k^i\}_{k=1}^{\infty}, \{\hat{z}_k^i\}_{k=1}^{\infty}$, denoted by the same symbols for convenience, and functions $\hat{w}, \hat{z}$ in $\Im$, such that $||\hat{w}_k^i - \hat{w}||_\Im \to 0, ||\hat{z}_k^i - \hat{z}||_\Im \to 0$ as $k \to \infty$, and $||\hat{w}||_\Im = ||\hat{z}||_\Im = 1$. From (2.13) and (2.14), we readily obtain $-\Delta\hat{w} = \lambda_0\hat{w}, -\Delta\hat{z} = \lambda_0\hat{z}$ in Ω. Since λ_0 is the first eigenvalue of $(-\Delta)^{-1}$, we must have $\hat{w} > 0$ or $\hat{w} < 0$ on Ω. The same is true for $\hat{z}$.

It follows that there exists $k_0 > 0$ such that w_k^i and z_k^i have the same sign as $\hat{w}$ and $\hat{z}$ respectively, for all $k \geq k_0$. Since the additive inverse of any solution of (2.13) or (2.14) is also a solution of the same equation, we may assume that $w_k^i > 0$ and $z_k^i > 0$ in Ω for all $k \geq k_0$. From hypothesis (2.10) and the limits of λ_k^i and $\hat{\lambda}_k^i$, we may also assume that $0 < \lambda_k^i \leq 1$ and $0 < \hat{\lambda}_k^i \leq 1$ for all $k \geq k_0$. Moreover, from the fact that $||w_k^i||_\Im \to 0$ and $||z_k^i||_\Im \to 0$ as $k \to \infty$, there exists an integer $k_1 \geq k_0$ such that

$$(2.15) \qquad w_k^i \leq \bar{w}_i, \quad i = 1, ..., m \quad \text{and} \quad z_k^i \leq \bar{z}_i, \quad i = 1, ..., n \text{ in } \Omega$$

for all $k \geq k_1$.

Define $\underline{w}_i = w_{k_1}^i, i = 1, ..., m$, $\underline{z}_i = z_{k_1}^i, i = 1, ..., n$, in $\bar{\Omega}$. From (2.13) and (2.14), we obtain

$$(2.16) \qquad \begin{aligned} \Delta\underline{w}_i &= -\underline{w}_i[\lambda_{k_1}^i \underline{p} + a_{ii}\underline{w}_i - c^*] \\ &\geq -\underline{w}_i[p_i + (A\underline{w})_i + (R\bar{z})_i] \quad \text{in } \Omega, \end{aligned}$$

for $i = 1, ..., m$; and

$$(2.17) \qquad \begin{aligned} \Delta\underline{z}_i &= -\underline{z}_i[\hat{\lambda}_{k_1}^i \underline{q} + b_{ii}\underline{z}_i - d^*] \\ &\geq -\underline{z}_i[q_i + (B\underline{z})_i + (S\bar{w})_i] \quad \text{in } \Omega, \end{aligned}$$

for $i = 1, ..., n$. Moreover, we have $0 \le \underline{w}_i \le \bar{w}_i$ for $i = 1, ..., m$ and $0 \le \underline{z}_i \le \bar{z}_i$ for $i = 1, ..., n$, for all $x \in \bar{\Omega}$. On the other hand, hypotheses (2.3), (2.8) and (2.9) imply that

$$\Delta \bar{w}_i \le -\bar{w}_i[p_i + (A\bar{w})_i + (R\underline{z})_i] \text{ in } \Omega, \text{ for } i = 1, ..., m, \tag{2.18}$$

and

$$\Delta \bar{z}_i \le -\bar{z}_i[q_i + (B\bar{z})_i + (S\underline{w})_i] \text{ in } \Omega, \text{ for } i = 1, ..., n. \tag{2.19}$$

Thus the pair of functions $(\underline{w}_1, ..., \underline{w}_m, \underline{z}_1, ..., \underline{z}_m)$ and $(\bar{w}_1, ..., \bar{w}_m, \bar{z}_1, ..., \bar{z}_n)$ form coupled lower and upper solutions for the system (2.1). Here we have used (2.16)-(2.19) and hypotheses (2.3), (2.6) and (2.7) concerning the signs of the matrices A, B, R and S. Consequently, by a well-known theorem (cf., e.g. Theorem 1.4-2 in [125] or Theorem 8-10.5 in [183], there is a solution $(u_1, u_2, ..., u_{m+n})$ of (2.1) with $0 < \underline{w}_i \le u_i \le \bar{w}_i, i = 1, ..., m$ and $0 < \underline{z}_i \le u_i \le \bar{w}_i, i = 1, ..., n$ in Ω. This proves the Theorem.

We next consider the situation when the species inside group I do not interact cooperatively as in (2.6). We assume the other relations between the groups I and II, and inside group II, remain unchanged. In group I, we assume that there is one predator (i.e. the m-th species) with $m - 1$ competing preys. More precisely, we suppose:

$$\begin{aligned} &a_{ij} < 0 \text{ for } 1 \le i \le m-1, \ 1 \le j \le m; \\ &a_{mj} > 0 \text{ for } 1 \le j \le m-1. \end{aligned} \tag{2.20}$$

In this case we will not assume as before that the subsystem (2.4), i.e. the subsystem for species inside group I, has an upper solution as given in (2.8). If we have the hypothesis:

$$\lambda_0 < \underline{p}, \tag{2.21}$$

we define positive functions $y = (y_1(x), ..., y_m(x))$ in Ω as follows:

$$\begin{cases} -\Delta y_i = y_i[p_i + a_{ii}y_i] \text{ in } \Omega, \\ y_i = 0 \text{ on } \partial\Omega, \quad \text{for } i = 1, ..., m-1; \\ -\Delta y_m = y_m[p_m + \Sigma_{j=1}^{m-1} a_{mj}y_j + a_{mm}y_m] \text{ in } \Omega \\ y_m = 0 \text{ on } \partial\Omega. \end{cases}$$

From hypotheses (2.21) and the assumption on a_{ii} in (2.3), we see that small positive multiples of an eigenfunction of the operator Δ and large constant functions are respectively lower and upper solutions of each of the above scalar equations successively. By means of Lemma 2.1 in Chapter 1 or Theorem 1.4-2 and Lemma 5.2-2 in [125], we can readily show that the above equations have a solution $y \in [C^{2+\mu}(\bar{\Omega})]^m$, and each y_i is a uniquely defined positive function in $\Omega, i = 1, ..., m$. For convenience, we define the matrix:

$$\tilde{A} = \{\tilde{a}_{ij}\}, \;\; 1 \leq i, j \leq m,$$

where

$$\tilde{a}_{ij} = a_{ij} \text{ if } i \neq j, \;\; \text{and } \tilde{a}_{ij} = 0 \text{ if } i = j$$

Theorem 2.2 (Synthesizing a Prey-Predator Group with a Cooperative Group). *Assume hypotheses (2.3), (2.7), (2.20) and (2.21). Suppose that there exists $\bar{z} = (\bar{z}_1(x), ..., \bar{z}_n(x))$ whose components are in $C^{2+\mu}(\bar{\Omega}), 0 < \mu < 1$, and are positive for all $x \in \Omega$ such that*

$$(2.22) \qquad \begin{cases} \Delta\bar{z}_i + \bar{z}_i[q_i + (B\bar{z})_i] \leq 0 \;\; \textit{in } \Omega, \\ \bar{z}_i = 0 \;\; \textit{on } \partial\Omega, \qquad \textit{for } i = 1, ..., n. \end{cases}$$

Further, assume

$$(2.23) \qquad \tilde{c} + \lambda_0 \leq \underline{p} \quad \textit{and} \quad \tilde{d} + \lambda_0 \leq \underline{q},$$

where

$$\tilde{c} := max\{\, sup\{(\tilde{A}y(x) - R\bar{z}(x))_i : x \in \bar{\Omega}\}, i = 1, ..., m-1;$$

$$sup\{(-R\bar{z}(x))_m : x \in \bar{\Omega}\}\},$$

$$\tilde{d} := max_{1 \leq i \leq n}\{sup(-Sy(x))_i : x \in \bar{\Omega}\}.$$

Then there exists a solution $(u_1, u_2, ..., u_{m+n}) \in [C^{2+\mu}(\bar{\Omega}]^{m+n}$ of (2.1) such that $u_i > 0$ in $\Omega, i = 1, ..., m+n$.

Remark 2.2. (a) Note that y is constructed independent of the off-diagonal entries of the first $m-1$ rows of the matrix A. On the other hand, the matrix $\tilde{A}$ is independent of the diagonal entries of A; and thus, the first inequality of (2.23) is valid provided that $-R$ and the absolute value of the off-diagonal entries of the first $m-1$ rows of A is sufficiently small.

(b) Hypothesis (2.21)is actually included in assumption (2.23).

Proof of Theorem 2.2. As in the proof of the last theorem, we consider the following equations:

$$(2.24) \qquad \begin{cases} -\Delta w + \tilde{c}w = w[\lambda \underline{p} + a_{ii}|w|] \;\; \text{in } \Omega, \\ w = 0 \;\; \text{on } \partial\Omega, \qquad \text{for } i = 1, ..., m; \end{cases}$$

and

$$(2.25)\qquad \begin{cases} -\Delta z + \tilde{d}z = z[\tilde{\lambda}\underline{\mathrm{q}} + b_{ii}|z|] & \text{in } \Omega, \\ z = 0 \text{ on } \partial\Omega, \quad \text{for } i = 1, ..., n. \end{cases}$$

As in Lemma 2.1, we can show that $(\lambda, w) = ([\lambda_0 + \tilde{c}]\underline{\mathrm{p}}^{-1}, 0)$ is a bifurcation point of (2.24), and $(\tilde{\lambda}, z) = ([\lambda_0 + \tilde{d}]\underline{\mathrm{q}}^{-1}, 0)$ is a bifurcation point of (2.25). There exist sequences $\{(\lambda_k^i, w_k^i)\}_{k=1}^{\infty}$ and $\{(\tilde{\lambda}_k^i, z_k^i)\}_{k=1}^{\infty}$ in $R \times \Im$ with $(\lambda_k^i, w_k^i) \to ([\lambda_0 + \tilde{c}]\underline{\mathrm{p}}^{-1}, 0)$ and $(\tilde{\lambda}_k^i, z_k^i) \to ([\lambda_0 + \tilde{d}]\underline{\mathrm{q}}^{-1}, 0)$ as $k \to \infty$, where $w_k^i > 0$ and $z_k^i > 0$ are solutions of:

$$(2.26)\qquad \begin{cases} -\Delta w_k^i + \tilde{c}w_k^i = w_k^i[\lambda_k^i\underline{\mathrm{p}} + a_{ii}|w_k^i|] & \text{in } \Omega, \\ w_k^i = 0 \text{ on } \partial\Omega, \quad \text{for } i = 1, ..., m; \end{cases}$$

and

$$(2.27)\qquad \begin{cases} -\Delta z_k^i + \tilde{d}z_k^i = z_k^i[\tilde{\lambda}_k^i\underline{\mathrm{q}} + b_{ii}|z_k^i|] & \text{in } \Omega, \\ z_k^i = 0 \text{ on } \partial\Omega, \quad \text{for } i = 1, ..., n. \end{cases}$$

By hypothesis (2.23), we may assume that $0 < \lambda_k^i \leq 1$ and $0 < \tilde{\lambda}_k^i \leq 1$ for all $k \geq k_0$, some k_0.

Since $||w_k^i||_{\Im} \to 0$, $||z_k^i||_{\Im} \to 0$ as $k \to \infty$, there exists a positive integer $k_1 \geq k_0$, such that $0 < w_k^i \leq y_i, i = 1, ..., m$, $0 < z_k^i \leq \bar{z}_i, i = 1, ..., n$, in Ω, for all $k \geq k_1$. Define $\underline{\mathrm{w}}_i = w_{k_1}^i$ for $i = 1, ..., m$, $\underline{\mathrm{z}}_i = z_{k_1}^i$ for $i = 1, ..., n$ in $\bar{\Omega}$; and let $\underline{\mathrm{w}} = (\underline{\mathrm{w}}_1, ..., \underline{\mathrm{w}}_m), \underline{\mathrm{z}} = (\underline{\mathrm{z}}_1, ..., \underline{\mathrm{z}}_n)$. From (2.26) and (2.27), we obtain:

$$(2.28)\qquad \begin{aligned} \Delta\underline{\mathrm{w}}_i &= -\underline{\mathrm{w}}_i[\lambda_{k_1}^i\underline{\mathrm{p}} + a_{ii}\underline{\mathrm{w}}_i - \tilde{c}] \\ &\geq -\underline{\mathrm{w}}_i[p_i + a_{ii}\underline{\mathrm{w}}_i + (\tilde{A}y)_i + (R\bar{z})_i] \quad \text{in } \Omega, \end{aligned}$$

for $i = 1, ..., m-1$; and

$$(2.29)\qquad \begin{aligned} \Delta\underline{\mathrm{w}}_m &= -\underline{\mathrm{w}}_m[\lambda_{k_1}^m\underline{\mathrm{p}} + a_{mm}\underline{\mathrm{w}}_m - \tilde{c}] \\ &\geq -\underline{\mathrm{w}}_m[p_i + a_{mm}\underline{\mathrm{w}}_m + (\tilde{A}\underline{\mathrm{w}})_m + (R\bar{z})_m] \quad \text{in } \Omega. \end{aligned}$$

Moreover, we have

$$(2.30)\qquad \begin{aligned} \Delta\underline{\mathrm{z}}_i &= -\underline{\mathrm{z}}_i[\hat{\lambda}_{k_1}^i\underline{\mathrm{q}} + b_{ii}\underline{\mathrm{z}}_i - \tilde{d}] \\ &\geq -\underline{\mathrm{z}}_i[q_i + b_{ii}\underline{\mathrm{z}}_i + (B\underline{\mathrm{z}})_i + (Sy)_i] \quad \text{in } \Omega, \end{aligned}$$

for $i = 1, ..., n$, and $0 < \underline{w}_i \le y_i, 0 < \underline{z}_i \le \bar{z}_i$ in Ω. On the other hand, from the construction of y, hypotheses (2.20) and (2.22), we deduce that:

$$\Delta y_i \le -y_i[p_i + a_{ii}y_i + (\tilde{A}\underline{w})_i + (R\underline{z})_i] \text{ in } \Omega, \text{ for } i = 1, ..., m-1,$$

$$\Delta y_m \le -y_m[p_m + a_{mm}y_m + (\tilde{A}y)_m + (R\underline{z})_m] \text{ in } \Omega, \tag{2.31}$$

$$\Delta \bar{z}_i \le -\bar{z}_i[q_i + (B\bar{z})_i + (S\underline{w})_i] \text{ in } \Omega, \text{ for } i = 1, ..., n.$$

Thus, the pair of functions $(\underline{w}_1, ..., \underline{w}_m, \underline{z}_1, ..., \underline{z}_n)$ and $(y_1, ..., y_m, \bar{z}_1, ..., \bar{z}_n)$ form lower and upper solutions for the system (2.1). Consequently, we conclude as in the last theorem that problem (2.1) has a positive solution in Ω, between the upper and lower solutions.

In the third case, we consider the situation when there is a competing relation inside the first group of species and a food-chain relation inside the second group. Again, the two groups compete against each other as before. More specifically, we assume:

$$a_{ij} < 0 \text{ for } 1 \le i, j \le m; \tag{2.32}$$

$$\begin{aligned} &b_{i,i+1} < 0 \text{ for } 1 \le i \le n-1, \\ &b_{i,i-1} > 0 \text{ for } 2 \le i \le n, \text{ and} \\ &b_{ij} = 0 \text{ for } |i-j| > 1. \end{aligned} \tag{2.33}$$

Positive solutions inside a food-chain group have been studied in López-Gómez and Pardo San Gil [163]. Here, we will not need to assume the existence of positive solution explicitly inside each group as in Theorem 2.1. For convenience, we define

$$C_1 = q_1/|b_{11}|, \; C_i = (q_i + C_{i-1}b_{i,i-1})/|b_{ii}| \text{ for } i = 2, ..., n; \tag{2.34}$$

$$\begin{aligned} &A_i^* = \Sigma_{j=1, j\ne i}^m |a_{ij}p_j/a_{jj}| + \Sigma_{j=1}^n |r_{ij}|C_j \text{ for } i = 1, ..., m; \text{ and} \\ &B_i^* = |b_{i,i+1}|C_{i+1} + \Sigma_{j=1}^m |s_{ij}p_j/a_{jj}| \text{ for } i = 1, ..., n. \end{aligned} \tag{2.35}$$

Theorem 2.3 (Synthesizing a Competing Group with a Food-Chain Group). *Assume hypotheses (2.3), (2.32) and (2.33). Suppose*

$$\begin{aligned} &A_i^* + \lambda_0 < p_i \textit{ for } i = 1, ..., m, \textit{ and} \\ &B_i^* + \lambda_0 < q_i \textit{ for } i = 1, ..., n. \end{aligned} \tag{2.36}$$

Then there exists a solution $(u_1, u_2, ..., u_{m+n}) \in [C^{2+\mu}(\bar{\Omega})]^{m+n}$ *of problem (2.1) such that* $u_i > 0$ *in* Ω, $i = 1, ..., m+n$.

Proof. First define $\bar{w}_i \in C^{2+\mu}(\bar{\Omega})$ for $i = 1, ..., m$ to be the unique positive solution of the problem:

$$\begin{cases} -\Delta \bar{w}_i = \bar{w}_i[p_i + a_{ii}\bar{w}_i] & \text{in } \Omega, \\ \bar{w}_i = 0 \text{ on } \partial\Omega, \quad \bar{w}_i > 0 & \text{in } \Omega. \end{cases}$$

It is clear that $\bar{w}_i \le p_i/a_{ii}$ in $\bar{\Omega}$.(See e.g. Chapter 5 in [125]). Further, define $\bar{z}_i \equiv C_i$ in Ω for $i = 1, ..., n$. From (2.34), we see that $\bar{z}_i$ satisfies

$$-\Delta \bar{z}_i = 0 \ge \bar{z}_i[q_i + b_{i,i-1}\bar{z}_{i-1} + b_{ii}\bar{z}_i] \text{ in } \Omega, \text{ for } i = 1, ..., n.$$

(Here, we denote $\bar{z}_0 = 0, b_{1,0} = 0$ for convenience.)

To construct a lower solution for system (2.1) in this theorem, we consider the auxiliary problems:

$$(2.37) \qquad \begin{cases} -\Delta w + A_i^* w = w[\lambda p_i + a_{ii}|w|] \text{ in } \Omega, \\ w = 0 \text{ on } \partial\Omega, \quad \text{for } i = 1, ..., m; \end{cases}$$

and

$$(2.38) \qquad \begin{cases} -\Delta z + B_i^* z = z[\hat{\lambda} q_i + b_{ii}|z|] \text{ in } \Omega, \\ z = 0 \text{ on } \partial\Omega, \quad \text{for } i = 1, ..., n. \end{cases}$$

As in Theorem 2.1, the points $(\lambda, w) = ([\lambda_0 + A_i^*]/p_i, 0) \in R \times \Im$ are bifurcation points for equation (2.37), and $(\hat{\lambda}, z) = ([\lambda_0 + B_i^*]/q_i, 0) \in R \times \Im$ are bifurcation points for equation (2.38). There exist sequences $\{(\lambda_k^i, w_k^i)\}_{k=1}^{\infty}, \{(\hat{\lambda}_k^i, z_k^i)\}_{k=1}^{\infty}$, with $w_k^i > 0$ and $z_k^i > 0$ in Ω, $(\lambda_k^i, w_k^i) \to ([\lambda_0 + A_i^*]/p_i, 0), (\hat{\lambda}_k^i, z_k^i) \to ([\lambda_0 + B_i^*]/q_i, 0)$ in $R \times \Im$, as $k \to \infty$. Moreover,

$$(2.39) \qquad \begin{cases} -\Delta w_k^i + A_i^* w_k^i = w_k^i[\lambda_k^i p_i + a_{ii}|w_k^i|] \text{ in } \Omega, \\ w_k^i = 0 \text{ on } \partial\Omega, \quad \text{for } i = 1, ..., m; \end{cases}$$

and

$$(2.40) \qquad \begin{cases} -\Delta z_k^i + B_i^* z_k^i = z_k^i[\hat{\lambda}_k^i q_i + b_{ii}|z_k^i|] \text{ in } \Omega, \\ z_k^i = 0 \text{ on } \partial\Omega, \quad \text{for } i = 1, ..., n. \end{cases}$$

By hypotheses (2.36), we have $0 < [\lambda_0 + A_i^*]/p_i < 1$ and $0 < [\lambda_0 + B_j^*]/q_j < 1$ for $i = 1, ..., m$ and $j = 1, ..., n$, respectively. It follows that there exists a positive integer k_0 such that $0 < \lambda^i_{k_0} < 1, 0 < \hat{\lambda}^j_{k_0} < 1$ and $0 < w^i_{k_0} \le \bar{w}_i, 0 < z^j_{k_0} \le \bar{z}_j$ in $\bar{\Omega}$ for $i = 1, ..., m$ and $j = 1, ..., n$, respectively.

Finally, define $\underline{w}_i = w^i_{k_0}$ and $\underline{z}_j = z^j_{k_0}$ in Ω in for $i = 1, ..., m$ and $j = 1, ..., n$, respectively. Set $\bar{w} = (\bar{w}_1, ..., \bar{w}_m)$, $\underline{w} = (\underline{w}_1, ..., \underline{w}_m), \bar{z} = (\bar{z}_1, ..., \bar{z}_n)$ and $\underline{z} = (\underline{z}_1, ..., \underline{z}_n)$ in $\bar{\Omega}$. From (2.39) and (2.35), we have for $i = 1, ..., m$,

$$
\begin{aligned}
-\Delta \underline{w}_i &= -A_i^* \underline{w}_i + \underline{w}_i[\lambda_{k_0} p_i + a_{ii} \underline{w}_i] \\
&\le \underline{w}_i[p_i + a_{ii}\underline{w}_i + (\tilde{A}\bar{w})_i + (R\bar{z})_i] \quad \text{in } \Omega.
\end{aligned} \tag{2.41}
$$

From (2.33), (2.40) and (2.35), we have for $i = 1, ..., n$,

$$
\begin{aligned}
-\Delta \underline{z}_i &= -B_i^* \underline{z}_i + \underline{z}_i[\hat{\lambda}_{k_0} q_i + b_{ii} \underline{z}_i] \\
&\le \underline{z}_i[q_i + b_{ii}\underline{z}_i + b_{i,i-1}\underline{z}_{i-1} + b_{i,i+1}\bar{z}_{i+1} + (S\bar{w})_i] \quad \text{in } \Omega.
\end{aligned} \tag{2.42}
$$

Moreover, from hypotheses (2.32) and (2.33), we obtain for $i = 1, ..., m$,

$$
\begin{aligned}
-\Delta \bar{w}_i &= \bar{w}_i[p_i + a_{ii}\bar{w}_i] \\
&\ge \bar{w}_i[p_i + a_{ii}\bar{w}_i + (\tilde{A}\underline{w})_i + (R\underline{z})_i] \quad \text{in } \Omega;
\end{aligned} \tag{2.43}
$$

and we obtain for $i = 1, ..., n$,

$$
\begin{aligned}
-\Delta \bar{z}_i &\ge \bar{z}_i[q_i + b_{i,i-1}\bar{z}_{i-1} + b_{ii}\bar{z}_i] \\
&\ge \bar{z}_i[q_i + b_{i.i-1}\bar{z}_{i-1} + b_{ii}\bar{z}_i + b_{i,i+1}\underline{z}_{i+1} + (S\underline{w})_i] \quad \text{in } \Omega.
\end{aligned} \tag{2.44}
$$

Consequently, we find that the pair $(\bar{w}_1, ..., \bar{w}_m, \bar{z}_1, ..., \bar{z}_n)$ and $(\underline{w}_1, ..., \underline{w}_m, \underline{z}_1, ..., \underline{z}_n)$ form upper and lower solutions for the system (2.1) under the hypotheses of this theorem. We then conclude that there exists a solution $(u_1, ..., u_{m+n})$ between the pair, and it has the properties as stated in the theorem.

In general, it is nontrivial to verify assumptions (2.8) to (2.10) in Theorem 2.1. However, in some cases, it is not difficult to express these hypotheses directly in terms of the growth and interaction coefficients of the components, as illustrated in the following example.

Example 2.1. Consider system (2.1) with $m = 3, n = 2$. Assume hypotheses (2.3), (2.6) and (2.7). Let

$$
D = det \begin{vmatrix} a_{11} & a_{12} & a_{13} \\ a_{21} & a_{22} & a_{23} \\ a_{31} & a_{32} & a_{33} \end{vmatrix}, \quad D_1 = det \begin{vmatrix} -1 & a_{12} & a_{13} \\ -1 & a_{22} & a_{23} \\ -1 & a_{32} & a_{33} \end{vmatrix},
$$

$$D_2 = det\begin{vmatrix} a_{11} & -1 & a_{13} \\ a_{21} & -1 & a_{23} \\ a_{31} & -1 & a_{33} \end{vmatrix}, \quad D_3 = det\begin{vmatrix} a_{11} & a_{12} & -1 \\ a_{21} & a_{22} & -1 \\ a_{31} & a_{32} & -1 \end{vmatrix},$$

$$D_4 = det\begin{vmatrix} b_{11} & b_{12} \\ b_{21} & b_{22} \end{vmatrix}, \quad D_5 = det\begin{vmatrix} -1 & b_{12} \\ -1 & b_{22} \end{vmatrix}, \quad D_6 = det\begin{vmatrix} b_{11} & -1 \\ b_{21} & -1 \end{vmatrix}.$$

Suppose that

(2.45) D, D_1, D_2 and D_3 have the same sign;

(2.46) D_4, D_5 and D_6 have the same sign; and

(2.47) $\underline{p} > \lambda_0, \quad \underline{q} > \lambda_0.$

Then there exists a solution $(u_1, ..., u_5) \in [C^{2+\mu}(\bar{\Omega})]^5$ of problem (2.1) such that $u_i > 0$ in Ω, for $i = 1, ..., 5$, provided that all the entries of the matrices R and S are sufficiently small in absolute value.

Remark 2.3. An example for a_{ij} and b_{ij} satisfying (2.45), (2.46), (2.6) and (2.7) is given by $a_{11} = -3, a_{12} = 2, a_{13} = 3, a_{21} = 1, a_{22} = -5, a_{23} = 2, a_{31} = 1, a_{32} = 1, a_{33} = -4$ and $b_{11} = -2, b_{12} = 1, b_{21} = 1, b_{22} = -1$.

Proof. Let θ and ξ be respectively the unique positive solution of the boundary value problems:

(2.48) $$-\Delta v = v[\bar{p} - v] \text{ in } \Omega, \quad v = 0 \text{ on } \partial\Omega,$$

and

(2.49) $$-\Delta \tilde{v} = \tilde{v}[\bar{q} - \tilde{v}] \text{ in } \Omega, \quad \tilde{v} = 0 \text{ on } \partial\Omega,$$

where $\bar{p} = max\{p_1, p_2, p_3\}$ and $\bar{q} = max\{q_1, q_2\}$. For $\alpha = D_1/D > 0, \beta = D_2/D > 0, \gamma = D_3/D > 0$, the function $w = (w_1, w_2, w_3) = (\alpha\theta, \beta\theta, \gamma\theta)$ is a solution of the problem

(2.50) $$\begin{cases} -\Delta w_i = w_i[\bar{p} + (Aw)_i] \text{ in } \Omega, \\ w_i = 0 \text{ on } \partial\Omega, \text{ for } i = 1, 2, 3. \end{cases}$$

For $\bar{\alpha} = D_5/D_4 > 0, \bar{\beta} = D_6/D_4 > 0$, the function $z = (z_1, z_2) = (\bar{\alpha}\xi, \bar{\beta}\xi)$ is a solution to the problem

(2.51) $$\begin{cases} -\Delta z_i = z_i[\bar{q} + (Bz)_i] \text{ in } \Omega, \\ z_i = 0 \text{ on } \partial\Omega, \text{ for } i = 1, 2. \end{cases}$$

Equations (2.50) and (2.51) imply that $w = (w_1, w_2, w_3) = (\alpha\theta, \beta\theta, \gamma\theta)$ satisfies

$$\begin{cases} \Delta w_i + w_i[p_i + (Aw)_i] \le 0 \text{ in } \Omega, \\ w_i = 0 \text{ on } \partial\Omega, \text{ for } i = 1, 2, 3, \end{cases} \tag{2.52}$$

and $z = (z_1, z_2) = (\bar{\alpha}\xi, \bar{\beta}\xi)$ satisfies

$$\begin{cases} \Delta z_i + z_i[q_i + (Bz)_i] \le 0 \text{ in } \Omega, \\ z_i = 0 \text{ on } \partial\Omega, \text{ for } i = 1, 2. \end{cases} \tag{2.53}$$

Inequality (2.47) and the assumptions on the matrices R and S imply that condition (2.10) in Theorem 2.1 is valid. Thus the existence of positive solution in this example follows from Theorem 2.1.

Remark 2.4. Since $1/\bar{p}$ and $1/\bar{q}$ are upper bounds for the positive solutions in (2.48) and (3.49), respectively, one can readily obtain an explicit condition on the matrices R and S such that (2.10) is satisfied.

Further extentions of the theories in this section can be found in [139].

2.3 Application to Epidemics of Many Interacting Infected Species

This section considers the application of systems of reaction-diffusion to epidemiology. Blat and Brown [12] model the spread of infections by a system of two reaction-diffusion equations. Capasso and Maddalena [23], [24] use similar models to investigate the spread of oro-faecal or other bacterial and viral diseases. In this section, we consider large systems modeling the spread of several bacterial infections among many interacting species, as described in Leung and Villa [142]. More precisely, we consider

(3.1)
$$\begin{cases} -\Delta u_i + a_i(x)u_i = \Sigma_{j=1}^m b_{ij}v_j & \text{for } x \in \Omega,\ i = 1, ..., n, \\ -\Delta v_k + \tilde{a}_k v_k = \Sigma_{j=1}^n f_{kj}(u_j) + v_k \Sigma_{j=1}^m c_{kj}v_j & \text{for } x \in \Omega,\ k = 1, ..., m, \\ u_i = v_k = 0 & \text{for } x \in \partial\Omega,\ i = 1, ..., n,\ k = 1, ..., m, \end{cases}$$

where $b_{ij} > 0$ and c_{kj} are any constant, $f_{kj} \in C^1(R)$, and Ω is a bounded domain in R^N, with boundary $\partial\Omega$ of class $C^{2+\alpha}, 0 < \alpha < 1$.

We will also consider the corresponding parabolic system, with $\partial u_i/\partial t$ and $\partial v_k/\partial t$ added to the first n and second m equations respectively on the left on

(3.1). The functions u_i represent n different kinds of bacterial population densities and the functions v_k represent m different-type infected-species population densities. The populations are assumed to diffuse in the space domain Ω. The functions $a_i(x)$ are assumed to be positive, because the bacterial populations tend to die in the absence of other factors; the term $b_{ij}v_j$ represent the growth of the number of bacterial due to infected species. The functions $\tilde{a}_k(x)$ are assumed to be positive, because a certain proportion of the infected species recover per unit time; the term $f_{kj}(u_j)$ describe the rate at which the kth species becomes infected by u_j, and the terms $v_k c_{kj} v_j$ describe the interaction between the kth and the jth infected species. The model can be more naturally interpreted in the form of the corresponding parabolic system, with the positive solutions of (3.1) considered as steady-states. The prototype form with $m = n = 1, c_{11} = 0$, is introduced in e.g. [12] and [23]. We will consider the case when all c_{jk} are zero as well as other cases.

For convenience, we will adopt the following conventions. Let B and K_0 be respectively $n \times m$ and $m \times n$ constant matrices as follows:

$$B = \begin{bmatrix} b_{11} & \cdots & b_{1m} \\ \cdot & & \cdot \\ \cdot & & \cdot \\ \cdot & & \cdot \\ b_{n1} & \cdots & b_{nm} \end{bmatrix}, \quad K_0 = \begin{bmatrix} f'_{11}(0) & \cdots & f'_{1n}(0) \\ \cdot & & \cdot \\ \cdot & & \cdot \\ \cdot & & \cdot \\ f'_{m1}(0) & \cdots & f'_{mn}(0) \end{bmatrix}. \tag{3.2}$$

For abbreviation, we write $\hat{F} = col.(F_1, ..., F_{n+m})$, where F_j are operators from $[C^1(\bar{\Omega})]^{n+m}$ into $C^1(\bar{\Omega})$ defined by

(3.3)

$$F_i[col.(u_1, ..., u_n, v_1, ..., v_m)] = \Sigma_{j=1}^m b_{ij} v_j \qquad \text{for } i = 1, ..., n,$$

$$F_{n+k}[col.(u_1, ..., u_n, v_1, ..., v_m)] = \Sigma_{j=1}^n f_{kj}(u_j) + v_k \Sigma_{j=1}^m c_{kj} v_j \quad \text{for } k = 1, ..., m.$$

We now label a few key assumptions, some or all of which will be used in various theorems in this section:

(3.4) The functions $a_i, \tilde{a}_k$ are members of $C^\alpha(\bar{\Omega}), 0 < \alpha < 1$ and satisfy $a_i(x) > 0, \tilde{a}_k(x) > 0$ for all $x \in \bar{\Omega}, i = 1, ..., n, k = 1, ..., m$.

(3.5)
The functions $f_{kj} \in C^1(R)$ satisfy $f'_{kj}(0) > 0$ and $f_{kj}(0) = 0$; $f_{kj}(s) \geq 0$ for all $s \geq 0, k = 1, ..., m,\ j = 1, ..., n$. For each k, there exists at least one j such that $0 < f_{kj}(s)$ for all $s > 0$.

(3.6) The functions f_{kj} satisfy $f_{kj}(s) \leq K_1 s$ for all $s > 0$, where K_1 is some positive constant.

(3.7) There exists a constant vector $\vec{d} = col.(d_1, ..., d_n)$, with $d_i > 0$, $i = 1, ..., n$, such that $BK_0\vec{d} > (\lambda_1 + a^*)^2\vec{d}$.

Here, the (strict) inequality between the two vectors are interpreted to be satisfied for each component. The quantity λ_1 is the principal eigenvalue for $-\Delta$ as defined by (1.5) in Section 1.1, with principal eigenfunction ϕ where $||\phi||_\infty = 1$. The symbols a^* and a^{**} are defined as $a^* = sup.\{a_i(x), \tilde{a}_k(x)|x \in \bar{\Omega}, i = 1, ..., n, k = 1, ..., m\}$ and $a^{**} = inf.\{a_i(x), \tilde{a}_k(x)|x \in \bar{\Omega}, i = 1, ..., n, k = 1, ..., m\}$.

Another assumption which will sometimes be used concerning the interaction of the infected species v_k is as follows:

(3.8) $$c_{kk} < 0, \text{ and } |c_{kk}| > \Sigma_{j=1, j\neq k}^{m}|c_{kj}|, \text{ for each } k = 1, ..., m.$$

In order to simplify writing, we introduce a few more notations. Let

$$E = \{w = col.(w_1, ..., w_{n+m})|w_i \in C^1(\bar{\Omega}), w_i = 0 \text{ on } \partial\Omega, i = 1, ..., m+n\},$$

with norm $||w||_E = max.\{||w_i||_{C^1(\bar{\Omega})}|i = 1, ..., m+n\}$; and $\mathcal{P}$ denotes the cone $\mathcal{P} = \{col.(w_1, ..., w_{n+m}) \in E|w_i \geq 0 \text{ in } \bar{\Omega}, i = 1, ..., n+m\}$. Also, let

$$Y = \{w = col.(w_1, ..., w_{n+m})|w_i \in C^{2+\alpha}(\bar{\Omega}), w_i = 0 \text{ on } \partial\Omega\}$$

with its norm denoted as $||w||_Y = max.\{||w_i||_{C^{2+\alpha}(\bar{\Omega})}|i = 1, ..., m+n\}$. As operators from $C^{2+\alpha}(\bar{\Omega})$ into $C^\alpha(\bar{\Omega})$, we write

$$L_i := -\Delta + a_i, \; i = 1, ..., n,$$

$$L_{n+k} := -\Delta + \tilde{a}_k, \; k = 1, ..., m.$$

As an operator from Y into $[C^\alpha(\bar{\Omega})]^{n+m}$, we write $L = col.(L_1, ..., L_{n+m})$.

The system (3.1) can be abbreviated as

(3.9) $$L[w] = \hat{F}[w], \quad \text{where } w \in Y,$$

To study this nonlinear problem, we consider the auxiliary problem:

(3.10) $$L[w] = \lambda\hat{F}[w], \; w \in Y,$$

where λ is a parameter, and investigate the bifurcation from the trivial solution $w = 0$ as the parameter λ passes through a certain value λ_0. Under conditions (3.4) and (3.5), we will see that this bifurcation actually occurs in Theorem 3.1. Moreover, in Theorem 3.2, we will see that hypotheses (3.5), (3.6) and (3.7) insure that $\lambda_0 < 1$; and hypothesis (3.8) insures that the bifurcation curve of nontrivial solutions connects to $\lambda = +\infty$. Thus (3.10) has a nontrivial solution when $\lambda = 1$, i.e. (3.9) has a nontrivial solution under appropriate conditions. For convenience, we let M_0 denotes the $(n+m)\times(n+m)$ square constant matrix:

$$M_0 = \begin{bmatrix} 0 & B \\ K_0 & 0 \end{bmatrix}$$

where the 0's along the diagonal are zero matrices with appropriate dimensions. Applying L^{-1} to both sides of (3.10), using zero Dirichlet boundary condition, we obtain: $w = \lambda L^{-1}\hat{F}[w]$. Thus, (3.10) can be written as:

$$Q(\lambda, w) = 0, \quad (\lambda, w) \in R \times E, \tag{3.11}$$

where $Q : R \times E \to E$ is an operator given by

$$Q(\lambda, w) := w - \lambda L^{-1}\hat{F}[w]$$

(for this entire section the inverse operators L^{-1} or L_i^{-1} will always mean finding the solution using zero Dirichlet boundary condition).

Theorem 3.1. *Under hypotheses (3.4) and (3.5), the point $(\lambda_0, 0)$ is a bifurcation point for problem (3.11). Here $\lambda = \lambda_0$ is the unique positive number so that the problem:*

$$L[w] = \lambda M_0 w \ \ in\, \Omega, \quad w = 0 \ \ on\ \partial\Omega, \tag{3.12}$$

has a non-negative eigenfunction in E. (The nullspace for $L^{-1}M_0 - \frac{1}{\lambda_0}I$ is one dimensional as described in Lemma 3.2 below.) Moreover, the component of $\bar{\mathcal{S}}$ containing the point $(\lambda_0, 0)$ is unbounded, where

$$\mathcal{S} := \{(\lambda, w) \in R^+ \times \mathcal{P}|\ Q(\lambda, w) = 0, \lambda > 0 \ \ and \ \ w \in \mathcal{P}\backslash\{0\}\};$$

and it also has the property that $\bar{\mathcal{S}} \cap (R^+ \times \partial\mathcal{P}) = (\lambda_0, 0)$.

We first state a sequence of lemmas which will lead to the proof of Theorem 3.1. The proof of the lemmas will be given afterwards.

Lemma 3.1. (Comparison) *Let $w, \hat{w} \in [C^2(\Omega)\cap C^1(\bar{\Omega})]^{n+m}, w \not\equiv 0, \hat{w}_i \geq 0, \hat{w}_i \not\equiv 0$ in Ω, for $i = 1, \ldots, n+m$, satisfy*

$$\begin{cases} L_i[w_i(x)] = \sum_{j=1}^{n+m} p_{ij}(x) w_j(x), \ \ for\ x \in \Omega, \ \ i = 1, \ldots, n+m \\ w|_{\partial\Omega} = 0, \quad w = col.(w_1, \ldots, w_{n+m}), \end{cases}$$

$$\begin{cases} L_i[\hat{w}_i(x)] \geq \sum_{j=1}^{n+m} q_{ij}(x) \hat{w}_j(x), \ \ for\ \ x \in \Omega, i = 1, \ldots, n+m \\ \hat{w} = col.(\hat{w}_1, \ldots, \hat{w}_{n+m}) \end{cases}$$

where p_{ij} and q_{ij} are bounded functions in Ω. Suppose that

$$q_{ij} \geq p_{ij} \ \text{in} \ \bar{\Omega} \ \text{for} \ i, j = 1, \ldots, n+m,$$

$$\text{and} \ q_{ij}, p_{ij} \geq 0 \ \text{in} \ \bar{\Omega} \ \text{for all} \ i \neq j,$$

then there exists an integer k, $1 \leq k \leq n+m$ and a real number δ such that

$$\hat{w}_k \equiv \delta w_k, \ \ p_{kj} \equiv q_{kj} \ \text{ in } \ \bar{\Omega} \ \text{ for all } \ j = 1, \ldots, n+m,$$

$$\text{and } \ \hat{w}_j - \delta w_j \geq 0 \ \text{ for all } \ j = 1, \ldots, n+m.$$

Lemma 3.2. *Under hypotheses (3.4) and (3.5), there exists $(\lambda_0, w^0) \in R \times Y$, $\lambda_0 > 0$, such that*

$$L[w^0] = \lambda_0 M_0 w^0 \ \ in \ \Omega, \ \ w^0 = 0 \ \ on \ \partial\Omega \tag{3.13}$$

with each component $w_i^0 > 0$ in Ω, $\partial w_i^0/\partial\nu < 0$ on $\partial\Omega$ for $i = 1, \ldots, n+m$. Furthermore, the eigenfunction corresponding to the eigenvalue $1/\lambda_0$ for the operator $L^{-1}M_0 : [C^1(\bar{\Omega})]^{n+m} \to [C^1(\bar{\Omega})]^{n+m}$ is unique up to a multiple. Also, the number $\lambda = \lambda_0$ is the unique positive number so that the problem $w = \lambda L^{-1}M_0 w$ has a nontrivial non-negative solution for $w \in \mathcal{P}$.

Lemma 3.3. *Let $G : E \to E$ be the operator defined by:*

$$G[w] = L^{-1}[\hat{F}(w) - M_0 w],$$

then $||G[w]||_E/||w||_E \to 0$ as $\|w\|_E \to 0$.

Proof of Theorem 3.1. The operator G described in Lemma 3.3 is completely continuous, and the operator $L^{-1}M_0$ described in Lemma 3.2 is compact and positive with respect to $\mathcal{P}$. Equation (3.11) can be written as:

$$w - \lambda L^{-1}[M_0 w] - \lambda G[w] = 0, \ \text{ for } (\lambda, w) \in R^+ \times E.$$

By means of Lemma 3.3 and the existence and uniqueness part of Lemma 3.2, we can apply Theorem 29.2 in Diemling [49] to conclude that $(\lambda_0, 0)$ is a bifurcation point for problem (3.11), and the component of $\mathcal{S}$ containing the point $(\lambda_0, 0)$ as described above is unbounded.

Let $(\lambda_i, w_i) \in \mathcal{S}, i = 1, 2, \ldots$, be a sequence tending to a limit point $(\bar{\lambda}, \bar{w})$ in $R^+ \times \partial\mathcal{P}$, and $(\bar{\lambda}, \bar{w}) \neq (\lambda_0, 0)$. We now show that $\bar{w} = col.(\bar{w}_1, \ldots, \bar{w}_{n+m})$ must satisfy $\bar{w}_i \equiv 0$ in $\bar{\Omega}$, for each $i = 1, \ldots, n+m$. Consider the first case when there exists some $x_0 \in \Omega$ and some $j \in \{1, \ldots, n\}$ such that $\bar{w}_j(x_0) = 0$. The equation $L_j[\bar{w}_j] = \bar{\lambda}\Sigma_{k=1}^m b_{jk}\bar{w}_{n+k} \geq 0$ implies $\bar{w}_j \equiv 0$ in $\bar{\Omega}$; and subsequently the right side of this equation implies that $\bar{w}_{n+k} \equiv 0$ in $\bar{\Omega}$ for each $k = 1, \ldots, m$. This leads to $L_i[\bar{w}_i] = 0$ in Ω, for $i = 1, ..., n$, and thus $\bar{w}_i \equiv 0$ for such i too. Hence $\bar{w} \equiv 0$ in this case.

Consider the second case when $\bar{w}_j > 0$ in Ω for all $j \in \{1, ..., n\}$. For each $k = 1, ..., m$, $\lambda > 0$, we introduce the problem

$$(3.14) \qquad \begin{cases} L_k[z] = \lambda \sum_{j=1}^{n} f_{kj}(\bar{w}_j) + \lambda z[\sum_{j=1, j\neq k}^{m} c_{kj}\bar{w}_{n+j} + c_{kk}z] & \text{in } \Omega, \\ \\ z = 0 \qquad \text{on } \partial\Omega. \end{cases}$$

The function $\omega_0 \equiv 0$ is a lower solution of (3.14) due to the sign of f_{kj}; and $\bar{w}_{n+k} \geq 0$ is a solution which is $\not\equiv 0$, due to (3.5). Since $f'_{kj}(0)\partial\bar{w}_j/\partial\nu < 0$ on $\partial\Omega$, there exists a small $\delta_1 > 0$ such that for $0 \leq \delta \leq \delta_1$, we have

$$(3.15) \quad \lambda_1\delta\phi + \tilde{a}_k(x)\delta\phi < \lambda\Sigma_{j=1}^{n} f_{kj}(\bar{w}_j) + \lambda\delta\phi[\Sigma_{j=1,j\neq k}^{m} c_{kj}\bar{w}_{n+j} + c_{kk}\delta\phi] \quad \text{in } \Omega.$$

Consequently, the functions $\omega_\delta := \delta\phi, 0 \leq \delta \leq \delta_1$, form a family of lower solutions for problem (3.14), and $\bar{w}_{n+k} \not\equiv \omega_\delta$ for all $\delta \in [0, \delta_1]$. By means of a sweeping principle (cf. Theorem 1.4-3 in [125]), it follows that $\bar{w}_{n+k} \geq \omega_{\delta_1} > 0$ in Ω. Moreover, for each $k = 1, ..., m$, the function $\bar{w}_{n+k}$ satisfies

$$L_k[\bar{w}_{n+k}] \;\geq\; \lambda\Sigma_{j=1}^{n} f_{kj}(\bar{w}_j) \;-\; \lambda\bar{w}_{n+k}|\Sigma_{j=1}^{m} c_{kj}\bar{w}_{n+j}| \quad \text{in } \Omega,$$

and thus

$$-\Delta\bar{w}_{n+k} + \tilde{a}_k\bar{w}_{n+k} + \lambda|\Sigma_{j=1}^{m} c_{kj}\bar{w}_{n+j}|\bar{w}_{n+k} \;\geq\; \lambda\Sigma_{j=1}^{n} f_{kj}(\bar{w}_j) \geq 0 \;\text{ in } \Omega.$$

It follows from the maximum principle that $\partial\bar{w}_{n+k}/\partial\nu < 0$ on $\partial\Omega$. We thus obtain in this second case that $\bar{w} \in Int(\mathcal{P})$, which is a contradiction. Hence, only the first case can happen, and we must have $\bar{w} \equiv 0$.

Next define $z_i := w_i/||w_i||_E, i = 1, 2, \ldots$; they satisfy

$$(3.16) \qquad z_i - \lambda_i L^{-1}[M_0 z_i] - \lambda_i G[w_i]/||w_i||_E = 0.$$

Since $L^{-1}M_0$ is compact, there exists a subsequence of $\{z_i\}$(again denoted by $\{z_i\}$ for convenience) such that $L^{-1}M_0 z_i$ converges in E. Since $G[w_i]/||w_i||_E$ tends to zero in E, as $||w_i||_E \to ||\bar{w}||_E = 0$, Eq. (3.16) implies that $\{z_i\}$ converges in E to a function z_0 say, and

$$z_0 = \bar{\lambda} L^{-1} M_0 z_0.$$

Moreover, we have $z_0 \in \mathcal{P}$ since $w_i \in \mathcal{P}$, and $z_0 \not\equiv 0$ since $||z_i|| = 1$ for all i. The uniqueness part of Lemma 3.2 thus implies that $\bar{\lambda} = \lambda_0$. Consequently, we must have $(\bar{\lambda}, \bar{w}) = (\lambda_0, 0)$. This completes the proof of Theorem 3.1.

For completeness, we will present the proofs of Lemmas 3.1 to 3.3.

Proof of Lemma 3.1. Let $K > 0$ be a positive constant such that $p_{ii}+K, q_{ii}+K > 0$ in $\bar{\Omega}$, for $i = 1, ..., n+m$. From the inequalities for $L_i[\hat{w}_i(x)]$ and q_{ij}, p_{ij}, we have

$$L_i[\hat{w}_i] + K\hat{w}_i \geq (q_{ii} + K)\hat{w}_i + \sum_{j=1,\neq i}^{m+n} q_{ij}\hat{w}_j \geq 0.$$

Thus the maximum principles imply the following properties

$$\hat{w}_i > 0 \text{ in } \Omega, \text{ and } (\partial \hat{w}_i/\partial \eta)(\bar{x}) < 0 \text{ if } \bar{x} \in \partial\Omega \text{ and } \hat{w}_i(\bar{x}) = 0.$$

Without loss of generality, we may assume that some component of w takes a positive value somewhere. Otherwise, replace w by $-w$. Since $w = 0$ on $\partial\Omega$, we can readily obtain from properties just obtained that $\hat{w}_i(x) - \epsilon w_i(x) > 0$ for some $\epsilon > 0$ and all $x \in \Omega, i = 1, ..., n+m$. Let $\delta_i = \sup\{a : \hat{w}_i - aw_i > 0 \text{ in } \Omega\}$ for those $i = 1, ..., n+m$ such that δ_i can be finitely defined. Define δ to be the minimum of such δ_i's. Thus $\delta = \delta_k$ for some k, and $0 < \delta_k < \infty, \hat{w}_i - \delta w_i \geq 0$ in Ω for $i = 1, ..., n+m$. From the inequalities for $L_i[w_i(x)]$ and $L_i[\hat{w}_i(x)]$, we find the inequality

$$(L_k + K)(\hat{w}_k - \delta_k w_k) \geq (K + p_{kk})(\hat{w}_k - \delta_k w_k) + \textstyle\sum_{i=1,\neq k}^{m+n} p_{ki}(\hat{w}_i - \delta_k w_i)$$

$$+ \textstyle\sum_{i=1}^{m+n}(q_{ki} - p_{ki})\hat{w}_i \geq 0 \quad \text{in } \Omega.$$

Consequently, the maximum principle implies that $\hat{w}_k - \delta w_k \equiv 0$ in Ω. Then, the last inequality above further implies that we must have $q_{ki} \equiv p_{ki}$ for all $i = 1, ..., n+m$.

Proof of Lemma 3.2. The operator $L^{-1}M_0$ as described in the Lemma is completely continuous and positive with respect to the cone $\mathcal{P}$. Let $z = col.(z_1, ..., z_{n+m}) = col.(L_1^{-1}(1), ..., L_{n+m}^{-1}(1))$. The functions satisfy $z_i > 0$ in Ω, $z_i|_{\partial\Omega} = 0$ for $i = 1, ..., n+m$. Define $v = L^{-1}M_0 z$. The positivity of B, K_0 and the maximum principle imply that the components satisfy $v_i > 0$ and $\partial v_i/\partial \nu < 0$ on $\partial\Omega$ for $i = 1, ..., n+m$. Thus, there exists $\delta > 0$ such that $L^{-1}M_0 z \geq \delta z$ with $z \in \mathcal{P}$. Theorem 2.5 in Krasnosel'skii [108] or Theorem A3-10 in Chapter 6 asserts that there exists a nontrivial $w^0 = col.(w_1^0, ..., w_{n+m}^0) \in \mathcal{P}$ and $\rho_0 \geq \delta > 0$ such that $L^{-1}M_0 w^0 = \rho_0 w^0$ (i.e. (3.13) with $\lambda_0 = 1/\rho_0$). Suppose $w_i^0 \not\equiv 0$ for some $1 \leq i \leq n$ (or $n+1 \leq i \leq n+m$); using the maximum principle on each of the last m (or first n) equations in (3.13), we obtain by means of the positivity of K_0 (or B) that for each $j = n+1, ..., n+m$ (or $j = 1, ..., n$) we must have $w_j^0 > 0$ in Ω and $\partial w_j^0/\partial \nu < 0$ on $\partial\Omega$. Then, we use the first n (or last m) equations and the positivity of B (or K_0) to obtain by means of the maximum principle that $w_j^0 > 0$ in Ω and $\partial w_j^0/\partial \nu < 0$ on $\partial\Omega$ for each $i = 1, ..., n$ (or $j = n+1, ..., n+m$.)

Now, let $\hat{w}^0 = col.(\hat{w}_1^0, ..., \hat{w}_{n+m}) \not\equiv 0$ be such that $L\hat{w}^0 = \lambda_0 M_0 \hat{w}^0$. From Lemma 3.1, there must exist $\delta^* \in R$ and some $k, 1 \le k \le m+n$, such that

$$w_k^0 \equiv \delta^* \hat{w}_k^0 \text{ and } w_j^0 - \delta^* \hat{w}_j^0 \ge 0 \text{ in } \bar{\Omega} \quad \text{for all } j = 1, ..., m+n. \tag{3.17}$$

If there is an integer r such that

$$w_r^0(\bar{x}) - \delta^* \hat{w}_r^0(\bar{x}) > 0 \text{ for } \bar{x} \in \Omega, \tag{3.18}$$

then $L_r[w_r^0 - \delta^* \hat{w}_r^0] \ge 0$ implies that $w_r^0 - \delta^* \hat{w}_r^0 > 0$ in Ω. If $1 \le r \le n$ (or $n+1 \le r \le n+m$), consider the i-th equation in (3.13) for $n+1 \le i \le n+m$ (or $1 \le i \le n$). We obtain $L_i[w_i^0 - \delta^* \hat{w}_i^0] > 0$ in Ω, which implies $w_i^0 - \delta^* \hat{w}_i^0 > 0$ in Ω, for each $i = n+1, ..., n+m$ (or $i = 1, ..., n$). We then consider the other n (or m) equations to obtain $w_i^0 - \delta^* \hat{w}_i^0 > 0$ in Ω for each $i = 1, ..., n$ (or $i = n+1, ..., n+m$). This contradicts the existence of an integer k such that (3.17) holds. This means that if (3.17) holds, there cannot exist an r such that (3.18) holds. That is we have $w^0 \equiv \delta^* \hat{w}^0$. Finally, suppose that there is another $\lambda_1 > 0, \lambda_1 \ne \lambda_0$, such that $L\tilde{w} = \lambda_1 M_0 \tilde{w}$ for some $\tilde{w} \in \mathcal{P}\backslash\{0\}$. We can deduce as before that $\tilde{w}_i > 0$ in Ω, $\partial\tilde{w}_i/\partial\nu < 0$ on $\partial\Omega$ for $i = 1, ..., n+m$. Then we can obtain from Lemma 3.1 that $\lambda_1 = \lambda_0$. This completes the proof of Lemma 3.2.

Proof of Lemma 3.3. We use Schauder's theory to deduce

$$\begin{aligned} &||G[w]||_E/||w||_E \le \bar{k}||\,[\hat{F}(w) - M_0 w]\,||_\infty/||w||_\infty \\ &\quad \le \bar{k}\Sigma_{i=1}^m \textstyle\sum_{j=1}^n ||\,[f_{ij}(w_j) - f'_{ij}(0)w_j]\,||_\infty/||w_j||_\infty + \sum_{i=1}^m \sum_{j=1}^m |c_{ij}|||w_{n+j}||_\infty \\ &\quad \to 0, \qquad \text{as } ||w||_E \to 0. \end{aligned}$$

Theorem 3.2 (Coexistence of Bacteria and Infected Species). *Under hypotheses (3.4) to (3.8), the problem (3.9) has a solution* $w = col.(w_1, \ldots, w_{n+m}) \in Y$*, such that* $w_i \ge 0$ *in* $\bar{\Omega}$ *for each* i *and* $w \not\equiv 0$ *(i.e.* $w \in \mathcal{P}\backslash\{0\}$*).*

Remark 3.1. By the property: $\bar{\mathcal{S}} \cap (R^+ \times \partial\mathcal{P}) = (\lambda_0, 0)$, of the part $\bar{\mathcal{S}}$ of the bifurcation curve described in Theorem 3.1, we must have $w_i > 0$ in Ω, for $i = 1, ..., n+m$.

In order to prove Theorem 3.2, we first show the following two lemmas.

Lemma 3.4. *Under hypotheses (3.4), (3.5) and (3.7), the positive number where bifurcation occurs described in Theorem 3.1 satisfies* $\lambda_0 < 1$.

Proof. By hypothesis (3.7), there exists a small enough $\epsilon > 0$ such that $B[K_0(\lambda_1 + a^*)^{-1}\vec{d} - col.(\epsilon, ..., \epsilon)] > (\lambda_1 + a^*)\vec{d}$ and $K_0(\lambda_1 + a^*)^{-1}\vec{d} > col.(\epsilon, ..., \epsilon)$. Let $\vec{p} = col.(p_1, ..., p_m)$ be the positive constant vector defined by $\vec{p} := K_0(\lambda_1 + a^*)^{-1}d - col.(\epsilon, ..., \epsilon)$. We can readily verify that the positive constant vector

$\vec{g} := col.(d_1, ..., d_n,\ p_1, ..., p_m)$ satisfies $M_0\vec{g} > (\lambda_1 + a^*)\vec{g}$. Thus there exists $r > 1$ such that

$$M_0\vec{g} \geq r(\lambda_1 + a^*)\vec{g}. \tag{3.19}$$

Let $\hat{z} := \phi\vec{g}$. Since $L_i[\phi] \leq (\lambda_1 + a^*)\phi$ in Ω for each $i = 1, ..., n+m$, a simple calculation shows

$$\begin{aligned} L^{-1}[M_0\hat{z}] &\geq L^{-1}[r(\lambda_1 + a^*)\hat{z}] \\ &= rL^{-1}[(\lambda_1 + a^*)\phi]\vec{g} \geq r\phi\vec{g} = r\hat{z} \quad \text{in } \Omega, \end{aligned}$$

where $r > 1$. As in Lemma 3.2, we obtain by means of Theorem 2.5 in Krasnosel'skii [108] or Theorem A3-10 in Chapter 6 the existence of a nontrivial function $\hat{w} \in \mathcal{P}$ such that $L^{-1}M_0\hat{w} = \hat{\rho}\hat{w}$ with $\hat{\rho} \geq r$. The uniqueness part of Lemma 3.2 implies that $\hat{\rho} = 1/\lambda_0$. Thus we have $\lambda_0 \leq 1/r < 1$.

Lemma 3.5. *Under the hypotheses (3.4) and (3.5), let $(\bar{\lambda}, \bar{w}) \in \mathcal{S}$, where $\mathcal{S}$ is described in Theorem 3.1. Suppose R_k are positive constants such that $0 \leq \bar{w}_{n+k}(x) \leq R_k$ for all $x \in \bar{\Omega}$, $k = 1, \ldots, m$. Then for each $i = 1, \ldots, n$,*

$$0 \leq \bar{w}_i(x) \leq \bar{\lambda}(\inf . a_i)^{-1} \sum_{k=1}^{m} b_{ik} R_k \ \text{ for all } \ x \in \bar{\Omega}. \tag{3.20}$$

Proof. For each $i = 1, \ldots, n$, consider the scalar linear problem

$$L_i[z] = \bar{\lambda} \sum_{k=1}^{m} b_{ik}\bar{w}_{n+k} \quad \text{in } \Omega,\ z = 0 \text{ on } \partial\Omega. \tag{3.21}$$

Since the trivial function and $\hat{z} = \bar{\lambda}(\inf . a_i)^{-1} \sum_{k=1}^{m} b_{ik}R_k$ are respectively lower and upper solution of (3.21), the unique solution $z = \bar{w}_i$ must satisfy (3.20).

Proof of Theorem 3.2. Let the component of $\bar{\mathcal{S}}$ containing the point $(\lambda_0, 0)$, described in Theorem 3.1, be denoted by $\mathcal{S}^+$. Since $\lambda_0 < 1$, by Lemma 3.4, it suffices to show that the set $I := \{\lambda \in R^+ | (\lambda, w) \in \mathcal{S}^+ \text{ for some } w\}$ is unbounded. For convenience, let

$$\alpha_k := \sum_{i=1}^{n} b_{ik}(\inf . a_i)^{-1} > 0 \quad \text{for } k = 1, \ldots, m,$$

$$\beta_k := |c_{kk}| - \sum_{j=1, j\neq k}^{m} |c_{kj}| \quad \text{for } k = 1, \ldots, m.$$

Hypothesis (3.8) implies each $\beta_k > 0$. Suppose that the set I is contained in the interval $[0, C]$, we will deduce a contradiction. Let $\hat{N} > 0$ be large enough such that

$$CK_1(\sum_{j=1}^{m} \alpha_j)(\theta\hat{N}) - \beta_k(\theta\hat{N})^2 < 0 \quad \text{for all } \theta \geq 1, \text{ each } k = 1, \ldots, m. \tag{3.22}$$

Here, the constant K_1 is described in (3.6). We claim that if there exists λ such that $(\lambda, w) \in \mathcal{S}^+$, then $w = \text{col.}(w_1, \ldots, w_{n+m})$ must satisfy:

$$0 \leq w_i(x) \leq \hat{N} \quad \text{for all } x \in \bar{\Omega},\ i = n+1, \ldots, n+m. \tag{3.23}$$

Suppose (3.23) is not true. Let $(\tilde{\lambda}, \tilde{w}) \in \mathcal{S}^+$, such that $\tilde{w}$ does not satisfy the corresponding (3.23). For each $i = n+1, \ldots, n+m$, let $x_i^* \in \bar{\Omega}$ be the point where $\tilde{w}_i$ attains its maximum value on $\bar{\Omega}$. Let $r \in \{n+1, \ldots, n+m\}$ where $\tilde{w}_r(x_r^*) = \max\{\tilde{w}_{n+1}(x_{n+1}^*) \ldots, \tilde{w}_{n+m}(x_{n+m}^*)\}$. By assumption on $\tilde{w}$, we must have $\alpha^* := \tilde{w}_r(x_r^*)/\hat{N} > 1$; moreover, we have $0 \leq \tilde{w}_i(x) \leq \tilde{w}_r(x_r^*) = \alpha^*\hat{N}$ for all $x \in \bar{\Omega}, i = n+1, \ldots, n+m$. Suppose we have $x_r^* \in \Omega$. Then (3.6), Lemma 3.5 and (3.22) imply that $\tilde{w}_r(x_r^*)$ satisfies:

$$\begin{aligned}
&-\Delta\tilde{w}_r(x_r^*) + \tilde{a}_{r-n}(x_r^*)\tilde{w}_r(x_r^*)\\
&= \tilde{\lambda}\textstyle\sum_{j=1}^{n} f_{r-n,j}(\tilde{w}_j(x_r^*)) + \tilde{\lambda}\tilde{w}_r(x_r^*)\sum_{j=1}^{m} c_{r-n,j}\tilde{w}_{n+j}(x_r^*)\\
&\leq \tilde{\lambda}[K_1\textstyle\sum_{j=1}^{n} \tilde{w}_j(x_r^*)] + \tilde{\lambda}(\alpha^*\hat{N})^2\ [-|c_{r-n,r-n}| + \sum_{j=1, j\neq r-n}^{m} |c_{r-n,j}|]\\
&\leq \tilde{\lambda}[K_1 C(\textstyle\sum_{k=1}^{m} \alpha_k)(\alpha^*\hat{N}) - (\alpha^*\hat{N})^2\beta_{r-n} < 0.
\end{aligned}$$

Note that we have used the assumption that I is contained in $[0, C]$ and $\alpha^* > 1$ in obtaining the last two inequalities above. This contradicts the fact that x_r^* is an interior maximum. Thus we must have $x_r^* \in \partial\Omega$. Then the boundary conditions and Lemma 3.5 imply that $\tilde{w} \equiv 0$; and we can conclude that (3.23) must be true.

Finally, inequality (3.23), Lemma 3.5 and gradient estimates by means of equation (3.10) imply that $\mathcal{S}^+$ cannot be unbounded if I is bounded. Consequently, I must be unbounded, and this completes the proof of Theorem 3.2.

We next consider system (3.1), under the special situation when all $c_{kj} = 0$, for $k, j = 1, \ldots, m$; that is, we consider the problem:

$$\begin{cases}
-\Delta u_i + a_i(x)u_i = \sum_{j=1}^{m} b_{ij}v_j & \text{for } x \in \Omega, i = 1, ..., n,\\
-\Delta v_k + \tilde{a}_k(x)v_k = \sum_{j=1}^{n} f_{kj}(u_j) & \text{for } x \in \Omega, k = 1, ..., m,\\
u_i = v_k = 0 \quad \text{for } x \in \partial\Omega,\ i = 1, ..., n,\ k = 1, ..., m.
\end{cases} \tag{3.24}$$

In other words, the infected species v_k will not interact among themselves. This situation is a direct generalization of the theory in Blat and Brown [12]. Under additional assumptions on $f_{kj}(u_j)$ for large u_j (see (3.28)), Theorem 3.3 shows that problem (3.24) has a positive solution. Letting $w = col.(w_1, \ldots, w_{n+m}) = col.(u_1, \ldots, u_n, v_1, \ldots, v_m)$, system (3.24) can be written as

$$L[w] = \tilde{F}[w], \quad w \in E, \tag{3.25}$$

where $\tilde{F}$ is the same as $\hat{F}$ described in (3.3) with the special restriction $c_{kj} = 0$, for all $k, j = 1, \ldots, m$. For convenience, we define

$$\hat{f}_{ij}(\eta) := \begin{cases} f_{ij}(\eta)/\eta & \text{if} \quad \eta \neq 0 \\ f'_{ij}(0) & \text{if} \quad \eta = 0 \end{cases} \qquad \text{for } i = 1, \ldots, m, \ j = 1, \ldots, n. \tag{3.26}$$

Also define the $m \times n$ matrix

$$\left[\hat{f}_{ij}(\eta_{ij})\right]_{i,j=1}^{m,n} := \begin{bmatrix} \hat{f}_{11}(\eta_{11}) & \cdots & \hat{f}_{1n}(\eta_{1n}) \\ \vdots & & \vdots \\ \hat{f}_{m1}(\eta_{m1}) & \cdots & \hat{f}_{mn}(\eta_{mn}) \end{bmatrix} \tag{3.27}$$

where η_{ij} are real numbers for $i = 1, \ldots, m$, $j = 1, \ldots, n$. We will use the following hypothesis:

(3.28)
There exist a real number $\eta_0 > 0$, and a constant vector $\vec{q} = col.(q_1, \ldots, q_n)$, with $q_i > 0, i = 1, \ldots, n$, such that:

$$\vec{q}^T B \left[\hat{f}_{ij}(\eta_{ij})\right]_{i,j=1}^{m,n} < (\lambda_1 + a^{**})^2 \vec{q}^T \ \text{ for all } \ \eta_{ij} \geq \eta_0.$$

Under hypothesis (3.28), one can always choose a number ρ_1 with $0 < \rho_1 < (\lambda_1 + a^{**})$ such that:

$$\vec{q}^T B \left[\hat{f}_{ij}(\eta_{ij})\right]_{i,j=1}^{m,n} < \rho_1^2 \vec{q}^T < (\lambda_1 + a^{**})^2 \vec{q}^T \ \text{ for all } \ \eta_{ij} \geq \eta_0. \tag{3.29}$$

The following theorem is the main result for case when the infected species do not interact among each other.

Theorem 3.3 (Coexistence When Infected Species Do Not Interact). *Under hypotheses (3.4) to (3.7), and (3.28), the problem (3.25) (alternatively, problem (3.24) with $w = col.(u_1, \ldots, u_n, v_1, \ldots, v_m)$) has a solution $w = col.(w_1, \ldots, w_{n+m}) \in Y$, such that $w_i \geq 0$ in $\bar{\Omega}$ for each i and $w \not\equiv 0$ (i.e. $w \in \mathcal{P}\backslash\{0\}$).*

In this theorem, $\hat{F}$ is considered with the special restriction $c_{kj} = 0$, and assumption (3.8) concerning c_{kj} is not assumed. We are thus led to the problem:

$$L[w] = \lambda \tilde{F}[w], \quad w \in E, \tag{3.30}$$

with $\tilde{F}$ as described above. Since $\tilde{F}$ is a special case of $\hat{F}$, Theorem 3.1 applies. Under the assumptions of Theorem 3.3, let $\mathcal{S}$ be as defined in Theorem 3.1, and $\mathcal{S}^+$ be the component of $\bar{\mathcal{S}}$ containing the point $(\lambda_0, 0)$. Recall that $\mathcal{S}^+$ is proved to be unbounded in Theorem 3.1. The following Lemma will be needed in the proof of Theorem 3.3.

Lemma 3.6. *Assume all the hypotheses of Theorem 3.3. Suppose* $\{(\tilde{\lambda}_r, \tilde{w}^r)\}$, $r = 1, 2, \ldots$ *is a sequence in* $\mathcal{S}^+$ *with the property:* $\tilde{\lambda}_r \to \hat{\lambda}$, $0 < \hat{\lambda} < \infty$, *and* $\|\tilde{w}^r\|_E \to \infty$, *as* $r \to \infty$. *Then there exists a subsequence* $\{(\tilde{\lambda}_{r(j)}, \tilde{w}^{r(j)})\}$ *such that the first* n *components of* $\tilde{w}^{r(j)}$ *tend to* $+\infty$ *uniformly in compact subsets of* Ω, *as* $r(j) \to \infty$.

Proof of Lemma 3.6. Let $z_i^r = \tilde{u}_i^r / ||\tilde{w}^r||_E$, $y_k^r = \tilde{v}_k^r / ||\tilde{w}^r||_E$, $\hat{w}^r = \tilde{w}^r / ||\tilde{w}^r||_E = col.(z_1^r, \ldots, z_n^r, y_1^r, \ldots, y_m^r)$, for $r = 1, 2, \ldots$. Dividing equation (3.30) for $\tilde{w}^r$ by $||\tilde{w}^r||_E$, we obtain

$$(3.31)\qquad \begin{cases} -\Delta z_i^r + a_i(x) z_i^r = \tilde{\lambda}_r \sum_{j=1}^m b_{ij} y_j^r & \text{for } x \in \Omega, i = 1, ..., n, \\ -\Delta y_k^r + \tilde{a}_k(x) y_k^r = \tilde{\lambda}_r \sum_{j=1}^n \hat{f}_{kj}(\tilde{u}_j^r) z_j^r & \text{for } x \in \Omega, k = 1, ..., m, \\ z_i^r = y_k^r = 0 \quad \text{for } x \in \partial\Omega,\ i = 1, ..., n,\ k = 1, ..., m. \end{cases}$$

Since the right-hand side of (3.31) is bounded in $|| \cdot ||_\infty$- norm, using Schauder's estimates we obtain a subsequence of $\{\hat{w}^r\}_{r=1}^\infty$, denoted again by $\{\hat{w}^r\}_{r=1}^\infty$, which converges in the $\|.\|_E$ - norm to a function $\hat{w} = col.(z_1, \ldots, z_n, y_1, \ldots, y_m) \in E$. Here, from the definition of $\mathcal{S}$, we must have $z_i \geq 0, y_k \geq 0$ in Ω for $i = 1, \ldots, n, k = 1, \ldots, m$, and $||\hat{w}||_E = 1$. Using Schauder's theory, we see that $\hat{w}$ also satisfy (3.31) by taking limit as $r \to \infty$.

If $y_k \equiv 0$ in Ω for all $k = 1, \ldots, m$, the first n equations in (3.31) imply by maximum principle that $z_i \equiv 0$ in Ω, for each $i = 1, \ldots, n$. This contradicts the fact that $||\hat{w}||_E = 1$. Thus, there must exists some $l \in \{1, \ldots, m\}$, such that $y_l \not\equiv 0$ in Ω. The first n equations in (3.31) then imply that $z_i(x) > 0$ in Ω for all $i = 1, \ldots, n$. The conclusion of Lemma 3.6 follows since $u_i^r(x) = z_i^r(x)||\tilde{w}^r||_E$, $||\tilde{w}^r||_E \to \infty$, and $z_i^r(x) \to z_i(x) > 0$ in Ω (in $C^1(\Omega)$-norm) as $r \to \infty$.

Proof of Theorem 3.3. Recall that Lemma 3.4 implies that $\lambda_0 < 1$. To prove this Theorem, it suffices to show that if there exists a sequence in $\mathcal{S}^+$ with property as described in Lemma 3.6, then we must have $\hat{\lambda} > 1$. This leads to the existence of $(\tilde{\lambda}, \tilde{w})$ in $\mathcal{S}^+$ with $\tilde{\lambda} = 1$; that is, we obtain a solution of problem (3.24).

If $\{\lambda : (\lambda, w) \in \mathcal{S}^+\}$ is contained in a bounded interval, then there must exists a sequence $\{(\tilde{\lambda}_r, \tilde{w}^r)\}_{r=1}^\infty$ in $\mathcal{S}^+$ with property as described in Lemma 3.6; and $(\tilde{\lambda}_r, \tilde{w}^r)$ satisfies (3.30) for each r. Denoting $\tilde{w}^r = col.(\tilde{w}_1^r, \ldots, \tilde{w}_{n+m}^r) =$

$col.(\tilde{u}_1^r, \ldots, \tilde{u}_n^r, \tilde{v}_1^r, \ldots, \tilde{v}_m^r)$, the conclusion of Lemma 3.6 implies that we may assume that $\tilde{u}_i^r(x) \to \infty$ uniformly in compact subsets of Ω, as $r \to \infty, i = 1, \ldots, n$. Let $\psi_i^r = \tilde{u}_i^r/||\tilde{w}^r||_{L_2(\Omega)}$, $\xi_k^r = \tilde{v}_k^r/||\tilde{w}^r||_{L_2(\Omega)}$, $\Psi_r = col.(\psi_1^r, \ldots, \psi_n^r, \xi_1^r, \ldots, \xi_m^r)$. Dividing the equation for $(\tilde{\lambda}_r, \tilde{w}^r)$ by $||\tilde{w}^r||_{L_2(\Omega)}$, we obtain

$$(3.32) \qquad \begin{cases} -\Delta\psi_i^r + a_i(x)\psi_i^r = \tilde{\lambda}_r \sum_{j=1}^m b_{ij}\xi_j^r & \text{for } x \in \Omega,\ i = 1, ..., n, \\ \\ -\Delta\xi_k^r + \tilde{a}_k(x)\xi_k^r = \tilde{\lambda}_r \sum_{j=1}^n \hat{f}_{kj}(\tilde{u}_j^r)\psi_j^r & \text{for } x \in \Omega,\ k = 1, ..., m. \end{cases}$$

Since the right side of (3.32) is bounded in $|| \cdot ||_{L_2(\Omega)}$- norm, using Agmon-Douglis-Nirenberg estimate, we obtain a subsequence of $\{\Psi_r\}_{r=1}^\infty$, also denoted by $\{\Psi_r\}_{r=1}^\infty$, which converges to a function $\Psi = col.(\psi_1, \ldots, \psi_n, \xi_1, \ldots, \xi_m)$ in $L_2(\Omega)$ at each component. Moreover we have $||\Psi||_{L_2(\Omega)} = 1$, and $\psi_i \geq 0, \xi_k \geq 0$ a.e. in Ω, for $i = 1, \ldots, n$, $k = 1, \ldots, m$.

Let $(q_1, \ldots, q_n)$ be as described in hypothesis (3.28), and ρ_1 be as chosen in (3.29). Define ρ_2 such that $0 < \rho_1 < \rho_2 < (\lambda_1 + a^{**})$, and let $\hat{q}^T = (q_1, \ldots, q_n, q_{n+1}, ..., q_{n+m})$, where $q_{n+k} = (1/\rho_1)\sum_{j=1}^n q_j b_{jk}$, $k = 1, \ldots, m$. (Note that q_{n+k} is the k-th component of $(1/\rho_1)\vec{q}^T B$). By the first inequality in (3.29), we readily obtain from the definition of $\hat{q}^T$ and ρ_2 that

$$(3.33) \qquad \hat{q}^T \begin{bmatrix} 0 & B \\ \left[\hat{f}_{ij}(\eta_{ij})\right]_{i,j=1}^{m,n} & 0 \end{bmatrix} < \rho_2 \hat{q}^T \qquad \text{for all } \eta_{ij} \geq \eta_0.$$

To simplify writing, for $i = 1, \ldots, n$, denote $h_{ij}(\eta) = 0$ when $j = 1, .., n$ and $h_{i,j+n}(\eta) = b_{ij}$ when $j = 1, \ldots, m$. For $i = 1, \ldots, m$, denote $h_{i+n,j}(\eta) = \hat{f}_{ij}(\eta)$ when $j = 1, \ldots, n$ and $h_{i+n,j+n}(\eta) = 0$ when $j = 1, \ldots, m$. Inequality (3.33) can then be rewritten as:

$$(3.34) \qquad \hat{q}^T([h_{ij}(\eta_{ij})]_{i,j=1}^{n+m,n+m}) < \rho_2 \hat{q}^T \qquad \text{for all } \eta_{ij} \geq \eta_0.$$

(Note that h_{ij} are really constants for $j > n, i = 1, \ldots, n+m$.)

Multiplying the first n equations of (3.32) by $q_i\phi, i = 1, \ldots, n$, and the last m equations of (3.32) by $q_{n+k}\phi, k = 1, \ldots, m$, we obtain, after integrating by parts:

$$(3.35) \qquad 0 = \int_\Omega [-\psi_i^r(\lambda_1 + a_i) + \tilde{\lambda}_r \sum_{j=1}^m b_{ij}\xi_j^r] q_i\phi\, dx, \quad \text{for } i = 1 \ldots, n,$$

$$(3.36) \quad 0 = \int_\Omega [-\xi_k^r(\lambda_1 + \tilde{a}_k) + \tilde{\lambda}_r \sum_{j=1}^n \hat{f}_{kj}(\tilde{u}_j^r)\psi_j^r]\, q_{n+k}\phi\, dx, \qquad \text{for } k = 1, \ldots, m.$$

Adding the last $n+m$ equations, and using the definitions of a^{**} and h_{ij}, we obtain:

$$0 \le \int_{\Omega} \{-(\lambda_1 + a^{**})\hat{q}^T \Psi_r \phi + \tilde{\lambda}_r \hat{q}^T [h_{ij}(\tilde{w}_j^r)]_{i,j=1}^{n+m,n+m} \Psi_r \phi\}\, dx. \tag{3.37}$$

(Recall that h_{ij} are constants for $j > n$, and $h_{ij}(\tilde{w}_j^r)$ above actually depends only on $\tilde{u}_j^r, j = 1, \ldots, n$.)

Let $\{\Omega_k\}_{k=1}^{\infty}$ be a sequence of open subsets of Ω, such that $\Omega_k \subset \bar{\Omega}_k \subset \Omega_{k+1} \subset \Omega$, for all k with $\Omega = \cup_{k=1}^{\infty} \Omega_k$; and let $\epsilon > 0$ be an arbitrary small constant. Since the quantities $\tilde{\lambda}_r, \hat{q}^T, h_{ij}$ are bounded, and $\Psi_r \to \Psi$ in $L_2(\Omega)$, we may assume that for l and r sufficiently large, we have:

$$\int_{\Omega \setminus \Omega_l} \tilde{\lambda}_r \hat{q}^T [h_{ij}(\tilde{w}_j^r)]_{i,j=1}^{n+m,n+m} \Psi_r \phi\, dx < \epsilon. \tag{3.38}$$

We can thus obtain from (3.37) that:

$$0 \le \int_{\Omega_l} \{-(\lambda_1 + a^{**})\hat{q}^T + \tilde{\lambda}_r \hat{q}^T [(h_{ij}(\tilde{w}_j^r)]_{i,j=1}^{n+m,n+m}\} \Psi_r \phi\, dx + \epsilon. \tag{3.39}$$

By Lemma 3.6, we may assume that $\tilde{u}_j^r \to \infty$ uniformly in Ω_l for each fixed l, as $r \to \infty$, for $j = 1, \ldots, n$.

Consequently, from (3.34) and (3.39), we obtain

$$0 \le \int_{\Omega_l} \{-(\lambda_1 + a^{**}) + \tilde{\lambda}_r \rho_2\}\, \hat{q}^T \Psi_r \phi\, dx + \epsilon \tag{3.40}$$

for all sufficiently large r. Letting $r \to \infty$, we find that for all sufficiently large l,

$$0 \le \int_{\Omega_l} \{-(\lambda_1 + a^{**}) + \hat{\lambda} \rho_2\}\, \hat{q}^T \Psi \phi\, dx + \epsilon. \tag{3.41}$$

Since ϵ is arbitrary, we conclude

$$0 \le \int_{\Omega} \{-(\lambda_1 + a^{**}) + \hat{\lambda} \rho_2\}\, \hat{q}^T \Psi \phi\, dx.$$

This leads to $\hat{\lambda} \ge (\lambda_1 + a^{**})/\rho_2 > 1$. The proof of Theorem 3.3 is complete.

Remark 3.2. The global continuum of positive solutions of (3.30) and (3.10) cannot cross $\lambda = 0$, because when $\lambda = 0$ the trivial solution is the only solution by the maximum principle.

In the remaining part of this section, we will consider the stability of the steady state solutions found in Theorem 3.2 as a solution of the corresponding parabolic system. It will be seen in Theorem 3.5 that if hypotheses (3.5) and (3.8) are strengthened, then the bifurcating steady-states near the bifurcation point are asymptotically stable in time. Before obtaining further results with additional hypotheses, we first deduce a few more consequences of hypotheses (3.4) and (3.5).

Lemma 3.7. *Under hypotheses (3.4) and (3.5), the problem*

$$L[w] = \lambda M_0^T w \quad \text{in } \Omega,\ w = 0 \ \text{ on } \partial\Omega, \tag{3.42}$$

has a solution $(\lambda, w) = (\lambda_0, \hat{w}^0)$, $\hat{w}^0 \in Y$ *with each component* $\hat{w}_i^0 > 0$ *in* Ω, $\partial\hat{w}_i^0/\partial\nu < 0$ *on* $\partial\Omega$ *for* $i = 1, ..., n+m$. *(Here,* λ_0 *is exactly the same number as in Lemma 3.2.) Moreover, any solution of (3.42) with* $\lambda = \lambda_0$ *is a multiple of* $\hat{w}^0$.

Proof. The existence of a positive solution and the simplicity of the corresponding eigenvalue is proved in exactly the same way as Lemma 3.2 with the role of B and K_0 interchanged. The fact that λ_0 is exactly the same as in Lemma 3.2 follows exactly the same procedure as in the proof of Lemma 2.4 in Leung and Villa [141] or Lemma 6.4 in Section 2.6, and will thus be postponed.

For convenience, we will define two operators L_0 and $L_1 : E \to E$ as follow:

$$L_0 := I - \lambda_0 L^{-1} M_0, \tag{3.43}$$

$$L_1 := -L^{-1} M_0. \tag{3.44}$$

Lemma 3.8. *Under hypotheses (3.4) and (3.5), the null space and range of* L_0*, denoted respectively by* $N(L_0)$ *and* $R(L_0)$*, satisfy:*

(i) $N(L_0)$ *is one-dimensional, spanned by* w^0,

(ii) $dim[E/R(L_0)] = 1$, *and*

(iii) $L_1 w^0 \notin R(L_0)$.

Proof. Part (i) was proved in Lemma 3.2. The remaining parts are proved in the same way as in Lemma 2.5 in [141] or Lemma 6.5 in Section 2.6. For the proof of part (iii), the positivity property of w^0 and $\hat{w}^0$ are used.

Theorem 3.4 (Local Bifurcation). *Assume hypotheses (3.4), (3.5) and the additional condition that* $f_{kj} \in C^2(R)$ *for* $k = 1, ..., n, j = 1, ..., m$. *Then there exists* $\delta > 0$ *and a* C^1*-curve* $(\hat{\lambda}(s), \hat{\phi}(s)) : (-\delta, \delta) \to R \times E$ *with* $\hat{\lambda}(0) = \lambda_0, \hat{\phi}(0) = 0$, *such that in a neighborhood of* $(\lambda_0, 0)$, *any solution of (3.11) is either of the form* $(\lambda, 0)$ *or on the curve* $(\hat{\lambda}(s), s[w^0 + \hat{\phi}(s)])$ *for* $|s| < \delta$. *Furthermore, the set* $\mathcal{S}^+ \backslash \{(\lambda_0, 0)\}$ *is contained in* $R^+ \times (Int\, \mathcal{P})$, *where* $\mathcal{S}^+$ *is the component of the closure of* $\mathcal{S}$ *(described in Theorem 3.1) containing the point* $(\lambda_0, 0)$ *in* $R^+ \times E$.

Proof. Equation (3.11) can be written as

$$Q(\lambda, w) := L_0[w] + (\lambda - \lambda_0)L_1[w] - \lambda G[w] = 0, \tag{3.45}$$

where the operator $G : E \to E$ is defined in Lemma 3.3. Under the additional smoothness condition on f_{kj}, we can show that $Q \in C^2(R^+ \times E, E)$ and the Fréchet derivative of G is continuous on E. Moreover, we can readily deduce as in Lemma 3.3 that $L_0 = D_2Q(\lambda_0, 0)$, $L_1 = D_{12}Q(\lambda_0, 0)$, and $G[0] = D_2G[0] = 0$. Hence we can apply the local bifurcation theorem in Crandall and Rabinowitz [34] to obtain the C^1-curve $(\hat{\lambda}(s), \hat{\phi}(s))$ describing the nontrivial solution of (3.11) near $(\lambda_0, 0)$ as stated above.

For $s > 0$ sufficiently small, the function $s[w^0 + \hat{\phi}(s)]$ is clearly in $Int\,\mathcal{P}$. Suppose $(\lambda_i, w_i) \in \mathcal{S}$, $i = 1, 2\ldots$ is a sequence tending to a limit point $(\bar{\lambda}, \bar{w})$ in $R^+ \times \partial\mathcal{P}$; we can show as in Theorem 3.1 that $(\bar{\lambda}, \bar{w}) = (\lambda_0, 0)$. By means of the local behavior of all the solutions (3.11) near $(\lambda_0, 0)$, we conclude that $\mathcal{S}^+ \backslash \{(\lambda_0, 0)\} \subset R^+ \times (Int\,\mathcal{P})$. This completes the proof of Theorem 3.4.

We next consider the linearized and then the asymptotic stability of the positive bifurcating solution described in Theorem 3.3, near $(\lambda_0, 0)$. Applying the bifurcation theory in [34], and the fact that

$$\int_\Omega \hat{w}^0 \cdot \Delta^{-1} w^0 \, dx \neq 0, \tag{3.46}$$

(note that each component of both $\hat{w}^0$ and w^0 is strictly positive in Ω), we can assert that there exist $\delta_1 \in (0, \delta)$ and two functions: $(\gamma(\cdot), z(\cdot)) : (\lambda_0 - \delta_1, \lambda_0 + \delta_1) \to R \times E$, $(\eta(\cdot), h(\cdot)) : [0, \delta_1) \to R \times E$, with $(\gamma(\lambda_0), z(\lambda_0)) = (\eta(0), h(0)) = (0, w^0)$, such that

$$D_2Q(\lambda, 0)z(\lambda) = \gamma(\lambda)(-L)^{-1}(z(\lambda)), \tag{3.47}$$

and

$$D_2Q(\hat{\lambda}(s), s(w^0 + \hat{\phi}(s))h(s) = \eta(s)(-L)^{-1}(h(s)). \tag{3.48}$$

Here, (3.46) and the theory in [34] imply that $\gamma(\lambda)$ and $\eta(s)$ are respectively L^{-1}-simple eigenvalues of $D_2Q(\lambda, 0)$ and $D_2Q(\hat{\lambda}(s), s(w^0 + \hat{\phi}(s))$ with eigenfunctions $z(\lambda)$ and $h(s)$. Moreover, the theory in [34] further lead to the following lemmas.

Lemma 3.9. *Assume all the hypotheses in Theorem 3.4. There exists $\rho > 0$ such that for each $s \in [0, \delta_1)$, there is a unique (real) eigenvalue $\eta(s)$ for the linear operator*

$$Q_s^* \equiv -LD_2Q(\hat{\lambda}(s), s(w^0 + \hat{\phi}(s))) : Y \to [C^\alpha(\bar{\Omega})]^{m+n} \tag{3.49}$$

satisfying $|\eta(s)| < \rho$ *with eigenfunction* $h(s) \in Y$. *That is,*

$$Q_s^* h(s) \equiv -L[h(s)] + \hat{\lambda}(s) F_w[s(w^0 + \hat{\phi}(s))]h(s) = \eta(s)h(s). \tag{3.50}$$

The next few lemmas study the behavior of the eigenvalues $\hat{\lambda}(s)$ and $\eta(s)$ for small $s \geq 0$, and $\gamma(\lambda)$ near $\lambda = \lambda_0$.

Lemma 3.10. *Assume all the hypotheses of Theorem 3.4. Suppose further that*

$$f''_{kj}(0) < 0 \qquad \text{for } k = 1, ..., m, j = 1, ..., n; \tag{3.51}$$

$$c_{kj} < 0 \qquad \text{for all } k, j = 1, ..., m. \tag{3.52}$$

Then the function $\hat{\lambda}(s)$ *satisfies* $\hat{\lambda}'(0) > 0$.

Remark 3.3. Note that hypotheses (3.51) and (3.52) are respectively modifications of (3.6) and (3.8). Hypothesis (3.52) means the infected species compete among each other.

Proof. Theorem 3.4 asserts that $\hat{\lambda}'(0)$ exists. Equation (3.11) implies that $s(w^0 + \hat{\phi}(s))$ is in Y; and for $s \in [0, \delta)$, we have

$$L[s(w^0 + \hat{\phi}(s))] = \hat{\lambda}(s)\hat{F}[s(w^0 + \hat{\phi}(s))].$$

Dividing by s and then differentiating with respect to s for $s > 0$, we find

$$L[\hat{\phi}'(s)] = \hat{\lambda}'(s)(1/s)\hat{F}[s(w^0 + \hat{\phi}(s))] + \hat{\lambda}(s)\frac{d}{ds}\{(1/s)\hat{F}[s(w^0 + \hat{\phi}(s))]\}.$$

Letting $s \to 0^+$, we obtain

$$L[\hat{\phi}'(0)] = \hat{\lambda}'(0)M_0 w^0 + \hat{\lambda}(0)M_0\hat{\phi}'(0) + \hat{\lambda}(0)R_0 w^0, \tag{3.53}$$

where

$$R_0 = \begin{bmatrix} 0_{n\times n} & 0_{n\times m} \\ H_0 & Z_0 \end{bmatrix}. \tag{3.54}$$

Here, H_0 is the $m \times n$ matrix whose (i,j)-th entry is $(1/2)f''_{ij}(0)w_j^0$ for $i = 1, \dots, m, j = 1, \dots, n$; and Z_0 is the diagonal $m \times m$ matrix $col.(\sum_{j=1}^m c_{1j}w_{n+j}^0, \dots, \sum_{j=1}^m c_{mj}w_{n+j}^0)$. We multiply (3.53) by $(\hat{w}^0)^T$ and integrate over Ω. Then, after integrating by parts on the left and canceling with the second term on the right, we obtain

$$\hat{\lambda}'(0) = \frac{-\hat{\lambda}(0)\int_\Omega (\hat{w}^0)^T R_0 w^0\, dx}{\int_\Omega (\hat{w}^0)^T M_0 w^0\, dx} > 0.$$

The last inequality is due to the additional assumptions (3.51) and (3.52).

Lemma 3.11. *Under all the hypotheses in Theorem 3.4, the function $\gamma(\lambda)$ satisfies $\gamma'(\lambda_0) > 0$.*

Proof. Note that $D_2Q(\lambda, 0) = I - \lambda L^{-1}M_0$. From (3.47), we have

$$(I - \lambda L^{-1}M_0)z(\lambda) = \gamma(\lambda)(-L)^{-1}z(\lambda), \text{ for } \lambda \in (\lambda_0 - \delta_1, \lambda_0 + \delta_1).$$

Applying L, multiplying by $(\hat{w}^0)^T$, and integrating over Ω, we obtain

$$\int_\Omega (\hat{w}^0)^T L[z(\lambda)] - \lambda(\hat{w}^0)^T M_0 z(\lambda)\, dx = [\gamma(\lambda_0) - \gamma(\lambda)] \int_\Omega (\hat{w}^0)^T z(\lambda)\, dx,$$

since $\gamma(\lambda_0) = 0$. Integrating by parts, using the equation for $(\hat{w}^0)^T$, and letting λ tends to λ_0 after transposing, we obtain

$$\gamma'(\lambda_0) = \frac{\int_\Omega (\hat{w}^0)^T M_0 w^0\, dx}{\int_\Omega (\hat{w}^0)^T w^0\, dx} > 0.$$

Lemma 3.12. *Under the hypotheses of Theorem 3.4, (3.51) and (3.52), there exists $\delta_2 \in (0, \delta_1)$ such that $\eta(s) < 0$ for all $s \in (0, \delta_2)$.*

Proof. From Theorem 1.16 in [34] or exposition in Smoller [209], we find $-s\hat{\lambda}'(s)\gamma'(\lambda_0)$ and $\eta(s)$ have the same sign for $s > 0$ near 0. Hence Lemma 3.10 and Lemma 3.11 imply $\eta(s) < 0$ for small positive s.

The linearized eigenvalue problem for (3.10), (3.11) at the bifurcating solution $w = s(w^0 + \hat{\phi}(s))$ is precisely (3.50). When $\lambda = \hat{\lambda}(0) = \lambda_0$, the eigenvalue problem corresponding to (3.50) becomes

$$-L[h] + \lambda_0 M_0 h = \eta h \qquad h \in E, \tag{3.55}$$

where η is the eigenvalue. Under hypotheses (3.4) and (3.5), Theorem 3.1 asserts that $\eta = 0$ is an eigenvalue for (3.55), with positive eigenfunction. Using this property and the fact that the off diagonal terms of M_0 are all non-negative, we can proceed to show the following.

Lemma 3.13. *Under hypotheses (3.4) and (3.5), all eigenvalues in equation (3.55) except $\eta = 0$ satisfies $Re(\eta) < -r$ for some positive number r.*

The proof of this lemma is essentially the same as the proof of Lemma 2.8 in Leung and Ortega [138]. The details will be omitted here

Lemmas 3.9 to 3.12 above essentially shows the eigenvalue $\eta = 0$ corresponding to (3.50) at $s = 0$ moves to the left as s increases. As to the other eigenvalues with $Re(\eta) < -r$ described in Lemma 3.13, one can show by perturbation arguments that they should still be in the left open half plane for $s > 0$ sufficiently small. More precisely, we have the following.

Lemma 3.14. *Under the hypotheses of Lemma 3.10, there exists a number* $\delta^* \in (0, \delta)$ *and a positive function* $\eta(s)$ *for* $s \in (0, \delta^*)$ *such that the real parts of all the numbers in the point spectrum of the linear operator* Q_s^* *are contained in the interval* $(-\infty, -\eta(s))$, *for* $s \in (0, \delta^*)$. *(Here,* δ *is described in Theorem 3.4 and* Q_s^* *is described in (3.49) in Lemma 3.9.)*

The proof of Lemma 3.14, using the assertions in Lemmas 3.9 to 3.13 are exactly the same as the proof of Theorem 2.2 in [138]. The details are thus omitted here.

For each $s \in (0, \delta^*)$, the function $\bar{w}_s := s[w^0 + \hat{\phi}(s)]$ described in Theorem 3.4 can be considered as a steady-state solution of (3.10) with L replaced by $\partial/\partial t + L$ and homogeneous boundary condition. We now consider the time asymptotic stability of this steady-state as a solution of the corresponding parabolic problem. In order to obtain a precise statement, we let B_1 and B_2 be Banach spaces as follow:

$$B_1 = \{u : u \in [C(\bar{\Omega})]^{n+m}, u = 0 \quad \text{on } \partial\Omega\},$$

and

$$B_2 = \{u : u \in [L_p(\Omega)]^{n+m}\} \quad \text{for } p \text{ large enough such that } N/(2p) < 1.$$

Let A_1 be the operator L on B_1 with domain $D(A_1) = \{u : u \in [W^{2,p}(\Omega)]^{n+m}$ for all p, $\Delta u \in [C(\bar{\Omega})]^{n+m}, u = 0$ and $\Delta u = 0$ on $\partial\Omega\}$; and A_2 be the operator L on B_2 with domain $D(A_2) = \{u \in B_2 : u \in [W^{2,p}(\Omega) \cap W_0^{1,p}(\Omega)]^{n+m}\}$. For $w = col.(w_1, \ldots, w_{n+m})$, we consider the following nonlinear initial-boundary value problem for each i = 1,2:

$$\frac{dw}{dt} + A_i w(t) = \lambda \hat{F}(w(t)), \; w(0) = w_0 \quad \text{for } t \in (0, T]. \tag{3.56}$$

A solution of (3.56) in B_i is a function $w \in C([0, T], B_i) \cap C^1((0, T], B_i)$, with $w(0) = w_0, w(t) \in D(A_i)$ for all $t \in (0, T]$; and $w(t)$ satisfies (3.56) for $t \in (0, T]$. The operator A_2 is an infinitesimal generator of an analytic semigroup, say $S(t)$, on B_2 for $t \geq 0$. It is well known that for $\alpha > 0$,

$$(-A_2)^{-\alpha} = \frac{1}{\Gamma(\alpha)} \int_0^\infty \tau^{\alpha - 1} S(\tau) d\tau$$

defines a bounded linear operator on B_2 (called fractional power of the operator A_2). Moreover, $[(-A_2)^{-\alpha}]^{-1} := (-A_2)^{\alpha}$ is a closed operator on B_2 with dense domain $D((-A_2)^{\alpha}) = (-A_2)^{-\alpha}(B_2)$. We denote by X^{α} the Banach space $(D(-A_2)^{\alpha}, \|\cdot\|_{\alpha})$ where $\|w\|_{\alpha} = \|(-A_2)^{\alpha} w\|_{L^p}$ for all $w \in D((-A_2)^{\alpha})$.

Using Lemma 3.14, we can use the theory by Mora [176] (Theorem A4-9 in Chapter 6) or apply the stability Theorem 5.1.1 in Henry [84] (Theorem A4-11 in Chapter 6) for sectorial operators to obtain the following asymptotic stability theorem for the steady-state solution $\bar{w}_s$.

Theorem 3.5 (Local Stability of Coexistence Steady-State when the Infected Species Compete). *Assume all the hypotheses of Theorem 3.4, (3.51) and (3.52). For each fixed $s \in (0, \delta^*)$, let $\lambda = \hat{\lambda}(s)$, $\bar{w} = s[w^0 + \hat{\phi}(s)]$. Then, for each $i = 1, 2$, there exists $\rho > 0, \beta > 0$ and $M > 1$ such that equation (3.56) has a unique solution in B_i for all $t > 0$ if $w_0 \in B_1$ and $||w_0 - \bar{w}||_\infty \leq \rho/(2M)$ for $i = 1$ (or $w_0 \in X^\alpha$ and $||w_0 - \bar{w}||_\alpha \leq \rho/(2M)$ for $i = 2$; here we assume $\alpha \in (N/(2p), 1)$ for the space X^α). Moreover, the solution satisfies*

$$(3.57) \qquad ||w(t) - \bar{w}||_\infty \leq 2Me^{-\beta t}\, ||w_0 - \bar{w}||_\infty \quad \text{for all } t \geq 0,\ i = 1,$$

or

$$(3.58) \qquad ||w(t) - \bar{w}||_\alpha \leq 2Me^{-\beta t}\, ||w_0 - \bar{w}||_\alpha \quad \text{for all } t \geq 0, i = 2.$$

2.4 Conditions for Coexistence in Terms of Signs of Principal Eigenvalues of Related Single Equations, Mixed Boundary Data

In the last two sections, the conditions for coexistence solutions are expressed in terms of relationship of the coefficients with the principal eigenvalue of the operator $-\Delta$. In Theorem 2.5 for prey-predator and Theorem 3.4 for competing species studies in the last chapter, the conditions are expressed in terms of the signs of the principal eigenvalues of appropriate operators on scalar equations deduced from the original system. Such results are more direct and convenient. In this section, we investigate extentions of such results to the case of a system of three equations. Moreover, we consider the situation of mixed boundary condition rather than Dirichlet type. The results of this section are mainly adopted from Li and Liu [150] and Liu [160].

As described in Section 2.1, the study of three species diffusion-reaction systems is applicable to many practical problems. As a further example from immunology, we consider the interaction of viruses with different types of antibodies. An external antigen (e.g viruses, called V) stimulates a large increase in a type of immunoglobulin molecules (called antibody 1, *Ab*1), which then induce the production of another type of antibody (called *Ab*2β). By binding with V, *Ab*1 will act as a predator with V as its prey. Induced by *Ab*1 and then neutralizing *Ab*1, *Ab*2β will be a predator with *Ab*1 as prey. Since some of the idiotopes on *Ab*2β resemble the shape of the antigenic determinant on V, the interaction between V and $Ab2\beta$ is competing for the binding sites on $Ab1$. (See e.g. Linthicum and Faird [156] or Roitt, Brostoff and Male [193].)

Let Ω be a bounded domain in $R^N, N \geq 2$, and assume $\partial\Omega$ belongs to $C^{2+\alpha}$ for some $0 < \alpha < 1$. We define a boundary operator B by

$$Bu = a(x)u + b(x)\partial u/\partial\eta \tag{4.1}$$

where $a, b \in C^{1+\alpha}(\partial\Omega)$. We assume that either $b(x) > 0$ and $0 \not\equiv a(x) \geq 0$ on $\partial\Omega$, or $b(x) \equiv 0$ and $a(x) > 0$ on $\partial\Omega$. Here, η is the outward unit normal at the boundary. For a linear operator L, with $+L$ (or $-L$) elliptic, we denote by $\lambda_1(L, B)$, the largest (or smallest) eigenvalue for the problem:

$$Lu = \lambda u \text{ in } \Omega, \quad Bu = 0 \text{ on } \partial\Omega.$$

When the boundary condition is clear from the context, the eigenvalue is abbreviated as $\lambda_1(L)$. The spectral radius of a compact operator T is denoted by $r(T)$. For a given boundary operator B as described above, we define for convenience a class of function $\mathcal{F}(B) \subset C(\bar{\Omega} \times \bar{R}^+)$ as follows:

Definition 4.1. A function $f : \bar{\Omega} \times [0, \infty) \to R$ belongs to the class $\mathcal{F}(B)$, denoted as $f \in \mathcal{F}(B)$, if

(i) $f(x, z) \in C^\alpha(\bar{\Omega} \times [0, \infty)), 0 < \alpha < 1$, and its partial derivative with respect to the second component $f_z(x, z)$ are continuous in $\bar{\Omega} \times [0, \infty)$;

(ii) $-C < f_z(x, z) < 0$ in $\Omega \times [0, \infty)$ for some $C > 0$;

(iii) $limsup_{z \to +\infty} f(x, z) < \lambda_1(-\Delta, B)$ uniformly for $x \in \bar{\Omega}$.

We first deduce an extention of Lemma 2.1 of Chapter 1 concerning the existence of positive solution for scalar equations.

Lemma 4.1. *Suppose $f \in \mathcal{F}(B)$. Consider the boundary value problem:*

$$\begin{cases} -\Delta u = uf(x, u) & \text{in } \Omega, \\ Bu = 0 & \text{on } \partial\Omega. \end{cases} \tag{4.2}$$

(i) If $\lambda_1(\Delta + f(x, 0)) > 0$, then the problem (4.2) has a unique positive solution.
(ii) If $\lambda_1(\Delta + f(x, 0)) \leq 0$, the $u \equiv 0$ is the only non-negative solution of (4.2).

Proof. (i) Assume $\lambda_1 = \lambda_1(\Delta + f(x, 0)) > 0$. Let $\phi(x) \geq 0$ be the corresponding eigenfunction. We have $\Delta\phi - P\phi \leq 0$ in Ω, for sufficiently large $P > 0$, and $B\phi = 0$ on $\partial\Omega$. The boundary condition (4.1), (4.2) implies ϕ is not a constant function, and thus the Hopf Lemma implies that at the boundary minimum point x_0, we must have $\partial\phi/\partial\eta < 0$. The boundary condition (4.1), (4.2) again implies that $\phi(x) \geq \phi(x_0) > 0$ for all $x \in \bar{\Omega}$, for the case $b(x) \not\equiv 0$. Let $\delta \in (0, \lambda_1)$, and $\epsilon > 0$ be small enough such that $|f(x, \epsilon\phi(x)) - f(x, 0)| < \delta$ for all $x \in \bar{\Omega}$. We have

$$(4.3) \qquad \begin{cases} \Delta\epsilon\phi + f(x,\epsilon\phi)\epsilon\phi \\ = \Delta\epsilon\phi + f(x,0)\epsilon\phi + [f(x,\epsilon\phi) - f(x,0)]\epsilon\phi \\ \geq \epsilon\phi(x)(\lambda_1 - \delta) > 0 \qquad \text{in } \Omega, \\ B(\epsilon\phi) \leq 0 \qquad \text{on } \partial\Omega. \end{cases}$$

Let $\tilde{b}(x) = b(x) + \sigma$ with $\sigma > 0$, and $\tilde{\lambda}_1 = \lambda_1(-\Delta, \tilde{B})$ where $\tilde{B}u = a(x)u + \tilde{b}(x)\partial u/\partial\eta$ on $\partial\Omega$. By the continuous dependence of eigenvalues on the boundary conditions, we have $\tilde{\lambda}_1$ is close enough to $\lambda_1(-\Delta, B)$ for sufficiently small σ so that

$$limsup_{z\to+\infty} f(x,z) < \tilde{\lambda}_1 \;\text{ uniformly for } x \in \Omega.$$

Let $\omega(x) \geq 0$ be an eigenfunction for the eigenvalue $\tilde{\lambda}_1 = \lambda_1(-\Delta, \tilde{B})$. By the same argument as in the last paragraph, we obtain $\omega(x) > 0$ for all $x \in \bar{\Omega}$. Thus, for $K > 0$ sufficiently large, we have $\bar{u} := K\omega > \underline{u} := \epsilon\phi$ in $\bar{\Omega}$, where ϵ was chosen in the last paragraph to satisfy (4.3), and $f(x,\bar{u}) < \tilde{\lambda}_1$ for all $x \in \Omega$. We thus have

$$(4.4) \qquad \begin{cases} \Delta\bar{u} + \bar{u}f(x,\bar{u}) = \Delta K\omega + K\omega f(x,\bar{u}) \\ \qquad\qquad = K\omega(f(x,\bar{u}) - \tilde{\lambda}_1) < 0 \quad \text{in } \Omega, \\ B\bar{u} \geq 0 \qquad \text{on } \partial\Omega. \end{cases}$$

The last inequality above is true because $\partial\bar{u}/\partial\eta = (-a/\tilde{b})\bar{u} \geq (-a/b)\bar{u}$ on $\partial\Omega$. Define $u_0 := \bar{u}$ in $\bar{\Omega}$, and inductively, for $i = 1, 2, ..., u_i$ to be the solution of

$$(4.5) \qquad -\Delta u_i + Pu_i = u_{i-1}f(x, u_{i-1}) + Pu_{i-1} \text{ in } \Omega, \;\; Bu_i = 0 \text{ on } \partial\Omega.$$

Here $P > 0$ is chosen sufficiently large so that the function $z \to zf(x,z) + Pz$ is nondecreasing in z for $x \in \bar{\Omega}, 0 \leq z \leq max_{\bar{\Omega}}\bar{u}$. Using (4.3), (4.4) and (4.5), one can show by means of Hopf Lemma and maximum principle (cf. Section 5.1 in [125] for the case of Dirichlet boundary condition) that

$$(4.6) \qquad \underline{u} \leq \cdots \leq u_i \leq \cdots \leq u_2 \leq u_1 \leq \bar{u} \;\text{ in } \bar{\Omega}.$$

The sequence $\{u_k\}$ is bounded in $C(\bar{\Omega})$. Using the mapping $(-\Delta + P)^{-1}(uf(x,u) + Pu)$ and Theorem 4.2 in Amann [3], we can choose subsequence $\{u_{k(i)}\}$ so that it converges in $C^1(\bar{\Omega})$. Then, we can use Lemma 9.2 in [3] to assert that the sequence $\{u_{k(i)}(x)f(x, u_{k(i)}(x))\}$ is bounded in $C^\alpha(\bar{\Omega})$. Thus by Schauder's theory and equations (4.5), the sequence $\{u_{k(i)+1}\}$ is bounded in $C^{2+\alpha}(\bar{\Omega})$, and hence relatively compact in $C^2(\bar{\Omega})$. We obtain another subsequence of $\{u_{k(i)+1}\}$ converging to a $C^2(\bar{\Omega})$ solution of (4.2).

The uniqueness part can be proved by modifying the proof for the case of Dirichlet boundary condition as in Lemma 5.2-2 in [125]. We construct a solution larger than or equal to two given ones by iterating from a large upper solution as in the last paragraph. We then show by means of Green's identity that both given solutions are identically equal to this large solution.

(ii) Suppose $\lambda_1(\Delta+f(x,0)) \leq 0$, and there is a positive solution u of problem (4.2). Then by Theorem 3.2 in Amann [3], we must have $\lambda_1(\Delta + f(x,u)) = 0$. Since $f(x,0) > f(x,u)$ in Ω, we obtain by variational characterization of principal eigenvalue that $\lambda_1(\Delta+f(x,0)) > \lambda_1(\Delta+f(x,u)) = 0$. This contradicts our assumption.

For a given $f \in \mathcal{F}(B)$, we will use u_f to denote the unique positive solution of problem (4.2) in case (i) of Lemma 4.1 ; and set $u_f \equiv 0$ in case (ii) of Lemma 4.1. The following continuity and comparison properties will be needed for further considerations.

Lemma 4.2. *(i) Let f and f_n be functions in $\mathcal{F}(B), n = 1,2,\ldots,$ and $f_n \to f$ uniformly in each compact subset of $\bar{\Omega} \times [0,\infty)$ as $n \to \infty$, then $u_{f_n} \to u_f$ in $C^{1+\nu}(\bar{\Omega})$, for any $\nu \in (0,1)$.*
(ii) Let $f_1, f_2 \in \mathcal{F}(B)$ and $f_1 \geq f_2$, then $u_{f_1} \geq u_{f_2}$.

Proof. (i) First assume $\lambda_1(\Delta+f(x,0)) > 0$. Then by means of variational characterization of principal eigenvalue, we have $\lambda_1(\Delta+f_n(x,0)) > 0$ for sufficiently large n. Let $\phi(x) > 0$ be the eigenfunction corresponding to $\lambda_1(\Delta+f(x,0)) > 0$. As described in Lemma 4.1, we obtain for sufficiently small $\epsilon > 0$:

$$\Delta\epsilon\phi + f(x,\epsilon\phi)\epsilon\phi \geq p\phi(x) > 0 \text{ in } \Omega \text{ for some constant } p > 0,$$

and $B(\epsilon\phi) \leq 0$ on $\partial\Omega$. Thus for large n, we also have

$$\Delta\epsilon\phi + f_n(x,\epsilon\phi)\epsilon\phi > 0 \text{ in } \Omega.$$

Thus the function $\epsilon\phi$ is a lower solution for problem (4.2) and all such problems if f is replaced by f_n for sufficiently large n. As in the proof of Lemma 4.1, we find the function $K\omega$ is an upper solution for problem (4.2). By the uniform convergence of f_n on compact subsets, we can readily see that $K\omega$ is also an upper solution for problem (4.2) if f is replaced with f_n for sufficiently large n. Consequently, we have

$$\epsilon\phi \leq u_f \leq K\omega, \ \epsilon\phi \leq u_{f_n} \leq K\omega,$$

for all large n, $\epsilon > 0$ sufficiently small. As in the proof of Lemma 4.1, we use Theorem 4.2 and Lemma 9.2 in [3] to choose a subsequence of $\{u_{f_n}\}$ convergent in $C^1(\bar{\Omega})$, and a subsequence of $\{u_{f_n} f_n(x, u_{f_n})\}$ bounded in $C^\alpha(\bar{\Omega})$. This will lead to a subsequence of $\{u_{f_n}\}$ convergent in $C^2(\bar{\Omega})$ to a function $\tilde{u} \geq \epsilon\phi > 0$, and $\tilde{u}$ is a positive solution of problem (4.2). By uniqueness part of Lemma 4.1, we must have $\tilde{u} \equiv u_f$. Since this can be done for any subsequence of $\{u_{f_n}\}$, we conclude that the full sequence $\{u_{f_n}\}$ converges to u_f in $C^2(\bar{\Omega})$, and hence in $C^{1+\nu}(\bar{\Omega})$. In case $\lambda_1(\Delta + f(x,0)) \leq 0$, we can argue similarly that $u_{f_n} \to u_f = 0$ in $C^{1+\nu}(\bar{\Omega})$.

(ii) Assume that $f_1 \geq f_2$. If $\lambda_1(\Delta + f_1(x,0)) \leq 0$, then $u_{f_1} \equiv u_{f_2} \equiv 0$ by Lemma 4.1, part (ii). If $\lambda_1(\Delta + f_1(x,0)) > 0$, then $u_{f_1} > 0$. The usual upper-lower solution argument and uniqueness of positive solution of part (i) of Lemma 4.1 leads to $u_{f_1} \geq u_{f_2}$.

For characterization of the principal eigenvalue and the property that the principal eigenfunction does not change sign involving homongeneous boundary condition for (4.1), the reader is referred to Chapter 11 in Smoller [209]. The following is an extention of Theorem A2-6 in Chapter 6.

Lemma 4.3. *Let $c(x) \in C(\bar{\Omega})$ and P is a positive constant such that $P+c(x) > 0$ in Ω. Then*

(i) $\lambda_1(\Delta + c) > 0$ if and only if $r[(-\Delta + P)^{-1}(P + c)] > 1$,

(ii) $\lambda_1(\Delta + c) < 0$ if and only if $r[(-\Delta + P)^{-1}(P + c)] < 1$,

(iii) $\lambda_1(\Delta + c) = 0$ if and only if $r[(-\Delta + P)^{-1}(P + c)] = 1$.

(Here, $(-\Delta + P)^{-1}(P + c)$ denotes the compact operator from $C(\bar{\Omega})$ into itself subject to boundary condition $Bu = 0$ for the inverse, and $r[(-\Delta+P)^{-1}(P+c)]$ denotes the spectral radius of the operator.)

Proof. Let $\phi > 0$ be the corresponding eigenfunction of $\lambda_1 = \lambda_1(\Delta + c)$. Consider the expression $(-\Delta + P)\phi = (P + c)\phi - \lambda_1\phi$ which leads to

$$\phi + \lambda_1(-\Delta + P)^{-1}\phi = (-\Delta + P)^{-1}(P + c)\phi.$$

Let T be the compact operator $(-\Delta + P)^{-1}(P + c(x))$. Then $T\phi$ is $>, <$ or $=$ ϕ according to λ_1 is $>, <$ or $= 0$ respectively. From Theorem A2-5, we obtain the forward implication in all three cases. Conversely, suppose that the spectral radius satisfies $r[-\Delta + P)^{-1}(P + c)] > 1$, then from the forward implication part of cases (ii) and (iii) we must have $\lambda_1(\Delta + c) > 0$ in order not to have any contradiction on the sign of the spectral radius. The other parts of (ii) and (iii) are proved the same way.

We consider the existence of positive solutions for the following nonlinear elliptic homogeneous boundary value problem. (Here, positive means positive in each component; the boundary condition is sometimes known as Robin type.)

$$
(4.7) \qquad \begin{cases} \Delta u + u f_1(x,u,v,w) = 0 & \text{in } \Omega, \\ \Delta v + v f_2(x,u,v,w) = 0 & \text{in } \Omega, \\ \Delta w + w f_3(x,u,v,w) = 0 & \text{in } \Omega, \\ B_1 u = B_2 v = B_3 w = 0 & \text{on } \partial\Omega. \end{cases}
$$

The $B_i, i = 1,2,3$ are boundary operators as described in (4.1) with B, a, b respectively replaced by B_i, a_i, b_i with corresponding properties. The following hypotheses will be assumed:
(4.8)
The functions f_1, f_2, f_3 and all their first partial derivatives with respect to u, v, w are continuous on $\bar{\Omega} \times [0,\infty) \times [0,\infty) \times [0,\infty)$.

We shall use index method to investigate this problem. One main difficulty we encounter is that the number of semi-trivial non-negative solutions of the form $(u,v,0), (u,0,w)$ or $(0,v,w)$ is unknown. In order to circumvent this handicap, we evaluate the fixed point index of certain operator on a slice of the positive cone. This slice contains all non-negative solutions of the form, say, $(u,v,0)$. We will show that under appropriate situations, the homotopy invariance principle can be applied on the slice so that index calculations can be readily simplified. The existence of positive solutions will be characterized by the signs of the principal eigenvalues of certain scalar linear operators. These operators are determined by a-priori bound, the marginal densities, and some other functions derived from marginal densities. Here, the marginal densities of a species means the densities of the species when all the other species are absent.

Let $C(\bar{\Omega})$ be the ordered Banach space of continuous functions on $\bar{\Omega}$ with the maximum norm. Let $X = C_{B_1}(\bar{\Omega}) \oplus C_{B_2}(\bar{\Omega}) \oplus C_{B_3}(\bar{\Omega})$, where $C_{B_i}(\bar{\Omega})$ denotes the subspace of $C(\bar{\Omega})$ subject to boundary condition $B_i u = 0$. For $\theta_i \in [0,1], i = 1,2,3$ and $P > 0$, define the operator $A_{\theta_1,\theta_2,\theta_3} : C(\bar{\Omega}) \oplus C(\bar{\Omega}) \oplus C(\bar{\Omega}) \to X \subset C(\bar{\Omega}) \oplus C(\bar{\Omega}) \oplus C(\bar{\Omega})$ by

$$
(4.9) \qquad A_{\theta_1,\theta_2,\theta_3}(u,v,w) = (-\Delta + P)^{-1} \begin{bmatrix} \theta_1 u f_1(x,u,v,w) + Pu \\ \theta_2 v f_2(x,u,v,w) + Pv \\ \theta_3 w f_3(x,u,v,w) + Pw \end{bmatrix}^T,
$$

where $(-\Delta + P)^{-1}$ is taken componentwise. (Here, the boundary conditions are taken as in (4.7)). Further, define $A = A_{1,1,1}$.

Let $K = C^+_{B_1}(\bar{\Omega}) \oplus C^+_{B_2}(\bar{\Omega}) \oplus C^+_{B_3}(\bar{\Omega})$, where $C^+_{B_i}(\bar{\Omega})$ denotes the subset of non-negative functions in $C_{B_i}(\bar{\Omega})$. As in Section 1.2, we will be studying positive operators which map the cone K into K. For any $C > 0$, define $[[0, C]] := \{u \in C(\bar{\Omega}) : 0 \leq u \leq C \text{ in } \bar{\Omega}\}$. Suppose (4.8) is satisfied, it can be shown (cf. Theorem 4.2 in Amann [3]) that the operator $A_{\theta_1,\theta_2,\theta_3}$ restricted to $[[0,C]]\oplus[[0,C]]\oplus[[0,C]]$ defined above is positive, compact, and Fréchet differentiable for large P,where P does not depend on $\theta_i, i = 1, 2, 3$. In order to use the theory of positive cone mapping, we will modifying the functions f_i for large u, v, w so that $A_{\theta_1,\theta_2,\theta_3}$ maps K into K. If we know that the non-negative solutions of problem (4.7), with f_i changed to $\theta_i f_i, \theta_i \in [0, 1]$ must have all components in $[[0, M]]$ for some large $M > 0$, then we can further impose that the modifications of $f_i(u, v, w)$ will not change their values for u, v, w in $[0, M]$. Thus the fixed points of the altered map for $A_{\theta_1,\theta_2,\theta_3}$ in $[[0, M]] \oplus [[0, M]] \oplus [[0, M]]$ will not be changed. Such fixed points will be non-negative solutions of problem (4.7),with f_i changed to $\theta_i f_i$. Let $g \in C^\infty[0, \infty)$ such that $g = 1$ on $[0, M], g \in (0, 1)$ on $(M, M + 1)$ and $g = 0$ on $[M + 1, \infty)$. We may for instance change f_i in the definition of $A_{\theta_1,\theta_2,\theta_3}$ to $\tilde{f}_i(x, u, v, w) = f_i(x, ug(u), vg(v), wg(w))$. Therefore, we will explicitly assume that the non-negative solutions of (4.7), with f_i changed to $\theta_i f_i$, will possess a-priori bound $M > 0$; and thus without loss of generality, we may assume that $A_{\theta_1,\theta_2,\theta_3}$ is an operator mapping K into K.

For convenience, for any $t > 0$, define $E(t) := \{u \in C(\bar{\Omega}) : |u(x)| < t$ for all $x \in \bar{\Omega}\}$, and define

$$D := K \cap [E(M)]^3,$$

where M is the a-priori bound described in the last paragraph. As explained above, we assume that the operator $A_{\theta_1,\theta_2,\theta_3} : K \to K$ satisfies:

$$A_{\theta_1,\theta_2,\theta_3} \text{ has no fixed point on } \partial D \text{ for all } \theta_i \in [0, 1]. \tag{4.10}$$

For the mapping A_{111}, we will search in D for fixed points which has no identically zero component. These will be the coexistence solutions of the original problem (4.7).

Remark 4.1. As an example, we can show readily that the following system has an a-priori bound for all positive solutions:

$$\begin{cases} \Delta u + u[a - \frac{p_1 u}{1+q_1 u} - \frac{p_2 v}{1+q_2 v} - \frac{p_3 w}{1+q_3 w}] = 0 & \text{in } \Omega, \quad Bu = 0 \text{ on } \partial\Omega, \\ \Delta v + v[\alpha_1 + \beta_1 u - c_{11} v + c_{12} w] = 0 & \text{in } \Omega, \quad v = 0 \text{ on } \partial\Omega, \\ \Delta w + w[\alpha_2 + \beta_2 u + c_{21} v - c_{22} w] = 0 & \text{in } \Omega, \quad w = 0, \text{ on } \partial\Omega, \end{cases}$$

where $a > \lambda_1(-\Delta, B) > a - \frac{p_1}{q_1}$, $c_{11}c_{22} > c_{21}c_{12}$, and all parameter constants $a, p_i, q_i, \alpha_i, \beta_i, c_{ij}$ are positive. We first deduce a bound for u in the first equation, and then use the method in Section 1.4 to obtain a bound for v and w.

For a compact operator $\mathcal{A} : K \to K$, and an open subset $U \subset K$, define the $index(\mathcal{A}, U, K) := i_K(\mathcal{A}, U) = deg(I - \mathcal{A} \circ r, r^{-1}(U), 0)$, where I is the identity map and r is a retraction of K into $C(\bar{\Omega}) \oplus C(\bar{\Omega}) \oplus C(\bar{\Omega})$. (Recall the definition of such concepts in part B of Section 1.2). If y is an isolated fixed point of $\mathcal{A}$, then the fixed point index of $\mathcal{A}$ at y in K is defined by $index(\mathcal{A}, y, K) := index(\mathcal{A}, U(y), K) = i_K(\mathcal{A}, U(y))$, where $U(y)$ is a small open neighborhood of y in K. For convenience, we abbreviate $i(\mathcal{A}, D) = index(\mathcal{A}, D, K), i(\mathcal{A}, y) = index(\mathcal{A}, y, K)$. The following lemma will be needed for our main results.

Lemma 4.4. *Suppose hypotheses (4.8) and (4.10) are satisfied, then $i(A, D) = 1$. (Here $A = A_{111}$ as defined in (4.9).)*

Proof. From hypothesis (4.10), we can use homotopy invariance of indices of the mappings to obtain $i(A, D) := i(A_{111}, D) = i(A_{000}, D)$. From definition, the i-th component of $A_{000}(u_1, u_2, u_3)$ is $(-\Delta + P)^{-1}(Pu_i)$. If $A_{000}(u, v, w) = \lambda(u, v, w)$ for $\lambda \geq 1$, then $-\Delta u + P(\lambda - 1)\lambda^{-1}u = 0$ in Ω, and $B_1 u = 0$ on $\partial\Omega$. Thus if $\lambda \geq 1$ we must have $u = 0$, since $\lambda_1(-\Delta, B_1) > 0$. Similarly, we also deduce $v = w = 0$. This means $A_{000}(u, v, w) \neq \lambda(u, v, w)$ for all $(u, v, w) \in \partial D$ and every $\lambda \geq 1$. Hence, by Theorem A2-4, we assert that $i(A_{000}) = 1$.

For further study of problem (4.7) we need more hypotheses on the functions f_i.

(4.11)
For fixed non-negative functions $u(x), v(x), w(x)$ in $C^\alpha(\bar{\Omega})$, the functions $f_1(x, \cdot, v(x), w(x)), f_2(x, u(x), \cdot, w(x)), f_3(u(x), v(x), \cdot)$ belong respectively to the set $\mathcal{F}(B_i)$ of functions $i = 1, 2, 3$.

Before we state some of the main results, we define three operators $S_i : C^\alpha(\bar{\Omega}) \oplus C^\alpha(\bar{\Omega}) \to C^2(\bar{\Omega}), i = 1, 2, 3$ for convenience as follows. For given $(p, q) \in C^\alpha(\bar{\Omega}) \oplus C^\alpha(\bar{\Omega})$, $u = S_1(p, q)$ is the unique positive solution of the problem:

$$\Delta u + u f_1(x, u, p(x), q(x)) = 0 \quad \text{in } \Omega, \quad B_1 u = 0 \text{ on } \partial\Omega,$$

if $\lambda_1(\Delta + f_1(x, 0, p, q), B_1) > 0$ and $S_1(p, q) \equiv 0$ otherwise. Similarly, $v = S_2(p, q)$ or $w = S_3(p, q)$ are the unique positive solutions of:

$$\Delta v + v f_2(x, p(x), v, q(x)) = 0 \quad \text{in } \Omega, \quad B_2 v = 0 \text{ on } \partial\Omega, \text{ or}$$

$$\Delta w + w f_3(x, p(x), q(x), w) = 0 \quad \text{in } \Omega, \quad B_3 w = 0 \text{ on } \partial\Omega,$$

if $\lambda_1(\Delta + f_2(x, p, 0, q), B_2) > 0$ or $\lambda_1(\Delta + f_3(x, p, q, 0), B_3) > 0$ respectively. Otherwise $S_2(p, q)$ or $S_3(p, q)$ are the trivial functions. In particular we denote $u_0 = S_1(0, 0), v_0 = S_2(0, 0)$ and $w_0 = S_3(0, 0)$. For $i = 1, 2, 3$, denote the principal eigenvalues by

$$\mu_i := \lambda_1(\Delta + f_i(0, 0, 0), B_i).$$

By Lemma 4.1, we have u_0 is non-trivial if and only if $\mu_1 > 0$, v_0 is non-trivial if and only if $\mu_2 > 0$, and w_0 is non-trivial if and only if $\mu_3 > 0$.

Theorem 4.1 and 4.2 deduce the indices for the trivial solution and those solutions when two of the species are absent. For $y \in K$, as in Section 1.2, we define $K_y = \{p \in C_{B_1}(\bar{\Omega}) \oplus C_{B_2}(\bar{\Omega}) \oplus C_{B_3}(\bar{\Omega}) : y + sp \in K \text{ for some } s > 0\}$; $\bar{K}_y$ denotes the closure of K_y, and $S_y = \{p \in \bar{K}_y : -p \in \bar{K}_y\}$.

Theorem 4.1 (Index of the Trivial Solution). *Assume hypotheses (4.8), (4.10) and (4.11).*

(i) If $max(\mu_1, \mu_2, \mu_3) > 0$ and at most one of the μ_i's is zero, then $i(A, (0,0,0)) = 0$.

(ii) If $\mu_i \leq 0$ for $i = 1, 2, 3$ and at most one of them is zero, then $i(A, (0,0,0)) = 1$.

Proof. (i) Without loss of generality, we only have to consider two cases:

(a) Suppose $\mu_1 > 0, \mu_2 \neq 0, \mu_3 \neq 0$. We have $\bar{K}_{(0,0,0)} = K$, $S_{(0,0,0)} = \{(0,0,0)\}$. Denote the Fréchet derivative of A at $(0,0,0)$ by:

$$\begin{aligned} L &:= A'(0,0,0) \\ &= (-\Delta + P)^{-1} \circ \begin{bmatrix} f_1(0,0,0) + P & 0 & 0 \\ 0 & f_2(0,0,0) + P & 0 \\ 0 & 0 & f_3(0,0,0) + P \end{bmatrix}, \end{aligned}$$

with boundary condition B_i on the i-th component. To show $I - L$ is invertible on $\bar{K}_{(0,0,0)}$, it suffices to show that the only fixed point of L in K is zero. Let $[I - L](u, v, w)] = 0$ with $(u, v, w) \in K$. Then $[\Delta + f_1(0,0,0)]u = 0$, with $u(x) \geq 0$ in $\bar{\Omega}$. Hence the assumption $\mu_1 > 0$ implies $u = 0$. Similarly, $\mu_2 \neq 0$ and $\mu_3 \neq 0$ imply that $v = w = 0$. We next show the operator $L = A'(0,0,0)$ has property (α) on K, and apply Lemma 2.7(i) to conclude $i(A, (0,0,0)) = 0$. For $s \in R$, let $\phi(s) := \lambda_1(\Delta + sf_1(x,0,0,0) + (1-s)(-P), B_1)$. Since $\phi(0) = \lambda_1(\Delta - P) < 0$, (by the characterization of principal eigenvalue), $\phi(1) = \mu_1 > 0$ and ϕ is continuous in s, there must exist $t \in (0,1)$ such that $\phi(t) = \lambda_1(\Delta + tf_1(0,0,0) + (1 - t)(-P)) = 0$. The corresponding eigenfunction u satisfies $(u,0,0) = tL(u,0,0)$, with $(u,0,0) \in \bar{K}_{(0,0,0)} \backslash S_{(0,0,0)}$. Thus L has property (α), and $i(A, (0,0,0)) = 0$.

(b) Suppose $\mu_1 > 0, \mu_2 = 0$ and $\mu_3 \neq 0$. We modify f_2 by letting $f_2^t(x, u, v, w) := f_2(x, u, v, w) - t$ and define the operator A^t in the same way as A with f_2 replaced by f_2^t. Then we have $A = A^0$ and $\lambda_1(\Delta + f_2^t(x,0,0,0), B_2) < \lambda_1(\Delta + f_2(x,0,0,0), B_2) = \mu_2 = 0$ for $t > 0$. From part (a), we have $i(A^t, (0,0,0)) = 0$ for $t > 0$. We next deduce that $(0,0,0)$ is an isolated fixed point of A^t for all $t \geq 0$ by the following argument. Suppose there exist $t_n \geq 0, (u_n, v_n, w_n) \to (0,0,0)$, and for each n,(u_n, v_n, w_n) is a fixed point of A^{t_n} in K. Then $\Delta u_n + u_n f_1(x, u_n, v_n, w_n) = 0$. Since $f_1(x, u_n, v_n, w_n) \to f_1(x,0,0,0)$ and $\mu_1 = \lambda_1(\Delta +$

$f_1(x,0,0,0),B_1) > 0$, we find $\lambda_1(\Delta + f_1(x,u_n,v_n,w_n),B_1) > 0$ for all large n. It follows that $u_n = 0$. Since $\mu_3 \neq 0$, a similar argument shows that $w_n = 0$ for all large n. We next observe that for large n, v_n is a non-negative solution of

$$\Delta v + v f_2^{t_n}(x,0,v,0) = 0 \text{ in } \Omega, \quad B_2 v = 0 \text{ on } \partial\Omega.$$

From $\lambda_1(\Delta + f_2^{t_n}(x,0,0,0),B_2) \leq 0$ and hypothesis (4.11), we conclude by Lemma 4.1 that $v_n = 0$ also, for large n. Hence $(0,0,0)$ is an isolated fixed point as claimed. Finally, by homotopy invariance, we obtain for some $t > 0$ that $i(A,(0,0,0)) = i(A^0,(0,0,0)) = i(A^t,(0,0,0)) = 0$.
(ii) First assume $\mu_i < 0$ for $i = 1,2,3$. We readily obtain $r(L) < 1$. Therefore $I - L$ is invertible on $\bar{K}_{(0,0,0)}$. By comparison of principal eigenvalues and the choice of large P we deduce that L does not have property (α). By Lemma 2.7 in Chapter 1 and Theorem A2-3 in Chapter 6, we obtain $i(A,(0,0,0)) = (-1)^0 = 1$. If one of the $\mu_i's$ is zero, we use a similar argument by modifying the corresponding f_i as in part(i)(b) above to deduce by homotopy that $i(A,(0,0,0)) = 1$.

Theorem 4.2 (Indices of Semi-Trivial Solutions). *Assume hypotheses (4.8), (4.10), (4.11) and $\mu_1 > 0$.*

(i) If $\lambda_1(\Delta+f_2(x,u_0,0,0),B_2) \neq 0$ and $\lambda_1(\Delta+f_3(x,u_0,0,0),B_3) > 0$, or $\lambda_1(\Delta+f_2(x,u_0,0,0),B_2) > 0$ and $\lambda_1(\Delta+f_3(x,u_0,0,0),B_3) \neq 0$, then $i(A,(u_0,0,0)) = 0$.

(ii) If $\lambda_1(\Delta+f_2(x,u_0,0,0),B_2) < 0$ and $\lambda_1(\Delta+f_3(x,u_0,0,0),B_3) < 0$, then $i(A,(u_0,0,0)) = 1$.
(Note that $\mu_1 > 0$ implies that u_0 is non-trivial.)

Proof. (i) Without loss of generality, we assume $\lambda_1(\Delta+f_3(x,u_0,0,0),B_3) > 0$ and $\lambda_1(\Delta+f_2(x,u_0,0,0),B_2) \neq 0$. Let $y = (u_0,0,0)$. Then

$$L := A'(y) = (-\Delta+P)^{-1}\circ$$

$$\begin{bmatrix} u_0 D_u f_1(x,y) + f_1(x,y) + P & u_0 D_v f_1(x,y) & u_0 D_w f_1(x,y) \\ 0 & f_2(x,y)+P & 0 \\ 0 & 0 & f_3(x,y)+P \end{bmatrix}.$$

One readily verifies that $\bar{K}_y = C_{B_1}(\bar{\Omega}) \oplus C^+_{B_2}(\bar{\Omega}) \oplus C^+_{B_3}(\bar{\Omega})$. and $S_y = C_{B_1}(\bar{\Omega}) \oplus \{0\} \oplus \{0\}$. Let $L(u,v,w) = (u,v,w) \in \bar{K}_y$. As in the proof of Theorem 4.1(i), we have $v = w = 0$. Therefore $[\Delta + u_0 D_u f_1(x,u_0,0,0) + f_1(x,u_0,0,0)]u = 0$ in Ω, $B_1 u = 0$. From the sign of $D_u f_1$ and comparison of principal eigenvalues, we obtain $u \equiv 0$. Thus $I - L$ is invertible in $\bar{K}_y$.

Since $\lambda_1(\Delta+f_3(x,u_0,0,0),B_3) > 0$, as in the proof of Theorem 4.1(i), we obtain a number $t \in (0,1)$ and a non-trivial $w \in C^+_{B_3}(\bar{\Omega})$ such that $(-\Delta+P)w = t(f_3(x,u_0,0,0)+P)w$ in Ω. Then $(u,0,w) \in \bar{K}_y \backslash S_y$ and $(u,0,w) - tL(u,0,w) \in$

S_y for any $u \in C_{B_1}(\bar{\Omega})$. This implies that L has property (α) on $\bar{K}_y$. Thus $i(A,(u_0,0,0)) = 0$, by Lemma 2.7 in Chapter 1.
(ii) Let L be as defined above, we first show that no eigenvalue of L is greater than one. Suppose $L(u,v,w) = \mu(u,v,w)$ for some $(u,v,w) \in X$ with $\mu \geq 1$. Then $(-\Delta + P)^{-1}[f_2(x,u_0,0,0) + P]v = \mu v$. However, from Lemma 4.3 and assumption $\lambda_1(\Delta + f_2(x,u_0,0,0)) < 0$, we know that $r(H) < 1$ where $H := (-\Delta + P)^{-1}[f_2(x,u_0,0,0) + P]$. Therefore we obtain $v = 0$. Similarly, we deduce $w = 0$. Thus $(-\Delta + P)^{-1}[u_0 D_u f_1(x,u_0,0,0) + f_1(u_0,0,0) + P]u = \mu u$. This implies $u = 0$ again by Lemma 4.3 because $\lambda_1(\Delta + u_0 D_u f_1(x,u_0,0,0) + f_1(u_0,0,0) < \lambda_1(\Delta + f_1(u_0,0,0) = 0$. Therefore we must have $(u,v,w) = (0,0,0)$ and such μ cannot be an eigenvalue of L.

Suppose that the operator L has property (α) on $\bar{K}_y$. Then there exist $(u,v,w) \in \bar{K}_y \backslash S_y$ and $t \in (0,1)$ such that $(u,v,w) - tL(u,v,w) \in S_y$. Thus v and w are non-negative solutions of $(-\Delta + P)^{-1}[f_2(x,u_0,0,0) + P]v = (1/t)v$ and $(-\Delta+P)^{-1}[f_3(x,u_0,0,0)+P]w = (1/t)w$. If v is non-trivial, then $(1/t) > 1$ is an eigenvalue of the operator $(-\Delta + P)^{-1}[f_2(x,u_0,0,0) + P]$, contradicting Lemma 4.3. Thus we must have $v = 0$. Similarly, we deduce $w = 0$. However, this means $(u,v,w) \in S_y$, leading to another contradiction. Consequently, the operator L cannot have property (α). We can thus use Lemma 2.7 in Chapter 1 and Theorem A2-3 in Chapter 6 to conclude that $i(A,(u_0,0,0) = (-1)^0 = 1$.

Let M be the a-priori bound for (4.7) defined above, and $\hat{E}_i(t) = \{u \in C(\bar{\Omega}) : B_i u = 0, 0 \leq u < t \text{ in } \bar{\Omega}\}$. Define $D_1^\epsilon = \hat{E}_1(t) \oplus \hat{E}_2(M) \oplus \hat{E}_3(M)$, $D_2^\epsilon = \hat{E}_1(M) \oplus \hat{E}_2(t) \oplus \hat{E}_3(M)$, and $D_3^\epsilon = \hat{E}_1(M) \oplus \hat{E}_2(M) \oplus \hat{E}_3(t)$. D_1^ϵ is a "slice" in D containing all fixed points of A of the form $(0,v,w)$; similarly, D_2^ϵ and D_3^ϵ are different slices. The following theorem is concerned with the fixed point index of A on such a slice.

Theorem 4.3 (Indices of Thin Slices of the Positive Cone). *Assume hypotheses (4.8), (4.10) and (4.11). Define:*

$$t := inf.\{||u|| : u \neq 0, (u,v,w) \text{ is a fixed point of } A_{1,\theta,\theta} \text{ in } D \text{ for some } v,w \geq 0 \text{ and some } \theta \in [0,1]\}.$$

Suppose that $t > 0$ and ϵ is a positive number with $\epsilon \in (0,t)$. Then $i(A, D_1^\epsilon) = 0$ if $\mu_1 > 0$, and $i(A, D_1^\epsilon) = 1$ if $\mu_1 \leq 0$.

Proof. First note that the assumption $t > 0$ means that if $\epsilon \in (0,t)$, all fixed points of $A_{1,\theta,\theta}$, for all $\theta \in [0,1]$ in D_1^ϵ are of the form $(0,u,w)$ and are thus not on the boundary ∂D_1^ϵ in D. Hence we can use homotopy invariance in the 'slice' D_1^ϵ to obtain $i(A, D_1^\epsilon) = i(A_{1,1,1}, D_1^\epsilon) = i(A_{1,0,0}, D_1^\epsilon)$. However, by definition $A_{1,0,0}(u,v,w) = (-\Delta + P)^{-1}[uf_1(x,u,v,w) + Pu), Pv, Pw]$. We see that $A_{1,0,0}(u,v,w) = (u,v,w) \in D_1^\epsilon$ implies $v = w = 0$ and $||u|| < \epsilon$. Thus by the choice of $\epsilon < t$, we must have $u = 0$. This means the only fixed point of $A_{1,0,0}$ in

D_1^ϵ is $(0,0,0)$. Therefore $i(A_{1,0,0}, D_1^\epsilon) = i(A_{1,0,0}, (0,0,0))$. We can readily verify from the proof of Theorem 4.1 that it still applies if we change f_2 and f_3 to be the trivial functions with the new $\mu_2 = \lambda_1(\Delta, B_2) < 0$, $\mu_3 = \lambda_1(\Delta, B_3) < 0$, and $\mu_1 = \lambda_1(f_1(x,0,0,0), B_1)$ unchanged. That is the hypotheses of Theorem 4.1 can be modified so that it is applicable to the mapping $A_{1,0,0}$. Thus we conclude that $i(A_{1,0,0}, (0,0,0)) = 0$ if $\mu_1 > 0$, and $i(A_{1,0,0}, (0,0,0)) = 1$ if $\mu_1 \leq 0$.

We next define $D_{1,2}^\epsilon = D_1^\epsilon \cup D_2^\epsilon$, $D_{1,3}^\epsilon = D_1^\epsilon \cup D_3^\epsilon$, and $D_{2,3}^\epsilon = D_2^\epsilon \cup D_3^\epsilon$. Note that the set D_{12}^ϵ contains all fixed points of A in D of the form $(0,v,w)$ and $(u,0,w)$, and $\partial D_{12}^\epsilon = \{(u,v,w) \in D : min\{||u||, ||v||\} = \epsilon$ or $min\{||u||, ||v||\} \leq \epsilon$ and $||w|| = M\}$. The following theorem consider the indices of these sets.

Theorem 4.4 (Indices for Unions of Thin Slices). *Assume hypotheses (4.8), (4.10) and (4.11). Define*

$$t := inf.\{||u||, ||v|| : (u,v,w) \textit{ is a fixed point of } A_{1,1,\theta} \textit{ in } D \textit{ for some } w \geq 0$$
$$\textit{with both } u \neq 0,\ v \neq 0 \textit{ and some } \theta \in [0,1]\}.$$

Suppose $t > 0$ and ϵ is a number satisfying $\epsilon \in (0,t)$, then $i(A, D_{1,2}^\epsilon)$ has the following properties:

(i) Suppose $\mu_1 > 0$ and $\mu_2 > 0$.

If $\lambda_1(\Delta + f_1(x,0,v_0,0), B_1) > 0$ and $\lambda_1(\Delta + f_2(x,u_0,0,0), B_2) > 0$, then $i(A, D_{1,2}^\epsilon) = 0$.

If $\lambda_1(\Delta + f_1(x,0,v_0,0), B_1) > 0$ and $\lambda_1(\Delta + f_2(x,u_0,0,0), B_2) < 0$, or vice versa, then $i(A, D_{1,2}^\epsilon) = 1$.

If $\lambda_1(\Delta + f_1(x,0,v_0,0), B_1) < 0$ and $\lambda_1(\Delta + f_2(x,u_0,0,0), B_2) < 0$, then $i(A, D_{1,2}^\epsilon) = 2$.

(ii) Suppose $\mu_1 > 0$ and $\mu_2 \leq 0$.

If $\lambda_1(\Delta + f_2(x,u_0,0,0), B_2) > 0$, then $i(A, D_{1,2}^\epsilon) = 0$.
If $\lambda_1(\Delta + f_2(x,u_0,0,0), B_2) < 0$, then $i(A, D_{1,2}^\epsilon) = 1$.

(iii) Suppose $\mu_1 \leq 0$ and $\mu_2 > 0$.

If $\lambda_1(\Delta + f_1(x,0,v_0,0), B_2) > 0$, then $i(A, D_{1,2}^\epsilon) = 0$.
If $\lambda_1(\Delta + f_1(x,0,v_0,0), B_2) < 0$, then $i(A, D_{1,2}^\epsilon) = 1$.

(iv) Suppose $\mu_1 < 0$ and $\mu_2 \leq 0$, or $\mu_1 \leq 0$ and $\mu_2 < 0$. Then $i(A, D_{1,2}^\epsilon) = 1$.

Proof. (i) Assume $\mu_1 > 0$ and $\mu_2 > 0$. By the choice of $\epsilon < t$, we see that the mapping $A_{1,1,\theta}$ has no fixed point on $\partial D_{1,2}^\epsilon$. Hence, we can use homotopic invariance to obtain $i(A, D_{1,2}^\epsilon) = i(A_{1,1,0}, D_{1,2}^\epsilon)$. As in the proof of Theorem 4.3, we see that if $A_{1,1,0}(u,v,w) = (u,v,w) \in D_{1,2}^\epsilon$, then $w = 0$ and either $||u|| < \epsilon$ or $||v|| < \epsilon$. Again, by the choice of ϵ, we must further have either $u = 0$ or $v = 0$. Consequently, we have $u = 0$ or u_0, $v = 0$ or v_0. This means that $(0,0,0), (u_0,0,0)$ and $(0,v_0,0)$ are the only three fixed points of the mapping $A_{1,0,0}$ in $D^{\epsilon}_{1,2}$. Hence, we obtain $i(A_{1,1,0}, D^{\epsilon}_{1,2}) = i(A_{1,1,0}, (0,0,0)) + i(A_{1,1,0}, (u_0, 0, 0)) + i(A_{1,1,0}, (0, v_0, 0))$. As described in the proof of Theorem 4.3, we can apply a modification of Theorem 4.1(i) to find $i(A_{1,1,0}, (0,0,0)) = 0$. If we review the proof of Theorem 4.2(i), in the case when $\lambda_1(\Delta + f_2(x, u_0, 0, 0), B_2) > 0$ and $(\lambda_1(\Delta + f_3(x, u_0, 0, 0), B_3) \neq 0$, the proof still applies if f_3 is changed to the trivial function. Thus we obtain $i(A_{1,1,0}, (u_0, 0, 0)) = 0$. By similar arguments, we apply a modification of Theorem 4.2(i) to deduce $i(A_{1,1,0}, (0, v_0, 0)) = 0$. The conclusion is obtained by adding the three indices.

Parts (ii), (iii) and (iv) of this theorem can be proved similarly.

Remark 4.2. Results analogous to Theorems 4.2, 4.3 and 4.4 are valid for all their natural symmetric variations. For example, we can calculate the fixed point indices of A on $D^{\epsilon}_{1,3}$ and $D^{\epsilon}_{2,3}$ in the same way as in Theorem 4.4.

Motivated by application to biological problems, we will assume some monotone properties concerning the functions $f_i, i = 1, 2, 3$.

(4.12)
The partial derivatives $D_v f_1, D_w f_1, D_u f_2, D_w f_2, D_u f_3, D_v f_3$ are all non-negative or non-positive for all $(x, u, v, w) \in \bar{\Omega} \times [0, \infty) \times [0, \infty) \times [0, \infty)$.

For convenience, we use the following symbols:

$u \to v$ if u preys on v, i.e. $D_v f_1 \geq 0$ and $D_u f_2 \leq 0$;

$u \sim v$ if u cooperates with v, i.e. $D_v f_1 \geq 0$ and $D_u f_2 \geq 0$;

$u \leftrightarrow v$ if u competes with v, i.e. $D_v f_1 \leq 0$ and $D_u f_2 \leq 0$.

For example, the diagram $u \sim v \leftrightarrow w \leftarrow u$ represent a model in which u, v cooperate, v, w compete, and w, u is a prey-predator pair with w as the prey. Consequently, each such model or diagram represents some hypotheses on the signs of the partial derivatives of the various $f_i' s$.

Recall a solution of problem (4.7) is called positive if each component is not identically zero and non-negative in $\bar{\Omega}$). We now give sufficient conditions for the existence of positive solutions of problem (4.7) under various combinations of predation, competition and symbiosis (described by diagram models above). The conditions are in terms of the signs of the principal eigenvalues of related elliptic operators for single equation. They are thus convenient for applications.

Theorem 4.5 (Main Theorem for Coexistence of Three Species). *Assume hypotheses (4.8), (4.10), (4.11) and (4.12) for $f_i, i = 1, 2, 3$. Suppose that for each of the following models, the corresponding conditions are satisfied. Then the problem (4.7) has a positive solution. (Here, M is the a-priori bound used in the definition of the set D.)*

Model 1. $u \to v \leftarrow w \sim u$. *Suppose* $\lambda_1(\Delta + f_1(x,0,0,w_0), B_1) > 0$, $\lambda_1(\Delta + f_2(x,M,0,M), B_2) > 0$, *and* $\lambda_1(\Delta + f_3(x,u_0,0,0), B_3) > 0$.

Model 2. $u \leftarrow v \leftarrow w \sim u$. *Let* $\mu_i > 0$ *for* $i = 1,2,3$. *Suppose either both* $\lambda_1(\Delta + f_1(x,0,v_0,0), B_1) > 0$ *and* $\lambda_1(\Delta + f_2(x,u_0,0,M), B_2) > 0$, *or both* $\lambda_1(\Delta + f_1(x,0,v_0,w_0), B_1) > 0$ *and* $\lambda_1(\Delta + f_2(x,0,0,M), B_2) > 0$.

Model 3. $u \leftarrow v \to w \sim u$. *Let* $\mu_1 > 0$ *and* $\mu_3 > 0$. *Suppose one of* $\lambda_1(\Delta + f_2(x,u_0,0,0), B_2) > 0$ *and* $\lambda_1(\Delta + f_2(x,0,0,w_0), B_2) > 0$ *is positive. Moreover, suppose* $\lambda_1(\Delta + f_1(x,0,v_2,0), B_1) > 0$ *and* $\lambda_1(\Delta + f_3(x,0,v_1,0), B_3) > 0$, *where* $v_1 := S_2(u_0,0)$ *and* $v_2 := S_2(0,w_0)$.

Model 4. $u \leftrightarrow v \leftarrow w \sim u$. *Let* $\mu_2 > 0$ *and* $\mu_3 > 0$. *Suppose* $\lambda_1(\Delta + f_1(x,0,v_0,w_0), B_1) > 0$, *and* $\lambda_1(\Delta + f_2(x,M,0,M), B_2) > 0$.

Model 5. $u \leftarrow v \leftrightarrow w \sim u$. *Let* $\mu_i > 0$ *for* $i = 1,2,3$. *Suppose* $\lambda_1(\Delta + f_1(x,0,v_0,0), B_1) > 0$. *Define* $v_1 := S_2(u_0,0)$. *Moreover, suppose either both* $\lambda_1(\Delta + f_2(x,u_0,0,M), B_2) > 0$ *and* $\lambda_1(\Delta + f_3(x,0,v_1,0), B_3) > 0$, *or both* $\lambda_1(\Delta + f_2(x,M,0,w_0), B_2) < 0$ *and* $\lambda_1(\Delta + f_3(x,M,v_0,0), B_3) < 0$.

Model 6. $u \leftrightarrow v \leftrightarrow w \sim u$. *Let* $\mu_i > 0$ *for* $i = 1,2,3$. *Suppose* $\lambda_1(\Delta + f_1(x,0,v_0,0), B_1) > 0$, $\lambda_1(\Delta + f_2(x,M,0,M), B_2) > 0$, *and* $\lambda_1(\Delta + f_3(x,0,v_0,0), B_3) > 0$.

Proof. For each model described in the statement, it suffices to show that $i(A,\tilde{D}) \neq 0$, where $\tilde{D} = \{(u,v,w) \in D : u,v,w > 0 \text{ in } \Omega\}$.

Model 1. The assumption $\lambda_1(\Delta + f_2(x,M,0,M)) > 0$ implies $\mu_2 > 0$. Suppose that both μ_1 and μ_3 are non-positive, then $u_0 \equiv w_0 \equiv 0$, contradicting the assumptions $\lambda_1(\Delta + f_1(x,0,0,w_0)) > 0$ and $\lambda_1(\Delta + f_3(x,u_0,0,0)) > 0$. Hence, we must have $\mu_2 > 0$ and at least one μ_1, μ_3 is positive. There are three cases as follows.

(i) Let $\mu_1 > 0, \mu_3 \leq 0$. Consider the mappings $A_{\theta,1,\theta}$, for $\theta \in [0,1]$. Since $\lambda_1(\Delta + f_2(x,M,0,M)) > 0$, the problem:

$$\Delta v + v f_2(x,M,v,M) = 0 \text{ in } \Omega, \quad B_2 v = 0 \text{ on } \partial\Omega, \tag{4.13}$$

has a unique positive solution $\bar{v}$ by Lemma 4.1. We have $||\bar{v}|| \geq \delta > 0$ for some δ. Let (u,v,w) be a fixed point of $A_{\theta,1,\theta}$ in D with $v \neq 0$. We have $\Delta v + v f_2(x,u,v,w) = 0$ in Ω. Since $f_2(x,u,v.w) \geq f_2(x,M,v,M)$, Lemma 4.2(ii) implies that $v \geq \bar{v}$. Choose $\epsilon \in (0,\delta)$, then by Theorem 4.3 (with th role of first and second components interchanged) we obtain $i(A, D_2^\epsilon) = 0$. Here, we have used hypothesis (4.12) and the interaction relation in model 1 to find that the hypothesis $t > 0$ is satisfied in Theorem 4.3.

Next, consider the mappings $A_{1,\theta,1}$. Let (u,v,w) be a fixed point of $A_{1,\theta,1}$ in D with $u \neq 0$ and $w \neq 0$. Since $\Delta u + u f_1(x,u,v,w) = 0$ in Ω and $f_1(x,u,v,w) \geq$

$f_1(x,u,0,0)$, by Lemma 4.2(ii), we have $u \geq u_0 \neq 0$. Let $\bar{w}$ be the unique positive solution of:

$$\Delta w + wf_3(x,u_0,0,w) = 0 \text{ in } \Omega, \quad B_3 w = 0 \text{ on } \partial\Omega. \tag{4.14}$$

Since $\Delta w + f_3(x,u,v,w) = 0$ and $f_3(x,u,v,w) \geq f_3(x,u_0,0,w)$, we obtain from Lemma 4.2(ii) again that $w \geq \bar{w}$. Choose a positive $\epsilon < min\{||u_0||, ||\bar{w}||\}$. We obtain from Theorem 4.4, with the second and third components interchanged, that $i(A, D^{\epsilon}_{1,3}) = 0$.

We next use Theorem 4.1(i) and Theorem 4.2(i) to deduce that $i(A,(0,0,0)) = i(A,(u_0,0,0)) = 0$. Then, unless there is already positive solution to the problem, we can obtain from the additivity of indices, Theorem A2-1(ii), that

$$i(A,\tilde{D}) = i(A,D) - i(A,D^{\epsilon}_{1,3}) - i(A,D^{\epsilon}_{2}) + i(A,(u_0,0,0)) + i(A,(0,0,0))$$

$$= 1 - 0 - 0 + 0 + 0 = 1.$$

Note that $(u_0,0,0)$ and $(0,0,0)$ are members of both $D^{\epsilon}_{1,3}$ and D^{ϵ}_{2}.

(ii) $\mu_1 \leq 0, \mu_3 > 0$. This is symmetric to case (i). The proof is the same with the role of the first and third components interchanged.

(iii) Both $\mu_1 > 0$ and $\mu_3 > 0$, with both u_0 and w_0 non-trivial. As in the proof above for case (i), we can show that $i(A,D^{\epsilon}_{i}) = 0$ for sufficiently small positive ϵ, $i = 1,2,3$, by means of Theorem 4.3. We can also similarly show by means of Theorem 4.1 and Theorem 4.2 that $i(A,(0,0,0)) = i(A,(u_0,0,0)) = i(A,(0,v_0,0)) = i(A,(0,0,w_0)) = 0$. Again, by the additivity of indices, we deduce for our problem that $i(A,\tilde{D}) = i(A,D) - i(A,D^{\epsilon}_{1}) - i(A,D^{\epsilon}_{2}) - i(A,D^{\epsilon}_{3}) + i(A,(u_0,0,0)) + i(A,(0,v_0,0)) + i(A,(0,0,w_0)) + 2i(A,(0,0,0)) = 1 - 0 - 0 - 0 + 0 + 0 + 0 + 0 = 1$.

Model 2. As in the proof in Model 1, we can use Theorem 4.3 to deduce $i(A,D^{\epsilon}_{3}) = 0$. We can also use Theorem 4.1(i) to deduce $i(A,(0,0,0) = 0$, and use Theorem 4.2(i) to deduce $i(A,(u_0,0,0)) = i(A,(0,v_0,0)) = i(A,(0,0,w_0) = 0$

(i) Suppose both $\lambda_1(\Delta + f_1(x,0,v_0,0), B_1) > 0$ and $\lambda_1(\Delta + f_2(x,u_0,0,M), B_2) > 0$. We consider the mapping $A_{1,1,\theta}$ in D; and assume that there exist fixed points (u_n,v_n,w_n) of $A_{1,1,\theta}$ with $\theta_n \in [0,1], u_n \neq 0, v_n \neq 0$ and either $u_n \to 0$ or $v_n \to 0$. First suppose $u_n \to 0$. From the relation in the model, we have for $v \geq 0$, $f_2(x,u_n,v,w_n) \leq f_2(x,u_n,v,0) \to f_2(x,0,v,0)$. Thus by Lemma 4.2, we have $v_n \leq \bar{v}_n \to v_0$ where $\bar{v}_n$ is the unique positive solution of

$$\Delta v + vf_2(x,u_n,v,0) = 0 \text{ in } \Omega, \quad B_2 v = 0 \text{ on } \partial\Omega.$$

Therefore the model relation and hypothesis (4.12) imply that $\lambda_1(\Delta + f_1(x,0,v_n,0), B_1) \geq \lambda_1(\Delta + f_1(x,0,\bar{v}_n,0), B_1) \to \lambda_1(\Delta + f_1(x,0,v_0,0), B_1) > 0$. From Lemma 4.2, we obtain the contradiction that u_n cannot tend to 0.

Next, suppose $v_n \to 0$. Then for $u \geq 0$, we have $f_1(x, u, v_n, w_n) \geq f_1(x, u, v_n, 0) \to f_1(x, u, 0, 0)$. By Lemma 4.2, we have $u_n \geq \bar{u}_n \to u_0$, where $\bar{u}_n$ is the unique positive solution of

$$\Delta u + uf_1(x, u, v_n, 0) = 0 \text{ in } \Omega, \quad B_1 u = 0 \text{ on } \partial\Omega.$$

Therefore the model relation and hypothesis (4.12) imply that $\lambda_1(\Delta + f_2(x, u_n, 0, w_n), B_2) \geq \lambda_1(\Delta + f_2(x, \bar{u}_n, 0, M), B_2) \to \lambda_1(\Delta + f_2(x, u_0, 0, M), B_2) > 0$. From Lemma 4.2, we obtain the contradiction that v_n cannot tend to 0.

From the arguments above, we see that for small $\epsilon > 0$, $A_{1,1,\theta}$ has no fixed point on $\partial D^{\epsilon}_{1,2}$; and we obtain from Theorem 4.4 that $i(A, D^{\epsilon}_{1,2}) = 0$. Again, we obtain from the additivity of indices, Theorem A2-1(ii), that $i(A, \tilde{D}) = i(A, D) - i(A, D^{\epsilon}_{1,2}) - i(A, D^{\epsilon}_{3}) + i(A, (u_0, 0, 0)) + i(A, (0, v_0, 0)) + i(A, (0, 0, 0)) = 1 - 0 - 0 + 0 + 0 + 0 = 1$.

(ii) Suppose both $\lambda_1(\Delta + f_1(x, 0, v_0, w_0), B_1) > 0$ and $\lambda_1(\Delta + f_2(x, 0, 0, M), B_2) > 0$. Using similar arguments as in case (i), we can consider the mappings $A_{1,\theta,1}$ and $A_{\theta,1,1}$ to show that $i(A, D^{\epsilon}_{1,3}) = i(A, D^{\epsilon}_{2,3}) = 0$ for a small $\epsilon > 0$. Then, we use additivity of indices to obtain $i(A, \tilde{D}) = i(A, D) - i(A, D^{\epsilon}_{1,3}) - i(A, D^{\epsilon}_{2,3}) + i(A, D^{\epsilon}_{3}) + i(A, (0, 0, w_0)) = 1 - 0 - 0 + 0 + 0 = 1$.

Model 3. Using $\mu_1 > 0, \mu_3 > 0$ and the cooperating relation between u and w, we obtain from Theorem 4.1(i) and 4.2(i) that $i(A, (0, 0, 0)) = i(A, (u_0, 0, 0)) = i(A, (0, 0, w_0)) = 0$. We consider the following three cases.

(i) Suppose $\mu_2 > 0$. Assume there exist fixed points (u_n, v_n, w_n) of $A_{1,\theta_n,1}$ in D with $\theta_n \in [0, 1], u_n \neq 0, w_n \neq 0$ and $u_n \to 0$ or $w_n \to 0$. First suppose $u_n \to 0$. We can show as in the proof of Model 2(i) above that there are some $\bar{w}_n$ such that $w_n \leq \bar{w}_n \to w_0$. This would imply that there are $\bar{v}_n$ such that $v_n \leq \bar{v}_n \to v_2$. Therefore from the model relation, we have $\lambda_1(\Delta + f_1(x, 0, v_n, w_n), B_1) \geq \lambda_1(\Delta + f_1(x, 0, \bar{v}_n, 0), B_1) \to \lambda_1(\Delta + f_1(x, 0, v_2, 0), B_1) > 0$. This leads to the contradiction that u_n cannot tend to 0. In the same way, the assumption that $w_n \to 0$ leads to a contradiction. This shows that there exists $\epsilon > 0$ such that $A_{1,\theta,1}$ has no fixed point on $\partial D^{\epsilon}_{1,3}$ and we obtain from Theorem 4.4 that $i(A, D^{\epsilon}_{1,3}) = 0$. We can also readily show as in the proof of Model 1 that $i(A, D^{\epsilon}_{2}) = 0$. Then using additivity of indices, we obtain $i(A, \tilde{D}) = i(A, D) - i(A, D^{\epsilon}_{1,3}) - i(A, D^{\epsilon}_{2}) + i(A, (u_0, 0, 0)) + i(A, (0, 0, w_0)) + i(A, (0, 0, 0)) = 1 - 0 - 0 + 0 + 0 + 0 = 1$.

(ii) Suppose $\mu_2 \leq 0$ and $\lambda_1(\Delta + f_2(x, u_0, 0, 0), B_2) > 0$. Assume there exist fixed points (u_n, v_n, w_n) of $A_{1,1,\theta_n}$ in D with $\theta_n \in [0, 1], u_n \neq 0, v_n \neq 0$ and $u_n \to 0$ or $v_n \to 0$. Then we can show that $u_n \to 0$ implies that there are some $\bar{v}_n$ such that $v_n \leq \bar{v}_n \to v_2$. This in turn implies from the assumption $\lambda_1(\Delta + f_1(x, 0, v_2, 0), B_1) > 0$ that u_n cannot really tend to zero. On the other hand, we can show that $v_n \to 0$ implies that there are some $\bar{u}_n$ such that $u_n \geq \bar{u}_n \to u_0$. This in turn implies from the assumption $\lambda_1(\Delta + f_2(x, u_0, 0, 0), B_1) > 0$

that v_n cannot really tend to zero. Therefore we obtain from Theorem 4.4 that $i(A, D^\epsilon_{1,2}) = 0$ for some $\epsilon > 0$.

Assume there exist fixed points (u_n, v_n, w_n) of $A_{\theta_n,\theta_n,1}$ in D with $\theta_n \in [0,1], w_n \neq 0$, and $w_n \to 0$. We can show that $w_n \to 0$ implies that there are some $\bar{u}_n$ such that $u_n \leq \bar{u}_n \to u_0$. This implies that there are some $\bar{v}_n$ such that $v_n \leq \bar{v}_n \to v_1$. This is turn implies by means of the assumption $\lambda_1(\Delta + f_3(x, 0, v_1, 0), B_3) > 0$ that w_n cannot really tend to zero. Therefore we obtain from Theorem 4.3 that $i(A, D^\epsilon_3) = 0$ for some $\epsilon > 0$. Thus we obtain $i(A, \tilde{D}) = i(A, D) - i(A, D^\epsilon_{1,2}) - i(A, D^\epsilon_3) + i(A, (u_0, 0, 0)) + i(A, (0, 0, 0)) = 1 - 0 - 0 + 0 + 0 = 1$.

(iii) Suppose $\mu_2 \leq 0$ and $\lambda_1(\Delta + f_2(x, 0, 0, w_0), B_2) > 0$. This is symmetric to case (ii) above with the role of u_0 and w_0 interchanged. We show that $i(A, D^\epsilon_{2,3}) = i(A, D^\epsilon_1) = 0$ for some $\epsilon > 0$. This leads to $i(A, \tilde{D}) = 1$, as above.

Model 4. Using the assumptions $\mu_2 > 0$ and $\lambda_1(\Delta + f_2(x, M, 0, M), B_2) > 0$ we can show as in the proof of Model 1 that $i(A, D^\epsilon_2) = 0$. Since we also assume $\mu_3 > 0$, we obtain $i(A, (0, 0, w_0)) = i(A, (0, 0, 0)) = 0$.

We next show that $i(A, D^\epsilon_{1,3}) = 0$ for some small $\epsilon > 0$. Assume (u, v, w) is a fixed point of $A_{1,\theta,1}$ in D with $u \neq 0$ and $w \neq 0$. Using the model relation, we obtain by comparison that $v \leq v_0$ and $w \geq w_0$, and therefore $f_1(x, u, v, w) \geq f_1(x, u, v_0, w_0)$. Let $\bar{u}$ be the unique positive solution of

$$\Delta u + u f_1(x, u, v_0, w_0) = 0 \text{ in } \Omega, \quad B_1 u = 0 \text{ on } \partial\Omega.$$

By Lemma 4.2, we have $u \geq \bar{u}$. Choosing a positive $\epsilon < min\{||w_0||, ||\bar{u}||\}$, we can apply Theorem 4.4 to obtain $i(A, D^\epsilon_{1,3}) = 0$. Now if $\mu_1 \leq 0$, then $i(A, \tilde{D}) = i(A, D) - i(A, D^\epsilon_{1,3}) - i(A, D^\epsilon_2) + i(A, (0, 0, w_0)) + i(A, (0, 0, 0)) = 1-0-0+0+0 = 1$. If $\mu_1 > 0$, then Theorem 4.2 also implies $i(A, (u_0, 0, 0)) = 0$. Consequently, we have $i(A, \tilde{D}) = i(A, D) - i(A, D^\epsilon_{1,3}) - i(A, D^\epsilon_2) + i(A, (u_0, 0, 0)) + i(A, (0, 0, w_0)) + i(A, (0, 0, 0)) = 1 - 0 - 0 + 0 + 0 + 0 = 1$.

Model 5. We only outline the procedures. As in the proof of the models above, we use the model relations and assumptions here to first prove $i(A, D^\epsilon_1) = i(A, (u_0, 0, 0)) = i(A, (0, v_0, 0)) = i(A, (0, 0, 0)) = 0$.

(i) Suppose both $\lambda_1(\Delta + f_2(x, u_0, 0, M), B_2) > 0$ and $\lambda_1(\Delta + f_3(x, 0, v_1, 0), B_3) > 0$. Then we show that $i(A, D^\epsilon_{1,2}) = i(A, D^\epsilon_{1,3}) = i(A, D^\epsilon_1) = i(A, (u_0, 0, 0)) = 0$. Then, by the additivity of indices, we obtain $i(A, \tilde{D}) = i(A, D) - i(A, D^\epsilon_{1,2}) - i(A, D^\epsilon_{1,3}) + i(A, D^\epsilon_1) + i(A, (u_0, 0, 0)) = 1 - 0 - 0 + 0 + 0 = 1$.

(ii) Suppose that both $\lambda_1(\Delta + f_2(x, M, 0, w_0), B_2) < 0$ and $\lambda_1(\Delta + f_3(x, M, v_0, 0), B_3) < 0$. We use Theorem 4.4 to obtain $i(A, D^\epsilon_{2,3}) = 2$. We obtain $i(A, \tilde{D}) = i(A, D) - i(A, D^\epsilon_{2,3}) - i(A, D^\epsilon_1) + i(A, (0, v_0, 0)) + i(A, (0, 0, w_0)) + i(A, (0, 0, 0)) = 1 - 2 - 0 + 0 + 0 + 0 = -1$.

Model 6. We first show that $i(A, D^\epsilon_1) = i(A, D^\epsilon_2) = i(A, D^\epsilon_3) = i(A, (u_0, 0, 0)) =$

$i(A,(0,v_0,0)) = i(A,(0,0,w_0)) = i(A,(0,0,0)) = 0$. We thus obtain $i(A,\tilde{D}) = i(A,D) - i(A,D_1^\epsilon) - i(A,D_2^\epsilon) - i(A,D_3^\epsilon) + i(A,(u_0,0,0)) + i(A,(0,v_0,0)) + i(A,(0,0,w_0)) + 2i(A,(0,0,0)) = 1 - 0 - 0 - 0 + 0 + 0 + 0 + 0 = 1$.

Remark 4.3. The results above are obtained from Liu [160]. In [160], the following models are considered, and results are also obtained with similar proof. Assume hypotheses (4.8), (4.10), (4.11) and (4.12) for $f_i, i = 1,2,3$. Suppose that for each of the following models, the corresponding conditions are satisfied. Then the problem (4.7) has a positive solution. (Here, M is the a-priori bound used in the definition of the set D.)

Model 7. $u \sim v \to w \sim u$. Suppose $\lambda_1(\Delta + f_1(x,0,v_0,0), B_1) > 0$, $\lambda_1(\Delta + f_2(x,u_0,0,0), B_2) > 0$, and $\lambda_1(\Delta + f_3(x,u_0,M,0), B_3) > 0$.

Model 8. $u \sim v \leftrightarrow w \sim u$. Let $\mu_i > 0$ for $i = 1,2,3$. Suppose either both $\lambda_1(\Delta + f_2(x,u_0,0,M), B_2) > 0$ and $\lambda_1(\Delta + f_3(x,u_0,M,0), B_3) > 0$, or both $\lambda_1(\Delta + f_2(x,M,0,w_0), B_2) < 0$ and $\lambda_1(\Delta + f_3(x,M,v_0,0), B_3) < 0$.

Model 9. $u \sim v \sim w \sim u$. Suppose at least one of $\lambda_1(\Delta + f_1(x,0,v_0,0), B_1)$ and $\lambda_1(\Delta + f_1(x,0,0,w_0), B_1)$ is positive, at least one of $\lambda_1(\Delta + f_2(x,u_0,0,0), B_2)$ and $\lambda_1(\Delta + f_2(x,0,0,w_0), B_2)$ is positive, and at least one of $\lambda_1(\Delta + f_3(x,u_0,0,0), B_3)$ and $\lambda_1(\Delta + f_3(x,0,v_0,0), B_3)$ is positive.

Model 10. $u \leftarrow v \leftarrow w \leftarrow u$. Let $\mu_i > 0$ for $i = 1,2,3$. Suppose $\lambda_1(\Delta + f_1(x,0,v_0,0), B_1) > 0$, $\lambda_1(\Delta + f_2(x,0,0,w_0), B_2) > 0$, and $\lambda_1(\Delta + f_3(x,u_0,v_0,0), B_3) > 0$.

Model 11. $u \to v \leftarrow w \leftarrow u$. Let $\mu_2 > 0$ and $\mu_3 > 0$. Define $u_1 = S_1(v_0,0)$ and $u_2 = S_1(0,w_0)$. Suppose $\lambda_1(\Delta + f_1(x,0,0,w_0), B_1) > 0$, $\lambda_1(\Delta + f_2(x,u_2,0,w_0), B_2) > 0$ and $\lambda_1(\Delta + f_3(x,u_1,0,0), B_3) > 0$.

Model 12. $u \leftrightarrow v \leftarrow w \leftarrow u$. Let $\mu_2 > 0$ and $\mu_3 > 0$. Define $u_1 = S_1(0,w_0)$. Suppose $\lambda_1(\Delta + f_1(x,0,v_0,w_0), B_1) > 0$, $\lambda_1(\Delta + f_2(x,u_1,0,w_0), B_2) > 0$, and $\lambda_1(\Delta + f_3(x,u_0,0,0), B_3) > 0$.

Model 13. $u \leftrightarrow v \to w \leftarrow u$. Let $\mu_i > 0$ for $i = 1,2,3$. Suppose $\lambda_1(\Delta + f_3(x,u_0,v_0,0), B_3) > 0$. Define $u_1 = S_1(0,w_0)$ and $v_1 = S_2(0,w_0)$. Suppose either $\lambda_1(\Delta + f_1(x,0,v_1,0), B_1) > 0$ and $\lambda_1(\Delta + f_2(x,u_1,0,0), B_2) > 0$, or $\lambda_1(\Delta + f_1(x,0,v_0,w_0), B_1) < 0$ and $\lambda_1(\Delta + f_2(x,u_0,0,w_0), B_2) < 0$.

Model 14. $u \to v \leftrightarrow w \leftarrow u$. Let $\mu_i > 0$ for $i = 1,2,3$. Define $u_1 = S_1(v_0,0)$ and $u_2 = S_1(0,w_0)$. Suppose $\lambda_1(\Delta + f_2(x,u_2,0,w_0), B_2) > 0$, and $\lambda_1(\Delta + f_3(x,u_1,v_0,0), B_3) > 0$.

Model 15. $u \leftrightarrow v \leftrightarrow w \leftarrow u$. Let $\mu_i > 0$ for $i = 1,2,3$. Define $u_1 = S_1(0,w_0)$. Suppose $\lambda_1(\Delta + f_1(x,0,v_0,0), B_1) > 0$, $\lambda_1(\Delta + f_2(x,u_1,0,w_0), B_2) > 0$, and $\lambda_1(\Delta + f_3(x,u_0,v_0,0), B_3) > 0$.

Model 16. $u \leftrightarrow v \leftrightarrow w \leftrightarrow u$. Let $\mu_i > 0$ for $i = 1, 2, 3$. Suppose $\lambda_1(\Delta + f_1(x, 0, v_0, w_0), B_1) > 0$, $\lambda_1(\Delta + f_2(x, u_0, 0, w_0), B_2) > 0$, and $\lambda_1(\Delta + f_3(x, u_0, v_0, 0), B_3) > 0$.

Remark 4.4. In some of the models in Theorem 4.5 and the above remark, the conditions for coexistence involve the knowledge of the a-priori bound M. They are thus sometimes not very readily applicable. In the next section, we consider many cases involving even more species when the a-priori bound M can be obtained from the coefficients of the equations. Consequently, the conditions for coexistence can be readily verified.

2.5 Positive Steady-States for Large Systems by Index Method

This section considers the existence of positive solutions for various systems of four nonlinear coupled elliptic partial differential equations subjected to zero Dirichlet boundary condition. In applications, the four components can be interpreted as concentrations of four interacting chemicals or populations. The four species can interact nonlinearly in many different ways, leading to various cases below. The positive solutions represent coexistence of all four species in equilibrium with each other.

The species are divided into two groups, with a pair of species within each group. In Theorems 5.1 and 5.2, we assume that the two groups interact with a predator-prey relation. Each species in the first group is a predator for the prey-species in the second group. Within each group, the pair of species interact with each other in a competitive (or cooperative) manner. On the microscopic scale of immunology, for example, killer and helper T lymphocytes stimulate each others growth and proliferation through chemical mediators. They both directly or indirectly eliminate bacteria or viruses, which may compete for resources such as host cellular products and proteins (cf. DeLisi [44] and Kuby [111]). We prove Theorems 5.1 and 5.2 using index calculation methods as in the previous section for three species. The method for calculating the cone indices of the mappings for three components in the last section does not apply immediately for the case of four components in this present section. The methods of calculation of the cone indices are extended to the case for four components in this section. The extended methods are then applied to study the various cases. Sufficient conditions are found for the existence of solutions with each component positive. In other theorems, we assume that the two groups interact with a cooperating relation. Within each group, the pair of species interact in a competing or cooperating manner. For simplicity in this section, we assume the interaction terms are of Volterra-Lotka type, which is common in many biological applications.

A systematic investigation of the existence of positive solutions by means of cone index method for systems with more than three equations should be of value in future research of complex biological models. The results give elegant conditions, in terms of the spectral property of simpler appropriately related operators on only one component, for the existence of positive coexistence states for the full system. Furthermore, the method of analysis here can be extended to include other boundary conditions and reactions more general than Volterra-Lotka type as in the previous section. We also deduce the a-priori bounds of the solutions directly in terms of the coefficients of the equations. Consequently, the conditions of coexistence are expressed directly in terms of the coefficients of the equations, and are thus much easier to apply. The presentation in this section follows the results in Leung [129].

More precisely, we consider the system of elliptic equations:

$$\Delta u_i + u_i[\,e_i + \sum_{i=1}^{4} a_{ij}u_j] = 0 \quad \text{in } \Omega; \qquad u_i = 0 \quad \text{on } \partial\Omega, \tag{5.1}$$

where e_i and a_{ij} are constants satisfying:

$$a_{ii} < 0 \qquad \text{and} \qquad e_i > 0 \qquad \text{for} \qquad i = 1, 2, 3, 4. \tag{5.2}$$

The constants e_i and $a_{ii}, i = 1,..,4$ represent the intrinsic growth rates and crowding effects of the corresponding species. The constants $a_{ij}, i \neq j$ are the interaction rates, whose signs will satisfying various assumptions according to the cases considered by the particular theorems. Ω is a bounded domain in R^N with smooth boundary. A solution of problem (1.1) is called positive if each component is not identically zero and non-negative in $\bar{\Omega}$.

We divide the species into two groups. Group I consists of the species $m = 1, 2$ and group II consists of species $n = 3$ and 4. In the first two theorems, we assume that Groups I and II have a predator-prey relationship, with species in I as predators and species in II as prey. More precisely, we assume that:

[C1] $\quad a_{m3}$ and a_{m4} are ≥ 0, for $m = 1, 2$,
$\quad a_{n1}$ and a_{n2} are ≤ 0, for $n = 3, 4$.

Within the two groups, we will consider 4 different cases. In the first case, the species in group I form a cooperating pair, and in group II also form a cooperating pair. More precisely, we assume

[A1] $\quad a_{12}$ and a_{21} are ≥ 0; $\quad a_{34}$ and a_{43} are ≥ 0.

In the second case, we assume species in group I form a cooperating pair, while in group II form a competing pair. That is

[A2] $\quad a_{12}$ and a_{21} are ≥ 0; $\quad a_{34}$ and a_{43} are ≤ 0.

In the third case, we assume species in group I form a competing pair, while in group II form a cooperating pair. That is

[A3] $\qquad a_{12}$ and a_{21} are ≤ 0; $\quad a_{34}$ and a_{43} are ≥ 0.

Finally in the fourth case, we assume species in group I form a competing pair, while in group II also form a competing pair. That is

[A4] $\qquad a_{12}$ and a_{21} are ≤ 0; $\quad a_{34}$ and a_{43} are ≤ 0.

Let c be a function defined on $\bar{\Omega}$, we will use the symbol $\lambda_1(\Delta + c)$ to denote the first eigenvalue for the eigenvalue problem: $\Delta u + cu = \lambda u$ in Ω, $u = 0$ on $\partial\Omega$. For each $i = 1, ..., 4$, if $\lambda_1(\Delta + e_i) > 0$, we will use u_i^0 to denote the unique positive solution of the problem: $\Delta u_i^0 + u_i^0[e_i + a_{ii}u_i^0] = 0$ in Ω, $u_i^0 = 0$ on $\partial\Omega$. Moreover, let $y_i := (0, ..., u_i^0, \ldots, 0)$ where each of the four component is zero except the i-th component as shown. For convenience, we define the following expressions:

$$B_4^1 = [a_{11}a_{22} - a_{12}a_{21}]^{-1}[(e_1 + a_{14}e_4/|a_{44}|)|a_{22}| + (e_2 + a_{24}e_4/|a_{44}|)a_{12}],$$

$$B_4^2 = [a_{11}a_{22} - a_{12}a_{21}]^{-1}[(e_1 + a_{14}e_4/|a_{44}|)a_{21} + (e_2 + a_{24}e_4/|a_{44}|)|a_{11}|],$$

$$B_3^1 = [a_{11}a_{22} - a_{12}a_{21}]^{-1}[(e_1 + a_{13}e_3/|a_{33}|)|a_{22}| + (e_2 + a_{23}e_3/|a_{33}|)a_{12}],$$

$$B_3^2 = [a_{11}a_{22} - a_{12}a_{21}]^{-1}[(e_1 + a_{13}e_3/|a_{33}|)a_{21} + (e_2 + a_{23}e_3/|a_{33}|)|a_{11}|].$$

The following theorem gives sufficient conditions for the coexistence of positive solution of problem (5.1) when the predators cooperate while the preys may either cooperate or compete. In Theorem 5.1, we will see in the proof that for each predator i = 1,2, B_4^i or B_3^i respectively represents a bound for the predator u_i when the prey u_4 or u_3 is the only one present.

Theorem 5.1 (Cooperating Predators with Preys which Cooperate or Compete). *(i) Assume interaction relations [C1] and [A1]. Suppose that*

$$a_{11}a_{22} > a_{12}a_{21} \qquad \textit{and} \qquad a_{33}a_{44} > a_{34}a_{43}, \tag{5.3}$$

then problem (5.1) has a positive solution if the following conditions are satisfied:

$$\lambda_1(\Delta + e_1) > 0, \qquad \lambda_1(\Delta + e_2) > 0, \tag{5.4}$$

$$\lambda_1(\Delta + e_3 - (|a_{31}|B_4^1 + |a_{32}|B_4^2) > 0, \qquad \textit{and} \tag{5.5}$$

$$\lambda_1(\Delta + e_4 - (|a_{41}|B_3^1 + |a_{42}|B_3^2) > 0. \tag{5.6}$$

(ii) Assume interaction relations [C1] and [A2]. Suppose that

$$a_{11}a_{22} > a_{12}a_{21}, \tag{5.7}$$

then problem (5.1) has a positive solution if the following conditions are satisfied:

$$\lambda_1(\Delta + e_1) > 0, \qquad \lambda_1(\Delta + e_2) > 0, \tag{5.8}$$

$$\lambda_1(\Delta + e_3 - (|a_{31}|B_4^1 + |a_{32}|B_4^2) - |a_{34}|e_4/|a_{44}|) > 0, \quad \textit{and} \tag{5.9}$$

$$\lambda_1(\Delta + e_4 - (|a_{41}|B_3^1 + |a_{42}|B_3^2) - |a_{43}|e_3/|a_{33}|) > 0. \tag{5.10}$$

The next theorem give sufficient conditions for the coexistence of positive solution for problem (5.1) when the predators compete while the prey may co-operate or compete. For convenience of stating the theorem, we define the following expressions:

$$\hat{B}_4^1 = |a_{11}|^{-1}[e_1 + a_{14}e_4/|a_{44}|], \qquad \hat{B}_4^2 = |a_{22}|^{-1}[e_2 + a_{24}e_4/|a_{44}|],$$
$$\hat{B}_3^1 = |a_{11}|^{-1}[e_1 + a_{13}e_3/|a_{33}|], \qquad \hat{B}_3^2 = |a_{22}|^{-1}[e_2 + a_{23}e_3/|a_{33}|],$$
$$K_3 = [a_{33}a_{44} - a_{34}a_{43}]^{-1}[e_3|a_{44}| + e_4a_{34}],$$
$$K_4 = [a_{33}a_{44} - a_{34}a_{43}]^{-1}[e_3a_{43} + e_4|a_{33}|].$$

Theorem 5.2 (Competing Predators with Preys which Cooperate or Compete). *(i) Assume interaction relations [C1] and [A3]. Suppose that*

$$a_{33}a_{44} > a_{34}a_{43}, \tag{5.11}$$

then problem (5.1) has a positive solution if the following conditions are satisfied:

$$\lambda_1(\Delta + e_1 - |a_{12}a_{22}^{-1}|[e_2 + a_{23}K_3 + a_{24}K_4]) > 0, \tag{5.12}$$

$$\lambda_1(\Delta + e_2 - |a_{21}a_{11}^{-1}|[e_1 + a_{13}K_3 + a_{14}K_4]) > 0, \tag{5.13}$$

$$\lambda_1(\Delta + e_3 - (|a_{31}|\hat{B}_4^1 + |a_{32}|\hat{B}_4^2)) > 0, \quad \textit{and} \tag{5.14}$$

$$\lambda_1(\Delta + e_4 - (|a_{41}|\hat{B}_3^1 + |a_{42}|\hat{B}_3^2)) > 0. \tag{5.15}$$

(ii) Assume interaction relations [C1] and [A4]. Then problem (5.1) has a positive solution if the following conditions are satisfied:

$$\lambda_1(\Delta + e_1 - |a_{12}a_{22}^{-1}|[e_2 + a_{23}e_3/|a_{33}| + a_{24}e_4/|a_{44}|]) > 0, \tag{5.16}$$

$$\lambda_1(\Delta + e_2 - |a_{21}a_{11}^{-1}|[e_1 + a_{13}e_3/|a_{33}| + a_{14}e_4/|a_{44}|]) > 0, \tag{5.17}$$

$$\lambda_1(\Delta + e_3 - (|a_{31}|\hat{B}_4^1 + |a_{32}|\hat{B}_4^2) - |a_{34}|e_4/|a_{44}|) > 0, \quad \textit{and} \tag{5.18}$$

$$\lambda_1(\Delta + e_4 - (|a_{41}|\hat{B}_3^1 + |a_{42}|\hat{B}_3^2) - |a_{43}|e_3/|a_{33}|) > 0. \tag{5.19}$$

In the proofs of the two theorems stated above, we will use indices of various mappings from the cone of non-negative functions into itself. In order to emphasize the main ideas of the proof of the theorems, the details for calculating these indices are explained in later lemmas. However, in order to start proving Theorem 5.1, we need the following lemma which gives a-priori bounds for two cooperative species under appropriate conditions.

Lemma 5.1. *Consider the following Dirichlet problem:*

$$\Delta v_i + v_i(b_i + g_i(x) + \sum_{j=1}^{2} c_{ij}v_j) = 0 \quad \text{in } \Omega, \quad v_i = 0 \quad \text{on } \partial\Omega, \quad \text{for } i = 1, 2, \tag{5.20}$$

where $g_i(x)$ are non-positive continuous functions on Ω, and b_i, c_{ij} are constants satisfying $b_i > 0, c_{ii} < 0$ for $i = 1, 2$, $c_{12} \geq 0, c_{21} \geq 0$. Suppose that $c_{11}c_{22} > c_{12}c_{21}$. Then any positive solution (v_1, v_2), with $v_i \in C^2(\bar{\Omega}), i = 1, 2$, must satisfy:

$$v_1 \leq [b_1|c_{22}| + b_2c_{12}]/[c_{11}c_{22} - c_{12}c_{21}], \quad \text{and}$$
$$v_2 \leq [b_1c_{21} + b_2|c_{11}|]/[c_{11}c_{22} - c_{12}c_{21}] \quad \text{in } \Omega.$$

Proof. On the $x - y$ plane, the two lines $b_i + c_{i1}x + c_{i2}y = 0, i = 1, 2$, intersect at (x_0, y_0) where $x_0 := (-b_1c_{22} + b_2c_{12})/(c_{11}c_{22} - c_{12}c_{21}), y_0 := (b_1c_{21} - b_2c_{11})/(c_{11}c_{22} - c_{12}c_{21})$. The assumptions on b_i and c_{ij} of this lemma implies that x_0 and y_0 are positive. Let k be a positive number satisfying, $c_{21}/|c_{22}| < k < |c_{11}|/c_{12}$. (Here, $|c_{11}|/c_{12}$ is replaced with $+\infty$ if $c_{12} = 0$). For each $\delta > 0$, the pair of constant functions $v_1^\delta := x_0 + \delta, v_2^\delta := y_0 + k\delta$ on $\bar{\Omega}$ satisfy: $\Delta v_i^\delta + v_i^\delta[b_i + g_i(x) + c_{i1}v_1^\delta + c_{i2}v_2^\delta] < 0$ in Ω, $v_i^\delta > 0$ on $\partial\Omega$, for i = 1, 2. That is, they form a family of coupled upper solutions for problem (5.20). For $M > 0$ sufficiently large, the positive solution (v_1, v_2) of problem (5.20) satisfies $v_i(x) < v_i^M, i = 1, 2$. Let $J := \{\delta \in (0, M]|$ for both $i = 1, 2$, $v_i(x) < v_i^\delta$ for all

$x \in \bar{\Omega}\}$. Suppose the set J has a positive glb $\bar{\delta} > 0$; and let there be a point $x \in \Omega$ where $v_i(x) = v_i^{\bar{\delta}}$ for some i. We may assume, without loss of generality, that i $= 1$. For $x \in \Omega, u \geq 0$, define $f_1(x,u) = u(b_1 + g_1(x) + c_{11}u)$; and let P be a large positive constant such that$|\partial f_1/\partial u| < P$ for all $x \in \bar{\Omega}, 0 \leq u \leq v_1^M$. Consider the expression:

$$
\begin{aligned}
&\Delta(v_1(x) - v_1^{\bar{\delta}}) - P(v_1(x) - v_1^{\bar{\delta}}) \\
&\quad = \Delta v_1(x) + v_1(x)[b_1 + g_1(x) + c_{11}v_1(x) + c_{12}v_2(x)] \\
&\qquad - \{\Delta v_1^{\bar{\delta}} + v_1^{\bar{\delta}}[b_1 + g_1(x) + c_{11}v_1^{\bar{\delta}} + c_{12}v_2^{\bar{\delta}}\} + f_1(x, v_1^{\bar{\delta}}) + v_1^{\bar{\delta}}c_{12}v_2^{\bar{\delta}} \\
&\qquad - f_1(x, v_1(x)) - v_1(x)c_{12}v_2(x) - P(v_1(x) - v_1^{\bar{\delta}}) > 0 \qquad \text{for all} \quad x \in \Omega.
\end{aligned}
$$

The last inequality is true due to the facts that $(v_1^{\bar{\delta}}, v_2^{\bar{\delta}})$ is a strict upper solution, $c_{12} \geq 0$, and the choice of large P. The maximum principle asserts that $v_1(x) \equiv v_1^{\bar{\delta}}$ on $\bar{\Omega}$. This contradiction implies that the positive glb $\bar{\delta}$ can be reduced, and cannot be positive. Thus, we must have $v_1(x) \leq x_0$ and $v_2(x) \leq y_0$ in $\bar{\Omega}$.

Remark 5.1. Note that the proof of the above lemma uses an extension of a sweeping principle to quasimonotone nondecreasing elliptic systems (cf. Theorem A3-9 in Chapter 6).

We will next prove Theorem 5.1(i) by the following procedure. Under the hypotheses of this part, we use Lemma 5.1 to obtain a bound for all non-negative solutions of (5.21) below. Let $C_0^+(\bar{\Omega})$ denotes the set of non-negative continuous functions on $\bar{\Omega}$, vanishing on the boundary $\partial\Omega$. We will construct a bounded set D in $[C_0^+(\bar{\Omega})]^4$ containing all solutions of (5.21). Then we define various subsets of D containing solutions with certain components identically zero. The solutions will be fixed points of appropriate positive compact mappings on D. We will show that the index of the mapping on D is equal to one, by homotopy invariance and normalization property (cf. Theorem A2-1 in Chapter 6). By appropriate deformations and homotopy invariance principle again we will show that the indices are zero on the various subsets of D described above. By the additive property of the indices of the maps on disjoint open subsets (cf. Theorem A2-1 in Chapter 6), we will conclude by index formula (5.24) below that there must exist a solution of (5.1) with each component positive.

Proof of Theorem 5.1(i). Assume [C1], [A1] and (5.3) to (5.6). Consider any non-negative solutions of the problem:

$$(5.21) \quad -\Delta u_i = \theta_i u_i[e_i + \sum_{i=1}^{4} a_{ij}u_j] \text{ in } \Omega, \quad u_i = 0 \text{ on } \partial\Omega, \text{ for } \quad i = 1, ..., 4,$$

for $\theta_i \in (0, 1]$. By Lemma 5.1, the second part of [C1] and second part of (5.3),

we have

$$(5.22)\qquad \begin{cases} u_3 \le [e_3|a_{44}| + e_4 a_{34}]/[a_{33}a_{44} - a_{34}a_{43}] := \bar{u}_3, \\ u_4 \le [e_3 a_{43} + e_4|a_{33}|]/[a_{33}a_{44} - a_{34}a_{43}] := \bar{u}_4. \end{cases}$$

For $i = 1, 2$, let $b_i = \theta_i[e_i + a_{i3}\bar{u}_3 + a_{i4}\bar{u}_4]$, $g_i(x) \equiv \theta_i[a_{i3}u_3 + a_{i4}u_4 - a_{i3}\bar{u}_3 - a_{i4}\bar{u}_4] \le 0$ and $c_{ij} = a_{ij}$, we apply Lemma 5.1 again to obtain a uniform bound for u_1 and u_2, for $\theta_i \in (0,1]$. Note that we have made use of the second condition of [C1] and first condition of (5.3). In case θ_3(or θ_4) is equal to zero, u_3(or u_4) will be the trivial function, and u_4(or u_3) will be bounded above by $e_4/|a_{44}|$(or $e_3/|a_{33}|$). The subsequent bound on u_1 and u_2 will be again established by Lemma 5.1 as before if both θ_1 and θ_2 are positive. If one of the θ_1 or θ_2 is zero, the corresponding u_1 or u_2 will be the trivial function, and the bound on the other component can be established by the corresponding scalar equation uniformly for $\theta_i \in [0,1]$. In any case, there is a constant $M > 0$ such that all components of all non-negative solutions of (5.21) must have values in $[0, M)$, uniformly for $\theta_i \in [0,1], i = 1,2,3,4$.

For any $t > 0$, let $E(t) := \{u \in C(\bar{\Omega}) : |u| < t\}$, and $\bar{E}(t)$ denotes its closure. For $\theta_i \in [0,1], i = 1,2,3,4$ and $P > 0$, define the operator $A_{\theta_1\theta_2\theta_3\theta_4}$: $[C(\bar{\Omega})]^4 \cap [\bar{E}(M)]^4 \to [C_0(\bar{\Omega})]^4$ by

$$A_{\theta_1\theta_2\theta_3\theta_4}(u_1, .., u_4) = (v_1, .., v_4)$$

where $v_i := (-\Delta + P)^{-1}(\theta_i u_i[e_i + \sum_{i=1}^4 a_{ij}u_j] + Pu_i)$. Here, the inverse operator is taken with homogeneous Dirichlet boundary condition on $\partial\Omega$. We can take P sufficiently large so that the operator $A_{\theta_1\theta_2\theta_3\theta_4}$ is positive, compact and Fréchet differentiable on $[C^+(\bar{\Omega})]^4 \cap [\bar{E}(M)]^4$. Let $D := [C_0^+(\bar{\Omega})]^4 \cap [E(M)]^4$, the bound on the solutions implies that these operators has no fixed point on ∂D (in the relative topology). We can further use a familiar cut-off procedure to extend $A_{\theta_1\theta_2\theta_3\theta_4}$ to be defined outside D as a compact positive mapping from the cone $K := [C_0^+(\bar{\Omega})]^4$ into itself (cf. Ladde, Lakshmikanthm and Vatsala [112]). For convenience, we will denote $i(A_{\theta_1\theta_2\theta_3\theta_4}, y) = index(A_{\theta_1\theta_2\theta_3\theta_4}, y, K)$ for a fixed point y of the map in K, and denote $i(A_{\theta_1\theta_2\theta_3\theta_4}, D) = index(A_{\theta_1\theta_2\theta_3\theta_4}, D, K) = i_K(A_{\theta_1\theta_2\theta_3\theta_4}, D)$ as in Section 2.4 (cf. Section 6.2 in Chapter 6). Let $A := A_{1111}$, we obtain by homotopy invariance that the cone indices of the mappings satisfy $i(A, D) = i(A_{1111}, D) = i(A_{0000}, D)$. From definition, the i-th component of $A_{0000}(u_1, .., u_4)$ is $(-\Delta + P)^{-1}(Pu_i)$. One readily verifies that $A_{0000}(u) \ne \lambda u$ for every $u \in \partial D$ and every $\lambda \ge 1$. Hence by Theorem A2-4 in Chapter 6, we conclude that $i(A, D) = i(A_{0000}, D) = 1$.

We will next define various subsets of D containing solutions with some components identically zero, and then proceed to calculate the indices of the mapping A on these subsets. For $i = 1, .., 4, \epsilon \in (0, M)$, let

$$D_i^\epsilon := \{u = (u_1, .., u_4) : u \in K, 0 \le u_i < \epsilon, 0 \le u_j < M \ for \ j \ne i\},$$

which is a "slice" in D containing all fixed point of A with small i-th component. Let

$$D^{\epsilon}_{i,j} := D^{\epsilon}_i \cup D^{\epsilon}_j, \text{ and } \hat{D}^{\epsilon}_{i,j} := D^{\epsilon}_i \cap D^{\epsilon}_j.$$

Note that $\partial D^{\epsilon}_{i,j} = \{u \in D : \min\{\|u_i\|, \|u_j\|\} = \epsilon,$ or $\min\{\|u_i\|, \|u_j\|\} \le \epsilon$ and $\max\{\|u_k\| : 1 \le k \le 4\} = M\}$, and $\partial \hat{D}^{\epsilon}_{i,j} = \{u \in D : \max\{\|u_i\|, \|u_j\|\} = \epsilon,$ or $\max\{\|u_i\|, \|u_j\|\} \le \epsilon$ and $\max\{\|u_k\| : k \ne i \text{ and } j\} = M\}$. For convenience, we will use the notation:

$$f_i(u_1, u_2, u_3, u_4) = e_i + \sum_{i=1}^{4} a_{ij} u_j \quad \text{for } i = 1, 2, 3, 4. \tag{5.23}$$

Consider the mapping $A_{\theta 11 \theta}$ in D. Suppose there exists a sequence of fixed points $(u^n_1, u^n_2, u^n_3, u^n_4), n = 1, 2, 3...$, of the map $A_{\theta_n 11 \theta_n}$ in $D, \theta_n \in [0,1]$, with $u^n_2 \not\equiv 0$ or $u^n_3 \not\equiv 0$. We have $\Delta u^n_3 + u^n_3 f_3(u^n_1, ..., u^n_4) = 0$ in $\Omega, u^n_i = 0$ on $\partial\Omega, i = 1, ..., 4$. If both u^n_2 and $u^n_3 \to 0$ in $C(\bar{\Omega})$, the equation for u^n_4 implies that $u^n_4 \le [e_4/|a_{44}|] + \delta$ for any small $\delta > 0$, provided n is sufficiently large. (Note that a_{41} and a_{42} are ≤ 0). The equations for u^n_1, u^n_2 and Lemma 5.1 then imply that $u^n_1 \le B^1_4 + \delta$, $u^n_2 \le B^2_4 + \delta$ for any small $\delta > 0$, provided n is sufficiently large. Thus for $\delta > 0$ sufficiently small, we have $f_3(u^n_1, \dots, u^n_4) > e_3 - (|a_{31}|B^1_4 + |a_{32}|B^2_4) - \delta$ for n sufficiently large. By assumption (5.5), we find $\lambda_1(\Delta + f_3(u^n_1, u^n_2, u^n_3, u^n_4)) > 0$, and the equation for u^n_3, which is in K, implies that $u^n_3 \equiv 0$ for all large n. (Note that (5.5) also implies that $e_3 > |a_{31}|B^1_4 + |a_{32}|B^2_4$). Then, we have $\Delta u^n_2 + u^n_2 f_2(u^n_1, u^n_2, 0, u^n_4) = 0$ in Ω, and the second condition in (5.4) implies that $\lambda_1(\Delta + f_2(u_1, u^n_2, 0, u^n_4)) > 0$ for n sufficiently large. Thus we have $u^n_2 \equiv 0$ for large n too. This contradicts the assumptions above on u^n_2 and u^n_3. Consequently, the number t:= inf.$\{$max.$\{\|u_2\|, \|u_3\|\} : (u_2, u_3) \not\equiv (0,0)$, where (u_1, u_2, u_3, u_4) is a fixed point of $A_{\theta 11 \theta}$ in D for some $\theta \in [0,1]\}$ must satisfy $t > 0$. Choosing $\epsilon \in (0, t)$, we obtain by Lemma 5.4 below that $i(A, \hat{D}^{\epsilon}_{2,3}) = 0$. (The lemma indicates that all the fixed points of $A_{\theta 11 \theta}$ in $\hat{D}^{\epsilon}_{2,3}$ have both their 2nd and 3rd components identically zero, and $A = A_{1111}$ can be deformed into A_{0110} by homotopy in $\hat{D}^{\epsilon}_{2,3}$. It then shows that the only fixed point of A_{0110} in $\hat{D}^{\epsilon}_{2,3}$ is (0,0,0,0), whose index is 0.)

Consider the mapping $A_{11\theta\theta}$ in D. Suppose that there exists a sequence of fixed points $(v^n_1, v^n_2, v^n_3, v^n_4), n = 1, 2, 3...$, of the map $A_{11\theta_n\theta_n}$ in $D, \theta_n \in [0,1]$, with both $v^n_1 \not\equiv 0$ and $v^n_2 \not\equiv 0$. The signs of a_{1j} implies that $\lambda_1(\Delta + f_1(0, v^n_2, v^n_3, v^n_4)) > \lambda_1(\Delta + e_1)$ which is > 0 by assumption (5.4). We can obtain by comparison from the equation satisfied by v^n_1 that they are uniformly bounded away from zero by a positive function, and thus cannot tend to zero as n tends to infinity. Similarly, we find $\lambda_1(\Delta + f_2(v^n_1, 0, v^n_3, v^n_4)) > \lambda_1(\Delta + e_2) > 0$, and deduce v^n_2 also cannot tend to zero as n tends to infinity. Consequently, the number $t^* := inf.\{\|u_1\|, \|u_2\|$: both $u_1 \not\equiv 0$ and $u_2 \not\equiv 0$, where (u_1, u_2, u_3, u_4) is a fixed point of $A_{11\theta\theta}$ in D, some $\theta \in [0,1]\}$ must satisfy $t^* > 0$. Further, the signs of a_{1j}, a_{2j} and assumptions (5.4) imply that $\lambda_1(\Delta + f_1(0, u_2^0, 0, 0)) > 0$ and $\lambda_1(\Delta + f_2(u_1^0, 0, 0, 0)) > 0$. Choosing $\epsilon \in (0, t^*)$, we can thus obtain by Lemma 5.3 below that $i(A, D_{1,2}^\epsilon) = 0$. (Lemma 5.3 shows that A can be deformed in $D_{1,2}^\epsilon$ to A_{1100} which has in $D_{1,2}^\epsilon$ three fixed points, all with index zero.)

Consider the mapping $A_{\theta\theta 11}$ in D. Suppose there exists a sequence of fixed points $(w_1^n, w_2^n, w_3^n, w_4^n), n = 1, 2, 3, \ldots$, of the map $A_{\theta_n\theta_n 11}$ in $D, \theta_n \in [0,1]$, with both $w_3^n \not\equiv 0$ and $w_4^n \not\equiv 0$. If $w_3^n \to 0$, the equation for w_4^n implies that $w_4^n \le e_4/|a_{44}| + \delta$ for any small $\delta > 0$ provided that n is large enough. The equations for w_1^n and w_2^n and Lemma 5.1 then imply that $w_1^n \le B_4^1 + \delta, w_2^n \le B_4^2 + \delta$ for any small δ provided n is large enough. (Note that B_4^i is a bound for $u_i, i = 1, 2$, with u_4 as the only prey, i.e. with u_3 absent). Thus for $\delta > 0$ sufficiently small, we have $f_3(w_1^n, \ldots, w_4^n) > e_3 - (|a_{31}|B_4^1 + |a_{32}|B_4^2) - \delta > 0$, for n sufficiently large. Hence we obtain $\lambda_1(\Delta + f_3(w_1^n, w_2^n, w_3^n, w_4^n)) > 0$ by assumption (5.5), and the equation for w_3^n implies that $w_3^n \equiv 0$ for all large n. This contradicts $w_3^n \not\equiv 0$, and thus w_3^n cannot tend to zero as n tends to infinity. On the other hand, if $w_4^n \to 0$, the equation for w_3^n implies that $w_3^n \le e_3/|a_{33}| + \delta$ for small $\delta > 0$ and n large enough. We continue to deduce in a symmetric way that $\lambda_1(\Delta + f_4(w_1^n, w_2^n, w_3^n, w_4^n)) > 0$ by assumption (5.6), leading to $w_4^n \equiv 0$ for all large n. We again conclude by contradiction that w_4^n cannot tend to zero. Consequently, the number $t^{**} := \inf.\{\|u_3\|, \|u_4\|$:both $u_3 \not\equiv 0$ and $u_4 \not\equiv 0$, where (u_1, u_2, u_3, u_4) is a fixed point of $A_{\theta\theta 11}$ in D, some $\theta \in [0,1]\}$ must satisfy $t^{**} > 0$. Further, the signs of a_{3j}, a_{4j} and assumptions (5.5), (5.6) imply that $\lambda_1(\Delta + f_3(0,0,0,u_4^0)) > 0$ and $\lambda_1(\Delta + f_4(0,0,u_3^0,0)) > 0$. Choosing $\epsilon \in (0, t^{**})$, we obtain by Lemma 5.3 again that $i(A, D_{3,4}^\epsilon) = 0$.

Under the conditions of this part, the above paragraphs show that for sufficiently small $\epsilon > 0$, all the fixed points of A in $\hat{D}_{2,3}^\epsilon$ must have both the 2nd and 3rd components identically zero. Because of the symmetry of relations between u_1 and u_2 with respect to u_3 and u_4, analogous property can be obtained for fixed points of A in $\hat{D}_{1,3}^\epsilon, \hat{D}_{1,4}^\epsilon$ and $\hat{D}_{2,4}^\epsilon$ (Here 1, 2 can be interchanged and 3, 4 can be interchanged). Also, the above paragraphs show that for sufficiently small $\epsilon > 0$, all the fixed points of A in $D_{1,2}^\epsilon$ must have either the 1st or 2nd components identically zero; and those in $D_{3,4}^\epsilon$ must have either the 3rd or 4th component identically zero. Let $\tilde{D} := \{u \in D$: each component of u is $\not\equiv 0\}$. Thus we obtain $i(A, \tilde{D}) = i(A, D) - i(A, D_{3,4}^\epsilon \cup D_{1,2}^\epsilon)$ for sufficiently small $\epsilon > 0$. Moreover, $i(A, D_{3,4}^\epsilon \cup D_{1,2}^\epsilon) = i(A, D_{3,4}^\epsilon) + i(A, D_{1,2}^\epsilon) - i(A, D_{3,4}^\epsilon \cap D_{1,2}^\epsilon)$. Deducing from Venn diagrams, using the additive property of the indices of a map on disjoint open sets (cf. Theorem A2-1 in Chapter 6), we also have:

$$\begin{aligned} i(A, D_{3,4}^\epsilon \cap D_{1,2}^\epsilon) &= i(A, \hat{D}_{1,3}^\epsilon \cup \hat{D}_{2,3}^\epsilon \cup \hat{D}_{1,4}^\epsilon \cup \hat{D}_{2,4}^\epsilon) \\ &= i(A, \hat{D}_{1,3}^\epsilon) + i(A, \hat{D}_{2,3}^\epsilon) + i(A, \hat{D}_{1,4}^\epsilon) + i(A, \hat{D}_{2,4}^\epsilon) \\ &\quad - i(A, y_4) - i(A, y_2) - i(A, y_1) - i(A, y_3) - 3\, i(A, (0,0,0,0)). \end{aligned}$$

Note that each $y_i, i = 1,2,3,4$ is inside exactly two of the sets $\hat{D}^\epsilon_{1,3}$, $\hat{D}^\epsilon_{2,3}$, $\hat{D}^\epsilon_{1,4}$, $\hat{D}^\epsilon_{2,4}$ and (0,0,0,0) is inside all the four sets.

Combining the above formulas we find

(5.24)
$$i(A,\tilde{D}) = i(A,D) - i(A,D^\epsilon_{3,4}) - i(A,D^\epsilon_{1,2}) + i(A,\hat{D}^\epsilon_{1,3}) + i(A,\hat{D}^\epsilon_{2,3})$$
$$+ i(A,\hat{D}^\epsilon_{1,4}) + i(A,\hat{D}^\epsilon_{2,4}) - \sum_{i=1}^4 i(A,y_i) - 3i(A,(0,0,0,0)).$$

We have proved $i(A,\hat{D}^\epsilon_{2,3}) = 0$ above. Using the symmetry of the relations between u_1 and u_2 with respect to u_3 and u_4, we can interchange the role of u_1 with u_2 and the role of u_3 with u_4 to deduce $i(A,\hat{D}^\epsilon_{1,3}) = i(A,\hat{D}^\epsilon_{1,4}) = i(A,\hat{D}^\epsilon_{2,4}) = 0$.

In order to use (5.24), it remains to find $i(A,y_i)$ and $i(A,(0,0,0,0))$. For convenience, we let $\mu_i = \lambda_1(\Delta + f_i(0,0,0,0)), i = 1,..,4$. The first part of (5.4) asserts that μ_1 is positive. The equation for u_1^0 implies $u_1^0 \leq e_1/|a_{11}| < B_4^1$; thus (5.5) implies $\lambda_1(\Delta + f_3(y_1)) > 0$. Similarly, we have $u_1^0 < B_3^1$ and (5.6) implies $\lambda_1(\Delta + f_4(y_1)) > 0$. The second part of (5.4) and $a_{21} \geq 0$ lead to $\lambda_1(\Delta + f_2(y_1)) > 0$. By part (ii) of Lemma 5.2 below, we obtain $i(A,y_1) = 0$. The symmetry of the relations among u_1 and u_2 would readily lead to $i(A,y_2) = 0$.

Condition (5.5) and comparison show that μ_3 is positive. Condition (5.4) and $a_{i3} \geq 0$ for $i = 1,2$ imply that $\lambda_1(\Delta + f_i(y_3)) > 0$ for $i = 1,2$. Condition (5.6) and $a_{43} \geq 0$ lead to $\lambda_1(\Delta + f_4(y_3)) > 0$. By Lemma 5.2, part (ii), we obtain $i(A,y_3) = 0$. We deduce in a symmetric way that $i(A,y_4) = 0$. Finally, conditions (5.4) to (5.6) imply $\mu_i > 0$ for each i. We conclude that $i(A,(0,0,0,0)) = 0$ from part (i) of Lemma 5.2 below.

Finally, we apply index formula (5.24). The above paragraphs show that every term on the right of the formula is equal to zero, except $i(A,D) = 1$. Consequently, we obtain $i(A,\tilde{D}) = 1$. That is there must exist at least one positive solution for problem (5.1). This completes the proof of part (i) of Theorem 5.1.

The proof of Theorem 5.1(ii) is similar to that of Theorem 5.1(i). The details will thus be omitted.

Proof of Theorem 5.2(i). Assume [C1], [A3], and (5.11) to (5.15). Consider the non-negative solutions of problem (5.21) in the present conditions. Since (5.11) is the same as the second part of (5.3), we can apply Lemma 5.1 to obtain inequalities (5.22). Note that $\bar{u}_3 = K_3$ and $\bar{u}_4 = K_4$ by definition. We then compare the first equation of (5.22) with the scalar problem $-\Delta z = \theta_1 z[e_1 + a_{13}K_3 + a_{14}K_4]$ in Ω, $z = 0$ on $\partial\Omega$, (using a_{11} and a_{12} are ≤ 0), we readily obtain a bound for u_1 on $\bar{\Omega}$. Similarly, we deduce a bound for u_2. Thus, as in the proof of Theorem 5.1, we obtain a constant $M > 0$ such that all components of all non-negative solutions of (5.21) must have values in $[0, M)$, uniformly for $\theta_i \in [0,1], i = 1,2,3,4$. We define sets $E(t), D, D_i^\epsilon, D_{i,j}^\epsilon, \hat{D}_{i,j}^\epsilon$ and the operators $A_{\theta_1\theta_2\theta_3\theta_4}$ as compact positive mappings from the cone $K := [C_0^+(\bar{\Omega})]^4$ into itself, with no fixed point on ∂D, exactly as in the proof of Theorem 5.1. Also we obtain in the same way that $i(A, D) = 1$.

Consider the mapping $A_{\theta 11\theta}$ in D. Suppose there exists a sequence of fixed points $(u_1^n, u_2^n, u_3^n, u_4^n), n = 1,2,3...,$ of the map $A_{\theta_n 11\theta_n}$ in $D, \theta_n \in [0,1]$, with $u_2^n \not\equiv 0$ or $u_3^n \not\equiv 0$. We have $\Delta u_3^n + u_3^n f_3(u_1^n, ..., u_4^n) = 0$ in $\Omega, u_i^n = 0$ on $\partial\Omega, i = 1, ..., 4$. If both u_2^n and $u_3^n \to 0$ in $C(\bar{\Omega})$, the equation for u_4^n implies that $u_4^n \leq [e_4/|a_{44}|] + \delta$ for any small $\delta > 0$, provided n is sufficiently large. The equation for u_1^n then implies that $u_1^n \leq \hat{B}_4^1 + \delta$ any small $\delta > 0$, provided n is sufficiently large. Similarly we obtain $u_2^n \leq \hat{B}_4^2 + \delta$. Thus for $\delta > 0$ sufficiently small, we have $f_3(u_1^n, \ldots, u_4^n) > e_3 - (|a_{31}| \hat{B}_4^1 + |a_{32}| \hat{B}_4^2) - \delta > 0$, for n sufficiently large. Thus we have $\lambda_1(\Delta + f_3(u_1^n, u_2^n, u_3^n, u_4^n)) > 0$ by assumption (5.14), and the equation for u_3^n implies that $u_3^n \equiv 0$ for all large n. Then, we have $\Delta u_2^n + u_2^n f_2(u_1^n, u_2^n, 0, u_4^n) = 0$ in Ω. For any $\delta > 0$, we verify that $\lambda_1(\Delta + f_2(u_1^n, u_2^n, 0, u_4^n)) > \lambda_1(\Delta + e_2 - |a_{21}a_{11}^{-1}|][e_1 + a_{14}K_4] + \delta)$ for n sufficiently large. Thus using (5.13) we obtain by comparison that $u_2^n \equiv 0$ for large n too. This contradicts the assumptions above on u_2^n and u_3^n. We deduce by contradiction as in the proof of Theorem 5.1 that $i(A, \hat{D}_{2,3}^\epsilon) = 0$ for $\epsilon > 0$ sufficiently small. By symmetry, we also obtain $i(A, \hat{D}_{1,3}^\epsilon) = i(A, \hat{D}_{1,4}^\epsilon) = i(A, \hat{D}_{2,4}^\epsilon) = 0$, as in the proof of Theorem 5.1.

Consider the mapping $A_{11\theta\theta}$ in D. Suppose there exists a sequence of fixed points $(v_1^n, v_2^n, v_3^n, v_4^n), n = 1,2,3...,$ of the map $A_{11\theta_n\theta_n}$ in $D, \theta_n \in [0,1]$, with both $v_1^n \not\equiv 0$ and $v_2^n \not\equiv 0$. Using (5.22) to estimate v_3^n and v_4^n, we then deduce from the second equation for v_2^n that $v_2^n \leq | a_{22}^{-1} | [e_2 + a_{23}K_3 + a_{24}K_4]$. Hence, we have $\lambda_1(\Delta + f_1(0, v_2^n, v_3^n, v_4^n)) > \lambda_1(\Delta + e_1 - | a_{12}a_{22}^{-1} | [e_2 + a_{23}K_3 + a_{24}K_4]) > 0$, by assumption (5.12). We can then compare with the equation satisfied by v_1^n to deduce that v_1^n cannot tend to zero as n tends to infinity. Similarly, we deduce that $\lambda_1(\Delta + f_2(v_1^n, 0, v_3^n, v_4^n)) > 0$ by assumption (5.13), and find v_2^n also cannot tend to zero as n tends to infinity. Consequently, the number $t^* :=$ inf.$\{\|u_1\|, \|u_2\|$: both $u_1 \not\equiv 0$ and $u_2 \not\equiv 0$, where (u_1, u_2, u_3, u_4) is a fixed point of $A_{11\theta\theta}$ in D, some $\theta \in [0,1]\}$ must satisfy $t^* > 0$. Further, assumptions (5.12) and (5.13) imply that $\lambda_1(\Delta + f_1(0, u_2^0, 0, 0)) > 0$ and $\lambda_1(\Delta + f_2(u_1^0, 0, 0, 0)) > 0$. Choosing $\epsilon \in (0, t^*)$, we can thus obtain by Lemma 5.3 below that $i(A, D_{1,2}^\epsilon) = 0$.

Consider the mapping $A_{\theta\theta 11}$in D. Suppose there exists a sequence of fixed points $(w_1^n, w_2^n, w_3^n, w_4^n), n = 1,2,3...,$ of the map $A_{\theta_n\theta_n 11}$ in $D, \theta \in [0,1]$, with both $w_3^n \not\equiv 0$ and $w_4^n \not\equiv 0$. If $w_3^n \to 0$, the equation for w_4^n implies that $w_4^n \leq e_4/|a_{44}| + \delta$ for any small $\delta > 0$ provided that n is large enough. The equation for w_1^n then imply that $w_1^n \leq \hat{B}_4^1 + \delta$ for any small δ provided n is large enough.

Similarly, we have $w_2^n \leq \hat{B}_4^2 + \delta$. Thus for $\delta > 0$ sufficiently small, we have $f_3(w_1^n, \ldots, w_4^n) > e_3 - (|a_{31}| \hat{B}_4^1 + |a_{32}| \hat{B}_4^2) - \delta > 0$, for n sufficiently large. Hence we obtain $\lambda_1(\Delta + f_3(w_1^n, w_2^n, w_3^n, w_4^n)) > 0$ by assumption (5.14), and the equation for w_3^n implies that $w_3^n \equiv 0$ for all large n. This contradicts $w_3^n \not\equiv 0$, and thus w_3^n cannot tend to zero as n tends to infinity. On the other hand, if $w_4^n \to 0$, the equation for w_3^n implies that $w_3^n \leq e_3/|a_{33}| + \delta$ for small $\delta > 0$ and n large enough. We continue to deduce in a symmetric way that $\lambda_1(\Delta + f_4(w_1^n, w_2^n, w_3^n, w_4^n)) > 0$ by assumption (5.15), leading to $w_4^n \equiv 0$ for all large n. We again conclude by contradiction that w_4^n cannot tend to zero. Further, the assumptions (5.14), (5.15) imply that $\lambda_1(\Delta + f_3(0, 0, 0, u_4^0)) > 0$ and $\lambda_1(\Delta + f_4(0, 0, u_3^0, 0)) > 0$. Consequently, we obtain by Lemma 5.3 again that $i(A, D_{3,4}^\epsilon) = 0$, for $\epsilon > 0$ sufficiently small.

As in the proof of Theorem 5.1, we next use index formula (5.24). It remains to show that under the conditions of the present theorem, we still have $i(A, (0, 0, 0, 0)) = i(A, y_i) = 0$, for $i = 1, \ldots, 4$. They are all readily verified by applying Lemma 5.2, using (5.12) to (5.15). For evaluating $i(A, y_1)$, we observe $\lambda_1(\Delta + f_2(y_1)) > \lambda_1(\Delta + e_2 - \mid a_{21} a_{11}^{-1} \mid e_1) > 0$ by (5.13), $\lambda_1(\Delta + f_3(y_1)) > \lambda_1(\Delta + e_3 - \mid a_{31} | | e_1 a_{11}^{-1} \mid) > \lambda_1(\Delta + e_3 - \mid a_{31} \mid \hat{B}_4^1) > 0$ by (5.14) and $\lambda_1(\Delta + f_4(y_1)) > \lambda_1(\Delta + e_4 - \mid a_{41} e_1 a_{11}^{-1} \mid) > \lambda_1(\Delta + e_4 - \mid a_{41} \mid \hat{B}_3^1) > 0$ by (5.15). Then we apply Lemma 5.2(ii) to verify that $i(A, y_1) = 0$. The other cases are similar or easier. Finally, applying formula (5.24), we obtain $i(A, \tilde{D}) = 1$, and complete the proof of part (i) of Theorem 5.2.

The proof of Theorem 5.2(ii) is similar to the proof of the theorems above. The details will thus be omitted here.

Remark 5.2. In order to complete the proof of Theorems 5.1 and 5.2, we now carefully justify the method for calculating the indices of the fixed points and the mappings over various sets used in the proofs above. They are described in detail in Lemmas 5.2 to 5.4 below. Lemma 5.2 and Lemma 5.3 are generalizations of results given in the last section, and Lemma 5.4 is different.

Consider the problem:

$$-\Delta u_i = \theta_i u_i f_i(u_1, u_2, u_3, u_4) \quad \text{in } \Omega, \qquad u_i = 0 \quad \text{on } \partial\Omega,$$

for $i = 1, \ldots, 4$, $\theta_i \in [0, 1]$. For $\theta_i \in [0, 1]$, $i = 1, \ldots, 4$ and $P > 0$, define the operator $A_{\theta_1\theta_2\theta_3\theta_4} : [C(\bar{\Omega})]^4 \to [C_0(\bar{\Omega})]^4$ by: $A_{\theta_1\theta_2\theta_3\theta_4}(u_1, \ldots, u_4) = (v_1, v_2, v_3, v_4)$, where $v_i := (-\Delta + P)^{-1}(\theta_i u_i f_i(u_1, \ldots, u_4) + P u_i)$. Here, the inverse operator is taken with homogeneous Dirichlet boundary condition on $\partial\Omega$. For simplicity, we use the abbreviations:

$$A = A_{1111},$$

$$A_\theta^{ij} = A_{\theta_1\theta_2\theta_3\theta_4} \text{ where } \theta_i = \theta_j = 1,\ \theta_k = \theta \text{ for } k \neq i \text{ or } j.$$

For convenience, we let $\lambda = \mu_i$ represent the first eigenvalue of the problem:

$$\Delta u + f_i(0,0,0,0)u = \lambda u \ \text{ in } \Omega,\ \ u = 0 \text{ on } \partial\Omega. \tag{5.25}$$

Remark 5.3. In the proofs of the main Theorems 5.1 and 5.2 above, it is shown that under the hypotheses of the theorems, the positive solutions of (5.1), or the fixed points of $A_{\theta_1\theta_2\theta_3\theta_4}$, are uniformly bounded for all $\theta_i \in [0,1]$. Let M be the uniform bound, $E(M) = \{ u \in C(\bar{\Omega}) : |u| < M.\}$ and $D = [C_0^+(\bar{\Omega})]^4 \cap [E(M)]^4$. It is shown that $A_{\theta_1\theta_2\theta_3\theta_4}$ has no fixed point on ∂D, and can be extended to be defined as a compact, Fréchet differentiable mapping from the cone $K = [C_0^+(\bar{\Omega})]^4$ into itself. We will assume these properties for all such operators in Lemmas 5.2 to 5.4 below.

Let j be an integer between 1 to 4. For simplicity, we denote $y_j = (0,..,u_j^0,..0)$ where every component is the zero function except the j-th component. Also, recall the definitions of the sets $D_i^\epsilon, D_{i,j}^\epsilon, \hat{D}_{i,j}^\epsilon$ given in the proof of Theorem 5.1(i) above.

Lemma 5.2 (Indices for the Trivial and Semi-trivial Solutions). *(i) If $max.\{\mu_1,...,\mu_4\} > 0$, and at most one of μ_i, $i = 1,...,4$, is zero, then $i(A,(0,0,0,0)) = 0$.*

(ii) Suppose that $\mu_j > 0$. If there exists $k \neq j$ such that $\lambda_1(\Delta + f_k(y_j)) > 0$, and $\lambda_1(\Delta + f_r(y_j)) \neq 0$ for all $r \neq j$ and k, then $i(A, y_j) = 0$.

Proof of (i). The proof is the same as Theorem 4.1 in Section 2.4. We outline the main idea again for convenience.

Suppose $\mu_j > 0$ and $\mu_i \neq 0$ for all other i's. For $y \in K$, define $K_y := \{p \in [C(\bar{\Omega})]^4 : y + sp \in K \text{ for some } s > 0\}$, and $S_y := \{p \in \bar{K}_y : -p \in \bar{K}_y\}$, as in Section 2.4. We have $\bar{K}_{(0,0,0,0)} = K$, $S_{(0,0,0,0)} = \{(0,0,0,0)\}$. The k-th component of $A'(0,0,0,0)u$ is $(-\Delta+P)^{-1}[f_k(0,0,0,0)+P]u_k$. Hence $[I - A'(0,0,0,0)]u = 0$ for $u \in K$ implies that $[\Delta + f_j(0,0,0,0)]u_j = 0$, $u_j \in C_0^+(\bar{\Omega})$. Thus the assumption $\mu_j > 0$ implies that $u_j = 0$. Similarly the assumption $\mu_i \neq 0$ implies that $u_i = 0$ for all other i's. Further, the assumption $\mu_j > 0$ and the continuity in $t \in [0,1]$ for $\lambda_1(\Delta + tf_j(0,...,0) + (t-1)P)$ imply that there exists a nontrivial function $\bar{u} \in C_0^+(\bar{\Omega})$ such that $(-\Delta+P)\bar{u} = t(f_j(0,0,0,0)+P)\bar{u}$, or $\bar{u} - t(-\Delta+P)^{-1}(f_j(0,0,0,0)+P)\bar{u} = 0 \in S_{(0,0,0,0)}$, for some $t \in (0,1)$. Thus it follows from Lemma 2.7(i) in Section 1.2 of Chapter 1 that $i(A,(0,0,0,0)) = 0$.

Next, suppose $\mu_j > 0$, $\mu_r = 0$ for some $1 \leq j, r \leq 4$, and $\mu_i \neq 0$ for all other i's. Define an operator A_t by modifying the r-th component of the operator A by changing $f_r(u_1, u_2, u_3, u_4)$ to $f_r(u_1, u_2, u_3, u_4) - t$. From the last paragraph, we obtain $i(A_t, (0, 0, 0, 0)) = 0$ for $t > 0$. Then by means of comparison and the nonincreasing property of f_r with respect to the r-th component, we show that (0,0,0,0) is an isolated fixed point of A_t uniformly for $t \geq 0$. We finally conclude that $i(A, (0, 0, 0, 0)) = i(A_t, (0, 0, 0, 0)) = 0$ by homotopy invariance of degree. See Theorem 4.1 in Section 2.4 for more details. This proves part (i)

Proof of (ii). Let $u = (u_1, \ldots, u_4)$ be any element in $[C(\bar{\Omega})]^4$. For $i \neq j$, the i-th component of $A'(y_j)u$ is $(-\Delta+P)^{-1}\{[f_i(y_j)+P]u_i\}$; the j-th component is $(-\Delta+P)^{-1}\{\sum_{i\neq j}[u_j^0(\partial f_j/\partial u_i)(y_j)u_i] + [u_j^0(\partial f_j/\partial u_j)(y_j) + f_j(y_j) + P]u_j\}$. One readily checks that $\bar{K}_{y_j} = C_0^+(\bar{\Omega})\oplus..\oplus C_0(\bar{\Omega})..\oplus C_0^+(\bar{\Omega})$ and $S_{y_j} = \{0\}\oplus..\oplus C_0(\bar{\Omega})..\oplus\{0\}$ where $C_0(\bar{\Omega})$ appear in the j-th component in both cases. Let $[I - A'(y_j)]\hat{u} = 0$ for $\hat{u} \in \bar{K}_{y_j}$. As in part (i), the assumption that $\lambda_1(\Delta + f_i(y_j)) \neq 0$ for $i \neq j$ implies that $\hat{u}_i = 0$ for $i \neq j$. Thus the j-th component can be written as $\Delta\hat{u}_j + (e_j + 2a_{jj}u_j^0)\hat{u}_j = 0$. Since $\lambda_1(\Delta + (e_j + 2a_{jj}u_j^0)) < 0$ by comparison, we must have $\hat{u}_j = 0$. Thus we have $\hat{u} = 0$.

As in the proof of part (i), the assumption $\lambda_1(\Delta + f_k(y_j)) > 0$ implies that there exists a nontrivial function $\bar{u}_k \in C_0^+(\bar{\Omega})$ satisfying $(-\Delta + P)\bar{u}_k - t(f_k(y) + P)\bar{u}_k = 0$. Let w be the column vector function on $\bar{\Omega}$ with $\bar{u}_k$ as its k-th component and zero function as all other components. Then w has the properties: $w \in \bar{K}_{y_j}\backslash S_{y_j}$ and $[I - tA'(y_j)]w \in S_{y_j}$. Thus the operator $A'(y_j)$ has the properties as described in Lemma 2.7(i) in Section 1.2, and we assert that $i(A, y_j) = 0$. This completes the proof of part (ii).

Lemma 5.3 (Indices for Unions of Two Thin Slices). *Suppose $i \neq j$ are integers with $1 \leq i, j \leq 4$ with $\mu_i > 0$ and $\mu_j > 0$. Let $t :=$ inf. $\{\|u_i\|, \|u_j\|$: both $u_i \not\equiv 0$ and $u_j \not\equiv 0$, where col. $u = (u_1, u_2, u_3, u_4)$ is a fixed point of A_θ^{ij} in D with $u_k \geq 0$ for $k = 1, 2, 3, 4$, some $\theta \in [0, 1]\}$. Let $y_k \in [C_0^+(\bar{\Omega})]^4$ with the k-th component as u_k^0 and all other components as the trivial function. Assume $t > 0$, and further that*

$$\lambda_1(\Delta + f_i(y_j)) > 0, \quad \lambda_1(\Delta + f_j(y_i)) > 0;$$

then for any $\epsilon \in (0, t)$, we have $i(A, D_{ij}^\epsilon) = 0$. (Recall that D_{ij}^ϵ is defined in the proof of Theorem 5.1(i)).

Proof. Since $0 < \epsilon < t$, the operator A_θ^{ij} in D has no fixed point on ∂D_{ij}^ϵ. By homotopy invariance, $i(A, D_{ij}^\epsilon) = i(A_0^{ij}, D_{ij}^\epsilon)$. Let $(\hat{u}_1, \hat{u}_2, \hat{u}_3, \hat{u}_4)$ be a nontrivial fixed point of A_0^{ij} in D_{ij}^ϵ. Then $(-\Delta + P)^{-1}[P\hat{u}_k] = 0$, for $k \neq i$ and j, implies that such $\hat{u}_k = 0$. The condition on ϵ implies either $\hat{u}_i = 0$ or $\hat{u}_j = 0$. Thus A_0^{ij} has three fixed points in D_{ij}^ϵ, namely, $(0, 0, 0, 0)$, y_i and y_j. Moreover, we find $i(A_0^{ij}, D_{ij}^\epsilon) = i(A_0^{ij}, (0, 0, 0, 0)) + i(A_0^{ij}, y_i) + i(A_0^{ij}, y_j)$. Applying a natural modification of Lemma 5.2(i) for the operator A_0^{ij}, we find $i(A_0^{ij}, (0, 0, 0, 0)) = 0$.

Also, applying a natural modification of Lemma 5.2(ii) for the operator A_0^{ij}, we obtain $i(A_0^{ij}, y_j) = 0$ and $i(A_0^{ij}, y_i) = 0$. This proves Lemma 5.3.

Lemma 5.4 (Indices for Intersections of Two Thin Slices). *Let $i \neq j$ be integers, $1 \leq i, j \leq 4$, and $t := inf.\{max.\{\|u_i\|, \|u_j\|\}, (u_i, u_j) \not\equiv (0,0),$, where (u_1, u_2, u_3, u_4) is a fixed point of A_θ^{ij} in D with $u_k \geq 0$ for $k = 1,2,3,4$, for some $\theta \in [0,1]\}$. Suppose that $t > 0$, and further assume: either $\mu_i > 0$ or $\mu_j > 0$. Then for any $\epsilon \in (0,t)$, we have $i(A, \hat{D}_{ij}^\epsilon) = 0$. (Recall that $\hat{D}_{ij}^\epsilon$ is defined in the proof of Theorem 5.1(i)).*

Proof. The assumption that $0 < \epsilon < t$ implies that all fixed points $\hat{u} = (\hat{u}_1, \hat{u}_2, \hat{u}_3, \hat{u}_4)$ of A_θ^{ij} in D contained in $\hat{D}_{ij}^\epsilon$ must satisfy $\hat{u}_i = \hat{u}_j = 0$; and they are bounded away from the other non-negative fixed points in D by $\partial \hat{D}_{ij}^\epsilon$. Applying homotopy invariance principle on the set $\hat{D}_{ij}^\epsilon$, we find $i(A, \hat{D}_{ij}^\epsilon) = i(A_1^{ij}, \hat{D}_{ij}^\epsilon) = i(A_0^{ij}, \hat{D}_{ij}^\epsilon)$. Note that for $k \neq i$ and j, the k-th component of $A_0^{ij}(u_1, .., u_4)$ is of the form $(-\Delta + P)^{-1}[Pu_k]$; thus any fixed point of A_0^{ij} in $\hat{D}_{ij}^\epsilon$ must have all the k-th component identically zero for $k \neq i$ and j too. Hence (0,0,0,0) is the only fixed point of A_0^{ij} in $\hat{D}_{ij}^\epsilon$, and $i(A_0^{ij}, \hat{D}_{ij}^\epsilon) = i(A_0^{ij}, (0,0,0,0))$. We then apply a modification of Lemma 5.2(i) to the operator A_0^{ij} with μ_k changed to $\hat{\mu}_k$, where $\hat{\mu}_k = \lambda_1(\Delta) < 0$ for $k \neq i$ and j, and $\hat{\mu}_i = \mu_i, \hat{\mu}_j = \mu_j$. Since at least one of $\hat{\mu}_i$ and $\hat{\mu}_j$ is positive, we conclude as in Lemma 5.2(i) that $i(A_0^{ij}, (0,0,0,0)) = 0$.

Theorems 5.1 and 5.2 only consider the special situation when the two groups of species interact in prey-predator relationship. The two groups can of course interact in many different ways. We next consider the coexistence of positive solutions for problem (5.1) when the two groups of species cooperate with each other. As before, there are two groups with a pair of species within each group. The cooperation between the groups means that the assumption [C1]is replaced by:

[C2] $\qquad a_{m3}$ and a_{m4} are ≥ 0 for $m = 1, 2;$

$\qquad a_{n1}$ and a_{n2} are ≥ 0 for $n = 3, 4.$

This will always be assumed in the next Theorem 5.3. Within each group, the species may compete or cooperate as expressed in [A4] and [A2] in Theorems 5.1 and 5.2. The case of [A3] is the same as [A2] if we interchange species 1 and 2 with 3 and 4. The case of [C2] together with [A1] would mean all species cooperate. It will then be unnecessary to classify the species into two groups for study.

Theorem 5.3 (Cooperating Groups with Competition or Cooperation within Each Group). *(i) Assume interaction relations [C2] and [A4].* Suppose that

$$|a_{mm}| > a_{m3} + a_{m4} \text{ for } m = 1, 2, \text{ and } |a_{nn}| > a_{n1} + a_{n2} \text{ for } n = 3, 4. \tag{5.26}$$

then the problem (5.1) has a positive solution if the following conditions are satisfied:

(5.27)
$$\lambda_1(\Delta + e_1 + a_{12}\hat{Q}_2) > 0, \text{ where } \hat{Q}_2 = max.\{|e_2/k_2|, |e_3/k_2|, |e_4/k_2|\},$$
$$k_2 = min.\{|a_{22}| - a_{23} - a_{24}, |a_{33}| - a_{32}, |a_{44}| - a_{42}\},$$

(5.28)
$$\lambda_1(\Delta + e_2 + a_{21}\hat{Q}_1) > 0, \text{ where } \hat{Q}_1 = max.\{|e_1/k_1|, |e_3/k_1|, |e_4/k_1|\},$$
$$k_1 = min.\{|a_{11}| - a_{13} - a_{14}, |a_{33}| - a_{31}, |a_{44}| - a_{41}\},$$

(5.29)
$$\lambda_1(\Delta + e_3 + a_{34}\hat{Q}_4) > 0, \text{ where } \hat{Q}_4 = max.\{|e_4/k_4|, |e_1/k_4|, |e_2/k_4|\},$$
$$k_4 = min.\{|a_{44}| - a_{41} - a_{42}, |a_{11}| - a_{14}, |a_{22}| - a_{24}\},$$

(5.30)
$$\lambda_1(\Delta + e_4 + a_{43}\hat{Q}_3) > 0, \text{ where } \hat{Q}_3 = max.\{|e_3/k_3|, |e_1/k_3|, |e_2/k_3|\},$$
$$k_3 = min.\{|a_{33}| - a_{31} - a_{32}, |a_{11}| - a_{13}, |a_{22}| - a_{23}\},$$

(ii) Assume interaction relations [C2] and [A2]. Suppose

$$\begin{aligned} &|a_{11}| > a_{12} + a_{13} + a_{14}, && |a_{22}| > a_{21} + a_{23} + a_{24}, \\ &|a_{nn}| > a_{n1} + a_{n2} && \text{for } n = 3, 4, \end{aligned} \tag{5.31}$$

then problem (5.1) has a positive solution if the following conditions are satisfied:

$$\lambda_1(\Delta + e_1) > 0, \qquad \lambda_1(\Delta + e_2) > 0, \tag{5.32}$$

(5.33)
$$\lambda_1(\Delta + e_3 + a_{34}R_4) > 0 \text{ where } R_4 = max.\{|e_4/\rho_4|, |e_1/\rho_4|, |e_2/\rho_4|\},$$
$$\rho_4 = min.\{|a_{44}| - a_{41} - a_{42}, |a_{11}| - a_{12} - a_{14}, |a_{22}| - a_{21} - a_{24}\},$$

(5.34)
$$\lambda_1(\Delta + e_4 + a_{43}R_3) > 0 \text{ where } R_3 = max.\{|e_3/\rho_3|, |e_1/\rho_3|, |e_2/\rho_3|\},$$
$$\rho_3 = min.\{|a_{33}| - a_{31} - a_{32}, |a_{11}| - a_{12} - a_{13}, |a_{22}| - a_{21} - a_{23}\}.$$

The proof of Theorem 5.3 uses the methods as described in the proof of Theorems 5.1 and 5.2. It can be found in [129], and thus the details will be omitted.

Remark 5.4. In Theorems 5.1 and 5.2 we consider the situation when species in group I consists of predators with species in group II as preys. The species within each group are assumed to cooperate or compete with each other. We have omitted the case when there may be further prey-predator relationship within one group. For example, species in group I are cooperative and the species in group II form a prey-predator pair. More generalizations of Theorems 5.1 and 5.2 can also be done for prey-predator groups when prey-predator relations occur within each group, or prey-predator within one group and cooperating relation within another. Other cases can be treated similarly. Some of the theorems may conceivably be proved by other methods. However, if we consider the first case, i.e. Theorem 5.1(i), it does not seem that one can readily prove the theorem by other methods. Note that B_3^i or B_4^i may not be a bound for predator species i when all the prey species 3 and 4 are present. Thus the condition in (5.5) and (5.6) may not be strong enough for proving the result by using other methods. Generalization of Theorem 5.3 is also possible for cooperative groups with prey-predator relations within each group. Since the methods are similar for these cases, the details will be omitted here. There is also the situation of a group of 3 interacting with a fourth species in the same way.

When there is a large number of m species in group I, each of which competes with n species in group II, existence of positive solutions is studied in Section 2.2 with bifurcation and upper-lower solutions methods. Within each group, there may be various types of structures. There are, however, limitations to the amount of interactions between the groups in order to prove the existence of positive solutions in Section 2.2. In order to use the technique of this section when there are groups of large numbers of m and n species, the methods in this section have to be extended more systematically. More interesting results remain to be found.

Remark 5.5. Further research should also address the issue of time stability and persistence of the systems. Some stability problems are considered in earlier sections. The problem of persistence will be considered in Chapter 4. Under the hypotheses that the various related principal eigenvalues are positive, it should be possible to obtain some information about the dynamics when the boundary equilibria are repellers relative to the positive cone. Some conclusions concerning persistence are possible, as explained in Chapter 4. It would also be interesting to treat the cases where some of the principal eigenvalues are negative.

2.6 Application to Reactor Dynamics with Temperature Feedback

In this section, we consider applications to a physical and engineering problem. Our methods of analysis will readily lead to useful results, as in the previous biological problems. We study a system of reaction-diffusion equations describing the dynamics of fission reactor with temperature feedback. There are m equations for the neutrons in m energy groups and a last temperature equation. Basic theory of reactor dynamics can be found in Duderstat and Hamilton [52] and Kasterberg [101]. We use the bifurcation method to find positive steady-states for the system which is not symmetric. We then analyze the linearized stability of the steady-state as a solution of the full system of $m+1$ parabolic equations. The asymptotic stability of the steady-state solution is proved by means of a stability theorem for sectorial operators. In the study of steady states and dynamics of nuclear fission reactors, it is crucial to understand the effect of temperature-dependent feedback, fission rates and reactor size on the behavior of the system. In Leung and Chen [131], [132] and Ortega [180], various nonlinear models concerning multigroup neutron fission reaction-diffusion with temperature feedback are investigated. In these articles, it is always assumed that the scattering and reaction rates are in some sense larger than the principal eigenvalue of the domain representing the reactor core. In Leung and Ortega [137], the scattering and reaction rates are only assumed to be positive, and the emphasis is on the bifurcation of a positive steady state at certain critical size of the reactor core. The stability of the bifurcating solution has also been investigated in [137]. The only important drawback in [137] is that the temperature is first expressed in terms of neutron-fluxes, and then substituted back into the first m equations for the m energy groups of neutron-fluxes. The model is thus implicitly assuming that the temperature is changing in a faster time-scale. This section treats the $m+1$ equations simultaneously, without eliminating the last temperature equation as in [137]. Here, we follow the presentation in Leung and Villa [141]. The stability of the positive steady-state here is considered for the full system of $m+1$ equations. Thus the theory is more elegant and less restrictive. It will be found that the decreasing property of the reaction coefficients with respect to temperature is crucial for the stability of the system (cf. hypothesis [H6] below). Such property may be achieved by means of control rods in the reactor.

More precisely, we first consider the following elliptic system with Dirichlet boundary conditions on a domain with various sizes:

(6.1)
$$\begin{cases} \Delta \hat{u}_i(x) + \sum_{j=1}^{m+1} \hat{H}_{ij}(x, \hat{u}_{m+1})\hat{u}_j(x) = 0 & \text{for all } x \in k\Omega, i = 1, ..., m+1, \\ \hat{u}_i(x) = 0 & \text{for all } x \in \partial(k\Omega), i = 1, ..., m+1, \end{cases}$$

where $k\Omega = \{x = ky : y \in \Omega\}, k > 0, \Omega$ is a fixed domain in R^N, and $m \geq 2$. The domain $k\Omega$ represents the reactor core; $\hat{u}_j(x), j = 1, ..., m$ is the neutron-flux of the j-th energy group; and $\hat{u}_{m+1}(x)$ denotes the temperature. $\hat{H}_{ij}, i, j = 1, ..., m$ represent the temperature-dependent fission and scattering rates of various energy groups; Δ is the Laplacian operator with x as independent variable. The function $\hat{H}_{m+1,m+1}$ denotes the cooling coefficient, and $\hat{H}_{m+1,j}$, $j = 1, ..., m$, denotes the rate of temperature increase due to neutrons in group j. Consequently, we should have $\hat{H}_{m+1,m+1} \leq 0$ and $\hat{H}_{ij} \geq 0$ for all $(i, j) \neq (m+1, m+1)$. We will determine the parameter k when positive steady-state will bifurcate from the trivial solution, and will thus find the critical size of the reactor core.

With the change of variable $x = ky$, the problem (6.1) is transformed into

(6.2)
$$\begin{cases} \Delta_y u_i(y) + \lambda \sum_{i=1}^{m+1} H_{ij}(y, u_{m+1}(y)) u_j(y) = 0, & \text{for } y \in \Omega, i = 1, \ldots, m+1. \\ u_i(y) = 0, & \text{for } y \in \partial\Omega, i = 1, \ldots, m+1, \end{cases}$$

which is the Dirichlet problem for a fixed domain Ω, where $\lambda = k^2 > 0$. Here, $u_i(y) = \hat{u}_i(x) = \hat{u}_i(ky)$, and

$$H_{ij}(y, u_{m+1}(y)) = \hat{H}_{ij}(x, \hat{u}_{m+1}(x)) = \hat{H}_{ij}(ky, \hat{u}_{m+1}(ky));$$

Δ_y is the Laplacian on the y-variable; and for convenience, we will not display this variable in the following context. We will obtain a positive steady state for (6.2) for certain value of λ; and consider the stability of this steady state as a solution of the nonlinear parabolic system:

(6.3)
$$\begin{cases} \partial u_i / \partial t = \Delta u_i + \lambda \sum_{j=1}^{m+1} H_{ij}(y, u_{m+1}(y,t)) u_j(y,t) & \text{for } (y,t) \in \Omega \times (0, \infty), \\ u_i(y,t) = 0 & \text{for } (y,t) \in \partial\Omega \times (0, \infty), i = 1, ..., m+1, \\ u_i(y, 0) = u_0^i(y) & \text{for } y \in \bar{\Omega}, i = 1, ..., m+1. \end{cases}$$

To fixed ideas, we assume Ω is a bounded domain in $R^N, N \geq 1$, with boundary $\partial\Omega$ of class $C^{2+\mu}$ for some $\mu \in (0, 1)$. For convenience, we denote

$$J = \{1, \ldots, m+1\}, \ \ B_{ij}(x) := H_{i,j}(x, 0), \ \ \text{for } x \in \bar{\Omega}, i, j \in J.$$

We will assume

[H1] $H_{ij}(., \eta) \in C^\mu(\bar{\Omega})$ uniformly for η in bounded subsets of R^1, $i, j \in J$; $B_{ij}(x) \geq 0$ in $\bar{\Omega}$, for all $(i, j) \neq (m+1, m+1)$;

[H2] $B_{ij} \not\equiv 0$, for $i \neq j$, $i, j = 1, ..., m$; $B_{m+1,j} \not\equiv 0, j = 1, \ldots, m$; $B_{i,m+1} \equiv 0$ for $i = 1, ..., m$.

Note that in (6.3), the coefficients H_{ij} depends on temperature u_{m+1}. Moreover, the last part in [H2], ($B_{i,m+1} \equiv 0, i = 1, \ldots, m$), means that the effect of temperature on the system enters only through changes in the fission and scattering coefficients $B_{ij}, j < m+1$. Also, note that so far, the crucial assumptions are made only at temperature $u_{m+1} = 0$, which can be normalized as the exterior temperature. More hypotheses will be added later when stability is investigated.

For convenience, we will use the following notations:

$$\mathcal{E} = \{u = col.(u_1, \ldots, u_{m+1}) | u_i \in C^{2+\mu}(\bar{\Omega}), u_i|_{\partial\Omega} = 0, i = 1, \ldots, m+1\},$$

$$\mathcal{F} = \{u = col.(u_1, \ldots, u_{m+1}) | u_i \in C^{\mu}(\bar{\Omega}), u_i|_{\partial\Omega} = 0, i = 1, \ldots, m+1\},$$

$$\mathcal{F}_1 = \{u = col.(u_1, \ldots, u_{m+1}) | u_i \in C^{1}(\bar{\Omega}), u_i|_{\partial\Omega} = 0, i = 1, \ldots, m+1\},$$

$$||\cdot||_{\mathcal{E}} = ||\cdot||_{C^{2+\mu}(\bar{\Omega})}, \quad ||\cdot||_{\mathcal{F}} = ||\cdot||_{C^{\mu}(\bar{\Omega})}, \quad ||\cdot||_{\mathcal{F}_1} = ||\cdot||_{C^{1}(\bar{\Omega})};$$

$H(x, \eta) = [H_{ij}(x, \eta)]$ is a $(m+1) \times (m+1)$ matrix for (x, η) in $\bar{\Omega} \times R^1$, and $\mathcal{P} = \{u = col.(u_1, \ldots, u_{m+1}) \in \mathcal{F}_1 | u_i \geq 0$ on $\bar{\Omega}\}$. Throughout this section we will use the symbol $\Delta_D^{-1} f$, for $f \in \mathcal{F}$ or $\mathcal{F}_1$, to denote the function $w \in \mathcal{E}$ such that $\Delta w = f$ in Ω. Applying Δ_D^{-1} to (6.2), it can be written as:

$$u + \lambda \Delta_D^{-1}[H(., u_{m+1})u] = 0 \tag{6.4}$$

where $u = \text{col.}(u_1, \ldots, u_{m+1})$. We will use the bifurcation method to find a positive solution to (6.4). Then, we analyze the linearized stability of this solution as a steady-state solution of (6.3). Finally, the asymptotic stability of the steady-state solution will be proved.

In order to find a positive solution bifurcating from zero for equation (6.4), we will first consider the corresponding linearized eigenvalue problem. We will need the following comparison Lemma 6.1 in order to prove the main theorems in this section. For convenience we let $J = \{1, 2, .., m+1\}$, and define the operator:

$$L_i = -\Delta + c_i(x)$$

for $i \in J$ where $c_i(x) \geq 0$ are functions in $C^{\mu}(\bar{\Omega}), 0 < \mu < 1$.

Lemma 6.1 (Comparison). *Let* $u, v \in [C^2(\Omega) \bigcap C^1(\bar{\Omega})]^{m+1}, u \not\equiv 0, v_i \geq 0, \not\equiv 0$ *in* Ω, *for* $i \in J$, *satisfy:*

$$\begin{cases} L_i[u_i(x)] = \sum_{j=1}^{m+1} p_{ij}(x) u_j(x), & \textit{for } x \in \Omega, i \in J, \\ u|_{\partial\Omega} = 0, \quad u = col.(u_1, \ldots, u_{m+1}); \end{cases} \tag{6.5}$$

$$(6.6)\qquad \begin{cases} L_i[v_i(x)] \geq \sum_{j=1}^{m+1} q_{ij}(x)v_j(x), & \textit{for } x \in \Omega, i \in J, \\ v = col.(v_1, ..., v_{m+1}), \end{cases}$$

where p_{ij} *and* q_{ij} *are bounded functions in* Ω. *Suppose that*

$$(6.7)\qquad \begin{array}{ll} q_{ij} \geq p_{ij} & \textit{in } \bar{\Omega} \ \textit{ for } i,j \in J, \textit{ and} \\ q_{ij}, p_{ij} \geq 0 & \textit{in } \bar{\Omega} \ \textit{ for all } i \neq j; \end{array}$$

then there exists $k \in J$ *and a real number* δ *such that:*

$$(6.8)\qquad \begin{array}{ll} v_k \equiv \delta u_k, \ p_{kj} \equiv q_{kj} & \text{in } \bar{\Omega} \text{ for all } j \in J, \\ \text{and } v_j - \delta u_j \geq 0 & \text{for all } j \in J. \end{array}$$

Proof. Let $K > 0$ be a positive constant such that $p_{ii} + K, q_{ii} + K > 0$ in $\bar{\Omega}$, for $i \in$ J. From (6.6) and (6.3), we have

$$L_i[v_i] + Kv_i \geq (q_{ii} + K)v_i + \sum_{j=1, j\neq i}^{m+1} q_{ij}v_j \geq 0.$$

Thus the maximum principles imply that

$$(6.9)\qquad v_i > 0 \text{ in } \Omega, \text{ and } (\partial v_i/\partial \eta)(\bar{x}) < 0 \text{ if } \bar{x} \in \partial\Omega \text{ and } v_i(\bar{x}) = 0.$$

Without loss of generality, we may assume that some component of u takes a positive value somewhere. Otherwise, replace u by $-u$. Since $u = 0$ on $\partial\Omega$, we can readily obtain from properties (6.9) that $v_i(x) - \epsilon u_i(x) > 0$ for some $\epsilon > 0$ and all $x \in \Omega, i \in J$. Let $\delta_i = \sup\{a : v_i - au_i > 0 \text{ in } \Omega\}$ for those $i \in J$ such that δ_i can be finitely defined. Define δ to be the minimum of such δ_i's. Thus $\delta = \delta_k$ for some k, and $0 < \delta_k < \infty, v_i - \delta u_i \geq 0$ in Ω for $i \in$ J. From (6.5) and (6.6), we find

$$(6.10)\qquad \begin{aligned} (L_k + K)(v_k - \delta_k u_k) \geq (K + p_{kk})(v_k - \delta_k u_k) + \textstyle\sum_{i=1,\neq k}^{m+1} p_{ki}(v_i - \delta_k u_i) \\ + \textstyle\sum_{i=1}^{m+1}(q_{ki} - p_{ki})v_i \geq 0 \quad \text{in } \Omega. \end{aligned}$$

Consequently, the maximum principle implies that $v_k - \delta u_k \equiv 0$ in Ω. Then, (6.10) further implies that we must have $q_{ki} \equiv p_{ki}$ for all $i \in J$.

For convenience, we define the following operators with $m + 1$ components:

$\tilde{L}_q \equiv (-\Delta, \ldots, -\Delta, -\Delta + q(x))$, where $q(x) \geq 0$ is a function in $C^{\mu}(\bar{\Omega})$;

$T \equiv \tilde{L}_q^{-1}(B) : [C^1(\bar{\Omega})]^{m+1} \to [C^1(\bar{\Omega})]^{m+1}$, so that for $u \in [C^1(\bar{\Omega})]^{m+1}$,

$w = Tu$ is the function which satisfies $\tilde{L}_q w = Bu$ and $w|_{\partial\Omega} = 0$.

We first prove the existence of positive eigenfunction for an appropriate linear system related to (6.4), under the restrictive condition $B_{m+1,m+1} \equiv 0$. This restriction will be modified later in Theorem 6.2.

Theorem 6.1. *Suppose that B satisfies $[H1], [H2]$ and $B_{m+1,m+1} \equiv 0$ then there exists $(\lambda_0, u^0) \in R \times \mathcal{E}, \lambda_0 > 0$ such that*

$$\tilde{L}_q[u^0] = \lambda_0 B u^0 \quad in\ \Omega,\ u^0 = 0 \quad on\ \partial\Omega, \tag{6.11}$$

with each component $u_i^0 > 0$ in Ω and $\partial u_i^0/\partial\eta < 0$ on $\partial\Omega$ for $i \in J$. Furthermore, the eigenfunction corresponding to the eigenvalue $1/\lambda_0$ for the operator T is unique up to a multiple. Also, The number $\lambda = \lambda_0$ is the unique positive number so that the problem $u = \lambda T u$ has a nontrivial non-negative solution for $u \in \mathcal{P}$. (Recall that B is the matrix whose (i, j) entries are defined as $B_{ij}(x) := H_{ij}(x, 0)$ for $i, j = 1, ..., m + 1$).

Proof. The operator $T : [C^1(\bar{\Omega})]^{m+1} \to [C^1(\bar{\Omega})]^{m+1}$ is completely continuous and positive with respect to the cone $\mathcal{P}$. Let $z = col.(z_1, \ldots, z_{m+1}) = col.((-\Delta_D)^{-1}(1), \ldots, (-\Delta_D)^{-1}(1))$. The functions z_i satisfies $z_i(x) > 0$ in $\Omega, z_i|_{\partial\Omega} = 0$ for $i \in J$. Define $v = T(z)$. Hypotheses $[H1], [H2]$ and the maximum principles imply that the components $v_i > 0$ in $\Omega, \partial v_i/\partial\eta < 0$ on $\partial\Omega$ for $i \in J$. Thus, there exists $\delta > 0$ such that $T(z) \geq \delta z$ with $z \in \mathcal{P}$. Theorem A3-10 in Chapter 6 asserts that there exists a nontrivial $u^0 = col.(u_1^0, \ldots, u_{m+1}^0) \in \mathcal{P}$ and $\rho_0 \geq \delta > 0$ such that $Tu^0 = \rho_0 u^0$ (i.e. (6.11) with $\lambda_0 = 1/\rho_0$.) The last component of (6.11) implies that we cannot have $u_i^0 \equiv 0$ in $\bar{\Omega}$ for all $i = 1, \ldots, m$. The maximum principle further implies that if $u_j^0 \not\equiv 0$, for $j \in J$, then $u_j^0(x) > 0$ for all $x \in \Omega$. We can then obtain from hypotheses $[H1], [H2]$ and the maximum principle that $u_i^0 > 0$ in Ω and $\partial u_i^0/\partial\eta < 0$ on $\partial\Omega$ for all $i \in J$.

Now, let $w = col.(w_1, \ldots, w_{m+1}) \not\equiv 0$ be such that $\tilde{L}_q[w] = \lambda_0 B w$. From Lemma 6.1, there must exist $\delta^* \in R$ and some $k \in J$ such that

$$u_k^0 \equiv \delta^* w_k \text{ and } u_j^0 - \delta^* w_j \geq 0 \text{ in } \bar{\Omega} \text{ for all } j \in J. \tag{6.12}$$

If there is an integer $r \in J$ such that

$$u_r^0(\bar{x}) - \delta^* w_r(\bar{x}) > 0 \text{ for some } \bar{x} \in \Omega, \tag{6.13}$$

then $-\Delta[u_r^0 - \delta^* w_r] = \lambda_0 \sum_{j=1}^{m+1} B_{rj}(u_j^0 - \delta^* w_j) \geq 0$ implies that $u_r^0 - \delta^* w_r > 0$ in Ω for the case $r \neq m+1$. For the case $r = m+1$, we have $(-\Delta + q)(u_{m+1}^0 - \delta^* w_{m+1}) \geq 0$, which also implies that $u_r^0 - \delta^* w_r > 0$ in Ω. We then consider the i-th equation in (6.11), for $i \neq r$; the hypothesis $[H2]$ implies that $-\Delta(u_i^0 - \delta^* w_i) \not\equiv 0$ if $i \neq m+1$ also, or $(-\Delta + q)(u_i^0 - \delta^* w_i) \not\equiv 0$ if $i = m+1$. Consequently, $u_i^0 - \delta^* w_i \not\equiv 0$ for each $i \neq r$. This contradicts the existence of an integer $k \in J$ such that (6.12) holds. This means that if (6.12) holds, there cannot exist an

$r \in J$ such that (6.13) holds. That is, we have $u^0 \equiv \delta^* w$. Finally, suppose that there is another $\lambda_1 > 0, \lambda_1 \neq \lambda_0$ so that $\tilde{u} = \lambda_1 T\tilde{u}$ for some $\tilde{u} \in \mathcal{P}, \tilde{u} \not\equiv 0$. We can deduce as before that $\tilde{u}_i > 0$ in $\Omega, \partial\tilde{u}_i/\partial\eta < 0$ on $\partial\Omega$ for $i \in J$. Then we can obtain from comparison Lemma 6.1 that $\lambda_1 = \lambda_0$. This completes the proof of the Theorem.

As described in the beginning of this section, the last component u_{m+1} denotes the temperature in the reactor, and the term $H_{m+1,m+1}(x,0) = B_{m+1,m+1}$ denotes the cooling coefficient. It is therefore physically reasonable to impose the hypothesis:

[H3] $H_{m+1,m+1}(x,0) = B_{m+1,m+1}(x) \leq 0$ for all $x \in \bar{\Omega}$.

To insure the existence of positive eigenfunction, we further assume that:

[H4] There exists some $k \in J$ such that $B_{kk}(x) > 0$ for some $x \in \Omega$.

The following theorem remove the restrictive assumption that $B_{m+1,m+1} \equiv 0$ in Theorem 6.1, and the entries of B are no longer all non-negative.

Theorem 6.2 (Positive Eigenvalue for the Linear Part). *Suppose B satisfies all the hypotheses [H1] to [H4]; then there exists $(\hat{\lambda}_0, v^0) \in R\times\mathcal{E}, \hat{\lambda}_0 > 0$, such that*

$$-\Delta[v^0] = \hat{\lambda}_0 B v^0 \ \ in \ \Omega, v^0 = 0 \quad on \ \partial\Omega, \tag{6.14}$$

with each component $v_i^0 > 0$ in $\Omega, \partial v_i^0/\partial\eta < 0$ on $\partial\Omega$ for $i \in J$. Furthermore, $1/\hat{\lambda}_0$ is an eigenvalue of the operator $(-\Delta_D)^{-1}B : [C^1(\bar{\Omega})]^{m+1} \to [C^1(\bar{\Omega})]^{m+1}$, with an one-dimensional nullspace for $(-\Delta_D)^{-1}B - (1/\hat{\lambda}_0)I$.

We will use Theorem 6.1 to prove this Theorem. For convenience, define $\tilde{B}$ to be the $(m+1)\times(m+1)$ matrix function on $\bar{\Omega}$ as follows:

$$\begin{aligned} &\tilde{B}_{ij}(x) = B_{ij}(x) \qquad \text{for } (i,j) \in J, (i,j) \neq (m+1,m+1), x \in \bar{\Omega}, \\ &\tilde{B}_{m+1,m+1} \equiv 0. \end{aligned} \tag{6.15}$$

For each $\lambda \geq 0$, define the $m+1$ component vector operator:

$$\tilde{L}_\lambda \equiv (-\Delta, \ldots, -\Delta, -\Delta - \lambda B_{m+1,m+1});$$

and consider the eigenvalue problem

$$\tilde{L}_\lambda u = \rho\tilde{B}u \ \text{ in } \Omega, \qquad u|_{\partial\Omega} = 0, \tag{6.16}$$

with eigenvalue ρ. Since $\tilde{B}$ satisfies the conditions in Theorem 6.1, problem (6.16) has a unique positive eigenvalue $\rho = \hat{\rho}(\lambda)$ with corresponding eigenfunction $u = u_\lambda = col.((u_\lambda)_1, \ldots, (u_\lambda)_{m+1})$, with $(u_\lambda)_i > 0$ in $\Omega, \partial(u_\lambda)_i/\partial\eta < 0$ on

$\partial\Omega$, for $i \in J$. The proof of Theorem 6.2 will follow readily from the next two lemmas.

Lemma 6.2. *Under the hypotheses of Theorem 6.2, the function $\hat{\rho}(\lambda)$ is bounded for all $\lambda \in [0, \infty)$.*

Proof. Let G be an open bounded set in Ω with its closure contained in Ω, and $B_{kk}(x) > 0$ for all $x \in G$. (Here, k is the integer described in $[H4]$). Let $\Phi \not\equiv 0$ be a C^∞ function with compact support contained in G. We clearly have $\int_G B_{kk}\Phi^2 dx > 0$. Let u_λ be as described above, and set $w_\lambda(x) = \ln(u_\lambda)_k(x)$ for $x \in \Omega$. Thus, we have in G that:

$$(6.17) \qquad -\Delta w_\lambda - \sum_{i=1}^{N}(\frac{\partial(w_\lambda)}{\partial x_i})^2 = \frac{\hat{\rho}(\lambda)}{(u_\lambda)_k}\sum_{j=1}^{m}\tilde{B}_{kj}(u_\lambda)_j \geq B_{kk}\hat{\rho}(\lambda).$$

Multiplying by Φ^2 and integrating over G, we obtain

$$(6.18) \qquad \int_G[-\Delta w_\lambda - \sum_{i=1}^{N}(\frac{\partial(w_\lambda)}{\partial x_i})^2]\Phi^2 dx \geq \hat{\rho}(\lambda)\int_G B_{kk}(x)\Phi^2 dx.$$

Integrating by parts gives

$$(6.19) \qquad \int_G < \Phi\nabla w_\lambda, 2\nabla\Phi - \Phi\nabla w_\lambda > dx \;\; \geq \;\; \hat{\rho}(\lambda)\int_G B_{kk}\Phi^2 dx.$$

Note that in G, we have

$$\begin{aligned} < \Phi\nabla w_\lambda, 2\nabla\Phi - \Phi\nabla w_\lambda > &= - < \nabla\Phi - \Phi\nabla w_\lambda, \nabla\Phi - \Phi\nabla w_\lambda > + < \nabla\Phi, \nabla\Phi > \\ &\leq < \nabla\Phi, \nabla\Phi > . \end{aligned}$$

Hence, (6.19) gives

$$(6.20) \qquad 0 < \hat{\rho}(\lambda) \leq \frac{\int_G < \nabla\Phi, \nabla\Phi > dx}{\int_G B_{kk}\Phi^2 dx}$$

for all $\lambda \in [0, \infty)$.

Lemma 6.3. *Under the hypotheses of Theorem 6.2, the function $\hat{\rho}(\lambda)$ is continuous on $\lambda \in [0, \infty)$.*

Proof. Let $\lambda^* \geq 0$ and λ_i be a sequence with $\lambda_i \to \lambda^*$. By Lemma 6.2, we may assume without loss of generality that $\hat{\rho}(\lambda_i) \to d$ for some $d \geq 0$. From Theorem 6.1, for each i there exists an eigenfunction $u_i \geq 0$, normalized to $\| u_i \|_\infty = 1$ satisfying

$$(6.21) \qquad \tilde{L}_{\lambda_i}u_i = \hat{\rho}(\lambda_i)\tilde{B}u_i \text{ in } \Omega, \; u|_{\partial\Omega} = 0.$$

Schauder's theory implies that $\{u_i\}$ is bounded in the $C^{1+\mu}(\bar{\Omega})$ norm. Using an embedding theorem we obtain without loss of generality that $u_i \to v$ for some $v \in [C^1(\bar{\Omega})]^{m+1}$ and $v \not\equiv 0, \geq 0$ in $\bar{\Omega}$. In the limit, we obtain

$$\tilde{L}_{\lambda^*} v = d\tilde{B}v \text{ in } \Omega, \; v|_{\partial\Omega} = 0. \tag{6.22}$$

From the maximum principle, we must have $v > 0$ in Ω. On the other hand, for $\lambda = \lambda^*$, Theorem 6.1 implies that there exists $v^* \in [C^{2+\mu}(\bar{\Omega})]^{m+1}, v^* > 0$ in Ω, and a number $\hat{\rho}(\lambda^*)$ satisfying

$$\tilde{L}_{\lambda^*} v^* = \hat{\rho}(\lambda^*)\tilde{B}v^* \text{ in } \Omega, \; v^*|_{\partial\Omega} = 0. \tag{6.23}$$

Using the comparison Lemma 6.1, (6.22) and (6.23), we can readily deduce by contradiction that $\hat{\rho}(\lambda^*) = d$.

Proof of Theorem 6.2. By Lemma 6.3, the function $\omega(\lambda) := \hat{\rho}(\lambda) - \lambda$ is continuous on $[0, \infty)$. From Theorem 6.1, we have $\omega(0) > 0$; and from Lemma 6.2, we have $\omega(\lambda) < 0$ for large $\lambda > 0$. Thus, there exists $\bar{\lambda}$ such that $\hat{\rho}(\bar{\lambda}) = \bar{\lambda}$ and (6.16) becomes

$$\tilde{L}_{\bar{\lambda}} v^0 = \hat{\rho}(\bar{\lambda})\tilde{B}v^0 \text{ in } \Omega, \; v^0|_{\partial\Omega} = 0 \tag{6.24}$$

for some v^0, with $v_i^0 > 0$ in Ω and $\partial v_i^0/\partial\eta < 0$ on $\partial\Omega, i \in J$. Comparing with (6.14), we clearly see that it is the same as (6.24) with $\hat{\lambda}_0 = \hat{\rho}(\bar{\lambda}) = \bar{\lambda}$. Finally, the dimension of the nullspace of $(-\Delta_D)^{-1}B - (1/\hat{\lambda}_0)I$ follows from (6.24) and Theorem 6.1. This completes the proof of Theorem 6.2.

In order to apply Crandall-Rabinowitz's bifurcation theorem to obtain positive solution for the nonlinear problem (6.4) we will have to analyze the range of the operator: $I - \hat{\lambda}_0(-\Delta_D)^{-1}$B. This leads to the study of the adjoint problem:

$$(-\Delta)v = \lambda B^T v \text{ in } \Omega, \; v|_{\partial\Omega} = 0. \tag{6.25}$$

Lemma 6.4. *Let B satisfies $[H1]$ to $[H4]$. Then problem (6.25) has a nontrivial solution when $\lambda = \hat{\lambda}_0$ with the corresponding eigenfunction $v(\hat{\lambda}_0) \equiv \hat{v}^0$satisfying $\hat{v}_i^0 > 0$ for $i = 1, ..., m$, and $\hat{v}_{m+1}^0 \equiv 0$ in Ω. Moreover, the solution $v(\hat{\lambda}_0)$ is unique up to a multiple. (Here, $\hat{\lambda}_0$ is the same as that defined in Theorem 6.2).*

Proof. Let $\bar{k}$ be a positive constant such that $\bar{k} + H_{m+1,m+1} > 0$ in Ω. Let $D^* = B^T + \bar{k}I$ and consider the problem $(-\Delta + \hat{\lambda}_0\bar{k})u = \lambda D^* u$ in $\Omega, u|_{\partial\Omega} = 0$. As in the proof of Theorem 6.1, we can show that there exists $(\tilde{\lambda}_0, u^*)$ such that

$$(-\Delta + \hat{\lambda}_0\bar{k}I)u^* = \tilde{\lambda}_0 D^* u^* = \tilde{\lambda}_0(B^T + \bar{k}I)u^* \quad \text{in } \Omega, \; u^*|_{\partial\Omega} = 0, \tag{6.26}$$

with $u^* \geq 0, \not\equiv 0$ in $\Omega, \tilde{\lambda}_0 > 0$. Hypotheses $[H1]$ and $[H2]$ imply that $u_i^* > 0$ in Ω for $i = 1, \ldots, m$. Recall that from (6.14), v^0 satisfies:

$$(6.27) \qquad (-\Delta + \hat{\lambda}_0 \bar{k} I) v^0 = \hat{\lambda}_0 (B + \bar{k} I) v^0 \text{ in } \Omega, \quad v^0|_{\partial\Omega} = 0.$$

Multiplying both sides of (6.27) by u^* and integrating over Ω, we obtain

$$(6.28) \qquad \begin{aligned} \hat{\lambda}_0 < (B + \bar{k}I)v^0, u^* > &= < (-\Delta + \hat{\lambda}_0 \bar{k} I) v^0, u^* > \\ &= < v^0, (-\Delta + \hat{\lambda}_0 \bar{k} I) u^* > \\ &= \tilde{\lambda}_0 < v^0, (B^T + \bar{k} I) u^* > \\ &= \tilde{\lambda}_0 < (B + \bar{k} I) v^0, u^* > . \end{aligned}$$

This implies that $\tilde{\lambda}_0 = \hat{\lambda}_0$, and (6.26) becomes

$$(6.29) \qquad (-\Delta)\hat{v}^0 = \hat{\lambda}_0 B^T \hat{v}^0 \text{ in } \Omega, \qquad \hat{v}^0|_{\partial\Omega} = 0,$$

where we label $\hat{v}_0 = u^*$. The $(m+1)$-th equation in (6.29) clearly implies that $\hat{v}^0_{m+1} \equiv 0$ in Ω, since the last row of B^T is $\equiv 0$ except the diagonal entry, which is ≤ 0. Applying Lemma 6.1 to the first m equations of (6.29), (i.e with m replacing $m+1$), we obtain the uniqueness of $u^* = \hat{v}^0 = v(\hat{\lambda}_0)$ up to a multiple.

As described above, we can consider a solution of problem (6.2) as a solution of (6.4). For convenience, we define an operator $F : R^+ \times \mathcal{F}_1 \to \mathcal{F}_1$ by

$$(6.30) \qquad F(\lambda, u) = u - \lambda(-\Delta_D)^{-1} H(., u_{m+1}) u \text{ for } (\lambda, u) \in R^+ \times \mathcal{F}_1.$$

Problem (6.4) can be written in the form

$$(6.31) \qquad F(\lambda, u) = 0.$$

Defining

$$(6.32) \qquad \begin{aligned} &L_0 : \mathcal{F}_1 \to \mathcal{F}_1 \text{ by } L_0 = I - \hat{\lambda}_0(-\Delta_D)^{-1} B, \\ &L_1 : \mathcal{F}_1 \to \mathcal{F}_1 \text{ by } L_1 = (\Delta_D)^{-1} B, \text{ and} \\ &G : R^+ \times \mathcal{F}_1 \to \mathcal{F}_1 \text{ by } G(\lambda, u) = -\lambda(-\Delta_D)^{-1}([H(., u_{m+1}) - B]u), \end{aligned}$$

equation (6.31) becomes

$$(6.33) \quad F(\lambda, u) := L_0 u + (\lambda - \hat{\lambda}_0) L_1 u + G(\lambda, u) = 0, \quad \text{for} \quad (\lambda, u) \in R^+ \times \mathcal{F}_1.$$

We clearly have $F(\lambda, 0) = 0$, for all $\lambda \in R^+$. Let $N(L_0)$ and $R(L_0)$ respectively denote the null space and range of L_0. We will show that they have the following properties:

Lemma 6.5. *Assume that B satisfies hypotheses $[H1]$ to $[H4]$ then*

(i) $N(L_0)$ *is one-dimensional, spanned by* v^0*;*

(ii) $\dim[\mathcal{F}_1/R(L_0)] = 1$*;*

(iii) $L_1 v^0 \notin R(L_0)$.

Proof. Part (i) was proved in Theorem 6.2. The operator L_0 can be extended naturally to $\tilde{L}_0$ with the set $[L^2(\Omega)]^{m+1}$ as its domain. The range of $\tilde{L}_0$ can be described by

$$\{z \in [L^2(\Omega)]^{m+1} : \textstyle\int_\Omega g \cdot z dx = 0, \quad \text{for all } g \text{ satisfying } \Delta g + \hat{\lambda}_0 B^T g = 0 \text{ in } \Omega, \ g|_{\partial\Omega} = 0\}.$$

By Lemma 6.4, all the g described above have to be a multiple of $\hat{v}^0$. Thus, by means of the mapping $u \to \int_\Omega \hat{v}^0 \cdot u dx$ from $\mathcal{F}_1$ onto R^1, we conclude that (ii) must be true.

To prove (iii), we first assume the contrary that $L_1 v^0 \in R(L_0)$. Then there exists $w \in [\mathcal{F}_1]^{m+1}$ such that

$$[I - \hat{\lambda}_0(-\Delta_D)^{-1}B]w = (\Delta_D)^{-1}Bv^0, \quad \text{i.e.} \ -\Delta w - \hat{\lambda}_0 Bw = -Bv^0.$$

Multiplying both sides by $\hat{v}^0$ (cf. equation (6.9)), and integrate over Ω, we obtain

$$< -\Delta w - \hat{\lambda}_0 Bw, \hat{v}^0 > \ = \ < -Bv^0, \hat{v}^0 >,$$

and thus

$$0 \ = < w, 0 > = \ < w, (-\Delta - \hat{\lambda}_0 B^T)\hat{v}^0 > \ = \ < -Bv^0, \hat{v}^0 > .$$

However, the assumptions on B, and the fact that $v_i^0 > 0$ for $i = 1, \ldots, m+1$, and $\hat{v}_i^0 > 0$ for $i = 1, \ldots, m$ imply that the expression on the right above is strictly negative. This contradiction implies that (iii) is valid.

In order to have enough smoothness for the function G given in (6.32), we now impose further hypotheses on the smoothness of $H(., \eta)$:

$[H5]$ $(\partial H_{ij}/\partial\eta)(., \eta)$ and $(\partial^2 H_{ij}/\partial\eta^2)(., \eta)$ are in $C^\mu(\bar{\Omega})$ uniformly in η in bounded subsets of R^1, for each $i, j \in J$.

Lemma 6.6. *(i) Assume hypotheses $[H1]$ to $[H5]$; then the Fréchet derivatives $D_2G, D_1G, D_{12}G$ exist and are continuous on $R^1 \times \mathcal{F}_1$. Moreover, we have*

$$D_2G(\lambda, u)w = \lambda(\Delta_D)^{-1}[H(., u_{m+1})w + \tilde{H}(., u_{m+1})uw_{m+1} - Bw] \tag{6.34}$$

for all $(\lambda, u) \in R^+ \times \mathcal{F}_1, w \in \mathcal{F}_1$ where $\tilde{H}(., u_{m+1}) = (\partial H_{ij}/\partial\eta)(., u_{m+1})$ is an $(m+1) \times (m+1)$ matrix function;

(ii) $||G(\lambda,u)||_{\mathcal{F}_1}/||u||_{\mathcal{F}_1} \to 0$ *as* $||u||_{\mathcal{F}_1} \to 0$ *uniformly in* λ *near* λ_0.

Proof. Direct calculation and estimating the range of $(-\Delta)^{-1}$ gives

$$\begin{aligned}
&||G(\lambda,w) - G(\lambda,u) - \lambda(\Delta_D)^{-1}[H(.,u_{m+1})(w-u) \\
&\qquad\qquad + \tilde{H}(.,u_{m+1})u(w_{m+1}-u_{m+1}) - B(w-u)]||_{\mathcal{F}_1} \\
&\le k\lambda||H(.,w_{m+1})w - H(.,u_{m+1})u - H(.,u_{m+1})(w-u) \\
&\qquad\qquad - \tilde{H}(.,u_{m+1})u(w_{m+1}-u_{m+1})||_\infty \\
&\le k\lambda \textstyle\sum_{i,j=1}^{m+1} ||H_{ij}(.,w_{m+1})w_j - H_{ij}(.,u_{m+1})w_j \\
&\qquad\qquad - \partial H_{ij}/\partial\eta(.,u_{m+1})u_j(w_{m+1}-u_{m+1})||_\infty \\
&\le k\lambda \textstyle\sum_{i,j=1}^{m+1}\{||[H_{ij}(.,w_{m+1}) - H_{ij}(.,u_{m+1}) \\
&\qquad\qquad - \partial H_{ij}/\partial\eta(.,u_{m+1})(w_{m+1}-u_{m+1})]u_j||_\infty \\
&\qquad\qquad + ||[H_{ij}(.,w_{m+1}) - H_{ij}(.,u_{m+1})][w_j-u_j]||_\infty\} \\
&\le k\lambda \textstyle\sum_{i,j=1}^{m+1}\{||\partial H_{ij}/\partial\eta(.,u_{m+1}+s_{ij}(w_{m+1}-u_{m+1})) \\
&\qquad\qquad - \partial H_{ij}/\partial\eta(.,u_{m+1})||_\infty||w_{m+1}-u_{m+1}||_\infty||u||_\infty \\
&\qquad\qquad + ||H_{ij}(.,w_{m+1}) - H_{ij}(.,u_{m+1})||_\infty||w-u||_\infty\} \\
&\le c(\lambda,\rho)||w-u||_\infty^2
\end{aligned}$$

for $||w||_{\mathcal{F}_1}, ||u||_{\mathcal{F}_1} \le \rho$, where s_{ij} above is a number between 0 and 1 and $c(\lambda,\rho)$ is a constant which depends on (λ,ρ). This inequality proves (6.34) and $G \in C^1(R^+ \times \mathcal{F}_1, \mathcal{F}_1)$. One can then similarly show the existence of $D_{12}G$ and D_1G.

For part (ii), we use Schauder's theory to obtain

$$\begin{aligned}
\tfrac{||G(\lambda,u)||_{\mathcal{F}_1}}{||u||_{\mathcal{F}_1}} &\le \bar{k}\tfrac{||[H(.,u_{m+1})-B]u||_\infty}{||u||_\infty} \\
&\le \bar{k}\textstyle\sum_{i,j=1}^{\infty} ||H_{ij}(.,u_{m+1}) - H_{ij}(.,0)||_\infty \to 0, \text{ as } ||u||_{\mathcal{F}_1} \to 0.
\end{aligned}$$

Continuing as in Lemma 6.6, one can show $F \in C^2(R^+ \times \mathcal{F}_1, \mathcal{F}_1)$. Using (6.34) and Lemma 6.6 again, we obtain

$$L_0 = D_2F(\hat{\lambda}_0, 0), \quad L_1 = D_{12}F(\hat{\lambda}_0, 0),$$

$$G(\lambda, 0) \equiv 0, \quad D_2G(\hat{\lambda}_0, 0) = D_{12}G(\hat{\lambda}_0, 0) = 0.$$

Consequently, by means of Lemma 6.5, we can apply a local bifurcation theorem of Crandall-Rabinowitz (Theorem A1-3 in Chapter 6) to the equation $F(\lambda,u) = 0$ to obtain a C^1 curve $(\lambda(s), \theta(s))$ of solutions as described in the following theorem. (See also Diemling [49]).

Theorem 6.3 (Bifurcating Positive Steady-State). *Under the hypotheses* $[H1]$ *to* $[H5]$, *the point* $(\hat{\lambda}_0, 0)$ *is a bifurcation point for the problem (6.33) (or equivalently for the problem (6.2)). Here,* $\hat{\lambda}_0 > 0$, *with corresponding eigenfunction* v^0, *is the eigenvalue described in Theorem 6.2. More precisely, there exists an interval* $[0, \delta), \delta > 0$, *and a* C^1*-curve* $(\lambda(s), \theta(s)) : [0, \delta) \to R \times \mathcal{F}_1$ *such that:* $\lambda(0) = \hat{\lambda}_0$, $\theta(0) = 0$, *and the solution* $\hat{u}(x) = u(y) = u(x/k)$ *of (6.33) is of the form*

$$u(x/k) = s(v^0 + \theta(s))(x/k) \quad \textit{for } x \in k\Omega, \ k = \sqrt{\lambda(s)}.$$

The corresponding solution $\hat{u}(x) = u(x/k)$ *of (6.1) is positive in* $k\Omega$ *and is in* $C^{2+\mu}(k\bar{\Omega})$.

Note that $\lambda(s)$ and $u(x/k)$ satisfy

$$F(\lambda(s), s(v^0 + \theta(s)) = 0 \ \text{ for } \ s \in [0, \delta). \tag{6.35}$$

Remark 6.1. The above number $\hat{\lambda}_0$, which is first defined in the statement of Theorem 6.2, is unique.

Proof of Remark. Suppose $\hat{\lambda}_1$ is any number so that there exists $v^1 \in \mathcal{E}$ (with each component $v_i^1 > 0$ in $\Omega, i \in J$) satisfying

$$(-\Delta)v^1 = \hat{\lambda}_1 B v^1, \ \text{ in } \ \Omega, \ \ v^1 = 0 \ \text{ on } \ \partial\Omega, \tag{6.36}$$

where B satisfies hypotheses $[H1]$ to $[H4]$. Let $\hat{v}^0$ be as defined in Lemma 6.4. Taking inner product on both sides of (6.36) with $\hat{v}^0$, we obtain

$$\begin{aligned} \hat{\lambda}_1 < Bv^1, \hat{v}^0 > &= < -\Delta v^1, \hat{v}^0 > = < v^1, -\Delta \hat{v}^0 > \\ &= \hat{\lambda}_0 < v^1, B^T \hat{v}^0 > = \ \hat{\lambda}_0 < Bv^1, \hat{v}^0 > . \end{aligned}$$

Since $\hat{v}_{m+1}^0 \equiv 0$, the hypotheses $[H1]$ to $[H4]$ on B implies that $< Bv^1, \hat{v}^0 > \neq 0$. This shows that $\hat{\lambda}_1 = \hat{\lambda}_0$.

For the rest of this section, we will always assume hypotheses $[H1]$ to $[H5]$. We will investigate the linearized and asymptotic stability of the positive bifurcating solution found in Theorem 6.3. For this purpose, we will introduce more assumptions on the derivative of H_{ij} below. Applying the theory in [34] (Theorem A1-3 in Chapter 6), and the fact that

$$\int_\Omega \hat{v}^0 \cdot \Delta_D^{-1} v^0 dx \neq 0 \tag{6.37}$$

(note that $\hat{v}_i^0 > 0$ for $i = 1, \ldots, m, \hat{v}_{m+1}^0 = 0$, and $v_i^0 > 0$ for $i = 1, \ldots, m+1$), we can assert that there exist $\delta_1 \in (0, \delta)$ and two functions

$$(\gamma(.), z(.)) : (\hat{\lambda}_0 - \delta_1, \hat{\lambda}_0 + \delta_1) \to R \times \mathcal{F}_1,$$

$$(\eta(.), w(.)) : [0, \delta_1) \to R \times \mathcal{F}_1,$$

with $(\gamma(\hat{\lambda}_0), z(\hat{\lambda}_0)) = (\eta(0), w(0)) = (0, v^0)$, such that

$$D_2F(\lambda, 0)z(\lambda) = \gamma(\lambda)\Delta_D^{-1}(z(\lambda)), \quad \text{and} \tag{6.38}$$

$$D_2F(\lambda(s), s(v^0 + \theta(s)))w(s) = \eta(s)\Delta_D^{-1}(w(s)). \tag{6.39}$$

Here, (6.37) and the theory in [34] implies that $\gamma(\lambda)$ and $\eta(s)$ are respectively Δ_D^{-1}-simple eigenvalues of $D_2F(\lambda, 0)$ and $D_2F(\lambda(s), s(v^0+\theta(s))$, with eigenfunctions $z(\lambda)$ and $w(s)$. Moreover, the theory in [34] further leads to the following lemmas.

Lemma 6.7. *Assume hypotheses $[H1]$ to $[H5]$. There exists $\rho > 0$ such that for each $s \in [0, \delta_1)$ there is a unique (real) eigenvalue $\eta(s)$ for the linear operator*

$$F_s^* := \Delta D_2F(\lambda(s), s(v^0 + \theta(s))) : \ \mathcal{E} \to \mathcal{F} \tag{6.40}$$

satisfying $|\eta(s)| < \rho$ with eigenfunction $w(s) \in \mathcal{E}$. That is

$$\begin{aligned} F_s^* w(s) &:= \Delta w(s) + \lambda(s)H(., (u_s)_{m+1})w(s) + \lambda(s)\tilde{H}(., (u_s)_{m+1})u_s w_{m+1} \\ &= \eta(s)w(s), \end{aligned} \tag{6.41}$$

where $u_s := s(v^0 + \theta(s))$.

The next few lemmas study the behavior of the eigenvalues $\lambda(s), \eta(s)$ for small $s \geq 0$, and $\gamma(\lambda)$ near $\lambda = \lambda_0$. In order to obtain stability we will need the following additional hypotheses:

$$\text{[H6]} \quad \begin{aligned} &(\partial H_{ij}/\partial \eta)(., 0) \leq 0 \ \text{ in } \Omega, \quad \text{for all } i = 1, .., m, j = 1, .., m+1; \\ &(\partial H_{ij}/\partial \eta)(\bar{x}, 0) < 0 \ \text{ for some } \bar{x} \in \Omega, \ \text{ some } i = 1, \ldots, m, \\ &\qquad\qquad\qquad\qquad \text{and some } j = 1, \ldots, m+1. \end{aligned}$$

For all the remaining part of this section, we will always assume hypotheses $[H1]$ to $[H6]$.

Lemma 6.8. *The function $\lambda(s)$ defined in Theorem 6.3 satisfies $\lambda'(0) > 0$.*

Proof. Theorem 6.3 asserts that $\lambda'(0)$ exists. Equation (6.35) implies that $s(v^0 + \theta(s))$ is in $\mathcal{E}$; and for $s \in [0, \delta)$, we have

$$\Delta(s(v^0 + \theta(s)) + \lambda(s)H(., s((v^0)_{m+1} + \theta_{m+1}(s)))s(v^0 + \theta(s)) = 0.$$

Dividing by s, then differentiating with respect to s and setting $s = 0$, we obtain

$$\Delta(\theta'(0)) + \lambda'(0)H(., 0)v^0 + \hat{\lambda}_0 H(., 0)\theta'(0) + \hat{\lambda}_0 \tilde{H}(., 0)(v^0)_{m+1}v^0 = 0.$$

Multiplying by $(\hat{v}^0)^T$ and integrating over Ω, we obtain

$$\int_\Omega \{(\hat{v}^0)^T \Delta\theta'(0) + \lambda'(0)(\hat{v}^0)^T B v^0 + \hat{\lambda}_0 (\hat{v}^0)^T B\theta'(0)$$
$$+ \hat{\lambda}_0 (\hat{v}^0)^T \tilde{H}(.,0)(v^0)_{m+1} v^0\} dx = 0.$$

Integrating by parts, we find

$$\int_\Omega \lambda'(0)(\hat{v}^0)^T B v^0 dx = -\hat{\lambda}_0 \int_\Omega (\hat{v}^0)^T \tilde{H}(.,0) v^0 (v^0)_{m+1} dx.$$

Consequently, we obtain

$$\lambda'(0) = \frac{-\hat{\lambda}_0 \int_\Omega (\hat{v}^0)^T \tilde{H}(.,0) v^0 (v^0)_{m+1} dx}{\int_\Omega (\hat{v}^0)^T B v^0 dx} > 0. \tag{6.42}$$

Note that the sign of the numerator above is determined by hypotheses [$H6$].

Lemma 6.9. *The function $\gamma(\lambda)$ in (6.38) satisfies $\gamma'(\hat{\lambda}_0) > 0$.*

Proof. Note that $D_2F(\lambda, 0) = I + \lambda\Delta_D^{-1}B$. From (6.38), we have

$$(I + \lambda\Delta_D^{-1}B)z(\lambda) = \gamma(\lambda)\Delta_D^{-1}z(\lambda), \text{ for } \lambda \in (\hat{\lambda}_0 - \delta_1, \hat{\lambda}_0 + \delta_1).$$

We can thus readily obtain

$$\Delta z(\lambda) + \lambda B z(\lambda) = \gamma(\lambda) z(\lambda),$$

$$\textstyle\int_\Omega (\hat{v}^0)^T \Delta z(\lambda) + \lambda(\hat{v}^0)^T B z(\lambda) dx = [\gamma(\lambda) - \gamma(\hat{\lambda}_0)] \int_\Omega (\hat{v}^0)^T z(\lambda) dx,$$

since $\gamma(\hat{\lambda}_0) = 0$. Integrating the first term on the left above by parts, using the equation satisfied by $\hat{v}^0$, factoring and cross multiplying, we deduce

$$\frac{\gamma(\lambda) - \gamma(\hat{\lambda}_0)}{\lambda - \hat{\lambda}_0} = \frac{\int_\Omega (\hat{v}^0)^T B z(\lambda) dx}{\int_\Omega (\hat{v}^0)^T z(\lambda) dx}.$$

Taking limit as λ tends to $\hat{\lambda}_0$, we obtain

$$\gamma'(\hat{\lambda}_0) = \frac{\int_\Omega (\hat{v}^0)^T B v^0 dx}{\int_\Omega (\hat{v}^0)^T v^0 dx} > 0.$$

The strict inequality above is due to hypothesis [$H4$].

Lemma 6.10. *There exists $\delta_2 \in (0, \delta_1)$, such that $\eta(s) < 0$ and $u_s \equiv s(v^0 + \theta(s)) > 0$ in Ω for all $s \in (0, \delta_2)$. Here, $\eta(s)$ and u_s are defined in Lemma 6.7.*

Proof. From Theorem 1.16 in [34], we find $-s\lambda'(s)\gamma'(\hat{\lambda}_0)$ and $\eta(s)$ have the same sign for $s > 0$ near 0. Hence Lemmas 6.8 and 6.9 imply that $\eta(s) < 0$ for small positive s. Since v^0 is positive in Ω with negative outward normal derivative on the part of $\partial\Omega$ where it is zero, and $\theta(s) \to 0$ in C^1 as $s \to 0$, we must have $u_s \equiv s(v^0 + \theta(s)) > 0$ in Ω for $s > 0$ sufficiently small.

It remains to investigate the other eigenvalues of F_s^*. Let $b > 0$ be a large enough constant such that

$$\sum_{j=1}^{m+1} \hat{\lambda}_0 H_{ij}(x, 0) - b < 0 \quad \text{for all } x \in \Omega,\ i = 1, \ldots, m+1. \tag{6.43}$$

For convenience, let M_0 be the complex extension of the operator from $\mathcal{E}$ into $\mathcal{F}$ defined by

$$M_0 w = \Delta(w) + \hat{\lambda}_0 B(w) - bI(w), \tag{6.44}$$

for $w \in \mathcal{E}$, where I is the identity operator.

Lemma 6.11. *(i) The inverse of M_0 can be defined as $M_0^{-1} \in \mathcal{L}(\mathcal{F})$, i.e. a bounded linear operator $\mathcal{F} \to \mathcal{F}$, and it is compact.*

(ii) If $\lambda \neq 0$ is an eigenvalue of $M_0 + bI$ (i.e. $\Delta + \hat{\lambda}_0 B$), then $Re(\lambda) < -r$ for some positive number r.

The proof uses the sign of the off-diagonal terms of B, (6.43) and the maximum principle for the corresponding systems. It is essentially the same as the proof of Lemma 2.8 in [137]. The details will be omitted here.

For convenience, we let σ_s denote the point spectrum of F_s^*. The next linearized stability theorem follows from Lemmas 6.7 to 6.11.

Theorem 6.4 (Linearized Stability of Bifurcating Positive Steady-State). *Under hypotheses [H1] to [H6], there exists a number $\delta^* \in (0, \delta)$ where δ is described in Theorem 6.3, and a positive function $\eta(s)$ for $s \in (0, \delta^*)$ such that the point spectrum σ_s satisfies:*

$$Re\,\sigma_s \subset \{w \in R^1 | w \leq -\eta(s)\} \quad \text{for } s \in (0, \delta^*). \tag{6.45}$$

Here, $Re\,\sigma_s$ denotes the set of real numbers which are real parts of numbers in σ_s.

The proof of Lemmas 6.8 to 6.10 are given in detail above, and they are the consequences of [H1] to [H6]. They are different from the model in [137]. However, the proof of Theorem 6.4, using assertions in Lemmas 6.7 to 6.11 are exactly the same as the proof of Theorem 2.2 in [137]. The details are thus omitted here.

For each fixed $\bar{s} \in (0, \delta^*)$, the function $u_{\bar{s}} = \bar{s}(v^0 + \theta(\bar{s}))$ is a steady- state solution of problem (6.3) with $\lambda = \lambda(\bar{s})$ given by Theorem 6.3. We now proceed to investigate the time asymptotic stability of this steady state as a solution of the parabolic system (6.3). Let B_1 and B_2 be Banach spaces as follows:

$$\begin{aligned} &B_1 = \{u : u \in [C(\bar{\Omega})]^{m+1}, u = 0 \text{ on } \partial\Omega\}, \text{ and} \\ &B_2 = \{u : u \in [L_p(\Omega)]^{m+1}\} \text{ for } p \text{ large enough such that } N/(2p) < 1. \end{aligned} \tag{6.46}$$

Let A_1 be the Δ operator on B_1 with domain $D(A_1) = \{u : u \in [W^{2,p}(\Omega)]^{m+1}$ for all p, $\Delta u \in [C(\bar{\Omega})]^{m+1}$, $u = 0$ and $\Delta u = 0$ on $\partial\Omega\}$; and A_2 be the Δ operator on B_2 with domain $D(A_2) = \{u \in B_2 : u \in [W^{2,p}(\Omega) \cap W_0^{1,p}(\Omega)]^{m+1}\}$. For $u = \text{col.}(u_1, \ldots, u_{m+1})$, and $f(u) = \lambda H(\cdot, u_{m+1})u$, we can consider the following nonlinear initial-boundary value problem for each i = 1, 2 corresponding to (6.3):

$$\frac{du}{dt} - A_i u(t) = f(u(t)) \quad \text{for } t \in (0, T] \tag{6.47}$$

with $u(t) \in D(A_i), t > 0$, respectively for $i = 1, 2$. Here, we suppress writing the dependence of f on $\lambda(\bar{s})$, since it is fixed for some $\bar{s} \in (0, \delta^*)$.

Definition 6.1. *A solution of an initial value problem corresponding to (6.47) in B_i is a function*

$$u(t) \in C([0, T], B_i) \cap C^1((0, T], B_i),$$

with $u(0) = u_0, u(t) \in D(A_i)$ for all $t \in (0, T]$; and $u(t)$ satisfies (6.47) for $t \in (0, T]$.

The operator A_2 is an infinitesimal generator of an analytic semigroup $M(t)$, $t \geq 0$, on B_2. It is well-known that for $\alpha > 0$

$$(-A_2)^{-\alpha} = \frac{1}{\Gamma(\alpha)} \int_0^\infty \tau^{\alpha-1} M(\tau) d\tau$$

defines a bounded linear operator on B_2. Moreover, $[(-A_2)^{-\alpha}]^{-1} = (-A_2)^\alpha$ is a closed linear operator on B_2 with dense domain $D((-A_2)^\alpha) = (-A_2)^{-\alpha}(B_2)$. We denote by X^α the Banach space $(D(-A_2)^\alpha), ||\cdot||_\alpha)$, where $||u||_\alpha = ||(-A_2)^\alpha u||_{L^p}$ for all $u \in D((-A)^\alpha)$. Moreover, for $N/(2p) < \alpha < 1$, there exists a constant $C(\alpha) > 0$ such that $||u||_\infty \leq C(\alpha)||u||_\alpha$ for all $u \in X^\alpha$. (See Definition A4-6 to A4-8 in Chapter 6 or Pazy [184]).

For further discussion of solutions in these spaces, we will make the following additional assumptions for $i, j = 1, \ldots, m+1$:

$$\begin{aligned} &H_{ij}(x, \eta) \text{ and } (\partial H_{ij}/\partial \eta)(x, \eta) \quad \text{are bounded for all } (x, \eta) \in \Omega \times R^1, \\ &|\tfrac{\partial H_{ij}}{\partial \eta}(x, \eta_1) - \tfrac{\partial H_{ij}}{\partial \eta}(x, \eta_2| \leq C|\eta_1 - \eta_2)| \text{ for all } \eta_1, \eta_2 \in R^1, \\ &\qquad\qquad\qquad\qquad\qquad\qquad \text{for some constant } C > 0. \end{aligned} \tag{H7}$$

For the function f described above, the hypotheses $[H1]$ and $[H7]$ lead to the following Lipschitz properties in B_1 and B_2: There exists $\bar{C} > 0$ and $\rho > 0$, such that

(6.48)
$$\| f(u) - f(w) \|_\infty \leq \bar{C} \| u - w \|_\infty \text{ for all } u, w \in B_1 \text{ with } \| u \|_\infty, \| w \|_\infty \leq \rho,$$

and

(6.49)
$$\| f(u) - f(w) \|_{L^p} \leq \bar{C} \| u - w \|_\alpha \text{ for all } u, w \in X^\alpha \text{ with } \| u \|_\alpha, \| w \|_\alpha \leq \rho.$$

Here $\bar{C}$ depends on α and ρ. From the Lipschitz properties (6.48) and (6.49), we obtain local existence for solutions of (6.47) in B_1 and B_2 respectively. (See e.g. [184]). Moreover, Corollary 3.3.5 in Henry [84] implies that solutions of (6.47) are global. From hypotheses $[H1]$ and $[H7]$, we can further deduce:

$$\| f(w) - f(u) - df_u(w - u) \|_\infty = o(\| w - u \|_\infty) \tag{6.50}$$

for all $u, w \in B_1, ||u||_\infty \leq \rho$, as $||w - u||_\infty \to 0$, and

$$\| f(w) - f(u) - df_u(w - u) \|_{L^p} = o(\| w - u \|_\alpha) \tag{6.51}$$

for all $u, w \in X^\alpha, ||u||_\alpha \leq \rho$, as $||w - u||_\alpha \to 0$. Here,

$$df_u z = \lambda H(\cdot, u_{m+1})z + \lambda[\frac{\partial H}{\partial \eta}(\cdot, u_{m+1})]u z_{m+1}.$$

Note that the operator $A_i + df_{\bar{u}}, \bar{u} := u_{\bar{s}}$, on B_i can be written as $F^*_{\bar{s}}$ as in (6.41). Thus the point spectrum of $A_i + df_{\bar{u}}$ lies in $\{\lambda \in C : Re\,\lambda < -\eta(\bar{s})\}$ by Theorem 6.4. Moreover, the spectrum of $F^*_{\bar{s}}$ is the set $\{1/\mu : \mu$ is in the spectrum of $(F^*_{\bar{s}})^{-1}\}$, and thus consists only of eigenvalues. (See e.g. pp. 51 and 79 in [184]). By means of (6.50), (6.51) and Theorem 6.4, we can apply the stability Theorem 5.1.1 in [84] or Theorem A4-11 in Chapter 6 for sectorial operators to obtain (6.53) in the following theorem for the asymptotic stability of the steady state solution.

Theorem 6.5 (Local Asymptotic Stability of Positive Steady-State). *Assume hypotheses $[H1]$ to $[H7]$, and let $\bar{u} := u_{\bar{s}}$, $\lambda = \lambda(\bar{s})$ for a fixed $\bar{s} \in (0, \delta^*), \alpha \in (N/(2p), 1)$. Then, for each $i = 1, 2$, there exist $\rho > 0, \beta > 0$ and $M > 1$ such that equation (6.47) has a unique solution in B_i for all $t > 0$ if $u_0 \in B_1$ and $\| u_0 - \bar{u} \|_\infty \leq \rho/(2M)$ for $i = 1$ (or $u_0 \in X^\alpha$ and $\| u_0 - \bar{u} \|_\alpha \leq \rho/(2M)$ for $i = 2$). Moreover, the solution satisfies:*

$$\| u(t) - \bar{u} \|_\infty \leq 2Me^{-\beta t} \| u_0 - \bar{u} \|_\infty \quad \textit{for all } t \geq 0,\ i = 1, \ \textit{or} \tag{6.52}$$

$$\| u(t) - \bar{u} \|_\alpha \leq 2Me^{-\beta t} \| u_0 - \bar{u} \|_\alpha \quad \textit{for all } t \geq 0,\ i = 2. \tag{6.53}$$

Note: the condition on α is only assumed for solutions in B_2.

The details of the proof are the same as that for Theorem 3.1 in [137], and will be omitted here. For (6.52), we can use the theories in Theorems A4-8 and A4-9 in Chapter 6 by Mora [176] or results in Stewart [211].

Note that hypotheses $[H1]$ to $[H5]$ are sufficient to insure the existence of the steady-state $\bar{u}$, by Theorem 6.3. The addition of hypotheses $[H6]$ and $[H7]$ leads to asymptotic stability.

Notes.

Theorem 2.1 to Theorem 2.3 are due to Leung and Ortega [139]. Theorems 3.1 to 3.5 are obtained from Leung and Villa [142]. Theorems 4.1 to 4.5 are found in Liu [160], and Theorems 5.1 to 5.3 are due to Leung [129]. Theorems 6.1 to 6.5 are obtained from Leung and Villa [141].

Chapter 3

Optimal Control for Nonlinear Systems of Partial Differential Equations

3.1 Introduction and Preliminary Results for Scalar Equations

The last two chapters study the basic properties concerning the existence and stability of positive solutions of reaction-diffusion systems. This chapter is concerned with the control of these systems with various definite purposes in mind. We summarize some recent results concerning the optimal control of nonlinear partial differential equations related to those studied in previous chapters. We control the interaction parameters or boundary conditions in order to optimize expressions involving the solutions of the systems. For example, in Section 3.2 we consider the optimal control of harvesting effort for a prey-predator system in an environment. Such problems arise naturally in fisheries and agriculture when various species are harvested for economic profit. We first assume that the species are in steady-state under diffusion and Volterra-Lotka type interaction. They are harvested for economic return, leading to reduction of growth rate. The problem is to control the costly spatial distribution of harvesting effort in the habitat so as to maximize the profit. In Section 3.5, we assume the control is imposed on the boundary of the habitat. The study leads to better understanding of the relationship between the growth of the species and the economic cost of maintaining an ecologically favorable boundary environment. In many ecological systems, progressive deterioration of the surrounding of the habitat may lead to pattern of species extinction in remnant patches, as described in Bierregarrd et al. [9]. It is also suggested in Angelstam [4] that as human activities make the habitat surrounding more dissimilar, there is an increase in the severity of predation impact on Swedish forest birds by predators (e.g. corvids and foxes) residing in the habitat. Application of similar model is also made in wildlife damage management for controlling population of diffusive small mammal species such as beavers, raccoons and muskrats (see Lenhart and Bhat [116]). Since there are natural seasonal environmental variations, we consider time-periodic optimal harvesting control of competing populations in Section 3.3. In Section 3.4, we consider the optimal control of nuclear fission reactors modeled by parabolic differential equations. The neutrons are divided into fast and thermal groups with two equations describing their interaction and fission, while the third equation describes the temperature in the reactor. The coefficient for fission and absorption of the thermal neutron is assumed to be controlled by a function through the use of control rods in the reactor. The object is to maintain a target neutron flux shape, while a desired power level and adjustment costs are taken into consideration. For other applications to medical and physical sciences, one can find examples in e.g. Kirschner, Lenhart and Serbin [104], and Lasiecka and Triggiani [114].

In this section, we first consider a simple preliminary problem concerning the steady-state control of one species, whose growth is governed by the diffusive Volterra-Lotka equation with no-flux boundary condition:

$$\Delta u + u[(a(x) - f(x)) - bu] = 0 \text{ in } \Omega, \quad \frac{\partial u}{\partial \nu} = 0 \text{ on } \partial\Omega. \tag{1.1}$$

Here, u is the species concentration. The function $a(x)$ describes spatially dependent intrinsic growth rate, and b designates crowding effect which is assumed to be constant for simplicity. The function $f(x)$ denotes spatially dependent control harvesting effort on the biological species. The optimal control criteria is to maximize the difference between economic revenue and cost. This is expressed by the payoff functional

$$J(f) = \int_\Omega \{Kuf - Mf^2\}dx, \tag{1.2}$$

where K and M are constants describing the price of the species and the cost of the control. Here $\int_\Omega uf dx$ is a measure of the total harvest, and u itself depends on f through (1.1). Such a model is certainly only a prototype, and various other variations can be made. In this section, we follow the presentations in Leung and Stojanovic [140].

To fixed ideas, we assume Ω is a bounded domain in R^n with $\partial\Omega \in C^2$; Δ and $\frac{\partial}{\partial \nu}$ respectively denote the Laplacian and outward normal derivative. K, M and b are positive constants. We also assume that

$$\begin{aligned} a(x) \geq 0, \;\; f(x) \geq 0 \text{ a.e. in } \Omega, \text{ and} \\ a \in L^\infty(\Omega), \;\; f \in L^\infty(\Omega). \end{aligned}$$

For convenience, we denote

$$L^\infty_+(\Omega) = \{f | f \in L^\infty(\Omega), \ \ f \geq 0 \ \text{ a.e. in } \Omega\};$$

and for $\delta > 0$

$$C_\delta = \{f \in L^\infty_+(\Omega) | 0 \leq f \leq \delta \ \text{ a.e. in } \Omega\}. \tag{1.3}$$

We define an optimal control (if it exists) to be an $f^* \in C_\delta$ such that

$$J(f^*) = \sup_{f \in C_\delta} J(f). \tag{1.4}$$

Before investigating the optimal control, we first note that for each fixed $f \in C_\delta$, problem (1.1) has a unique positive solution under appropriate assumptions.

Theorem 1.1 (Positive Solution for a Given Control). *Suppose that $a(x)$ and δ satisfy hypothesis*

[H1] $$0 < \delta < \inf_\Omega a(x).$$

Then for each $f \in C_\delta$, the problem (1.1) has a unique strictly positive solution $u := u(f) \in W^{2,p}(\Omega)$, for any $p \in [1, \infty)$. Furthermore, the estimate

$$\|u(f)\|_{2,p} \leq const. \tag{1.5}$$

is valid, uniformly for all $f \in C_\delta$.

Proof. Let $C > 0$ be a constant large enough so that

$$||a||_\infty C - bC^2 \leq 0. \tag{1.6}$$

Then choose a constant $P > 0$ so that

$$\delta + 2bC - P < 0. \tag{1.7}$$

Define an initial iterate

$$u_0 \equiv C; \tag{1.8}$$

and then inductively define u_k, $k = 1, 2, \ldots$ as solutions in $W^{2,p}(\Omega)$ of the linear problem

$$\begin{cases} \Delta u_k - P u_k = -(a(x) - f(x)) u_{k-1} + b u_{k-1}^2 - P u_{k-1} \ \text{ in } \Omega, \\ \frac{\partial u_k}{\partial \nu} = 0 \ \text{ on } \partial\Omega. \end{cases} \tag{1.9}$$

One readily verifies that (1.9) implies

$$\begin{cases} \Delta(C-u_1) - P(C-u_1) = (a-f)C - bC^2 \leq 0 \text{ in } \Omega, \\ \frac{\partial}{\partial \nu}(C-u_1) = 0 \text{ on } \partial\Omega. \end{cases} \tag{1.10}$$

By means of maximum principle and boundary condition, we can use e.g. Theorem 9.6 and arguments in Lemma 3.4 in Gilbarg and Trudinger [71] to deduce

$$u_1 \leq C = u_0 \text{ in } \Omega. \tag{1.11}$$

The choice of P in (1.7) implies that the function

$$h(x,u) \equiv -(a(x) - f(x))u + bu^2 - Pu$$

is decreasing in $u \in [0, C]$ a.e. in Ω. Using the same argument as above, we readily obtain

$$u_k \leq u_{k-1} \text{ in } \Omega \tag{1.12}$$

for each $k = 2, 3, \ldots$ On the other hand, hypothesis [H1] implies that for $\epsilon > 0$ sufficiently small

$$\begin{aligned} \Delta(u_k - \epsilon) - P(u_k - \epsilon) &= h(x, u_{k-1}) + P\epsilon \\ &\leq h(x, u_{k-1}) + P\epsilon + (a(x) - f(x))\epsilon - b\epsilon^2 \\ &= h(x, u_{k-1}) - h(x, \epsilon) \qquad \text{in } \Omega. \end{aligned} \tag{1.13}$$

Thus assuming $\epsilon < C = u_0$, and starting from $k = 0$, we can inductively deduce that $\Delta(u_k - \epsilon) - P(u_k - \epsilon) \leq 0$ in Ω for each k. Consequently, we have

$$u_k(x) \geq \epsilon > 0 \quad \text{a.e. in } \Omega \tag{1.14}$$

for each $k = 0, 1, 2, \ldots$

From the uniform bounds for u_k, and $W^{2,p}(\Omega)$ estimate, we conclude from (1.9) that for any $p \in [1, \infty)$, we have

$$||u_k||_{2,p} \leq const.,$$

uniformly for $f \in C_\delta$. Passing to limit as $k \to \infty$ in (1.9), and using $W^{2,p}(\Omega)$ estimate again, we obtain a solution $u = u(f)$ of (1.1) in $W^{2,p}(\Omega)$.

Moreover, we have

$$0 < \epsilon \leq u(f) \leq C, \tag{1.15}$$

and (1.5) holds for all $f \in C_\delta$.

It remains to prove uniqueness. Suppose that w is another bounded strictly positive solution of (1.1). Let

$$l = max\{C, ||w||_\infty\}; \tag{1.16}$$

and start iterating from an initial constant function l. We can construct as in the first part of the proof another solution of (1.1), say v, such that

$$0 < u \leq v, 0 < w \leq v \quad a.e \text{ in } \Omega. \tag{1.17}$$

(Note that (1.17) is deduced by comparing u and w with successive iterates starting from l.) However, we have from the equations for u and v

$$\begin{aligned} 0 &= \int_\Omega (u\Delta v - v\Delta u)dx \\ &= \int_\Omega [uv^2 b - uv(a-f) - vu^2 b + vu(a-f)]dx \\ &= \int_\Omega buv(v-u)dx. \end{aligned} \tag{1.18}$$

Thus from the first inequality in (1.17), we obtain from the last equation $u = v$ in Ω. Similarly, replacing the role of u by w in (1.18), we deduce that $w = v$ in Ω.

Remark 1.1. From the proof of the above theorem we see that the unique strictly positive solution satisfies (1.15). From (1.13), one sees that the lower bound ϵ can be chosen as

$$\epsilon = \frac{1}{b}\Big[\inf_\Omega a(x) - \delta\Big] > 0. \tag{1.19}$$

From (1.6) in the proof, we also see that we can choose

$$C = \sup_\Omega \{a(x)/b\} \tag{1.20}$$

to be the upper bound for all $u = u(f), f \in C_\delta$.

Since there are upper and lower bounds for all solutions of (1.1) uniformly for all $f \in C_\delta$, we next show that we can find a maximizing sequence $f_n \in C_\delta$ and prove in the usual manner that there exist a subsequence which converges weakly in $L^2(\Omega)$ to an optimal control $f^* \in C_\delta$.

Theorem 1.2 (Existence of Optimal Control). *Let δ and $a(x)$ satisfy hypothesis [H1], then an optimal control does exist in the sense of (1.4).*

Proof. From (1.15), it follows that

$$\sup_{f \in C_\delta} J(f) < \infty.$$

Let $f_n \in C_\delta$ be a maximizing sequence. Then, there exists a subsequence, again denoted by f_n for convenience, so that

$$f_n \to f^* \in C_\delta \quad \text{weakly in } L^2(\Omega),$$

$$u_n \equiv u(f_n) \to u^* \quad \text{strongly in } W^{1,2}(\Omega)$$

(by using (1.5)). Passing to the limit as $n \to \infty$ in

$$\int_\Omega \{\nabla u_n \cdot \nabla \phi - (a - f_n)u_n\phi + bu_n^2\phi\}dx = 0 \ \ \forall \phi \in W^{2,p}(\Omega) \cap L^\infty(\Omega),$$

and noting that

$$\int_\Omega f_n u_n \phi dx \to \int_\Omega f^* u^* \phi dx, \ \forall \phi \in L^\infty(\Omega),$$

we conclude from the two limits as $n \to \infty$ above that

$$u^* = u(f^*).$$

But then

$$\begin{aligned} J(f^*) &= \textstyle\int_\Omega \{Ku^* f^* - M(f^*)^2\}dx \\ &\geq \lim_{n\to\infty} \textstyle\int_\Omega Ku_n f_n dx - \liminf_{n\to\infty} \int_\Omega M f_n^2 dx \\ &= \limsup_{n\to\infty} J(f_n) = \sup_{f \in C_\delta} J(f), \end{aligned}$$

and consequently, f^* is an optimal control in C_δ.

In order to characterize the optimal control, we next find a slightly stronger condition than [H1] to obtain differentiability of $u(f)$ with respect to f as described in the following lemma.

Lemma 1.1 (Differentiability with Respect to Control). *Suppose δ and $a(x)$ satisfy*

[H2] $$0 < \delta \leq \tfrac{1}{3}\{2 \inf_\Omega a(x) - \sup_\Omega a(x)\}.$$

Then, the mapping

$$C_\delta \ni f \mapsto u(f) \in W^{1,2}(\Omega)$$

is differentiable in the following sense:

$$\frac{u(f + \beta \bar{f}) - u(f)}{\beta} \to \xi \ \ \textit{weakly in } W^{1,2}(\Omega) \tag{1.21}$$

as $\beta \to 0$, for any $f \in C_\delta$ and $\bar{f} \in L^\infty(\Omega)$ such that $f + \beta\bar{f} \in C_\delta$. Further, ξ is the unique solution of

$$\Delta\xi + (a - f - 2bu(f))\xi = \bar{f}u(f) \text{ in } \Omega, \quad \frac{\partial\xi}{\partial\nu} = 0 \text{ on } \partial\Omega. \tag{1.22}$$

(Note that [H2] implies that [H1] is satisfied.)

Proof. From (1.1), we deduce that

$$\xi_\beta := \frac{u(f + \beta\bar{f}) - u(f)}{\beta} \tag{1.23}$$

satisfies

$$\begin{cases} \Delta\xi_\beta + (a - f)\xi_\beta - b(u(f + \beta\bar{f}) + u(f))\xi_\beta = \bar{f}u(f + \beta\bar{f}) \text{ in } \Omega, \\ \frac{\partial\xi_\beta}{\partial\nu} = 0 \text{ on } \partial\Omega. \end{cases} \tag{1.24}$$

Consequently, we have

$$\begin{aligned} &\int_\Omega \{|\nabla\,\xi_\beta|^2 + [b(u(f + \beta\bar{f}) + u(f)) - a + f]\,\xi_\beta^2\}\,dx \\ &\quad = \int_\Omega -\bar{f}u(f + \beta\bar{f})\xi_\beta\,dx. \end{aligned}$$

From Remark 1.1, we obtain

$$b(u(f + \beta\bar{f}) + u(f)) - a + f \geq 2\Big[\inf_\Omega a - \delta\Big] - \sup_\Omega a \geq \delta, \tag{1.25}$$

where the last inequality is due to [H2]. Consequently

$$||\xi_\beta||_{1,2}^2 \leq const.||\bar{f}||_\infty||u(f + \beta\bar{f})||_2||\xi_\beta||_2.$$

and thus

$$||\xi_\beta||_{1,2} \leq const., \tag{1.26}$$

where the constant is independent of β.

Thus, using (1.26), we choose a sequence $\beta \to 0$ and deduce, as in Theorem 1.2, that the weak limit satisfies (1.22). From the uniqueness of solution to (1.22) we conclude that $\xi_\beta \to \xi$ weakly in $W^{1,2}(\Omega)$ as $\beta \to 0$, for a full sequence.

Remark 1.2. Hypothesis [H2] ensures the positivity of the expression on the left of (1.25). If

$$2b\inf_\Omega \{u(f)(x)|f \in C_\delta\} - a(x) \geq \sigma > 0, \tag{1.27}$$

for all $x \in \Omega$, then Lemma 1.1 can be proved in the same way. Hence if $\delta > 0$ is such that (1.27) holds, then Lemma 1.1 is true without assuming [H2].

We now obtain a characterization of the optimal control as follows

Theorem 1.3 (Characterization of the Optimal Control). *Suppose $a(x)$ and δ satisfy hypothesis [H2]; and the constant K, M have the property:*

$$M \geq [K \sup_{\Omega} a]/(2b\delta). \tag{1.28}$$

Then for any optimal control $f \in C_\delta$, there exists (u, p) with $b^{-1}[\inf_\Omega a - \delta] \leq u \leq b^{-1} \sup_\Omega a$, $0 \leq p \leq K$, such that

$$f = \frac{u}{2M}(K - p) \ \textit{in}\ \Omega, \tag{1.29}$$

and (u, p) is a solution of the optimality system

$$\begin{cases} \Delta u + au - (b + \frac{K-p}{2M})u^2 = 0 \\ \Delta p + (a - 2bu)p + \frac{(K-p)^2 u}{2M} = 0 \ \text{in} \ \ \Omega, \\ \frac{\partial u}{\partial \nu} = \frac{\partial p}{\partial \nu} = 0 \ \text{on} \ \partial\Omega. \end{cases} \tag{1.30}$$

Proof. The existence of an optimal control in C_δ has been justified by Theorem 1.2. Let $f \in C_\delta$ be an optimal control. For $\bar{g} \in L^\infty_+(\Omega)$, $\epsilon > 0$, set

$$\bar{f} = \bar{f}_\epsilon = \begin{cases} \bar{g} & \text{if } f \leq \delta - \epsilon \|\bar{g}\|_\infty \\ 0 & \text{elsewhere.} \end{cases}$$

Then, for $\beta > 0$ small enough, we have

$$J(f) \geq J(f + \beta\bar{f}). \tag{1.31}$$

Dividing by β, we obtain

$$\int_\Omega \{K[\frac{u(f + \beta\bar{f}) - u(f)}{\beta}(f + \beta\bar{f}) + u(f)\bar{f}] - M\bar{f}(2f + \beta\bar{f})\}\, dx \leq 0. \tag{1.32}$$

Letting $\beta \to 0$, we use Lemma 1.1 and (1.32) to obtain

$$\int_\Omega Kf\xi + Ku(f)\bar{f} - 2Mf\bar{f}dx \leq 0. \tag{1.33}$$

Now, define p to be the solution of

$$\Delta p + [a - f - 2bu(f)]p = -Kf \ \text{in} \ \Omega, \ \frac{\partial p}{\partial \nu} = 0 \ \text{on} \ \partial\Omega. \tag{1.34}$$

Note that from the lower bound for $u(f)$ described in Remark 1.1, we have $a - f - 2bu(f) \le \sup_\Omega a - 2\inf_\Omega a + 2\delta \le -3\delta + 2\delta = -\delta$. Hence, the problem (1.34) has a unique solution p, and moreover

$$0 \le p \le K. \tag{1.35}$$

Combining (1.33) and (1.34), we integrate by parts and use Lemma 1.1 to obtain

$$\int_\Omega \bar{f}_\epsilon [u(f)(K-p) - 2Mf] dx \le 0. \tag{1.36}$$

Letting $\epsilon \to 0$, (1.36) leads to

$$\int_{\Omega \cap \{x \in \Omega | f(x) < \delta\}} \bar{g}[u(f)(K-p) - 2Mf] dx \le 0 \quad \forall \bar{g} \in L^\infty_+(\Omega). \tag{1.37}$$

Consequently, we must have

$$f \ge \frac{u(f)}{2M}(K-p) \text{ in } \Omega \cap \{x \in \Omega | f(x) < \delta\}. \tag{1.38}$$

On the other hand, for $-\bar{g} \in L^\infty_+(\Omega)$, $\epsilon > 0$, we set

$$\bar{f} = \bar{f}_\epsilon = \begin{cases} \bar{g} & \text{if } f \ge \epsilon \|\bar{g}\|_\infty \\ 0 & \text{elsewhere.} \end{cases}$$

We deduce in the same way as above that (1.36) holds for such $\bar{f}_\epsilon$. Passing to the limit as $\epsilon \to 0$, we obtain

$$\int_{\Omega \cap \{x \in \Omega | f(x) > 0\}} -\bar{g}[u(f)(K-p) - 2Mf] dx \ge 0 \quad \forall (-\bar{g}) \in L^\infty_+(\Omega). \tag{1.39}$$

Consequently we must have

$$f \le \frac{u(f)}{2M}(K-p) \text{ in } \Omega \cap \{x \in \Omega | f(x) > 0\}. \tag{1.40}$$

Hence, combining (1.38) and (1.40), we conclude that

$$f = \frac{u(f)}{2M}(K-p) \text{ in } \Omega \cap \{x \in \Omega | 0 < f(x) < \delta\}. \tag{1.41}$$

Since by (1.38) we have

$$f \ge \frac{u(f)}{2M}(K-p) \text{ in } \Omega \cap \{x \in \Omega | f(x) = 0\}; \tag{1.42}$$

and by (1.40) we have
(1.43)

$$f \leq \frac{u(f)}{2M}(K-p) \leq \frac{u(f)}{2M}K \leq (\sup_{\Omega} a\ K)/(2Mb) \leq \delta \quad \text{in } \Omega \cap \{x \in \Omega | f(x) = \delta\}.$$

Consequently, from (1.41), (1.42), (1.35), and (1.43) we obtain

$$f = \frac{u(f)}{2M}(K-p) \quad \text{in } \Omega. \tag{1.44}$$

From (1.1), (1.34), and (1.44) we easily derive (1.30). This completes the proof of Theorem 1.3.

Theorem 1.3 expresses an optimal control in terms of (u, p) which is a solution of the system (1.30) with the property that

$$\begin{aligned} & b^{-1}[\inf_{\Omega} a(x) - \delta] \leq u(x) \leq b^{-1} \sup_{\Omega} a(x) \\ & \qquad\qquad\qquad\qquad\qquad\qquad\qquad\qquad\qquad \text{in } \Omega. \\ & 0 \leq p(x) \leq K. \end{aligned}$$

We now provide a much better approximation for (u, p). We construct monotone sequences converging from above and below to upper and lower estimates for (u, p). These sequences are found by solving scalar equations rather than the larger system (1.30). In case where the limits of the upper and lower iterates agree, then the optimal control problem is completely solved.

For convenience we denote

$$\tilde{a} = \inf_{\Omega} a(x), \quad \bar{a} = \sup_{\Omega} a(x).$$

In the remaining part of this section, we will always assume [H2] and (1.28).

Let R_1, R_2 be positive constants so that the expressions

$$\begin{aligned} & -p(a(x) - 2bu) - R_1 p, \quad \text{and} \\ & \tfrac{-(K-p)^2 u}{2M} - R_2 p \end{aligned} \tag{1.45}$$

are decreasing in p in the interval $0 \leq p \leq K$ for all $x \in \Omega, (\tilde{a} - \delta)/b \leq u \leq \bar{a}/b$. Choose $R \geq R_1 + R_2$ so that the expression

$$-u[a(x) - (b + \frac{K}{2M})u + \frac{pu}{2M}] - Ru \tag{1.46}$$

is also decreasing in u in the interval $(\tilde{a} - \delta)/b \leq u \leq \bar{a}/b$, for all $x \in \Omega, 0 \leq p \leq K$.

For convenience, we define the following constant functions in Ω:

$$u_{-1} \equiv \bar{a}/b, \qquad u_0 \equiv \tfrac{\tilde{a}-\delta}{b}, \qquad p_{-1} \equiv K, \qquad p_0 \equiv 0. \tag{1.47}$$

We can readily verify that they satisfy

$$\Delta u_{-1} - Ru_{-1} \le -u_{-1}[a - (b + \frac{K}{2M})u_{-1} + \frac{p_{-1}u_{-1}}{2M}] - Ru_{-1} \quad \text{in } \Omega, \tag{1.48}$$

$$\Delta u_0 - Ru_0 \ge -u_0[a - (b + \frac{K}{2M})u_0 + \frac{p_{-1}u_0}{2M}] - Ru_0 \quad \text{in } \Omega; \tag{1.49}$$

and that

$$\Delta p_{-1} - Rp_{-1} \le -p_{-1}(a - 2bu_0) - \frac{(K - p_{-1})^2 u}{2M} - Rp_{-1} \quad \text{in } \Omega, \tag{1.50}$$

$$\Delta p_0 - Rp_0 \ge -p_0(a - 2bu_0) - \frac{(K - p_0)^2 u}{2M} - Rp_0 \quad \text{in } \Omega \tag{1.51}$$

for each u in the interval $[u_0, u_{-1}]$.

We now inductively define sequences of functions $u_i(x), p_i(x)$ in $\Omega, i = 1, 2, \ldots$ as solutions of the following scalar problems:

(1.52)

$$\Delta u_i - Ru_i = -u_{i-2}[a - (b + \tfrac{K}{2M})u_{i-2} + \tfrac{p_{i-2}u_{i-2}}{2M}] - Ru_{i-2} \text{ in } \Omega,\ \tfrac{\partial u_i}{\partial \nu} = 0 \text{ on } \partial\Omega;$$

$$\Delta p_i - Rp_i = -p_{i-2}(a - 2bu_{i-1}) - \tfrac{(K-p_{i-2})^2 u_i}{2M} - Rp_{i-2} \text{ in } \Omega,\ \tfrac{\partial p_i}{\partial \nu} = 0 \text{ on } \partial\Omega.$$

Theorem 1.4 (Approximation Scheme for the Optimal Control). *Assume [H2] and (1.28). The sequences of functions $u_i(x), p_i(x)$, defined above satisfy the order relation:*

$$u_0 \le u_2 \le \cdots \le u_{2r} \le u_{2r-1} \le \cdots \le u_1 \le u_{-1}, \qquad p_0 \le p_2 \le \cdots \le p_{2r} \le p_{2r-1} \le \cdots \le p_1 \le p_{-1} \tag{1.53}$$

for all $x \in \Omega$. Moreover, any solution (u, p) of problem (1.30) with the property

$$u_0 \le u \le u_{-1}, \quad p_0 \le p \le p_{-1} \quad \textit{in } \Omega \tag{1.54}$$

must satisfy

$$u_{2r} \le u \le u_{2r-1}, \quad p_{2r} \le p \le p_{2r-1} \quad \textit{in } \Omega, \tag{1.55}$$

for all positive integer r.

Proof. Using the equation satisfied by u_i and inequality (1.48), we obtain $\Delta(u_{-1}-u_1)-R(u_{-1}-u_1) \leq 0$ in Ω, $(\partial/\partial\nu)(u_{-1}-u_1) = 0$ on $\partial\Omega$. Hence $u_1 \leq u_{-1}$ in Ω. Similarly using (1.49) and the decreasing property of the expression (1.46), we deduce that $\Delta(u_0 - u_1) - R(u_0 - u_1) \geq 0$ in Ω, $(\partial/\partial\nu)(u_0 - u_1) = 0$ on $\partial\Omega$. Thus we have $u_0 \leq u_1$ in Ω.

Inequality (1.50) is satisfied with u replaced by u_1 on the right hand side. Then subtracting equation (1.52) for p_1, we obtain $\Delta(p_{-1}-p_1)-R(p_{-1}-p_1) \leq 0$ in Ω, $(\partial/\partial\nu)(p_{-1} - p_1) = 0$ on $\partial\Omega$. We therefore have $p_1 \leq p_{-1}$ in Ω. Inequality (1.51) is satisfied with u replaced by u_1 on the right. Then subtracting (1.52) for p_1 and using the decreasing property of the second expression in (1.45), we obtain $\Delta(p_1-p_0)-R(p_1-p_0) \leq 0$ in Ω, $(\partial/\partial\nu)(p_1-p_0) = 0$ on $\partial\Omega$. We therefore have $p_0 \leq p_1$ in Ω.

From (1.52) we obtain

$$\Delta u_2 - Ru_2 \geq -u_0[a - (b + \frac{K}{2M})u_0 + \frac{p_{-1}u_0}{2M}] - Ru_0 \quad \text{in } \Omega.$$

Subtracting the equation for u_1, and using the decreasing property of (1.46), we deduce as above that $u_2 \leq u_1$ in Ω. From the property $M \geq K\bar{a}/(2b\delta)$, we find that

$$\Delta u_0 - Ru_0 \geq -u_0[a - (b + \frac{K}{2M})u_0 + \frac{p_0 u_0}{2M}] - Ru_0 \quad \text{in } \Omega.$$

Thus, subtracting the equation for u_2, we obtain $\Delta(u_0 - u_2) - R(u_0 - u_2) \geq 0$ in Ω and deduce that $u_0 \leq u_2$ in Ω. So far, we have

$$u_0 \leq u_2 \leq u_1 \leq u_{-1} \quad \text{in } \Omega. \tag{1.56}$$

From (1.52), we obtain

$$\Delta p_1 - Rp_1 \leq -p_{-1}(a - 2bu_1) - \frac{(K - p_{-1})^2 u_2}{2M} - Rp_{-1} \quad \text{in } \Omega.$$

Subtracting the equation for p_2, and using the decreasing property of expressions in (1.45), we deduce that $\Delta(p_1 - p_2) - R(p_1 - p_2) \leq 0$ in Ω and hence $p_2 \leq p_1$ in Ω. From (1.51), we have

$$\Delta p_0 - Rp_0 \geq -p_0(a - 2bu_1) - \frac{(K - p_0)^2 u_2}{2M} - Rp_0 \quad \text{in } \Omega.$$

Subtracting the equation for p_2 in (1.52), we thus deduce that $p_0 \leq p_2$ in Ω. So far, we have

$$p_0 \leq p_2 \leq p_1 \leq p_{-1} \quad \text{in } \Omega. \tag{1.57}$$

Inequalities (1.56) and (1.57) show that

$$\begin{aligned} &u_{2n} \le u_{2n+2} \le u_{2n+1} \le u_{2n-1} \\ &p_{2n} \le p_{2n+2} \le p_{2n+1} \le p_{2n-1} \end{aligned} \qquad \text{in } \Omega \tag{1.58}$$

are true for $n = 0$.

For convenience, let the expression in (1.46) be denoted as $h(u, p)$. From the equation in (1.52), we deduce that

$$\Delta(u_1 - u_3) - R(u_1 - u_3)$$
$$= [h(u_{-1}, p_{-1}) - h(u_1, p_{-1})] + [h(u_1, p_{-1}) - h(u_1, p_1)] \quad \text{in } \Omega.$$

From the decreasing property of h in u and p, we find that

$$\Delta(u_1 - u_3) - R(u_1 - u_3) \le 0,$$

and thus $u_3 \le u_1$ in Ω. Similarly,

$$\Delta(u_2 - u_3) - R(u_2 - u_3)$$
$$= [h(u_0, p_0) - h(u_1, p_0)] + [h(u_1, p_0) - h(u_1, p_1)] \ge 0 \quad \text{in } \Omega,$$

and thus $u_2 \le u_3$ in Ω. We have now obtained $u_2 \le u_3 \le u_1$ in Ω.

We then follow the arguments as in the above paragraphs to deduce in turn that $p_2 \le p_3 \le p_1$, $u_2 \le u_4 \le u_3$, and $p_2 \le p_4 \le p_3$ in Ω. Therefore we obtain the validity of (1.58) for $n = 1$. Following the same procedures as in the above paragraphs, we prove by induction that (1.58) is true for all positive integers n. Consequently, we obtain the order relation (1.53).

To prove the second part of the theorem, we first assume the validity of (1.54). From the equations for u and u_1, we deduce that

$$\Delta(u - u_1) - R(u - u_1)$$
$$= [h(u, p) - h(u_{-1}, p)] + [h(u_{-1}, p) - h(u_{-1}, p_{-1})] \ge 0 \quad \text{in } \Omega,$$

and hence $u \le u_1$ in Ω. We next prove in turn in the same way that $p \le p_1$, $u_2 \le u$, and $p_2 \le p$ in Ω. We thus obtain (1.55) for $r = 1$.

Following the same arguments, we prove (1.55) by induction.

Remark 1.3. From Theorems 1.3 and 1.4, we find that if

$$\lim_{r\to\infty} u_{2r} = \lim_{r\to\infty} u_{2r-1} \quad \text{and} \quad \lim_{r\to\infty} p_{2r} = \lim_{r\to\infty} p_{2r-1}, \tag{1.59}$$

then the optimal control problem is determined. In other words, if the upper and lower bounds resulting from the limits of the two monotone sequences of the functions are the same, the optimality system has a unique solution with the property $(\tilde{a}-\delta)/b \le u \le \bar{a}/b, 0 \le p \le K$, and hence the unique optimal control is determined.

Remark 1.4. On the other hand, if the optimality system has more than one solution satisfying $(\tilde{a}-\delta)/b \le u \le \bar{a}/b, 0 \le p \le K$, some solution(s) might not provide optimal control(s). However, at least one solution gives the optimal control.

The following examples illustrate some practical values for the application of Theorem 1.3 and Theorem 1.4.

Example 1.1. Consider the optimal control problem (1.1) to (1.4), with

$$(x,y) \in \Omega := (0,1) \times (0,1), \quad a = 7 + sin(2\pi xy),$$

$$b = 1, \quad K = 4, \quad M = 13, \quad \delta = 4/3.$$

One readily verifies that hypotheses [H2] and (1.28) are satisfied; thus Theorems 1.3 and 1.4 are both applicable. Numerical experiments indicate that we have the situation as described in Remark 1.3. The sequences constructed by (1.52) practically provides a complete solution of the optimal control problem.

Example 1.2. Consider the optimal control problem (1.1) to (1.4), with Ω, b and K as given in Example 1.1, while a, M and δ are modified to

$$a = 7 + 4\,sin(2\pi xy), \quad M = 5, \quad \delta = 2.5.$$

Both hypotheses [H2] and (1.28) are violated. However, numerical experiments indicate that (1.59) in Remark 1.3 still holds. The proof that the optimal control can be described by Remark 1.4 remains open.

3.2 Optimal Harvesting-Coefficient Control of Steady-State Prey-Predator Diffusive Volterra-Lotka Systems

This section considers the optimal harvesting control of two interacting populations. The species concentrations satisfy a prey-predator Volterra- Lotka system under diffusion. When they are in steady state situation, they are assumed to satisfy the following system, with no-flux boundary conditions.

$$
(2.1) \qquad \begin{cases} \Delta u + u[(a_1(x) - f_1(x)) - b_1 u - c_1 v] = 0 \text{ in } \Omega, \\ \Delta v + v[(a_2(x) - f_2(x)) + c_2 u - b_2 v] = 0 \text{ in } \Omega, \\ \frac{\partial u}{\partial \nu} = \frac{\partial v}{\partial \nu} = 0 \text{ on } \partial\Omega. \end{cases}
$$

The functions $u(x), v(x)$ respectively describe prey, predator population concentrations with intrinsic growth rates $a_1(x), a_2(x)$. The functions $f_1(x), f_2(x)$ respectively denote distribution of control harvesting effect on the biological species. Such problem arises naturally in ecological systems, e.g. fisheries and agriculture, when various species are harvested for economic return. The parameters $b_i, c_i, i = 1, 2$ designate crowding and interaction effects which are assumed constant for simplicity. The optimal control criteria is to maximize profit, which is the difference between economic revenue and cost. This is expressed by the payoff functional

$$
(2.2) \qquad J(f_1, f_2) = \int_\Omega \{K_1 u f_1 + K_2 v f_2 - M_1 f_1^2 - M_2 f_2^2\}\, dx
$$

where K_1, K_2 are constants describing the price of the prey and predator species, and M_1, M_2 are constants describing the costs of the controls f_1, f_2. Here $\int_\Omega u f_1 dx$ and $\int_\Omega v f_2 dx$ represent the total harvest of respectively u, v which depends on f_i through (2.1). Roughly speaking, the object is to maximize (2.2) through controlling f_1, f_2 which determine the solution u, v of (2.1). Many analogous models of this nature had appeared in the literature, see e.g. Clark [30], and Okubo and Levin [179].

The case of competing species under control will be considered in Section 3.3. Here, the species are under prey-predator type of interaction which are usually more difficult to analyze than the competing or cooperative case, because the relation between the species is not symmetric. Nevertheless, Section 3.3 considers the time-dependent parabolic case which gives rise to other concerns. In this section for the prey-predator steady-state case, we will find explicit conditions for rigorous characterization of the optimal control. We also justify the existence of solution for the resulting nonlinear optimality system of four equations. The conditions on the various coefficients are elaborate, and some of them seem incompatible with each other. However, Example 2.1 shows that they can all be simultaneously satisfied. Our results will provide framework for further investigation to consider whether some of the hypotheses can be successively relaxed for more practical applications. In the last part of this section, we will further solve the optimality system by an iterative scheme. Due to the fact that the nonlinear terms for prey-predator case are not really monotonic in the same direction in each component, it requires special treatment to find a particular scheme so that an oscillatory sequence is obtained for approximating each component of the solution. We have assumed that the cost in the payoff functional depends quadratically on the control in the form $M_i f_i^2$ in a customary way in (2.2). One can certainly modify the condition to obtain a new payoff functional for $J(f_1, f_2)$, leading to a new system for investigation. Our prototype problem here can provide a guideline for the analysis of many other concrete similar problems which are beyond the scope of this present study. The materials in this section are mainly obtained from Leung [126].

To fix ideas, we assume Ω is a bounded domain in R^N with $\delta\Omega \in C^2; \Delta$ and $\frac{\partial}{\partial \nu}$ respectively denote the Laplacian and outward normal derivative. K_i, M_i, b_i and $c_i, i = 1, 2$ are positive constants. We also assume that

$$\begin{aligned} &a_i(x) \geq 0, f_i(x) \geq 0 \text{ a.e. in } \Omega, \\ &a_i \in L^\infty(\Omega), f_i \in L^\infty(\Omega), i = 1, 2. \end{aligned} \tag{2.3}$$

For convenience, we denote

$$L^\infty_+(\Omega) := \{f | f \in L^\infty(\Omega), f \geq 0 \text{ a.e. in} \, \Omega\}, \tag{2.4}$$

and for $\delta_i > 0, i = 1, 2$

$$\mathcal{C}(\delta_1, \delta_2) := \{(f_1, f_2) | 0 \leq f_i \leq \delta_i \text{ a.e. in } \Omega, i = 1, 2\}. \tag{2.5}$$

Finally, we denote an optimal control (if it exists) to be an $(f^*, f^*) \in \mathcal{C}(\delta_1, \delta_2)$ such that

$$J(f_1^*, f_2^*) = \sup\{J(f_1, f_2) | (f_1, f_2) \in \mathcal{C}(\delta_1, \delta_2)\}. \tag{2.6}$$

We discuss the existence and uniqueness of positive solutions (2.1). Then we show the existence of optimal control for our problem (2.1), (2.2), (2.6). We find stronger conditions which enables the characterization of an optimal control in terms of solution of an elliptic optimality system of 4 equations. Several theorems and corollaries are given with increasingly more stringent hypothesis and consequently giving rise to increasingly simpler optimality systems and results. In last part, we construct monotone sequences closing in to all appropriate solutions of an optimality system by methods analogous to that described in Chapter V of Leung [125]. If the monotone increasing and decreasing sequences (i.e. oscillatory sequence) converge to the same function, then the optimal control is unique. An example satisfying all the hypotheses is given at the end of the section. For our prey-predator control problem, we thus obtain very precise conditions on the interspecies relation and control constraints etc. so that the problem can be rigorously solved, together with a method for approximating the optimal control. For convenience, we will denote

$$\bar{a}_i = ess \sup_{x\in\Omega} a_i(x), \quad \tilde{a}_i = ess \inf_{x\in\Omega} a_i(x), \quad for \ i = 1, 2.$$

As a start, we first consider the existence of positive solution of (2.1) for an arbitrary fixed control $(f_1, f_2) \in \mathcal{C}(\delta_1, \delta_2)$. This is established in Theorem 2.1 under hypotheses $[H1]$ and $[H2]$. Furthermore, a uniform estimate is given for all solutions under such controls. However, in Theorem 2.1, the solution may not be unique. Under the further hypothesis $[H3]$, Theorem 2.2 shows the uniqueness of solution in the appropriate range, under each given control in $\mathcal{C}(\delta_1, \delta_2)$. Theorem 2.3 shows the existence of optimal control when solutions are uniquely defined for each fixed given control.

Theorem 2.1 (Positive Solution for Given Control). *Suppose that $a_i(x), b_i$, c_i and δ_i satisfy the hypothesis:*

[H1] $\quad \tilde{a}_1 - \frac{c_1}{b_2}(\bar{a}_2 + c_2\bar{a}_1/b_1) > \delta_1 > 0,$

[H2] $\quad \tilde{a}_2 > \delta_2 > 0.$

Then for each pair $(f_1, f_2) \in \mathcal{C}(\delta_1, \delta_2)$ problem (2.1) has a strictly positive solution $(u, v) = (u(f_1, f_2), v(f_1, f_2))$, i.e. $u, v > 0$ in $\bar{\Omega}$, and with each component in $W^{2,p}(\Omega)$ for any $p \in (1, \infty)$. Moreover, the estimate

$$\| u(f_1, f_2) \|_{2,p}, \| v(f_1, f_2) \|_{2,p} \le constant \tag{2.7}$$

is valid uniformly for $(f_1, f_2) \in \mathcal{C}(\delta_1, \delta_2)$.

Proof. Define constant functions

$$\psi_1(x) \equiv \tfrac{\bar{a}_1}{b_1}, \quad \psi_2(x) \equiv \tfrac{1}{b_2}(\bar{a}_2 + \tfrac{c_2\bar{a}_1}{b_1}),$$

$$\phi_1(x) \equiv \tfrac{1}{b_1}[\tilde{a}_1 - \tfrac{c_1}{b_2}(\bar{a}_2 + \tfrac{c_2\bar{a}_1}{b_1}) - \delta_1], \text{ and} \tag{2.8}$$

$$\phi_2(x) \equiv \tfrac{1}{b_2}[\tilde{a}_2 - \delta_2 + \tfrac{c_2}{b_1}\{\tilde{a}_1 - \tfrac{c_1}{b_2}(\bar{a}_2 + \tfrac{c_2\bar{a}_1}{b_1}) - \delta_1\}]$$

for all x in $\bar{\Omega}$. It is clear from [H1], [H2] that $\psi_i, \phi_i, i = 1, 2,$ are strictly positive in $\bar{\Omega}$. One can readily see that

$$\begin{aligned} &\Delta\psi_1 + \psi_1[(a_1(x) - f_1(x))) - b_1\psi_1 - c_1 v] \\ &= \tfrac{\bar{a}_1}{b_1}[(a_1(x) - f_1(x)) - \bar{a}_1 - c_1 v] \le 0, \end{aligned} \tag{2.9}$$

for all $\phi_2 \le v \le \psi_2, z \in \Omega$. Moreover, we have

$$\begin{aligned} &\Delta\phi_1 + \phi_1[(a_1(x) - f_1(x)) - b_1\phi_1 - c_1 v] \\ &= \phi_1[(a_1(x) - \tilde{a}_1) + (\delta_1 - f_1(x)) + c_1 b_2^{-1}(\bar{a}_2 + c_2\bar{a}_1 b_1^{-1}) - c_1 v] \ge 0 \end{aligned} \tag{2.10}$$

for all $\phi_2 \le v \le \psi_2, 0 \le f_1 \le \delta_1, x \in \Omega$. Similarly, we obtain

$$\Delta\psi_2 + \psi_2[(a_2(x) - f_2(x)) - b_2\psi_2 + c_2u] \le 0, \tag{2.11}$$

$$\Delta\phi_2 + \phi_2[(a_2(x) - f_2(x)) - b_2\phi_2 + c_2u] \ge 0, \tag{2.12}$$

for all $\phi_1 \le u \le \psi_1, 0 \le f_2 \le \delta_2, x \in \Omega$.

Let $X_i = \{w \in C(\bar{\Omega}), \phi_i \le w \le \psi_i\}, i = 1,2$. Define the map $T : X_1 \times X_2 \to X_1 \times X_2$ as $T(y_1, y_2) = (z_1, z_2)$ for $(y_1, y_2) \in X_1 \times X_2$, where $z_1, z_2 \in W^{2,p}(\Omega), p > 1$ and (z_1, z_2) is determined uniquely as the solution of the decoupled linear system

$$\Delta z_1 - Qz_1 + y_1[a_1(x) - f_1(x) - b_1y_1 - c_1y_2] + Qy_1 = 0 \text{ in } \Omega,$$

$$\Delta z_2 - Qz_2 + y_2[a_2(x) - f_2(x) + c_2y_1 - b_2y_2] + Qy_2 = 0 \text{ in } \Omega,$$

$$\frac{\partial z_1}{\partial \nu} = \frac{\partial z_2}{\partial \nu} = 0 \text{ on } \delta\Omega.$$

Here $Q > 0$ is a fixed constant. Using (2.9) to (2.12) and the maximum principle for $W^{2,p}(\Omega)$ solution with Neumann boundary condition we can show as in Theorem 3.1 in Leung and Fan [135] (or Theorem 4.2 in Chapter 4) that indeed $(z_1, z_2) \in X_1 \times X_2$. Using Theorem 15.1 in Agmon, Douglas and Nirenberg [1], we can obtain a bound for the $W^{2,p}(\Omega)$ norms of z_1, z_2 in terms of the $L^P(\Omega)$ norms of $y_1[a_1(x) - f_1(x) - b_1y_1 - c_1y_2] + Qy_1, y_2[a_2(x) - f_2(x) + c_2y_1 - b_2y_2] + Qy_2$. Since y_1, y_2 are all bounded functions in X_1 or X_2, we can obtain a uniform bound for the $W^{2,p}(\Omega)$ norm of z_1, z_2. Following the proof of Theorem 4.2 in Chapter 4, we can then use such bound to show that the mapping T is compact and eventually obtain a fixed point. Such fixed point is a solution of (1.1) in $X_1 \times X_2$, and the uniform bound for the $W^{2,p}(\Omega)$ norm gives precisely (2.7). For more details, see Theorem 4.2 in Chapter 4; the later part is completely analogous and is thus omitted.

Theorem 2.2 (Unique Positive Solution for Given Control). *Assume hypotheses [H1] and [H2]. Define*

$$S := min.\{\frac{b_2}{b_1}[\tilde{a}_1 - \frac{c_1}{b_2}(\bar{a}_2 + \frac{c_2\bar{a}_1}{b_1}) - \delta_1](\bar{a}_2 + \frac{c_2\bar{a}_1}{b_1})^{-1}, \ (\tilde{a}_2 - \delta_2)\frac{b_1}{\bar{a}_1 b_2}\}. \tag{2.13}$$

Suppose further that

[H3] $\qquad \frac{c_1c_2}{b_1b_2} < S^4$

is satisfied. Then for each pair $(f_1, f_2) \in \mathcal{C}(\delta_1, \delta_2)$, problem (2.1) has a unique solution $(u, v), u, v \in W^{2,p}(\Omega)$ for any $p \in (1, \infty)$, with the property that

$$\phi_1 \le u(x) \le \psi_1, \ \ \phi_2 \le v(x) \le \psi_2 \ \ in \ \bar{\Omega}. \tag{2.14}$$

Here ϕ_i, ψ_i, $i = 1, 2$ is given in (2.8).

Remark 2.1. Hypotheses $[H1]$ and $[H2]$ imply that S is positive. Hypothesis $[H3]$ is readily satisfied if c_1 or c_2 is reduced to a sufficiently small positive number.

Proof. For convenience, let

$$G_1 = b_1 b_2^{-1}[\tilde{a}_1 - \frac{c_1}{b_2}(\bar{a}_2 + \frac{c_2\bar{a}_1}{b_1}) - \delta_1]^{-1}(\bar{a}_2 + \frac{c_2\bar{a}_1}{b_1})$$

and

$$G_2 = (\tilde{a}_2 - \delta_2)^{-1} b_1^{-1} \bar{a}_1 b_2.$$

We now define $\hat{U}(x), \hat{V}(x), \tilde{U}(x)$ and $\tilde{V}(x)$ to be respectively solutions of the scalar problems:

$$\Delta\hat{U} + \hat{U}[(a_1(x) - f_1(x)) - b_1\hat{U}] = 0 \text{ in } \Omega, \ \frac{\partial \hat{U}}{\partial \nu} = 0 \quad \text{on } \partial\Omega, \tag{2.15}$$

$$\Delta\hat{V} + \hat{V}[(a_2(x) - f_2(x)) + \frac{c_2\bar{a}_1}{b_1} - b_2\hat{V}] = 0 \text{ in } \Omega, \ \frac{\partial \hat{V}}{\partial \nu} = 0 \text{ on } \partial\Omega, \tag{2.16}$$

$$\Delta\tilde{U} + \tilde{U}[(a_1(x) - f_1(x)) - b_1\tilde{U} - c_1\hat{V}(x)] = 0, \ \text{ in } \Omega, \ \frac{\partial \tilde{U}}{\partial \nu} = 0 \quad \text{on } \partial\Omega, \tag{2.17}$$

$$\Delta\tilde{V} + \tilde{V}[(a_2(x) - f_2(x)) - b_2\tilde{V}] = 0 \text{ in } \Omega, \ \frac{\partial \tilde{V}}{\partial \nu} = 0 \text{ on } \partial\Omega. \tag{2.18}$$

Using the constant functions $\bar{a}_1/b_1$ and $(1/b_1)[\tilde{a}_1 - \delta_1]$ as upper and lower solutions for (2.15), we can readily obtain by means of monotone iterations from the upper solution as in Section 5.1 in [125] that a unique solution $\hat{U}$ of (2.15) exists in $W^{2,p}(\Omega)$ for any $p \in (1, \infty)$; and $(1/b_1)(\tilde{a}_1 - \delta_1) \le \hat{U}(x) \le \bar{a}_1/b_1$ for all $x \in \bar{\Omega}$. Similarly, we obtain the unique positive solutions in $W^{2,p}(\Omega)$,

$$b_2^{-1}(\tilde{a}_2 - \delta_2) \le \hat{V}(x) \le b_2^{-1}(\bar{a}_2 + \frac{c_2\bar{a}_1}{b_1}),$$

$$b_1^{-1}[\tilde{a}_1 - \delta_1 - \frac{c_1}{b_2}(\bar{a}_2 + \frac{c_2\bar{a}_1}{b_1})] \le \tilde{U}(x) \le b_1^{-1}(\tilde{a}_1 - \delta_1),$$

$$b_2^{-1}(\tilde{a}_2 - \delta_2) \le \tilde{V}(x) \le b_2^{-1}\bar{a}_2,$$

respectively for (2.16), (2.17) and (2.18). We thus have the comparison,

$$\hat{U}(x) \le G_2\tilde{V}(x), \ \ \hat{V}(x) \le G_1\tilde{U}(x), \ \text{ in } \bar{\Omega}. \tag{2.19}$$

We next inductively define u_i, v_i to be strictly positive functions in $W^{2,p}(\Omega)$, starting with $u_1 = \hat{U}, v_1$ satisfying

$$\Delta v_1 + v_1[(a_2(x) - f_2(x)) + c_2\hat{U}(x) - b_2 v_1] = 0 \ \text{ in } \Omega, \ \frac{\partial v_1}{\partial \nu} = 0 \quad \text{on} \quad \partial\Omega,$$

and $u_i, v_i, i = 2, 3, \ldots$ satisfying
(2.20)

$$\begin{cases} \Delta u_i + u_i[(a_1(x) - f_1(x)) - b_1 u_i - c_1 v_{i-1}] = 0 \ \text{ in } \Omega, \ \frac{\partial u_i}{\partial \nu} = 0 \quad \text{on} \quad \partial\Omega, \\ \Delta v_i + v_i[(a_2(x) - f_2(x)) + c_2 u_i - b_2 v_i] = 0 \ \text{ in } \Omega, \ \frac{\partial v_i}{\partial \nu} = 0 \quad \text{on} \quad \partial\Omega. \end{cases}$$

Using the maximum principle that a function $u \in W^{2,p}(\Omega), p \geq N$, satisfying

$$\Delta u - cu \leq 0 \ \text{ in } \Omega, \ \frac{\partial u}{\partial \nu} = 0 \ \text{ on } \partial\Omega,$$

where $c > 0$ is a constant, must have the property that $u > 0$ in $\bar{\Omega}$ or $u \equiv 0$, we can deduce as in Sections 5.2 and 5.3 in [125] that

$$\begin{aligned} &\tilde{U} \leq u_2 \leq u_4 \leq u_6 \cdots \leq u_5 \leq u_3 \leq u_1 \leq \hat{U}, \\ &\tilde{V} \leq v_2 \leq v_4 \leq v_6 \cdots \leq v_5 \leq v_3 \leq v_1 \leq \hat{V}. \end{aligned} \tag{2.21}$$

for all $x \in \bar{\Omega}$. Using the Green's identity and the equations (2.20), we obtain for $i \geq 1$

$$\begin{aligned} 0 &= \textstyle\int_\Omega (u_{2i+2}\Delta u_{2i+1} - u_{2i+1}\Delta u_{2i+2})\, dx \\ &= -\textstyle\int_\Omega u_{2i+1}u_{2i+2}[b_1(u_{2i+2} - u_{2i+1}) + c_1(v_{2i+1} - v_{2i})]\, dx, \end{aligned} \tag{2.22}$$

$$0 = \int_\Omega v_{2i}v_{2i+1}[c_2(u_{2i} - u_{2i+1}) + b_2(v_{2i+1} - v_{2i})]\, dx, \tag{2.23}$$

$$0 = \int_\Omega u_{2i}u_{2i+1}[b_1(u_{2i} - u_{2i+1}) + c_1(v_{2i-1} - v_{2i})]\, dx, \tag{2.24}$$

$$0 = \int_\Omega v_{2i-1}v_{2i}[c_2(u_{2i} - u_{2i-1}) + b_2(v_{2i-1} - v_{2i})]\, dx. \tag{2.25}$$

Using (2.22), (2.23), and (2.19), (2.21) we deduce that

$$\begin{aligned} \textstyle\int_\Omega (u_{2i+1} - u_{2i+2})u_{2i+1}u_{2i+2} dx &= \tfrac{c_1}{b_1} \textstyle\int_\Omega (v_{2i+1} - v_{2i})u_{2i+1}u_{2i+2}\, dx \\ &\leq \tfrac{c_1}{b_1} \textstyle\int_\Omega G_2^2 (v_{2i+1} - v_{2i})v_{2i}v_{2i+1}\, dx \\ &= G_2^2 \tfrac{c_1 c_2}{b_1 b_2} \textstyle\int_\Omega (u_{2i+1} - u_{2i})v_{2i}v_{2i+1}\, dx. \end{aligned} \tag{2.26}$$

Then, we use (2.24), (2.25) and (2.19), (2.21) again to obtain

$$\begin{aligned}\int_\Omega (u_{2i-1}-u_{2i})u_{2i}u_{2i+1}dx &= \tfrac{c_1}{b_1}\int_\Omega (v_{2i-1}-v_{2i})u_{2i}u_{2i+1}\,dx\\ &\leq \tfrac{c_1}{b_1}\int_\Omega G_2^2(v_{2i-1}-v_{2i})v_{2i-1}v_{2i}\,dx\\ &= G_2^2\tfrac{c_1c_2}{b_1b_2}\int_\Omega (u_{2i-1}-u_{2i})v_{2i-1}v_{2i}\,dx.\end{aligned} \tag{2.27}$$

Combining (2.26), (2.27) and using (2.19), (2.21) once more, we obtain

$$\begin{aligned}&\int_\Omega (u_{2i+1}-u_{2i+2})u_{2i+1}u_{2i+2}\,dx\\ &\quad\leq G_2^4G_1^4(\tfrac{c_1}{b_1})^2(\tfrac{c_2}{b_2})^2\int_\Omega (u_{2i-1}-u_{2i})u_{2i-1}u_{2i}\,dx\end{aligned} \tag{2.28}$$

for each integer $i \geq 1$.

By means of (2.28), we conclude that if $[H3]$ is satisfied then $\lim_{i\to\infty}\int_\Omega (u_{2i+1} - u_{2i+2})u_{2i+1}u_{2i+2}\,dx = 0$. By (2.21), the limits

$$\lim_{i\to\infty} u_{2i+1} := u^* > 0 \text{ and } \lim_{i\to\infty} u_{2i+2} := u_* > 0$$

must exist. The argument above shows that $u^* = u_*$ a.e. in Ω. Further, using the maximum principle described above and the subsequently modified comparison theorem as in Theorem 5.2-1 in [125], with homogeneous Neumann boundary condition, we can show as in Section 5.2 in [125] that any solution (u,v) of (2.1) with $\tilde{U} \leq u \leq \hat{U}, \tilde{V} \leq v \leq \hat{V}$ in $\bar{\Omega}, u, v \in W^{2,p}(\Omega)$ must satisfy

$$u_* \leq u \leq u*, \ \lim_{i\to\infty} v_{2i} := v_* \leq v \leq v^* := \lim_{i\to\infty} v_{2i+1}, \ x \in \bar{\Omega}.$$

(For more details, see Theorem 5.2-4 in [125]). Since $u^* = u_*$, we can show that v^* and v_* satisfy the same equation and use comparison again as above to conclude that $v^* = v_*$ (see Theorem 5.2-3 in [125]). Comparing $\phi_i, \psi_i, i = 1, 2$ with the estimates for $\tilde{U}, \hat{U}, \tilde{V}, \hat{V}$, we conclude that any solution (u, v) of (2.1) satisfying (2.14) must have $u = u^* = u_*, v = v^* = v_*$ in $\bar{\Omega}$. The existence part follows from Theorem 2.1.

Remark 2.2. In Theorem 2.2, uniform $\|\cdot\|_{2,p}$ bound for u, v can be obtained for all $(f_1, f_2) \in \mathcal{C}(\delta_1, \delta_2), p > 1$, as in Theorem 2.1.

Remark 2.3. Under the hypotheses of Theorem 2.2, the payoff functional $J(f_1, f_2)$ is uniquely defined, if (u, v) is chosen as the one solution satisfying (2.14).

Theorem 2.3 (Existence of Optimal Control). *Assume hypotheses* $[H1]$, $[H2]$ *and that* $(u(f_1, f_2), v(f_1, f_2))$ *is defined uniquely so that (2.14) and (2.7)* *are satisfied uniformly for all* $(f_1, f_2) \in \mathcal{C}(\delta_1, \delta_2)$. *Then* $(f_1^*, f_2^*) \in \mathcal{C}(\delta_1, \delta_2)$ *exists such that* $J(f_1^*, f_2^*)$ *is the optimal control for all* $(f_1, f_2) \in \mathcal{C}(\delta_1, \delta_2)$.

Remark 2.4. By theorem 2.2 and Remark 2.3, the addition of hypothesis $[H3]$ to $[H1]$ and $[H2]$ ensures that $(u(f_1, f_2), v(f_1, f_2))$ can be chosen uniquely in a way as described in Theorem 2.3. Hence, under hypotheses $[H1]$ to $[H3]$, an optimal control does exist.

Proof. The uniform boundedness of $(u(f_1, f_2), v(f_1, f_2))$ for all $(f_1, f_2) \in \mathcal{C}(\delta_1, \delta_2)$ implies that $\sup\{J(f_1, f_2)|(f_1, f_2) \in \mathcal{C}(\delta_1, \delta_2)\} < \infty$. Let $(f_{1n}, f_{2n}) \in \mathcal{C}(\delta_1, \delta_2)$ be a maximizing sequence. Then there exists a subsequence, again denoted as (f_{1n}, f_{2n}) for convenience, so that

$$f_{in} \to f_i^* \text{ weakly in } L^2(\Omega), \text{ with } (f_1^*, f_2^*) \in \mathcal{C}(\delta_1, \delta_2),$$

and

$$u_n \equiv u(f_{1n}, f_{2n}) \to \bar{u}^*, \ v_n \equiv v(f_{1n}, f_{2n}) \to \bar{v}^* \text{ strongly in } W^{1,2}(\Omega)$$

(by using (2.7)). Passing to the limit as $n \to \infty$ in

$$\int_\Omega (\nabla u_n \nabla \varphi - (a_1 - f_{1n}) u_n \varphi + b_1 u_n^2 \varphi + c_1 u_n v_n \varphi)\, dx = 0$$

and

$$\int_\Omega (\nabla v_n \nabla \varphi - (a_2 - f_{2n}) v_n \varphi - c_2 v_n u_n \varphi + b_2 v_n^2 \varphi)\, dx = 0,$$

for all $\varphi \in W^{1,2}(\Omega) \cap L^\infty(\Omega)$, and noting that, for example,

$$\int_\Omega f_{1n} u_n \varphi dx \to \int_\Omega f_1^* \bar{u}^* \varphi\, dx \quad \text{for all } \varphi \in L^\infty(\Omega),$$

we conclude that $(\bar{u}^*, \bar{v}^*)$ is a solution of (2.1) with (f_1, f_2) replaced by (f_1^*, f_2^*). Since (u_n, v_n) are uniquely defined in a certain range of values, hence its limit $(\bar{u}^*, \bar{v}^*)$ is within the same bounds. Consequently, (2.1) implies that $\| u \|_{2,p}, \| v \|_{2,p}$ is bounded by the same constant as in (2.7). By assumption, $u(f_1^*, f_2^*)$ is uniquely defined so that such properties are satisfied. We thus conclude that $(\bar{u}^*, \bar{v}^*) = (u(f_1^*, f_2^*), v(f_1^*, f_2^*))$. Finally, the conclusion follows from semicontinuity of J; that is we have $J(f_1^*, f_2^*) = sup\{J(f_1, f_2)|(f_1, f_2) \in \mathcal{C}(\delta_1, \delta_2)\}$.

In order to further describe the optimal control, we will need stronger assumptions on the intrinsic growth rate functions $a_i(x), i = 1, 2$. When $[H1], [H2]$ are respectively strengthened to $[H1^*], [H2^*]$ and additional assumptions are made on the interaction rates between the species, Lemma 2.1 below shows the differentiability of $u(f_1, f_2)$ and $v(f_1, f_2)$ with respect to (f_1, f_2). The additional assumptions are satisfied, for instance, when the interspecies interactions are small compared with the intraspecies interactions. Theorem 2.4 gives a characterization of an optimal control in terms of solutions of an elliptic system of 4 equations. The optimal control is related to the solution of the systems in terms of various inequalities. Corollary 2.5 shows that under further assumptions on the cost and price parameters $M_i, K_i, i = 1, 2$, the optimal control can be exactly characterized by a solution of the optimality system of 4 equations. Under the additional assumption $[H5]$, Corollary 2.6 shows that the optimality system has solutions with all four components non-negative. In this last case, the solutions can be more readily found or approximated, and this last topic will be described at the end of the section.

Lemma 2.1 (Differentiability with Respect to Control). *Assume that there exist δ_1, δ_2 such that*

$[H1^*]$ $\qquad 0 < \delta_1 \leq (1/3)\{2\tilde{a}_1 - \bar{a}_1 - (2c_1/b_2)(\bar{a}_2 + c_2\bar{a}_1/b_1)\},$

$[H2^*]$ $\qquad 0 < \delta_2 \leq (1/3)\{2\tilde{a}_2 - \bar{a}_2 - c_2\bar{a}_1/b_1\},$

and that $u(f_1, f_2), v(f_1, f_2)$ is uniquely defined for all $(f_1, f_2) \in \mathcal{C}(\delta_1, \delta_2)$ in the sense described in Theorem 2.3. Further suppose

$[H4]$ $\qquad c_1\bar{a}_1/b_1 + (c_2/b_2)(\bar{a}_2 + c_2\bar{a}_1/b_1) < 2\min\{\delta_1, \delta_2, 1\}.$

Then the mappings $\mathcal{C}(\delta_1, \delta_2) \ni (f_1, f_2) \to u(f_1, f_2), v(f_1, f_2) \in W^{1,2}(\Omega)$ are differentiable in the following sense:

$$(2.29) \qquad \begin{aligned} &\left(\tfrac{u(f_1+\beta_i\bar{f}_1, f_2)-u(f_1,f_2)}{\beta_i}, \tfrac{v(f_1+\beta_i\bar{f}_1, f_2)-v(f_1,f_2)}{\beta_i}\right) \to (\xi, \eta), \\ &\left(\tfrac{u(f_1, f_2+\beta_i\bar{f}_2)-u(f_1,f_2)}{\beta_i}, \tfrac{v(f_1, f_2+\beta_i\bar{f}_2)-v(f_1,f_2)}{\beta_i}\right) \to (\tilde{\xi}, \tilde{\eta}), \end{aligned}$$

componentwise weakly in $W^{1,2}(\Omega)$ for some $\beta_i \to 0$, for any given $(f_1, f_2) \in \mathcal{C}(\delta_1, \delta_2)$ and $\bar{f}_1, \bar{f}_2 \in L^\infty(\Omega)$ such that $(f_1 + \beta_i\bar{f}_1, f_2 + \beta_i\bar{f}_2) \in \mathcal{C}(\delta_1, \delta_2)$. Further, (ξ, η) is a solution of

$$(2.30) \quad \begin{cases} \Delta\xi + [(a_1 - f_1) - 2b_1u(f_1, f_2) - c_1v(f_1, f_2)]\xi - c_1u(f_1, f_2)\eta = u(f_1, f_2)\bar{f}_1 \\ \hfill \text{in } \Omega, \\ \Delta\eta + c_2v(f_1, f_2)\xi + [(a_2 - f_2) - 2b_2v(f_1, f_2) + c_2u(f_1, f_2)]\eta = 0 \\ \frac{\partial\xi}{\partial\nu} = \frac{\partial\eta}{\partial\nu} = 0 \quad \text{on } \partial\Omega; \end{cases}$$

and $(\tilde{\xi}, \tilde{\eta})$ is a solution in Ω of

(2.31)
$$\begin{cases} \Delta\tilde{\xi} + [(a_1 - f_1) - 2b_1 u(f_1, f_2) - c_1 v(f_1, f_2)]\tilde{\xi} - c_1 u(f_1, f_2)\tilde{\eta} = 0 \\ \Delta\tilde{\eta} + c_2 v(f_1, f_2)\tilde{\xi} + [(a_2 - f_2) - 2b_2 v(f_1, f_2) + c_2 u(f_1, f_2)]\tilde{\eta} = v(f_1, f_2)\bar{f}_2 \\ \frac{\partial\tilde{\xi}}{\partial\nu} = \frac{\partial\tilde{\eta}}{\partial\nu} = 0 \quad on\ \partial\Omega. \end{cases}$$

Here, $\xi, \eta, \tilde{\xi}, \tilde{\eta}$ *are in* $W^{2,2}(\Omega)$.

Proof. From (2.1), we deduce that

$$\xi_\beta = \frac{u(f_1 + \beta\bar{f}_1, f_2) - u(f_1, f_2)}{\beta}, \quad \eta_\beta = \frac{v(f_1 + \beta\bar{f}_1, f_2) - v(f_1, f_2)}{\beta} \tag{2.32}$$

satisfy

(2.33)
$$\begin{cases} \Delta\xi_\beta + [(a_1 - f_1) - b_1 u(f_1 + \beta\bar{f}_1, f_2) - b_1 u(f_1, f_2) - c_1 v(f_1, f_2)]\xi_\beta \\ \qquad\qquad -c_1 u(f_1 + \beta\bar{f}_1, f_2)\eta_\beta = u(f_1 + \beta\bar{f}_1, f_2)\bar{f}_1 \quad \text{in } \Omega, \\ \Delta\eta_\beta + c_2 v(f_1 + \beta\bar{f}_1, f_2)\xi_\beta + [(a_2 - f_2) - b_2 v(f_1 + \beta\bar{f}_1, f_2) - b_2 v(f_1, f_2) \\ \qquad\qquad + c_2 u(f_1, f_2)]\eta_\beta = 0 \quad \text{in } \Omega, \\ \frac{\partial\xi_\beta}{\partial\nu} = \frac{\partial\eta_\beta}{\partial\nu} = 0 \text{ on } \partial\Omega, \end{cases}$$

if (f_1, f_2) and $(f_1 + \beta\bar{f}_1, f_2) \in \mathcal{C}(\delta_1, \delta_2)$. The lower bounds in (2.14) imply that

$$\begin{aligned} &b_1[u(f_1 + \beta\bar{f}_1, f_2) + u(f_1, f_2)] - a_1 + f_1 + c_1 v(f_1, f_2) \\ &\geq 2\tilde{a}_1 - \tfrac{2c_1}{b_2}(\bar{a}_2 + \tfrac{c_2\bar{a}_1}{b_1}) - 2\delta_1 - a_1 + c_1\phi_2 \geq \delta_1 \quad \text{in } \bar{\Omega}, \end{aligned} \tag{2.34}$$

where the last inequality follows from $[H1^*]$. The first equation in (2.33), and inequality (2.34) give

$$\min\{\delta_1, 1\} \parallel \xi_\beta \parallel_{1,2}^2 \leq \parallel u(f_1 + \beta\bar{f}_1, f_2) \parallel_\infty \parallel \xi_\beta \parallel_2 [c_1 \parallel \eta_\beta \parallel_2 + \parallel \bar{f}_1 \parallel_2]. \tag{2.35}$$

The bounds in (2.14) and $[H2^*]$ also imply that

$$b_2[v(f_1 + \beta\bar{f}_1, f_2) + v(f_1, f_2)] - a_2 + f_2 - c_2 u(f_1, f_2) \geq \delta_2 \text{ in } \bar{\Omega}.$$

The last inequality and the second equation in (2.33) give

$$\min\{\delta_2, 1\} \parallel \eta_\beta \parallel_{1,2}^2 \leq c_2 \parallel v(f_1 + \beta\bar{f}_1, f_2) \parallel_\infty \parallel \xi_\beta \parallel_2 \parallel \eta_\beta \parallel_2 . \tag{2.36}$$

Since $\| u(f_1+\beta\bar{f}_1, f_2) \|_\infty \leq \bar{a}_1/b_1$ and $\| v(f_1+\beta\bar{f}_1, f_2) \|_\infty \leq (1/b_2)(\bar{a}_2+c_2\bar{a}_1/b_1)$, inequalities (2.35), (2.36) and $[H4]$ yield

$$\begin{aligned}\min\{\delta_1, 1\} \| \xi_\beta \|_{1,2}^2 + \min\{\delta_2, 1\} \| \eta_\beta \|_{1,2}^2 \\ \leq k \min\{\delta_1, \delta_2, 1\} \| \xi_\beta \|_2 \| \eta_\beta \|_2 + \text{ const. } \| \xi_\beta \|_2\end{aligned} \tag{2.37}$$

for some $k \in (0, 2)$. Inequality (2.37) hence leads to

$$\hat{k}(\| \xi_\beta \|_{1,2}^2 + \| \eta_\beta \|_{1,2}^2) \leq \text{const. } \| \xi_\beta \|_{1,2} \tag{2.38}$$

for some $\hat{k} > 0$. This gives a uniform bound for $\| \xi_\beta \|_{1,2}$ and $\| \eta_\beta \|_{1,2}$ for all $(f_1, f_2), (f_1 + \beta\bar{f}_1, f_2) \in \mathcal{C}(\delta_1, \delta_2)$ with $\bar{f}_1$ fixed. We can therefore choose a sequence $\beta_i \to 0$ such that we have a weakly convergent sequence as described in (2.29).

Since $c_1 u(f_1, f_2)\eta + u(f_1, f_2)\bar{f}_1$ is a function in $L^2(\Omega)$, the first equation in (2.30) and the results in Agmon, Douglis and Nirenberg [1] imply that $\xi \in W^{2,2}(\Omega)$. Similarly, the second equation in (2.30) implies that $\eta \in W^{2,2}(\Omega)$.

Analogously, we obtain the result involving the second part of (2.29) and the solution $(\tilde{\xi}, \tilde{\eta})$ of (2.31). This proves the lemma.

For convenience, we next denote constants

$$E_1 = E_1(\bar{a}_1, \tilde{a}_1, \bar{a}_2, b_1, b_2, c_1, c_2) = 2\tilde{a}_1 - \bar{a}_1 - \frac{2c_1}{b_2}(\bar{a}_2 + \frac{c_2\bar{a}_1}{b_1}), \tag{2.39}$$

$$E_2 = E_2(\bar{a}_2, \tilde{a}_2, \bar{a}_1, b_1, c_2) = 2\tilde{a}_2 - \bar{a}_2 - \frac{c_2\bar{a}_1}{b_1}. \tag{2.40}$$

In order to obtain better characterization of the optimal control of the problem, we need to strengthen hypotheses $[H1^*]$ and $[H2^*]$ to

$[H1^{**}]$ $\quad 0 < \delta_1 \leq \min\{(1/4)E_1, (1/3)[E_1 + \frac{c_1\bar{a}_2}{b_2 2} - \frac{c_2}{b_2}(\bar{a}_2 + \frac{c_2\bar{a}_1}{b_1})\frac{K_2}{2K_1}]\}$,

$[H2^{**}]$ $\quad 0 < \delta_2 \leq (1/4)[E_2 - \frac{8c_1^2 c_2 K_1}{b_1^2 b_2 K_2}\bar{a}_1^2(\bar{a}_2 + \frac{c_2\bar{a}_1}{b_1})E_1^{-1}E_2^{-1}]$.

Note that the right hand sides of $[H1^*]$ and $[H2^*]$ are respectively $\frac{1}{3}E_1$ and $\frac{1}{3}E_2$. We will not be looking for the best possible sufficient condition for the characterization; the hypotheses $[H1^{**}]$ and $[H2^{**}]$ will be used because they can be readily satisfied if c_1 and/or c_2 are sufficiently small.

Theorem 2.4 (Characterization of the Optimal Control). *Assume hypotheses* $[H1^{**}], [H2^{**}], [H4]$ *and that* $(u(f_1, f_2), v(f_1, f_2))$ *is uniquely defined for all* $(f_1, f_2) \in \mathcal{C}(\delta_1, \delta_2)$ *in the sense described in Theorem 2.3. Suppose*

$(f_1^*, f_2^*) \in \mathcal{C}(\delta_1, \delta_2)$ is an optimal control. Then, let (u, v, p_1, p_2) be any solution of

$$\begin{cases} \Delta u + (a_1 - f_1^*)u - b_1 u^2 - c_1 uv = 0 \quad \text{in } \Omega, \\ \Delta v + (a_2 - f_2^*)v + c_2 uv - b_2 v^2 = 0 \quad \text{in } \Omega, \\ \Delta p_1 + (a_1 - f_1^*)p_1 - (2b_1 u + c_1 v)p_1 + c_2 v p_2 = -K_1 f_1^* \quad \text{in } \Omega, \\ \Delta p_2 + (a_2 - f_2^*)p_2 - (2b_2 v - c_2 u)p_2 - c_1 u p_1 = -K_2 f_2^* \quad \text{in } \Omega, \\ \frac{\partial u}{\partial \nu} = \frac{\partial v}{\partial \nu} = \frac{\partial p_1}{\partial \nu} = \frac{\partial p_2}{\partial \nu} = 0 \quad \text{on } \Omega, \end{cases} \tag{2.41}$$

with $u, v, p_1, p_2 \in H^{2,2}(\Omega)$, satisfying

$$\begin{aligned} &\phi_1 \le u \le \psi_1, \ \ \phi_2 \le v \le \psi_2 \ \ \text{in } \Omega, \\ &-4\bar{a}_1 c_1 c_2 K_1 (\bar{a}_2 + \tfrac{c_2 \bar{a}_1}{b_1})(b_1 b_2 E_1 E_2)^{-1} \le p_1 \le K_1 \quad \text{in } \Omega, \\ &-2\bar{a}_1 c_1 K_1 (b_1 E_2)^{-1} \le p_2 \le \tfrac{K_2}{2} \quad \text{in } \Omega, \end{aligned} \tag{2.42}$$

then the control (f_1^*, f_2^*) must satisfy, for $i = 1, 2$,

$$f_i^*(x) \ge \frac{(K_i - p_i(x))u_i(x)}{2M_i} \quad \text{in } \{x \in \Omega | f_i^*(x) < \delta_i\} \ \ a.e., \tag{2.43}$$

$$f_i^*(x) \le \frac{(K_i - p_i(x))u_i(x)}{2M_i} \quad \text{in } \{x \in \Omega | f_i^*(x) > 0\} \ \ a.e., \tag{2.44}$$

$$f_i^*(x) = \frac{(K_i - p_i(x))u_i(x)}{2M_i} \quad \text{in } \{x \in \Omega | 0 < f_i^*(x) < \delta_i\} \ \ a.e., \tag{2.45}$$

(Here, we denote $(u, v) = (u_1, u_2)$ for convenience. Recall $\phi_i, \psi_i, i = 1, 2$ are defined in (2.8).)

Proof. Since hypotheses $[H1^{**}]$ and $[H2^{**}]$ imply $[H1]$ and $[H2]$, Theorem 2.3 insures the existence of an optimal control $(f_1^*, f_2^*) \in \mathcal{C}(\delta_1, \delta_2)$.

For $\bar{f}_1 \in L_+^\infty(\Omega), \epsilon > 0$, define

$$\bar{f}_1^\epsilon = \begin{cases} \bar{f}_1 & \text{if } f_1^* \le \delta_1 - \epsilon ||\bar{f}_1||_\infty, \\ 0 & \text{elsewhere.} \end{cases} \tag{2.46}$$

Then for $\beta > 0$ small enough, we have $J(f_1^*, f_2^*) \geq J(f_1^* + \beta \bar{f}_1^\epsilon, \bar{f}_2^*)$. Dividing by β, and letting β tend to zero appropriately as in Lemma 2.1, we obtain

$$\int_\Omega K_1 f_1^* \xi + K_1 u(f_1^*, f_2^*) \bar{f}_1^\epsilon + K_2 f_2^* \eta - 2M_1 f_1^* \bar{f}_1^\epsilon \, dx \leq 0, \tag{2.47}$$

where (ξ, η) is a solution of (2.30) with $f_1, f_2, \bar{f}_1$, respectively replaced by $f_1^*, f_2^*, \bar{f}_1^\epsilon$.

Let (p_1, p_2) be any solution in Ω of

$$\begin{cases} -\Delta p_1 - (a_1 - f_1^*)p_1 + (2b_1 u(f_1^*, f_2^*) + c_1 v(f_1^*, f_2^*))p_1 - c_2 v(f_1^*, f_2^*)p_2 = K_1 f_1^*, \\ -\Delta p_2 - (a_2 - f_2^*)p_2 + (2b_2 v(f_1^*, f_2^*) - c_2 u(f_1^*, f_2^*))p_2 + c_1 u(f_1^*, f_2^*)p_1 = K_2 f_2^*, \\ \frac{\partial p_1}{\partial \nu} = \frac{\partial p_2}{\partial \nu} = 0 \quad \text{on } \partial\Omega. \end{cases} \tag{2.48}$$

Replacing $K_1 f_1^*$ and $K_2 f_2^*$ in (2.47) by the left-hand side of (2.48) and integrating by parts, we obtain by means of the equation for (ξ, η) that

$$\int_\Omega \bar{f}_1^\epsilon [p_1 u(f_1^*, f_2^*) - K_1 u(f_1^*, f_2^*) + 2M_1 f_1^*] \, dx \geq 0. \tag{2.49}$$

Letting $\epsilon \to 0^+$, (2.49) leads to

$$f_1^*(x) \geq \frac{(K_1 - p_1(x))u(f_1^*, f_2^*)}{2M_1} \quad \text{in } \{x \in \Omega | f_1^*(x) < \delta_1\}. \tag{2.50}$$

This proves (2.43) for $i = 1$. The rest of the proof for (2.44) and (2.45) is analogous to that of Theorem 1.3 in Section 3.1, the details are thus omitted here. Comparing (2.48) with (2.41) and noting the definition of $u(f_1^*, f_2^*), v(f_1^*, f_2^*)$, we see that it remains to show that (u, v, p_1, p_2) as described in the statement of the theorem actually exists.

The proof of the existence of solution with range of values as described in (2.42) is carried out as in Theorem 2.1 by using upper and lower solutions for system. For convenience, we denote

$$\begin{aligned} &\psi_3(x) = K_1, \qquad \phi_3(x) = -4\bar{a}_1 c_1 c_2 K_1 (\bar{a}_2 + \tfrac{c_2 \bar{a}_1}{b_1})(b_1 b_2 E_1 E_2)^{-1}, \\ &\psi_4(x) = K_2/2, \quad \phi_4(x) = -2\bar{a}_1 c_1 K_1 (b_1 E_2)^{-1} \end{aligned} \tag{2.51}$$

for $x \in \bar{\Omega}$. Consider for all $\phi_1 \leq u \leq \psi_1, \phi_2 \leq v \leq \psi_2, \phi_4 \leq p_2 \leq \psi_4$, that the expression

$$\Delta\psi_3 + (a_1 - f_1^*)\psi_3 - (2b_1 u + c_1 v)\psi_3 + c_2 v p_2 + K_1 f_1^*$$

$$\leq \bar{a}_1 K_1 - 2[\tilde{a}_1 - \tfrac{c_1}{b_2}(\bar{a}_2 + \tfrac{c_2\bar{a}_1}{b_1}) - \delta_1]K_1 - c_1\tfrac{1}{b_2}(\tilde{a}_2 - \delta_2)K_1$$

$$+\tfrac{c_2}{b_2}(\bar{a}_2 + \tfrac{c_2\bar{a}_1}{b_1})\tfrac{K_2}{2} + K_1\delta_1$$

$$\leq K_1[\bar{a}_1 - 2\tilde{a}_1 + \tfrac{2c_1}{b_2}(\bar{a}_2 + \tfrac{c_2\bar{a}_1}{b_1})] - \tfrac{c_1\bar{a}_2}{b_2 2}K_1 + \tfrac{c_2}{b_2}(\bar{a}_2 + \tfrac{c_2\bar{a}_1}{b_1})\tfrac{K_2}{2} + 3K_1\delta_1$$

is true in Ω, since $\delta_2 \leq (1/2)E_2$ implies $\tilde{a}_2 - \delta_2 \geq \bar{a}_2/2$. Thus we have for such situations,

$$\Delta\psi_3 + (a_1 - f_1^*)\psi_3 - (2b_1 u + c_1 v)\psi_3 + c_2 v p_2 + K_1 f_1^* \leq 0 \tag{2.52}$$

provided

$$\delta_1 \leq \frac{1}{3}[E_1 + \frac{c_1\bar{a}_2}{b_2 2} - \frac{c_2}{b_2}(\bar{a}_2 + \frac{c_2\bar{a}_1}{b_1})\frac{K_2}{2K_1}],$$

which is assumed in $[H1^{**}]$.

For all $\phi_1 \leq u \leq \psi_1, \phi_2 \leq v \leq \psi_2, \phi_3 \leq p_1 \leq \psi_3, x \in \Omega$, consider the expression

$$\Delta\psi_4 + (a_2 - f_2^*)\psi_4 - (2b_2 v - c_2 u)\psi_4 - c_1 u p_1 + K_2 f_2^*$$

$$\leq \tfrac{\bar{a}_2 K_2}{2} - 2(\tilde{a}_2 - \delta_2)\tfrac{K_2}{2} + c_2\tfrac{\bar{a}_1 K_2}{b_1 2}$$

$$+ c_1\tfrac{\bar{a}_1}{b_1}4\bar{a}_1 c_1 c_2 K_1(\bar{a}_2 + \tfrac{c_2\bar{a}_1}{b_1})(b_1 b_2 E_1 E_2)^{-1} + K_2\delta_2.$$

The above expression is ≤ 0 provided that

$$[\frac{\bar{a}_2}{2} - \tilde{a}_2 + \frac{c_2\bar{a}_1}{2b_1} + 2\delta_2]K_2 + K_1 4\bar{a}_1^2 c_1^2 c_2(\bar{a}_2 + \frac{c_2\bar{a}_1}{b_1})(b_1^2 b_2 E_1 E_2)^{-1} \leq 0.$$

Consequently, hypothesis $[H2^{**}]$ implies that

$$\Delta\psi_4 + (a_2 - f_2^*)\psi_4 - (2b_2 v - c_2 u)\psi_4 - c_1 u p_1 + K_2 f_2^* \leq 0 \tag{2.53}$$

in Ω for the appropriate u, v, p_1 described above.

For $\phi_1 \leq u \leq \psi_1, \phi_2 \leq v \leq \psi_2, \phi_4 \leq p_2 \leq \psi_4, x \in \Omega$, we have

$$\Delta\phi_3 + (a_1 - f_1^*)\phi_3 - (2b_1 u + c_1 v)\phi_3 + c_2 v p_2 + K_1 f_1^*$$

$$\geq -4\bar{a}_1 c_1 c_2 K_1(\bar{a}_2 + \tfrac{c_2\bar{a}_1}{b_1})(b_1 b_2 E_1 E_2)^{-1}$$

$$\times\{\bar{a}_1 - 2[\tilde{a}_1 - \tfrac{c_1}{b_2}(\bar{a}_2 + \tfrac{c_2\bar{a}_1}{b_1}) - \delta_1] - \tfrac{c_1\bar{a}_2}{2b_2}\}$$

$$- \tfrac{c_2}{b_2}(\bar{a}_2 + \tfrac{c_2\bar{a}_1}{b_1}) \cdot 2\bar{a}_1 c_1 K_1(b_1 E_2)^{-1}$$

is valid because $\delta_2 \leq (1/2)E_2$ implies $v \geq (\tilde{a}_2 - \delta_2)/b_2 \geq \bar{a}_2/(2b_2)$. Consequently

$$\Delta\phi_3 + (a_1 - f_1^*)\phi_3 - (2b_1 u + c_1 v)\phi_3 + c_2 v p_2 + K_1 f_1^* \geq 0 \tag{2.54}$$

in the described region provided that

$$2\delta_1 \leq 2\tilde{a}_1 - \bar{a}_1 - 2\frac{c_1}{b_2}(\bar{a}_1 + \frac{c_2\bar{a}_2}{b_1}) + \frac{\bar{a}_2 c_1}{2b_2} - \frac{E_1}{2} = \frac{E_1}{2} + \frac{\bar{a}_2 c_1}{2b_2},$$

which is clearly true due to $[H1^{**}]$.

For $\phi_1 \leq u \leq \psi_1, \phi_2 \leq v \leq \psi_2, \phi_3 \leq p_1 \leq \psi_3, x \in \Omega$, we have

$$\begin{aligned}&\Delta\phi_4 + (a_2 - f_2^*)\phi_4 - (2b_2 v - c_2 u)\phi_4 - c_1 u p_1 + K_2 f_2^* \\ &\geq 2\bar{a}_1 c_1 K_1 (b_1 E_2)^{-1}\{-\bar{a}_2 + 2[\tilde{a}_2 - \delta_2] - (\tfrac{c_2\bar{a}_1}{b_1})\} - c_1 \tfrac{\bar{a}_1}{b_1} K_1.\end{aligned}$$

Hypotheses $[H2^{**}]$ implies that $\delta_2 < E_2/4$, and thus

$$-\bar{a}_2 + 2[\tilde{a}_2 - \delta_2] - \frac{c_2\bar{a}_1}{b_1} > \frac{1}{2}[2\tilde{a}_2 - \bar{a}_2 - \frac{c_2\bar{a}_1}{b_1}] = \frac{1}{2}E_2.$$

Consequently, we have

$$\begin{aligned}&\Delta\phi_4 + (a_2 - f_2^*)\phi_4 - (2b_2 v - c_2 u)\phi_4 - c_1 u p_1 + K_2 f_2^* \\ &> 2\bar{a}_1 c_1 K_1 (b_1 E_2)^{-1} \cdot \tfrac{1}{2}E_2 - \tfrac{c_1\bar{a}_1}{b_1}K_1 = 0\end{aligned} \tag{2.55}$$

in the region described.

Since the first two equations of (2.41) are independent of p_1, p_2, we can show $\Delta\psi_1 + (a_1 - f_1^*)\psi_1 - b_1\psi_1^2 - c_1 v\psi_1 \geq 0, \Delta\phi_1 + (a_1 - f_1^*)\phi_1 - b_1\phi_1^2 - c_1 v\phi_1 \leq 0$ for all $\phi_2 \leq v \leq \psi_2, \phi_3 \leq p_1 \leq \psi_3, \phi_4 \leq p_2 \leq \psi_4, x \in \Omega$ in exactly the same way as in Theorem 2.1. Similarly, we show $\Delta\psi_2 + (a_2 - f_2^*)\psi_2 + c_2 u\psi_2 - b_2\psi_2^2 \leq 0, \Delta\phi_2 + (a_2 - f_2^*)\phi_2 + c_2 u\phi_2 - b_2\phi_2^2 \geq 0$ for all $\phi_1 \leq u \leq \psi_1, \phi_3 \leq p_1 \leq \psi_3, \phi_4 \leq p_2 \leq \psi_4, x \in \Omega$. Then we follow the same method in the last part of the proof of Theorem 2.1 to construct a natural mapping T from $X_1 \times X_2 \times X_3 \times X_4$ into itself, where $X_i = \{w \in C(\bar{\Omega}) : \phi_i \leq w \leq \psi_i\}, i = 1, 2, 3, 4$. Following the same arguments, the fixed point of T gives a solution of (2.41) with each component in $H^{2,2}(\Omega)$.

Corollary 2.5. *Assume all the hypotheses of Theorem 2.4; and moreover*

$$M_1 > \frac{K_1\bar{a}_1}{2b_1\delta_1}[1 + 4\bar{a}_1 c_1 c_2(\bar{a}_2 + \frac{c_2\bar{a}_1}{b_1})(b_1 b_2 E_1 E_2)^{-1}], \tag{2.56}$$

$$M_2 > \frac{1}{2b_2\delta_2}(\bar{a}_2 + \frac{c_2\bar{a}_1}{b_1})[K_2 + 2\bar{a}_1 c_1 K_1 (b_1 E_2)^{-1}], \tag{2.57}$$

where

$$E_1 = 2\tilde{a}_1 - \bar{a}_1 - \frac{2c_1}{b_2}(\bar{a}_2 + \frac{c_2\bar{a}_1}{b_1}), \quad E_2 = 2\tilde{a}_2 - \bar{a}_2 - \frac{c_2\bar{a}_1}{b_1}.$$

Let $(f_1^*, f_2^*) \in \mathcal{C}(\delta_1, \delta_2)$ *be an optimal control. Then*

$$f_1^* = \frac{(K_1 - p_1)u}{2M_1}, \quad f_2^* = \frac{(K_2 - p_2)v}{2M_2}, \tag{2.58}$$

where (u, v, p_1, p_2) *is a solution of the optimality system:*

$$\begin{cases} \Delta u + [(a_1 - \frac{(K_1-p_1)u}{2M_1}) - b_1 u - c_1 v]u = 0 \quad \text{in } \Omega, \\ \Delta v + [(a_2 - \frac{(K_2-p_2)v}{2M_2}) + c_2 u - b_2 v]v = 0 \quad \text{in } \Omega, \\ \Delta p_1 + a_1 p_1 + \frac{(K_1-p_1)^2 u}{2M_1} - (2b_1 u + c_1 v)p_1 + c_2 v p_2 = 0 \quad \text{in } \Omega, \\ \Delta p_2 + a_2 p_2 + \frac{(K_2-p_2)^2 v}{2M_2} - (2b_2 v - c_2 u)p_2 - c_1 u p_1 = 0 \quad \text{in } \Omega, \\ \frac{\partial u}{\partial \nu} = \frac{\partial v}{\partial \nu} = \frac{\partial p_1}{\partial \nu} = \frac{\partial p_2}{\partial \nu} = 0 \quad \text{on } \Omega, \end{cases} \tag{2.59}$$

with $u, v, p_1, p_2 \in H^{2,2}(\Omega)$, *satisfying (2.42).*

Proof. By Theorem 2.4, (f_1^*, f_2^*) must satisfy (2.43) to (2.45), where (u, v, p_1, p_2) is a solution of (2.41) with conditions (2.42).

From (2.43), in the set $\{x \in \Omega | f_1^*(x) = 0\}$ we have

$$0 = f_1^*(x) \geq \frac{(K_1 - p_1(x))u(x)}{2M_1} \geq 0. \tag{2.60}$$

Thus the first inequality in (2.58) must hold in this set. From (2.44), in the set $\{x \in \Omega | f_1^*(x) = \delta_1\}$, we have

$$\begin{aligned} \delta_1 = f_1^*(x) &\leq \frac{(K_1 - p_1(x))}{2M_1} u(x) \\ &\leq \frac{\bar{a}_1}{2M_1 b_1}[K_1 + 4\bar{a}_1 c_1 c_2 K_1(\bar{a}_2 + \frac{c_2\bar{a}_1}{b_1})(b_1 b_2 E_1 E_2)^{-1}] < \delta_1 \end{aligned} \tag{2.61}$$

due to (2.42) and (2.56). Thus there are no $x \in \Omega$ where $f_1^*(x) = \delta_1$. From (2.45) and (2.60), the first equality in (2.58) holds for all $x \in \Omega$.

From (2.43), in the set $\{x \in \Omega | f_2^*(x) = 0\}$, we have

$$0 = f_2^*(x) \geq \frac{(K_2 - p_2(x))v(x)}{2M_2} \geq 0. \tag{2.62}$$

Thus the second inequality in (2.58) must hold in this set. From (2.44), in the set $\{x \in \Omega | f_2^*(x) = \delta_2\}$, we have

$$\begin{aligned}\delta_2 = f_2^*(x) &\leq \tfrac{(K_2 - p_2(x))}{2M_2} v(x) \\ &\leq \tfrac{1}{2M_2 b_2}(\bar{a}_2 + \tfrac{c_2 \bar{a}_1}{b_1})[K_2 + 2\bar{a}_1 c_1 K_1 (b_1 E_2)^{-1}] < \delta_2\end{aligned} \tag{2.63}$$

due to (2.42) and (2.57). Thus there are no $x \in \Omega$ where $f_2^*(x) = \delta_2$. From (2.45) and (2.62), the second inequality in (2.58) holds for all $x \in \Omega$. Combining (2.41) and (2.58), we conclude that (u, v, p_1, p_2) satisfies (2.59). This completes the proof.

Remark 2.5. If some additional conditions are made, we can show with a little more care that the solution (u, v, p_1, p_2) of (2.59), which characterizes the optimal control, is actually positive. More precisely, assume all the hypotheses of Corollary 2.5, and further

$$[H5] \qquad c_1 \leq \tfrac{K_2^2 b_1 \bar{a}_2}{8K_1 b_2 \bar{a}_1 M_2}.$$

Let $(f_1^*, f_2^*) \in \mathcal{C}(\delta_1, \delta_2)$ be an optimal control. Then (f_1^*, f_2^*) satisfies (2.58), where (u, v, p_1, p_2) is a solution of the optimality system (2.59) with each component in $H^{2,2}(\Omega)$ satisfying

$$\begin{aligned}&\phi_1 \leq u \leq \psi_1, \;\; \phi_2 \leq v \leq \psi_2 && \text{in } \Omega, \\ &0 \leq p_1 \leq K_1, \;\; 0 \leq p_2 \leq K_2/2 && \text{in } \Omega.\end{aligned} \tag{2.64}$$

The arguments are the same as in Theorem 2.4 and Corollary 2.5, redefining ϕ_3, ϕ_4 in (2.51) as the trivial function. Details can be found in Corollary 3.2 in Leung [126].

In the remaining part of this section, we further deduce a constructive method of approximating or computing the positive solution (u, v, p_1, p_2) of the optimality system. We will now always assume all the hypotheses of Corollary 2.5 together with [H5] as in Remark 2.5, so that a positive solution of the optimality system (2.59) can be found with the property (2.64). We proceed to construct monotone sequences converging from above and below to give upper and lower estimates for (u, v, p_1, p_2). In the case when the limits of the upper and lower estimates agree, then the optimal control problem is completely solved.

Choose a large constant $R > 0$ so that the following four expressions:

$$-[a_1(x) - \tfrac{(K_1-p_1)u}{2M_1} - b_1 u - c_1 v]u - Ru,$$

$$-[a_2(x) - \tfrac{(K_2-p_2)v}{2M_2} + c_2 u - b_2 v]v - Rv,$$

$$-[a_1(x)p_1 + \tfrac{(K_1-p_1)^2 u}{2M_1} - (2b_1 u + c_1 v)p_1 + c_2 v p_2] - Rp_1, \text{ and}$$

$$-[a_2(x)p_2 + \tfrac{(K_2-p_2)^2 v}{2M_2} - (2b_2 v - c_2 u)p_2 - c_1 v p_1] - Rp_2$$

are decreasing respectively in the four corresponding variables u, v, p_1, p_2 for all $x \in \Omega, \phi_1 \le u \le \psi_1, \phi_2 \le v \le \psi_2, 0 \le p_1 \le K_1, 0 \le p_2 \le K_2/2$, when the other three variables are all fixed. (Recall the definitions of $\phi_i, \psi_i, i = 1,2$ in (2.8)). For convenience, let

$$u_0 \equiv \phi_1, \quad u_{-1} \equiv \psi_1; \quad v_0 \equiv \phi_2, \quad v_{-1} \equiv \psi_2;$$
$$(2.65)$$
$$p_{1,0} \equiv 0, \quad p_{1,-1} \equiv K_1; \quad p_{2,0} \equiv 0, \quad p_{2,-1} \equiv K_2/2.$$

We can readily verify that these constant functions satisfy
(2.66)

$$\Delta u_{-1} - Ru_{-1} \le -u_{-1}[a_1(x) - \frac{(K_1 - p_{1,-1})u_{-1}}{2M_1} - b_1 u_{-1} - c_1 v_0] - Ru_{-1} \quad \text{in } \Omega$$

and

$$(2.67) \qquad \Delta u_0 - Ru_0 \ge -u_0[a_1(x) - \frac{(K_1 - p_1)u_0}{2M_1} - b_1 u_0 - c_1 v] - Ru_0 \quad \text{in } \Omega$$

for each v, p_1 respectively in the intervals $[v_0, v_{-1}], [p_{1,0}, p_{1,-1}]$. The last inequality is true because

$$a_1(x) - \tfrac{(K_1-p_1)u_0}{2M_1} - b_1 u_0 - c_1 v \ge a_1 - \tfrac{K_1}{2M_1}\phi_1 - b_1\phi_1 - c_1\psi_2$$

$$\ge a_1 - \tfrac{\delta_1}{\tilde{a}_1}\tilde{a}_1 - \tilde{a}_1 + \delta_1 > 0$$

by using (2.56). Moreover, we have

$$(2.68)\ \Delta v_{-1} - Rv_{-1} \ge -v_{-1}[a_2(x) - \frac{(K_2 - p_2)v_{-1}}{2M_2} + c_2 u - b_2 v_{-1}] - Rv_{-1} \quad \text{in } \Omega$$

for each u, p_2 respectively in the interval $[u_0, u_{-1}], [p_{2,0}, p_{2,-1}]$, since

$$a_2(x) - \tfrac{(K_2-p_2)v_{-1}}{2M_2} + c_2 u - b_2 v_{-1} \le a_2(x) + c_2\psi_1 - b_2\psi_2$$

$$= a_2(x) - \bar{a}_2 \le 0.$$

On the other hand,

$$\Delta v_0 - Rv_0 \geq -v_0[a_2(x) - \frac{(K_2-p_2)v_0}{2M_2} + c_2 u - b_2 v_0] - Rv_0 \quad \text{in } \Omega \tag{2.69}$$

for each $u \in [u_0, u_{-1}], p_2 \in [p_{2,0}, p_{2,-1}]$, since

$$\begin{aligned} a_2(x) - \tfrac{(K_2-p_2)v_0}{2M_2} + c_2 u - b_2 v_0 &\geq a_2(x) - \tfrac{K_2}{2M_2}\phi_2 - b_2\phi_2 + c_2\phi_1 \\ &> a_2(x) - \tfrac{K_2}{2M_2 b_2}(\bar{a}_2 + \tfrac{c_1\bar{a}_1}{b_1}) - \tilde{a}_2 + \delta_2 \\ &> a_2(x) - \delta_2 - \tilde{a}_2 + \delta_2 \ \geq 0. \end{aligned}$$

(Here, we have used the hypothesis (2.57).) Further, we verify

$$\begin{aligned} &\Delta p_{1,-1} - Rp_{1,-1} \\ &\leq -[a_1(x)p_{1,-1} + \tfrac{(K_1-p_{1,-1})^2 u}{2M_1} - (2b_1u_0 + c_1v_0)p_{1,-1} + c_2vp_2] - Rp_{1,-1} \quad \text{in } \Omega \end{aligned} \tag{2.70}$$

for each $u \in [u_0, u_{-1}], v \in [v_0, v_{-1}], p_2 \in [p_{2,0}, p_{2,-1}]$, since

$$\begin{aligned} &a_1(x)p_{1,-1} + \tfrac{(K_1-p_{1,-1})^2 u}{2M_1} - (2b_1u_0 + c_1v_0)p_{1,-1} + c_2vp_2 \\ &\leq K_1[\bar{a}_1 - 2\tilde{a}_1 + 2\delta_1 + \tfrac{2c_1}{b_2}(\bar{a}_2 + \tfrac{c_2\bar{a}_1}{b_1}) - \tfrac{\bar{a}_2c_1}{2b_2}] + \tfrac{c_2}{b_2}(\bar{a}_2 + \tfrac{c_2\bar{a}_1}{b_1})\tfrac{K_2}{2} \leq 0. \end{aligned}$$

The last inequality is true due to hypothesis $[H1^{**}]$. We also have

$$\begin{aligned} &\Delta p_{1,0} - Rp_{1,0} \\ &\geq -[a_1(x)p_{1,0} + \tfrac{(K_1-p_{1,0})^2 u}{2M_1} - (2b_1u + c_1v)p_{1,0} + c_2vp_2] - Rp_{1,0} \quad \text{in } \Omega \end{aligned} \tag{2.71}$$

for each $u \in [u_0, u_{-1}], v \in [v_0, v_{-1}], p_2 \in [p_{2,0}, p_{2,-1}]$, since

$$a_1(x)p_{1,0} + \frac{(K_1-p_{1,0})^2 u}{2M_1} - (2b_1u_0 + c_1v_0)p_{1,0} + c_2vp_2 \geq \frac{K_1^2 u_0}{2M_1} > 0 \ \text{ in } \Omega.$$

For the last component, we have

$$\begin{aligned} &\Delta p_{2,-1} - Rp_{2,-1} \\ &\leq -[a_2(x)p_{2,-1} + \tfrac{(K_2-p_{2,-1})^2 v}{2M_2} - (2b_2v_0 - c_2u)p_{2,-1} - c_1u_0p_{1,0}] - Rp_{2,-1} \quad \text{in } \Omega \end{aligned} \tag{2.72}$$

for each $u \in [u_0, u_{-1}], v \in [v_0, v_{-1}]$, since

$$\begin{aligned} &a_2(x)p_{2,-1} + \tfrac{(K_2-p_{2,-1})^2 v}{2M_2} - (2b_2v_0 - c_2u)p_{2,-1} - c_1u_0p_{1,0} \\ &\leq \tfrac{\bar{a}_2K_2}{2} + \tfrac{K_2}{2}\tfrac{\delta_2}{2} - K_2\tilde{a}_2 + K_2\delta_2 + c_2\tfrac{\bar{a}_1}{b_1}\tfrac{K_2}{2} < 0. \end{aligned}$$

In the last line, we use hypotheses (2.57) and $[H2^{**}]$. Moreover, we have

$$\begin{aligned}(2.73)\quad &\Delta p_{2,0} - Rp_{2,0}\\ &\geq -[a_2(x)p_{2,0} + \tfrac{(K_2-p_{2,0})^2 v}{2M_2} - (2b_2 v - c_2 u)p_{2,0} - c_1 u p_1] - Rp_{2,0} \quad \text{in } \Omega\end{aligned}$$

for each $u \in [u_0, u_{-1}], v \in [v_0, v_{-1}], p_1 \in [p_{1,0}, p_{1,-1}]$, since

$$\begin{aligned}&a_2(x)p_{2,0} + \tfrac{(K_2-p_{2,0})^2 v}{2M_2} - (2b_2 v - c_2 u)p_{2,0} - c_1 u p_1\\ &\geq \tfrac{K_2^2}{2M_2}\phi_2 - c_1\psi_1 K_1 > \tfrac{2K_1\bar{a}_1 c_1}{b_1\phi_2}\phi_2 - c_1\psi_1 K_1 = \tfrac{K_1\bar{a}_1}{b_1}c_1 > 0 \quad \text{in } \Omega.\end{aligned}$$

(Here, we use $[H1^{**}], [H2^{**}]$ and [H5]).

We now inductively define sequences of functions $u_k(x), v_k(x), p_{1,k}(x), p_{2,k}(x)$ in $\Omega, k = 1, 2, \ldots,$ as solutions of scalar problems in Ω as follows:

(2.74)

$$\begin{aligned}\Delta u_k - Ru_k &= -u_{k-2}[a_1 - \tfrac{(K_1-p_{1,k-2})u_{k-2}}{2M_1} - b_1 u_{k-2} - c_1 v_{k-1}] - Ru_{k-2},\\ \Delta v_k - Rv_k &= -v_{k-2}[a_2 - \tfrac{(K_2-p_{2,k-2})v_{k-2}}{2M_2} + c_2 u_k - b_2 v_{k-2}] - Rv_{k-2},\\ \Delta p_{1,k} - Rp_{1,k} &= -[a_1 p_{1,k-2} + \tfrac{(K_1-p_{1,k-2})^2 u_k}{2M_1} - (2b_1 u_{k-1} + c_1 v_{k-1})p_{1,k-2}\\ &\quad + c_2 v_k p_{2,k-2}] - Rp_{1,k-2},\\ \Delta p_{2,k} - Rp_{2,k} &= -[a_2 p_{2,k-2} + \tfrac{(K_2-p_{2,k-2})^2 v_k}{2M_2} - (2b_2 v_{k-1} - c_2 u_k)p_{2,k-2}\\ &\quad - c_1 u_{k-1} p_{1,k-1}] - Rp_{2,k-2},\\ \tfrac{\partial u_k}{\partial\nu} = \tfrac{\partial v_k}{\partial\nu} &= \tfrac{\partial p_{1,k}}{\partial\nu} = \tfrac{\partial p_{2,k}}{\partial\nu} = 0 \quad \text{on } \Omega.\end{aligned}$$

Theorem 2.6 (Approximation Scheme for the Optimal Control). *Assume all the hypotheses of Corollary 2.5 and [H5]. The sequences of functions $u_k(x)$, $v_k(x), p_{1,k}(x), p_{2,k}(x)$, defined above satisfy the order relations:*

$$\begin{aligned}&u_0 \leq u_2 \leq \cdots \leq u_{2r} \leq u_{2r-1} \leq \cdots \leq u_1 \leq u_{-1},\\ (2.75)\quad &v_0 \leq v_2 \leq \cdots \leq v_{2r} \leq v_{2r-1} \leq \cdots \leq v_1 \leq v_{-1},\\ &p_{i,0} \leq p_{i,2} \leq \cdots \leq p_{i,2r} \leq p_{i,2r-1} \leq \cdots \leq p_{i,1} \leq p_{i,-1}, \quad i = 1, 2,\end{aligned}$$

for all $x \in \Omega$. *Moreover, any solution* (u, v, p_1, p_2) *of problem (2.59) with the property*

$$u_0 \le u \le u_{-1},\ v_0 \le v \le v_{-1},\ p_{i,0} \le p_i \le p_{i,-1},\ \ i = 1,2 \ \ \textit{in } \Omega, \tag{2.76}$$

must satisfy

$$u_{2r} \le u \le u_{2r-1},\ \ v_{2r}, \le v \le v_{2r-1},\ \ p_{i,2r} \le p_i \le p_{i,2r-1},\ \ i = 1,2 \ \textit{in } \Omega, \tag{2.77}$$

for all positive integer r.

Proof. Using the equation satisfied by u_1 and inequality (2.66), we obtain $\Delta(u_{-1} - u_1) - R(u_{-1} - u_1) \le 0$ in $\Omega, (\partial/\partial\nu)(u_{-1} - u_1) = 0$ on $\partial\Omega$. Hence $u_1 \le u_{-1}$ in Ω. Similarly, using (2.67) and the choice of R, we deduce that $\Delta(u_0 - u_1) - R(u_0 - u_1) \ge 0$ in $\Omega, (\partial/\partial\nu)(u_0 - u_1) = 0$ on $\partial\Omega$. Thus, we have

$$u_0 \le u_1 \le u_{-1} \quad \text{in } \Omega. \tag{2.78}$$

Using the equation for v_1, inequalities (2.68), (2.69), (2.78), and the choice of R, we obtain

$$\begin{aligned} &\Delta(v_{-1} - v_1) - R(v_{-1} - v_1) \le 0 \quad \text{in } \Omega, \quad \frac{\partial}{\partial\nu}(v_{-1} - v_1) = 0 \quad \text{on } \partial\Omega, \\ &\Delta(v_0 - v_1) - R(v_0 - v_1) \ge 0 \quad \text{in } \Omega, \quad \frac{\partial}{\partial\nu}(v_0 - v_1) = 0 \quad \text{on } \partial\Omega, \end{aligned} \tag{2.79}$$

and

$$v_0 \le v_1 \le v_{-1} \text{ in } \Omega. \tag{2.80}$$

Again, using the equations for p_1 and p_2, the inequalities above, and the choice of R, we deduce similarly that

$$p_{i,0} \le p_{i,1} \le p_{i,-1}, \quad i = 1,2 \text{ in } \Omega. \tag{2.81}$$

Next, we show

$$\begin{aligned} &\Delta(u_1 - u_2) - R(u_1 - u_2) \le 0 \text{ in } \Omega, \ \frac{\partial}{\partial\nu}(u_1 - u_2) = 0 \quad \text{on } \partial\Omega, \\ &\Delta(u_0 - u_2) - R(u_0 - u_2) \ge 0 \text{ in } \Omega, \ \frac{\partial}{\partial\nu}(u_0 - u_2) = 0 \quad \text{on } \partial\Omega, \end{aligned} \tag{2.82}$$

(Here, we use (2.67) at $v = v_1, p_1 = p_{1,0}$.) Thus we obtain

$$u_0 \le u_2 \le u_1 \text{ in } \Omega. \tag{2.83}$$

Similarly, comparing the related equations and using the inequalities above, we deduce successively that

$$v_0 \le v_2 \le v_1 \text{ in } \Omega, \tag{2.84}$$

$$p_{1,0} \leq p_{1,2} \leq p_{1,1} \text{ and } p_{2,0} \leq p_{2,2} \leq p_{2,1} \text{ in } \Omega. \tag{2.85}$$

From the above inequalities we have the validity of the inequalities

(2.86)

$$u_{2n} \leq u_{2n+2} \leq u_{2n+1} \leq u_{2n-1}, \quad v_{2n} \leq v_{2n+2} \leq v_{2n+1} \leq v_{2n-1} \text{ in } \Omega,$$

$$p_{1,2n} \leq p_{1,2n+2} \leq p_{1,2n+1} \leq p_{1,2n-1}, \quad p_{2,2n} \leq p_{2,2n+2} \leq p_{2,2n+1} \leq p_{2,2n-1} \text{ in } \Omega,$$

for $n = 0$. We then use the comparison method as above to prove the validity of (2.86) for any positive integer n by induction. This proves (2.75).

To prove the second part of the theorem, we use the comparison method on the appropriate equations as above, and proceed by induction on r in proving inequality (2.77). For more details of analogous procedures, see Theorem 5.5-1 in [125].

The following example indicates that when the intrinsic growth rates $a_i(x,y)$ various only moderately in different locations and the interaction parameters are relatively small, $i = 1, 2$, then all the theorems in this section are applicable.

Example 2.1. Let Ω be any bounded domain on the (x,y) plane with C^2 boundary, with $\{(x,y) | \, 0 \leq x \leq 1, 0 \leq y \leq 1\} \subset \Omega$. Consider the system (2.1) with

$$a_i(x,y) = 16 + 2\,sin(2\pi xy),\ \bar{a}_i = 18,\ \tilde{a}_i = 14 \text{ for } i = 1,2,\ b_1 = 12,\ b_2 = 7.8,$$
$$c_1 = 0.35,\ c_2 = 0.5.$$

The problem is to maximize (2.2) with

$$K_1 = 0.5,\ K_2 = 1,\ M_1 = 0.5,\ M_2 = 1$$

for all $f_i \in L^\infty(\Omega)$ in (2.5), where $\delta_1 = 2.079, \delta_2 = 2.308$.

We can verify that the hypotheses $[H1^{**}]$, $[H2^{**}]$, $[H3]$ and $[H4]$ are all satisfied, thus Theorem 2.4 applies. Moreover, conditions (2.56), (2.57) and $[H5]$ are also valid; hence Corollary 2.5, Remark 2.5 and Theorem 2.6 are all applicable to this example. The optimal solution can be characterized by positive solutions of the elliptic system of four equations (2.59). Moreover, the approximation Theorem 2.6 applies.

3.3 Time-Periodic Optimal Control for Competing Parabolic Systems

Optimal control models for reaction-diffusion systems are used for agriculture and environmental problems. Since the growth rates of life species change seasonally, one needs to analyze such models whose coefficients are periodic in time.

In this section, we consider time-periodic Voterra-Lotka diffusive competing systems whose growth and interaction rates are periodic functions of time. We will show that such systems has a unique positive periodic solution. We then search for optimal harvesting control of the systems in order to maximize certain pay-off.

Let Ω be a bounded domain in R^N with C^2 boundary, $G := \Omega \times [0,T)$, $S := \partial\Omega \times [0,T)$ for some $T > 0$, and b_i, c_i some positive constants, $i = 1, 2$. Throughout this section we will always assume that $f(x,t), g(x,t)$ and $a_i(x,t), i = 1,2$ are functions in $L^\infty_+(\Omega \times (-\infty,\infty)) := \{h \in L^\infty(\Omega \times (-\infty,\infty)| h \geq 0 \text{ in } G\}$, and they are periodic functions of t with period T for $(x,t) \in \Omega \times (-\infty,\infty)$. For any constant $\delta = (\delta_1, \delta_2), \delta_i > 0, i = 1, 2$, we let

$$B_{\delta,T} = \{(f,g)|f,g \in L^\infty_+(\Omega \times (-\infty,\infty)),$$

$$f \text{ and } g \text{ are periodic functions of } t \text{ with period } T \text{ and } f \leq \delta_1, g \leq \delta_2\}.$$

For any $(f,g) \in B_{\delta,T}$, we define $(u,v) = (u(f,g), v(f,g))$ as a solution of the problem:

$$(3.1)\quad \begin{cases} u_t - \Delta u - u[(a_1(x,t) - f(x,t)) - b_1 u - c_1 v] = 0 & \text{in } G, \\ v_t - \Delta v - v[(a_2(x,t) - g(x,t)) - b_2 v - c_2 u] = 0 & \text{in } G, \\ \frac{\partial u}{\partial \nu} = \frac{\partial v}{\partial \nu} = 0 & \text{on } S, \\ u(x,0) = u(x,T), \ v(x,0) = v(x,T), & \text{for } x \in \Omega. \end{cases}$$

We will show that such $(u(f,g), v(f,g))$ is uniquely defined when $\delta_i, a_i, b_i, c_i, i = 1, 2$ satisfy appropriate conditions (cf. [H1], [H2] below in this section).

Let $K_i, M_i, i = 1, 2$ be positive constants; we define the pay-off functional by

$$J(f,g) = \int_G [K_1 u(f,g) f + K_2 v(f,g) g - M_1 f^2 - M_2 g^2]\, dx dt,$$

which describes the economic profit of harvesting the competing species u, v. The problem is to find periodic control $(f^*, g^*) \in B_{\delta,T}$ such that

$$J(f^*, g^*) = sup_{(f,g)\in B_{\delta,T}} J(f,g).$$

In practical terms, we are searching for optimal harvesting of two competing biological species whose growth are governed by diffusive Volterra-Lotka system (3.1). The functions f, g denote the distributions of control harvesting effort on the biological species. The optimal control criterion is to maximize the pay-off functional, where K_1 and M_1 are constants describing the market price of species

u and the cost of control f, and similarly K_2 and M_2 are constants related to v anf g. The results of this section are obtained from He, Leung and Stojanovic in [82] and [83].

We first consider the basic preliminary problem of existence of t-periodic solution of scalar linear parabolic problem with L^∞ coefficients rather than the usual C^α coefficients.

Theorem 3.1. *Let $f \in L^p_+(G)$ and $c \in L^\infty(\bar{G})$ with $min_G c(x,t) \geq \lambda > 0$. Suppose further that f and c are periodic functions of t with period T in $\Omega \times [0, \infty)$. Then the problem*

$$(3.2) \qquad \begin{cases} u_t - \Delta u + c(x,t)u = f & \text{in } G, \\ \frac{\partial u}{\partial \nu} = 0 & \text{on } S, \\ u(x,0) = u(x,T) & \text{for all } x \in \Omega \end{cases}$$

has a unique non-negative solution u in $W^{2,1}_p(G)$, for $p > 1$, satisfying the estimate

$$(3.3) \qquad ||u||_{W^{2,1}_p(G)} \leq C||f||_{p,G}.$$

Here, C is a constant independent of f and dependent on $||c||_{\infty,G}$; λ is any positive constant.

Proof. First suppose that $c \in C^\alpha(\bar{G}), 0 < \alpha < 1$. Let $f_m \in L^p_+(G) \cap C^\alpha(\bar{G})$ be periodic about t with period T such that

$$f_m \to f \ \text{ in } L^p(G).$$

(The existence of such sequence $\{f_m\}$ can be found in e.g. Gilbarg and Trudinger [71].) Then by Fife's results in [58] (cf. Theorem 5.6.1 in Leung [125]), there exists a unique solution $u_m \in C^{2+\alpha}(\bar{G})$ of problem (3.2) with f replaced by f_m. Multiplying both sides of the equation

$$\frac{\partial u_m}{\partial t} - \Delta u_m + cu_m = f_m,$$

by u_m^{p-1} and integrating on G, we obtain

$$\int_G \frac{1}{p}\frac{d}{dt}u_m^p dxdt + (p-1)\int_G u_m^{p-2}|\nabla u_m|^2 dxdt + \int_G cu_m^p dxdt = \int_G f u_m^{m-1} dxdt.$$

Since u_m is periodic with period T, we have

$$\int_G \frac{1}{p}\frac{d}{dt}u_m^p dxdt = 0.$$

Using Young's inequality (see e.g. p. 622 in [57]) and the fact that

$$\int_G u_m^{p-2}|\nabla u_m|^2 dxdt = (2/p)^2 \int_G |\nabla u_m^{p/2}|^2 dxdt \geq 0,$$

we deduce that

$$\lambda||u_m||^p_{L^p(G)} \leq \epsilon^{-[1/(p-1)]}||f_m||^p_{L^p(G)} + \epsilon||u_m||^p_{L^p(\Omega)} \text{ for all } \epsilon > 0. \tag{3.4}$$

Hence, by choosing ϵ small enough, we obtain

$$||u_m||_{L^p(G)} \leq C||f_m||_{L^p(G)}. \tag{3.5}$$

Here, C is a constant independent of m and f_m, and $u_m \in C^{2+\alpha}(G)$. Since $f_m \geq 0$, the maximum principle implies that $u_m \geq 0$. Since $f_m \to f$ in $L^p(G)$, we obtain by (3.5) that

$$||u_m||_{p,G} \leq C. \tag{3.6}$$

Here C is independent of m. Using Theorem A5-1 in Chapter 6 (with $T_1 = T/2$ and $T_2 = 2T$), and the periodic property of u_m, we obtain

$$||u_m||_{W_p^{2,1}(G)} \leq C_1||f_m||_{p,G} + C_2||u_m||_{p,G} \leq C||f_m||_{p,G}. \tag{3.7}$$

We thus readily deduce that

$$u_m \to u_0 \text{ strongly in } W_p^{2,1}(G),$$

where $u_0 \in W_p^{2,1}(G)$ is a solution of (3.2) and

$$||u_0||_{W_p^{2,1}(G)} \leq C||f||_{p,G}. \tag{3.8}$$

In order to complete the proof of this theorem when $c \in L^\infty(\bar{G})$ in general, we need the following lemma.

Lemma 3.1. *Assume all the hypotheses of Theorem 3.1, with f, c replaced by f_i, c_i, here $f_i \in L^p(\Omega \times [0,\infty))$ and $c_i \in C^\alpha(\Omega \times [0,\infty)), i = 1,2$ and $p > 1$. Let $u_i, i = 1,2$ be the non-negative solutions of problem (3.2) with f, c replaced by f_i, c_i, respectively. If $f_1 \geq f_2$ and $c_2 \geq c_1$ in G, then $u_1 \geq u_2$ in G.*

Proof. By using mollifiers, we may assume without loss of generality that $f_i \in C^\alpha(G)$ for $i = 1,2$. The function $w = u_1 - u_2$ satisfies

$$\begin{cases} w_t - \Delta w + c_1 w = f_1 - f_2 + (c_2 - c_1)u_2 & \text{in } G, \\ \frac{\partial w}{\partial \nu} = 0 & \text{on } S, \\ w(x,0) = w(x,T) & \text{for all } x \in \Omega. \end{cases}$$

Suppose $w(x_0, t_0)$ is a negative minimum of w, then the fact that $(c_2 - c_1)u_2 \geq 0$ and $f_1 - f_2 \geq 0$, implies that (x_0, t_0) cannot be in the interior of G. Moreover, $w(x, 0) = w(x, T)$ and the maximum principle implies that $t_0 \neq 0, T$; unless w is a negative constant, which is a contradiction to the equation. Thus, we have $x_0 \in \partial\Omega$, and $\partial w/\partial\nu > 0$ at x_0, by maximum principle. This again contradicts the boundary condition. Hence, we must have $w \geq 0$, i.e. $u_1 \geq u_2$ in G.

We now continue to prove Theorem 3.1 when $c \in L^\infty(\Omega \times [0, \infty))$. Choose a constant P so that $P \geq ||c||_{\infty,G}$. Let u_0 and u_* be, respectively, the non-negative solution of the problems $(3.2)_\lambda$ and $(3.2)_P$ which are (3.2) with c replaced by λ and P respectively. Lemma 3.1 implies that

$$0 \leq u_* \leq u_0 \quad \text{in } G. \tag{3.9}$$

We then define u_k as the non-negative solution of the problems

$$\begin{cases} \frac{\partial u_k}{\partial t} - \Delta u_k + P u_k = f - c u_{k-1} + P u_{k-1} & \text{in } G, \\ \frac{\partial u_k}{\partial \nu} = 0 & \text{on } S, \\ u_k(x, 0) = u_k(x, T) & \text{for all } x \in \Omega. \end{cases} \tag{3.10}$$

The first part of the proof of Theorem 3.1 implies that $u_1 \geq 0$ is uniquely defined in G. Using the equations satisfied by u_*, u_1 and u_0, we obtain readily from Lemma 3.1 that

$$0 \leq u_* \leq u_1 \leq u_0 \quad \text{in } G.$$

Denoting $h(u) = f - cu + Pu$, we find h is increasing since $P \geq c$. Suppose that we have proved

$$0 \leq u_* \leq u_k \leq u_{k-1} \leq \cdots \leq u_1 \leq u_0 \quad \text{in } G.$$

Since u_{k+1} satisfies problem (3.10) with k replaced with $k+1$, and $h(0) \leq h(u_k) \leq h(u_{k-1})$, Lemma 3.1 implies that

$$u_* \leq u_{k+1} \leq u_k.$$

We thus obtain a monotone increasing sequence $\{u_k\}$ with $u_0 \geq u_k \geq u_*$ for all $k = 1, 2, 3, \ldots$. Using the first part of the proof (cf.(3.7)), and applying estimates (3.8) to (3.10) we obtain the estimate

$$||u_k||_{W_p^{2,1}(G)} \leq C||f||_{p,G}(1 + ||c||_{\infty,G}).$$

Applying (3.8) to (3.10) again, we obtain a subsequence u_{n_k} converging to a solution in $W_p^{2,1}(G)$. Further, applying the above estimate to u_{n_k}, and taking limit as $k \to \infty$, we obtain (3.3).

It remains to prove the uniqueness of solution. Suppose that (3.2) has two solutions u and v, then $u - v$ is a solution of problem (3.2) with f replaced by 0. Multiplying both sides of the equation by $u - v$ and integrating over G, we obtain

$$\int_G (u-v)_t(u-v)dxdt + \int_G |\nabla(u-v)|^2 dxdt + \int_G c(u-v)^2 dxdt = 0. \tag{3.11}$$

From the periodic property of u, v, we deduce

$$\int_G (u-v)_t(u-v)dxdt = \int_G \int_0^T \frac{1}{2}\frac{d}{dt}(u-v)^2 dtdx = 0.$$

Consequently, we can complete the proof of Theorem 3.1 by mean of (3.11).

In order to prepare for the consideration of the competitive system (3.1), we next consider the following scalar periodic nonlinear problem:

$$\begin{cases} u_t - \Delta u - (a(x,t) - f(x,t))u + bu^2 = 0 & \text{in } G, \\ \frac{\partial u}{\partial \nu} = 0 & \text{on } S, \\ u(x,0) = u(x,T) & \text{for all } x \in \Omega. \end{cases} \tag{3.12}$$

Theorem 3.2 (Periodic Solution for Scalar Nonlinear Problem). *Let $a(x,t), f(x,t)$ be both in $L_+^\infty(\Omega \times [0,\infty))$ and $f(x,t) \in B_{\delta,T}$, where δ satisfies*

$$0 < \delta \le inf_G\, a(x,t). \tag{3.13}$$

Suppose $a(x,t)$ is also a t-periodic function with period T in $\Omega \times [0,\infty)$, then the problem (3.12) has a positive t-periodic solution $u(x,t) \in W_p^{2,1}(G)$ of period T, for all $p \in (1,\infty)$, satisfying the estimate

$$||u||_{W_p^{2,1}(G)} \le \tilde{C}, \tag{3.14}$$

where $\tilde{C}$ is a constant determined by $||a||_{\infty,G}$. Moreover, if δ further satisfies the condition

$$0 < \delta < \frac{1}{3}\{2\; inf_G\, a(x,t) - sup_G\, a(x,t)\}, \tag{3.15}$$

then the positive solution is unique.

Proof. Let $C_1 > 0$ be a constant such that

$$||a||_{\infty,G} C_1 - bC_1^2 \le 0, \tag{3.16}$$

and choose another positive constant P so that

$$\delta + 2bC_1 - P < 0. \tag{3.17}$$

Define an initial iterate

$$u_0 \equiv C_1; \tag{3.18}$$

and then inductively define u_k, $k = 1, 2, \dots$, as the unique non-negative solutions in $W_p^{2,1}(G)$ of the problem:

$$\begin{cases} \frac{\partial u_k}{\partial t} - \Delta u_k + Pu_k = (a-f)u_{k-1} + Pu_{k-1} - bu_{k-1}^2 & \text{in } G, \\ \frac{\partial u_k}{\partial \nu} = 0 & \text{on } S, \\ u_k(x,0) = u_k(x,T) & \text{for all } x \in \Omega. \end{cases} \tag{3.19}$$

The existence of u_k has been proved by Theorem 3.1.

By (3.16), (3.18) and Lemma 3.1, one can easily verify that

$$u_1 \le C_1 \equiv u_0 \quad \text{in } G. \tag{3.20}$$

By the choice of P in (3.17), the function

$$h(x,t,u) = [a(x,t) - f(x,t)]u - bu^2 + Pu \tag{3.21}$$

is increasing in u for $0 < u \le C_1$. Again, Lemma 3.1 implies

$$u_k \le u_{k-1} \quad \text{in } G \quad \text{for each } k = 2, 3, \dots \tag{3.22}$$

On the other hand, hypothesis (3.13) implies that for $\epsilon > 0$ small enough, we have

(3.23)

$$\begin{aligned} \frac{\partial}{\partial t}(u_k - \epsilon) - \Delta(u_k - \epsilon) + P(u_k - \epsilon) &= h(x,t,u_{k-1}) - P\epsilon \\ &\ge h(x,t,u_{k-1}) - P\epsilon - (a-f)\epsilon + b\epsilon^2 \\ &= h(x,t,u_{k-1}) - h(x,t,\epsilon). \end{aligned}$$

Thus, assuming $\epsilon < C_1$ is sufficiently small, we can deduce inductively starting from $k = 1$ that

$$\frac{\partial}{\partial t}(u_k - \epsilon) - \Delta(u_k - \epsilon) + P(u_k - \epsilon) > 0.$$

Consequently, we have

$$C_1 = u_0 \ge u_1 \ge u_2 \ge \cdots u_k \ge \cdots \ge \epsilon > 0.$$

From Theorem 3.1, we have

$$||u_k||_{W_p^{2,1}(G)} \leq C(||(a-f)u_{k-1} + Pu_{k-1} - bu_{k-1}^2||_{p,G}). \tag{3.24}$$

Hence, combining with the uniform bound for u_k, we obtain

$$||u_k||_{W_p^{2,1}(G)} \leq \tilde{C}, \tag{3.25}$$

where $\tilde{C}$ is determined by C_1 and $||a||_{\infty,G}$, independent of k. Passing to the limit as $k \to \infty$ in (3.19) and using the a-priori estimate (3.24) again, we obtain a solution of problem (3.12) in $W_p^{2,1}(G)$ and the estimate (3.14).

We next prove the uniqueness part of this theorem. First, we note that from (3.23) we can choose the lower bound ϵ as

$$\epsilon = (1/b)(inf_G\, a(x,t) - \delta). \tag{3.26}$$

From (3.16) above, we also see that we can choose C_1 as

$$C_1 = \{sup_G\, a(x,t)\}/b. \tag{3.27}$$

Suppose w is another positive solution, let $\bar{k} = max\{C_1, ||w||_\infty\}$ and start iterating from an initial constant function $\bar{k}$. We construct as in the proof of existence above another solution, say v, such that

$$0 < \epsilon \leq u \leq v \quad \text{and} \quad 0 \leq w \leq v \quad \text{a.e. in } G. \tag{3.28}$$

By the choice of ϵ and using the same argument as above for proving $u \geq \epsilon$, we can prove $w \geq \epsilon$. Thus we have

$$0 < \epsilon \leq w \leq v.$$

We now prove $u \equiv v$ in G. Since u and v both satisfy (3.12), the function $U := v - u$ satisfies

$$\begin{cases} U_t - \Delta U - (a(x,t) - f(x,t))U + b(u+v)U = 0 & \text{in } G, \\ \frac{\partial U}{\partial \nu} = 0 & \text{on } S, \\ U(x,0) = U(x,T) & \text{for all } x \in \Omega. \end{cases}$$

From (3.28) we have

$$u + v > 2\epsilon. \tag{3.29}$$

Using (3.15), (3.26) and (3.29), we deduce that

$$a - f \leq 2\,inf_G\, a(x,t) - 3\delta$$

$$< 2\,inf_G\, a(x,t) - 2\delta \leq 2\epsilon b \leq b(u+v).$$

That is, we have

$$(a-f) - b(u+v) < 0 \text{ in } G.$$

We can then use the same methods as in the proof of uniqueness as in Theorem 3.1 to prove $U \equiv 0$, i.e. $u \equiv v$ in G. (Here, the role of c is replaced by $b(u+v) - (a-f)$.) Similarly, we can show $w \equiv v$ in G. This completes the proof of Theorem 3.2.

We now return to consider the existence of periodic positive solution for the competitive system (3.1), for a fixed given $(f,g) \in B_{\delta,,T}$. This will be established in Theorem 3.3 with the hypotheses

$[H1]$ $\quad 0 < \delta_i < \frac{1}{3}\{2\,\tilde{a}_i - \hat{a}_i - 2\,c_i\frac{\hat{a}_j}{b_j}\},\ i = j = 1,2$ and $i \neq j$.

Here $\tilde{a}_i = inf_G\, a_i(x,t)$ and $\hat{a}_i = sup_G\, a_i(x,t)$ for $i = 1,2$. To obtain the existence theorem (Theorem 3.3), we will construct two sequences by means of iteration. Let u_0 be the solution of the problem

$$(3.30)\qquad \begin{cases} u_t - \Delta u - (a_1(x,t) - f(x,t))u + b_1 u^2 = 0 & \text{in } G, \\ \frac{\partial u}{\partial \nu} = 0 & \text{on } S, \\ u(x,0) = u(x,T) & \text{for all } x \in \Omega; \end{cases}$$

and let v_0 be the solution of the problem

$$(3.31)\qquad \begin{cases} v_t - \Delta v - (a_2(x,t) - g(x,t) - c_2 u_0)v + b_2 v^2 = 0 & \text{in } G, \\ \frac{\partial v}{\partial \nu} = 0 & \text{on } S, \\ v(x,0) = v(x,T) & \text{for all } x \in \Omega. \end{cases}$$

From Theorem 3.2 and its proof, we see that hypotheses [H1] imply that u_0 and v_0 exist in $W_p^{2,1}(G)$ for $p > 1$, and

$$\frac{\hat{a}_1}{b_1} \geq u_0 \geq \frac{\tilde{a}_1 - \delta_1}{b_1} > 0, \quad \frac{\hat{a}_2}{b_2} \geq v_0 \geq \frac{\tilde{a}_2 - c_2\hat{a}_1/b_1 - \delta_2}{b_2} > 0.$$

Note that in (3.30) and (3.31), the derivatives are taken in the weak sense, and the equations are satisfied a.e. in G. For $i = 1,2,\ldots,$ we define u_i and v_i

as the solutions of the following problems (3.32) and (3.33) respectively:

$$(3.32) \quad \begin{cases} \frac{\partial u_i}{\partial t} - \Delta u_i - (a_1 - f - c_1 v_{i-1})u_i + b_1 u_i^2 = 0 & \text{in } G, \\ \frac{\partial u_i}{\partial \nu} = 0 & \text{on } S, \\ u_i(x,0) = u_i(x,T) & \text{for all } x \in \Omega; \end{cases}$$

and

$$(3.33) \quad \begin{cases} \frac{\partial v_i}{\partial t} - \Delta v_i - (a_2 - g - c_2 u_i)v_i + b_2 v_i^2 = 0 & \text{in } G, \\ \frac{\partial v_i}{\partial \nu} = 0 & \text{on } S, \\ v_i(x,0) = v_i(x,T) & \text{for all } x \in \Omega. \end{cases}$$

Inductively, from (3.32) and (3.33), and using Theorem 3.2 we obtain
(3.34)

$$0 < \frac{\tilde{a}_1 - c_1(\hat{a}_2/b_2) - \delta_1}{b_1} \leq u_k \leq \frac{\hat{a}_1}{b_1}, \quad 0 < \frac{\tilde{a}_2 - c_2(\hat{a}_1/b_1) - \delta_2}{b_2} \leq v_k \leq \frac{\hat{a}_2}{b_2}.$$

Moreover, u_k and v_k are in $W_p^{2,1}(G)$, for $p > 1, k = 1, 2, \ldots$. By means of a slight extension of the comparison Lemma 3.1, we can prove the following monotone properties.

Lemma 3.2. *Assume hypotheses [H1], then the sequences defined by (3.30) to (3.33) satisfy*

$$(3.35) \quad u_0 \geq u_1 \geq u_2 \geq \cdots \geq u_k \geq \cdots, \text{ in } G,$$

and

$$(3.36) \quad v_0 \leq v_1 \leq v_2 \leq \cdots \leq v_k \leq \cdots, \text{ in } G.$$

Proof. We first extend the comparison Lemma 3.1 to include the case for a fixed $c \in L_+^\infty(\Omega \times (-\infty, \infty))$, while all the other assumptions are unchanged. In fact, suppose that w_i is the solution of the problem

$$(3.37) \quad \begin{cases} w_t - \Delta w + c(x,t)w = f & \text{in } G, \\ \frac{\partial w}{\partial \nu} = 0 & \text{on } S, \\ w(x,0) = w(x,T) & \text{for all } x \in \Omega \end{cases}$$

for $i = 1, 2$, where $c \in L^\infty_+(\Omega \times (-\infty, \infty))$. We need to prove that $f_1 \geq f_2, f_i \in L^p(G)$ for $i = 1, 2$, implies $w_1 \geq w_2$ in G. Let $c_n \in C^\alpha(\Omega \times (-\infty, \infty)), n = 1, 2, ..,$ be positive functions periodic in t with period T such that

$$c_n \to c \text{ in } L^p(G), \quad \text{as } n \to \infty.$$

Moreover, let w_{in} be the solution of problem (3.37) with c replaced by c_n for $i = 1, 2$. Then, Lemma 3.1 asserts that

$$w_{1n} \geq w_{2n} \text{ in } G, \; n = 1, 2, 3, \ldots . \tag{3.38}$$

Moreover, Theorem 3.1 implies that $\{w_{in}\}$ is uniformly bounded in $W_p^{2,1}(G)$ for each $i = 1, 2$. Hence, by extracting appropriate subsequences, we can readily obtain $w_{in} \to w_i$ a.e. in G as $n \to \infty$ for $i = 1, 2$. Consequently, from (3.38), we obtain $w_1 \geq w_2$.

Now we are ready to prove (3.35) and (3.36). Let $w = u_0 - u_1$; then w satisfies the inequality

$$w_t - \Delta w + [b_1(u_0 + u_1) - (a_1 - f) + c_1 v_0]w \geq 0, \text{ in } G. \tag{3.39}$$

The hypothesis [H1] and (3.34) imply that

$$b_1(u_0 + u_1) - (a_1 - f) + c_1 v_0 \geq \delta > 0, \text{ in } G.$$

Then by the extension of Lemma 3.1 mentioned above, we conclude that $w \geq 0$, i.e.

$$u_0 \geq u_1, \text{ in } G.$$

From this we deduce that $(v_1 - v_0)$ satisfies the inequality

$$(v_1 - v_0)_t - \Delta(v_1 - v_0) + [b_2(v_1 + v_0) - (a_2 - g) + c_2 u_0](v_1 - v_0) \geq 0,$$

in G. The same argument as above implies that $v_1 \geq v_0$ in G. By iterating and induction in k, we deduce by the same argument that (3.35) and (3.36) hold.

By (3.34)-(3.36) and Theorem 3.2, we obtain the estimates

$$||u_k||_{W_p^{2,1}(G)} \leq R_1, \quad ||v_k||_{W_p^{2,1}(G)} \leq R_2, \tag{3.40}$$

where R_1 and R_2 are constants independent of k.

By a similar argument as in Theorem 3.2, taking limit as $i \to \infty$ in (3.35) and (3.36) and using a-priori estimates (3.40), we finally conclude that there exists a solution (u, v) of problem (3.1) in $W_p^{2,1}(G) \times W_p^{2,1}(G)$ and the estimates (3.40) for u and v hold. Hence, we have proved the following theorem.

Theorem 3.3 (Periodic Solution for Competitive Systems). *Suppose hypothesis [H1] holds. Let* $(f, g) \in B_{\delta,T}$, *then problem (3.1) has a solution in* $W_p^{2,1}(G) \times W_p^{2,1}(G)$ *for* $p > 1$ *with* u, v *satisfying*

$$0 < \epsilon_1 \le u \le \hat{a}_1/b_1, \quad 0 < \epsilon_2 \le v \le \hat{a}_2/b_2. \tag{3.41}$$

Here $\epsilon_i = [\tilde{a}_i - c_i(\hat{a}_j/b_j) - \delta_i]/b_i, i = 1, 2$ *and* $i \ne j$. *Moreover,* (u, v) *satisfies*

$$||u||_{W_p^{2,1}(G)} \le R_1, \quad ||v||_{W_p^{2,1}(G)} \le R_2. \tag{3.42}$$

Here, R_i *is a constant determined by* $||a_i||_{\infty,G}, i = 1, 2$ *respectively.*

In order to obtain uniqueness of solution to problem (3.1), we introduce the following hypothesis:

[$H2$] $\quad c_i\frac{\hat{a}_2}{b_2} + c_j\frac{\hat{a}_1}{b_1} \le 2\,min\{\delta_1, \delta_2\}$, for $i, j = 1, 2$ and $i \ne j$.

Theorem 3.4 (Uniqueness). *Let* δ_i, a_i, c_i *and* $b_i, i = 1, 2$, *satisfy [H1] and [H2]. Let* $(f, g) \in B_{\delta,T}$, *then the problem (3.1) has a unique solution* (u, v) *in* $W_p^{2,1}(G) \times W_p^{2,1}(G)$ *for* $p > 1$ *with* $u, v > 0$.

Proof. We first prove that if (u, v) is a solution of problem (3.1) with $u, v > 0$, then u and v satisfy (3.41). In fact, we can use the same comparison lemma described in the proof of Lemma 3.2 to prove $u_0 \ge u$, and then $v_0 \le v$. Similarly, we can show $u_k \ge u$ and $v_k \le v$ for all $k = 0, 1, 2, \ldots$. Finally, we obtain the inequalities $u \le lim_{k\to\infty}u_k \le \hat{a}_1/b_1$ and $v \ge lim_{k\to\infty}v_k \ge \epsilon_2$. Interchanging the role of u and v, we can show by means of symmetry that $u \ge \epsilon_1$ and $v \le \hat{a}_2/b_2$.

Suppose that there exist two solutions (u_1, v_1) and (u_2, v_2) of problem (3.1) with $u_i, v_i > 0$ for $i = 1, 2$; then $(u_1 - u_2, v_1 - v_2)$ satisfies

$$\begin{cases} (u_1 - u_2)_t - \Delta(u_1 - u_2) + [b_1(u_1 + u_2) - (a_1 - f)](u_1 - u_2) \\ \qquad + c_1v_1(u_1 - u_2) + c_1u_2(v_1 - v_2) = 0 \quad \text{in } G, \\ (v_1 - v_2)_t - \Delta(v_1 - v_2) + [b_2(v_1 + v_2) - (a_2 - g)](v_1 - v_2) \\ \qquad + c_2u_1(v_1 - v_2) + c_2v_2(u_1 - u_2) = 0 \quad \text{in } G. \end{cases} \tag{3.43}$$

By hypothesis [H1] and (3.41), we have

$$b_1(u_1 + u_2) - (a_1 - f) + c_1v_1 \ge \delta_1, \quad b_2(v_1 + v_2) - (a_2 - g) + c_2u_1 \ge \delta_2. \tag{3.44}$$

By hypothesis [H2] when $i = 1, j = 2$ and (3.41), we find

$$c_1u_2 + c_2v_2 \le 2\,min\{\delta_1, \delta_2\}. \tag{3.45}$$

From (3.43), (3.44), (3.45) and the periodic property of the solutions, we obtain

(3.46)
$$\int_G[|\nabla(u_1-u_2)|^2+|\nabla(v_1-v_2)|^2]\,dxdt+\int_G[\delta_1(u_1-u_2)^2+\delta_2(v_1-v_2)^2]\,dxdt$$
$$-2\,min\{\delta_1,\delta_2\}\int_G|(u_1-u_2)||(v_1-v_2)|\,dxdt\ \leq 0.$$

From (3.46), we finally obtain

$$u_1=u_2,\quad v_1=v_2,\quad \text{in } G.$$

Having proved the existence and uniqueness of problem (3.1), we can now consider the problem of existence of an optimal control $(f^*,g^*)\in B_{\delta,T}$, such that

$$J(f^*,g^*)=sup_{(f,g)\in B_{\delta,T}}J(f,g), \tag{3.47}$$

where

$$J(f,g)=\int_G[K_1u(f,g)f+K_2v(f,g)g-M_1f^2-M_2g^2]\,dxdt, \tag{3.48}$$

as described in the beginning of this section.

Theorem 3.5 (Existence of Periodic Optimal Control). *Let δ_i, a_i, c_i and $b_i, i=1,2$, satisfy the hypotheses $[H1]$ and $[H2]$. Then there exists an optimal control $(f^*,g^*)\in B_{\delta,T}$ for the problem (3.47), (3.48). Here, $(u(f,g),v(f,g))$ denotes the unique positive solution of (3.1) as described in Theorem 3.4.*

Proof. From (3.42), it follows that

$$sup_{(f,g)\in B_{\delta,T}}J(f,g)<\infty.$$

Let (f_n,g_n) be a maximizing sequence. Then, there exists a subsequence, again denoted by (f_n,g_n) for convenience, so that

$$f_n\to f^*,\ g_n\to g^*,\ \text{weakly in } L^2(G) \text{ with } (f^*,g^*)\in B_{\delta,T},$$

and

$$u_n(f_n,g_n)\to u^*,\ v_n(f_n,g_n)\to v^*,\ \text{ strongly in } W_p^{1,0}(G) \text{ and weakly in } W_p^{2,1}(G).$$

Since

$$\begin{cases} u_{nt}-\Delta u_n-u_n[(a_1-f_n)-b_1u_n-c_1v_n]=0 & \text{in } G,\\ v_{nt}-\Delta v_n-v_n[(a_2-g_n)-b_2v_n-c_2u_n]=0 & \text{in } G,\\ \frac{\partial u_n}{\partial\nu}=\frac{\partial v_n}{\partial\nu}=0 & \text{on } S,\\ u_n(x,0)=u_n(x,T),\ v_n(x,0)=v_n(x,T), & \text{for } x\in\Omega, \end{cases}$$

we have

$$\int_G \{-u_n\phi_t + \nabla u_n \cdot \nabla\phi - (a_1 - f_n)u_n\phi + b_1 u_n^2\phi + c_1 u_n v_n \phi\}\, dxdt = 0$$

and

$$\int_G \{-v_n\phi_t + \nabla v_n \cdot \nabla\phi - (a_2 - g_n)v_n\phi + b_2 v_n^2\phi + c_2 u_n v_n \phi\}\, dxdt = 0,$$

for any $\phi \in W_p^{1,1}(G) \cap L^\infty(G)$ with $\phi(x,T) = \phi(x,0)$. Passing to the limits as $n \to \infty$ in the two inequalities above, and noting that

$$\int_G f_n u_n \phi\, dxdt \to \int_G f^* u^* \phi\, dxdt \quad \text{and} \quad \int_G g_n v_n \phi\, dxdt \to \int_G g^* v^* \phi\, dxdt,$$

for all $\phi \in L^\infty(G)$, we find that (u^*, v^*) is a weak solution of (3.1) with (f,g) replaced by (f^*, g^*). Since $(u^*, v^*) \in W_p^{2,1}(G) \times W_p^{2,1}(G)$, the uniqueness of positive solution of problem (3.1) (Theorem 3.4) implies that

$$u^* = u^*(f^*, g^*) \quad \text{and} \quad v^* = v^*(f^*, g^*).$$

Moreover, we have

$$\begin{aligned} J(f^*, g^*) &= \textstyle\int_G \{K_1 u^* f^* + K_2 v^* g^* - M_1 (f^*)^2 - M_2 (g^*)^2\}\, dxdt \\ &\geq lim_{n\to\infty} \textstyle\int_G \{K_1 u_n f_n + K_2 v_n g_n\}\, dxdt \\ &\quad - lim\, inf_{n\to\infty} \textstyle\int_G \{M_1 f_n^2 + M_2 g_n^2\}\, dxdt \\ &= lim\, sup_{n\to\infty} J(f_n, g_n) = sup_{(f,g)\in B_{\delta,T}} J(f,g). \end{aligned}$$

Consequently, (f^*, g^*) is an optimal control in $B_{\delta,T}$, and this completes the proof.

In the remaining part of this section, we will deduce an optimal system which characterizes the optimal control described in Theorem 3.5. We first consider the differentiability property of the solutions of (3.1) with respect to the control.

Lemma 3.3 (Differentiability with Respect to Control). *Suppose that* δ_i, a_i, c_i *and* $b_i, i = 1, 2,$ *satisfy hypotheses* $[H1]$ *and* $[H2]$*; then we have*

$$\begin{aligned} &\tfrac{u(f+\beta\bar{f}, g+\beta\bar{g}) - u(f,g)}{\beta} \to \xi, \qquad \textit{weakly in } W_2^{2,1}(G), \\ &\tfrac{v(f+\beta\bar{f}, g+\beta\bar{g}) - u(f,g)}{\beta} \to \eta, \qquad \textit{weakly in } W_2^{2,1}(G), \end{aligned} \tag{3.49}$$

as $\beta \to 0$ for some subsequences, for any given $(f,g) \in B_{\delta,T}$, and $(\bar{f},\bar{g}) \in L^\infty(G)$ such that $(f+\beta\bar{f}, g+\beta\bar{g}) \in B_{\delta,T}$. Furthermore, (ξ,η) is a solution of the problem

(3.50)

$$\xi_t - \Delta\xi - [a_1 - f - 2b_1u(f,g) - c_1v(f,g)]\xi + c_1\eta u = -\bar{f}u(f,g) \quad \text{in } G,$$

$$\frac{\partial\xi}{\partial\nu} = 0 \quad \text{on } S,$$

$$\xi(x,0) = \xi(x,T) \quad \text{for all } x \in \Omega;$$

(3.51)

$$\eta_t - \Delta\eta - [a_2 - g - 2b_2v(f,g) - c_2u(f,g)]\eta + c_2\xi v = -\bar{g}v(f,g) \quad \text{in } G,$$

$$\frac{\partial\eta}{\partial\nu} = 0 \quad \text{on } S,$$

$$\eta(x,0) = \eta(x,T) \quad \text{for all } x \in \Omega.$$

(For the uniqueness of solution to problem (3.50) and (3.51), see Remark 3.1 below.)

Proof. Let

$$\xi_\beta = \frac{u(f+\beta\bar{f}, g+\beta\bar{g}) - u(f,g)}{\beta}, \quad \eta_\beta = \frac{u(f+\beta\bar{f}, g+\beta\bar{g}) - u(f,g)}{\beta};$$

then by (3.1), (ξ_β, η_β) satisfies

(3.52)

$$\begin{cases} \xi_{\beta t} - \Delta\xi_\beta - (a_1 - f)\xi_\beta + b_1(\bar{u}+u)\xi_\beta + c_1\bar{v}\xi_\beta + c_1u\eta_\beta = -\bar{f}\bar{u} & \text{in } G, \\ \eta_{\beta t} - \Delta\eta_\beta - (a_2 - g)\eta_\beta + b_2(\bar{v}+v)\eta_\beta + c_2\bar{u}\eta_\beta + c_2v\xi_\beta = -\bar{g}\bar{v} & \text{in } G, \\ \frac{\partial\xi_\beta}{\partial\nu} = \frac{\partial\eta_\beta}{\partial\nu} = 0 & \text{on } S, \\ \xi_\beta(x,0) = \xi_\beta(x,T) \text{ and } \eta_\beta(x,0) = \eta_\beta(x,T), & \text{for all } x \in \Omega. \end{cases}$$

Here, we denote $\bar{u} = u(f+\beta\bar{f}, g+\beta\bar{g}), \bar{v} = v(f+\beta\bar{f}, g+\beta\bar{g})$. Since ξ_β and $\eta_\beta \in W_p^{2,1}(G)$ are periodic with period T, we can readily prove by approximation that

$$(3.53) \quad \int_G \xi_\beta\xi_{\beta t}\,dxdt = \int_G \eta_\beta\eta_{\beta t}\,dxdt = \int_G \xi_{\beta t}\Delta\xi_\beta\,dxdt = \int_G \eta_{\beta t}\Delta\eta_\beta\,dxdt = 0.$$

In fact, the inequalities are all proved similarly, and we will only prove one of them here. Let $\{r_m\}$ be a sequence of $C^\infty(G)$-periodic functions of t with period

T, and $\partial r_m/\partial\nu = 0$ for all $(x,t) \in \partial\Omega \times [0,T), m = 1,2,\ldots$ such that

$$r_{m,t} \to \xi_{\beta,t}, \quad \Delta r_m \to \Delta\xi_\beta, \quad \text{in } \mathrm{L}^2(G), \quad \text{as } m \to \infty.$$

(Since $\xi_\beta \in W_p^{2,1}(G)$, the existence of such $\{r_m\}$ is given in Chapter 7 of [113]). Then we have

$$\begin{aligned}\textstyle\int_G \xi_{\beta,t}\Delta\xi_\beta\, dxdt &= lim_{m\to\infty} \textstyle\int_G r_{m,t}\Delta r_m\, dxdt \\ &= lim_{m\to\infty} \textstyle\int_G -\frac{1}{2}\frac{d}{dt}|\nabla r_m|^2\, dxdt = 0.\end{aligned}$$

Here, the second equality is due to the divergence theorem; and the last one is due to the periodic property of r_m.

Multiplying the first and second equations of (3.52) by ξ_β, η_β respectively and integrating both over G, we obtain

$$\begin{aligned}&\textstyle\int_G |\nabla\xi_\beta|^2\, dxdt + \int_G [b_1(\bar{u}+u) + c_1\bar{v} - (a_1 - f)]\xi_\beta^2\, dxdt + \int_G c_1 u\xi_\beta\eta_\beta\, dxdt \\ &\quad = -\textstyle\int_G \bar{f}\bar{u}\xi_\beta\, dxdt,\end{aligned}$$

and

$$\begin{aligned}&\textstyle\int_G |\nabla\eta_\beta|^2\, dxdt + \int_G [b_2(\bar{v}+v) + c_2\bar{u} - (a_2 - g)]\eta_\beta^2\, dxdt + \int_G c_2 v\eta_\beta\xi_\beta\, dxdt \\ &\quad = -\textstyle\int_G \bar{g}\bar{v}\eta_\beta\, dxdt.\end{aligned}$$

From Theorem 3.3 and $[H1]$, we find

$$b_1(\bar{u}+u) - (a_1 - f) \geq \delta_1, \quad b_2(\bar{v}+v) - (a_2 - \bar{g}) \geq \delta_2. \tag{3.54}$$

Moreover, by $[H2]$ when $i = 1, j = 2$, we have

$$c_1 u + c_2 v \leq 2\,min\{\delta_1, \delta_2\}. \tag{3.55}$$

Consequently we obtain the inequality

$$\int_G [\nabla\xi_\beta|^2 + |\nabla\eta_\beta|^2 + \xi_\beta^2 + \eta_\beta^2]\, dxdt \leq const. \tag{3.56}$$

where the constant is independent of β.

By transposing all the terms of the first equation in (3.52) except $\xi_{\beta t} - \Delta\xi_\beta$ to the right-hand side and using (3.56), we can use parabolic estimates to obtain the following inequality

$$||\xi_\beta||_{W_2^{2,1}(G)} \leq C,$$

where the constant C is independent of β. Similarly, from the second equation of (3.52) and (3.56), we obtain

$$||\eta_\beta||_{W_2^{2,1}(G)} \leq C,$$

where C is independent of β. Consequently, there exist subsequences (denoted again by ξ_β and η_β), such that

$$\xi_\beta \to \xi \quad \text{and} \quad \eta_\beta \to \eta,$$

strongly in $W_2^{1,0}(G)$ and weakly in $W_2^{2,1}(G)$. Moreover, taking limits as $\beta \to \infty$ in (3.52), we conclude that the limit (ξ, η) satisfies (3.50) and (3.51). This completes the proof of the Lemma.

Remark 3.1. Under the same hypotheses as Lemma 3.3, we can prove as in Theorem 3.4 that problem (3.50), (3.51) has only one solution, which is in $W_2^{2,1}(G) \times W_2^{2,1}(G)$. Therefore we can actually conclude that (3.49) holds for the full sequence. The proof of uniqueness under hypotheses [H1] and [H2] is nearly the same as that for Theorem 3.4, and will be omitted.

Theorem 3.6 (Characterization of the Optimal Control). *Let $p > 1$ be any positive number. Suppose $a_i(x,t), \delta_i, b_i, c_i, i = 1, 2$, satisfy hypotheses [H1] and [H2]; and the positive constants $K_i, M_i, i = 1, 2$, satisfy the hypotheses*

[H3] $M_i \geq \frac{K_i \sup_G a_i(x,t)}{2b_i\delta_i}$, *for $i = 1, 2$.*

For any optimal control $(f, g) \in B_{\delta,T}$ of problem (3.1), (3.47), (3.48), let (u, v) be the solution of problem (3.1) with

$$0 < \epsilon_1 \leq u \leq \hat{a}_1/b_1, \quad 0 < \epsilon_2 \leq v \leq \hat{a}_2/b_2,$$

and suppose (z, w) is a solution of

$$(3.57) \quad \begin{cases} z_t + \Delta z - [2ub_1 + c_1 v - a_1 + f]z - c_2 vw = -K_1 f & \text{in } G, \\ w_t + \Delta w - [2vb_2 + c_2 u - a_2 + g]w - c_1 uz = -K_2 g & \text{in } G, \\ \frac{\partial z}{\partial \nu} = \frac{\partial w}{\partial \nu} = 0 & \text{on } S, \\ z(x,0) = z(x,T) \text{ and } w(x,0) = w(x,T), & \text{for all } x \in \Omega. \end{cases}$$

satisfying

$$(3.58) \qquad -\frac{c_2 K_2 \hat{a}_2}{\delta_1 b_2 + c_1 \hat{a}_2} \leq z \leq K_1, \qquad -\frac{c_1 K_1 \hat{a}_1}{\delta_2 b_1 + c_2 \hat{a}_1} \leq w \leq K_2.$$

Then the optimal control (f, g) *satisfies*

$$f = \frac{u(x,t)}{2M_1}(K_1 - z) \quad and \quad g = \frac{v(x,t)}{2M_2}(K_2 - w), \quad in\ G. \tag{3.59}$$

Here u, v, z *and* w *are in* $W_p^{2,1}(G)$ *and* $\epsilon_i, i = 1, 2$ *are defined in (3.41).*

Proof. Theorem 3.5 implies that the conditions of this theorem suffice to ensure the existence of an optimal control in $B_{\delta,T}$. Let $(f, g) \in B_{\delta,T}$ be an optimal control, i.e. there exists a solution (u, v) of the problem (3.1) for (f, g) such that

$$J(f, g) = sup_{(\hat{f},\hat{g}) \in B_{\delta,T}} J(\hat{f}, \hat{g}).$$

For arbitrary $\tilde{f}, \tilde{g} \in L_+^\infty(G), \epsilon > 0$, set

$$\bar{f} = \bar{f}_\epsilon = \begin{cases} \tilde{f}, & \text{if } f \leq \delta_1 - \epsilon ||\tilde{f}||_{\infty,G}, \\ 0, & \text{elsewhere;} \end{cases}$$

similarly we define $g = \bar{g}_\epsilon$.

For $\beta > 0$ small enough (say $\beta < \epsilon$), such that $(f + \beta\bar{f}, g + \beta\bar{g}) \in B_{\beta,T}$, the optimality of (f, g) implies that

$$J(f, g) \geq J(f + \beta\bar{f}, g + \beta\bar{g}), \tag{3.60}$$

that is

$$\begin{aligned} &\textstyle\int_G (K_1 uf + K_2 vg - M_1 f^2 - M_2 g^2)\, dxdt \\ &\quad \geq \textstyle\int_G [K_1 u(f + \beta\bar{f}, g + \beta\bar{g})(f + \beta\bar{f}) + K_2 v(f + \beta\bar{f}, g + \beta\bar{g})(g + \beta\bar{g}) \\ &\quad -M_1(f + \beta\bar{f})^2 - M_2(g + \beta\bar{g})^2]\, dxdt. \end{aligned}$$

Dividing by β and letting $\beta \to 0$, we obtain from Lemma 3.3,

$$\int_G [K_1\xi f + K_1 u\bar{f} + K_2\eta g + K_2 v\bar{g} - 2M_1 f\bar{f} - 2M_2 g\bar{g}]\, dxdt \leq 0. \tag{3.61}$$

Since (z, w) is a solution of problem (3.57) satisfying (3.58), we deduce from (3.61), (3.57), (3.50), (3.51) and integrating by parts that

$$\int_G \{\bar{f}_\epsilon[(K_1 - z)u - 2M_1 f] + \bar{g}_\epsilon[(K_2 - w)v - 2M_2 g]\}\, dxdt \leq 0.$$

Now, letting $\tilde{g} = 0, \epsilon \to 0^+$, and using argument as in the proof of Theorem 2.4 in Section 3.2, we deduce from hypothesis [H3] and the above inequality that

$$f = \frac{K_1 - z}{2M_1} u(x,t), \quad \text{in } G.$$

Similarly, letting $\tilde{f} = 0$, we obtain

$$g = \frac{K_2 - w}{2M_2} v(x,t), \quad \text{in } G.$$

This completes the proof of the theorem.

Remark 3.2. Suppose $(f,g) \in B_{\delta,T}$ is any optimal control, we see from Theorem 3.6 that if (u,v) and (z,w) are the unique solutions of problem (3.1) and (3.57), respectively, then (u,v,z,w) is a solution of the following optimal system:

(3.62)
$$\begin{cases} u_t - \Delta u - a_1 u + (b_1 + \frac{K_1 - z}{2M_1})u^2 + c_1 uv = 0 & \text{in } G, \\ v_t - \Delta v - a_2 v + (b_2 + \frac{K_2 - w}{2M_2})v^2 + c_2 uv = 0 & \text{in } G, \\ z_t + \Delta z - [2b_1 u - a_1 + c_1 v]z + \frac{(K_1 - z)^2}{2M_1} u - c_2 vw = 0 & \text{in } G, \\ w_t + \Delta w - [2b_2 v - a_2 + c_2 u]w + \frac{(K_2 - w)^2}{2M_2} v - c_1 uz = 0 & \text{in } G, \\ \frac{\partial u}{\partial \nu} = \frac{\partial v}{\partial \nu} = \frac{\partial z}{\partial \nu} = \frac{\partial w}{\partial \nu} = 0 & \text{on } S, \\ u(x,0) = u(x,T),\ v(x,0) = v(x,T), & \text{for all } x \in \Omega, \\ z(x,0) = z(x,T),\ w(x,0) = w(x,T), & \text{for all } x \in \Omega. \end{cases}$$

Thus if (3.62) can be solved for (u,v,w,z), then the optimal control (f,g) can be found by using (3.59).

We next prove problem (3.57), described in Theorem 3.6, indeed has a unique solution satisfying (3.58)

Theorem 3.7. *Under the assumptions of Theorem 3.6, problem (3.57) has a unique solution* $(z,w) \in W_p^{2,1}(G) \times W_p^{2,1}(G)$ *with*

$$-D_1 \equiv -\frac{c_2 K_2 \hat{a}_2}{\delta_1 b_2 + c_1 \hat{a}_2} \le z \le K_1 \ \ and \ \ -D_2 \equiv -\frac{c_1 K_1 \hat{a}_1}{\delta_2 b_1 + c_2 \hat{a}_1} \le w \le K_2.$$

Here (u,v) *satisfies (3.1) and (3.41).*

Proof. We can readily show that $(-D_1, -D_2), (K_1, K_2)$ are respectively lower and upper solutions of problem (3.57) in the region $-D_1 \le z \le K_1, -D_2 \le w \le$

K_2, i.e.,

$$\begin{cases} (-D_1)_t + \Delta(-D_1) + [2b_1u + c_1v - a_1 + f]D_1 - c_2vK_2 \geq -K_1f & \text{in } G, \\ (-D_2)_t + \Delta(-D_2) + [2b_2v + c_2u - a_2 + g]D_2 - c_1uK_1 \geq -K_2g & \text{in } G, \\ K_{1t} + \Delta K_1 - [2b_1u + c_1v - a_1 + f]K_1 - c_2v(-D_2) \leq -K_1f & \text{in } G, \\ K_{2t} + \Delta K_2 - [2b_2v + c_2u - a_2 + g]K_2 - c_1u(-D_1) \leq -K_2g & \text{in } G. \end{cases}$$

To prove the existence of solution for (3.57), we first define

$$(p_0, q_0) = (-D_1, -D_2), \quad (p_{-1}, q_{-1}) = (K_1, K_2),$$

and $p_i, q_i, i = 1, 2, 3, ...$, to be solutions of

(3.63)
$$\begin{cases} -p_{it} - \Delta p_i + [2b_1u - a_1 + f + c_1v]p_{i-2} = K_1f - c_2vq_{i-1} & \text{in } G, \\ -q_{it} - \Delta q_i + [2b_2v - a_2 + g + c_2u]q_{i-2} = K_2g - c_1up_{i-1} & \text{in } G, \\ \frac{\partial p_i}{\partial \nu} = \frac{\partial q_i}{\partial \nu} = 0 & \text{on } S, \\ p_i(x,0) = p_i(x,T) \text{ and } q_i(x,0) = q_i(x,T), & \text{for all } x \in \Omega. \end{cases}$$

The existence of solutions for each of the scalar problems in (3.63) are ensured by Theorem 3.1. In fact, if we denote $\phi^0(x,s) = \phi(x,-s)$ for any function $\phi(x,t)$, then p_i^0 satisfies the parabolic problem

(3.64)
$$\begin{cases} p_{it}^0 - \Delta p_i^0 + [2b_1u^0 - a_1^0 + f^0 + c_1v^0]p_{i-2}^0 = K_1f^0 - c_2v^0q_{i-1}^0 & \text{in } \Omega \times [-T,0], \\ \frac{\partial p_i^0}{\partial \nu} = 0, & \text{on } \tilde{S} = \partial\Omega \times [-T,0], \\ p^0(x,-T) = p^0(x,0), & \text{for all } x \in \Omega. \end{cases}$$

Theorem 3.1 implies that the above parabolic problem has a unique solution. Therefore problem (3.63) has a unique solution $p_i(x,s) = p_i^0(x,-s)$. The same argument applies to q_i.

For a given positive number $R, (x,t) \in G$ we define two functions:

(3.65)
$$\begin{aligned} h_1(x,t,p,q) &= K_1f(x,t) - c_2v(x,t)q - [2b_1u(x,t) - a_1(x,t) + f(x,t) \\ &\quad + c_1v(x,t)]p + Rp, \\ h_2(x,t,p,q) &= K_2g(x,t) - c_1u(x,t)p - [2b_2v(x,t) - a_2(x,t) + g(x,t) \\ &\quad + c_2u(x,t)]q + Rq. \end{aligned}$$

We choose R to be sufficiently large such that h_1 and h_2 are increasing in p and q respectively in the domain $\epsilon_1 \le u(x,t) \le \hat{a}_1/b_1, \epsilon_2 \le v(x,t) \le \hat{a}_2/b_2, -D_1 \le p \le K_1$ and $-D_2 \le q \le K_2$. Moreover, it is obvious that h_1 and h_2 are decreasing in q and p respectively in the above domain.

Using h_1 and h_2 we modify (3.63) into

$$(3.66) \qquad \begin{cases} -p_{it} - \Delta p_i + Rp_i = h_1(x,t,p_{i-2},q_{i-1}) & \text{in } G, \\ -q_{it} - \Delta q_i + Rq_i = h_2(x,t,q_{i-2},p_{i-1}) & \text{in } G, \\ \frac{\partial p_i}{\partial \nu} = \frac{\partial q_i}{\partial \nu} = 0 & \text{on } S, \\ p_i(x,0) = p_i(x,T) \text{ and } q_i(x,0) = q_i(x,T), & \text{for all } x \in \Omega. \end{cases}$$

From the monotone properties of h_1 and h_2, the maximum principle of linear parabolic equations and the fact that $(p_0,q_0),(p_{-1},q_{-1})$ are lower and upper solutions, we can prove by means of (3.66) and induction that

$$p_0 \le p_2 \le \cdots \le p_{2i} \le \cdots \le p_{2i+1} \le \cdots \le p_1 \le p_{-1} \quad \text{in } G$$

and

$$q_0 \le q_2 \le \cdots \le q_{2i} \le \cdots \le q_{2i+1} \le \cdots \le q_1 \le q_{-1} \quad \text{in } G.$$

(See Section 5.5 in [125] for details of the scheme). Moreover, we have

$$||p_i||_{W_p^{2,1}(G)} \le C \text{ and } ||q_i||_{W_p^{2,1}(G)} \le C,$$

where C is a constant independent of i. Consequently, (3.66) implies that the limits

$$lim_{r\to\infty} q_{2r}, \quad lim_{r\to\infty} q_{2r-1}, \quad lim_{r\to\infty} p_{2r}, \quad lim_{r\to\infty} p_{2r-1}$$

exist in $W_p^{2,1}(G)$, say q_*, q^*, p_* and p^*, respectively. Moreover, we have $q_* \le q^*$ and $p_* \le p^*$. It remains to prove that $q_* = q^*$ and $p_* = p^*$. Taking limit as $i \to \infty$ in (3.66), we find (q_*, q^*, p_*, p^*) satisfies the problem

(3.67)

$$\begin{cases} -p_{*t} - \Delta p_* + [2b_1u - a_1 + f + c_1v]p_* + c_2vq^* = K_1f & \text{in } G, \\ -p^*_t - \Delta p^* + [2b_1u - a_1 + f + c_1v]p^* + c_2vq_* = K_1f & \text{in } G, \\ -q_{*t} - \Delta q_* + [2b_2v - a_2 + g + c_2u]q_* + c_1up^* = K_2g & \text{in } G, \\ -q^*_t - \Delta q^* + [2b_2v - a_2 + g + c_2u]q^* + c_1up_* = K_2g & \text{in } G, \\ \frac{\partial p_*}{\partial \nu} = \frac{\partial p^*}{\partial \nu} = \frac{\partial q_*}{\partial \nu} = \frac{\partial q^*}{\partial \nu} = 0 & \text{on } S, \\ p_*(x,0) = p_*(x,T), p^*(x,0) = p^*(x,T), q_*(x,0) = q_*(x,T), \text{ and} & \\ q^*(x,0) = q^*(x,T), & x \in \Omega. \end{cases}$$

Equation (3.67) consists of actually two separate systems, each with two equations. Moreover, (p_*, q^*) and (p^*, q_*) satisfy the same system of two equations. From hypotheses [H1] and [H2] and the fact that

$$2b_1u - a_1 + f + c_1v \geq \delta_1 + c_1v \geq \delta_1, \quad 2b_2v - a_2 + g + c_2u \geq \delta_2 + c_2u \geq \delta_2,$$

we can prove as in Theorem 3.4 that

$$(p^*, q_*) = (p_*, q^*).$$

Hence, we have proved the existence part. The uniqueness of solution in the prescribed range is proved by using the property that $(-D_1, -D_2)$ and (K_1, K_2) are lower and upper solutions of problem (3.63) and by showing

$$p_{2r} \leq z \leq p_{2r+1}, \quad \text{and } q_{2r} \leq w \leq q_{2r+1}, \quad \text{for } r = 1, 2, ..,$$

with similar arguments. This proves Theorem 3.7.

In the final part of this section we describe a method to find approximations for the solution (u, v, z, w) of problem (3.62). We construct monotone sequences converging from above and below, providing upper and lower estimates for (u, v, z, w). In the case when the limits of upper and lower iterates agree, then the optimal control problem is completely solved. That is, the optimal control is given by (3.59) in terms of (u, v, z, w), which is calculated iteratively. We will need the following additional conditions:

[H4] $\frac{\epsilon_i K_i}{2M_i} \leq \delta_i$ for $i = 1, 2$ and

[H5] $\frac{c_j \hat{a}_j b_i}{b_j} \leq \frac{K_i^2 \delta_i}{K_j M_i}$ for $i, j = 1, 2$ and $i \neq j$,

where ϵ_1, ϵ_2 are positive numbers defined by (3.41).

Remark 3.3. Under the additional conditions [H4] and [H5], together with [H1]-[H3], we can prove that $(u_0, v_0, p_0, q_0) := (\epsilon_1, \epsilon_2, 0, 0)$ is a lower solution of (3.62). This implies that the proofs of Theorems 3.6 and 3.7 still hold if we replace $(-D_1, -D_2)$ with $(0, 0)$. Then we can use the same arguments to show that the conclusions of Theorems 3.6 and 3.7 are still true. Consequently, there exists one solution (u, v, z, w) of problem (3.59) such that the functions u, v, z and w are positive.

Assume all the hypotheses [H1] to [H5] in the remaining part of this section; define

$$(u_0, v_0, p_0, q_0) := (\epsilon_1, \epsilon_2, 0, 0) \quad \text{and} \quad (u_{-1}, v_{-1}, p_{-1}, q_{-1}) := (\hat{a}_1/b_1, \hat{a}_2/b_2, K_1, K_2).$$

For $(x, t) \in G$, and a given positive number Q, we define four functions as follows:

$$\hat{h}_1(x, t, p, u_1, u_2, v) = p[a_1(x, t) - 2b_1 u_1 - c_1 v] + \frac{(K_1 - p)^2}{2M_1} u_2 + Qp,$$

$$\hat{h}_2(x, t, q, v_1, v_2, u) = q[a_2(x, t) - 2b_2 v_1 - c_2 u] + \frac{(K_2 - q)^2}{2M_2} v_2 + Qq,$$

$$\hat{h}_3(x, t, u, v, p) = u[a_1(x, t) - (b_1 + \frac{K_1}{2M_1})u + \frac{pu}{2M_1} - c_1 v] + Qu,$$

$$\hat{h}_4(x, t, v, u, q) = v[a_2(x, t) - (b_2 + \frac{K_2}{2M_2})v + \frac{qv}{2M_2} - c_2 u] + Qv.$$

We choose Q large enough such that $\hat{h}_i, i = 1, 2, 3, 4$ have the following corresponding properties.

(S1) $\hat{h}_1$ is increasing in p for $p \in [p_0, p_{-1}]$ with fixed $u_1, u_2 \in [u_0, u_{-1}]$ and $v \in [v_0, v_{-1}]$; moreover, $\hat{h}_1$ is increasing in u_2 but decreasing in u_1, v with the other variables fixed in the same intervals.

(S2) The properties of $\hat{h}_2$ in terms of q, v_1, v_2, u are the same as $\hat{h}_1$ in terms of p, u_1, u_2, v respectively.

(S3) $\hat{h}_3$ is increasing in u for $u \in [u_0, u_{-1}]$ with fixed $p \in [p_0, p_{-1}]$ and $v \in [v_0, v_{-1}]$; moreover, $\hat{h}_3$ is increasing in p but decreasing in v with the other variables fixed in the same intervals.

(S4) The properties of $\hat{h}_4$ in terms of v, u, q are the same as $\hat{h}_3$ in terms of u, v, p respectively.

We can readily verify that (u_0, v_0, p_0, q_0) and $(u_{-1}, v_{-1}, p_{-1}, q_{-1})$ satisfy

$$u_{-1t} - \Delta u_{-1} + Q u_{-1} \geq \hat{h}_3(x, t, u_{-1}, v_0, p_{-1}) \qquad \text{in } G, \tag{3.68}$$

(3.69) $$v_{-1t} - \Delta v_{-1} + Qv_{-1} \geq \hat{h}_4(x,t,v_{-1},u_0,q_{-1}) \quad \text{in } G,$$

(3.70) $$p_{-1t} + \Delta p_{-1} - Qp_{-1} \leq -\hat{h}_1(x,t,p_{-1},u_0,u_{-1},v_0) + c_2 v_0 q_0 \quad \text{in } G,$$

(3.71) $$q_{-1t} + \Delta q_{-1} - Qq_{-1} \leq -\hat{h}_2(x,t,q_{-1},v_0,v_{-1},u_0) + c_1 u_0 p_0 \quad \text{in } G,$$

(3.72) $$p_{0t} + \Delta p_0 - Qp_0 \geq -\hat{h}_1(x,t,p_0,u_{-1},u_0,v_{-1}) + c_2 v_{-1} q_{-1} \quad \text{in } G,$$

(3.73) $$q_{0t} + \Delta q_0 - Qq_0 \geq -\hat{h}_2(x,t,q_0,v_{-1},v_0,u_{-1}) + c_1 u_{-1} p_{-1} \quad \text{in } G,$$

(3.74) $$u_{0t} - \Delta u_0 + Qu_0 \leq \hat{h}_3(x,t,u_0,v_{-1},p_0) \quad \text{in } G,$$

(3.75) $$v_{0t} - \Delta v_0 + Qv_0 \leq \hat{h}_4(x,t,v_0,u_{-1},q_0) \quad \text{in } G,$$

Inequalities (3.68)-(3.71) can be readily verified using [H1]. We next show that (3.72) holds. Since $p_0 = 0$, proving (3.72) is equivalent to proving the inequality

$$0 \geq -(\frac{K_1^2}{2M_1})\epsilon_1 + c_2(\frac{\hat{a}_2}{b_2})K_2.$$

From [H1], we have

$$\epsilon_1 = \frac{\tilde{a}_1 - c_1(\hat{a}_2/b_2) - \delta_1}{b_1} \geq \frac{2\delta_1}{b_1}.$$

Thus, in order to prove (3.72), we only need to show

$$\frac{K_1^2 \delta_1}{M_1 K_2} \geq \frac{c_2 \hat{a}_2 b_1}{b_2},$$

which is our hypothesis [H5]. Inequality (3.73) is completely analogous to (3.72). Similarly, using [H1] and [H4], we can prove (3.74) and (3.75).

Now, we inductively define sequences of functions u_i, v_i, p_i and q_i for $i = 1, 2, \ldots$ as solutions of the following problems:

(3.76) $$\begin{cases} u_{it} - \Delta u_i + Qu_i = \hat{h}_3(x,t,u_{i-2},v_{i-1},p_{i-2}) & \text{in } G, \\ \frac{\partial u_i}{\partial \nu} = 0 & \text{on } S, \\ u_i(x,0) = u_i(x,T) & \text{for all } x \in \Omega; \end{cases}$$

$$(3.77)\qquad \begin{cases} v_{it} - \Delta v_i + Q v_i = \hat{h}_4(x, t, v_{i-2}, u_{i-1}, q_{i-2}) & \text{in } G, \\ \frac{\partial v_i}{\partial \nu} = 0 & \text{on } S, \\ v_i(x, 0) = v_i(x, T) & \text{for all } x \in \Omega; \end{cases}$$

(3.78)

$$\begin{cases} p_{it} + \Delta p_i - Q p_i = -\hat{h}_1(x, t, p_{i-2}, u_{i-1}, u_{i-2}, v_{i-1}) + c_2 v_{i-1} q_{i-1} & \text{in } G, \\ \frac{\partial p_i}{\partial \nu} = 0 & \text{on } S, \\ p_i(x, 0) = p_i(x, T) & \text{for all } x \in \Omega; \end{cases}$$

(3.79)

$$\begin{cases} q_{it} + \Delta q_i - Q q_i = -\hat{h}_2(x, t, q_{i-2}, v_{i-1}, v_{i-2}, u_{i-1}) + c_1 u_{i-1} p_{i-1} & \text{in } G, \\ \frac{\partial q_i}{\partial \nu} = 0 & \text{on } S, \\ q_i(x, 0) = q_i(x, T) & \text{for all } x \in \Omega. \end{cases}$$

The existence of solutions follows from Theorem 3.1. By using the induction argument, the monotone properties of $\hat{h}_i, i = 1, 2, 3, 4$ and the maximum principle, (cf. Section 5.5 in [125]), we can show that

$$(3.80)\qquad \begin{aligned} & u_0 \le u_2 \le \cdots \le u_{2i} \le \cdots \le u_{2i-1} \le \cdots \le u_1 \le u_{-1} && \text{in } G, \\ & v_0 \le v_2 \le \cdots \le v_{2i} \le \cdots \le v_{2i-1} \le \cdots \le v_1 \le v_{-1} && \text{in } G, \\ & p_0 \le p_2 \le \cdots \le p_{2i} \le \cdots \le p_{2i-1} \le \cdots \le p_1 \le p_{-1} && \text{in } G, \\ & q_0 \le q_2 \le \cdots \le q_{2i} \le \cdots \le q_{2i-1} \le \cdots \le q_1 \le q_{-1} && \text{in } G. \end{aligned}$$

Furthermore, we can readily prove the following theorem by repeated applications of Lemma 3.1.

Theorem 3.8 (Approximation Scheme for the Optimal Control). *Assume hypotheses [H1] – [H5]. The sequences u_i, v_i, p_i and q_i defined above by (3.76)-(3.79) satisfy relation (3.80) for all positive integer i and $(x, t) \in G$. Moreover, any solution (u, v, z, w) of problem (3.62) with properties*

$$(3.81)\qquad u_0 \le u \le u_{-1},\ v_0 \le v \le v_{-1},\ p_0 \le z \le p_{-1},\ q_0 \le w \le q_{-1},\ \text{in } G,$$

must satisfy the inequalities

(3.82)

$$u_{2i} \le u \le u_{2i-1},\ v_{2i} \le v \le v_{2i-1},\ p_{2i} \le z \le p_{2i-1},\ q_{2i} \le w \le q_{2i-1},\ \text{in } G,$$

for any positive integer i.

Remark 3.4. From Theorems 3.7 and 3.8, we find that if

$$lim_{i\to\infty} u_{2i} = lim_{i\to\infty} u_{2i-1}, \; lim_{i\to\infty} v_{2i} = lim_{i\to\infty} v_{2i-1},$$

$$lim_{i\to\infty} p_{2i} = lim_{i\to\infty} p_{2i-1}, \; lim_{i\to\infty} q_{2i} = lim_{i\to\infty} q_{2i-1},$$

then the optimal control problem (3.47), (3.48) for the competitive system (3.1) is completely solved (cf. Remark 3.2).

For an example of the control problem: (3.1), (3.47) and (3.48), which satisfies all the hypotheses [H1] to [H5], we consider the following:

Example 3.1. Let $\Omega = \{(x,y)|x^2+y^2<1\}$ and $\bar{G} = \bar{\Omega}\times[0,2\pi]$. Define

$$a_1(x,y,t) = [\tfrac{1}{4}(x^2+y^2)]cos\, t + 16, \;\; b_1 = 4, \;\; c_1 = 0.4,$$

$$a_2(x,y,t) = sin(\pi x)sin(\pi y)sin\, t + 25, \;\; b_2 = 6, \;\; c_2 = 0.5,$$

$$K_1 = 8, \; K_2 = 7, \; M_1 = 4, \; M_2 = 5.$$

We thus have

$$\tilde{a}_1 = 16 - \frac{1}{4}, \;\; \hat{a}_1 = 16 + \frac{1}{4}, \;\; \tilde{a}_2 = 24, \;\; \hat{a}_2 = 26.$$

Choosing $\delta = (\delta_1, \delta_2) = (11, 17)$, we can easily verify that the hypotheses [H1]-[H5] are all satisfied. For instance, we have

$$\epsilon_1 = \frac{\tilde{a}_1 - c_1(\hat{a}_2/b_2) - \delta_1}{b_1} = \frac{19}{16} - \frac{13}{30} < 2;$$

thus [H4] holds for $i = 1$ because

$$\frac{\epsilon_1 K_1}{2M_1} < 2 < \delta_1 = 11.$$

Similarly, [H4] holds for $i = 2$. Moreover, [H5] holds for $i = 1, j = 2$ or $i = 2, j = 1$.

Remark 3.5. Let $A_1(x,t)$ and $A_2(x,t)$ be given continuous t-periodic functions for $x \in \Omega, t \in (-\infty, \infty)$, where Ω is any bounded domain with C^2 boundary. Consider problem (3.1), (3.47) and (3.48) with fixed c_i, b_i, M_i and K_i for $i = 1, 2$. From the previous example, we see that we can always find large enough constants B and δ_i such that if we define $a_i = A_i + B, i = 1, 2$, then the hypotheses [H1]-[H5] are readily satisfied. Consequently, our results are applicable to a large family of problems.

3.4 Optimal Control of an Initial-Boundary Value Problem for Fission Reactor Systems

To illustrate the diversity of applications of the method in the last sections, we consider a different type of problem related to engineering physics. We will be concerned with the mathematical theory for the optimal control of nuclear fission reactors modeled by parabolic differential equations. The neutrons are divided into fast and thermal groups with two equations describing their interaction and fission, while a third equation describes the temperature in the reactor. The coefficient for the fission and absorption of the thermal neutron is assumed to be controlled by a function through the use of control rods in the reactor. The object is to maintain a target neutron flux shape, while a desired power level and adjustment costs are taken into consideration. Dividing the neutrons into two energy groups and taking into account of the temperature feedback, the two-group neutron diffusion reactor equations become:

$$(4.1)\qquad \begin{cases} \nu_1^{-1}\partial u_1/\partial t - \sigma_1 \Delta u_1 = a_{11}u_1 + a_{12}u_2 \\ \nu_2^{-1}\partial u_2/\partial t - \sigma_2 \Delta u_2 = a_{21}u_1 + (a_{22} - f)u_2 \qquad \text{for } x \in \Omega,\ t > 0, \\ \partial u_3/\partial t - \sigma_3 \Delta u_3 = a_{31}u_1 + a_{32}u_2 + a_{33}u_3, \end{cases}$$

$$(4.2)\qquad \begin{cases} u_1(x,t) = u_2(x,t) = u_3(x,t) = 0 \qquad \text{for } x \in \partial\Omega,\ t > 0, \\ u_i(x,0) = r_i(x) \qquad \text{for } i = 1,2,3,\ \ x \in \Omega. \end{cases}$$

Here, $u_1(x,t)$ and $u_2(x,t)$ are the fast and thermal neutron fluxes respectively, and $u_3(x,t)$ is the temperature function. Ω represents the reactor domain, and t is the time variable. The coefficients a_{12} and a_{21} are positive constants, representing group-transfer "cross-sections" between the two neutron groups. The parameter a_{33} is negative, representing the cooling constant; the constants a_{31}, a_{32} are positive, related to rate of energy released due to fission in the two neutron groups. The constants a_{11} and a_{22} can be any sign, and are related to fission and absorption rates of each neutron group. The coefficient for the thermal neutron is assumed to be controlled by the function f through the use of control rods in the reactor. The parameter $\sigma_i, \nu_i, i = 1, 2$ are positive constants related to diffusion rate and average number of fast neutrons released during fission. (See e.g. Christensen, Soliman and Nieva [29], Lewins [146], Lin, Lin and Jiang [154], and Terney and Wade [218] for more details.) We are thus making the following assumptions concerning the signs of the constants a_{ij}:

$$(4.3)\qquad \begin{aligned} &a_{12},\ a_{21},\ a_{31},\ a_{32} \ \text{ are positive and } a_{33} \text{ is negative;} \\ &a_{11} \text{ and } a_{22} \ \text{ can be any sign.} \end{aligned}$$

We will assume Ω to be a bounded domain in $R^N, N \geq 2$, with C^2 smooth boundary $\partial\Omega$. The control function $f = f(x,t)$ will be assumed to be in the set

$$C_\delta = \{f \in L_2(\Omega \times (0,T))| - \delta \leq f \leq \delta\}$$

where T and δ are some positive constants. One method for the control of such a reactor is to maintain a target neutron flux shape, so that a desired power level and adjustment cost are taken into consideration. For simplicity, we thus define the cost functional J by

$$J(f) = \int_{\Omega_T} \{\sum_{i=1}^{3} K_i[u_i(f) - e_i]^2 + Mf^2\}dxdt \tag{4.4}$$

for each $f \in C_\delta$. Here, $\Omega_T = \Omega \times [0,T], e_i = e_i(x,t)$ are the target profiles for the two neutron fluxes and temperature with corresponding weight factor K_i, and M is related to the cost of the control f. Moreover, K_i and M are constants; and $(u_1(f), u_2(f), u_3(f))$ denotes the solution of the problem (4.1), (4.2) with the corresponding control function $f = f(x,t)$ in Ω_T. The target functions e_i are all assumed to be in the class $L_2(\Omega \times (0,T))$, and the initial functions are in $W_p^2(\Omega), p = 2$ with $r_i|_{\partial\Omega} = 0$ throughout this section. The solutions of (4.1), (4.2) will be considered in $W_2^{2,1}(\Omega_T)$ for each component; and the equations in (4.1) are satisfied almost everywhere. Here, $W_2^{2,1}(\Omega_T)$ denotes the set of measurable functions in Ω_T with weak partial derivatives with respect to x up to second order, and weak first order partial derivative with respect to t in $L_2(\Omega_T)$.

An optimal control (if it exists) for the problem (4.1), (4.2) corresponding to the cost functional (4.4) is a function $f^* \in C_\delta$ such that

$$J(f^*) = \inf_{f \in C_\delta} J(f). \tag{4.5}$$

We will consider the existence of an optimal control and the differentiability of the cost functional with respect to the control. We then characterize the optimal control and derive an optimality system of equations. The optimality system is solved by an iterative procedure, which can be implemented into a computer algorithm. The iterative procedure is shown to be a contraction for small enough time interval. The optimal control can then be synthesized from the solution of the optimality system. The theory in this section is obtained from Leung and Chen [133], it can be extended to include the effects of delayed neutrons, e.g. xenon and iodide. It is known in practice that these delayed neutrons do have significant effect in the dynamics of the reactor. More details can be found in [133].

We first show the existence of optimal control for our problem in the set C_δ. Then, we consider the differentiability of the mapping

$$C_\delta \ni f \to (u_1(f), u_2(f), u_3(f)).$$

We characterize the derivative of the mapping as the solution of the linear system (4.6), (4.7) below.

Lemma 4.1. *There exists an optimal control satisfying (4.5) for the control problem (4.1), (4.2) with cost functional J defined by (4.4).*

From the boundedness of the set of admissible controls and the subsequent a-priori $W_2^{2,1}(\Omega_T)$ bound on the solutions, we can find a subsequence f_n of the minimizing sequence such that:

$$f_n \to f^* \quad \text{weakly in} \quad L_2(\Omega_T),$$

and

$$u_i(f_n) \to u_i^*, \;\; i = 1, 2, 3, \;\; \text{strongly in } W_2^{1,0}(\Omega_T) \;\; \text{and weakly in } W_2^{2,1}(\Omega_T),$$

as $n \to \infty$. We can then deduce $J(f^*)$ satisfies (1.5) by the usual semicontinuity argument. See e.g. Theorem 1.2 in Section 3.1, Theorem 2.3 in Section 3.2, and Theorem 3.5 in Section 3.3 for details of similar problem.

Lemma 4.2 (Differentiability with Respect to Control). *For $i = 1, 2, 3$, the mapping*

$$C_\delta \ni f \to u_i(f) \in W_2^{2,1}(\Omega_T)$$

is differentiable in the following sense:

$$[u_i(f + \beta\bar{f}) - u_i(f)]/\beta \to \xi_i \;\; \textit{weakly in } W_2^{2,1}(\Omega_T)$$

as $\beta \to 0$, for any $f \in C_\delta$ and $\bar{f} \in L^\infty(\Omega_T)$ such that $f + \beta\bar{f} \in C_\delta$. Further, (ξ_1, ξ_2, ξ_3) is the unique solution of

$$\text{(4.6)} \quad \begin{cases} \nu_1^{-1}\partial\xi_1/\partial t - \sigma_1\Delta\xi_1 = a_{11}\xi_1 + a_{12}\xi_2, \\ \nu_2^{-1}\partial\xi_2/\partial t - \sigma_2\Delta\xi_2 = a_{21}\xi_1 + (a_{22} - f)\xi_2 - \bar{f}u_2(f) \quad \text{in } \Omega \times (0,T), \\ \partial\xi_3/\partial t - \sigma_3\Delta\xi_3 = a_{31}\xi_1 + a_{32}\xi_2 + a_{33}\xi_3, \end{cases}$$

$$\text{(4.7)} \quad \xi_i(x,t) = 0 \qquad \text{in } (\Omega \times \{0\}) \cup (\partial\Omega \times (0,T)), \;\; \text{for } i = 1, 2, 3.$$

Proof. Let

$$\xi_i^\beta = [u_i(f + \beta\bar{f}) - u_i(f)]/\beta \;\; \text{for } i = 1, 2, 3.$$

By (4.1), $(\xi_1^\beta, \xi_2^\beta, \xi_3^\beta)$ satisfies

$$(4.8)\quad \begin{cases} \nu_1^{-1}\partial\xi_1^\beta/\partial t - \sigma_1\Delta\xi_1^\beta = a_{11}\xi_1^\beta + a_{12}\xi_2^\beta, \\ \nu_2^{-1}\partial\xi_2^\beta/\partial t - \sigma_2\Delta\xi_2^\beta = a_{21}\xi_1^\beta + (a_{22} - f)\xi_2^\beta - \bar{f}u_2(f+\beta\bar{f}), \\ \partial\xi_3^\beta/\partial t - \sigma_3\Delta\xi_3^\beta = a_{31}\xi_1^\beta + a_{32}\xi_2^\beta + a_{33}\xi_3^\beta, \qquad \text{in } \Omega\times(0,T), \\ \xi_i^\beta(x,t) = 0 \qquad \text{in } (\Omega\times\{0\})\cup(\partial\Omega\times(0,T)), \quad \text{for } i=1,2,3. \end{cases}$$

Let $\tilde{\xi}_i^\beta = \xi_i^\beta e^{-\lambda t}, i = 1,2,3; (\tilde{\xi}_1^\beta, .., \tilde{\xi}_3^\beta)$ satisfies the same equations (4.8), with ξ_i^β replaced by $\tilde{\xi}_i^\beta$ and diagonal coefficients a_{ii} replaced by $a_{ii} - \lambda, i = 1,2,3$. By choosing $\lambda > 0$ large enough, the diagonal coefficients in the equations for $\tilde{\xi}_i^\beta$ are all negative and large in absolute value compared with $a_{ij}, i \neq$ j. Multiplying the i-th equation by $\tilde{\xi}_i^\beta$ and integrating over Ω_T, we obtain the following estimates after adding the three equations and using the Green's identity:

$$(4.9)\quad \begin{aligned} &\textstyle\sum_{i=1}^3\{||\nabla\tilde{\xi}_i^\beta||^2_{L_2(\Omega_T)} + ||\tilde{\xi}_i^\beta||^2_{L_2(\Omega_T)}\} \\ &\qquad \leq const.||\bar{f}||_\infty||u_2(f+\beta\bar{f})||_{L_2(\Omega_T)}||\tilde{\xi}_2^\beta||_{L_2(\Omega_T)}. \end{aligned}$$

This implies that the expression on the left above must be bounded for all small $\beta > 0$. Applying parabolic estimates to the equations satisfied by $\tilde{\xi}_i^\beta$, we obtain:

$$(4.10)\quad \sum_{i=1}^3 ||\tilde{\xi}_i^\beta||_{W_2^{2,1}(\Omega_T)} \leq constant$$

for all small $\beta > 0$. The inequality (4.10) is readily satisfied also with $\tilde{\xi}_i^\beta$ replaced by ξ_i^β, since λ is chosen fixed.

Consequently, there exist subsequences (for convenience denoted again by $\xi_i^\beta, i = 1,2,3$) such that

$$\xi_i^\beta \to \xi_i, \quad i = 1,2,3, \quad \text{as } \beta \to 0,$$

strongly in $W_2^{1,0}(\Omega_T)$ and weakly in $W_2^{2,1}(\Omega_T)$. Taking limit as $\beta \to 0$, we conclude that the limit $\xi_i's$ satisfy (4.6) and (4.7). From the uniqueness of solution ξ_i to (4.6), (4.7) we conclude that $\xi_i^\beta \to \xi_i, i = 1,2,3$, weakly in $W_2^{2,1}(\Omega_T)$ as $\beta \to 0$, for a full sequence.

We now proceed to characterize the optimal control in Theorem 4.1 in terms of solutions of (4.1), (4.2) and its adjoint problem (4.13), (4.14) below. From this, we will obtain an optimality system (4.23) together with initial and terminal conditions (4.2), (4.14) respectively. In Theorem 4.2, we show that for small enough time interval, the solution of the optimality system can be obtained by a contractive iterative procedure. The solution of the optimality system within a certain range is also shown to be unique because the solution is the fixed point of a contraction. After the optimality system is solved, an optimal control can be readily constructed by means of formula (4.12).

For convenience, we adopt the following notations for the differential operators:

$$
\begin{aligned}
&L_1 := \nu_1^{-1}\partial/\partial t - \sigma_1\Delta, \quad L_1^* := -\nu_1^{-1}\partial/\partial t - \sigma_1\Delta; \\
&L_2 := \nu_2^{-1}\partial/\partial t - \sigma_2\Delta, \quad L_2^* := -\nu_2^{-1}\partial/\partial t - \sigma_2\Delta; \\
&L_3 := \partial/\partial t - \sigma_3\Delta, \quad L_3^* := -\partial/\partial t - \sigma_3\Delta.
\end{aligned}
\tag{4.11}
$$

The following theorem gives a characterization of any optimal control in C_δ.

Theorem 4.1 (Characterization of the Optimal Control). *For $M > 0$ sufficiently large, any optimal control $f \in C_\delta$ must satisfy:*

$$
f = qu_2/M \quad \text{in } \Omega \times [0,T], \tag{4.12}
$$

where (u_1, u_2, u_3) is the solution of (4.1), and (p, q, w) is the solution of

$$
\begin{cases}
L_1^*[p] = a_{11}p + a_{21}q + a_{31}w + K_1(u_1 - e_1), \\
L_2^*[q] = a_{12}p + (a_{22} - f)q + a_{32}w + K_2(u_2 - e_2), \\
L_3^*[w] = a_{33}w + K_3(u_3 - e_3)
\end{cases}
\tag{4.13}
$$

in $\Omega \times (0,T)$, with terminal and boundary conditions:

$$
\begin{cases}
p(x,T) = q(x,T) = w(x,T) = 0 \quad \text{for } x \in \Omega, \\
p(x,t) = q(x,t) = w(x,t) = 0 \quad \text{for } (x,t) \in \partial\Omega \times [0,T].
\end{cases}
\tag{4.14}
$$

Proof. Let $f \in C_\delta$ be an optimal control, i.e., the corresponding solution of the problem (4.1) for f satisfies the property:

$$
J(f) = \inf_{\tilde{f} \in C_\delta} J(\tilde{f}).
$$

For arbitrary $\bar{g} \in L_+^\infty(\Omega_T), \epsilon > 0$, set

$$
\bar{f} = \bar{f}_\epsilon = \begin{cases} \bar{g} & \text{if } f \le \delta - \epsilon\|\bar{g}\|_\infty, \\ 0 & \text{otherwise.} \end{cases}
$$

For $\beta > 0$ small enough (say $\beta < \epsilon$), such that $f + \beta \bar{f}_\epsilon \in C_\delta$, the optimality of f implies that $J(f) \le J(f + \beta \bar{f}_\epsilon)$, that is

$$0 \le \tfrac{1}{2} \int_{\Omega_T} \sum_{i=1}^{3} \{K_i[u_i(f + \beta \bar{f}_\epsilon) - e_i]^2 - K_i[u_i(f) - e_i]^2\}$$

$$+ M[(f + \beta \bar{f}_\epsilon)^2 - f^2]\, dxdt.$$

Dividing by β and letting $\beta \to 0$, we obtain

$$0 \le \int_{\Omega_T} \{\sum_{i=1}^{3} K_i[u_i(f) - e_i]\xi_i + M \bar{f}_\epsilon f\}\, dxdt, \tag{4.15}$$

where $\xi_i, i = 1, 2, 3$, are defined in (4.6), (4.7). For convenience, we define the matrix function:

$$A = \begin{bmatrix} a_{11} & a_{12} & 0 \\ a_{21} & (a_{22} - f) & 0 \\ a_{31} & a_{32} & a_{33} \end{bmatrix}, \tag{4.16}$$

and let A^T denotes the transpose of A. Inequality (4.15) can be written as:

$$\begin{aligned} 0 &\le \int_{\Omega_T} \{[\xi_1, \xi_2, \xi_3] \cdot (\begin{bmatrix} L_1^* p \\ L_2^* q \\ L_3^* w \end{bmatrix} - A^T \begin{bmatrix} p \\ q \\ w \end{bmatrix}) + M f \bar{f}_\epsilon\}\, dxdt \\ &= \int_{\Omega_T} \{[p, q, w] \cdot (\begin{bmatrix} L_1 \xi_1 \\ L_2 \xi_2 \\ L_3 \xi_3 \end{bmatrix} - A \begin{bmatrix} \xi_1 \\ \xi_2 \\ \xi_3 \end{bmatrix}) + M f \bar{f}_\epsilon\}\, dxdt. \end{aligned} \tag{4.17}$$

From Lemma 4.2, (4.17) leads to:

$$0 \le \int_{\Omega_T} \{q(-\bar{f}_\epsilon u_2) + M \bar{f}_\epsilon f\}\, dxdt = \int_{\Omega_T} \bar{f}_\epsilon[-q u_2 + M f]\, dxdt.$$

Letting $\epsilon \to 0$, we obtain:

$$0 \le \int_{\Omega_T \cap \{(x,t) | f < \delta\}} \bar{g}[-q u_2 + M f]\, dxdt.$$

This implies that:

$$f \geq qu_2/M \quad \text{in} \quad \Omega_T \cap \{(x,t)|f < \delta\}. \tag{4.18}$$

On the other hand, for arbitrary $-\bar{h} \in L^\infty_+(\Omega_T), \epsilon > 0$, set

$$\bar{f} = f_\epsilon = \begin{cases} \bar{h} & \text{if } f \geq -\delta + \epsilon||\bar{h}||_\infty, \\ 0 & \text{otherwise.} \end{cases}$$

Using the same argument as above, and letting $\epsilon \to 0$, we obtain

$$0 \leq \int_{\Omega_T \cap \{(x,t)|-\delta < f\}} (\bar{h})[-qu_2 + Mf]\, dxdt.$$

This implies that

$$f \leq qu_2/M \qquad \text{in} \quad \Omega_T \cap \{(x,t)| -\delta < f\}. \tag{4.19}$$

Hence, from (4.18) and (4.19) we obtain

$$f = qu_2/M \qquad \text{in} \quad \Omega_T \cap \{(x,t)| -\delta < f < \delta\}. \tag{4.20}$$

In the set $\Omega_T \cap \{(x,t)|f = \delta\}$, we find from (4.19) that

$$\delta = f \leq qu_2/M \leq \delta. \quad \text{provided that } M \text{ is sufficiently large.} \tag{4.21}$$

This means $f = qu_2/M$ there, under these conditions. (Note that q and u_2 are uniformly bounded in $[0,T]$, for all $M > 0$.) Similarly, in the set $\Omega_T \cap \{(x,t)|f = -\delta\}$, we find from (4.18) that

$$-\delta = f \geq qu_2/M \geq -\delta \quad \text{provided that } M \text{ is sufficiently large.} \tag{4.22}$$

Combining (4.20), (4.21) and (4.22), we conclude that (4.12) must hold for $f_\delta \in C$. This proves the Theorem.

Remark 4.1. Note again that since $f \in C_\delta$, the functions q and u_2 in the above theorem are uniformly bounded in $[0,T]$, independent of $M > 0$. This leads to the last inequality in (4.21), (4.22) and the conclusion of the theorem.

Remark 4.2 (The Optimality System). Combining equations (4.1), (4.13) and equality (4.12), we obtain the following optimality system of six equations, from which the optimal control f in C_δ can be found by means of the formula

(4.12), when M is sufficiently large:

$$
(4.23)\quad \begin{cases}
\nu_1^{-1}\partial u_1/\partial t - \sigma_1\Delta u_1 = a_{11}u_1 + a_{12}u_2, \\
\nu_2^{-1}\partial u_2/\partial t - \sigma_2\Delta u_2 = a_{21}u_1 + (a_{22} - u_2 q/M)u_2, \\
\partial u_3/\partial t - \sigma_3\Delta u_3 = a_{31}u_1 + a_{32}u_2 + a_{33}u_3, \\
-\nu_1^{-1}\partial p/\partial t - \sigma_1\Delta p = a_{11}p + a_{21}q + a_{31}w + K_1(u_1 - e_1), \\
-\nu_2^{-1}\partial q/\partial t - \sigma_2\Delta q = a_{12}p + (a_{22} - u_2 q/M)q + a_{32}w + K_2(u_2 - e_2), \\
-\partial w/\partial t - \sigma_3\Delta w = a_{33}w + K_3(u_3 - e_3)
\end{cases}
$$

in $\Omega\times(0,T)$. The components (u_1,u_2,u_3) satisfy initial conditions (4.2) at $t=0$, and the components (p,q,w) satisfy terminal conditions (4.14) at $t=T$. Here all $a'_{ij}s$ are constants, with all off-diagonal constants $a_{12},a_{21},a_{31},a_{32}>0$, the cooling constant $a_{33}<0$, and a_{11},a_{22} arbitrary.

For simplicity, we assume the following additional hypotheses on the smoothness on the target and initial functions. The assumptions are convenient for the use of Hölder space theory and estimates by mean's of Green's function.

[H] Each $e_i(x,t)$ is in $C^{\alpha,\alpha/2}(\bar{\Omega}\times[0,T])$, for some $0<\alpha<1, e_i(x,T)|_{\partial\Omega} = 0, i=1,2,3; r_i(x)$ is in $C^{2+\alpha}(\bar{\Omega}), r_i(x)|_{\partial\Omega}=0$, and $\Delta r_i(x)|_{\partial\Omega}=0$, for $i=1,2,3$.

Theorem 4.2 (Solution of the Optimality System). *Assume hypotheses [H]. For $T:=t^*>0$ sufficiently small, problem (4.23) with boundary, initial and terminal conditions (4.2) and (4.14) respectively, has a unique solution with the property that each component of the solution is in the space $C^{2+\alpha,1+\alpha/2}(\bar{\Omega}\times[0,t^*]), 0<\alpha<1$.*

Remark 4.3. The existence part is not really needed because it can be obtained by the existence of optimal control. However, the proof further gives a constructive method to approximate the optimal control. It can also be adapted to obtain bounds of the solution. An iterative scheme is described in Remark 4.4. The convergence of the scheme is ensured by contraction argument in the proof of this theorem.

Proof. For convenience, we will denote the operators $L_i^*, i=1,2,3$ above by:

$$L_1^* = L_4, \quad L_2^* = L_5, \quad L_3^* = L_6.$$

We will also denote the expressions on the right hand side of equations (4.23) successively by $f_i(u_1,u_2,u_3,p,q,w), i=1,\dots,6$. Let $\mathcal{C}$ be the set of continuous functions $y = (y_1(x,t), \ldots, y_6(x,t))$ for $(x,t) \in \bar{\Omega} \times [0,t^*], (t^* > 0$ to be determined); and let $G_i(x,t;x_0,t_0); i = 1,2,3$, be Green's functions for the equations $L_i[u] = \hat{f}$. For functions $y \in \mathcal{C}$, define a mapping H for functions $y \in \mathcal{C} \to \mathcal{C}$ by:

(4.24)
$$\begin{cases} z = H[y], \quad \text{where} \quad z = (z_1(x,t), \ldots, z_6(x,t)), \quad (x,t) \in \Omega \times [0,t^*], \\ z_i(x,t) = \int_0^t d\tau \int_\Omega G_i(x,t,\xi,\tau) f_i(y_1(\xi,\tau), \ldots, y_6(\xi,\tau)) d\xi \\ \qquad + \int_\Omega G_i(x,t,\xi,0) r_i(\xi) d\xi \quad \text{for} \quad i = 1,2,3, \\ z_i(x,t) = \int_0^{t^*-t} d\tau \int_\Omega G_{i-3}(x,t^*-t,\xi,\tau) f_i(y_1(\xi,t^*-\tau), \ldots, y_6(\xi,t^*-\tau)) d\xi \\ \qquad \text{for } i = 4,5,6. \end{cases}$$

It is well known that the functions z will be in $\mathcal{C}$. If the integrand f_i as a function of $(\xi,\tau) \in \Omega \times (0,t^*)$ is Hölder continuous in ξ uniformly in τ, then z_i satisfies $L_i[z_i] = f_i$ in $\Omega \times (0,t^*)$, (see e.g. Friedman [63]). The functions (z_1,z_2,z_3) satisfy corresponding initial boundary conditions (4.2); and the functions (z_4,z_5,z_6) satisfy corresponding terminal boundary condition (4.14) with $t^* = T$. Moreover, for $x,\xi \in \Omega, 0 \le \tau \le t \le t^*$, we have

$$|G_i(x,t;\xi,\tau)| \le \frac{K}{(t-\tau)^\mu |x-\xi|^{n-2+\mu}}, \quad i = 1,2,3. \tag{4.25}$$

for some constant $K > 0, 0 < \mu < 1$.

We next show that the mapping H is a contraction in $\mathcal{C}$, with t^* chosen sufficiently small. Let $\hat{y} = (\hat{y}_1, ..., \hat{y}_6)$ and $\tilde{y} = (\tilde{y}_1, ..., \tilde{y}_6) \in \mathcal{C}$, then for $i = 1,2,3$, we obtain from (4.24),

$$\hat{z}_i(x,t) - \tilde{z}_i(x,t) = \int_0^t \int_\Omega G_i(x,t;\xi,\tau)[f_i(\hat{y}_1, ..., \hat{y}_6) - f_i(\tilde{y}_1, ..., \tilde{y}_6)]\, d\xi d\tau \tag{4.26}$$

for $(x,t) \in \Omega \times [0,t^*]$. Since the functions f_i are Lipschitz for $\hat{y}$ and $\tilde{y}$ in $\mathcal{C}$, one can thus readily obtain, from (4.25) and (4.26)

$$|\hat{z}_i(x,t) - \tilde{z}_i(x,t)| \le \bar{K} t^{1-\mu} \sum_{i=1}^{6} max_{\Omega \times [0,t^*]} |\hat{y}_i - \tilde{y}_i| \tag{4.27}$$

for all $(x,t) \in \bar{\Omega} \times [0,t^*], i = 1,2,3$, some $\bar{K} > 0$. For $i = 4,5,6$ in (4.24), the corresponding differential equation is changed to a parabolic type by using the variable $\tau = t^* - t, 0 \le t \le t^*$, and one can deduce in the same way that

$$|\hat{z}_i(x,t) - \tilde{z}_i(x,t)| \le \hat{K}(t^*-t)^{1-\mu} \sum_{i=1}^{6} max_{\Omega \times [0,t^*]} |\hat{y}_i - \tilde{y}_i| \tag{4.28}$$

for all $(x,t) \in \bar{\Omega} \times [0,t^*], i = 4,5,6$, some $\hat{K} > 0$. By (4.26) and (4.27), one can readily see that if $t^* > 0$ is chosen to be sufficiently small, then the mapping $H : \mathcal{C} \to \mathcal{C}$ is a contraction with respect to the uniform norm. The contraction mapping H has a unique fixed point in $\mathcal{C}$. As an image of the mapping H, the fixed point in $\mathcal{C}$ is Hölder continuous in $\bar{\Omega} \times [0,t^*]$. Applying the mapping H again, we find that the fixed-point function is actually a classical solution of (4.23). By the smoothness and compatibility assumptions in hypotheses [H], we conclude that the fixed point is a solution to problem (4.23), (4.2), and (4.14), with components in $C^{2+\alpha,1+\alpha/2}(\bar{\Omega} \times [0,t^*])$. Since any solution of problem (4.23), (4.2), and (4.14) with components in $C^{2+\alpha,1+\alpha/2}(\bar{\Omega} \times [0,t^*])$ is a fixed point of the mapping H in $\mathcal{C}$, we conclude that such a kind of solution must be unique.

Remark 4.4. We may define $(y_1^0, ..., y_6^0) = (r_1(x), r_2(x), r_3(x), 0, 0, 0)$ for $(x,t) \in \bar{\Omega} \times [0,t^*]$, and define iteratively $(y_1^j, ..., y_6^j) = H[(y_1^{j-1}, ..., y_6^{j-1})]$ as functions in $\mathcal{C}$ as described in (4.24), $j = 1, 2, \ldots$; that is

$$\begin{cases} L_i[y_i^j] = f_i(y_1^{j-1}, ..., y_6^{j-1}) \quad \text{in } \Omega \times (0,t^*), i = 1, ..., 6, \ j = 1, 2, \ldots \\ \\ y_i^j(x,0) = r_i(x) \quad \text{in } \Omega, \ \ i = 1,2,3, \\ \\ y_i^j(x,t^*) = 0 \quad \text{in } i = 4,5,6. \end{cases}$$

Then the functions $(y_1^j, ..., y_6^j)$ will tend to the unique solution (u_1, u_2, u_3, p, q, w) as described in Theorem 4.2, as j tends to infinity.

Remark 4.5. Due to the signs of the coefficients of the optimality system (4.23), its solutions has a tendency to blow up quickly. More specifically, the coefficient a_{22} may be positive; moreover, q can possibly be negative and thus the nonlinear term $-u_2^2 q/M$ in the second equation in (4.23) can be positive. This causes the solution u_2 to grow quickly and makes it difficult to define an iterative scheme for functions on the whole interval $[0,T]$. The nature of the opposite time orientation of the last three equations in (4.23) also gives rise to additional difficulty. Thus we are able to obtain a constructive scheme for calculating a solution only for small intervals of time. Such restrictive condition is partly alleviated by Corollary 4.4 and system (4.32) below. Similar optimal control problem for other applications are studied in Section 3.3 and Stojanovic [213]. However, these other kinds of systems do not tend to blow up as in our present situation. The signs of the coefficients in these other problems ensure that the problems can be conveniently treated once in the entire interval [0,T].

In Theorem 4.1 the constant M is assumed to be large. If no assumption is made on the size of the positive constant M, we can follow exactly the same argument as in the proof of the theorem up to (4.20), to obtain the following result.

Theorem 4.3. *For a given fixed $\delta > 0$, any optimal control $f \in C_\delta$ must satisfy*

$$f(x,t) = \frac{q(x,t)u_2(x,t)}{M} \quad in \quad \Omega_T \cap \{(x,t) | -\delta < f(x,t) < \delta\}, \tag{4.29}$$

where (u_1, u_2, u_3) is the solution of (4.1) and (4.2), while (p, q, w) is the solution of (4.13)and (4.14).

Let $f \in C_\delta$ as described in Theorem 4.3 or 4.1. At those (x,t) in Ω_T where $f(x,t) = \delta$, the arguments in the proof of Theorem 4.1 show that we must also have $f(x,t) \le q(x,t)u_2(x,t)/M$. That is, at such $(x,t), f = \delta$ also satisfies

$$f = max\,\{-\delta,\ min\{qu_2/M, \delta\}\}. \tag{4.30}$$

Here, q and u_2 is as described in Theorem 4.3. At those (x,t) in Ω_T where $f(x,t) = -\delta$, the arguments in the proof of Theorem 4.1 show that $f(x,t) \ge q(x,t)u_2(x,t)/M$. That is, at such $(x,t), f = -\delta$ also satisfies (4.30). Combining with the statement of Theorem 4.3, we see that (4.30) is satisfied for all (x,t) in Ω_T. We can summarize as follows.

Corollary 4.4 (General Formula for the Optimal Control). *Let $f \in C_\delta$ be an optimal control and let q, u_2 be as described in Theorem 4.3, then*

$$f(x,t) = max\{-\delta,\ min\{q(x,t)u_2(x,t)/M, \delta\}\} \quad \text{for all} \quad (x,t) \in \Omega_T. \tag{4.31}$$

Consequently, the function (u_1, u_2, u_3, p, q, w) in Ω_T, described in Theorem 4.1 or 4.3, also satisfies system

(4.32)
$$\begin{cases} \nu_1^{-1}\partial u_1/\partial t - \sigma_1 \Delta u_1 = a_{11}u_1 + a_{12}u_2, \\ \nu_2^{-1}\partial u_2/\partial t - \sigma_2 \Delta u_2 = a_{21}u_1 + (a_{22} - max\{-\delta, min\{qu_2/M, \delta\}\})u_2, \\ \partial u_3/\partial t - \sigma_3 \Delta u_3 = a_{31}u_1 + a_{32}u_2 + a_{33}u_3, \\ -\nu_1^{-1}\partial p/\partial t - \sigma_1 \Delta p = a_{11}p + a_{21}q + a_{31}w + K_1(u_1 - e_1), \\ -\nu_2^{-1}\partial q/\partial t - \sigma_2 \Delta q = a_{12}p + (a_{22} - max\{-\delta, min\{qu_2/M, \delta\}\})q + a_{32}w \\ \qquad\qquad + K_2\,(u_2 - e_2), \\ -\partial w/\partial t - \sigma_3 \Delta w = a_{33}w + K_3(u_3 - e_3) \end{cases}$$

in $\Omega \times (0, T)$. The components (u_1, u_2, u_3) satisfy initial-boundary conditions (4.2), and the components (p, q, w) satisfy terminal-boundary conditions (4.14). One readily sees that if $f \in C_\delta$ is an optimal control, then there exists a solution (u_1, u_2, u_3, p, q, w) of (4.32) in $\Omega \times (0, T)$ satisfying initial and terminal conditions (4.2) and (4.14) such that f satisfies (4.31) for the functions $u_2(x, t)$ and $q(x, t)$. Letting [H] be as described before Theorem 4.2, we can prove the following theorem by exactly the same method as Theorem 4.2.

Theorem 4.5. *Assume hypotheses [H]. For $T := t^* > 0$ sufficiently small, problem (4.32) with boundary, initial, and terminal conditions (4.2) and (4.14), respectively, has a unique solution with the property that each component of the solution is in the space $C^{2+\alpha, 1+\alpha/2}(\bar{\Omega} \times [0, t^*]), 0 < \alpha < 1$.*

Although the existence part of the theorem is ensured by the existence of optimal control f, the usefulness of this theorem is justified by Remark 4.3. The convenience of hypotheses [H] is also explained before.

3.5 Optimal Boundary Control of a Parabolic Problem

In the previous sections, we analyze elliptic and parabolic systems with control in the interior of the domain. We now consider a simple optimal control of a parabolic equation with the control of coefficient at the boundary. Such problem is related to natural applications to environmental boundary preservations (cf. Friesen, Eagles and MacKay [65] , and Doa [51]), and to heat transfer studies (cf. Lenhart and Wilson [119]). We first consider the heat equation with convective boundary condition, and use the boundary heat transfer coefficient as the control. The objective is to maintain a target interior temperature profile and reduce the cost of the control, as in the last section. However, in our present case of boundary control, it will be convenient to obtain rigorous theory by using different type of function spaces and solutions. Our intention is to illustrate how the proofs can be adapted for a simple problem. More elaborate and larger systems can then be considered as in previous applications. In this section, our development for the control of scalar temperature problem (5.1)-(5.2) is obtained from [119], and for the more complex prey-predator system (5.21)-(5.21) is obtained from Lenhart, Liang and Protopopescu [117].

Part A: Scalar Problem.

We consider the heat equation in a bounded domain Ω with smooth boundary in R^N, and a finite time interval $[0, T]$. In order to avoid restricting to a small interval, we define a dependent variable $u(x, t) = e^{-\lambda t}\tilde{T}(x, t)$, where $\tilde{T}(x, t)$ is the temperature function and $\lambda > 0$ is a large constant to be chosen later. The initial-boundary value problem for the variable u becomes

$$
(5.1) \qquad \begin{cases} u_t - \alpha\Delta u + \lambda u = 0 & \text{in} \quad \Omega \times (0,T), \\ u(x,0) = u_0(x) & \text{in} \quad \Omega, \\ \frac{\partial u}{\partial \nu} = -hu & \text{on} \quad \partial\Omega \times (0,T), \end{cases}
$$

where $h \in L^\infty(\partial\Omega \times (0,T))$ is a non-negative heat transfer coefficient under control, and α is a positive constant. The boundary relations above is known as Newton's law of cooling. (Note that $\tilde{T}(x,t)$ satisfies (5.1) with $\lambda = 0$). In this section, we assume that the initial function satisfies $u_0 \in L^\infty(\Omega)$, and $u_0 \geq 0$ a.e. in Ω. We define the cost functional J by:

$$
(5.2) \qquad J(h) = \frac{1}{2}\{\beta \int_{\Omega_T} [u(h) - Z_d]^2 \, dxdt + \gamma \int_{\partial\Omega_T} h^2 dsdt\}.
$$

Here, $\Omega_T = \Omega \times (0,T), \partial\Omega_T = \partial\Omega \times (0,T)$, and $Z_d \in L^\infty(\Omega_T)$ is the desired temperature distribution. The symbol $u(h)$ denotes the weak solution of the initial-boundary value problem (5.1) with corresponding control $h = h(x,t)$ on $\partial\Omega_T$ described below. The positive constants β and γ are per unit costs associated with deviation from desired temperature profile and with controlling the heat transfer coefficient. The control functions will be restricted to the set:

$$
C_M = \{h | h \in L^2(\partial\Omega \times (0,T)); 0 \leq h(x,t) \leq M \, a.e.\}.
$$

The underlying state space is given by

$$
V = H^1(\Omega);
$$

and define

$$
W = L^2(0,T; H^1(\Omega)).
$$

For Φ and $\Psi \in V$, we define a time dependent functional a_h by

$$
a_h(t; \Phi, \Psi) = \int_\Omega [\alpha \nabla\Phi\nabla\Psi + \lambda\Phi\Psi] dx + \int_{\partial\Omega} h\Phi\Psi ds.
$$

Definition 5.1. A function u is a weak solution of problem (5.1) provided that:

$$
(5.3) \qquad \begin{cases} u \in W \cap L^\infty(0,T; L^2(\Omega)), \\ \int_0^T [< u'(t), v(t) > + a_h(t; u(t), v(t))] \, dt = 0 \quad \text{for all} \quad v \in W, \\ \frac{du}{dt} \in L^2(0,T; V'), \quad u(0) = u_0, \end{cases}
$$

where V' is the dual space of $H^1(\Omega)$, and the bracket < > in the integral denotes $V' - V$ duality.

Remark 5.1. Under the assumptions on the domain Ω, the initial function u_0, and the control set C_M described above, we readily obtain:

$$k_0||v||_V^2 \leq \int_0^T a_h(t; v(t), v(t))dt \leq K_0||v||_V^2 \text{ for all } v \in W,$$

where k_0 and K_0 are positive constants independent of $h \in C_M$. Consequently, for each $h \in C_M$, the existence of a unique solution $u(h) \in W \cap L^\infty(0, T; L^2(\Omega))$ for (5.3) is given by standard theory from Chapter 3 of Lions and Magenes [159] (cf. Theorem A5-2 in Chapter 6). Note that by choosing suitable test functions in (5.3), we can deduce that

$$du/dt \in L^2(0, T; V'), \text{ and thus } u \in C([0, T], L^2(\Omega)).$$

Hence it makes sense for the initial condition in (5.3) (see e.g. Section 5.9 in [57]).

We will assume that in (5.2) the function Z_d is in $L^2(\Omega)$. An optimal control (if it exists) for problem (5.1) (or (5.3)) corresponding to the functional (5.2) is a function $h^* \in C_M$ such that

$$J(h^*) = inf_{h \in C_M} J(h). \tag{5.4}$$

Theorem 5.1 (Existence of Optimal Control). *There exists an optimal control satisfying (5.4) for the problem (5.1) (or (5.3)) corresponding to the functional (5.2).*

Proof. Let $\{h_n\}$ be a minimizing sequence in C_M. From the above remark, the functions $u_n := u(h_n)$ are uniquely defined for each n. Using (5.3) with $u = v = u_n$ and the time variable from 0 to t, $t \in (0, T)$ we obtain:

$$\begin{aligned} &\tfrac{1}{2} \textstyle\int_\Omega [(u_n(x, t))^2 - (u_0(x))^2]\, dx \\ &\qquad + \textstyle\int_0^t \{\int_\Omega \alpha |\nabla u_n|^2 + |\lambda u_n|^2 dx + \int_{\partial\Omega \times (0,t)} h_n |u_n|^2 ds\} dt = 0. \end{aligned} \tag{5.5}$$

Since $a_n(t, u_n(\cdot, t), u_n(\cdot, t)) \geq 0$, we have

$$\int_\Omega (u_n(x, s))^2 dx \leq \int_\Omega (u_0(x))^2 dx \text{ for all } s \in (0, T).$$

Hence, from above and (5.5) we find

$$\begin{aligned} &\textstyle\int_0^t \int_\Omega (u_n(x, s))^2\, dxds \leq T \int_\Omega (u_0(x))^2 dx, \quad \text{and} \\ &\alpha \textstyle\int_0^t \int_\Omega |\nabla u_n(x, s)|^2\, dxds \leq \frac{1}{2} \int_\Omega (u_0(x))^2 dx, \end{aligned} \tag{5.6}$$

for $t \in (0,T]$. Thus, u_n ranges in a bounded subset of W. By using appropriate test functions for u_n in (5.3), we deduce that $du_n/dt = a_h(t, u_n(t), v) \in V'$ for each t. Consequently, the boundedness of u_n in W implies that du_n/dt ranges in a bounded subset of $L^2(0,T,V')$. Since $H^1(\Omega) \subset H^{\frac{1}{2}+\epsilon}(\Omega) \subset V'$, for $0 \leq \epsilon < 1/2$, we can apply a compact embedding theorem of the space $H_1(0,T;H^1(\Omega),V')$ into $L^2(0,T;H^{\frac{1}{2}+\epsilon}(\Omega))$ (see Theorem A5-3 in Chapter 6), to extract a subsequence,(again labeled as $\{u_n\}$), with the following properties:

$$
\begin{aligned}
&u_n \to u^* && \text{weakly in } W,\\
&\frac{du_n}{dt} \to \frac{du^*}{dt} && \text{weakly in } L^2(0,T;V'),\\
&u_n \to u^* && \text{strongly in } L^2(0,T;H^{\frac{1}{2}+\epsilon}(\Omega)).
\end{aligned} \tag{5.7}
$$

We can also obtain a subsequence $\{h_n\}$ such that

$$
h_n \to h^* \qquad \text{weakly in } L^\infty(\partial\Omega_T); \tag{5.8}
$$

and $h^* \in C_M$, since C_M is closed in this topology. From the continuity of the restriction mapping of $H^{\frac{1}{2}+\epsilon}(\Omega)$ into $L^2(\partial\Omega)$, we find

$$
u_n \to u^* \qquad \text{strongly in } L^2(0,T;L^2(\partial\Omega)). \tag{5.9}
$$

Finally, passing to the limit with u replaced with u_n, $n \to \infty$ in (5.3), we obtain a weak solution u^* of (5.1) with h^* as the heat transfer coefficient. By the usual lower semi-continuity property of the cost functional as in the earlier sections or [159], we conclude by means of (5.8) that h^* is the optimal control in C_M.

Remark 5.2. Note that by (5.6), the function $u = u(h)$ for $h \in C_M$, satisfies $||u||_W \leq C$, for some constant C independent of M.

We will proceed to characterize the optimal control by an optimality system which consists of the coupling of the state equation (5.1) with an adjoint equation. For this purpose, we first consider the differentiation of the solution $u(h)$ of (5.3) with respect to the control h in C_M.

Lemma 5.1 (Differentiability of Optimal Control). *The mapping*

$$
C_M \ni h \to u(h) \in W := L^2(0,T;H^1(\Omega))
$$

is differentiable in the following sense:

$$
\frac{u(h+\rho\bar{h}) - u(h)}{\rho} \to U \text{ weakly in } W
$$

as $\rho \to 0$, *for any* $h \in C_M$ *and* $\bar{h} \in C_M$ *such that* $h + \rho\bar{h} \in C_M$. *Further,* $U = U(h, \bar{h})$ *is the unique weak solution of the problem*

$$(5.10) \qquad \begin{cases} U_t - \alpha\Delta U + \lambda U = 0 & \text{in } \Omega_T, \\ U(x,0) = 0 & \text{in } \Omega, \\ \frac{\partial U}{\partial \nu} = -hU - \bar{h}u(h) & \text{in } \partial\Omega_T. \end{cases}$$

Proof. From (5.1) and (5.3), we deduce that

$$(5.11) \qquad \xi_\rho \equiv \frac{u(h + \rho\bar{h}) - u(h)}{\rho}$$

satisfies

$$\int_{\Omega\times\{T\}} \frac{1}{2}\xi_\rho^2 dx + \int_{\Omega_T} \alpha|\nabla\xi_\rho|^2 + \lambda\xi_\rho^2 \, dxdt + \int_{\partial\Omega_T} h\xi_\rho^2 + \bar{h}u(h+\rho\bar{h})\xi_\rho \, dsdt = 0.$$

This implies that

$$(5.12) \qquad \int_{\Omega_T} \alpha|\nabla\xi_\rho|^2 + \lambda\xi_\rho^2 \, dxdt \le \int_{\partial\Omega_T} \bar{h}|u(h+\rho\bar{h})\xi_\rho| \, dsdt.$$

This leads to

$$||\xi_\rho||_W^2 \le C||\bar{h}||_{L^\infty(\partial\Omega_T)}||u(h+\rho\bar{h})||_W||\xi_\rho||_W,$$

and thus

$$(5.13) \qquad ||\xi_\rho||_W \le const,$$

where the constant is independent of ρ. Thus, we can choose a sequence $\rho \to 0$ and deduce that the weak limit satisfies (5.10). From the uniqueness of solution to (5.10), we conclude that $\xi_\rho \to U$ weakly in W as $\rho \to 0$, for a full sequence.

Theorem 5.2 (Characterization of the Optimal Control for the Single Heat Equation). *Let* h^* *be an optimal control for problem (5.1), (5.2), (5.4), and* $u^* = u^*(h^*)$ *be the corresponding solution of (5.1). Then there exists a solution* p^* *of the adjoint equation:*

$$(5.14) \qquad \begin{cases} -p_t^* - \alpha\Delta p^* + \lambda p^* = (u^* - Z_d) & \textit{in } \Omega_T, \\ p^*(x,T) = 0 & \textit{in } \Omega, \\ \frac{\partial p^*}{\partial \nu} = -h^*p^* & \textit{on } \partial\Omega_T, \end{cases}$$

with h^* satisfying

$$h^* = min\,\{max\{0, \frac{\alpha\beta}{\gamma}p^*u^*\}, M\} \qquad \text{on } \partial\Omega_T. \tag{5.15}$$

Proof. Let $h^* + \rho\bar{h}$ be another control in C_M with the corresponding solution $u_\rho := u(h^* + \rho\bar{h})$. We obtain

$$0 \leq lim_{\rho\to 0}\frac{J(h^*+\rho\bar{h})-J(h^*)}{\rho}$$

$$= lim_{\rho\to 0}\beta\int_{\Omega_T}(\tfrac{u_\rho - u^*}{\rho})(\tfrac{u_\rho+u^*}{2} - Z_d)dxdt + \gamma\int_{\partial_T}(\bar{h}h^* + \rho\tfrac{\bar{h}^2}{2})dsdt \tag{5.16}$$

$$= \beta\int_{\Omega_T}U(u^* - Z_d)dxdt + \gamma\int_{\partial_T}\bar{h}h^*dsdt.$$

Here, $U = U(h,\bar{h})$ is the weak solution of (5.10) as described in Lemma 5.1. Let p^* be the solution of the adjoint problem (5.14). Then, integrating by parts in the last line and using (5.10), we obtain from above

$$0 \leq \beta\int_{\Omega_T}U(-p_t^* - \alpha\Delta p^* + \lambda p^*)dxdt + \gamma\int_{\partial\Omega_T}\bar{h}h^*dsdt$$

$$= \int_{\partial\Omega_T}[\beta\alpha(p^*\tfrac{\partial U}{\partial\nu} - U\tfrac{\partial p^*}{\partial\nu}) + \gamma\bar{h}h^*]dsdt$$

$$= \int_{\partial\Omega_T}\bar{h}(\gamma h^* - \alpha\beta p^*u^*)dsdt.$$

Standard arguments as in the previous sections in this chapter shows that h^* satisfies equation (5.15).

Note that since the initial function in (5.1) is non-negative and bounded, the zero and large positive constant functions are respectively lower and upper solutions for problem (5.1) for all $h \in C_M$ independent of M. We can compare different solutions by maximum principle of parabolic equations by smoothing out the coefficient h and take limit as in Section 3.3. Thus we deduce that there exists a constant B independent of M such that all solutions of (5.1) with $h \in C_M$ satisfies

$$0 \leq u(h) \leq B \quad \text{a. e. in } \Omega_T.$$

Choose

$$\tilde{B} = max\{B, ||Z_d||_\infty\}.$$

Let p be the solution of

$$\begin{cases} -p_t - \alpha\Delta p + \lambda p = \tilde{B} - Z_d & \text{in } \Omega_T, \\ \frac{\partial p}{\partial\nu} = 0 & \text{on } \partial\Omega_T, \\ p|_T = 0 & \text{in } \Omega. \end{cases}$$

By comparison, we obtain $p \geq 0$ in Ω_T. Moreover, p is an upper solution to problem (5.14) in the sense that it satisfies (5.14) with all the three equalities replaced by $\geq$. Again, by comparison we have

$$p \geq p^* \text{ in } \partial\Omega_T,$$

where $p^* = p^*(h^*), h^* \in C_M$, independent of M. Since by comparison there exists a constant K independent of M such that $K \geq p$ in Ω_T, we obtain a constant K such that

$$p^* \leq K,$$

$h^* \in C_M$, independent of M. Consequently, we can choose M sufficiently large such that

$$\frac{\alpha\beta}{\gamma} p^* u^* \leq M,$$

and h^* is independent of M. Since $u^* > 0$ in Ω_T, we have

$$h^* = \frac{\alpha\beta}{\gamma}(p^*)^+ u^*.$$

From (5.14) and (5.15), we find that provided that M is sufficiently large for the set of admissible control C_M, the optimality system satisfied by u^* and p^* is:

$$(5.17) \qquad \begin{cases} u_t - \alpha\Delta u + \lambda u = 0 & \text{in } \Omega_T, \\ u(x,0) = u_0(x) & \text{in } \Omega, \\ \frac{\partial u}{\partial \nu} = -\frac{\alpha\beta}{\gamma} p^+ u^2 & \text{on } \partial\Omega_T, \\ -p_t - \alpha\Delta p + \lambda p = (u - Z_d) & \text{in } \Omega_T, \\ p(x,T) = 0 & \text{in } \Omega, \\ \frac{\partial p}{\partial \nu} = -\frac{\alpha\beta}{\gamma} u (p^+)^2 & \text{on } \partial\Omega_T. \end{cases}$$

Remark 5.3 (The Optimality System). In case the optimality system (5.17) has a unique solution, the optimal control problems (5.1), (5.2), (5.4) is completely solved by finding the solution $(u,p) = (u^*, p^*)$ of problem (5.17). Then, we set the optimal control h^* as

$$(5.18) \qquad h^* = \frac{\alpha\beta}{\gamma} p^+ u = \frac{\alpha\beta}{\gamma}(p^*)^+ u^*.$$

Note that the existence of solution for (5.17) is ensured by means of Theorems 5.1, 5.2 and the uniform bound arguments above. The question of uniqueness is considered in the next theorem.

Theorem 5.3 (Uniqueness of Solution for the Optimality System). *For $\lambda > 0$ sufficiently large, the optimality problem (5.17) has at most one solution.*

Proof. Suppose (u_1, p_1) and (u_2, p_2) are two solution pairs for the system (5.17). Using the test functions:

$$\phi_1 = u_1 - u_2, \text{ in the equation for } u_1,$$

$$\phi_2 = p_1 - p_2, \text{ in the equation for } p_1,$$

$$\phi_3 = u_2 - u_1, \text{ in the equation for } u_2, \text{ and}$$

$$\phi_4 = p_2 - p_1, \text{ in the equation for } p_2;$$

and adding all the four equations together, we obtain after integrating and transposing:

(5.19)

$$\begin{aligned}
&\int\int_{\Omega_T}(u_1 - u_2)(p_1 - p_2)dxdt \\
&= \tfrac{1}{2}\int_\Omega (u_1 - u_2)^2(x,T)dx + \tfrac{1}{2}\int_\Omega (p_1 - p_2)^2(x,0)dx + \alpha\int\int_{\Omega_T}|\nabla(u_1 - u_2)|^2 dxdt \\
&+ \alpha\int\int_{\Omega_T}|\nabla(p_1 - p_2)|^2 dxdt + \lambda\int\int_{\Omega_T}(p_1 - p_2)^2 + (u_1 - u_2)^2)dxdt \\
&+ \int_{\partial\Omega_T}\tfrac{\alpha\beta}{\gamma}(p_1^+ u_1^2 - p_2^+ u_2^2)(u_1 - u_2)dsdt \\
&+ \int_{\partial\Omega_T}\tfrac{\alpha\beta}{\gamma}(u_1(p_1^+)^2 - u_2(p_2^+)^2(p_1 - p_2)dsdt.
\end{aligned}$$

We estimate one of the boundary integral on the right above as follows:

$$\begin{aligned}
&\int_{\partial\Omega_T}\tfrac{\alpha\beta}{\gamma}(p_1^+ u_1^2 - p_2^+ u_2^2)(u_1 - u_2)dsdt \\
&= \int_{\partial\Omega_T}\tfrac{\alpha\beta}{\gamma}(p_1^+ u_1^2 - p_1^+ u_2^2 + p_1^+ u_2^2 - p_2^+ u_2^2)(u_1 - u_2)dsdt \\
&\le C\int_{\partial\Omega_T}(u_1 - u_2)^2 + (p_1^+ - p_2^+)^2 dsdt \\
&\le C\int_{\partial\Omega_T}(u_1 - u_2)^2 + (p_1 - p_2)^2 dsdt \\
&\le C_1\int\int_{\Omega_T}(u_1 - u_2)^2 + (p_1 - p_2)^2 dxdt \\
&\quad + \tfrac{\alpha}{2}\int\int_{\Omega_T}|\nabla(u_1 - u_2)|^2 + |\nabla(p_1 - p_2)|^2 dxdt,
\end{aligned}$$

for some large constant C_1, using continuous embedding of $H^1(\Omega)$ into $L^2(\partial\Omega)$ and Young's inequality with ϵ (cf. p. 258 in [57]) in the last line. The other boundary integral term can be estimated in the same way. We also have

$$\int\int_{\Omega_T}(u_1-u_2)(p_1-p_2)dxdt \leq \frac{1}{2}\int\int_{\Omega_T}[(u_1-u_2)^2+(p_1-p_2)^2]dxdt.$$

Rearranging all the terms in (5.19) and using the above estimates, we obtain

$$\begin{aligned} &\tfrac{1}{2}\int_\Omega (u_1-u_2)^2(x,T)dx + \tfrac{1}{2}\int_\Omega (p_1-p_2)^2(x,0)dx \\ &\quad +(\lambda - C_1)\int\int_{\Omega_T}[(u_1-u_2)^2+(p_1-p_2)^2]dxdt \leq 0. \end{aligned} \tag{5.20}$$

Substituting a large enough λ in (5.20), we conclude

$$u_1 = u_2, \text{ and } p_1 = p_2 \text{ in } \Omega_T.$$

Remark 5.4. By means of an iterative scheme, we can construct sequences of functions by solving scalar problems, converging to upper and lower estimates of the solution of (5.17), as in Sections 3.2 and 3.3. The upper and lower estimates are solutions of a larger modified system. However, since this is an initial-boundary value problem, we can show that the upper and lower estimates are equal by means of large λ in a variable change as above. Consequently, the sequences actually converge monotonically to the solution of the original problem (5.17).

Part B: Prey-Predator System.

The second half of this section extends the method in the first half to study the optimal boundary control for the following prey-predator system:

$$\begin{cases} u_t = \alpha_1 \Delta u + u[a_1(x,t) - b_1(x,t)u - c_1(x,t)v] & \text{in } \Omega_T, \\ v_t = \alpha_2 \Delta v + v[-a_2(x,t) + c_2(x,t)u] & \text{in } \Omega_T, \\ u(x,0) = u_0(x), \ v(x,0) = v_0(x) & \text{for } x \in \Omega, \\ \alpha_1 \frac{\partial u}{\partial \nu} + \beta(x,t)u = 0, \ \ \alpha_2 \frac{\partial v}{\partial \nu} + \beta(x,t)v = 0 & \text{on } \partial\Omega_T. \end{cases} \tag{5.21}$$

As in the first half of this section, we set the class of admissible control as:

$$C_M = \{\beta | \beta \in L^2(\partial\Omega \times (0,T)); 0 \leq \beta(x,t) \leq M \, a.e.\}.$$

We prescribe our payoff functional as

$$J(\beta) = \int\int_{\Omega_T}(A_1 u + A_2 v)dxdt - \frac{1}{2}\int_{\partial\Omega_T}(M-\beta)^2 dsdt, \tag{5.22}$$

with positive constants A_1, A_2. This functional combines the control on the boundary and its effect on the populations. It reflects the relationship between the growth of the species and the economic cost of maintaining an ecologically favorable boundary environment. The cost could include a loss of profit to the neighboring industries. To make the most favorable boundary environment, we would take $\beta = 0$ and incur the largest cost, $(M-0)^2$. We seek to maximize the functional over the admissible class above, i.e. to find β^* such that

$$J(\beta^*) = max_{\beta \in C_M} J(\beta). \tag{5.23}$$

We clarify the assumptions for our problem (5.21) to (5.23) for the remainder of this section. The set Ω is a bounded domain in R^N with smooth boundary $\partial\Omega$. The coefficient functions $a_i(x,t), c_i(x,t), i = 1,2$ and $b_1(x,t)$ are non-negative in Ω_T, and are all in $L^\infty(\Omega_T)$. The parameters $\alpha_i > 0, i = 1,2$, are constants. The initial functions $u_0(x), v_0(x) \in L^\infty(\Omega)$ satisfy $0 \le u_0(x) \le B, 0 \le v_0(x) \le B$ a.e. in Ω for some constant $B > 0$. As before, we let

$$V = H^1(\Omega),\ W = L^2(0,T;V),\ V' = H^{-1}(\Omega)$$

is the dual space of V; and $<,>$ denotes the duality between V' and V.

Definition 5.2. A pair of functions (u,v) is a weak solution of problem (5.21) provided that:

$$\begin{cases} u, v \in W,\ u_t, v_t \in L^2(0,T;V'), \\ \int_0^T < u_t, \phi > dt + \alpha_1 \int_0^T \int_\Omega \nabla u \nabla \phi \, dxdt \\ \quad = \int_0^T \int_\Omega u[a_1 - b_1 u - c_1 v]\phi \, dxdt - \int_0^T \int_{\partial\Omega} \beta u \phi \, dsdt, \\ \int_0^T < v_t, \psi > dt + \alpha_2 \int_0^T \int_\Omega \nabla v \nabla \psi \, dxdt \\ \quad = \int_0^T \int_\Omega v[-a_2 + c_2 u]\psi \, dxdt - \int_0^T \int_{\partial\Omega} \beta v \psi \, dsdt \\ \text{for all } \phi, \psi \in W, \text{ and} \\ u(x,0) = u_0(x),\ v(x,0) = v_0(x). \end{cases} \tag{5.24}$$

Theorem 5.4 (Solution of Initial-Boundary Value Problem). *For each $\beta \in C_M$, there exists a unique solution (u,v) for problem (5.21) (i.e. (5.24)).*

Remark 5.5. Note that there are much fewer restrictions on the relative sizes of the coefficients a_i, c_i and b_1 here compared with those on Sections 3.2 and 3.3

concerning positive steady states and time periodic solutions, because we are now only concerned with an initial value problem.

Proof. Let $\bar{U}$ and $\bar{V}$ be respectively the solutions of the following linear problems:

$$(5.25)\qquad \begin{cases} u_t - \alpha_1 \Delta u = a_1 u & \text{in } \Omega_T, \\ \alpha_1 \frac{\partial u}{\partial \nu} + \beta u = 0 & \text{on } \partial\Omega \times (0,T), \\ u(x,0) = u_0(x) & \text{for } x \in \Omega; \quad \text{and} \end{cases}$$

$$(5.26)\qquad \begin{cases} v_t - \alpha_2 \Delta v = c_2 \bar{U} v & \text{in } \Omega_T, \\ \alpha_2 \frac{\partial v}{\partial \nu} + \beta v = 0 & \text{on } \partial\Omega \times (0,T), \\ v(x,0) = v_0(x) & \text{for } x \in \Omega. \end{cases}$$

By comparison, the functions $\bar{U}$ and $\bar{V}$ are in $L^\infty(\Omega_T)$, with:

$$(5.27)\qquad \begin{aligned} 0 \le \bar{U} \le ||u_0||_{L^\infty(\Omega_T)} e^{\gamma T}, \\ 0 \le \bar{V} \le ||v_0||_{L^\infty(\Omega_T)} e^{\eta T}, \end{aligned}$$

where $\gamma = ||a_1||_{L^\infty(\Omega)}$ and $\eta = ||c_2||_{L^\infty(\Omega)} ||u_0||_{L^\infty(\Omega)} e^{\gamma T}$. Define $u^1 = \bar{U}, v^2 = \bar{V}, u^0 = 0$ and $v^1 = 0$, where the superscript denotes the iterative step which we will set up presently. Choose a large constant R such that

$$R > sup_{\Omega_T}\{a_2 + 2b_1\bar{U} + c_1\bar{V}\},$$

and for convenience, write

$$f_1(x,t,u,v) = Ru + u[a_1(x,t) - b_1(x,t)u - c_1(x,t)v],$$

$$f_2(x,t,u,v) = Rv + v[-a_2(x,t) + c_2(x,t)u].$$

For $i = 2, 3, \ldots$, define u^i, v^{i+1} inductively as the solutions of the following problems:

$$(5.28)\qquad \begin{cases} u^i_t - \alpha_1 \Delta u^i + Ru^i = f_1(x,t,u^{i-2},v^i) & \text{in } \Omega_T, \\ \alpha_1 \frac{\partial u^i}{\partial \nu} + \beta u^i = 0 & \text{on } \partial\Omega \times (0,T), \\ u^i(x,0) = u_0(x) & \text{for } x \in \Omega; \quad \text{and} \end{cases}$$

$$(5.29)\quad \begin{cases} v_t^i - \alpha_2 \Delta v^i + R v^i = f_2(x,t,u^{i-1},v^{i-2}) & \text{in } \Omega_T, \\ \alpha_2 \frac{\partial v^i}{\partial \nu} + \beta v^i = 0 & \text{on } \partial\Omega \times (0,T), \\ v^i(x,0) = v_0(x) & \text{for } x \in \Omega. \end{cases}$$

Since the problems (5.28) and (5.29) are linear, solutions $u^i, v^i, i = 0,1,2\ldots$ exist. By comparison, for $i = 1,2,\ldots,$ we have $0 \le u^i \le \bar{U}, 0 \le v^i \le V$ in Ω_T. (See e.g. p. 54 [183] for comparison of classical solutions and extension to weak solutions in Section 3.3). Also note that for $(x,t) \in \Omega_T, i = 1,2,\ldots$

$$f_1(x,t,u^{i-2},v^i) \text{ is increasing in } u^{i-2}, \text{ decreasing in } v^i,$$

$$f_2(x,t,u^{i-1},v^{i-2}) \text{ is increasing in } u^{i-1} \text{ and } v^{i-2}.$$

We have $v^1 = 0$, and from (5.29)

$$(v^3 - v^1)_t - \alpha_2 \Delta(v^3 - v^1) + R(v^3 - v^1) = c_2 u^2 v^1 \ge 0,$$

we find $v^1 \le v^3$. Similarly, we deduce $u^3 \le u^1, v^2 \ge v^4, u^0 \le u^2$. Moreover, we obtain $u^2 \le u^3$ and $v^4 \ge v^3$. We can then inductive show that

$$0 = u^0 \le u^2 \le \cdots \le u^{2i} \cdots \le u^{2i+1} \le \cdots \le u^3 \le u^1 = \bar{U}$$

$$0 = v^1 \le v^3 \le \cdots \le v^{2i+1} \cdots \le v^{2i} \le \cdots \le v^4 \le v^2 = \bar{V}$$

in Ω_T as in Sections 3.2, 3.3 or Chapter 5 in [125]. From the boundedness of u^i, v^i and the monotone properties, we obtain the pointwise convergence of the odd and even iterates. Since the right-hand side of (5.28) is bounded in $L^\infty(\Omega_T)$, we can use u^k as its own test function to obtain:

$$(5.30)\quad \begin{aligned} &sup_{0\le t\le T}(\textstyle\int_\Omega (u^k)^2(x,t)dx) + \int_0^T \int_\Omega |\nabla u^k|^2 dxdt + \int_0^T \int_{\partial\Omega} \beta (u^k)^2 dsdt \\ &\le C_3 \textstyle\int_\Omega u_0^2 dx + C_4 \int_0^T \int_\Omega [f_1(x,t,u^{k-2},v^k)]^2 + (u^k)^2 dxdt \\ &\le C_3 \textstyle\int_\Omega u_0^2 dx + C_5 T \end{aligned}$$

for some constants C_3, C_4 and C_5. Thus, the functions u^{2i}, u^{2i+1} are uniformly bounded in W. Similarly, we deduce that v^{2i}, v^{2i+1} are also uniformly bounded in W. Consequently, without loss of generality, or by relabeling, we can select subsequences so that:

$$u^{2j} \to \underline{u},\ u^{2j+1} \to \bar{u},\ v^{2j} \to \bar{v},\ v^{2j+1} \to \underline{v} \text{ weakly in } W.$$

From the W boundedness of $u^{2i}, u^{2i+1}, v^{2i}, v^{2i+1}$ and the equations (5.28), (5.29), we deduce that $u_t^{2i}, u_t^{2i+1}, v_t^{2i}$ and v_t^{2i+1} are bounded in $L^2(0,T;V')$. Thus by weak compactness again, we may assume by relabeling

$$u_t^{2j} \to \underline{u}_t,\ u_t^{2j+1} \to \bar{u}_t,\ v_t^{2j} \to \bar{v}_t,\ v_t^{2j+1} \to \underline{v}_t \ \text{ weakly in } L^2(0,T,V').$$

Since $L^2(0,T;H_0^1(\Omega))$ compactly embeds into $L^2(0,T;H^{1/2+\epsilon}(\Omega)), 0 \le \epsilon < \frac{1}{2}$, we may extract subsequences and assume

$$u^{2j} \to \underline{u},\ u^{2j+1} \to \bar{u},\ v^{2j} \to \bar{v},\ v^{2j+1} \to \underline{v} \ \text{ strongly in } L^2(0,T;H^{1/2+\epsilon}(\Omega)).$$

The above assertion can also be obtained by applying Theorem A5-3 in Chapter 6. Using the continuous mapping from $H^{1/2+\epsilon}(\Omega)$ into $L^2(\partial\Omega)$, we also have

$$u^{2j} \to \underline{u},\ u^{2j+1} \to \bar{u},\ v^{2j} \to \bar{v},\ v^{2j+1} \to \underline{v} \ \text{ strongly in } L^2(0,T;L^2(\partial\Omega)).$$

Passing to the limit with u^{2i+1}, u^{2i}, v^{2i} and v^{2i+1} in (5.28) and (5.29), we find that the limits satisfy:

(5.31)

$$\begin{cases} \bar{u}_t - \alpha_1\Delta\bar{u} + R\bar{u} = f_1(x,t,\bar{u},\underline{v}), \quad \underline{u}_t - \alpha_1\Delta\underline{u} + R\underline{u} = f_1(x,t,\underline{u},\bar{v}) & \text{in } \Omega_T, \\ \alpha_1\frac{\partial\bar{u}}{\partial\nu} + \beta\bar{u} = \alpha_1\frac{\partial\underline{u}}{\partial\nu} + \beta\underline{u} = 0 \quad \text{on } \partial\Omega_T, \\ \bar{u}(x,0) = \underline{u}(x,0) = u_0(x) \qquad \text{for } x \in \Omega; \\ \bar{v}_t - \alpha_2\Delta\bar{v} + R\bar{v} = f_2(x,t,\bar{u},\bar{v}), \quad \underline{v}_t - \alpha_2\Delta\underline{v} + R\underline{v} = f_2(x,t,\underline{u},\underline{v}) & \text{in } \Omega_T, \\ \alpha_2\frac{\partial\bar{v}}{\partial\nu} + \beta\bar{v} = \alpha_2\frac{\partial\underline{v}}{\partial\nu} + \beta\underline{v} = 0 \quad \text{on } \partial\Omega_T, \\ \bar{v}(x,0) = \underline{v}(x,0) = v_0(x) \qquad \text{for } x \in \Omega. \end{cases}$$

In order to show $\bar{u} = \underline{u}$ and $\bar{v} = \underline{v}$, we substitute $\bar{u} = e^{\lambda t}\bar{w}, \underline{u} = e^{\lambda t}\underline{w}, \bar{v} = e^{\lambda t}\bar{z}, \underline{v} = e^{\lambda t}\underline{z}$ where $\lambda > 0$ is to be chosen. Note that, we have

$$\begin{cases} \bar{w}_t - \alpha_1\Delta\bar{w} + \lambda\bar{w} = a_1\bar{w} - b_1e^{\lambda t}\bar{w}^2 - c_1e^{\lambda t}\bar{w}\underline{z} \quad \text{in } \Omega_T, \\ \underline{w}_t - \alpha_1\Delta\underline{w} + \lambda\underline{w} = a_1\underline{w} - b_1e^{\lambda t}\underline{w}^2 - c_1e^{\lambda t}\underline{w}\bar{z} \quad \text{in } \Omega_T, \\ \bar{z}_t - \alpha_2\Delta\bar{z} + \lambda\bar{z} = -a_2\bar{z} + c_2\bar{w}\bar{z} \quad \text{in } \Omega_T, \\ \underline{z}_t - \alpha_2\Delta\underline{z} + \lambda\underline{z} = -a_2\underline{z} + c_2\underline{wz} \quad \text{in } \Omega_T. \end{cases}$$

By the weak formulations of the equations for $\bar{w}, \underline{w}, \bar{z}$ and $\underline{z}$, we find

$$\int_{\Omega_T}\{(\bar{w}-\underline{w})_t(\bar{w}-\underline{w})+\alpha_1|\nabla(\bar{w}-\underline{w})|^2+\lambda(\bar{w}-\underline{w})^2$$

$$+(\underline{z}-\bar{z})_t(\underline{z}-\bar{z})+\alpha_2|\nabla(\underline{z}-\bar{z})|^2+\lambda(\underline{z}-\bar{z})^2\}dxdt$$

$$+\int_{\partial\Omega_T}\{\beta(\bar{w}-\underline{w})^2+\beta(\underline{z}-\bar{z})^2\}dsdt$$

$$=\int_{\Omega_T}\{a_1(\bar{w}-\underline{w})^2-b_1e^{\lambda t}(\bar{w}^2-\underline{w}^2)(\bar{w}-\underline{w})-c_1e^{\lambda t}(\bar{w}\underline{z}-\underline{w}\bar{z})(\bar{w}-\underline{w})\}dxdt$$

$$+\int_{\Omega_T}\{-a_2(\underline{z}-\bar{z})^2-c_2e^{\lambda t}(\underline{wz}-\bar{w}\bar{z})(\underline{z}-\bar{z})\}dxdt.$$

Since $\bar{u}, \underline{u}, \bar{v}, \underline{v}, a_i, c_i, i=1,2$, and b_1 are all L^∞ bounded in Ω_T, we can use Cauchy's inequality to obtain

(5.32)

$$\frac{1}{2}\int_{\Omega}[\bar{w}(x,T)-\underline{w}(x,T)]^2+[\bar{z}(x,T)-\underline{z}(x,T)]^2dx$$

$$+\int_{\Omega_T}[\alpha_1|\nabla(\bar{w}-\underline{w})|^2+\alpha_2|\nabla(\bar{z}-\underline{z})|^2]dxdt$$

$$+(\lambda-C)\int_{\Omega_T}[(\bar{w}-\underline{w})^2+(\bar{z}-\underline{z})^2]dxdt+\int_{\partial\Omega_T}\beta(\bar{w}-\underline{w})^2+\beta(\bar{z}-\underline{z})^2dsdt\le 0,$$

where C depends only on the coefficients, T and $||u^1||_{L^\infty}, ||v^2||_{L^\infty}$. By choosing $\lambda > C$, we see that inequality (5.32) holds if and only if

$$\bar{w}=\underline{w} \quad \text{and } \bar{z}=\underline{z} \quad \text{a.e. in } \Omega_T.$$

Consequently, $\bar{u}=\underline{u}, \bar{v}=\underline{v}$ in Ω_T, and the solution of problem (5.31) becomes the solution of problem (5.21) with $(u,v)=(\bar{u},\bar{v})=(\underline{u},\underline{v})$.

As in the above paragraphs, we can show by comparison and iteration that a solution (u,v) of (5.21) must satisfy

$$\underline{u}\le u\le\bar{u}, \quad \underline{v}\le v\le\bar{v} \text{ in } \Omega_T.$$

Since $\bar{u}=\underline{u}$ and $\bar{v}=\underline{v}$, we must have uniqueness.

We have the following existence theorem for the optimal control problem (5.21)-(5.23).

Theorem 5.5 (Existence of Optimal Control). *There exists an optimal control satisfying (5.23) for the problem (5.21) corresponding to the functional (5.22).*

Proof. Since the controls are bounded, there exists a maximizing sequence $\{\beta_n\}\subset C_M$. From the last theorem, the corresponding solutions of problem

(5.21): $u_n = u(\beta_n), v_n = v(\beta_n)$ are uniquely defined for each n. From (5.21) for (u_n, v_n) we deduce readily

$$sup_{0\le t\le T}\{\int_\Omega [u_n^2(x,t)+v_n^2(x,t)]\,dx\} + \int_0^T\int_\Omega [|\nabla u_n|^2 + |\nabla v_n|^2]dxdt$$

$$\le C_1(\int_\Omega [u_0^2+v_0^2]dx+1),$$

for some constant C_1. Using this bound, we then select subsequence and pass to the limit, by means of the same compactness arguments as in the proof of Theorem 5.1, to obtain the optimal control β^*.

The following theorem is concerned with the differentiability of the solution of (5.21) with respect to the control.

Theorem 5.6 (Differentiability with Respect to Control). *The mapping* $C_M \ni \beta \to (u,v) = (u(\beta), v(\beta)) \in W \times W$ *is differentiable in the following sense:*

$$\frac{u(\beta+\rho\bar\beta)-u(\beta)}{\rho} \to \xi \qquad \textit{weakly in } W,$$

$$\frac{v(\beta+\rho\bar\beta)-v(\beta)}{\rho} \to \eta \qquad \textit{weakly in } W,$$

as $\rho \to 0$ *for any* $\beta \in C_M, \bar\beta \in L^\infty(\Omega_T)$ *such that* $\beta + \rho\bar\beta \in C_M$. *Also* $\xi = \xi(\beta,\bar\beta), \eta = \eta(\beta,\bar\beta)$ *satisfy*

$$(5.33)\quad \begin{cases} \xi_t - \alpha_1\Delta\xi = a_1\xi - 2b_1u\xi - c_1v\xi - c_1u\eta & \text{in } \Omega_T,\\ \eta_t - \alpha_2\Delta\eta = -a_2\eta + c_2u\eta + c_2v\xi & \text{in } \Omega_T,\\ \alpha_1\frac{\partial\xi}{\partial\nu} = -\beta\xi - \bar\beta u & \text{on } \partial\Omega_T,\\ \alpha_2\frac{\partial\eta}{\partial\nu} = -\beta\eta - \bar\beta v & \text{on } \partial\Omega_T,\\ \xi(x,0) = \eta(x,0) = 0 & \text{for } x \in \Omega. \end{cases}$$

Proof. Define $u^\rho = u(\beta+\rho\bar\beta), v^\rho = v(\beta+\rho\bar\beta)$. Let $u^\rho = e^{\lambda t}w^\rho, u = e^{\lambda t}w, v^\rho = e^{\lambda t}z^\rho, v = e^{\lambda t}z$ where $\lambda > 0$ is to be determined. Estimating by means of the differential equations satisfied by $(w^\rho - w)/\rho, (z^\rho - z)/\rho$, and using the bounds

(5.27) for the solutions, we obtain

$$
\begin{aligned}
&\tfrac{1}{2}\int_\Omega[(\tfrac{w^\rho(x,t)-w(x,t)}{\rho})^2+(\tfrac{z^\rho(x,t)-z(x,t)}{\rho})^2]dx \\
&\quad+\int_{\Omega\times(0,t)}[\alpha_1|\nabla(\tfrac{w^\rho-w}{\rho})|^2+\alpha_2|\nabla(\tfrac{z^\rho-z}{\rho})|^2]dxdt \\
&\quad+\lambda\int_{\Omega\times(0,t)}[(\tfrac{w^\rho-w}{\rho})^2+(\tfrac{z^\rho-z}{\rho})^2]dxdt \\
&\le C\int_{\Omega_T}[(\tfrac{w^\rho-w}{\rho})^2+(\tfrac{z^\rho-z}{\rho})^2]dxdt \\
&\quad+\int_{\partial\Omega_T}\{|\bar{\beta}w^\rho(\tfrac{w^\rho-w}{\rho})|+|\bar{\beta}z^\rho(\tfrac{z^\rho-z}{\rho})|\}dsdt
\end{aligned}
\tag{5.34}
$$

where C depends on the coefficients and the final time T. Continuing to estimate using the continuous mapping of $H^1(\Omega)$ into $L^2(\partial\Omega)$ and Cauchy's inequality with ϵ, we deduce

$$
\begin{aligned}
&\tfrac{1}{2}\int_\Omega[(\tfrac{w^\rho(x,t)-w(x,t)}{\rho})^2+(\tfrac{z^\rho(x,t)-z(x,t)}{\rho})^2]dx \\
&\quad+\int_{\Omega_T}[\tfrac{\alpha_1}{2}|\nabla(\tfrac{w^\rho-w}{\rho})|^2+\tfrac{\alpha_2}{2}|\nabla(\tfrac{z^\rho-z}{\rho})|^2]dxdt \\
&\quad+(\lambda-C_1)\int_{\Omega_T}[(\tfrac{w^\rho-w}{\rho})^2+(\tfrac{z^\rho-z}{\rho})^2]dxdt \\
&\le C_2\int_{\partial\Omega_T}\bar{\beta}^2dsdt,
\end{aligned}
$$

where C_1, C_2 are constants depending on the bounds of the coefficients and u, v, u^ρ and v^ρ. Choosing $\lambda > C_1$, we conclude

$$\|\frac{w^\rho-w}{\rho}\|_W^2+\|\frac{z^\rho-z}{\rho}\|_W^2\le C_2\int_{\partial\Omega_T}\bar{\beta}^2dsdt.$$

This bound leads to the existence of ξ, η in W, such that

$$\frac{u_\rho-u}{\rho}\to\xi,\ \frac{v^\rho-v}{\rho}\to\eta \text{ weakly in } W.$$

As in Theorem 5.4, we also have

$$(\frac{u_\rho-u}{\rho})_t\to\xi_t,\ (\frac{v^\rho-v}{\rho}\rho)_t\to\eta_t \text{ weakly in } L^2(0,T;V').$$

These lead to $(u^\rho-u)/\rho\to\xi$ and $(v^\rho-v)/\rho\to\eta$ strongly in $L^2(\Omega_T)$. Also, we have $(u^\rho-u)/\rho\to\xi$ and $(v^\rho-v)/\rho\to\eta$ strongly in $L^2(\partial\Omega_T)$. Taking limit in the differential equations satisfied by $(u^\rho-u)/\rho$ and $(v^\rho-v)/\rho$, we obtain (ξ,η) satisfy (5.33).

Using the above theorem, we can characterize the optimal control in terms of the adjoint system for (5.21) as in the following theorem. The proof is similar to that for Theorem 5.2. The details will thus be omitted.

Theorem 5.7 (Characterization of Optimal Control for the Prey-Predator System). *Let β^* be an optimal control satisfying (5.23) for problem (5.21), (5.22), and $(u^*, v^*) \in W \times W$ be the corresponding solution, then there exists a unique solution $(p, q) \in W \times W$ satisfying the adjoint system:*

$$(5.35)\quad \begin{cases} -p_t - \alpha_1 \Delta p = A_1 + a_1 p - 2b_1 u^* p - c_1 v^* p + c_2 v^* q & \textit{in } \Omega_T, \\ -q_t - \alpha_2 \Delta q = A_2 - c_1 u^* p - a_2 q + c_2 u^* q & \textit{in } \Omega_T, \\ \alpha_1 \frac{\partial p}{\partial \nu} + \beta^* p = 0, \quad \alpha_2 \frac{\partial q}{\partial \nu} + \beta^* q = 0 & \textit{on } \partial\Omega_T, \\ p(x,T) = q(x,T) = 0 & \textit{for } x \in \Omega, \end{cases}$$

with the property,

$$(5.36)\qquad \beta^* = min.\{(M - pu^* - qv^*)^+, M\}.$$

Combining (5.21), (5.35) and (5.36), we obtain the optimality system:

$$(5.37)\quad \begin{cases} u_t - \alpha_1 \Delta u = u[a_1 - b_1 u - c_1 v] & \text{in } \Omega_T, \\ v_t - \alpha_2 \Delta v = v[-a_2 + c_2 u] & \text{in } \Omega_T, \\ -p_t - \alpha_1 \Delta p = A_1 + a_1 p - 2b_1 up - c_1 vp + c_2 vq & \text{in } \Omega_T, \\ -q_t - \alpha_2 \Delta q = A_2 - c_1 up - a_2 q + c_2 uq & \text{in } \Omega_T, \\ u(x,0) = u_0(x), \ v(x,0) = v_0(x) & \text{for } x \in \Omega, \\ p(x,T) = q(x,T) = 0 & \text{for } x \in \Omega, \\ \alpha_1 \frac{\partial u}{\partial \nu} + min.\{(M - pu - qv)^+, M\}u = 0 & \text{on } \partial\Omega_T, \\ \alpha_2 \frac{\partial v}{\partial \nu} + min.\{(M - pu - qv)^+, M\}v = 0 & \text{on } \partial\Omega_T, \\ \alpha_1 \frac{\partial p}{\partial \nu} + min.\{(M - pu - qv)^+, M\}p = 0 & \text{on } \partial\Omega_T, \\ \alpha_2 \frac{\partial q}{\partial \nu} + min.\{(M - pu - qv)^+, M\}q = 0 & \text{on } \partial\Omega_T, \end{cases}$$

Here, A_1, A_2, M, α_1 and α_2 are arbitrary given positive constants; and the functions $a_i(x,t), c_i(x,t), i = 1, 2, b_1(x,t), u_0(x)$ and $v_0(x)$ satisfy the assumptions as in Theorems 5.4 to 5.7 above. We know that weak solution of (5.37) exists by Theorem 5.4 and Theorem 5.7. Once the solution of (5.37) is found, then the optimal control for (5.21)-(5.23) can be expressed by:

$$\beta^* = min.\{M - pu - qv)^+, M\}.$$

For $T > 0$ sufficiently small, we can further deduce the uniqueness of solution for problem (5.37) as follows.

Theorem 5.8 (Uniqueness of Optimal Control) *Consider the optimality system (5.37) with hypotheses on the coefficients and parameters described above. For T sufficiently small, the optimality system (5.37) has a unique solution $(u, v, p, q) \in W \times W \times W \times W$.*

Proof. From the proof of Theorem 5.4, the solution components u and v of the state system part of (5.37) are bounded in $L^\infty(\Omega_T)$. The bounds are independent of M, A_1 and A_2. Thus the adjoint system can be interpreted as linear inside the domain with bounded coefficients, and appropriate sign for the coefficients on the boundary. We can deduce by comparison that

$$||p||_{L^\infty(\Omega_T)} \leq C(A_1 + A_2),$$

$$||q||_{L^\infty(\Omega_T)} \leq C(A_1 + A_2).$$

The constant C may depend on M, the coefficients of the adjoint system, and the bounds of the solution (u, v) of the state system.

Suppose (u, v, p, q) and $(\tilde{u}, \tilde{v}, \tilde{p}, \tilde{q})$ are two weak solutions of (5.37), we define new functions by

$$u = e^{\lambda t} w,\ \tilde{u} = e^{\lambda t}\tilde{w},\ v = e^{\lambda t} z,\ \tilde{v} = e^{\lambda t}\tilde{z},$$

$$p = e^{\lambda t} y,\ \tilde{p} = e^{\lambda t}\tilde{y},\ q = e^{\lambda t}\xi,\ \tilde{q} = e^{\lambda t}\tilde{\xi},$$

$$\beta = min.\{(M - pu - qv)^+, M\},\ \ \tilde{\beta} = min.\{(M - \tilde{p}\tilde{u} - \tilde{q}\tilde{v})^+, M\},$$

where $\lambda > 0$ is to be determined. Subtracting the equations for the new variables, multiplying with the appropriate test functions as in the proof of Theorem 5.3 and integrate, we obtain:

(5.38)

$$
\begin{aligned}
&\tfrac{1}{2}\int_\Omega\{[w(x,T)-\tilde{w}(x,T)]^2+[z(x,T)-\tilde{z}(x,T)]^2 \\
&\quad+[y(x,0)-\tilde{y}(x,0)]^2+[\xi(x,0)-\tilde{\xi}(x,0)]^2\}dx \\
&\quad+\int_{\Omega_T}\{\alpha_1|\nabla(w-\tilde{w})|^2+\alpha_2|\nabla(z-\tilde{z})|^2+\alpha_1|\nabla(y-\tilde{y})|^2+\alpha_2|\nabla(\xi-\tilde{\xi})|^2\}dxdt \\
&\quad+\int_{\Omega_T}\lambda[(w-\tilde{w})^2+(z-\tilde{z})^2+(y-\tilde{y})^2+(\xi-\tilde{\xi})^2]\,dxdt \\
&\le\int_{\Omega_T}\{a_1[(w-\tilde{w})^2+(y-\tilde{y})^2]-a_2[(z-\tilde{z})^2+(\xi-\tilde{\xi})^2] \\
&\quad-(w-\tilde{w})c_1e^{\lambda t}(wz-\tilde{w}\tilde{z})+(y-\tilde{y})[-2b_1e^{\lambda t}(wy-\tilde{w}\tilde{y})-c_1e^{\lambda t}(zy-\tilde{z}\tilde{y}) \\
&\quad+c_2e^{\lambda t}(z\xi-\tilde{z}\tilde{\xi})]+(z-\tilde{z})c_2e^{\lambda t}(wz-\tilde{w}\tilde{z}) \\
&\quad+(\xi-\tilde{\xi})[c_1e^{\lambda t}(wy-\tilde{w}\tilde{y})+c_2e^{\lambda t}(w\xi-\tilde{w}\tilde{\xi})]\}\,dxdt \\
&\quad-\int_{\partial\Omega_T}e^{\lambda t}[(\beta w-\tilde{\beta}\tilde{w})(w-\tilde{w})+(\beta z-\tilde{\beta}\tilde{z})(z-\tilde{z}) \\
&\quad-(\beta y-\tilde{\beta}\tilde{y})(y-\tilde{y})-(\beta\xi-\tilde{\beta}\tilde{\xi})(\xi-\tilde{\xi})]\,dsdt.
\end{aligned}
$$

As an illustration, we can use the L^∞ bound of $z,\tilde{z},w$, and $\tilde{w}$ to obtain an estimate of a typical integral term on the right as follows:

$$
\begin{aligned}
&\int_{\Omega_T}(w-\tilde{w})c_1e^{\lambda t}(wz-\tilde{w}\tilde{z})dxdt \\
&\le C_1e^{\lambda T}\int_{\Omega_T}|wz-\tilde{w}z+\tilde{w}z-\tilde{w}\tilde{z}||w-\tilde{w}|dxdt \qquad (5.39)\\
&\le \hat{C}_1e^{\lambda T}\int_{\Omega_T}[(w-\tilde{w})^2+(z-\tilde{z})^2]dxdt,
\end{aligned}
$$

for some constants $C_1,\hat{C}_1$. We can use the fact that $|\beta-\tilde{\beta}|^2\le(\tilde{w}\tilde{y}-wy+\tilde{z}\tilde{\xi}-z\xi)^2$ and the trace theorem to estimate a typical boundary term on the right as in Theorem 5.3 to obtain:

(5.40)

$$
\begin{aligned}
&\int_{\partial\Omega_T}e^{\lambda t}(\beta y-\tilde{\beta}\tilde{y})(y-\tilde{y})\,dsdt \\
&\quad\le C_2e^{2\lambda T}\int_{\partial\Omega_T}[(w-\tilde{w})^2+(y-\tilde{y})^2+(z-\tilde{z})^2+(\xi-\tilde{\xi})^2]\,dsdt \\
&\quad\le C_2\epsilon e^{2\lambda T}\int_{\Omega_T}[|\nabla(w-\tilde{w})|^2+|\nabla(y-\tilde{y})|^2+|\nabla(z-\tilde{z})|^2+|\nabla(\xi-\tilde{\xi})|^2]\,dxdt \\
&\qquad+C_2C(\epsilon)e^{2\lambda T}\int_{\Omega_T}[(w-\tilde{w})^2+(z-\tilde{z})^2+(y-\tilde{y})^2+(\xi-\tilde{\xi})^2]\,dxdt,
\end{aligned}
$$

for some constant C_2, ϵ is a small constant to be chosen, and $C(\epsilon)$ is a large constant determined by ϵ. By means of (5.39) and (5.40), we deduce from (5.38) that

$$\begin{aligned}
&\tfrac{1}{2}\int_{\Omega}\{[w(x,T)-\tilde{w}(x,T)]^2+[z(x,T)-\tilde{z}(x,T)]^2\\
&+[y(x,0)-\tilde{y}(x,0)]^2+[\xi(x,0)-\tilde{\xi}(x,0)]^2\}dx\\
&+\int_{\Omega_T}\{(\alpha_1-C_3\epsilon e^{2\lambda T})|\nabla(w-\tilde{w})|^2+(\alpha_2-C_3\epsilon e^{2\lambda T})|\nabla(z-\tilde{z})|^2\\
&+(\alpha_1-C_3\epsilon e^{2\lambda T})|\nabla(y-\tilde{y})|^2+(\alpha_2-C_3\epsilon e^{2\lambda T})|\nabla(\xi-\tilde{\xi})|^2\}dxdt\\
&+\int_{\Omega_T}(\lambda-C_4e^{2\lambda T}-C_5C(\epsilon)e^{2\lambda T})[(w-\tilde{w})^2+(z-\tilde{z})^2\\
&+(y-\tilde{y})^2+(\xi-\tilde{\xi})^2]\,dxdt\ \le 0
\end{aligned} \tag{5.41}$$

where $C_i, i=1,2,...,5$ depend only on the coefficients and the L^∞ bounds of z, w, p and q. Let $\alpha = min.\{\alpha_1,\alpha_2\}$. If we choose ϵ, λ and T such that

$$\epsilon < \tfrac{\alpha}{C_3},\quad \lambda > C_4 + C_5C(\epsilon),\quad \text{and}$$

$$T < \tfrac{1}{2\lambda} min.\{ln(\tfrac{\lambda}{C_4+C_5C(\epsilon)}),\ ln(\tfrac{\alpha}{C_3\epsilon})\},$$

then we find from (5.41) that $w=\tilde{w}, z=\tilde{z}, y=\tilde{y}$ and $\xi=\tilde{\xi}$. Consequently, we have $u=\tilde{u}, v=\tilde{v}, p=\tilde{p}$ and $q=\tilde{q}$.

Remark 5.6. In Part A of this section, Theorem 5.1, Lemma 5.1 and Theorem 5.2 are all valid even if $\lambda=0$. However, for Theorem 5.3, if $\lambda=0$, then we can only show that the optimality system (5.17) has at most one solution for small time interval $[0,T]$, with T sufficiently small, as in Theorem 5.8.

Notes.

Theorem 1.1 to Theorem 1.4, for the control of scalar equations, are due to Leung and Stojanovic [140]. For the control and steady-state prey-predator systems, Theorems 2.1 to 2.6 are obtained from Leung [126]. For the control of periodic competitive systems, Theorems 3.1 to 3.8 are summarized from the two articles [82] and [83] by He, Leung and Stojanovic. Theorems 4.1 to 4.5, concerning the target profile control for fission reactors, are found in Leung and Chen [133]. For boundary controls, Theorems 5.1 to 5.3 are results in Lenhart and Wilson [119]. Theorems 5.4 to 5.8, concerning the boundary control of prey-predator systems, are due to Lenhart, Liang and Protopopescu [117].

Chapter 4

Persistence, Upper and Lower Estimates, Blowup, Cross-Diffusion and Degeneracy

4.1 Persistence

In Chapters 1 and 2, we are mostly concerned with finding positive coexistence steady-states for systems of reaction-diffusion equations. We also investigate whether these steady-states are stable locally or even globally as time changes. In other words, we study whether the steady-states attract other solutions of the parabolic systems as time tends to infinity. We developed very complicated theories, which are sometimes difficult to apply. Moreover, in many situations in population dynamics or environmental studies, we do not need to obtain such detailed information for the system. We may be only concerned whether all species under consideration will survive in the long term. Conditions for globally attracting steady-state will be too strong because it excludes possibilities of periodic solutions or other steady-states. In many occasions, it will be practical enough to find criteria to insure all components which are positive initially must eventually enter and remain inside a fixed set of positive states which are strictly bounded away from zero in each component. This property of the system is called persistence or permanence. It may not require to obtain excessive knowledge of the dynamics of the system, and is thus sometimes more mathematically tractable. An exposition of the idea of permanence in genetics, population dynamics and evolutionary theory is given in Hofbauer and Sigmund [87]. Related criteria for coexistence are discussed in Bhatia and Szego [8] and Butler, Freedman and Waltman [16]. In the context of reaction-diffusion systems, we will follow the rigorous methods in Hale and Waltman [81], Cantrell, Cosner and Hutson [21] and [22] in this section. In part of the next section, we will further enhance the theory of persistence by estimating, by means of comparison method, the location of the set which attracts solutions whose components are positive initially.

Part A: Chains and Uniform Persistence.

We first introduce some topological concepts which will be used to analyze the reaction-diffusion systems. Let X be a complete metric space (with metric d) and suppose that $T(t) : X \to X, t \geq 0$, is a C_0-semigroup on X; that is, $T(0) = I, T(t+s) = T(t)T(s)$ for $t, s \geq 0$, and $T(t)x$ is continuous in t, x. The positive orbit through x is defined as $\gamma^+(x) := \cup_{t \geq 0}\{T(t)x\}$. The ω-limit set is defined as

$$\omega(x) := \cap_{\tau \geq 0} Cl \cup_{t \geq \tau} \{T(t)x\}.$$

(Here Cl denotes the closure). This is equivalent to saying that $y \in \omega(x)$ if and only if there is a sequence $t_n \to \infty$ as $n \to \infty$ such that $T(t_n)x \to y$ as $n \to \infty$. If B is a subset of X, we define the ω-limit set of B as

$$\omega(B) := \cap_{\tau \geq 0} Cl \cup_{t \geq \tau} T(t)B, \quad \text{where } T(t)B = \cup_{x \in B}\{T(t)x\}.$$

If the point x or the sets B have negative orbits, we can define the α-limit set $\alpha(x)$ of x and α-limit set $\alpha(B)$ of B in a similar manner taking into account the possibility of multiple backward orbits. A set B in X is said to be invariant if $T(t)B = B$ for $t \geq 0$; that is, the mapping $T(t)$ takes B onto B for each $t \geq 0$. This implies, in particular, that there is a negative orbit through each point of an invariant set. When the points or sets belong to an invariant set A, we will restrict the backward orbits to those remaining in the invariant set and denote this by $\alpha_A(x)$. Sometimes it is convenient to have the alpha limit set of a specific full orbit, $\gamma(x)$ through a point x. We denote this by $\alpha_\gamma(x)$.

A nonempty invariant subset M of X is called an isolated invariant set if it is the maximal invariant set of a neighborhood of itself. The neighborhood is called an isolating neighborhood. The stable (or attracting) set of a compact invariant set A is denoted by W^s and is defined as

$$W^s(A) := \{x | x \in X, \omega(x) \neq \Phi, \omega(x) \subset A\}.$$

(Here, Φ denotes the empty set.) The unstable (or repelling) set, W^u is defined by

$$W^u(A) = \{x | x \in X, \text{there exists a backward orbit}\, \gamma^-(x) \\ \text{such that } \alpha_\gamma(x) \neq \Phi, \alpha_\gamma \subset A\}.$$

A set A in X is said to be a global attractor if it is compact, invariant and, for any bounded set B in X, $\delta(T(t)B, A) \to 0$ as $t \to \infty$, where $\delta(B, A)$ is the distance from the set B to the set A:

$$\delta(B, A) = sup_{y \in B} inf_{x \in A} \, d(y, x).$$

In particular, this implies $\omega(B)$ exists and belong to A. A global attractor is always a maximal compact invariant set. The semigroup $T(t)$ is said to be point dissipative in X if there is a bounded nonempty set B in X such that, for any $x \in X$, there is a $t_0 = t_0(x, B)$ such that $T(t)x \in B$ for $t \geq t_0$. By Theorem A4-13 in Chapter 6, if $T(t)$ is point dissipative in X and there is a $t_0 \geq 0$ such that $T(t)$ is compact for $t > t_0$, then there is a nonempty global attractor A in X.

In order to study the problem of persistence, we will now assume that the metric space X satisfies $X = X^0 \cup \partial X^0$, where X^0 is open in X, which is the closure of X^0. ∂X^0 (assumed to be nonempty) is the boundary of X^0. We suppose that the C_0-semigroup $T(t)$ on X satisfies

$$T(t) : X^0 \to X^0, \qquad T(t) : \partial X^0 \to \partial X^0 \tag{1.1}$$

and let $T_0(t) = T(t)|_{X^0}, T_\partial(t) = T(t)|_{\partial X^0}$. The set ∂X^0 is a complete metric space. If $T(t)$ satisfies the conditions of Theorem A4-13 in Chapter 6, then T_∂ will satisfy the same conditions in ∂X^0. Thus, there will be a global attractor A_∂ in ∂X^0. However, if $T(t)$ satisfies Theorem A4-13 in Chapter 6 in X, it does not follow that the semigroup $T_0(t)$ must have a maximal compact invariant set in X^0. There may be points x in X^0 for which $\omega(x) \cap \partial X^0 \neq \phi$. In order to analyze such problem, we give the following definition.

Definition 1.1. Let X be a metric space, and $X = X^0 \cup \partial X^0$ as described above. Let $T(t)$ be a C_0 semigroup on X with properties as described in (1.1). We say the semigroup $T(t)$ is persistent if $\lim inf_{t \to \infty} d(T(t)x, \partial X^0) > 0$ for any $x \in X^0$. The semigroup $T(t)$ is said to be uniformly persistent if there is an $\eta > 0$ such that for any $x \in X^0$, $\lim inf_{t \to \infty} d(T(t)x, \partial X^0) \geq \eta$.

In order study uniform persistence by means of the behavior of the semigroup on ∂X^0, we introduce a few more definitions. Let M, N be isolated invariant sets (not necessarily distinct). M is said to be chained to N, denoted by $M \to N$, if there exist an element $x, x \notin M \cup N$, such that $x \in W^u(M) \cap W^s(N)$. A finite sequence $M_1, M_2, ..., M_k$ of isolated invariant sets is called a chain if $M_1 \to M_2 \to \cdots \to M_k (M_1 \to M_1$, if $k = 1)$. The chain is called a cycle if $M_k = M_1$. A special invariant set of interest is the following:

$$\tilde{A}_\partial := \cup_{x \in A_\partial} \omega(x). \tag{1.2}$$

(Here, A_∂ is the global attractor in ∂X^0 described above.) The set $\tilde{A}_\partial$ is isolated if there exists a covering $M = \cup_{i=1}^k M_k$ of $\tilde{A}_\partial$ by pairwise disjoint, compact, isolated invariant sets $M_1, M_2, ..., M_k$ for T_∂ such that each M_i is also an isolated invariant set for T. M is called an isolated covering. $\tilde{A}_\partial$ is called acyclic if there exists some isolated covering $M = \cup_{i=1}^k M_i$ of $\tilde{A}_\partial$ such that no subset of the $M_i's$ form a cycle. An isolated covering satisfying this condition is called acyclic.

The following topological theorem is very important for analyzing the problem of persistence.

Theorem 1.1 (Abstract Persistence Theorem). *Let the C_0-semigroup $T(t)$ on $X = X^0 \cup \partial X^0$ satisfy conditions as described in (1.1). Suppose that*

(i) There exists $t_0 \geq 0$ such that $T(t)$ is compact for $t > t_0$,

(ii) $T(t)$ is point dissipative in X, and

(iii) The set $\tilde{A}_\partial$, defined by (1.2), is isolated and has an acyclic covering M as described above.

Then $T(t)$ is uniformly persistent if and only if each M_i of the covering has the property:

$$W^s(M_i) \cap X^0 = \Phi. \tag{1.3}$$

Roughly speaking, condition (iii) above states that there are no cycle in the boundary which links up various of the limits; and the final condition states that no solution orbit from the interior converges to one of the boundary limits in forward time. With the aid of the above abstract theorem, we are now ready to return to the analysis of reaction-diffusion systems. We consider the following reaction-diffusion system for three interacting populations involving one predator with two competing preys, for $(x,t) \in \Omega \times [0,\infty)$.

$$\begin{cases} \partial u_1/\partial t = \sigma_1 \Delta u_1 + u_1(a_1 - u_1 - c_{12}u_2 - c_{13}u_3), \\ \partial u_2/\partial t = \sigma_2 \Delta u_2 + u_2(a_2 - c_{21}u_1 - u_2 - c_{23}u_3), \\ \partial u_3/\partial t = \sigma_3 \Delta u_3 + u_3(a_3 + c_{31}u_1 + c_{32}u_2 - u_3), \\ u_i(x,t) = 0, \ i = 1,2,3 \qquad \text{for } (x,t) \in \partial\Omega \times [0,\infty) \end{cases} \tag{1.4}$$

where σ_i, a_i, c_{ij} are all constants, with $\sigma_i > 0, i = 1,2,3$; $c_{ij} > 0$, $i,j = 1,2,3$, $i \neq j$.

We will discuss solutions of (1.4) such that each component as a function of x, at a given t, is in $C^k(\bar{\Omega})$, for some integer k. For this section, we assume Ω is a bounded domain in R^N, with boundary $\partial\Omega$ uniformly $C^{3+\alpha}$ for some $\alpha > 0$. The norm in $[C^k(\bar{\Omega})]^3$ will be denoted by $||\cdot||_k$, and the closed subspaces of functions vanishing on $\partial\Omega$ by $[C_0^k(\bar{\Omega})]^3$. $[C_+^k(\bar{\Omega})]^3$ will denote the positive cones with respect to the usual ordering; and let $C_{0,+}^k(\bar{\Omega}) = C_0^k(\bar{\Omega}) \cap C_+^k(\bar{\Omega})$. From the theory of semigroup of solutions of reaction-diffusion equations in $C^k(\bar{\Omega})$, and the structure of the quadratic reaction terms in (1.4), we see that $[C_{0,+}^2(\bar{\Omega})]^3$ is invariant on the maximal interval of existence. From the signs of the coefficients in (1.4), we will show in Lemma 1.1 below that the solution of (1.4) satisfy:

[H1] (Uniform boundedness in $[C^0_{0,+}(\bar{\Omega})]^3$). For given $\beta > 0$, there exists a number $B(\beta)$ such that if $||(u_1(\cdot,0),.,u_3(\cdot,0)||_0 \leq \beta$, then $||u_1(\cdot,t),.,u_3(\cdot,t)||_0 \leq B(\beta)$ for $t > 0$.

[H2] (Point dissipativity in $[C^0_{0,+}(\bar{\Omega})]^3$). There exists γ such that for any $u_0 = (u_1(\cdot,0),..,u_3(\cdot,0)) \in [C^0_{0,+}(\bar{\Omega})]^3$, there exists a $t(u_0)$ such that the solution of (1.4) with initial condition u_0 satisfies $||u_1(\cdot,t),.,u_3(\cdot,t)||_0 \leq \gamma$ for $t \geq t(u_0)$.

By the smoothing properties of parabolic equations, it is known from general theory (cf. [113]) that the following is true.

Theorem 1.2. *Suppose the solutions of (1.4) satisfy [H1], [H2], then the solutions of (1.4) generates a semigroup on $[C^0_{0,+}(\bar{\Omega})]^3$, and its restriction to $[C^1_{0,+}(\bar{\Omega})]^3$ is also a semigroup. Point dissipativity in $[C^1_{0,+}(\bar{\Omega})]^3$ will also hold. Moreover, the solution operator $T(t)$ is compact on $[C^1_{0,+}(\bar{\Omega})]^3$ for every $t > 0$. There is a bounded set U_2 in $[C^2_+(\bar{\Omega})]^3$ such that if $U \subset [C^1_{0,+}(\bar{\Omega})]^3$ is bounded, then $T(t)U \subset U_2$ for $t \geq 1$.*

In view of Theorem A4-13 and Theorem 1.2, we will take X^0 to be functions in $[C^1_{0,+}(\bar{\Omega})]^3$ with each component non-negative and not identically zero in Ω. The functions in ∂X^0 are those with at least one component identically zero. We will show that under further conditions of the coefficients in (1.4), we can study the uniform persistence of the semigroup of solutions $T(t)$ by means of Theorem 1.1, choosing $X = [C^1_{0,+}(\bar{\Omega})]^3$. We now consider problem (1.4) under the following hypotheses:

[C1] $\hat{\rho}_1(\sigma_i\Delta + a_i) > 0$, for $i = 1, 2$. (From Chapter 1, for each $i = 1, 2$, the scalar problem $w_{it} = \sigma_i \Delta w_i + (a_i - w_i)w_i$ has a corresponding equilibrium $\bar{u}_i > 0$ in Ω, which is a stable global attractor of all non-negative nontrivial solutions.)

[C2] $\hat{\rho}_1(\sigma_1\Delta + a_1 - c_{12}\bar{u}_2) > 0$, and problem (1.4) has no non-negative equilibrium solution of the form $(u_1^*, u_2^*, 0)$ with both $u_i^* \not\equiv 0, i = 1, 2$ in $\bar{\Omega}$.

[C3] $\hat{\rho}_1(\sigma_2\Delta + a_2 - c_{21}\bar{u}_1) < 0$.

[C4] There are unique globally attracting equilibria $P_1(\hat{u}_1, 0, \hat{u}_3)$ in the interior of the $u_1 - u_3$ face and $P_2(0, \tilde{u}_2, \tilde{u}_3)$ in the interior of the $u_2 - u_3$ face. The equilibria P_1 and P_2 are also stable with respect to solutions in $C^0_{0,+}(\bar{\Omega})$ on the $u_1 - u_3$ and $u_2 - u_3$ face respectively, if initial conditions are close to the equilibria in $C^1_{0,+}(\bar{\Omega})$.

[C5] $\hat{\rho}_1(\sigma_1\Delta + a_1 - c_{12}\tilde{u}_2 - c_{13}\tilde{u}_3) > 0$, and $\hat{\rho}_1(\sigma_2\Delta + a_2 - c_{21}\hat{u}_1 - c_{23}\hat{u}_3) > 0$.

(Recall the definition of the first eigenvalue $\hat{\rho}_1$ for an elliptic operator in Section 1.1.)

Theorem 1.3 (Persistence Theorem for a Prey-Predator System of 3 Species). *Assume hypotheses [C1]-[C5], then the semigroup $T(t)$ of solutions of (1.4) in $X = X^0 \cup \partial X^0 = [C^1_{0,+}(\bar{\Omega})]^3$ is uniformly persistent.*

We will first prove the following Lemmas 1.1 to 1.4 and apply Theorems 1.1 and 1.2 in order to prove Theorem 1.3. The hypotheses [C1] to [C5] impose properties on solutions of subsystems of (1.4) on the boundary faces and on the dynamical behavior near the boundary. These properties will lead to the fact that the corresponding set $\tilde{A}_\delta$ in Theorem 1.1 will be isolated and has an acyclic covering satisfying (1.3).

Lemma 1.1. *Under hypothesis [C1], the semigroup of solutions of (1.4) is point dissipative in $[C^0_{0,+}(\bar{\Omega}]^3$. It is also uniformly bounded in $[C^0_{0,+}(\bar{\Omega}]^3$. As a consequence of Theorem 1.2, these properties are also valid in $[C^1_{0,+}(\bar{\Omega}]^3$.*

Proof. For each $i = 1, 2$, the i-th component of the solution of (1.4) is a lower solution of the scalar problem: $w_{it} = \sigma_i \Delta w_i + (a_i - w_i)w_i$ subject to the same initial and boundary conditions. Thus by comparison, we have $u_i \leq w_i$ as long as they exist. Since $w_i(\cdot, t)$ tends to the corresponding steady state solution in $C^1_0(\bar{\Omega})$, (see e.g. proof of Theorem 5.7 in Chapter 1), and the steady state is $< a_i$ in $\bar{\Omega}$ with strictly negative outward normal derivative on $\partial\Omega$, we obtain $w_i(x, t) < a_i$ in $\bar{\Omega}$ for t sufficiently large. Similarly, since u_3 is a lower solution for $w_{3t} = \sigma_3 \Delta w_3 + (a_3 + c_{31}a_1 + c_{32}a_2 - w_3)w_3$ for sufficiently large t, eventually we have $u_3(x, t) \leq w_3(x, t) < a_3 + c_{31}a_1 + c_{32}a_2$ in $\bar{\Omega}$ for t sufficiently large. The last statement follows from semigroup theory, integral representation formula and Theorem 1.2 above.

Lemma 1.2. *Under hypotheses [C1], [C2], any solution of (1.4) of the form $(u_1, u_2, 0)$, with $u_i(x, 0) \geq 0, i = 1, 2$ and $u_1(x, 0) \not\equiv 0$, must $\to (\bar{u}_1(x), 0, 0)$ as $t \to \infty$ in $\bar{\Omega}$.*

Proof. Let $(u_1, u_2, 0)$ be a solution of (1.4) with non-negative initial condition and $u_1(x, 0) \not\equiv 0$ in $\bar{\Omega}$. Then by maximum principle, we have $u_1 > 0$ in Ω and $\partial u_1/\partial\nu < 0$ on $\partial\Omega$ for any $t > 0$. Moreover, u_2 is a lower solution of $w_t = \sigma_2 \Delta w + (a_2 - w)$ with the same initial boundary condition; and for any $\epsilon > 0$, we have $u_2 \leq (1 + \epsilon)\bar{u}_2$ for t sufficiently large. By [C2], we can choose $\epsilon > 0$ sufficiently small such that the eigenvalue $\hat{\rho}_1(\sigma_1\Delta + a_1 - c_{12}(1 + \epsilon)\bar{u}_2) > 0$ with a positive principal eigenfunction ϕ_1; and let t_0 be sufficiently large such that $u_2 < (1 + \epsilon)\bar{u}_2$ for $t \geq t_0$. Next, let $\delta > 0$ be sufficiently small so that $u_1(x, t_0) \geq \delta\phi_1(x)$ in $\bar{\Omega}$; and define $\underline{w}_1 = \delta\phi_1$, $\bar{w}_2 = (1 + \epsilon)\bar{u}_2$. One can readily verify that

$$\text{(1.5)}\quad \begin{cases} \sigma_1 \Delta \underline{w}_1 + (a_1 - \underline{w}_1 - c_{12}\bar{w}_2)\underline{w}_1 \geq 0 & \text{in } \Omega, \\ \sigma_2 \Delta \bar{w}_2 + (a_2 - \bar{w}_2 - c_{21}\underline{w}_1)\bar{w}_2 \leq 0 & \text{in } \Omega, \\ u_1(x, t_0) \geq \underline{w}_1(x), \;\; u_2(x, t_0) \leq \bar{w}_2(x) & \text{in } \Omega. \end{cases}$$

Finally, let $\underline{w}_2 \equiv 0$ and $\bar{w}_1 = (1+C)\bar{u}_1$ for some large C so that $u_1(x, t_0) \leq \bar{w}_1(x)$ in $\bar{\Omega}$. We thus have

$$\text{(1.6)}\quad \begin{cases} \sigma_1 \Delta \bar{w}_1 + (a_1 - \bar{w}_1 - c_{12}\underline{w}_2)\bar{w}_1 \leq 0 & \text{in } \Omega, \\ \sigma_2 \Delta \underline{w}_2 + (a_2 - \underline{w}_2 - c_{21}\bar{w}_1)\underline{w}_2 \geq 0 & \text{in } \Omega, \\ \underline{w}_1(x) \leq u_1(x, t_0) \leq \bar{w}_1(x), \;\; \underline{w}_2(x) \leq u_2(x, t_0) \leq \bar{w}_2(x) & \text{in } \Omega. \end{cases}$$

Let $(v_1(x,t), v_2(x,t))$ be a solution of

$$\text{(1.7)}\quad \begin{cases} v_{it} = \sigma_1 \Delta v_i + (a_i - v_i - c_{ij}v_j)v_i, & j \neq i, \;\; i = 1, 2 \text{in } \Omega \times (0, \infty), \\ v_i(x,t) = 0 & (x,t) \in \partial\Omega \times (0, \infty), \\ v_1(x, 0) = \underline{w}_1(x), \;\; v_2(x, 0) = \bar{w}_2(x), & x \in \Omega. \end{cases}$$

By the special monotone property of the reaction term of the competing relation, we can deduce by comparison, as in [125] or [183] that

$$v_1 \leq u_1 \leq \bar{w}_1, \quad v_2 \geq u_2 \geq \underline{w}_2 \equiv 0, \quad \text{for } \; (x,t) \in \bar{\Omega} \times \infty;$$

and further $v_1 \uparrow v_1^*, v_2 \downarrow v_2^*$ as $t \to \infty$ where (v_1^*, v_2^*) is an equilibrium solution of (1.7) without initial condition. By hypothesis [C2], we must have $v_2^* \equiv 0$ and $v_1^* = \bar{u}_1$.

Lemma 1.3. *Assume hypotheses [C1]. Suppose that there exists a global attractor (u_1^*, u_3^*) for solutions (with each component nontrivial and non-negative) of*

$$\text{(1.8)}\quad \begin{cases} \partial u_1/\partial t = \sigma_1 \Delta u_1 + u_1(a_1 - u_1 - c_{13}u_3), \\ \partial u_3/\partial t = \sigma_3 \Delta u_3 + u_3(a_3 + c_{31}u_1 - u_3), \\ u_i(x,t) = 0, \;\; i = 1, 3 \qquad \textit{for } (x,t) \in \partial\Omega \times [0, \infty), \end{cases}$$

where σ_i, a_i, c_{ij} satisfy conditions for problem (1.4); and each $u_i^* \not\equiv 0$ in $\bar{\Omega}$, for $i = 1, 3$. Then, the eigenvalue with homogeneous Dirichlet boundary condition satisfies

$$\text{(1.9)}\qquad \hat{\rho}_1(\sigma_3 \Delta + (a_3 + c_{31}\bar{u}_1)) > 0.$$

If the system (1.8) has a positive equilibrium $(0, \bar{u}_3), \bar{u}_3 \not\equiv 0$ *in* $\bar{\Omega}$, *then we also have*

$$\hat{\rho}_1(\sigma_1\Delta + (a_1 - c_{13}\bar{u}_3)) > 0. \tag{1.10}$$

Proof. Observe that $\bar{u}_1$ is a strict upper solution of the scalar homogeneous boundary value problem for $\sigma_1\Delta u_1 + (a_1 - c_{13}u_3^* - u_1)u_1 = 0$, which has the solution u_1^*. By means of scalar upper-lower solution theory, we obtain $\bar{u}_1 > u_1^*$ in Ω and $\partial\bar{u}_1/\partial\nu < \partial u_1^*/\partial\nu$ on $\partial\Omega$. Since (u_1^*, u_3^*) is a global attractor for nontrivial non-negative solutions, we must have $u_1 \leq \bar{u}_1$ for large enough t for any solution (u_1, u_3) of (1.8) with each component non-negative and nontrivial. (Recall that we can consider the semigroup of solutions with components in $C^1_{0,+}(\bar{\Omega})$). Suppose contrary to (1.9), we have $\hat{\rho}_1(\sigma_3\Delta + (a_3 + c_{31}\bar{u}_1)) := \sigma_0 \leq 0$, with the corresponding eigenfunction $\phi_0 > 0$ in Ω. Let (u_1, u_3) be any non-negative nontrivial solution of (1.8), so that $u_i \to u_i^*$ as $t \to \infty$ for $i = 1, 3$. We consider the expression:

$$\begin{aligned}\tfrac{d}{dt}\textstyle\int_\Omega \phi_0 u_3 dx &= \textstyle\int_\Omega \phi_0[\sigma_3\Delta u_3 + (a_3 + c_{31}u_1 - u_3)u_3]\,dx \\ &= \textstyle\int_\Omega [\sigma_3\Delta\phi_0 + (a_3 + c_{31}\bar{u}_1)\phi_0]u_3\,dx + \int_\Omega [c_{31}(u_1 - \bar{u}_1) - u_3]\phi_0 u_3\,dx \\ &= \textstyle\int_\Omega \sigma_0\phi_0 u_3 + c_{31}(u_1 - \bar{u}_1)\phi_0 u_3 - \phi_0 u_3^2\,dx.\end{aligned}$$

We have $\sigma_0\phi_0 u_3 \leq 0$ in $\bar{\Omega}$. Moreover, since $u_i \to u_i^*$ as $t \to \infty$ and $u_1^* < \bar{u}_1$ on Ω, $\partial u_1^*/\partial\nu > \partial\bar{u}_1/\partial\nu$ on $\partial\Omega$, there must exist large $T_0 > 0$ such that we have $u_1 - \bar{u}_1 \leq 0$, and $u_3 \geq (1/2)u_3^*$ in $\bar{\Omega}$ for $t > T_0$. Consequently, for $t > T_0$, we find

$$\frac{d}{dt}\int_\Omega \phi_0 u_3\,dx \leq -\frac{1}{4}\int_\Omega \phi_0 (u_3^*)^2\,dx := -\delta_0 < 0. \tag{1.11}$$

From (1.11), we readily deduce the fact that $\int_\Omega \phi_0 u_3 dx < 0$ for all large t. However, this fact is impossible because both ϕ_0 and u_3 are non-negative in $\bar{\Omega}$. We thus conclude the validity of (1.9) in order to avoid contradiction.

Suppose the system (1.8) has a positive equilibrium $(0, \bar{u}_3), \bar{u}_3 \not\equiv 0$ in $\bar{\Omega}$. Then u_3^* is a strict upper solution for the problem $\sigma_3\Delta w + (a_3 - w)w = 0$ in Ω, $w = 0$ on $\partial\Omega$; and thus $u_3^* > \bar{u}_3$ in Ω. Hence $a_1 - c_{13}u_3^* < a_1 - c_{13}\bar{u}_3$ in Ω, and the characterization of principal eigenvalues gives $\hat{\rho}_1(\sigma_1\Delta + (a_1 - c_{13}u_3^*)) < \hat{\rho}_1(\sigma_1\Delta + (a_1 - c_{13}\bar{u}_3))$. However, the fact that u_1^* is a positive solution of $\sigma_1\Delta w + (a_1 - c_{13}u_3^* - w) = 0$ with homogeneous boundary condition implies that $\hat{\rho}_1(\sigma_1\Delta + (a_1 - c_{13}u_3^*)) > 0$. Consequently, (1.10) must hold.

Lemma 1.4. *Suppose* $f \in C^2(\bar{\Omega}, R)$ *and* $\sigma > 0$. *Let* $\rho = \hat{\rho}_1(\sigma\Delta + f(x))$ *be the principal eigenvalue for the problem:*

$$\sigma\Delta w + f(x)w = \rho w \text{ in } \Omega, \quad \phi = 0 \text{ on } \partial\Omega; \tag{1.12}$$

and assume $\hat{\rho}_1 > 0$ with corresponding positive eigenfunction ϕ. Suppose that there exist some constants $k > 0$ and $\epsilon \in (0, \hat{\rho}_1)$, so that $u(x,t)$ satisfies the following in a neighborhood U of 0 in $C^1_{0,+}(\bar{\Omega})$:

$$
\begin{cases} \frac{\partial u}{\partial t} \geq \sigma \Delta u + [f(x) - \epsilon]u & \text{for } x \in \Omega,\ t > 0, \\ u(x,t) = 0 \ \ \text{for } x \in \partial\Omega, & u(x,0) \geq k\phi(x) \ \ \text{for } x \in \bar{\Omega}, \end{cases} \tag{1.13}
$$

as long as it exist. Then as long as $u(x,t) \in U$, the following inequality is satisfied

$$
u(x,t) \geq k e^{(\hat{\rho}_1 - \epsilon)t} \phi(x) \tag{1.14}
$$

for $x \in \bar{\Omega}$.

Proof. For $(x,t) \in \bar{\Omega} \times [0, \infty)$, define $v(x,t) = k e^{(\hat{\rho}_1 - \epsilon)t}\phi(x)$. We readily verify

$$
\begin{cases} \frac{\partial v}{\partial t} = \sigma \Delta v + [f(x) - \epsilon]v & \text{for } x \in \Omega,\ t > 0, \\ v(x,t) = 0 \ \ \text{for } x \in \partial\Omega, & v(x,0) = k\phi(x) \ \ \text{for } x \in \bar{\Omega}. \end{cases} \tag{1.15}
$$

By comparison of scalar parabolic equations, we obtain (1.14) from (1.13) and (1.15). (We are considering a neighborhood in $C^1_{0,+}$, but the result is still valid in $C^0_{0,+}(\bar{\Omega})$.)

Proof of Theorem 1.3. Since point dissipativity in $C^1_{0,+}(\bar{\Omega})$ follows from Lemma 1.1, we can prove this theorem by applying Theorem 1.1 accordingly, by choosing $X = X^0 \cup \partial X^0 = [C^1_{0,+}(\bar{\Omega})]^3$. The hypotheses of this theorem imply that the ω-limit set of the boundary consists exactly of the equilibria: $\bigcirc = (0,0,0), A_1 = (\bar{u}_1, 0, 0), A_2 = (0, \bar{u}_2, 0), P_1 = (\hat{u}_1, 0, \hat{u}_3), P_2 = (0, \tilde{u}_2, \tilde{u}_3)$ and $Q = (0, 0, \bar{u}_3)$ (if this exists). We thus proceed to take the isolated covering $\cup M_i$ (as described above) to be these points themselves. In order to apply Theorem 1.1, we need to show that: (i) this covering is isolated, (ii) $W^s(M_i) \cap X^0 = \Phi$, and (iii) the covering is acyclic. We now consider properties (i) and (ii) for each equilibrium. The proof for each equilibrium is the same, and we illustrate the detail arguments for the point A_1. In order to show that A_1 is isolated, we must show that there is a neighborhood of A_1 which does not contain a full orbit (other than A_1) itself. Suppose the contrary; that is, every neighborhood of A_1 contains a full orbit.

First, assume such an orbit lies in the u_1 axis in a small neighborhood of A_1. The first half of hypothesis [C1] implies that A_1 is a stable equilibrium which is a global attractor for nontrivial positive orbits on the axis. The α-limit set of this orbit is nonempty by compactness; and the set is either A_1 itself or contains a point other than A_1. Since all points are attracted to A_1, the α-limit set must then include a forward orbit tending to A_1. Let $q \neq A_1$ be a point on the full orbit first considered, the stability of the solution A_1 implies that there is a small neighborhood N of A_1 such that any orbit starting inside N can never reach as far as q from A_1 in forward time. Choose a large enough negative time so that the orbit first considered is inside N, then the forward orbit cannot contain the point q. This shows that the full orbit on the u_1 axis cannot contain a point other than A_1. Next, by [C1],[C3] and $\hat{\rho}_1(\sigma_1\Delta + (a_1 - 2\bar{u}_1)) < 0$, we can deduce by the linearized system at A_1 and Theorem A4-11 in Chapter 6 that the equilibrium A_1 is stable in $C^1_{0,+}(\bar{\Omega})$ on the $u_1 - u_2$ face. Thus, [C1], [C2],[C3] and Lemma 1.2 together imply that on the $u_1 - u_2$ face, A_1 is a $C^1_{0,+}(\bar{\Omega})$ stable equilibrium which is a global attractor for non-negative solutions not on the u_2 axis. We can therefore use the same argument above to assert that the full orbit in a neighborhood of A_1 on the $u_1 - u_2$ face cannot contain a point other than A_1. Consequently, if the full orbit in the neighborhood contains a point other the A_1, we must have $u_3 > 0$ for $x \in \Omega$. We now show that such an orbit must leave every sufficiently small neighborhood U of A_1. By maximum principle, any orbit with $u_3 \geq 0$ and $u_3 \not\equiv 0$ must satisfy $u_3(x, t_0) \geq k\tilde{\phi}(x)$ for some $t_0 > 0$ and $k > 0$. (Here, $\tilde{\phi}$ is the positive eigenfunction for the operator and principal eigenvalue described in (1.9). Note that hypothesis [C4] and Lemma 1.3 imply that $\hat{\rho}_1(\sigma_3\Delta + a_3 + c_{31}\bar{u}_1) > 0$). From the third equation in (1.4), we see that for any $\epsilon > 0$, there is a neighborhood U of $A_1 = (\bar{u}_1, 0, 0)$ so that in U, $\partial u_3/\partial t - \sigma_3\Delta u_3 - u_3(a_3 + c_{31}\bar{u}_1 - \epsilon) \geq 0$. Choose $\epsilon \in (0, \hat{\rho}_1(\sigma_3\Delta + (a_3 + c_{31}\bar{u}_1))$, and U to be such a corresponding neighborhood. From Lemma 1.4, the component u_3 must increase until (u_1, u_2, u_3) leaves U, so that U cannot contain a full orbit. We thus conclude that A_1 is isolated. A similar argument shows that $W^s(A_1) \cap X^0 = \Phi$, since at any point of X^0 sufficiently close to A_1, the component u_3 must increase so that (u_1, u_2, u_3) cannot approach A_1 along $W^s(A_1)$. (Note that in order to show that $P_i, i = 1, 2$ is isolated and $W^s(P_i) \cap X^0 = \Phi$, we will need to use hypotheses [C5]).

It remains to show that there does not exist any cycle in the boundary as defined above. We first consider the case when the equilibrium Q exists. Since P_1 and P_2 are attracting and stable as described in [C4], they cannot form part of any cycle on the boundary faces. Also, by [C1]-[C4]and Lemma 1.2, the solutions starting near the origin on the faces are all attracted away to other equilibria, thus the origin cannot form part of a cycle too. Therefore we only need to consider the equilibria A_1, A_2 and Q. On the $u_1 - u_3$ face (apart from the axes), all solutions are attracted to P_1, thus A_1 cannot be chained to itself, Q to itself, nor can A_1 and Q be chained by an orbit on the $u_1 - u_3$ face. Similarly, on the $u_2 - u_3$ face, we find A_2, Q cannot be chained to themselves or each other. The only other possibilities are that A_1, A_2 are cyclic or A_i is chained to itself on the $u_1 - u_2$ face. However, by Lemma 1.2, all solutions on this face are attracted to A_1. Moreover, by [C3] and $\hat{\rho}_1(\sigma_1\Delta + (a_1 - 2\bar{u}_1)) < 0$, we can deduce by the linearized system at A_1 and Theorem A4-11 in Chapter 6 that the equilibrium A_1 is stable on this face. Thus, both possibilities cannot happen. We thus conclude that there cannot be any cycle on the boundary faces. Finally, in the situation when Q does not exist, there are even fewer possible cycles to consider. They can all be shown to be impossible by exactly the same arguments just given. This completes the proof of Theorem 1.3.

In order to see how Theorem 1.3 can be readily applied, we consider various situations when condition [C4] concerning dynamics for the reduced problem on the boundary faces can be satisfied. This will be studied in Theorems 1.4 and 1.5 below. Recall that in Theorem 5.4 of Chapter 1, we consider

$$\begin{cases} \Delta u + u(a - bu - cv) = 0 \\ \qquad\qquad\qquad\qquad\qquad\qquad \text{in } \Omega, \\ \Delta v + v(e + fu - gv) = 0 \\ u = v = 0 \quad \text{on } \partial\Omega. \end{cases} \tag{1.16}$$

We found that the boundary value problem (1.16) under hypotheses:

$$\begin{cases} a > \lambda_1, \ e > \lambda_1, \\ cf < gb, \qquad\qquad\qquad\qquad \text{and} \\ a > gb(gb - cf)^{-1}[\lambda_1 + ce/g] \end{cases} \tag{1.17}$$

has a unique coexistence solution with each component strictly positive in Ω, and in $C^{2+\alpha}(\bar{\Omega})$, provided that

$$cf < k(bg), \tag{1.18}$$

for some sufficiently small constant $k \in (0, 1)$. Note that by reducing cf while holding all other parameters fixed, all the inequalities in hypotheses (1.17) remain valid.

Let a, b, C, e, F, g be positive constants such that (1.17) holds with c, f respectively replaced by C, F. Define $\hat{U}, \tilde{U}, \hat{V}, \tilde{V} \in C^{2+\alpha}(\Omega)$ be strictly positive functions in Ω satisfying the following scalar problems:

$$\begin{aligned} &\Delta\hat{U} + \hat{U}(a - b\hat{U}) = 0 \text{ in } \Omega, \ \hat{U} = 0 \text{ on } \partial\Omega, \\ &\Delta\hat{V} + \hat{V}(e + \tfrac{Fa}{b} - g\hat{V}) = 0 \text{ in } \Omega, \ \hat{V} = 0 \text{ on } \partial\Omega, \\ &\Delta\tilde{U} + \tilde{U}(a - b\tilde{U} - C\hat{V}) = 0 \text{ in } \Omega, \ \tilde{U} = 0 \text{ on } \partial\Omega, \\ &\Delta\tilde{V} + \tilde{V}(e - g\tilde{V}) = 0 \text{ in } \Omega, \ \tilde{V} = 0 \text{ on } \partial\Omega. \end{aligned} \tag{1.19}$$

Note that $\hat{U}, \hat{V}, \tilde{V}$ exist because $a, e, e+Fa/b$ are $> \lambda_1$; and $\hat{U}, \hat{V}, \tilde{V}$ are $\geq \delta\phi > 0$ in Ω for sufficiently small $\delta > 0$. Here ϕ is a positive eigenfunction for the principal eigenvalue λ_1 of the Laplacian with zero Dirichlet boundary data. One can readily deduce by upper lower solutions method that $\hat{V}(x) \leq \frac{1}{g}(e + \frac{Fa}{b})$, hence $a - C\hat{V} \geq a - \frac{C}{g}(e + \frac{Fa}{b}) > \lambda_1$ for all $x \in \bar{\Omega}$. Consequently, we obtain

$$0 < \delta\phi \leq \tilde{U} \leq \hat{U}, \;\; 0 < \delta\phi \leq \tilde{V} \leq \hat{V}$$

for $x \in \Omega, \delta > 0$ sufficiently small. It is shown in the proof of Theorem 5.4 of Section 5.5 that if $c \in (0, C)$ and $f \in (0, F)$ and cf satisfies (1.18), then the unique positive solution $(u, v) = (u^*, v^*)$ for (1.16) described above must have the property:

$$\tilde{U}(x) < u^*(x) < \hat{U}(x), \;\; \tilde{V}(x) < v^*(x) < \hat{V}(x), \tag{1.20}$$

for all $x \in \Omega$. Note that since the outward normal derivatives of ϕ are negative on the boundary, there must exist a constant $\bar{K} > 0$ such that

$$\hat{U} \leq \bar{K}\tilde{U}, \;\; \hat{V} \leq \bar{K}\tilde{V}, \;\; \hat{U} \leq \bar{K}\tilde{V}, \;\; \hat{V} \leq \bar{K}\tilde{U} \tag{1.21}$$

for all $x \in \bar{\Omega}$. The proof of Theorem 5.4 uses contraction argument, and is accomplished by choosing

$$k = (\frac{1}{\bar{K}})^4. \tag{1.22}$$

We now consider the corresponding parabolic problem for (1.16)

$$\begin{cases} u_t = \Delta u + u(a - bu - cv) & \\ & \text{in } \Omega \times [0, \infty), \\ v_t = \Delta v + v(e + fu - gv) & \\ & \\ u(x,t) = v(x,t) = 0 & \text{on } \partial\Omega \times [0, \infty). \end{cases} \tag{1.23}$$

Theorem 1.4 (Attractor for Reduced Problem). *Let a, b, C, e, F, g satisfy the conditions in (1.17) described above with c, f replaced respectively by C, F; and let $\tilde{U}(x), \bar{U}(x), \tilde{V}(x), \bar{V}(x)$ be functions defined by (1.19). Suppose that*

$$c \in (0, C), \;\; f \in (0, F), \;\; \textit{and} \;\; cf < k(bg), \tag{1.24}$$

where k described in (1.24) satisfies (1.22), ensuring (1.16) has a unique positive equilibrium solution as described above. Let $(u(x,t), v(x,t))$ be a solution of (1.23) with $u(x,0) \geq 0, \not\equiv 0$ and $v(x,0) \geq 0, \not\equiv 0$ in $\bar{\Omega}$. Then for any $\epsilon > 0$, there exists a $t_\epsilon > 0$ such that for $t \geq t_\epsilon$, we have:

$$(1-\epsilon)\tilde{U}(x) \leq u(x,t) \leq (1+\epsilon)\hat{U}(x), \;\; (1-\epsilon)\tilde{V}(x) \leq v(x,t) \leq (1+\epsilon)\hat{V}(x), \tag{1.25}$$

for all $x \in \Omega$.

Proof. The first component of the solution of (1.23) is a lower solution of $w_t = \Delta w + w(a - bw)$ in Ω, with $w(x,0) = u(x,0)$ on $\bar{\Omega}$ and zero homogeneous boundary condition. Thus by the $C^1(\bar{\Omega})$ convergence of $w(\cdot,t)$ to $\hat{U}(x)$ and comparison, we find that for any $\epsilon > 0$, we have

$$u(x,t) \leq (1+\epsilon)\hat{U}(x)$$

in $\bar{\Omega}$ for t sufficiently large. Hence for t large, $z = v$ is a lower solution of $z_t = \Delta z + z[e + f(1+\epsilon)\hat{U}(x) - gz]$ in Ω, with homogeneous boundary conditions. Moreover for small enough $\epsilon > 0$, we have $f(1+\epsilon)\hat{U}(x) \leq Fa/b$ in $\bar{\Omega}$. Thus by comparison again, we obtain

$$v(x,t) \leq (1+\epsilon)\hat{V}(x)$$

in $\bar{\Omega}$ for t sufficiently large. For small enough $\epsilon > 0$ and for t large, $w = u$ is an upper solution of $w_t = \Delta w + w[a - bw - C\hat{V}]$ in Ω. Since $\tilde{U}$ is a global attractor, we obtain by comparison that

$$u(x,t) \geq (1-\epsilon)\tilde{U}(x)$$

in $\bar{\Omega}$ for t sufficiently large. Finally, $z = v$ is an upper solution of $z_t = \Delta z + z(e - gz$ in Ω for any $\epsilon > 0$, we obtain by comparison that

$$v(x,t) \geq (1-\epsilon)\tilde{V}(x)$$

in $\bar{\Omega}$ for a large t.

Theorem 1.5 (Globally Asymptotically Stable Equilibrium for Reduced Problem). *Let a, b, C, e, F, g be as described in Theorem 1.4. Then there exist $\delta_1 \in (0, C)$ and $\delta_2 \in (0, F)$ such that if c, f satisfies hypotheses (1.24) in Theorem 1.4, and furthermore*

$$c \in (0, \delta_1), \quad f \in (0, \delta_2), \tag{1.26}$$

then the unique coexistence positive equilibrium (u^*, v^*) of (1.16) is a global attractor for all nontrivial non-negative solutions of (1.23). Here, $(u(x,t), v(x,t)) \to (u^*, v^*)$ in $[C_0^1(\bar{\Omega})]^2$; and (u^*, v^*) is a stable equilibrium in $[C_0^0(\bar{\Omega})]^2$, if initial conditions are close to the equilibrium in $[C_0^1(\bar{\Omega})]^2$.

Proof. Recall that under conditions (1.24), the system (1.16) has a unique coexistence positive solution (u^*, v^*) which always satisfy (1.20) for any such small c and f. Moreover, for any solution $(u(x,t), v(x,t))$ of (1.23) with each component nontrivial non-negative in $C^0(\bar{\Omega})$ at $t = 0$, and any $\epsilon > 0$, it must satisfy (1.25) for large enough t. Let

$$p(x,t) := u(x,t) - u^*(x) \quad q(x,t) := v(x,t) - v^*(x).$$

We have

$$(1.27)\qquad \begin{cases} p_t = \Delta p + (a - bu^* - cv^*)p - bup - cuq \\ \\ q_t = \Delta q + (e + fu^* - gv^*)q + fvp - gvq \end{cases}$$

in $\Omega \times (0, \infty), p = q = 0$ on $\partial\Omega \times (0, \infty)$. Since $u^* > 0$ is a solution of the eigenvalue problem: $\Delta\psi + (a - bu^* - cv^*)\psi = \rho\psi$ with $\rho = 0$, we must have $\rho = 0$ as the principal eigenvalue. Thus by characterization of eigenvalues, we have

$$(1.28)\qquad \int_\Omega [|\nabla\psi|^2 + (a - bu^* - cv^*)\psi^2]dx \le 0$$

for any $\psi \in W_0^{1,2}(\Omega)$, and similarly,

$$(1.29)\qquad \int_\Omega [|\nabla\psi|^2 + (e + fu^* - gv^*)\psi^2]dx \le 0.$$

Multiplying the first and second equation in (1.27) respectively by p and q, adding and using (1.28) and (1.30) we deduce.

$$(1.30)\qquad \frac{d}{dt}(\frac{1}{2}\int_\Omega (p^2 + q^2)dx) \le -\int_\Omega [bup^2 + (cu - fv)pq + gvq^2]dx.$$

The quadratic expression inside the integral on the right above is positive definite if $(cu - fv)^2 - 4bugv < 0$, or equivalently

$$(1.31)\qquad c^2(\frac{u}{v})^2 - (2cf + 4bg)(\frac{u}{v}) + f^2 < 0.$$

Thus at any $t > 0$, if

(1.32)

$$\frac{(2cf + 4bg) - \sqrt{16bg(bg + cf)}}{2c^2} < \frac{u(x,t)}{v(x,t)} < \frac{(2cf + 4bg) + \sqrt{16bg(bg + cf)}}{2c^2}$$

for all $x \in \bar{\Omega}$, then the expression in (1.30) is negative, unless $p(x,t) \equiv q(x,t) \equiv 0$ for $x \in \bar{\Omega}$. For fixed $b > 0, g > 0$, denote

$$K_1(c,f) := \frac{(2cf+4bg)-\sqrt{16bg(bg+cf)}}{2c^2} = \frac{2f^2}{(2cf+4bg)+\sqrt{16bg(bg+cf)}},$$

$$K_2(c,f) := \frac{(2cf+4bg)+\sqrt{16bg(bg+cf)}}{2c^2}.$$

We see $K_1(c,f) \to 0$ as $f \to 0$, and $K_2(c,f) \to +\infty$ as $c \to 0$. Consequently, from (1.25) and (1.21), we must have (1.32) holds for all large enough t, provided

(1.26) is valid for some small enough δ_1, δ_2. By (1.30), we deduce that $(p,q) \to (0,0)$ in $[L^2(\Omega)]^2$ as $t \to \infty$. By $W^{2,p}$ estimates for parabolic equations, we obtain $(p,q) \to (0,0)$ in $[C_0^1(\bar{\Omega})]^2$. The last assertion concerning stability in $[C_0^0(\bar{\Omega})]^2$ follows from (1.30) again.

In order to find a simple example so that condition [C2] of Theorem 1.3 applies, we consider:

$$
\begin{cases}
\Delta u_1 + u_1(a - u_1 - c_{12}u_2) = 0 & \text{in } \Omega, \\
\Delta u_2 + u_2(a - c_{21}u_1 - u_2) = 0 & \\
u = v = 0 & \text{on } \partial\Omega
\end{cases}
\tag{1.33}
$$

where a, c_{12}, c_{21} are constants satisfying

$$
\begin{aligned}
& a > \lambda_1, \\
& 0 < c_{12} < 1 < c_{21}.
\end{aligned}
\tag{1.34}
$$

Lemma 1.5 (Non-Coexistence for Another Reduced Problem). *The problem (1.33) under hypotheses (1.34) does not have any coexistence positive solution such that both components are non-negative and $\not\equiv 0$ in $\bar{\Omega}$. Moreover, let $\bar{u}$ denote the solution of $\Delta\bar{u} + \bar{u}(a - \bar{u}) = 0$ in Ω with homogeneous Dirichlet boundary condition, then the principal eigenvalue for the operator $\Delta + a - c_{12}\bar{u}$ satisfies*

$$
\hat{\rho}_1(\Delta + a - c_{12}\bar{u}) > 0.
\tag{1.35}
$$

Proof. We have $\hat{\rho}_1(\Delta + a - \bar{u}) = 0$, because $\bar{u}$ is a positive solution of $\Delta\bar{u} + \bar{u}(a - \bar{u}) = 0$ in Ω with homogeneous Dirichlet boundary condition. By comparison of principal eigenvalues with Rayleigh's quotient and (1.34), we have $\hat{\rho}_1(\Delta + a - c_{12}\bar{u}) > 0$ and $\hat{\rho}_1(\Delta + a - c_{21}\bar{u}) < 0$. By Theorem 3.5 of Chapter 1, this difference in sign implies that problem (1.33) does not have any coexistence positive solution such that both components are non-negative and $\not\equiv 0$ in $\bar{\Omega}$.

Example 1.1 (Application of Persistence Theorem 1.3 to Full System). Consider the problem for $(x,t) \in \Omega \times (0,\infty)$:

$$
\begin{cases}
\partial u_1/\partial t = \Delta u_1 + u_1(a - u_1 - c_{12}u_2 - \epsilon_1 u_3), \\
\partial u_2/\partial t = \Delta u_2 + u_2(a - c_{21}u_1 - u_2 - \epsilon_2 u_3), \\
\partial u_3/\partial t = \Delta u_3 + u_3(e + \delta_1 u_1 + \delta_2 u_2 - u_3), \\
u_i(x,t) = 0, \ i = 1,2,3 \qquad \text{for } (x,t) \in \partial\Omega \times [0,\infty)
\end{cases}
\tag{1.36}
$$

where $a, c_{12}, c_{21}, \epsilon_1, \epsilon_2, \delta_1$ and δ_2 are positive constants satisfying

$$e > \lambda_1, \quad 0 < c_{12} < 1 < c_{21} \tag{1.37}$$

$$a > \frac{\lambda_1 + \sqrt{\lambda_1^2 + 4c_{21}}}{2}. \tag{1.38}$$

From Theorem 1.3, we can readily deduce that the non-negative solutions of this problem are uniformly persistent provided that ϵ_i and δ_i are sufficiently small, for $i = 1, 2$. We simply verify all the conditions [C1] to [C5] for Theorem 1.3. The assumptions here on a, e imply [C1]. The hypotheses in [C2] are valid due to the second part of (1.37) and Lemma 1.5. [C3] is true due to the assumption $1 < c_{21}$. Theorem 1.5 implies [C4] is valid for sufficiently small $\epsilon_i, \delta_i, i = 1, 2$. (Note that the last inequality in (1.17) will be satisfied if $a > \lambda_1$ and ϵ_i are small enough). From (1.36), we readily obtain the components $\hat{u}_1(x)$ and $\tilde{u}_2(x)$ defined in [C4], [C5] satisfy $\hat{u}_1(x) < \frac{1}{a}$ and $\tilde{u}_2(x) < \frac{1}{a}$ for all $x \in \bar{\Omega}$. Thus assumption (1.38) and the roots of quadratic equations imply that [C5] is valid for sufficiently small $\epsilon_i, i = 1, 2$ by comparing principal eigenvalues.

Part B: Average Liapunov Functions.

We now introduce another useful tool for analyzing persistence or permanence for reaction-diffusion systems. Recall we assume the metric space X satisfies $X = X^0 \cup \partial X^0$ where X^0 is open in X, which is the closure of X^0. ∂X^0 (assumed to be nonempty) is the boundary of X^0. We suppose that the C_0-semigroup $T(t)$ on X satisfies

$$T(t) : X^0 \to X^0, \qquad T(t) : \partial X^0 \to \partial X^0 \tag{1.39}$$

and let $T_0(t) = T(t)|_{X^0}, T_\partial(t) = T(t)|_{\partial X^0}$. Let U, V be subsets of X, we denote the distances

$$\bar{d}(U, V) = sup_{u \in U}\, d(u, V), \;\; \underline{d}(U, V) = inf_{u \in U}\, d(u, V).$$

The following terminology is commonly used for $T(t)$ with the above properties:

Definition 1.2. The semigroup $T(t)$ is said to be permanent if there exists a bounded set U with $\underline{d}(U, \partial X^0) > 0$ such that $\lim_{t \to \infty} d(T(t)v, U) = 0$ for all $v \in X^0$.

Note that if the boundedness condition on U above is removed, then $T(t)$ is uniformly persistent as defined earlier. Consequently, uniform persistence is less stringent than permanent. However, we have actually proved the following stronger result in the proof of Theorem 1.3.

Remark 1.1. *Assume hypotheses [C1]-[C5], then the semigroup $T(t)$ of solutions of (1.4) in $X = X^0 \cup \partial X^0 = [C^1_{0,+}(\bar{\Omega})]^3$ is permanent.*

We next introduce another abstract permanence theorem which is based on the constructive use of a so called 'average' Liapunov function P. The method can be applied to systems for which we cannot rule out the possibility of a cycle, and thus unable to find an acyclic cover of $\omega(\partial X^0)$ for application of Theorem 1.3. We construct a function P whose value depends on the size of the solution $u_i(x,t)$ of the system, and study its variations as t changes. We analyze the behavior of the function starting at different initial states, in particular near the attractor of the semigroup and the boundary of the positive cone of functions. The system will tend to be permanent if the boundary of the positive cone acts as a repeller for the system.

More precisely, let $X = X^0 \cup \partial X^0$ be a complete metric space, and a semigroup $T(t)$ is point dissipative in X with properties (1.39) as described. Assume there is a $t_0 \geq 0$ such that $T(t)$ is compact for $t > t_0$, then by Theorem A4-13 in Chapter 6, there is a nonempty global attractor A in X. Let $B(A, \epsilon)$ be an ϵ neighborhood of A, and define the closure $Y := Cl\ T(B(A,\epsilon), [1,\infty))$, $S := Y \cap \partial X^0$.

Theorem 1.6 (Abstract Permanence Theorem). *Let X, $T(t)$, Y and S be as described in the last paragraph. (Note the point dissipative and compactness condition on $T(t)$). Suppose that Y, S are compact and for $u \in S, t > 0$, we define*

$$\alpha(t,u) = \lim inf_{v \to u,\, v \in Y \setminus S} \frac{P(T(t)v)}{P(v)},$$

where $P : Y \setminus S \to R^+$ is a continuous, strictly positive and bounded function. If the function $\alpha(t,u)$ has the property:

$$\begin{cases} sup_{t>0}\, \alpha(t,u) > 1 & \text{if } u \in \omega(S), \\ sup_{t>0}\, \alpha(t,u) > 0 & \text{if } u \in S, \end{cases} \tag{1.40}$$

then the semigroup $T(t)$ is permanent.

We will apply this theorem in the context of parabolic systems with $X = [C^1_{0,+}(\bar{\Omega}]^n$. Note that the theorem assumes that the semigroup is dissipative. This can be established by showing the solutions satisfy uniform bound of the form $0 \leq u_i \leq K_i$ for large t. The smoothing properties of the semigroup imply dissipativity in $C^1(\bar{\Omega})$. That is, we will have Y and S are compact. The theorem do not require any special assumptions concerning monotonicity of the interaction terms, uniqueness of coexistence states or a globally defined Liapunov function. However to construct the average Liapunov function, we must have a detailed knowledge of the semigroup generated on the boundary of the positive cone of functions.

Example 1.2. Consider the following system for $(x,t) \in \Omega \times (0,\infty)$,

$$(1.41) \qquad \begin{cases} \partial u_1/\partial t = \sigma \Delta u_1 + u_1(1 - u_1 - \alpha u_2 - \beta u_3), \\ \partial u_2/\partial t = \sigma \Delta u_2 + u_2(1 - \beta u_1 - u_2 - \alpha u_3), \\ \partial u_3/\partial t = \sigma \Delta u_3 + u_3(1 - \alpha u_1 - \beta u_2 - u_3), \\ u_i(x,t) = 0, \ i = 1,2,3 \qquad \text{for } (x,t) \in \partial\Omega \times [0,\infty), \end{cases}$$

where σ, α, β are all positive constants, satisfying the following

$$(1.42) \qquad 0 < \sigma < \frac{1}{\lambda_1}, \quad 0 < \alpha < 1 < \beta,$$

where λ_1 is the principal eigenvalue for the operator $-\Delta$ on Ω. The domain $\Omega \subset R^N$ is bounded and open, with $\partial\Omega$ uniformly $C^{3+\nu}$ for some $\nu > 0$. Here, we have three competing species. Each species is more competitive to one and less competitive to another in a cyclic fashion, as indicated by the sizes of α and β. By the assumption on σ, there exists a unique positive function $\bar{u} > 0$ in Ω satisfying:

$$(1.43) \qquad \sigma \Delta \bar{u} + \bar{u}(1 - \bar{u}) = 0 \text{ in } \Omega, \quad \bar{u} = 0 \text{ on } \partial\Omega.$$

Let $\rho = \hat{\rho}_1$ and $\phi > 0$ be the principal eigenvalue and a corresponding positive eigenfunction to the eigenvalue problem

$$(1.44) \qquad \sigma \Delta \phi + (1 - \alpha \bar{u})\phi = \rho \phi \text{ in } \Omega, \quad \phi = 0 \text{ on } \partial\Omega.$$

From the assumption that $0 < \alpha < 1$ and comparison with problem (1.43), we must have $\hat{\rho}_1 = \hat{\rho}_1(\sigma\Delta + 1 - \alpha\bar{u}) > 0$. We will proceed to construct an average Liapunov function for the problem (1.41) and apply Theorem 1.6 to analyze the relative sizes of $\hat{\rho}_1, \alpha$, and β which can ensure permanence for the system. As in the analysis of problem (1.4), the problem (1.41) is considered in the first quadrant of functions $[C^1_{0,+}(\bar{\Omega})]^3$. By comparison, we can show that the solutions are bounded as described above. We may choose Y so that the set S for Theorem 1.6 consists essentially of part of the three 'faces' obtained by setting successively $u_1 = 0, u_2 = 0$ and $u_3 = 0$. For the average Liapunov function, we define for $v = (v_1, v_2, v_3)$:

$$(1.45) \qquad P(v) = \prod_{i=1}^{3} \int_\Omega \phi\, v_i \, dx = exp\{\sum_{i=1}^{3} log \int_\Omega \phi v_i \, dx\},$$

where ϕ is the eigenfunction defined in (1.44). Direct computation shows

$$\frac{d}{ds}(logP(v\cdot s)) = \sum_{i=1}^{3} \frac{\int_\Omega \phi v_{is}\, dx}{\int_\Omega \phi v_i\, dx},$$

$$\frac{P(v\cdot t)}{P(v)} = exp\{\int_0^t ds \sum_{i=1}^{3}[\int_\Omega (\phi v_{is} dx)/(\int_\Omega \phi v_i dx)]\}. \tag{1.46}$$

In order to apply Theorem 1.6, we have to take limits as v tends to points in $\omega(S)$. We will only perform the formal calculations here. In view of the smoothness of solutions of the parabolic systems, the rigorous analytic justification can be done readily. More details can be found in [21]. Since $\hat{\rho}_1(\sigma\Delta+1-\alpha\bar{u}) > 0$, and $\hat{\rho}_1(\sigma\Delta+1-\beta\bar{u}) < 0$ are of different sign, we see from Theorem 3.5 in Chapter 1 that on the $u_1 - u_2$ face there is no coexistence solution with both components positive. Similarly we have the same situation on the $u_2 - u_3$ and $u_3 - u_1$ faces. That is the set $\omega(S)$ consists only of the equilibrium at the origin, and the three other equilibria $(\bar{u},0,0)$, $(0,\bar{u},0)$, $(0,0,\bar{u})$ on the axes (cf. Lemma 1.2).

For $(\bar{u},0,0)$, we calculate the first limit in sum on the right of (1.46) by using the first equation in (1.41) and (1.43):

$$\lim inf_{(v_1,v_2,v_3)\to(\bar{u},0^+,0^+)}\{\int_\Omega \phi v_{1t} dx/\int_\Omega \phi v_1 dx\}$$

$$= \lim inf_{(v_1,v_2,v_3)\to(\bar{u},0^+,0^+)}\{\int_\Omega \phi[\sigma\Delta v_1 + v_1(1-v_1-\alpha v_2-\beta v_3)]dx/\int_\Omega \phi v_1 dx\}$$

$$= 0.$$

For the second term in the sum on the right of (1.46), we obtain by means of (1.44):

(1.47)

$$\lim inf_{(v_1,v_2,v_3)\to(\bar{u},0^+,0^+)}\{\int_\Omega \phi v_{2t} dx/\int_\Omega \phi v_2 dx\}$$

$$= \lim inf_{(v_1,v_2,v_3)\to(\bar{u},0^+,0^+)}\{\int_\Omega \phi[\sigma\Delta v_2 + v_2(1-v_2-\alpha v_3-\beta v_1)]dx/\int_\Omega \phi v_2 dx\}$$

$$= \lim inf_{(v_1,v_2,v_3)\to(\bar{u},0^+,0^+)}\{\int_\Omega v_2[\sigma\Delta\phi + \phi(1-v_2-\alpha v_3-\beta v_1)]dx/\int_\Omega \phi v_2 dx\}$$

$$= \lim inf_{v_2\to 0^+}\{\int_\Omega v_2[\sigma\Delta\phi + \phi(1-\beta\bar{u})]dx/\int_\Omega \phi v_2 dx\}$$

$$= \hat{\rho}_1 + \lim inf_{v_2\to 0^+}\{(\alpha-\beta)\int_\Omega \phi v_2\bar{u} dx/\int_\Omega \phi v_2 dx\}.$$

Here, we denote $\hat{\rho}_1 = \hat{\rho}_1(\sigma\Delta+1-\alpha\bar{u})$. The second term on the last line above is the same as

$$(\alpha-\beta)\lim sup_{v_2\to 0^+}\{\int_\Omega \phi v_2\bar{u} dx/\int_\Omega \phi v_2 dx\}.$$

The lim sup term cannot be larger than $||\bar{u}||_0$. Moreover, the maximum value can be attained by choosing v_2 to tend to 0^+ through a sequence of smooth functions with successively smaller supports in a neighborhood of a point where $\bar{u}$ achieves its maximum. Thus the expression in (1.47) is exactly $\hat{\rho}_1 - (\beta - \alpha)||\bar{u}||_0$. As in (1.47), we deduce

$$\lim inf_{(v_1,v_2,v_3)\to(\bar{u},0^+,0^+)}\{\int_\Omega \phi v_{3t}dx/\int_\Omega \phi v_3 dx\} = \hat{\rho}_1.$$

Consequently, we have

(1.48)

$$\lim inf_{(v_1,v_2,v_3)\to(\bar{u},0^+,0^+)}\sum_{i=1}^{3}[\int_\Omega (\phi v_{it}dx)/(\int_\Omega \phi v_i dx)] = 2\hat{\rho}_1 - (\beta - \alpha)||\bar{u}||_0.$$

From the symmetry of the system (1.41), we readily see that the calculations for the limits at the equilibria on the other two axes are the same. The origin $(0,0,0)$ also lies on the set $\omega(S)$, and direct calculations as above show that the limit there is positive. As mentioned above, we can show by means of comparison as in the analysis of system (1.4), that hypotheses [H1] and [H2] (for uniform boundedness and dissipativity) also hold for problem (1.41). Theorem 1.2 also apply with system (1.41) as well as (1.4). We can thus apply Theorem 1.6, using (1.46) and (1.48), to obtain permanence in the following.

Theorem 1.7 (Permanence for 3 Competing Species). *The semigroup $T(t)$ of solutions of problem (1.41) under hypothesis (1.42) is permanent in $X = X^0 \cup \partial X^0 = [C^1_{0,+}(\bar{\Omega})]^3$ provided that*

$$2\hat{\rho}_1 - (\beta - \alpha)||\bar{u}||_0 > 0, \qquad (1.49)$$

where $\bar{u}$ and $\hat{\rho}_1$ are defined in (1.43) and (1.44) respectively.

Finally, using some more topological results, we can obtain sharper description of the permanent semigroups for reaction diffusion systems (1.4) and (1.41) as described in Remark 1.2 below. The following definition and Theorem 1.8 can be found in [81]. A set $U \subset X^0$ is said to be strongly bounded if it is bounded and $\underline{d}(U, \partial X^0) > 0$. A_0 is said to be a global attractor relative to strongly bounded sets if it is a compact invariant subset of X^0 and $\bar{d}(T(t)U, A_0) = 0$ for all strongly bounded sets.

Theorem 1.8. *Let X be complete and $T(t)$ be point dissipative in X. Assume that there is a $t_0 \geq 0$ such that $T(t)$ is compact for $t > t_0$. Let $X^0, \partial X^0$ be as described above. Suppose $T(t)$ is permanent, then there are global attractors A, A_∂ for $T_\partial(t)$, and a global attractor A_0 relative to strongly bounded sets.*

The above Theorem is an extension of Theorem A4-13 in Chapter 6. If we apply Theorem 1.8 and the permanence in $[C^1_{0,+}(\bar{\Omega})]^3$, for the semigroup $T(t)$

in Theorem 1.3 or Theorem 1.7, we can readily show that the solutions of (1.4) under hypotheses [C1]-[C5] or solutions of (1.41) under hypotheses (1.42) have the properties in the following remark.

Remark 1.2. There exists an $\epsilon > 0$ and functions $v, w \in C^2(\Omega, R^3_+)$ with v_i, w_i in $C^1(\bar{\Omega}), > 0$ on Ω and outward normal derivatives $\partial v_i/\partial\nu < -\epsilon$ on $\partial\Omega$, and a corresponding region $Z = \{z \in C^1(\bar{\Omega}, R^3) : v_i \leq z_i \leq w_i, i = 1, 2, 3\}$ such that all solutions $u = (u_1(x,t), u_2(x,t), u_3(x,t))$ with each component initially non-negative and nontrivial are attracted to Z in $C^1(\bar{\Omega})$. That is $\lim_{t\to\infty} d(u(\cdot,t), Z) = 0$ where d is the metric in $[C^1(\bar{\Omega})]^3$.

In order to prove the remark, we first note that the set A_0 defined in Theorem 1.8 is invariant. Thus for each $\hat{u} \in A_0$, there is a $\tilde{u}$ such that $T(t)\tilde{u} = \hat{u}$. Using the maximum principle for parabolic equations and the boundedness of A_0 in $[C^1(\bar{\Omega})]^3$, we then show that there exist positive constants c_1, c_2, γ and a function $e \in C^2(\bar{\Omega})$ with $e(x) > 0$ in Ω and $\partial e/\partial\nu < -\gamma$ on $\partial\Omega$, such that for each $\hat{u} \in A_0$, we must have

$$c_1 e(x) \leq \hat{u}_i(x) \leq c_2 e(x), \quad \text{for all } x \in \bar{\Omega}, \ i = 1, 2, 3.$$

Then we use convergence in $C^1(\bar{\Omega})$ to show that there exists a $\beta > 0$ such that for all solutions $u(x,t)$ with each component non-negative and nontrivial initially, there is $t(u)$ with the property

$$u_i(x,t) \geq \beta e(x) \ \text{ for } t \geq t(u), \ x \in \bar{\Omega}, \text{ and } i = 1, 2, 3.$$

For more description of the history of the development of persistence theory and its relation with dynamical system method, the readers should refer to Hofbauer and Sigmund [88] and Hutson and Schmitt [94]. Further explanations concerning the application of such spatial theory to ecology can be found in Cantrell and Cosner [20].

4.2 Upper-Lower Estimates, Attractor Set, Blowup

Throughout many sections in this book, solutions of elliptic and parabolic systems are estimated by comparison method, using coupled upper and lower solutions. Such techniques are explained in various books, e.g. Leung [125] and Pao [183]. Moreover, they have been developed to very general forms including coupled nonlinear boundary conditions. We review some of these results here in Part A, since they can be applied to many related problems. In Part B, we will use comparison method as in the first part to enhance the persistence theory of the last section. We will further construct the set which attracts all non-negative solutions by comparison scheme. This set can be calculated numerically in practice. In Part C, we discuss a case when there is no upper estimate. We study the problem when the solution of the parabolic system blow up due to excessive nonlinear boundary inflow. Note that the theories in this section are not restricted to Dirichlet type of boundary condition

Part A: General Monotone Schemes.

We consider the elliptic problem:

$$(2.1)\qquad \begin{cases} -L_i u_i = f_i(x, u_1, ..., u_n) & \text{in } \Omega, \\ B_i u_i = g_i(x, u_1, ..., u_n) & \text{on } \partial\Omega, \quad (i = 1, ..., n); \end{cases}$$

and the corresponding parabolic problem:

$$(2.2)\qquad \begin{cases} (u_i)_t - L_i u_i = f_i(x, t, u_1, ..., u_n) & \text{in } \Omega \times (0, T], \\ B_i u_i = g_i(x, t, u_1, ..., u_n) & \text{on } \partial\Omega \times (0, T], \quad (i = 1, ..., n) \\ u_i(x, 0) = u_{i,0}(x) & \text{in } \Omega. \end{cases}$$

Here, L_i and B_i are linear operators of the form:

$$(2.3)\qquad \begin{aligned} L_i u_i &:= \textstyle\sum_{j,l=1}^{N} a_{j,l}^{(i)}(x) \partial^2 u_i / \partial x_j \partial x_l + \sum_{j=1}^{N} b_j^{(i)}(x) \partial u / \partial x_j, \\ B_i u_i &:= \alpha_i(x) \partial u / \partial \nu + \beta_i(x) u. \end{aligned}$$

Ω is a bounded domain in R^N with smooth boundary $\partial\Omega$. For each $i = 1, ..., n$, L_i is a uniformly elliptic operator in Ω with coefficients in $C^\mu(\Omega)$ for some $\mu \in (0, 1)$. The functions α_i and β_i are respectively in C^μ and $C^{1+\mu}$, and are non-negative functions satisfying $\alpha_i + \beta_i > 0$ on $\partial\Omega$. The functions $u_{i,0}(x)$ are in $C^\mu(\bar{\Omega})$. The functions f_i and g_i are assumed to be in $C^\mu(\bar{\Omega})$ or $C^{\mu,\mu/2}(\bar{\Omega} \times [0, T])$ as functions of x or (x, t), and are continuously differentiable in $\mathbf{u} = (u_1, ..., u_n)$. In order to construct a monotone iteration process, we express $\mathbf{u}$ in the form

$$\mathbf{u} \equiv (u_i, [\mathbf{u}]_{a_i}, [\mathbf{u}]_{b_i}) \text{ or } \mathbf{u} \equiv (u_i, [\mathbf{u}]_{c_i}, [\mathbf{u}]_{d_i}),$$

where a_i, b_i, c_i and d_i are non-negative integers satisfying

$$a_i + b_i = c_i + d_i = n - 1, \text{ for each } i = 1, ..., n.$$

A vector function $\mathbf{f} = (f_1, ..., f_n)$ is said to have a quasimonotone property with respect to $\mathbf{u} = (u_1, ..., u_n)$, if for each $i = 1, ..., n$, there exist non-negative integers a_i, b_i, with $a_i + b_i = n - 1$, such that $f_i(u_i, [\mathbf{u}]_{a_i}, [\mathbf{u}]_{b_i})$ is monotone nondecreasing in the $[\mathbf{u}]_{a_i}$ components, and monotone nonincreasing in the $[\mathbf{u}]_{b_i}$

components. When $a_i = 0$ or $b_i = 0$, $\mathbf{f}$ is quasimonotone nonincreasing or nondecreasing respectively. Similar definitions are made for the vector function $\mathbf{g}$ with respect to the components $[\mathbf{u}]_{c_i}$ and $[\mathbf{u}]_{d_i}$. When $\mathbf{f}$ and $\mathbf{g}$ have such quasimonotone properties, we write problems (2.1) and (2.2) respectively in the form:

$$\begin{cases} -L_i u_i = f_i(x, u_i, [\mathbf{u}]_{a_i}, [\mathbf{u}]_{b_i}) & \text{in } \Omega, \\ B_i u_i = g_i(x, u_i, [\mathbf{u}]_{c_i}, [\mathbf{u}]_{d_i}) & \text{on } \partial\Omega, \quad (i = 1, ..., n), \end{cases} \tag{2.4}$$

and

(2.5)
$$\begin{cases} (u_i)_t - L_i u_i = f_i(x, t, u_i, [\mathbf{u}]_{a_i}, [\mathbf{u}]_{b_i}) & \text{in } \Omega \times (0, T], \\ B_i u_i = g_i(x, t, u_i, [\mathbf{u}]_{c_i}, [\mathbf{u}]_{d_i}) & \text{on } \partial\Omega \times (0, T], \quad (i = 1, ..., n) \\ u_i(x, 0) = u_{i,0}(x) & \text{in } \Omega. \end{cases}$$

We will construct monotone convergent sequences to find approximate solutions for system (2.4) and (2.5).

Definition 2.1. Let $\mathbf{f} = (f_1, ..., f_n), \mathbf{g} = (g_1, ..., g_n)$ be quasimonotone functions with respect to $\mathbf{u} = (u_1, ..., u_n)$ as described above. A pair of functions $\hat{\mathbf{u}} = (\hat{u}_1, ..., \hat{u}_n), \tilde{\mathbf{u}} = (\tilde{u}_1, ..., \tilde{u}_n)$ in $[C^2(\bar{\Omega})]^n$ are called (coupled) upper-lower solutions of (2.4) if $\hat{u}_i \geq \tilde{u}_i$ in $\bar{\Omega}$, for $i = 1, ..., n$, and

$$\begin{cases} -L_i \hat{u}_i \geq f_i(x, \hat{u}_i, [\hat{\mathbf{u}}]_{a_i}, [\tilde{\mathbf{u}}]_{b_i}) & \text{in } \Omega, \\ -L_i \tilde{u}_i \leq f_i(x, \tilde{u}_i, [\tilde{\mathbf{u}}]_{a_i}, [\hat{\mathbf{u}}]_{b_i}) & \text{in } \Omega, \\ B_i \hat{u}_i \geq g_i(x, \hat{u}_i, [\hat{\mathbf{u}}]_{c_i}, [\tilde{\mathbf{u}}]_{d_i}) & \text{on } \partial\Omega, \quad (i = 1, ..., n) \\ B_i \tilde{u}_i \leq g_i(x, \tilde{u}_i, [\tilde{\mathbf{u}}]_{c_i}, [\hat{\mathbf{u}}]_{d_i}) & \text{on } \partial\Omega. \end{cases} \tag{2.6}$$

Let $\hat{\mathbf{u}}, \tilde{\mathbf{u}}$ be a pair of coupled upper-lower solutions for (2.4). Define

$$\begin{aligned} &< \tilde{\mathbf{u}}, \hat{\mathbf{u}} > \equiv \{\mathbf{u} \in [C(\bar{\Omega})]^n; \tilde{u}_i \leq u_i \leq \hat{u}_i, i = 1, ..., n\}, \\ &\bar{c}_i(x) \equiv max\{max\{-\tfrac{\partial f_i}{\partial u_i}(x, \mathbf{u}); \mathbf{u} \in < \tilde{\mathbf{u}}, \hat{\mathbf{u}} >\}, 0\}, \\ &\bar{b}_i(x) \equiv max\{max\{-\tfrac{\partial g_i}{\partial u_i}(x, \mathbf{u}); \mathbf{u} \in < \tilde{\mathbf{u}}, \hat{\mathbf{u}} >\}, 0\}, \end{aligned} \tag{2.7}$$

where $[C(\bar{\Omega}]^n$ denotes the set of continuous functions $\mathbf{u}$ in $\bar{\Omega}$. Letting

$$\begin{aligned} &F_i(x, u_i, [\mathbf{u}]_{a_i}, [\mathbf{u}]_{b_i}) = \bar{c}_i u_i + f_i(x, u_i, [\mathbf{u}]_{a_i}, [\mathbf{u}]_{b_i}), \\ &G_i(x, u_i, [\mathbf{u}]_{c_i}, [\mathbf{u}]_{d_i}) = \bar{b}_i u_i + g_i(x, u_i, [\mathbf{u}]_{c_i}, [\mathbf{u}]_{d_i}), \end{aligned} \tag{2.8}$$

we may write (2.4) in the form

$$\text{(2.9)} \qquad \begin{cases} -L_i u_i + \bar{c}_i u_i = F_i(x, u_i, [\mathbf{u}]_{a_i}, [\mathbf{u}]_{b_i}) & \text{in } \Omega, \\ B_i u_i + \bar{b}_i = G_i(x, u_i, [\mathbf{u}]_{c_i}, [\mathbf{u}]_{d_i}) & \text{on } \partial\Omega \ \ (i = 1, ..., n). \end{cases}$$

A monotone approximating sequence to the solution of (2.9) can be constructed by the following scheme:

(2.10)

$$\begin{cases} -L_i \bar{u}_i^{(k)} + \bar{c}_i \bar{u}_i^{(k)} = F_i(x, \bar{u}_i^{(k-1)}, [\bar{\mathbf{u}}^{(\mathbf{k-1})}]_{a_i}, [\underline{\mathbf{u}}^{\mathbf{k-1}}]_{b_i}) & \text{in } \Omega, \\ B_i \bar{u}_i^{(k)} + \bar{b}_i \bar{u}_i^{(k)} = G_i(x, \bar{u}_i^{(k-1)}, [\bar{\mathbf{u}}^{(\mathbf{k-1})}]_{c_i}, [\underline{\mathbf{u}}^{(\mathbf{k-1})}]_{d_i}) & \text{on } \partial\Omega, \ (i = 1, ..., n); \\ -L_i \underline{u}_i^{(k)} + \bar{c}_i \underline{u}_i^{(k)} = F_i(x, \underline{u}_i^{(k-1)}, [\underline{\mathbf{u}}^{(\mathbf{k-1})}]_{a_i}, [\bar{\mathbf{u}}^{\mathbf{k-1}}]_{b_i}) & \text{in } \Omega, \\ B_i u_i^{(k)} + \bar{b}_i^{(k)} \underline{u}_i^{(k)} = G_i(x, \underline{u}_i^{(k-1)}, [\underline{\mathbf{u}}^{(\mathbf{k-1})}]_{c_i}, [\bar{\mathbf{u}}^{(\mathbf{k-1})}]_{d_i}) & \text{on } \partial\Omega, \ (i = 1, ..., n) \end{cases}$$

for $k = 1, 2, \ldots$, where we choose $\bar{u}_i^{(0)} = \hat{u}_i$ and $\underline{u}_i^{(0)} = \tilde{u}_i, i = 1, .., n$. The following existence-comparison theorem is well-known.

Theorem 2.1. *Let* $\mathbf{f} = (f_1, ..., f_n), \mathbf{g} = (g_1, ..., g_n)$ *be quasimonotone functions with respect to* $\mathbf{u} = (u_1, ..., u_n)$ *with smoothness properties as described above, and* $\hat{\mathbf{u}} = (\hat{u}_1, ..., \hat{u}_n), \tilde{\mathbf{u}} = (\tilde{u}_1, ..., \tilde{u}_n)$ *are (coupled) upper-lower solutions of (2.4). Then the sequences constructed by the recursive relation (2.10) satisfies the following monotonic properties for each* $i = 1, ..., n$*:*

$$\text{(2.11)} \qquad \tilde{u}_i \leq \underline{u}_i^{(k)} \leq \underline{u}_i^{(k+1)} \leq \bar{u}_i^{(k+1)} \leq \bar{u}_i^{(k)} \leq \hat{u}_i \quad \textit{in } \bar{\Omega}, \quad \textit{for } \ k = 1, 2, \ldots .$$

Moreover, any solution $\mathbf{u}^* = (u_1^*, ..., u_1^*)$ *of (2.4) which satisfies* $\tilde{u}_i \leq u_i^* \leq \hat{u}_i$ *in* $\bar{\Omega}$ *for* $i = 1, ..., n$ *must also satisfy*

(2.12)

$$\tilde{u}_i \leq \underline{u}_i^{(k)} \leq \underline{u}_i^{(k+1)} \leq \underline{u}_i \leq u_i^* \leq \bar{u}_i \leq \bar{u}_i^{(k+1)} \leq \bar{u}_i^{(k)} \leq \hat{u}_i \quad \textit{in } \bar{\Omega}, \ \textit{for } k = 1, 2, \ldots ,$$

where for each $i = 1, ..., n, x \in \bar{\Omega}$*:*

$$\text{(2.13)} \qquad \bar{u}_i(x) := \lim_{k \to \infty} \bar{u}_i^{(k)}(x) \ \text{ and } \ \underline{u}_i(x) := \lim_{k \to \infty} \underline{u}_i^{(k)}(x).$$

From (2.10), it can be shown that $\bar{\mathbf{u}} = (\bar{u}_1, ..., \bar{u}_n)$, and $\underline{u} = (\underline{u}_1, ..., \underline{u}_n)$

satisfy:

$$
(2.14) \quad \begin{cases} -L_i\bar{u}_i = f_i(x, \bar{u}_i, [\bar{\mathbf{u}}]_{a_i}, [\underline{\mathrm{u}}]_{b_i}) & \text{in } \Omega, \\ -L_i\underline{\mathrm{u}}_i = f_i(x, \underline{\mathrm{u}}_i, [\underline{\mathrm{u}}]_{a_i}, [\bar{\mathbf{u}}]_{b_i}) & \text{in } \Omega, \\ B_i\bar{u}_i = g_i(x, \bar{u}_i, [\bar{\mathbf{u}}]_{c_i}, [\underline{\mathrm{u}}]_{d_i}) & \text{on } \partial\Omega, \quad (i = 1, ..., n) \\ B_i\underline{\mathrm{u}}_i = g_i(x, \underline{\mathrm{u}}_i, [\underline{\mathrm{u}}]_{c_i}, [\bar{\mathbf{u}}]_{d_i}) & \text{on } \partial\Omega. \end{cases}
$$

Thus, the limiting functions $\bar{\mathbf{u}}$ and $\underline{\mathrm{u}}$ are in general not always a solution of problem (2.4). However, if $\mathbf{f}$ and $\mathbf{g}$ are both quasimonotone nondecreasing C^1-functions in $< \tilde{\mathbf{u}}, \hat{\mathbf{u}} >$, then the $\bar{\mathbf{u}}$ and $\underline{\mathrm{u}}$ are indeed solutions. In fact, they are respectively maximal and minimal solutions of (2.4) in the sense that any solution $\mathbf{u}^* = (u_1^*, .., u_n^*)$ in $< \tilde{\mathbf{u}}, \hat{\mathbf{u}} >$, must satisfy $\underline{\mathrm{u}}_i \le u_i^* \le \bar{u}_i$ in $\bar{\Omega}$ for each component. (For more details, see e.g. Chapter 8 and 9 in [183] or Chapter 5 in [125] where the notations are slightly different).

In the general case when $\mathbf{f}$ or $\mathbf{g}$ is not quasimonotone, there is also a more general definition of upper-lower solutions for system (2.4).

Definition 2.2. A pair of functions $\hat{\mathbf{u}} = (\hat{u}_1, ..., \hat{u}_n), \tilde{\mathbf{u}} = (\tilde{u}_1, ..., \tilde{u}_n)$ in $[C^2(\bar{\Omega})]^n$ are called (coupled) upper-lower solutions of the system (2.4) if $\hat{u}_i \ge \tilde{u}_i$ in $\bar{\Omega}$, for $i = 1, ..., n$, and

$$
(2.15) \quad \begin{cases} -L_i\hat{u}_i \ge f_i(x, \mathbf{v}) & \text{in } \Omega, \\ & \quad \text{for all } \mathbf{v} \in < \tilde{\mathbf{u}}, \hat{\mathbf{u}} >, \text{ with } v_i = \hat{u}_i \\ B_i\hat{u}_i \ge g_i(x, \mathbf{v}) & \text{on } \partial\Omega, \\ & \quad (i = 1, ..., n) \\ -L_i\tilde{u}_i \le f_i(x, \mathbf{v}) & \text{in } \Omega, \\ & \quad \text{for all } \mathbf{v} \in < \tilde{\mathbf{u}}, \hat{\mathbf{u}} >, \text{ with } v_i = \tilde{u}_i \\ B_i\hat{u}_i \le g_i(x, \mathbf{v}) & \text{on } \partial\Omega. \end{cases}
$$

In case when $\mathbf{f}$ and $\mathbf{g}$ are quasimonotone, Definition 2.2 becomes Definition 2.1. The following theorem by Tsai [221] is well-known (see e.g. Chapter 1 in [125] or Chapter 9 in [183].

Theorem 2.2. *Suppose there exists a pair of (coupled) upper-lower solutions* $\hat{\mathbf{u}}$, $\tilde{\mathbf{u}}$ *for problem (2.4) as described in Definition 2.2, where* $\mathbf{f}$ *and* $\mathbf{g}$ *are* C^1-*functions in* $< \tilde{\mathbf{u}}, \hat{\mathbf{u}} >$ *and* $x \in \bar{\Omega}$. *Then there is at least one solution* $\mathbf{u}^*$ *of (2.4) satisfying* $\mathbf{u}^* \in < \tilde{\mathbf{u}}, \hat{\mathbf{u}} >$.

For problem (2.4) with $\mathbf{f}$ quasimonotone and $\mathbf{g} = (0, ..., 0)$, there are some variations to the scheme described in (2.10) found in the literature. For each component i of the quasimonotone function $\mathbf{f} = (f_1, ..., f_n)$, we divide those components in $[\mathbf{u}]_{a_i}$ into two groups $[\mathbf{u}]_{a_{i1}}$ and $[\mathbf{u}]_{a_{i2}}$, where the first group consists of those components before the i-th and the second consists of those after the i-th. The scheme can be up-dated as follows:

(2.16)
$$\begin{cases} -L_i\bar{u}_i^{(k)} + \bar{c}_i\bar{u}_i^{(k)} = F_i(x, \bar{u}_i^{(k-1)}, [\bar{\mathbf{u}}^{(\mathbf{k})}]_{a_{i1}}, [\bar{\mathbf{u}}^{(\mathbf{k-1})}]_{a_{i2}}, [\underline{\mathbf{u}}^{\mathbf{k-1}}]_{b_i}) & \text{in } \Omega, \\ B_i\bar{u}_i^{(k)} = 0 & \text{on } \partial\Omega, \\ -L_i\underline{u}_i^{(k)} + \bar{c}_i\underline{u}_i^{(k)} = F_i(x, \underline{u}_i^{(k-1)}, [\underline{\mathbf{u}}^{(\mathbf{k})}]_{a_{i1}}, [\underline{\mathbf{u}}^{(\mathbf{k-1})}]_{a_{i2}}, [\bar{\mathbf{u}}^{\mathbf{k-1}}]_{b_i}) & \text{in } \Omega, \\ B_iu_i^{(k)} = 0 & \text{on } \partial\Omega, \quad (i = 1, ..., n) \end{cases}$$

for $k = 1, 2, \ldots$, where we choose $\bar{u}_i^{(0)} = \hat{u}_i$ and $\underline{u}_i^{(0)} = \tilde{u}_i, i = 1, ..., n$.

Theorem 2.3. *Let* $\mathbf{f} = (f_1, ..., f_n)$ *be quasimonotone functions with respect to* $\mathbf{u} = (u_1, ..., u_n)$ *as described in Theorem 2.1, and* $\hat{\mathbf{u}} = (\hat{u}_1, ..., \hat{u}_n), \tilde{\mathbf{u}} = (\tilde{u}_1, ..., \tilde{u}_n)$ *are (coupled) upper-lower solutions of (2.4) with* $\mathbf{g} = (0, ..., 0)$*. Then the sequences constructed by the recursive relation (2.16) satisfies the following monotonic properties for each* $i = 1, ..., n$*:*

$$\tilde{u}_i \leq \underline{u}_i^{(k)} \leq \underline{u}_i^{(k+1)} \leq \bar{u}_i^{(k+1)} \leq \bar{u}_i^{(k)} \leq \hat{u}_i \quad \text{in } \bar{\Omega}, \quad \text{for } k = 1, 2, ..., . \tag{2.17}$$

Moreover, any solution $\mathbf{u}^* = (u_1^*, ..., u_1^*)$ *of (2.4) which satisfies* $\tilde{u}_i \leq u_i^* \leq \hat{u}_i$ *in* $\bar{\Omega}$ *for* $i = 1, ..., n$ *must also satisfy (2.12) and (2.13).*

In the study of population equations, the functions f_i in (2.1) and (2.2) are usually of the form:

$$f_i(\cdot, u_1, ..., u_n) = u_i\tilde{f}_i(\cdot, u_1, ..., u_n), \quad i = 1, ..., n, \tag{2.18}$$

where $\tilde{f}_i$ are quasimonotone with respect to $\mathbf{u}$ for $u_i \geq 0, i = 1, .., n$, and has the same smoothness properties as f_i described before. Assume there exist coupled upper-lower solutions with all components non-negative in Ω and $\mathbf{g} \equiv 0$.. In such instance, we can replace the linear scheme in (2.10) and (2.14) by the following nonlinear scheme:

$$\begin{cases} -L_i\bar{u}_i^{(k)} = \bar{u}_i^{(k)}\tilde{f}_i(x, \bar{u}_i^{(k)}, [\bar{\mathbf{u}}^{(\mathbf{k})}]_{a_{i1}}, [\bar{\mathbf{u}}^{(\mathbf{k-1})}]_{a_{i2}}, [\underline{\mathbf{u}}^{\mathbf{k-1}}]_{b_i}) & \text{in } \Omega, \\ B_i\bar{u}_i^{(k)} = 0 & \text{on } \partial\Omega, \\ -L_i\underline{u}_i^{(k)} = \underline{u}_i^{(k)}\tilde{f}_i(x, \underline{u}_i^{(k)}, [\underline{\mathbf{u}}^{(\mathbf{k})}]_{a_{i1}}, [\underline{\mathbf{u}}^{(\mathbf{k-1})}]_{a_{i2}}, [\bar{\mathbf{u}}^{\mathbf{k-1}}]_{b_i}) & \text{in } \Omega, \\ B_iu_i^{(k)} = 0 & \text{on } \partial\Omega, \quad (i = 1, ..., n) \end{cases} \tag{2.19}$$

for $k = 1, 2, \ldots$, where we choose $\bar{u}_i^{(0)} = \hat{u}_i$ and $\underline{u}_i^{(0)} = \tilde{u}_i, i = 1, .., n$. Here, at each step k, $\bar{u}_i^k$ is a uniquely defined positive function or is the trivial function depending on the principal eigenvalue ρ for

(2.20)
$$L_i w + w\tilde{f}_i(x, 0, [\bar{\mathbf{u}}^{(\mathbf{k})}]_{a_{i1}}, [\bar{\mathbf{u}}^{(\mathbf{k-1})}]_{a_{i2}}, [\underline{\mathbf{u}}^{\mathbf{k-1}}]_{b_i}) = \rho w \quad \text{in } \Omega, \quad B_i w = 0 \quad \text{on } \partial\Omega,$$

is > 0 or ≤ 0 respectively. Similarly, $\underline{u}_i^k$ is a uniquely defined positive function or is the trivial function depending on the principal eigenvalue ρ for

(2.21)
$$L_i w + w\tilde{f}_i(x, 0, [\underline{\mathbf{u}}^{(\mathbf{k})}]_{a_{i1}}, [\underline{\mathbf{u}}^{(\mathbf{k-1})}]_{a_{i2}}, [\bar{\mathbf{u}}^{\mathbf{k-1}}]_{b_i}) = \rho w \quad \text{in } \Omega, \quad B_i w = 0 \quad \text{on } \partial\Omega,$$

is > 0 or ≤ 0 respectively. These estimates have been used in many situations in the earlier sections and chapters in analyzing the uniqueness and stability of positive coexistence states for 2 or 3 species.

Theorem 2.4. *Let $\mathbf{f} = (u_1\tilde{f}_1, ..., u_n\tilde{f}_n)$ be quasimonotone functions as described above with respect to $\mathbf{u} = (u_1, ..., u_n)$, and $\hat{\mathbf{u}} = (\hat{u}_1, ..., \hat{u}_n), \tilde{\mathbf{u}} = (\tilde{u}_1, ..., \tilde{u}_n)$ are coupled upper-lower solutions of (2.4) with $\tilde{u}_i \geq 0$, $\mathbf{g} = (0, ..., 0)$ in $\bar{\Omega}$. Then the sequences constructed by the nonlinear recursive relation (2.19) satisfies the following monotonic properties for each $i = 1, ..., n$:*

$$\tilde{u}_i \leq \underline{u}_i^{(k)} \leq \underline{u}_i^{(k+1)} \leq \bar{u}_i^{(k+1)} \leq \bar{u}_i^{(k)} \leq \hat{u}_i \quad in\ \bar{\Omega}, \quad for\ \ k = 1, 2, \ldots. \tag{2.22}$$

Moreover, any solution $\mathbf{u}^* = (u_1^*, ..., u_1^*)$ of (2.4) which satisfies $\tilde{u}_i \leq u_i^* \leq \hat{u}_i$ in $\bar{\Omega}$ for $i = 1, ..., n$ must also satisfy (2.12) and (2.13).

In case the principal eigenvalue of (2.20) or (2.21) is positive, the positive equilibrium for the related nonlinear scalar problem is a global attractor for the corresponding scalar parabolic problem. For 2 or 3 species, we have used this property to study the time asymptotic behavior for the corresponding parabolic system in Chapter 1 and the last section. We construct a region which attracts all non-negative orbits of the corresponding system (2.5) by means of the approximates obtained from (2.19). We will describe another such application for a system of n equations in Theorem 2.7 below. Before we study this situation, we first state the Theorems for the parabolic case corresponding to Theorem 2.1.

Definition 2.3. Let $\mathbf{f} = (f_1, ..., f_n), \mathbf{g} = (g_1, ..., g_n)$ be quasimonotone functions with respect to $\mathbf{u} = (u_1, ..., u_n)$ as described above. A pair of functions $\hat{\mathbf{u}} = (\hat{u}_1, ..., \hat{u}_n), \tilde{\mathbf{u}} = (\tilde{u}_1, ..., \tilde{u}_n)$ in $C^{2,1}(\bar{\Omega} \times [0, T])$ are called (coupled) upper-lower solutions of (2.5) if $\hat{u}_i \geq \tilde{u}_i$ in $\bar{\Omega} \times [0,T]$, for $i = 1, ..., n$, and

(2.23)
$$\begin{cases} (\hat{u}_i)_t - L_i\hat{u}_i \geq f_i(x,t,\hat{u}_i,[\hat{\mathbf{u}}]_{a_i},[\tilde{\mathbf{u}}]_{b_i}) & \text{in } \Omega \times (0,T], \\ (\tilde{u}_i)_t - L_i\tilde{u}_i \leq f_i(x,t,\tilde{u}_i,[\tilde{\mathbf{u}}]_{a_i},[\hat{\mathbf{u}}]_{b_i}) & \text{in } \Omega \times (0,T], \\ B_i\hat{u}_i \geq g_i(x,t,\hat{u}_i,[\hat{\mathbf{u}}]_{c_i},[\tilde{\mathbf{u}}]_{d_i}) & \text{on } \partial\Omega \times (0,T], \quad (i = 1,...,n) \\ B_i\tilde{u}_i \leq g_i(x,t,\tilde{u}_i,[\tilde{\mathbf{u}}]_{c_i},[\hat{\mathbf{u}}]_{d_i}) & \text{on } \partial\Omega \times (0,T], \\ \hat{u}_i(x,0) \geq u_{i,0}(x) \geq \tilde{u}_i(x,0) & \text{in } \Omega. \end{cases}$$

We replace (2.7) and (2.8) with

$$< \tilde{\mathbf{u}}, \hat{\mathbf{u}} > \equiv \{\mathbf{u} \in [C(\bar{\Omega} \times [0,T])]^n; \tilde{u}_i \leq u_i \leq \hat{u}_i, i = 1,...,n\},$$

$$\tilde{c}_i(x,t) \equiv max\{-\tfrac{\partial f_i}{\partial u_i}(x,t,\mathbf{u}); \mathbf{u} \in <\tilde{\mathbf{u}},\hat{\mathbf{u}}>\}, \tag{2.24}$$

$$\tilde{b}_i(x,t) \equiv max\{-\tfrac{\partial g_i}{\partial u_i}(x,t,\mathbf{u}); \mathbf{u} \in <\tilde{\mathbf{u}},\hat{\mathbf{u}}>\},$$

$$\begin{aligned} F_i(x,t,u_i,[\mathbf{u}]_{a_i},[\mathbf{u}]_{b_i}) &\equiv \tilde{c}_i u_i + f_i(x,t,u_i,[\mathbf{u}]_{a_i},[\mathbf{u}]_{b_i}), \\ G_i(x,t,u_i,[\mathbf{u}]_{c_i},[\mathbf{u}]_{d_i}) &\equiv \tilde{b}_i u_i + g_i(x,t,u_i,[\mathbf{u}]_{c_i},[\mathbf{u}]_{d_i}). \end{aligned} \tag{2.25}$$

We may write (2.5) in the form

$$\begin{cases} \mathcal{L}_i u_i = F_i(x,t,u_i,[\mathbf{u}]_{a_i},[\mathbf{u}]_{b_i}) & \text{in } \Omega \times (0,T], \\ \mathcal{B}_i u_i = G_i(x,t,u_i,[\mathbf{u}]_{c_i},[\mathbf{u}]_{d_i}) & \text{on } \partial\Omega \times (0,T], \quad (i = 1,...,n) \\ u_i(x,0) = u_{i,0}(x) & \text{in } \Omega, \end{cases} \tag{2.26}$$

where

$$\mathcal{L}_i u_i \equiv (u_i)_t - L_i u_i + \tilde{c}_i u_i,$$

$$\mathcal{B}_i u_i \equiv B_i u_i + \tilde{b}_i \qquad (i = 1,...,n).$$

A monotone approximating sequence to the solution of (2.26) can be constructed

by the following scheme:
(2.27)

$$
\begin{aligned}
&\mathcal{L}_i\bar{u}_i^{(k)} = F_i(x,t,\bar{u}_i^{(k-1)},[\bar{\mathbf{u}}^{(\mathbf{k-1})}]_{a_i},[\underline{\mathbf{u}}^{\mathbf{k-1}}]_{b_i}) && \text{in } \Omega\times(0,T],\\
&\mathcal{B}_i\bar{u}_i^{(k)} = G_i(x,t,\bar{u}_i^{(k-1)},[\bar{\mathbf{u}}^{(\mathbf{k-1})}]_{c_i},[\underline{\mathbf{u}}^{(\mathbf{k-1})}]_{d_i}) && \text{on } \partial\Omega\times(0,T],\quad (i=1,...,n)\\
&\mathcal{L}_i\underline{u}_i^{(k)} = F_i(x,t,\underline{u}_i^{(k-1)},[\underline{\mathbf{u}}^{(\mathbf{k-1})}]_{a_i},[\bar{\mathbf{u}}^{\mathbf{k-1}}]_{b_i}) && \text{in } \Omega\times(0,T],\\
&\mathcal{B}_i u_i^{(k)} = G_i(x,t,\underline{u}_i^{(k-1)},[\underline{\mathbf{u}}^{(\mathbf{k-1})}]_{c_i},[\bar{\mathbf{u}}^{(\mathbf{k-1})}]_{d_i}) && \text{on } \partial\Omega\times(0,T],\quad (i=1,...,n)\\
&\bar{u}_i^{(k)}(x,0) = \underline{u}^{(k)}(x,0) = u_{i,0}(x) && \text{in } \Omega
\end{aligned}
$$

for $k = 1, 2, \ldots$, where we choose $\bar{u}_i^{(0)} = \hat{u}_i$ and $\underline{u}_i^{(0)} = \tilde{u}_i, i = 1, ..., n$. The following existence-comparison result is given in Chapter 9 of [183].

Theorem 2.5. *Let $\mathbf{f} = (f_1, ..., f_n), \mathbf{g} = (g_1, ..., g_n)$ be quasimonotone functions with respect to $\mathbf{u} = (u_1, ..., u_n)$, and $\hat{\mathbf{u}} = (\hat{u}_1, ..., \hat{u}_n), \tilde{\mathbf{u}} = (\tilde{u}_1, ..., \tilde{u}_n)$ are (coupled) upper-lower solutions of (2.5). Then the sequences $\{\bar{\mathbf{u}}^{\mathbf{k}}\} = \{(\bar{u}_1^k, ..., \bar{u}_n^k)\}$, $\{\underline{u}^{\mathbf{k}}\} = \{(\underline{u}_1^k, ..., \underline{u}_n^k)\}$, constructed by the recursive relation (2.27) with $\bar{\mathbf{u}}^{\mathbf{0}} = \hat{\mathbf{u}}$ and $\underline{u}^{\mathbf{0}} = \tilde{\mathbf{u}}$ converge monotonically to a unique solution $\mathbf{u}^* = (u_1^*, ..., u_n^*)$ of (2.5). Moreover, for each $i = 1, ..., n$:*
(2.28)

$$
\tilde{u}_i \le \underline{u}_i^{(k)} \le \underline{u}_i^{(k+1)} \le u_i^* \le \bar{u}_i^{(k+1)} \le \bar{u}_i^{(k)} \le \hat{u}_i \quad \text{in } \bar{\Omega}\times[0,T], \quad \text{for } k = 1, 2, \ldots.
$$

If $\mathbf{f}$ and $\mathbf{g}$ are not quasimonotone, we have the following more general definition of upper-lower solutions.

Definition 2.4. A pair of functions $\hat{\mathbf{u}} = (\hat{u}_1, ..., \hat{u}_n), \tilde{\mathbf{u}} = (\tilde{u}_1, ..., \tilde{u}_n)$ in $[C^{2,1}(\bar{\Omega})\times[0,T])]^n$ are called (coupled) upper-lower solutions of the system (2.5) if $\hat{u}_i \ge \tilde{u}_i$ in $\bar{\Omega}\times[0,T]$, for $i = 1, ..., n$, and
(2.29)

$$
\begin{aligned}
&(\hat{u}_i)_t - L_i\hat{u}_i \ge f_i(x,t,\mathbf{v}) && \text{in } \Omega\times(0,T],\\
& && \text{for all } \mathbf{v} \in <\tilde{\mathbf{u}},\hat{\mathbf{u}}>, \text{ with } v_i = \hat{u}_i\\
&B_i\hat{u}_i \ge g_i(x,t,\mathbf{v}) && \text{on } \partial\Omega\times(0,T],\\
& && (i=1,...,n)\\
&(\tilde{u}_i)_t - L_i\tilde{u}_i \le f_i(x,t,\mathbf{v}) && \text{in } \Omega\times(0,T],\\
& && \text{for all } \mathbf{v} \in <\tilde{\mathbf{u}},\hat{\mathbf{u}}>, \text{ with } v_i = \tilde{u}_i\\
&B_i\hat{u}_i \le g_i(x,t,\mathbf{v}) && \text{on } \partial\Omega.\\
&\tilde{u}_i(x,0) \le u_{i,0}(x) \le \bar{u}_i(x,0) && \text{in } \bar{\Omega}.
\end{aligned}
$$

We also have the following general existence Theorem, which is a direct extension of a theorem in Chapter 9 of [183], where $\mathbf{g}$ does not depend on $\mathbf{u}$.

Theorem 2.6. *Suppose there exists a pair of (coupled) upper-lower solutions* $\hat{\mathbf{u}}$, $\tilde{\mathbf{u}}$ *for problem (2.5) as described in Definition 2.4, where* $\mathbf{f}$ *and* $\mathbf{g}$ *are* C^1*-functions in* $<\tilde{\mathbf{u}}, \hat{\mathbf{u}}>, (x,t) \in \Omega \times [0,T]$*. Then there exists a unique solution* $\mathbf{u}^*$ *of (2.5) satisfying* $\mathbf{u}^* \in <\tilde{\mathbf{u}}, \hat{\mathbf{u}}>$.

Part B: Construction Scheme for Set of Attraction.

We now apply the constructive method described in Part A above to obtain preliminary estimates of the solutions of reaction-diffusion systems. They can enhance the description of the long-term behavior concerning persistence, survival and extinction of particular components. Motivated by studies in ecological problems, we consider parabolic systems with time periodic coefficients. We consider

(2.30)

$$\begin{cases} (u_i)_t - \hat{L}_i u_i = \hat{f}_i(x,t,u_1,...,u_n)u_i & \text{in } \Omega \times (0,\infty), \\ \hat{B}_i u_i = 0 & \text{on } \partial\Omega \times (0,\infty), \quad (i = 1,...,n). \end{cases}$$

Here, Ω is a bounded domain in R^N with smooth boundary; $\hat{L}_i$ are uniformly elliptic linear operators of the form:

(2.31)

$$\hat{L}_i u_i := \sum_{j,l=1}^{N} \hat{a}_{j,l}^{(i)}(x,t)\partial^2 u_i/\partial x_j \partial x_l + \sum_{j=1}^{N} \hat{b}_j^{(i)}(x,t)\partial u_i/\partial x_j + \hat{c}^i(x,t)u_i$$

with $\sum_{j,l=1}^{N} \hat{a}_{j,l}^{(i)}\xi_j\xi_l \geq a_0^{(i)}|\xi|^2$ for some constant $a_0^{(i)} > 0$, where all the coefficients are T-periodic in t and Hölder continuous in x, t, lying in $C^{\alpha,\alpha/2}(\bar{\Omega} \times R)$ for some $\alpha > 0$ with $\hat{a}_{j,l}^i = \hat{a}_{l,j}^i$ and $\hat{c}^i \leq 0$. (Note that there need not actually be any t-dependence). The boundary operator $\hat{B}_i$ is of the form

$$\hat{B}_i u = u \quad \text{or} \quad \hat{B}_i u = \frac{\partial u}{\partial \nu} + \beta(x)u,$$

with $\beta(x)$ of class $C^{1+\alpha}$ and $\beta \geq 0$. From the theory of second-order periodic parabolic equations, we know that for given $R(x,t) \in C^{\alpha,\alpha/2}(\bar{\Omega}, R)$, the problem

$$(2.32) \qquad \begin{cases} \hat{L}_i\phi + R(x,t)\phi - \phi_t = \rho\phi & \text{in } \Omega \times R, \\ \hat{B}_i\phi = 0 & \text{on } \partial\Omega \times R, \\ \phi \text{ is } T\text{-periodic}, \end{cases}$$

has a principal (largest) eigenvalue $\rho = \hat{\rho}_1$, with positive eigenfunction. (See Castro and Lazer [25], Lazer [115] and Hess [85] for more details).

For the purpose of comparison, we shall use the solutions to diffusive logistic equations of the form

$$(2.33) \qquad \begin{cases} u_t = \hat{L}_i u + u(R - Cu) & \text{in } \Omega \times (0, \infty), \\ \hat{B}_i u = 0 & \text{on } \partial\Omega \times (0, \Omega), \end{cases}$$

where $C = C(x,t)$ is T-periodic, $C(x,t) \geq C_0 > 0$, and $C \in C^{\alpha,\alpha/2}(\bar{\Omega}, R)$. One can use upper-lower solution method for periodic parabolic equations to obtain positive periodic solution for (2.33) as explained in [85] and [125]. If the principal eigenvalue $\hat{\rho}_1$ for (2.32) is positive, we can use the corresponding eigenfunction to construct lower solution. On the other hand, since C is strictly positive, any large enough constant is an upper solution of (2.33). Moreover, by standard arguments based on maximum principle, any solution with non-negative nontrivial initial data will be strictly positive in Ω for $t > 0$, with $u > 0$ on $\bar{\Omega}$ in the case for Neumann or mixed boundary data, and $\partial u / \partial \nu < 0$ on $\partial\Omega$ in the case of Dirichlet boundary data. The following lemma is presented in Theorem 28.1 in [85].

Lemma 2.1 *The problem (2.33) has a positive T-periodic solution (denoted by $\theta(\hat{L}_i, R, C)$) if and only if the principal eigenvalue $\hat{\rho}_1$ in (2.32) satisfies $\hat{\rho}_1 > 0$. In this case, the solution $\theta(\hat{L}_i, R, C)$ is a global attractor for solutions of (2.33) with non-negative nontrivial initial data.*

Remark 2.1. If $\hat{\rho}_1 \leq 0$, then the non-negative solutions of (2.33) tend to 0 as $t \to \infty$. By the smoothing properties of parabolic equations, the convergence to $\theta(\hat{L}_i, R, C)$ in the above theorem can be taken in $C^{1+\alpha,\alpha/2}(\hat{\Omega} \times (0, \infty))$ (cf. [85]).

We now assume that the interacting species form a food pyramid structure. Such problems are studied in William and Chow [231], Alikakos [2], and Leung [125]. Here, we present some results in Cantrell and Cosner [19]. We relabel the species so that u_1 represent the species at the bottom of the food pyramid, while u_n is the one at the top. Each species can only be the food supply or competitor of those species above (with higher index). That is , for each k, u_k can only be the prey or competitor for those species with higher index i.e. $u_j, j > k$; and u_k can only be predator or competitor for species with lower index i.e. $u_j, j < k$. The scheme first estimate the asymptotic upper bound for u_1, at the bottom of the food pyramid. Using this upper bound, we estimate the asymptotic upper bound for u_2. Continuing this way, we reach an asymptotic upper bound for u_n. Having upper bound for all the possible predators $u_2, ..., u_n$, we can estimate the asymptotic lower bound for u_1. Then we can use the lower bound for u_1 and the upper bounds for $u_3, ..., u_n$ to deduce the lower bound for u_2. Continuing in this fashion, we finally obtain an asymptotic lower bound for u_n.

More precisely, we make the following assumption for $\hat{f}_i(x,t,u_1,...,u_n)$. For $u_i \geq 0$, $i = 1,...,n$, and any positive constants $K_i > 0, i = 1,...,n$, we assume that

$$\begin{aligned} &\hat{f}_1(x,t,u_1,...,u_n) \leq R_1(x,t) - C_1(x,t)u_1; \text{ and for } k = 2,...,n, \\ &\hat{f}_k(x,t,u_1,...,u_n) \leq R_k(x,t,K_1,..,K_{k-1}) - C_k(x,t)u_k \text{ for } u_i \leq K_i, \end{aligned} \tag{2.34}$$

$i = 1,...,k-1$, for all $(x,t) \in \bar{\Omega} \times [0,T]$. Here $R_k(x,t,K_1,...,K_{k-1})$ and $C_k(x,t)$ are some T-periodic functions in t for $x \in \bar{\Omega}$ in the class $C^{\alpha,\alpha/2}$. Moreover R_k is monotone nondecreasing in $K_1,...,K_{k-1}$, and $C_k(x,t) \geq C_0 > 0$ in $\bar{\Omega} \times [0,T]$.

We assume that the principal eigenvalue for the problem:

$$\begin{cases} \hat{L}_1\phi + R_1\phi - \phi_t = \rho\phi, \\ \hat{B}_1\phi = 0, \quad \phi \text{ is } T\text{-periodic}, \end{cases} \tag{2.35}$$

is positive, and by Lemma 2.1 we define $\bar{\theta}_1 \equiv \theta(\hat{L}_1, R_1, C_1)$ to be the unique globally attracting positive T-periodic solution of

$$\begin{cases} w_t = \hat{L}_1 w + (R_1 - C_1 w)w, \\ \hat{B}_1 w = 0. \end{cases} \tag{2.36}$$

For any sufficiently small $\epsilon > 0$, define $\bar{\theta}_k$ (inductively) to be $\bar{\theta}_k \equiv \theta(\hat{L}_k, R_k(x,t, (1+\epsilon)\bar{\theta}_1,...,(1+\epsilon)\bar{\theta}_{k-1}), C_k)$, where we suppose that for $\epsilon > 0$ the principal eigenvalue of

$$\begin{cases} \hat{L}_k\phi + R_k(x,t,(1+\epsilon)\bar{\theta}_1,...,(1+\epsilon)\bar{\theta}_{k-1})\phi - \phi_t = \rho\phi, \\ \hat{B}_k\phi = 0, \quad \phi \text{ is } T\text{-periodic}, \end{cases} \tag{2.37}$$

is positive for each k so that $\bar{\theta}_k > 0$ inside Ω. Here $w = \bar{\theta}_k$ is the unique globally attracting positive T-periodic solution of (2.36) with $\hat{L}_1, C_1, \hat{B}_1, R_1$ respectively replaced by $\hat{L}_k, C_k, \hat{B}_k$ and $R_k(x,t,(1+\epsilon)\bar{\theta}_1,...,(1+\epsilon)\bar{\theta}_{k-1})$.

Theorem 2.7 (Construction of Set of Attraction). *Consider problem (2.30) where $\hat{f}_i, i = 1,...,n$ satisfy (2.34) to (2.37). Let $\mathbf{u} = (u_1,...,u_n)$ be a solution of (2.30) with $u_i(x,0) \geq 0, u_i(x,0) \not\equiv 0$ for $i = 1,...,n$. Then the solution to (2.30) exists globally in time. Moreover, for t sufficiently large, we have*

$$u_i(x,t) \leq (1+\epsilon)\bar{\theta}_i(x,t) \quad \text{for } x \in \bar{\Omega},\ i = 1,..,n. \tag{2.38}$$

Suppose further, for each $k = 1,...,n$, there is an index $j(k) \in \{0,1,...,k-1\}$ and a function $r_k(x,t,K_1,..,K_{k-1},K_{k+1},..,K_n)$ with the same periodicity and *smoothness hypotheses as R_k in (2.34), such that r_k is monotonic nondecreasing in K_j for $j \leq j(k)$ and monotonic nonincreasing in K_j for $j > j(k)$,($j \neq k$), and a T-periodic Hölder $C^{\alpha,\alpha/2}, 0 < \alpha < 1$, continuous function $c_k(x,t) \geq c_0 > 0$ so that*

$$\hat{f}_k(x,t,u_1,...,u_n) \geq r_k(x,t,K_1,..,K_{k-1},K_{k+1},...,K_n) - c_k(x,t)u_k \tag{2.39}$$

whenever $u_j \geq K_j$ for $j \leq j(k)$ and $u_j \leq K_j$ for $j > j(k)$, $j \neq k$. We assume that the principal eigenvalue for

$$\begin{cases} \hat{L}_1\phi + r_1(x,t,(1+\epsilon)\bar{\theta}_2,...,(1+\epsilon)\bar{\theta}_n)\phi - \phi_t = \rho\phi, \\ \\ \hat{B}_1\phi = 0, \quad \phi \text{ is } T\text{-periodic}, \end{cases} \tag{2.40}$$

is positive for $\epsilon > 0$ sufficiently small and define $\tilde{\theta}_1 \equiv \theta(\hat{L}_1, r_1(x,t,(1+\epsilon)\bar{\theta}_2,...,(1+\epsilon)\bar{\theta}_n), c_1)$; then define $\tilde{\theta}_k \equiv \theta(\hat{L}_k, r_k(x,t,(1-\epsilon)\tilde{\theta}_1,...,(1-\epsilon)\tilde{\theta}_{j(k)},(1+\epsilon)\bar{\theta}_{j(k)+1},...,(1+\epsilon)\bar{\theta}_{k-1},(1+\epsilon)\bar{\theta}_{k+1},...,(1+\epsilon)\bar{\theta}_n), c_k)$, where we suppose that for $\epsilon > 0$ sufficiently small, the principal eigenvalue of

$$\begin{cases} \hat{L}_k\phi + r_k(x,t,(1-\epsilon)\tilde{\theta}_1,...,(1-\epsilon)\tilde{\theta}_{j(k)},(1+\epsilon)\bar{\theta}_{j(k)+1},..., \\ \qquad\qquad (1+\epsilon)\bar{\theta}_{k-1},(1+\epsilon)\bar{\theta}_{k+1},...,(1+\epsilon)\bar{\theta}_n)\phi - \phi_t = \rho\phi, \\ \\ \hat{B}_k\phi = 0, \quad \phi \text{ is } T\text{-periodic}, \end{cases} \tag{2.41}$$

is positive. Here, $w = \tilde{\theta}_k$ is the unique globally attracting positive T-periodic solution of (2.36) with $\hat{L}_1, C_1, \hat{B}_1, R_1$ respectively replaced by $\hat{L}_k, c_k, \hat{B}_k$ and $r_k(x,t,(1-\epsilon)\tilde{\theta}_1,...,(1-\epsilon)\tilde{\theta}_{j(k)},(1+\epsilon)\bar{\theta}_{j(k)+1},...,(1+\epsilon)\bar{\theta}_{k-1},(1+\epsilon)\bar{\theta}_{k+1},...,(1+\epsilon)\bar{\theta}_n), c_k)$. Then, for any sufficiently small $\epsilon > 0$, we have

$$u_i(x,t) \geq (1-\epsilon)\tilde{\theta}_i(x,t) \quad x \in \bar{\Omega},\ i = 1,...,n \tag{2.42}$$

for t sufficiently large.

Proof. We first prove that any solution must be bounded in its interval of existence. This would imply by the general theory for parabolic systems (cf. e.g. Amann [3]) that the solution must exist globally for $t > 0$. Since the i-th component in (2.30) is zero when $u_i = 0$, it follows from maximum principles that the set $\{\mathbf{u} \in R^n : u_i \geq 0, i = 1,...,n\}$ is positively invariant (see e.g. Smoller [209]). More precisely, if $u_i \geq 0, u_i \not\equiv 0$ at $t = 0$, then $u_i > 0$ for all $t > 0$. Moreover, under Dirichlet boundary conditions, we have $\partial u_i/\partial\nu < 0$ on $\partial\Omega$, and under Neumann and mixed case, we have $u_i > 0$ on $\bar{\Omega}$. Assume that a solution of (2.30) exists on $[0,\tau)$. From (2.34), we have

$$\begin{aligned} (u_1)_t &= \hat{L}_1 u_1 + \hat{f}_1(x,t,u_1,...,u_n)u_1 \\ &\leq \hat{L}_1 u_1 + (R_1(x,t) - C_1(x,t)u_1)u_1 \qquad \text{in } \Omega \times (0,\tau); \end{aligned}$$

so that by comparison we find $u_1 \leq v_1$, where v_1 is the solution of

$$(2.43) \qquad \begin{cases} (v_1)_t = \hat{L}_1 v_1 + (R_1(x,t) - C_1(x,t)v_1)v_1 & \text{in } \Omega \times (0,\tau), \\ \hat{B}_1 v_1 = 0 & \text{on } \partial\Omega \times (0,\tau), \\ v_1(x,0) = u_1(x,0) & \text{in } \bar{\Omega}. \end{cases}$$

Since $C_1(x,t)$ is larger than a positive constant, any sufficiently large constant is an upper solution for problem (2.43). Thus v_1 exists globally and $v_1(x,t) \leq V_1$ on $[0,\tau)$ for some constant V_1 which is dependent on $u_1(x,0)$. We thus have $0 \leq u_1 \leq V_1$ on $[0,\tau)$. From (2.34) again, we have

$$\begin{aligned} (u_2)_t &= \hat{L}_2 u_2 + \hat{f}_2(x,t,u_1,...,u_n)u_2 \\ &\leq \hat{L}_2 u_2 + (R_2(x,t,V_1) - C_2(x,t)u_2)u_2 \end{aligned}$$

in $\Omega \times [0,\tau)$. By comparison, we find $u_2 \leq v_2$ on $[0,\tau)$, where v_2 is the solution of

$$\begin{cases} (v_2)_t = \hat{L}_2 v_1 + (R_2(x,t,V_1) - C_2(x,t)v_2)v_2 & \text{in } \Omega \times (0,\tau), \\ \hat{B}_2 v_2 = 0 & \text{on } \partial\Omega \times (0,\tau), \\ v_2(x,0) = u_2(x,0) & \text{in } \bar{\Omega}. \end{cases}$$

As in the case for v_1, we deduce that $v_2 \leq V_2$ where V_2 is a constant which depends on V_1 and $u_2(x,0)$. We thus obtain $0 \leq u_2 \leq V_2$ on $[0,\tau)$. Continuing with this argument, we deduce that $0 \leq u_k \leq V_k$ on $[0,\tau)$ for $k = 1,...,n$. Since $sup\,|\mathbf{u}|$ does not tend to infinity as $t \to \tau^-$, the interval $(0,\tau)$ cannot be the maximal interval of existence. Since $\tau > 0$ is arbitrary, we can conclude that the solution exists globally for all $t > 0$.

We next refine our estimate for the upper bound to deduce (2.38). Again, we observe that u_1 is a lower solution for (2.43); and by Lemma 2.1, $v_1 \to \bar{\theta}_1 \equiv \theta(\hat{L}_1, R_1, C_1)$ as $t \to \infty$, with convergence in $C^{1+\alpha,\alpha/2}(\Omega,(0,\infty))$. Thus for any $\epsilon > 0$, we have $u_1 \leq v_1 \leq (1+\epsilon)\theta(\hat{L}_1, R_1, C_1)$ for sufficiently large t. It follows that for sufficiently large t, we have

$$(u_2)_t \leq \hat{L}_2 u_2 + [R_2(x,t,(1+\epsilon)\bar{\theta}_1) - C_2 u_2]u_2,$$

under boundary condition $\hat{B}_2$. Hence, u_2 is a lower solution of

$$(\hat{v}_2)_t = \hat{L}_2 \hat{v}_2 + [R_2(x,t,(1+\epsilon)\bar{\theta}_1) - C_2 \hat{v}_2]\hat{v}_2,$$

with the same initial-boundary conditions. By assumption (2.37) with $k = 2$ and Lemma 2.1, we have $\hat{v}_2 \to \bar{\theta}_2 \equiv \theta(\hat{L}_2, R_2(x,t,(1+\epsilon)\bar{\theta}_1), C_2)$ as $t \to \infty$. Consequently, we have $u_2 \leq \hat{v}_2 \leq (1+\epsilon)\bar{\theta}_2$ for t sufficiently large. Continuing with this argument, we conclude that the we have $u_k \leq (1+\epsilon)\bar{\theta}_k$ for $k = 1, ..., n$. Here $\bar{\theta}_k$ is the unique positive globally attracting periodic solution for

$$\begin{cases} v_t = \hat{L}_k v + [R_k(x,t,(1+\epsilon)\bar{\theta}_1, ..., (1+\epsilon)\bar{\theta}_{k-1}) - C_k v]v, \\ \\ \hat{B}_k v = 0. \end{cases}$$

This proves inequalities (2.38).

We next proceed to prove the lower bounds (2.42). We define w_1 to be the solution of

$$\begin{cases} (w_1)_t = \hat{L}_1 w_1 + [r_1(x,t,(1+\epsilon)\bar{\theta}_2, ..., (1+\epsilon)\bar{\theta}_n) - c_1(x,t)w_1]w_1, \\ \\ \hat{B}_1 w_1 = 0, \\ \\ w_1(x,t_0) = u_1(x,t_0), \end{cases} \tag{2.44}$$

with t_0 to be determined. From (2.38) and hypothesis (2.39), we find that for t_0 sufficiently large, u_1 is an upper solution for problem (2.44) for $t \geq t_0$. Also, u_1 is positive on $\bar{\Omega}$ under Neumann and mixed boundary conditions, and positive on Ω with $\partial u/\partial \nu < 0$ on $\partial\Omega$ under Dirichlet boundary condition. From the hypothesis that the principal eigenvalue of (2.40) is positive, we have $\tilde{\theta}_1 \equiv \theta(\hat{L}_1, r_1(x,t,(1+\epsilon)\bar{\theta}_2, ..., (1+\epsilon)\bar{\theta}_n), c_1)$ as a global attractor for nontrivial non-negative solutions. Hence, $w_1 \to \tilde{\theta}_1$ as $t \to \infty$; and for $t > t_1$ sufficiently large, we have $u_1 \geq (1-\epsilon)\tilde{\theta}_1$. Next, define w_2 to be the solution of the problem:

$$\begin{cases} (w_2)_t = \hat{L}_2 w_2 + [r_2(x,t,(1-\epsilon)\tilde{\theta}_1, (1+\epsilon)\bar{\theta}_3, ..., (1+\epsilon)\bar{\theta}_n) - c_2(x,t)w_2]w_2, \\ \\ \hat{B}_1 w_2 = 0, \\ \\ w_2(x,t_1) = u_2(x,t_1), \end{cases} \tag{2.45}$$

if $j(2) = 1$; or

$$\begin{cases} (w_2)_t = \hat{L}_2 w_2 + [r_2(x,t,(1+\epsilon)\bar{\theta}_1, (1+\epsilon)\bar{\theta}_3, ..., (1+\epsilon)\bar{\theta}_n) - c_2(x,t)w_2]w_2, \\ \\ \hat{B}_1 w_2 = 0, \\ \\ w_2(x,t_1) = u_2(x,t_1), \end{cases} \tag{2.46}$$

if $j(2) = 0$. In either case, the assumption on the sign of the principal eigenvalue of (2.41) and Lemma 2.1 imply that w_2 is positive inside Ω for $t \geq t_1$ and $w_2 \to \tilde{\theta}_2$

as $t \to \infty$. Again, by (2.38), (2.39) and comparison, we obtain $u_2 \geq w_2 \geq (1-\epsilon)\tilde{\theta}_2$ for t sufficiently large. We can continue with this argument to obtain the lower bounds (2.42) for $u_3, ..., u_n$. The different choices of w_k for comparison, analogous to (2.45) or (2.46), depend on the value of $j(k)$, $k = 3, ..., n$. They are taken into consideration in the definition of $\tilde{\theta}_k$.

Note that in Theorem 2.7, the upper bounds are used to construct the lower bounds. It is therefore more convenient and simpler to replace the upper bounds with constant bounds, although the estimate may not be as sharp. This is the object of the following corollary.

Corollary 2.8. *(i) Consider problem (2.30). Suppose $\hat{f}_i$ satisfies (2.34) and (2.35), and define $\bar{\theta}_1 \equiv \theta(\hat{L}_1, R_1, C_1)$ to be the unique globally attracting positive T-periodic solution of (2.36) as in Theorem 2.7. Let M_1 be a constant such that $M_1 > \bar{\theta}$ (By maximum principle, we may choose any $M_1 > sup(R_1/C_1)$). Define M_k inductively to be a constant such that $M_k > \theta(\hat{L}_k, R_k(x, t, M_1, ..., M_{k-1}), C_k)$, where we assume that the principal eigenvalue of*

$$(2.47) \qquad \begin{cases} \hat{L}_k\phi + R_k(x, t, M_1, ..., M_{k-1})\phi - \phi_t = \rho\phi, \\ \hat{B}_k\phi = 0, \quad \phi \text{ is } T\text{-periodic}, \end{cases}$$

is positive for each $k = 1, ..., n$. Then, for t sufficiently large, we have

$$(2.48) \qquad u_i(x, t) \leq M_i \quad \text{for } x \in \bar{\Omega}, \ i = 1, ..., n.$$

(ii) Assume all the above hypotheses in this Corollary and further $\hat{f}_k, k = 1, ..., n$ satisfy (2.39). Suppose the principal eigenvalue in (2.40) remain positive with $(1+\epsilon)\bar{\theta}_i$ replaced with $M_i, i = 2, ..., n$. Let $\tilde{\theta}_1^* \equiv \theta(\hat{L}_1, r_1(x, t, M_2, ..., M_n), c_1)$; and define inductively:

$$\tilde{\theta}_k^* \equiv \theta(\hat{L}_k, r_k(x, t, (1-\epsilon)\tilde{\theta}_1^*, ..., (1-\epsilon)\tilde{\theta}_{j(k)}^*, M_{j(k)+1}, ..., M_{k-1}, M_{k+1}, ..., M_n), c_k),$$

where we assume that for $\epsilon > 0$ sufficiently small the principal eigenvalue of

$$(2.49) \qquad \begin{aligned} &\hat{L}_k\phi + r_k(x, t, (1-\epsilon)\tilde{\theta}_1^*, ..., (1-\epsilon)\tilde{\theta}_{j(k)}^*, M_{j(k)+1}, ..., \\ &\qquad\qquad M_{k-1}, M_{k+1}, ..., M_n)\phi - \phi_t = \rho\phi, \\ &\hat{B}_k\phi = 0, \quad \phi \text{ is } T\text{-periodic}, \end{aligned}$$

is positive, for $\epsilon > 0$ sufficiently small. Then for any sufficiently small $\epsilon > 0$ and t sufficiently large, we have

$$(2.50) \qquad u_i(x, t) \geq (1-\epsilon)\tilde{\theta}_i^*(x, t) \quad \text{for } x \in \bar{\Omega}, \ i = 1, ..., n.$$

Example 2.1. We consider the case when u_3 represents a predator preying on two competing preys u_1 and u_2 in $\Omega \times (0,\infty)$. More precisely, consider

$$(2.51)\qquad \begin{cases} (u_1)_t = \sigma_1 \Delta u_1 + (a_1 - b_{11}u_1 - b_{12}u_2 - b_{13}u_3)u_1, \\ (u_2)_t = \sigma_2 \Delta u_2 + (a_2 - b_{21}u_1 - b_{22}u_2 - b_{23}u_3)u_2, \\ (u_3)_t = \sigma_3 \Delta u_3 + (a_3 + b_{31}u_1 + b_{32}u_2 - b_{33}u_3)u_3, \\ u_1 = u_2 = u_3 = 0 \quad \text{on } \partial\Omega \times (0,\infty). \end{cases}$$

The coefficients a_i, b_{ij} may depend on x, t and $\mathbf{u}$. We assume that there are constants $\underline{a}_i, \bar{a}_i, \underline{b}_{ij}, \bar{b}_{ij}$ with $\underline{a}_i \leq a_i \leq \bar{a}_i$ and $\underline{b}_{ij} \leq b_{ij} \leq \bar{b}_{ij}$, where $\underline{a}_1, \underline{a}_2$ and $\underline{b}_{ii}, i = 1,2,3$ are strictly positive, and $b_{ij} \geq 0$ for all i, j. In order to apply Corollary 2.8 to problem (2.51), we may choose the following constant functions for the estimates (2.34):

$$R_1 = \bar{a}_1,\ R_2 = \bar{a}_2,\ R_3 = \bar{a}_3 + \bar{b}_{31}K_1 + \bar{b}_{32}K_2, \text{ and } C_i = \underline{b}_{ii}.$$

The conditions for $k = 1,2$ in (2.47) are satisfied if

$$(2.52)\qquad \bar{a}_1 > \sigma_1\lambda_1,\ \ \bar{a}_2 > \sigma_2\lambda_1,$$

where $\lambda = \lambda_1$ is the principal eigenvalue for the eigenvalue problem $-\Delta\phi = \lambda\phi$ on Ω, with $\phi = 0$ on $\partial\Omega$. Under conditions (2.52), we have $u_1 \leq M_1 := \bar{a}_1/\underline{b}_{11}, u_2 \leq M_2 := \bar{a}_2/\underline{b}_{22}$ for sufficiently large t. Hypothesis (2.47) with $k = 3$ is satisfied if

$$(2.53)\qquad \bar{a}_3 + \bar{b}_{31}\frac{\bar{a}_1}{\underline{b}_{11}} + \bar{b}_{32}\frac{\bar{a}_2}{\underline{b}_{22}} > \sigma_3\lambda_1.$$

Then, we may choose the upper bound for u_3 as

$$M_3 = \underline{b}_{33}^{-1}[\bar{a}_3 + \bar{b}_{31}\frac{\bar{a}_1}{\underline{b}_{11}} + \bar{b}_{32}\frac{\bar{a}_2}{\underline{b}_{22}}].$$

To obtain lower bounds, we choose for (2.39),

$$r_1 = \underline{a}_1 - \bar{b}_{12}K_2 - \bar{b}_{13}K_3,\ r_2 = \underline{a}_2 - \bar{b}_{21}K_1 - \bar{b}_{23}K_3,\ r_3 = \underline{a}_3 + \underline{b}_{31}K_1 + \underline{b}_{32}K_2.$$

Here, we have $j(1) = j(2) = 0$, and $j(3) = 2$. The hypotheses corresponding to (2.40) and (2.49) become

$$(2.54)\qquad \begin{aligned} &\underline{a}_1 - \bar{b}_{12}M_2 - \bar{b}_{13}M_3 > \sigma_1\lambda_1, \\ &\underline{a}_2 - \bar{b}_{21}M_1 - \bar{b}_{23}M_3 > \sigma_2\lambda_1. \end{aligned}$$

From Corollary 2.8, we find that for any $\epsilon > 0$ and t sufficiently large we have

$$(2.55)\qquad \begin{aligned} u_1 &\geq (1-\epsilon)\theta(\sigma_1\Delta, \underline{a}_1 - \bar{b}_{12}M_2 - \bar{b}_{13}M_3, \bar{b}_{11}) \equiv (1-\epsilon)\tilde{\theta}_1^*, \\ u_2 &\geq (1-\epsilon)\theta(\sigma_2\Delta, \underline{a}_2 - \bar{b}_{21}M_1 - \bar{b}_{23}M_3, \bar{b}_{22}) \equiv (1-\epsilon)\tilde{\theta}_2^*. \end{aligned}$$

From (2.55) we obtain the lower estimate

$$(2.56)\qquad u_3 \geq (1-\epsilon)\theta(\sigma_3\Delta, \underline{a}_3 + \underline{b}_{31}(1-\epsilon)\tilde{\theta}_1^* + \underline{b}_{32}(1-\epsilon)\tilde{\theta}_2^*, \bar{b}_{33})$$

for any $\epsilon > 0$ and t sufficiently large, provided the principal eigenvalue for

$$(2.57)\qquad \begin{cases} \sigma_3\Delta\phi + (\underline{a}_3 + \underline{b}_{31}(1-\epsilon)\tilde{\theta}_1^* + \underline{b}_{32}(1-\epsilon)\tilde{\theta}_2^*)\phi = \rho\phi & \text{in } \Omega, \\ \phi = 0 & \text{on } \partial\Omega \end{cases}$$

is positive. Note that conditions (2.54) imply conditions (2.52), which can thus be removed.

Example 2.2. Consider the case of n competing species.

$$(2.58)\qquad \begin{cases} (u_i)_t = \sigma_i\Delta u + (a_i - \sum_{j=1}^n b_{ij}u_j)u_j & \text{in } \Omega \times (0,\infty), \\ u_i = 0 & \text{on } \partial\Omega \times (0,\infty), \end{cases}$$

$i = 1, ..., n$, where $\underline{a}_i \leq a_i(x,t,\mathbf{u}) \leq \bar{a}_i$ and $\underline{b}_i \leq b_i(x,t,\mathbf{u}) \leq \bar{b}_i$, with all the constants $\underline{a}_i, \bar{a}_i, \underline{b}_i, \bar{b}_i$ positive. We can again apply Corollary 2.8 and choose $R_i = \bar{a}_i, C_i = \underline{b}_i, r_i = \underline{a}_i - \sum_{j\neq i}\bar{b}_{ij}K_j, c_i = \bar{b}_{ii}$. As before, we find that, for large t, the upper estimates $u_i \leq M_i = \bar{a}_i/\underline{b}_{ii}$. From (2.49) and (2.50), we find that if

$$(2.59)\qquad \underline{a}_i - \sum_{j\neq i}(\bar{b}_{ij}\bar{a}_j/\underline{b}_{jj}) > \sigma_i\lambda_1, \quad i = 1, ..., n.$$

then, we have the following lower estimates for $x \in \Omega$ as $t \to \infty$:

$$(2.60)\qquad \begin{aligned} u_i(x,t) &\geq (1-\epsilon)\theta(\sigma_i\Delta, \underline{a}_i - \textstyle\sum_{j\neq i}(\bar{b}_{ij}\bar{a}_j/\underline{b}_{jj}), \bar{b}_{ii}) \\ &\geq (1-\epsilon)[(\underline{a}_i - \textstyle\sum_{j\neq i}(\bar{b}_{ij}\bar{a}_j/\underline{b}_{jj}) - \sigma_i\lambda_1)/\bar{b}_{ii}]\phi_1(x), \end{aligned}$$

for $i = 1, ..., n$, $\epsilon > 0$ small enough. Here, λ_1 is the principal eigenvalue for the operator $-\Delta$ on Ω with homogeneous Dirichlet boundary data; and $\phi_1(x)$ is the corresponding positive eigenfunction with $sup_{x\in\bar{\Omega}}\phi(x) = 1$.

Since λ_1 and ϕ_1 can be explicitly computed for simple geometries and there are well-known methods for computing them in general situations, formula (2.60) provides very convenient lower estimate for practical purposes. More examples can be found in [19].

Part C: Finite-Time Blowup caused by Boundary Inflow.

We next consider reaction-diffusion systems when there does not exist a any simultaneous upper and lower estimates for the solution of the system. Moreover, when the interior reaction are dominated by boundary feedback in an appropriate way, one can construct lower solution of the problem blowing up in finite time. This implies that the solution of the problem blow up in finite time. We first consider a simple scalar problem:

$$(2.61)\qquad \begin{cases} u_t - \Delta u = f(u) & \text{in } \Omega \times (0,T), \\ \frac{\partial u}{\partial \nu} = g(u) & \text{on } \partial\Omega \times (0,T), \\ u(x,0) = u_0(x) & \text{in } \Omega, \end{cases}$$

where $f(u), g(u)$ are continuously differentiable functions for $u \geq 0$, and

$$(2.62)\qquad g(u) > 0 \ \text{ for all } u > 0.$$

Here the interior reaction may be negative, i.e. $f(u) \leq 0$ for $u > 0$, suppressing the growth of the species. The species is flowing into the region, as indicated by the sign of $g(u)$, and $\partial/\partial\nu$ is the outward normal derivative. The next theorem indicates that if the boundary inflow dominates in appropriate amount, the species will blowup in the interior in finite time, even the species may be decaying inside. In this part, we assume that Ω is a bounded domain in R^N with smooth boundary. Substantial amount of early work appeared in the literature for such kind of problem when $f(u) \equiv 0$. Levine and Payne [145] showed that if $g(u) = |u|^{1+\epsilon}h(u)$ with $\epsilon > 0$ and $h(u)$ increasing, then there cannot be any global solution. Later, Walter [225] generalized to the case when $g(u)$ and $g'(u)$ are continuous, positive and increasing, then the solution will blow up in finite time when $\int^{\infty} ds/gg' < \infty$. Moreover, the solution exists globally provided that $\int^{\infty} ds/gg' = \infty$. Hu and Yin [91] studied the blow up properties for $g(u) = u^p$ with $p > 1$. They obtained many estimates concerning the blow-up rate of the solution. Many other related work can be found in e.g. Filo [61], Sperb [210] and Wang and Wu [227]. When $f(u) \not\equiv 0$, Chipot, Fila and Quittner [27] and Quittner [189] studied some properties of global solutions when $f(u) = au^p$ and $g(u) = u^q$. Theorem 2.9 is a modification of a theorem in Leung and Zhang [143], which extends some of the results just mentioned above.

Theorem 2.9 (Scalar Case). *Let $u(x,t) \in C^{2,1}(\Omega \times (0,T)) \cap C^{1,0}(\bar{\Omega} \times [0,T])$ be a positive solution of (2.61), under assumptions (2.62).*

(i) Suppose there exists a continuously differentiable function $m(u)$ for $0 \le u < \infty$ such that $m(0) = 0, m'(u) \ge 0$ and

$$\frac{-1}{\epsilon} f(u) \le m(u) \le \left(\frac{|\partial\Omega|}{(1+\epsilon)|\Omega|}\right) g(u) \tag{2.63}$$

for some positive constant ϵ. Then the solution will blow up in finite time if $u_0(x) > 0$ is large enough for all $x \in \bar{\Omega}$, and

$$\int^{\infty} \frac{1}{m(s)} ds < \infty, \qquad \int_0^1 \frac{1}{m(s)} ds = \infty. \tag{2.64}$$

(Here, $|\partial\Omega|$ and $|\Omega|$ are respectively the measures of $\partial\Omega$ and Ω. Note that if $f(u) \ge 0$, then the first part of inequalities (2.63) is always satisfied.).
(ii) Suppose $f(u) \ge 0$ for $u \ge 0$; $g(u)$ is continuously differentiable for $0 \le u < \infty$ such that $g(0) = 0, g'(u) \ge 0$ and

$$\int^{\infty} \frac{1}{g(s)} ds < \infty.$$

Then the solution will blow up in finite time if $u_0(x) > 0$ is large enough for all $x \in \bar{\Omega}$.

Proof. We seek lower solution for problem (2.61) of the form $\underline{u}(x,t) = v(t + h(x))$ where $v = v(s), s > 0$ satisfies:

$$\frac{dv}{ds} = m(v) \quad \text{and} \quad v(0) = c_0.$$

Here, c_0 is some positive constant and $h(x)$ is an unknown positive function to be determined. Denoting $t + h(x)$ by s, we find

$$\frac{dv}{ds} > 0 \quad \text{and} \quad \frac{d^2v}{ds^2} = m'(v)\frac{dv}{ds} \ge 0$$

when v is > 0. We then compute the derivatives of $\underline{u}(x,t)$ as follows:

$$\underline{u}_t = \tfrac{dv}{ds}, \quad \underline{u}_{x_i} = \tfrac{dv}{ds} h_{x_i},$$

$$\underline{u}_{x_i x_i} = \tfrac{d^2v}{ds^2}(h_{x_i})^2 + \tfrac{dv}{ds} h_{x_i x_i}.$$

Hence, we have

$$\underline{u}_t - \Delta\underline{u} - f(\underline{u}) = \frac{dv}{ds}(1 - \Delta h) - \frac{d^2v}{ds^2}|\nabla h|^2 - f(v). \tag{2.65}$$

Choose h to be the solution of the Neumann problem

$$\Delta h = 1 + \epsilon \text{ for } x \in \Omega, \text{ and } \frac{\partial h}{\partial \nu} = \beta \text{ for } x \in \partial\Omega,$$

where $\beta = (1+\epsilon)|\Omega|/|\partial\Omega|$ and $min_{x\in\bar{\Omega}} h(x) = 0$. Note that such $h(x)$ is uniquely determined. Since $m(0) = 0$, it follows from the second condition in (2.64) that we can define c_0 small enough such that $v(max_{x\in\bar{\Omega}} h(x)) < \infty$. By choosing $h(x)$ in this manner, we find from (2.65) that

$$\text{(2.66)} \qquad \underline{u}_t - \Delta\underline{u} - f(\underline{u}) \leq \frac{dv}{ds}(-\epsilon) - f(v) = -\epsilon m(v) - f(v) \leq 0.$$

The last inequality is due to (2.63). On the boundary, we compute

$$\text{(2.67)} \qquad \frac{\partial \underline{u}}{\partial \nu} - g(\underline{u}) = \frac{dv}{ds}\frac{\partial h}{\partial \nu} - g(v) = m(v)\beta - g(v) \leq 0.$$

Combining (2.66) and (2.67), we find that $\underline{u}$ is a lower solution for problem (2.61) under the assumptions in part (i) provided $u_0(x) \geq v(h(x))$. Thus for large enough $u_0(x)$ (it suffices to have $u_0(x) \geq v(max_{x\in\bar{\Omega}} h(x))$ for all $x \in \bar{\Omega}$), then $\underline{u} = v(t + h(x))$ is a lower solution. Since $v(s)$ blow up in finite time (due to the first condition in (2.64)), we conclude from standard comparison theory (see e.g. [125] or [183]) that $u(x,t)$ blows up in finite time. This proves part (i).

For part (ii), we define $m(u) = (1/(1+\epsilon))(|\partial\Omega|/|\Omega|)g(u)$ for any $\epsilon > 0$. The inequalities in (2.63) are trivially satisfied with $m(u)$ replaced by $\rho m(u)$ with any $\rho \in (0, 1]$. Note that since $f(u) > 0$, the first part of (2.63) must be valid. We can then apply nearly the same proof as part (i) to obtain a lower solution and conclude with the finite time blowup property as before. The only change needed is to replace all $m(v)$ that appear in the proof by $\rho m(v)$ where ρ is to be chosen in $(0, 1)$. We finally choose $\rho > 0$ sufficiently small so that $v(max_{x\in\bar{\Omega}} h(x)) < \infty$. This is possible because it would take longer for the solution of

$$\frac{dv}{ds} = \rho m(v), \;\; v(0) = c_0,$$

to blow up by reducing ρ. Thus there is no need to assume $\int_0^1 (1/m(s))ds = \infty$ as in part (i).

We next study a system of reaction-diffusion equations with nonlinear boundary inflow. We obtain sufficient conditions for finite-time blowup when the inflow is sufficiently strong, even the species are depleted in the interior. Consider the problem:

$$
(2.68)\qquad \begin{cases} u_t - \sigma_1 \Delta u = f_1(u,v) & \text{for } (x,t) \in \Omega \times (0,T), \\ v_t - \sigma_2 \Delta v = f_2(u,v) & \text{for } (x,t) \in \Omega \times (0,T), \\ \frac{\partial u}{\partial \nu} = g_1(u,v) & \text{for } (x,t) \in \partial\Omega \times [0,T), \\ \frac{\partial v}{\partial \nu} = g_2(u,v) & \text{for } (x,t) \in \partial\Omega \times [0,T), \\ u(x,0) = u_0(x) > 0 & \text{for } x \in \Omega, \\ v(x,0) = v_0(x) > 0 & \text{for } x \in \Omega. \end{cases}
$$

Here σ_1, σ_2 are positive constants, and $f_i(u,v), g_i(u,v)$ are smooth functions for $u, v \geq 0, i = 1, 2$. We assume the following for any $u, v \geq 0$:

$$
(2.69)\qquad \begin{aligned} &(i)\ \ f_i(u,v) < 0,\ \ g_i(u,v) > 0\ \text{ for } i = 1,2 \text{ and } u, v > 0, \\ &(ii)\ \ f_1(0,v) = f_2(u,0) = 0, \\ &(iii)\ \ f_{1,v}(u,v) \geq 0 \text{ and } f_{2,u}(u,v) \geq 0, \\ &(iv)\ \ g_{i,u}(u,v) \geq 0 \text{ and } g_{i,v}(u,v) \geq 0 \text{ for } i = 1,2. \end{aligned}
$$

We say a solution of (2.68) blows up in finite time in the sense that there exists a finite positive time T such that the solution is classical for $0 < t < T$ and $lim_{t \to T^-} max_{x \in \bar{\Omega}} (|u(x,t)| + |v(x,t)|) = \infty$.

Theorem 2.10 (Finite-time Blowup for Systems with Nonlinear Boundary Conditions). *Let (u,v) be a positive solution of (2.68) under assumptions (2.69i-iv) with $u, v \in C^{2,1}(\Omega \times (0,T)) \cap C^{1,0}(\bar{\Omega} \times [0,T])$. Further, suppose there exists a positive number $\lambda > 1$ such that*

$$
(2.70)\qquad \lambda \frac{|\Omega|}{\sigma_1 |\partial\Omega|} |f_1(u,v)| \leq g_1(u,v) \ \ \textit{and} \ \ \lambda \frac{|\Omega|}{\sigma_2 |\partial\Omega|} |f_2(u,v)| \leq g_2(u,v)
$$

for all $u, v \geq 0$. Let $(A(s), B(s))$ be the solution of the following system of ordinary differential equations:

$$
(2.71)\qquad \begin{cases} \frac{dA}{ds} = c\sigma_1 g_1(A,B) \ \ \textit{for } s \geq 0, \\ \frac{dB}{ds} = c\sigma_2 g_2(A,B) \ \ \textit{for } s \geq 0, \\ A(0) = A_0 > 0 \ \ \textit{and} \ \ B(0) = B_0 > 0, \end{cases}
$$

where $c = ((\lambda - 1)/\lambda) \cdot |\partial\Omega|/|\Omega|$, and A_0, B_0 are some positive constants. Then if $(A(s), B(s))$ blows up in finite time, the solution (u, v) of (2.68) blows up in finite time for large initial data.

Proof. We construct a pair of lower solutions for the system, with finite time blow up property, and then use a comparison theorem. Let $(A(s), B(s))$ be the solution of (2.71) above. Define $\underline{u}(x,t) = A(t + \Psi(x))$ and $\underline{v} = B(t + \Phi(x))$ with $\Psi(x)$ and $\Phi(x)$ to be determined later. They will be chosen so that $\underline{u}$ and $\underline{v}$ form a pair of lower solutions for problem (2.68). As in Theorem 2.9, we first calculate

$$\underline{u}_t - \sigma_1 \Delta \underline{u} - f_1(\underline{u}, \underline{v}) = \tfrac{dA}{ds}(1 - \sigma_1 \Delta\Psi) - \sigma_1 \tfrac{d^2A}{ds^2}|\nabla\Psi|^2 - f_1(A, B),$$

$$\tfrac{\partial \underline{u}}{\partial \nu} = \tfrac{dA}{ds} \cdot \tfrac{d\Psi}{\partial \nu},$$

where $s = t + \Psi(x)$. We choose $\Psi(x)$ to be a solution of the following problem:

$$\sigma_1 \Delta\Psi = \alpha \quad \text{for } x \in \Omega, \quad \text{and} \quad \frac{\partial\Psi}{\partial\nu} = \beta \ \text{ for } x \in \partial\Omega,$$

with α to be determined later and $\beta = \alpha|\Omega|/(\sigma_1|\partial\Omega|)$. From linear theory we know such Ψ exists and it can be chosen such that $\Psi > 0$ in $\bar{\Omega}$. With such choice of Ψ, we have

$$(2.72) \qquad \begin{cases} \underline{u}_t - \sigma_1 \Delta \underline{u} - f_1(\underline{u}, \underline{v}) \leq 0 & \text{in } \Omega \times (0, T), \\ \tfrac{\partial \underline{u}}{\partial \nu} \leq g_1(u, v) & \text{in } \partial\Omega \times (0, T), \end{cases}$$

provided the following holds:

$$(2.73) \qquad \frac{dA}{ds}(\alpha - 1) \geq -f_1(A, B), \quad \text{and} \quad \frac{dA}{ds}\beta \leq g_1(A, B) \ \text{ for } s > 0.$$

Let $\lambda > 1$, such that

$$\lambda \frac{|\Omega|}{\sigma_1 |\partial\Omega|} |f_1(u, v)| \leq g_1(u, v),$$

for any $u, v \geq 0$ as described in (2.70), and choose $\alpha = \lambda/(\lambda - 1)$. We see that the two inequalities in (2.73) are true from the definitions of $A(s), B(s)$. Therefore (2.72) is valid. Similarly, we can choose Φ such that $\underline{v}$ satisfies corresponding inequalities so that $(\underline{u}, \underline{v})$ forms a lower solution pair for problem (2.68). Note that if (2.70) holds for certain $\lambda > 1$, then it still holds by reducing λ to be closer to 1, with $\lambda > 1$. Then the parameter c in (2.71) can be reduced to as small as possible, and the blow up time for the solution of (2.71) will be expanded. This insures that $A(\Psi(x)) = \underline{u}(x, 0)$ and $B(\Phi(x)) = \underline{v}(x, 0)$ are defined for all

$x \in \bar{\Omega}$. Since the solution $(A(s), B(s))$ blow up in finite time, we conclude by using comparison theorem (see e.g. [125]) that the solution of problem (2.68) blows up in finite time for large initial data. This completes the proof of the theorem.

For more specific systems of the form (2.68), we can readily verify that Theorem 2.10 can be applied to the following two examples.

Example 2.3.

$$f_1(u,v) = -\tfrac{u^k}{v+c}, \quad f_2(u,v) = -\tfrac{v^l}{u+c},$$

$$g_1(u,v) = u^m + uv^p, \quad g_2(u,v) = v^n + vu^q,$$

where k, l, c, m, n, p, q are positive constants with $m, n > 1$ and further $m > k > 0, n > l > 0$.

Example 2.4.

$$f_1(u,v) = -u, \quad f_2(u,v) = -v,$$

$$g_1(u,v) = ku(v+1), \quad g_2(u,v) = kv(u+1)$$

for some large positive constant k.

It is interesting to note that in Example 2.4, any single equation does not blow-up in finite time if we set the other variable to be a constant. However, the coupled system does have a finite time blow-up solution for large initial data. To apply Theorem 2.10, it may take some effort to verify that the corresponding ordinary differential equations (2.71) have finite time blow-up solutions. In some situations, the following theorem is more readily applicable because it is easier to verify the hypotheses.

Theorem 2.11. *Consider problem (2.68) under hypotheses (2.69i-iv), with* $\sigma_1 = \sigma_2$. *Let* (u,v) *be a solution with* $u, v \in C^3(\bar{\Omega} \times [0,T])$. *Further, suppose that there exists a continuously differentiable function* $m(s)$ *for* $0 \le s < \infty$ *such that* $m(0) = 0, m'(s) \ge 0, \int^\infty ds/m(s) < \infty, \int_0^1 ds/m(s) = \infty$; *and for any* $u, v \ge 0$, *there exists a constant* $\lambda > 1$ *such that*

$$g_1(u,v) + g_2(u,v) \ge m(u+v) \ge \lambda \frac{|\Omega|}{\sigma_1 |\partial\Omega|}(|f_1(u,v)| + |f_2(u,v)|). \tag{2.74}$$

Then the solution (u,v) *of (2.68) blows up in finite time for large initial conditions.*

Corollary 2.12. *Assume $\sigma_1, \sigma_2 > 0$ and all the other hypotheses of Theorem 2.11, except that σ_1 in (2.74) is replaced by $\min\{\sigma_1, \sigma_2\}$. Furthermore, we assume the initial data satisfy*

$$\begin{cases} -\sigma_1 \Delta u_0(x) < f_1(u_0(x), v_0(x)), \\ -\sigma_2 \Delta v_0(x) < f_2(u_0(x), v_0(x)), \end{cases} \tag{2.75}$$

for $x \in \bar{\Omega}$. Then if $u_0(x)$ and $v_0(x)$ are large enough, the positive solution of (2.68) as described in Theorem 2.11 blows up in finite time.

We can readily verify that Theorem 2.11 can be applied to the following example.

Example 2.5.

$$\begin{cases} u_t - \Delta u = -(u^p + c(u,v)) & \text{in } \Omega \times (0,T), \\ v_t - \Delta v = -(v^q + d(u,v)) & \text{in } \Omega \times (0,T), \\ \frac{\partial u}{\partial \nu} = v^m, \ \frac{\partial v}{\partial \nu} = u^l & \text{in } \partial\Omega \times (0,T), \end{cases}$$

with initial data $u_0(x)$ and $v_0(x)$ where $0 \le c(u,v), d(u,v) \le M$ for some constant M, and $c_v(u,v), d_u(u,v) \le 0$. If $0 < p < l, 0 < q < m$ and $m > 1, l > 1$, then the solution blows up in finite time for large initial data.

Proof of Theorem 2.11 and Corollary 2.12. Let $w = u + v$, we find from (2.68) and assumption (2.74) that w satisfies

$$\begin{cases} w_t - \sigma_1 \Delta w = f_1(u,v) + f_2(u,v) \ge -\frac{\sigma_1 |\partial\Omega|}{\lambda |\Omega|} g(w) & \text{in } \Omega \times (0,T), \\ \frac{\partial w}{\partial \nu} = g_1(u,v) + g_2(u,v) \ge g(w) & \text{in } \partial\Omega \times (0,T). \end{cases}$$

Applying a slight modification of Theorem 2.9 for $\sigma_1 \ne 1$ we obtain the blow-up result for Theorem 2.11.

For the proof of Corollary 2.12, we will first use assumptions (2.69) and the conditions on the initial data to show that $u_t, v_t > 0$ for all $0 \le t < T$. Differentiating the equations in (2.68), we find that $w := u_t$ and $z := v_t$ satisfy the following:

$$\begin{cases} w_t - \sigma_1 \Delta w = f_{1u}(u,v)w + f_{1v}(u,v)z & \text{in } \Omega \times (0,T), \\ z_t - \sigma_2 \Delta z = f_{2u}(u,v)w + f_{2v}(u,v)z & \text{in } \Omega \times (0,T), \\ \frac{\partial w}{\partial \nu} = g_{1u}(u,v)w + g_{1v}(u,v)z & \text{on } \partial\Omega \times (0,T), \\ \frac{\partial z}{\partial \nu} = g_{2u}(u,v)w + g_{2v}(u,v)z & \text{on } \partial\Omega \times (0,T), \\ w(x,0) = u_t(x,0) = \sigma_1 \Delta u_0 + f_1(u_0,v_0) > 0 & \text{for } x \in \Omega, \\ z(x,0) = v_t(x,0) = \sigma_2 \Delta u_0 + f_2(u_0,v_0) > 0 & \text{for } x \in \Omega. \end{cases}$$

From the signs of the off diagonal terms in the parabolic system above, we can use comparison or maximum principle (see e.g. Protter and Weinberger [188]) to obtain the positivity of $w = u_t$ and $z = v_t$. This leads to the fact that $\sigma_1 \Delta u = u_t - f_1(u,v) \geq 0$ and $\sigma_2 \Delta v = v_t - f_2(u,v) \geq 0$. Consequently, we obtain the following inequalities:

$$\begin{cases} u_t - \sigma_0 \Delta u \geq f_1(u,v) & \text{in } \Omega \times (0,T), \\ v_t - \sigma_0 \Delta v \geq f_2(u,v) & \text{in } \Omega \times (0,T), \end{cases}$$

where $\sigma_0 = min\{\sigma_1, \sigma_2\}$. By means of comparison theorem as before and the result of Theorem 2.11, we obtain the conclusion of Corollary 2.12.

4.3 Diffusion, Self and Cross-Diffusion with No-Flux Boundary Condition

In this section we study reaction-diffusion models when the diffusion rate of a particular species at a point may depend on the concentration and gradient of another species at the point. Such property is called cross-diffusion. For example, we expect a species tends to diffuse in the direction of lower concentration of its predator or competitor. Kerner [103] and Jorné [96] examined such prey-predator models when they have spatially constant solutions, and formally studied their linearized stability. Gurtin [76] developed some models that include cross-diffusion and self-diffusion, and showed that the effect of cross-diffusion may give rise to segregation of two species. For more explanation of such theories, the reader is referred to Okubo and Levin [179] for further reading. We will consider special cases of such models involving competing species proposed by Shigesada, Kawasaki and Teramoto [206]. Here, we will carefully describe some rigorous results developed in Lou and Ni [164]. We will only discuss the corresponding steady-state equation under no-flux boundary condition, and investigate the possibility of non-constant steady states. More precisely, we consider the following system:

$$\begin{cases} \Delta[(d_1+\alpha_{11}u_1+\alpha_{12}u_2)u_1]+u_1(a_1-b_1u_1-c_1u_2)=0 & \text{in } \Omega, \\ \Delta[(d_2+\alpha_{21}u_1+\alpha_{22}u_2)u_2]+u_2(a_2-b_2u_1-c_2u_2)=0 & \text{in } \Omega, \\ \frac{\partial u_1}{\partial \nu}=\frac{\partial u_2}{\partial \nu}=0 \quad \text{on } \partial\Omega,\ u_1>0,\ u_2>0 & \text{in } \Omega. \end{cases} \tag{3.1}$$

Here Ω is a bounded domain in $R^N, N \geq 1$, with smooth boundary; ν is the outward unit normal vector on the boundary $\partial\Omega$. The parameters a_i, b_i, c_i and d_i, $i=1,2$ are all positive constants; and $\alpha_{ij}, i,j=1,2$ are non-negative constants. For convenience, we denote

$$\begin{aligned} &u=(u_1,u_2),\ f(u)=(f_1(u_1,u_2),f_2(u_1,u_2)), \\ &f_1(u_1,u_2)=a_1-b_1u_1-c_1u_2,\ f_2(u_1,u_2)=a_2-b_2u_1-c_2u_2. \end{aligned} \tag{3.2}$$

The system (3.1) is the steady-state diffusive competing species system with no-flux boundary condition. The constants a_i are the intrinsic growth rates, b_1, c_2 are the crowding effect (intra-species competition) coefficients, and b_2, c_1 are the (inter-species) competition coefficients. The constants d_i are the usual diffusion rates. The parameters α_{11}, α_{22} are referred as self-diffusion pressures, and α_{12}, α_{21} are cross-diffusion pressures. The objective of this section is to indicate and prove how these self-diffusion and cross-diffusion pressures may give rise to spatial segregation of the species, while usual diffusion alone will not have such effect under Neumann boundary condition.

It is convenient to consider the relative configuration of the two straight lines $f_i(u_i,u_2)=0, i=1,2$ in the first quadrant by classifying into four cases:

(I) $b_1/b_2 > a_1/a_2 > c_1/c_2$, (Weak Competition),
(II) $b_1/b_2 < a_1/a_2 < c_1/c_2$, (Strong Competition),
(III) $a_1/a_2 > max\{b_1/b_2, c_1/c_2\}$,
(IV) $a_1/a_2 < min\{b_1/b_2, c_1/c_2\}$.

In cases (I) and (II) the two straight lines intersect at a point in the interior of the first quadrant; and in cases (III) and (IV), there is no intersection in the first quadrant. If one considers the system of ordinary differential equations:

$$\begin{cases} \frac{du_1}{dt}=u_1f_1(u_1,u_2), \\ \frac{du_2}{dt}=u_2f_2(u_1,u_2). \end{cases}$$

In case (I), the point of intersection in the first quadrant is a stable equilibrium; while in case (II) it is unstable. Consider the corresponding parabolic problem with the usual diffusion assumptions:

$$(3.3)\quad \begin{cases} \frac{\partial u_1}{\partial t} = d_1 \Delta u_1 + u_1(a_1 - b_1 u_1 - c_1 u_2) & \text{in } \Omega \times (0, \infty), \\ \frac{\partial u_2}{\partial t} = d_2 \Delta u_2 + u_2(a_2 - b_2 u_1 - c_2 u_2) & \text{in } \Omega \times (0, \infty), \\ \frac{\partial u_1}{\partial \nu} = \frac{\partial u_2}{\partial \nu} = 0 & \text{on } \partial\Omega \times (0, \infty), \\ u_1(x, 0) = u_{1,0}(x) > 0, \ \ u_2(x, 0) = u_{2,0}(x) > 0 & \text{in } \Omega. \end{cases}$$

In case (I), the solution of (3.3) $(u_1(x,t), u_2(x,t)) \to (u_1^*, u_2^*)$ which is the spatially constant positive steady-state, as $t \to \infty$. In case (II), (u_1^*, u_2^*) is unstable, and $(a_1/b_1, 0)$ and $(0, a_2/c_2)$ are both locally stable. However, Matano and Mimura [167], Mimura, Ei and Fang [171], and Kan-on and Yanagida [100] showed that for certain dumb-bell shaped domain, it is possible to have a stable non-constant positive steady-state. If the domain Ω is convex, Kishimoto and Weinberger [105] found that there cannot be any stable positive steady state. In cases (III) and (IV), the solutions respectively tends to $(a_1/b_1, 0)$ or $(0, a_2/c_2)$. More general studies concerning constant steady-state solutions and their stabilities for the Neumann problem with larger systems and more elaborate interactions were investigated by Redheffer and Zhou [192], using Lyapunov method and graph theory. The results are explained in Chapter 7 in Leung [125].

The theories in the above paragraph provide directions for investigating problem (3.1) under the various cases. Recall that we are presently concerned with the possibility of non-constant steady-state under various self-diffusion and cross-diffusion pressures. For case (I), early results are found by Mimura and Kawasaki [172] for $N = 1, \alpha_{11} = \alpha_{21} = \alpha_{22} = 0$. They showed that there exist small amplitude solutions of (3.1) bifurcating from the constant positive coexistence steady-state. Under similar assumptions as in [172], Mimura [170] established the existence of large amplitude solutions of (3.1) when α_{12} is large. Other early results concerning the existence of non-constant steady-states when $N = 1, \alpha_{11} = \alpha_{21} = \alpha_{22} = 0$ for cases (II), (III) and (IV) are obtained by Mimura, Nishiura, Tesei and Tsujikawa [173]. The stability of some of these solutions are obtained in Kan-on [97]. For Dirichlet boundary condition, (3.1) was investigated by Wu [236] for $N = 1$.

Part A: Non-existence of Spatially Inhomogeneous Solutions.

We first show that all solutions of (3.1) must be spatially constant when one of the self-diffusion pressures α_{11}, α_{22} is large or one of the diffusion rates d_1, d_2

is large (while the self-diffusion pressures are both positive), under all cases (I), (II), (III) or (IV).

Theorem 3.1 (Conditions Prohibiting Non-constant Solutions). *Consider problem (3.1) under the assumptions*

$$\frac{a_1}{a_2} \neq \frac{b_1}{b_2}, \text{ and } \frac{a_1}{a_2} \neq \frac{c_1}{c_2}. \tag{3.4}$$

(i) There exists a positive constant $C_1 = C_1(d_i, a_i, b_i, c_i, \alpha_{12}, \alpha_{21}), i = 1, 2,$ *such that if*

$$max\{\alpha_{11}, \alpha_{22}\} \geq C_1, \tag{3.5}$$

then problem (3.1) does not have any non-constant solution.

(ii) There exists a positive constant $C_2 = C_2(a_i, b_i, c_i, \alpha_{ij}), i, j = 1, 2,$ *such that if*

$$max\{d_1, d_2\} \geq C_2, \tag{3.6}$$

and both α_{11} *and* α_{22} *are positive, then problem (3.1) does not have any non-constant solution.*

Proof. We will use the abbreviations given in (3.2) for the proof of this theorem. Assumption (3.4) implies that the two lines $f_1(u_1, u_2) = 0$ and $f_2(u_1, u_2) = 0$ do not intersect at any point on the two axes $u_1 = 0$ or $u_2 = 0$. We will now proceed to prove part (i).

First, consider the case when the two lines intercept at a point $(\hat{u}_1, \hat{u}_2)$ in the first open quadrant. We will show that for every small $\epsilon > 0$, there exists a constant $K(\epsilon)$ such that if $max\{\alpha_{11}, \alpha_{22}\} \geq K(\epsilon)$, then for any solution $u = (u_1, u_2)$ of (3.1), we must have

$$||u_1 - \hat{u}_1||_\infty + ||u_2 - \hat{u}_2||_\infty \leq \epsilon. \tag{3.7}$$

Suppose not, then there exists a constant $\epsilon_0 > 0$ and a sequence $\{\alpha_{11,k}, \alpha_{22,k}\}_{k=1}^\infty$ with say $\alpha_{11,k} \to \infty$, such that

$$||u_{1,k} - \hat{u}_1||_\infty + ||u_{2,k} - \hat{u}_2||_\infty \geq \epsilon_0, \tag{3.8}$$

where $(u_{1,k}, u_{2,k})$ is a solution of

$$\begin{cases} \Delta[(d_1 + \alpha_{11,k}u_{1,k} + \alpha_{12}u_{2,k})u_{1,k}] + u_{1,k}f_1(u_{1,k}, u_{2,k}) = 0 \text{ in } \Omega, \\ \Delta[(d_2 + \alpha_{21}u_{1,k} + \alpha_{22,k}u_{2,k})u_{2,k}] + u_{2,k}f_2(u_{1,k}, u_{2,k}) = 0 \text{ in } \Omega, \\ \frac{\partial u_{1,k}}{\partial \nu} = \frac{\partial u_{2,k}}{\partial \nu} = 0 \text{ on } \partial\Omega, \ u_{1,k} > 0, \ u_{2,k} > 0 \text{ in } \Omega. \end{cases} \tag{3.9}$$

We now show that there exists a subsequence of $\{u_{1,k}\}_{k=1}^{\infty}$, denoted again by $\{u_{1,k}\}_{k=1}^{\infty}$ for convenience, which converges uniformly to a constant as $k \to \infty$. For this purpose, let

$$\psi_{1,k} = u_{1,k}(u_{1,k} + \frac{d_1}{\alpha_{11,k}} + \frac{\alpha_{12}}{\alpha_{11,k}} u_{2,k}), \tag{3.10}$$

which satisfies

$$\alpha_{11,k}\Delta\psi_{1,k} + u_{1,k} f_1(u_k) = 0 \text{ in } \Omega, \quad \frac{\partial \psi_{1,k}}{\partial \nu} = 0 \text{ on } \partial\Omega. \tag{3.11}$$

Let

$$\psi_{1,k}(x_{0,k}) = max_{\bar{\Omega}}\, \psi_{1,k}.$$

Suppose that $x_{0,k} \in \partial\Omega$, and $f_1(u_{1,k}(x_{0,k}), u_{2,k}(x_{0,k})) < 0$. We can deduce from Hopf boundary lemma that this contradicts $\frac{\partial \psi_{1,k}}{\partial \nu} = 0$ on $\partial\Omega$. Thus, we have

$$f_1(u_{1,k}(x_{0,k}), u_{2,k}(x_{0,k})) \geq 0. \tag{3.12}$$

Suppose $x_{0,k} \in \Omega$, then equation (3.11) also leads to (3.12). By (3.12), we have

$$a_1 = f_1(0,0) \geq f_1(0,0) - f_1(u_{1,k}(x_{0,k}), u_{2,k}(x_{0,k})) = b_1 u_{1,k}(x_{0,k}) + c_1 u_{2,k}(x_{0,k}).$$

Thus

$$u_{1,k}(x_{0,k}) \leq \frac{a_1}{b_1} \quad \text{and} \quad u_{2,k}(x_{0,k}) \leq \frac{a_1}{c_1}. \tag{3.13}$$

From (3.10) and (3.13), we obtain

$$(d_1 + \alpha_{11,k} max_{\bar{\Omega}} u_{1,k}) max_{\bar{\Omega}} u_{1,k} \leq \alpha_{11,k} max_{\bar{\Omega}} \psi_{1,k} \leq \frac{a_1}{b_1}(\alpha_{11,k}\frac{a_1}{b_1} + d_1 + \alpha_{12}\frac{a_1}{c_1}). \tag{3.14}$$

By considering each case when $\alpha_{11,k} \leq d_1$ or $\alpha_{11,k} \geq d_1$, we deduce from (3.14) that

$$max_{\bar{\Omega}}\, u_{1,k} \leq C(1 + \frac{\alpha_{12}}{d_1}) \tag{3.15}$$

for each k, where $C = C(a_1, b_1, c_1)$. Similarly, letting

$$\psi_{2,k} = u_{2,k}(u_{2,k} + \frac{d_2}{\alpha_{22,k}} + \frac{\alpha_{21}}{\alpha_{22,k}} u_{1,k}),$$

and using the equation satisfied by $\psi_{2,k}$, we deduce by means of (3.13) that

$$max_{\bar{\Omega}}\, u_{2,k} \leq C(1 + \frac{\alpha_{21}}{d_2}) \tag{3.16}$$

for each k, where $C = C(a_2, b_2, c_2)$. By (3.10), (3.15), (3.16) and the fact that $\alpha_{11,k} \to \infty$, we see that $||\psi_{1,k}||_\infty \le K$ for some constant K. Again, by (3.10), (3.11), standard L^p estimates and Sobolov embedding, we obtain a uniform bound for $||\psi_{1,k}||_{C^{1,\alpha}(\bar{\Omega})}$ for some $\alpha \in (0,1)$. Thus there exists a subsequence, denoted again by $\{\psi_{1,k}\}_{k=1}^\infty$, which converges to some non-negative function ψ_1 in $C^1(\bar{\Omega})$, and ψ_1 satisfy the following equation weakly

$$\Delta\psi_1 = 0 \ \text{ in } \Omega, \ \ \frac{\partial \psi_1}{\partial \nu} = 0 \ \text{ on } \partial\Omega, \tag{3.17}$$

since $\alpha_{11,k} \to \infty$. From standard elliptic regularity theory, we find $\psi_1 \in C^2(\bar{\Omega})$, and (3.17) implies that $\psi_1 \equiv \hat{\psi}$ is a non-negative constant. Setting $\tilde{u}_1 = \sqrt{\hat{\psi}_1}$, we obtain

$$u_{1,k}^2 - \tilde{u}_1^2 = (\psi_{1,k} - \hat{\psi}_1) - \frac{d_1}{\alpha_{11,k}} u_{1,k} - \frac{\alpha_{12}}{\alpha_{11,k}} u_{1.k} u_{2.k} \to 0$$

as $k \to \infty$. Consequently, we have $u_{1,k} \to \tilde{u}_1$ uniformly, where $\tilde{u}_1$ is a non-negative constant.

We next show that there exists a subsequence of $\{u_{2,k}\}_{k=1}^\infty$, denoted again by $\{u_{2,k}\}_{k=1}^\infty$ for convenience, such that $u_{2,k} \to \tilde{u}_2$ uniformly as $k \to \infty$, where $\tilde{u}_2$ is also a non-negative constant. Suppose $\{\alpha_{22,k}\}_{k=1}^\infty$ is unbounded. We can choose a subsequence denoted again by $\{\alpha_{22,k}\}_{k=1}^\infty$ tending to ∞, and prove in the same manner as before to show that $u_{2,k} \to \tilde{u}_2$ for some non-negative constant $\tilde{u}_2$. On the other hand, if $\{\alpha_{22,k}\}_{k=1}^\infty$ is bounded, we may assume without loss of generality that $\alpha_{22,k} \to \alpha_{22} \in [0, \infty)$. Set

$$\tilde{\psi}_{2,k} = (d_2 + \alpha_{21} u_{1,k} + \alpha_{22,k} u_{2,k}) u_{2,k}.$$

Since $\{\alpha_{22,k}\}_{k=1}^\infty$ is bounded, from (3.15) and (3.16), we see that $||\tilde{\psi}_{2,k}||_\infty \le K$. Furthermore, $\tilde{\psi}_{2,k}$ satisfies

$$\Delta\tilde{\psi}_{2,k} + u_{2,k} f_2(u_{1,k}, u_{2,k}) = 0 \ \text{ in } \Omega, \ \ \frac{\partial \tilde{\psi}_{2,k}}{\partial \nu} = 0 \ \text{ on } \partial\Omega. \tag{3.18}$$

By means of L^p estimate and Sobolev embedding, we obtain $||\tilde{\psi}_{2,k}||_{C^{1,\alpha}(\bar{\Omega})} \le K$ for some $\alpha \in (0,1)$. Then by selecting to a subsequence if necessary, we may assume that $\tilde{\psi}_{2,k} \to \psi_2 \ge 0$ in $C^1(\bar{\Omega})$. From the definition of $\tilde{\psi}_{2,k}$ and the fact that $u_{1,k} \to \tilde{u}_1$, we find

$$(d_2 + \alpha_{21} \tilde{u}_1 + \alpha_{22,k} u_{2,k}) u_{2,k} - \psi_2 \to 0 \tag{3.19}$$

in $C^1(\bar{\Omega})$.

We first consider the case when $\alpha_{22} > 0$. We readily see from the quadratic formula that $u_{2,k} \to \bar{u}_2$ in $C^1(\bar{\Omega})$, where

$$\bar{u}_2 = [-(d_2 + \alpha_{21}\tilde{u}_1) + \sqrt{(d_2 + \alpha_{21}\tilde{u}_1)^2 + 4\alpha_{22}\psi_2}]/2\alpha_{22} \ge 0.$$

Thus, by letting $k \to \infty$ in (3.18), we find ψ_2 satisfies the following equation weakly

$$\Delta\psi_2 + \bar{u}_2 f_2(\tilde{u}_1, \bar{u}_2) = 0 \text{ in } \Omega, \quad \frac{\partial \psi_2}{\partial \nu} = 0 \text{ on } \partial\Omega. \tag{3.20}$$

From regularity theory, we have $\psi_2 \in C^2(\bar{\Omega})$ and is a classical solution of (3.20). Note that $\psi_2 \geq 0$. If $\psi_2 \equiv 0$, then we have $u_{2,k} \to 0$ in $C(\bar{\Omega})$. From (3.9), we deduce $f(\tilde{u}_1, 0) = 0$ and $\tilde{u}_1 \geq 0$. This contradicts assumption (3.4), as explained in the beginning of the proof of this theorem. Therefore $\psi_2 \geq 0$ and is not identically zero in Ω. Since $\psi_2 = (d_2 + \alpha_{21}\tilde{u}_1 + \alpha_{22}\bar{u}_2)\bar{u}_2$, we may write (3.20) as

$$\Delta\psi_2 + [f_2(\tilde{u}_1, \bar{u}_2)/(d_2 + \alpha_{21}\tilde{u}_1 + \alpha_{22}\bar{u}_2)]\psi_2 = 0 \text{ in } \Omega, \quad \frac{\partial \psi_2}{\partial \nu} = 0 \text{ on } \partial\Omega. \tag{3.21}$$

From the maximum principle, (see e.g. [125]) we find $\psi_2 > 0$ and thus $\bar{u}_2 > 0$ in $\bar{\Omega}$. Since $\bar{u}_2$ is a solution of

$$\Delta[d_2 + \alpha_{21}\tilde{u}_1 + \alpha_{22}\bar{u}_2(x))\bar{u}_2(x)] + \bar{u}_2 f_2(\tilde{u}_1, \bar{u}_2) = 0 \text{ in } \Omega, \quad \frac{\partial \bar{u}_2}{\partial \nu} = 0 \text{ on } \partial\Omega, \tag{3.22}$$

we can deduce as before at the point of maximum, that we have $f_2(\tilde{u}_1, max_{\bar{\Omega}}\bar{u}_2) \geq 0$. Thus from the formula of f_2, we find

$$f_2(\tilde{u}_1, \bar{u}_2(x)) \geq f_2(\tilde{u}_1, max_{\bar{\Omega}}\bar{u}_2) \geq 0, \quad \text{for all } x \in \Omega.$$

Integrating (3.22), we obtain

$$\begin{aligned} 0 &= \textstyle\int_\Omega \bar{u}_2(x) f_2(\tilde{u}_1, \bar{u}_2(x))dx \geq \int_\Omega \bar{u}_2(x) f_2(\tilde{u}_1, max_{\bar{\Omega}}\bar{u}_2)dx \\ &= f_2(\tilde{u}_2, max_{\bar{\Omega}}\bar{u}_2) \textstyle\int_\Omega \bar{u}_2(x)dx \geq 0. \end{aligned} \tag{3.23}$$

From (3.23), we deduce $\bar{u}_2(x) \equiv max_{\bar{\Omega}}\bar{u}_2 := \tilde{u}_2$ must be a positive constant, and $f_2(\tilde{u}_1, \tilde{u}_2) = 0$. That is, if $\alpha_{22,k} \to \alpha_{22} > 0$, then there exists a subsequence of $\{u_{2,k}\}_{k=1}^{\infty}$ which converges to some positive constant $\tilde{u}_2$.

In case $\alpha_{22} = 0$, then by (3.19) we have

$$u_{2,k} \to \bar{u}_2 = \frac{\psi_2}{d_2 + \alpha_{21}\tilde{u}_1}$$

in $C^1(\bar{\Omega})$ as $k \to \infty$. The argument that a subsequence of $\{u_{2,k}\}_{k=1}^{\infty}$ converges to some positive constant denoted again by $\tilde{u}_2$ is the same as the case for $\alpha_{22} > 0$ above with obvious modifications.

We have now obtained a subsequence $\{(u_{1,k}, u_{2,k})\}_{k=1}^{\infty}$ such that $(u_{1,k}, u_{2,k}) \to (\tilde{u}_1, \tilde{u}_2)$ uniformly as $k \to \infty$, where $\tilde{u}_1$ and $\tilde{u}_2$ are non-negative constants.

Integrating the system (3.9) in Ω, we find

$$\int_\Omega u_{1,k} f_1(u_{1,k}, u_{2,k})dx = \int_\Omega u_{2,k} f_2(u_{1,k}, u_{2,k})dx = 0. \tag{3.24}$$

From this, we will obtain $f_i(\tilde{u}_1, \tilde{u}_2) = 0, i = 1, 2$ and $\tilde{u}_i = \hat{u}_i, i = 1, 2$. For, suppose $f_1(\tilde{u}_1, \tilde{u}_2) \neq 0$. Without loss of generality assume $f_1(\tilde{u}_1, \tilde{u}_2) > 0$. Since $u_{i,k} \to \tilde{u}_i, i = 1, 2$, uniformly as $k \to \infty$, we must have $f_1(u_{1,k,}, u_{2,k}) > 0$ for k large, and therefore

$$\int_\Omega u_{1,k} f_1(u_{1,k}, u_{2,k})dx > 0$$

for large k since $u_{1,k}$ is always positive. This contradicts (3.24). Similarly we deduce that $f_2(\tilde{u}_1, \tilde{u}_2) \neq 0$ leads to a contradiction. Thus from the assumption on $f(u)$, we obtain $(\tilde{u}_1, \tilde{u}_2) = (\hat{u}_1, \hat{u}_2)$ and $\hat{u}_i > 0, i = 1, 2$. This contradicts the existence of $\epsilon_0 > 0$ as described in (3.8). This leads to the validity of assertion (3.7) for every small $\epsilon > 0$, under conditions described.

We next use (3.7) to show that if $max\{\alpha_{11}, \alpha_{22}\}$ is large enough, then equations (3.1) imply that $\nabla u_i \equiv 0$ for $i = 1, 2$, in Ω. For convenience, let $\delta = min\{\hat{u}_1, \hat{u}_2\}$. By (3.7), there exists positive constant $K(\delta/2)$ such that if $max\{\alpha_{11}, \alpha_{22}\} \geq K(\delta/2)$, then any solution (u_1, u_2) of (3.1) satisfies

$$\delta/2 \leq u_1(x),\ u_2(x) \leq \hat{K}, \quad \text{for all } x \in \Omega, \tag{3.25}$$

for some positive constant $\hat{K}$. Without loss of generality, we may assume α_{11} is sufficiently large. Let u_i^0 be the average of u_i in Ω, and $u^0 = (u_1^0, u_2^0)$. Multiplying the first equation of (3.1) by $u_1 - u_1^0$ and integrating over Ω, we find

$$\begin{aligned} &\textstyle\int_\Omega (d_1 + 2\alpha_{11}u_1 + \alpha_{12}u_2)|\nabla u_1|^2\,dx \;+\; \alpha_{12}\int_\Omega u_1 \nabla u_1 \cdot \nabla u_2\,dx \\ &\quad = \textstyle\int_\Omega (u_1 - u_1^0) u_1 f_1(u)\,dx \\ &\quad = \textstyle\int_\Omega (u_1 - u_1^0)[u_1 f_1(u) - u_1^0 f_1(u^0)]\,dx \\ &\quad = \textstyle\int_\Omega (u_1 - u_1^0)[(u_1 - u_1^0) f_1(u) + u_1^0(f_1(u) - f_1(u^0))]\,dx \\ &\quad \leq \tfrac{\hat{K}_1}{\epsilon}\textstyle\int_\Omega |u_1 - u_1^0|^2 dx + \epsilon \int_\Omega |u_2 - u_2^0|^2 dx \end{aligned} \tag{3.26}$$

for some positive constant $\hat{K}_1$, and ϵ can be chosen arbitrarily small. Moreover, we have

$$|\alpha_{12} \int_\Omega u_1 \nabla u_1 \cdot \nabla u_2\,dx| \leq \frac{K}{\epsilon} \int_\Omega |\nabla u_1|^2\,dx + \epsilon \int_\Omega |\nabla u_2|^2\,dx. \tag{3.27}$$

for some positive constant K. From (3.26), (3.27) and Poincare's inequality, we deduce

$$(\alpha_{11}\delta - \frac{K_2}{\epsilon})\int_\Omega |\nabla u_1|^2\,dx \le \epsilon K_3 \int_\Omega |\nabla u_2|^2\,dx, \tag{3.28}$$

for some positive constants K_2, K_3. Here, ϵ is arbitrarily small, provided α_{11} is sufficiently large. (Note that u_1 and u_2 depend on α_{11}).

We next multiply the second equation of (3.1) by $u_2 - u_2^0$ and integrate over Ω to obtain as in (3.26)

$$\begin{aligned} &\alpha_{21}\textstyle\int_\Omega u_2\nabla u_1\nabla u_2\,dx + \int_\Omega (d_2 + \alpha_{21}u_1 + 2\alpha_{22}u_2)|\nabla u_2|^2\,dx \\ &= \textstyle\int_\Omega (u_2 - u_2^0)u_2 f_2(u)\,dx \\ &\le \tfrac{\hat{K}_2}{\epsilon}\textstyle\int_\Omega |u_1 - u_1^0|^2 dx + \epsilon\int_\Omega |u_2 - u_2^0|^2 dx, \end{aligned} \tag{3.29}$$

for some positive constant $\hat{K}_2$. Moreover, we have

$$|\alpha_{21}\int_\Omega u_2\nabla u_1\cdot\nabla u_2\,dx| \le \frac{\bar{K}}{\epsilon}\int_\Omega |\nabla u_1|^2\,dx + \epsilon\int_\Omega |\nabla u_2|^2\,dx, \tag{3.30}$$

for some positive constant $\bar{K}$. From (3.29), (3.30) and Poincare's inequality, we find

$$\begin{aligned} &d_2\textstyle\int_\Omega |\nabla u_2|^2\,dx \\ &\le \alpha_{21}\textstyle\int_\Omega u_2|\nabla u_1||\nabla u_2|\,dx + \tfrac{\hat{K}_2}{\epsilon}\int_\Omega |u_1 - u_1^0|^2\,dx + \epsilon\int_\Omega |u_2 - u_2^0|^2 dx \\ &\le K_4\epsilon\textstyle\int_\Omega |\nabla u_2|^2\,dx + \tfrac{K_5}{\epsilon}\int_\Omega |\nabla u_1|^2\,dx, \end{aligned} \tag{3.31}$$

for some positive constants K_4 and K_5. Summing (3.28) and (3.31) we obtain

$$(\alpha_{11}\delta - \frac{K_2 + K_5}{\epsilon})\int_\Omega |\nabla u_1|^2\,dx + (d_2 - \epsilon(K_3 + K_4))\int_\Omega |\nabla u_2|^2\,dx \le 0. \tag{3.32}$$

Choosing $\epsilon < d_2(K_3 + K_4)^{-1}$, and then α_{11} sufficiently large, we conclude from (3.32) that $\nabla u_i \equiv 0$, that is $u_i \equiv$ *constant* for $i = 1, 2$. This completes the proof of part (i) when $f(u) = 0$ has a root in the first open quadrant.

In the case when $f(u) = 0$ does not has a root in the first quadrant. Suppose there exists a sequence $\{u_{1,k}, u_{2,k}\}_{k=1}^\infty$ of solutions of (3.1) with say $\alpha_{11,k} \to \infty$, we follow the first part of the proof up to (3.24) to find a constant root $(\tilde{u}_1, \tilde{u}_2)$ of $f(u) = 0$, with both $\tilde{u}_i \ge 0, i = 1, 2$. This is thus a contradiction, and completes the proof of part (i).

For the proof of part (ii), first assume that the equation $f(u) = 0$ has a root $(\hat{u}_1, \hat{u}_2)$ in the first open quadrant. We then show that for every small $\epsilon > 0$, there exists a constant $\tilde{K}(\epsilon)$ such that if $max\{d_1, d_2\} \geq \tilde{K}(\epsilon)$, then for any solution $u = (u_1, u_2)$ of (3.1), we must have

$$||u_1 - \hat{u}_1||_\infty + ||u_2 - \hat{u}_2||_\infty \leq \epsilon, \tag{3.33}$$

as in (3.7) for the proof of part (i). The rest of the proof follows the same arguments as in the proof of part (i) above, with the natural modifications. The details can be found in [164], and will be omitted here.

In the last theorem we see that if only one diffusion rate or self-diffusion pressure is large, there cannot any non-constant solution in all cases (I) to (IV). In the next theorem involving only the weak competition case (I), we see that if self-diffusion and cross-diffusion pressures are relatively small in relation to diffusion rate, there still cannot be any non-constant solution.

Theorem 3.2 (Other Conditions Prohibiting Non-Constant Solutions for Weak Competition Case). *Suppose that*

$$\frac{b_1}{b_2} > \frac{a_1}{a_2} > \frac{c_1}{c_2}. \tag{3.34}$$

Then there exists a positive constant $C_3 = C_3(a_i, b_i, c_i), i = 1, 2$ such that the constant $u^* := ((a_1c_2 - a_2c_1)/(b_1c_2 - b_2c_1), (b_1a_2 - b_2a_1)/(b_1c_2 - b_2c_1))$ is the only solution of problem (3.1), provided

$$max\{\alpha_{ij}/d_i \,|\, i, j = 1, 2\} \leq C_3. \tag{3.35}$$

Proof. For convenience, we rewrite (3.1) as

$$\begin{cases} d_1\Delta[(1 + r_{11}u_1 + r_{12}u_2)u_1] + u_1f_1(u) = 0 & \text{in } \Omega, \\ d_2\Delta[(1 + r_{21}u_1 + r_{22}u_2)u_2] + u_2f_2(u) = 0 & \text{in } \Omega, \\ \frac{\partial u_1}{\partial \nu} = \frac{\partial u_2}{\partial \nu} = 0 \quad \text{on } \partial\Omega, \ u_1 > 0, \ u_2 > 0 & \text{in } \Omega, \end{cases} \tag{3.36}$$

where $f_1(u)$ and $f_2(u)$ are defined in (3.2), and

$$r_{ij} = \frac{\alpha_{ij}}{d_i}, \quad 1 \leq i, j \leq 2.$$

We now transform problem (3.36) into a semilinear elliptic system. Define a mapping G by

$$G(u) = (u_1(1 + r_{11}u_1 + r_{12}u_2), u_2(1 + r_{21}u_1 + r_{22}u_2)), \tag{3.37}$$

where $u = (u_1, u_2)$, and denote $R^2_+ := \{(u_1, u_2) \,|\, u_1 > 0, u_2 > 0\}$. The mapping $G \in C(R^2_+, R^2_+)$, and $G^{-1}(K)$ is compact in R^2_+ for any compact subset K in R^2_+. The Fréchet derivative of G is given by

$$DG(u) = \begin{bmatrix} 1 + 2r_{11}u_1 + r_{12}u_2 & r_{12}u_1 \\ r_{21}u_2 & 1 + r_{21}u_1 + 2r_{22}u_2 \end{bmatrix}. \tag{3.38}$$

Thus for any non-negative constants $r_{ij}, i, j = 1, 2$, the determinant of $DG(u)$ is positive for all $u \in R^2_+$. It follows from the implicit function theorem that G is locally invertible in R^2_+. Since R^2_+ is also arcwise and simply connected, by Theorem A5-4 in Chapter 6 we assert that G is a homeomorphism from R^2_+ onto itself. From the smoothness of G and the implicit function theorem again, we see that the inverse G^{-1} of G is also smooth in R^2_+. That is G is a smooth diffeomorphism from R^2_+ onto itself.

Define $\psi = (\psi_1, \psi_2) \equiv G(u)$ and

$$H(\psi) = (h_1(\psi), h_2(\psi)) \equiv G^{-1}(\psi),$$

system (3.36) can be written as

$$\begin{cases} d_1 \Delta \psi_1 + h_1(\psi) f_1(H(\psi)) = 0 & \text{in } \Omega, \\ d_2 \Delta \psi_2 + h_2(\psi) f_2(H(\psi)) = 0 & \text{in } \Omega, \\ \frac{\partial \psi_1}{\partial \nu} = \frac{\partial \psi_2}{\partial \nu} = 0 & \text{on } \partial\Omega, \ \psi_1 > 0, \ \psi_2 > 0 \text{ in } \Omega. \end{cases} \tag{3.39}$$

Since G is a diffeomorphism, we see that (3.36) has a non-constant solution if and only if (3.39) has a non-constant solution.

We next show that there exist positive constants K_1 and K_2 such that any solution $\psi = (\psi_1, \psi_2)$ of (3.39), must satisfy

$$max_{\bar{\Omega}}\, \psi_i \le K_2, \; i = 1, 2, \;\; \text{provided that } r := max\{r_{ij} | \, i, j = 1, 2\} \le K_1. \tag{3.40}$$

Let $\psi_1(x_0) = max_{\bar{\Omega}}\, \psi_1$, then from the first equation in (3.39), we have $f_1(u_1(x_0), u_2(x_0)) \ge 0$. Hence

$$a_1 = f_1(0,0) \ge [f_1(0,0) - f_1(u_1(x_0), 0)] + [f_1(u_1(x_0), 0) - f_1(u_1(x_0), u_2(x_0))]$$
$$\ge b_1 u_1(x_0).$$

Consequently, we have $u_1(x_0) \le a_1/b_1$, and

$$max_{\bar{\Omega}}\, \psi_1 \le \frac{a_1}{b_1}(1 + r_{11}\frac{a_1}{b_1} + r_{12}\, max_{\bar{\Omega}}\, \psi_2). \tag{3.41}$$

Similarly, we deduce

$$max_{\bar{\Omega}}\,\psi_2 \le \frac{a_2}{c_2}(1 + r_{21}\,max_{\bar{\Omega}}\,\psi_1 + r_{22}\frac{a_2}{c_2}). \tag{3.42}$$

From (3.41) and (3.42), we obtain

$$max_{\bar{\Omega}}\,\psi_1 \le \frac{a_1}{b_1}(1 + r_{11}\frac{a_1}{b_1} + r_{12}\frac{a_2}{c_2} + r_{12}r_{22}\frac{a_2^2}{c_2^2} + \frac{a_2}{c_2}r_{12}r_{21}\,max_{\bar{\Omega}}\,\psi_1),$$

which then implies

$$max_{\bar{\Omega}}\,\psi_1 \le \frac{(a_1/b_1)(1 + r_{11}(a_1/b_1) + r_{12}(a_2/c_2) + r_{12}r_{22}(a_2^2/c_2^2))}{1 - (a_1a_2/b_1c_2)r_{12}r_{21}}$$

provided that $r_{12}r_{21} < (b_1c_2)/(a_1a_2)$. Thus there exist constants K_1, K_2 such that the inequality in (3.40) is valid for $i = 1$. Similarly, we can deduce that (3.40) is also true for $i = 2$.

The remaining part of the proof uses a Lyapunov function for the parabolic problem corresponding to (3.39):

$$\begin{cases} \frac{\partial\hat{\psi}_1}{\partial t} = d_1\Delta\hat{\psi}_1 + h_1(\hat{\psi})f_1(H(\hat{\psi})) & \text{in } \Omega\times(0,\infty), \\ \frac{\partial\hat{\psi}_2}{\partial t} = d_2\Delta\hat{\psi}_2 + h_2(\hat{\psi})f_2(H(\hat{\psi})) & \text{in } \Omega\times(0,\infty), \\ \frac{\partial\hat{\psi}_1}{\partial\nu} = \frac{\partial\hat{\psi}_2}{\partial\nu} = 0 & \text{on } \partial\Omega\times(0,\infty), \\ \hat{\psi}_1(x,0) = \hat{\psi}_{1,0}(x), \quad \hat{\psi}_2(x,0) = \hat{\psi}_{2,0}(x) & \text{in } \Omega. \end{cases} \tag{3.43}$$

We use the same Lyapunov function as the usual one for the Volterra-Lotka model (see e.g. [125]). That is, for any positive continuous functions $\tilde{\psi}_1, \tilde{\psi}_2$, define

$$E(\tilde{\psi}) = \int_\Omega \{b_2(\tilde{\psi}_1 - \psi_1^* - \psi_1^*\,log\,\frac{\tilde{\psi}_1}{\psi_1^*}) + c_1(\tilde{\psi}_2 - \psi_2^* - \psi_2^*\,log\,\frac{\tilde{\psi}_2}{\psi_2^*})\}dx, \tag{3.44}$$

where $\tilde{\psi} = (\tilde{\psi}_1, \tilde{\psi}_2)$ and $\psi^* = (\psi_1^*, \psi_2^*) = G(u^*)$. If $\hat{\psi} = (\hat{\psi}_1(x,t), \hat{\psi}_2(x,t))$ is positive solution of (3.43), we obtain

$$\begin{aligned} &\tfrac{d}{dt}\textstyle\int_\Omega(\hat{\psi}_1 - \psi_1^* - \psi_1^*\,log\,\frac{\hat{\psi}_1}{\psi_1^*})\,dx \\ &\quad = -d_1\psi_1^*\textstyle\int_\Omega \frac{|\nabla\hat{\psi}_1|^2}{\hat{\psi}_1^2}\,dx + \int_\Omega(\hat{\psi}_1 - \psi_1^*)\frac{h_1(\hat{\psi})}{\hat{\psi}_1}f_1(H(\hat{\psi}))\,dx. \end{aligned}$$

Since $f_1(H(\psi^*)) = 0$, we have

$$f_1(H(\hat{\psi})) = [f_1(H(\hat{\psi})) - f_1(u_1^*, h_2(\hat{\psi}))] + [f_1(u_1^*, h_2(\hat{\psi})) - f_1(H(\psi^*))]$$

$$= \tfrac{\partial f_1}{\partial u_1} \cdot (h_1(\hat{\psi}) - h_1(\psi^*)) + \tfrac{\partial f_1}{\partial u_2} \cdot (h_2(\hat{\psi}) - h_2(\psi^*)),$$

and

$$f_1(H(\hat{\psi})) = \sum_{i,j=1}^{2} \frac{\partial f_1}{\partial u_i}\frac{\partial h_i}{\partial \psi_j}(\hat{\psi}_j - \psi_j^*),$$

where $\frac{\partial h_1}{\partial \psi_1}$ is evaluated at $(\eta_1, \hat{\psi}_2)$, with $\eta_1(x,t)$ between $\hat{\psi}_1(x,t)$ and ψ_1^*, and $\frac{\partial h_1}{\partial \psi_2}$ is evaluated at (ψ_1^*, η_2), with $\eta_2(x,t)$ between $\hat{\psi}_2(x,t)$ and ψ_2^*. Similar convention is used for $\frac{\partial h_2}{\partial \psi_j}$.

Thus we have

$$\begin{aligned}&\tfrac{d}{dt}\textstyle\int_\Omega(\hat{\psi}_1 - \psi_1^* - \psi_1^* \log \frac{\hat{\psi}_1}{\psi_1^*})dx \\ &= -d_1\psi_1^* \textstyle\int_\Omega \frac{|\nabla\hat{\psi}_1|^2}{\hat{\psi}_1^2}\, dx + \sum_{i,j=1}^2 \int_\Omega \frac{h_1(\hat{\psi})}{\hat{\psi}_1}\frac{\partial f_1}{\partial u_i}\frac{\partial h_i}{\partial \psi_j}(\hat{\psi}_1 - \psi_1^*)(\hat{\psi}_j - \psi_j^*)\, dx,\end{aligned} \tag{3.45}$$

and similarly

$$\begin{aligned}&\tfrac{d}{dt}\textstyle\int_\Omega(\hat{\psi}_2 - \psi_2^* - \psi_2^* \log \frac{\hat{\psi}_2}{\psi_1^*})dx \\ &= -d_2\psi_2^* \textstyle\int_\Omega \frac{|\nabla\hat{\psi}_2|^2}{\hat{\psi}_2^2}\, dx + \sum_{i,j=1}^2 \int_\Omega \frac{h_2(\hat{\psi})}{\hat{\psi}_2}\frac{\partial f_2}{\partial u_i}\frac{\partial h_i}{\partial \psi_j}(\hat{\psi}_2 - \psi_2^*)(\hat{\psi}_j - \psi_j^*)\, dx.\end{aligned}$$

From the above formula and (3.45), we obtain

$$\begin{aligned}\tfrac{dE}{dt} &\le b_2 \textstyle\sum_{i=1}^2 \int_\Omega \frac{h_1(\hat{\psi})}{\hat{\psi}_1}\frac{\partial f_1}{\partial u_i}\frac{\partial h_i}{\partial \psi_1}(\hat{\psi}_1 - \psi_1^*)^2\, dx \\ &\quad + c_1 \textstyle\sum_{i=1}^2 \int_\Omega \frac{h_2(\hat{\psi})}{\hat{\psi}_2}\frac{\partial f_2}{\partial u_i}\frac{\partial h_i}{\partial \psi_2}(\hat{\psi}_2 - \psi_2^*)^2\, dx \\ &\quad + \textstyle\sum_{i=1}^2 \int_\Omega (\hat{\psi}_1 - \psi_1^*)(\hat{\psi}_2 - \psi_2^*)(b_2 \frac{\partial f_1}{\partial u_i}\frac{\partial h_i}{\partial \psi_2}\frac{h_1(\hat{\psi})}{\hat{\psi}_1} + c_1 \frac{\partial f_2}{\partial u_i}\frac{\partial h_i}{\partial \psi_1}\frac{h_2(\hat{\psi})}{\hat{\psi}_2})\, dx.\end{aligned}$$

From (3.38), we find

$$DH(\psi) = (DG(u))^{-1} = \frac{1}{det\, DG(u)}\begin{bmatrix} 1 + r_{21}u_1 + 2r_{22}u_2 & -r_{12}u_1 \\ -r_{21}u_2 & 1 + 2r_{11}u_1 + r_{12}u_2 \end{bmatrix}.$$

If there is some $K > 0$ such that $max_{\bar{\Omega}}\, \hat{\psi}_i \le K$ for all $t > 0, i = 1, 2$, then for small r we have

$$\begin{bmatrix} \frac{\partial h_1}{\partial \psi_1} & \frac{\partial h_1}{\partial \psi_2} \\ \frac{\partial h_2}{\partial \psi_1} & \frac{\partial h_2}{\partial \psi_2} \end{bmatrix} = \begin{bmatrix} 1 + O(1)r & O(1)r \\ O(1)r & 1 + O(r) \end{bmatrix},$$

and $h_i/\hat{\psi}_i = 1 + O(1)r$. Moreover, we find that for such solution $\hat{\psi}$

$$\begin{aligned} \tfrac{dE}{dt}(\hat{\psi}) \leq \textstyle\int_\Omega \{&(-b_1b_2 + O(1)r)(\hat{\psi}_1 - \psi_1^*)^2 \\ &+ (2b_2c_1 + O(1)r)|\hat{\psi}_1 - \psi_1^*||\hat{\psi}_2 - \psi_2^*| \\ &+ (-c_1c_2 + O(1)r)(\hat{\psi}_2 - \psi_2^*)^2\}\, dx. \end{aligned} \tag{3.46}$$

From the assumption $b_1/b_2 > c_1/c_2$, we see that there exists a positive constant $\bar{r} = \bar{r}(K)$ such that if $r \leq \bar{r}$, then $dE(\hat{\psi}(t))/dt \leq 0$ for all $t > 0$. Furthermore, $dE(\hat{\psi}(t_0))/dt = 0$ if and only if $\hat{\psi}(x,t) \equiv \psi^*$ for all $t \geq t_0$. Consequently, inequality (3.46) and property (3.40) for all equilibrium solutions imply that u^* is the only solution of problem (3.1) for small enough r as described in (3.35).

The following corollary is a slight modification of Theorem 3.2. It shows that if cross-diffusion is weak relative to diffusion, there is no non-constant solution for problem (3.1) in case (I).

Corollary 3.3. *Suppose that*

$$\frac{b_1}{b_2} > \frac{a_1}{a_2} > \frac{c_1}{c_2}. \tag{3.47}$$

Then there exists a positive constant $C_4 = C_4(a_i, b_i, c_i), i = 1,2$ such that the constant $u^* := ((a_1c_2 - a_2c_1)/(b_1c_2 - b_2c_1), (b_1a_2 - b_2a_1)/(b_1c_2 - b_2c_1))$ is the only solution of problem (3.1), provided

$$max\{(\alpha_{21}/d_1)(1 + \alpha_{12}/d_1),\ (\alpha_{12}/d_2)(1 + \alpha_{21}/d_2)\} \leq C_4, \tag{3.48}$$

or

$$max\{(\alpha_{21}/\sqrt{d_1d_2})(1 + \alpha_{12}/d_1),\ (\alpha_{12}/\sqrt{d_1d_2})(1 + \alpha_{21}/d_2)\} \leq C_4. \tag{3.49}$$

Corollary 3.3 is proved by the same method as Theorem 3.2, with slight modifications. Note that it is known that Theorem 3.2 and Corollary 3.3 can fail if the weak competition condition $b_1/b_2 > a_1/a_2 > c_1/c_2$ is dropped (cf. [100], [167] and [171]).

Remark 3.1. Roughly speaking, Theorems 3.1, 3.2 and Corollary 3.3 assert that problem (3.1) has no non-constant solution if diffusion or self-diffusion is strong, or if cross-diffusion is weak.

The above theorems and methods lead to consequences concerning steady-state solutions of (3.3) as follows.

Theorem 3.4. *Consider problem (3.1) with $\alpha_{ij} \equiv 0, i, j = 1,2$. The constant function $u = u^* = ((a_1c_2 - a_2c_1)/(b_1c_2 - b_2c_1), (b_1a_2 - b_2a_1/(b_1c_2 - b_2c_1))$ is the only solution if*

(i) $b_1/b_2 > a_1/a_2 > c_1/c_2$, *or*
(ii) $b_1/b_2 < a_1/a_2 < c_1/c_2$ *and* $max\{d_1, d_2\} \geq C^*$ *for some constant* C^*.

Note that part (i) follows directly from Theorem 3.2. Part (ii) is proved in the same way as part (ii) in Theorem 3.1. The details can be found in [164], and will be omitted here. Also, note that in (ii) we only assume one diffusion rate to be large in order to insure there is no non-constant equilibrium; while earlier well-known results (cf. e.g. [80]) assume both diffusion rates are large.

Part B: Spatial Inhomogeneity caused by Cross-Diffusion.

We will next study the existence of positive non-constant solution of (3.1) when cross-diffusion is strong. For $f(u)$ as described in (3.2), we will assume $f(u) = 0$ has a solution in the first open quadrant. Thus both components of $u = u^* = (u_1^*, u_2^*) = ((a_1c_2 - a_2c_1)/(b_1c_2 - b_2c_1), (b_1a_2 - b_2a_1/(b_1c_2 - b_2c_1))$ are positive. We will then assume

$$\frac{a_1}{a_2} > \frac{1}{2}(\frac{b_1}{b_2} + \frac{c_1}{c_2}), \tag{3.50}$$

which is equivalent to

$$-c_2u_2^* + b_2u_1^* > 0.$$

For the purpose of the next theorem, we define

$$d^{(l)} = \frac{-c_2u_2^* + b_2u_1^*}{(1 + 2r_{22}u_2^*)\mu_l}$$

for $l \geq 1$, where $0 = \mu_0 < \mu_1 < \cdots < \mu_k < \cdots$ are the eigenvalues of the negative Laplace operator in Ω under homogeneous Neumann boundary condition on $\partial\Omega$. Under assumption (3.50), we thus have

$$d^{(1)} > d^{(2)} > \cdots > d^{(l)} \to 0^+ \tag{3.51}$$

as $l \to \infty$. Let m_l denote the algebraic multiplicity of μ_l, then the following result gives the existence of non-constant solution for problem (3.36) for large r_{12}.

Theorem 3.5 (Non-Constant Solution under Weak or Strong Competition). *Consider problem (3.36), with the assumption that equation* $f(u) = 0$ *having a unique root in the first open quadrant, and hypothesis (3.50). Let* r_{11}, r_{21} *and* r_{22} *be arbitrarily given non-negative constants,* $d_2 \in (d^{(k+1)}, d^{(k)})$ *for some* $k \geq 1$ *with* $\sum_{l=1}^{k} m_l$ *being odd. Then there exists a positive constant* $\Lambda = \Lambda(d_1, d_2, r_{11}, r_{21}, r_{22}, f)$ *such that if* $r_{12} \geq \Lambda$, *problem (3.36) has at least one non-constant solution provided that one of the following conditions hold:*
(i) $b_1/b_2 > a_1/a_2 > c_1/c_2$, *or*

(ii) $b_1/b_2 < a_1/a_2 < c_1/c_2$ and $d_1 \geq \hat{C}$ where $\hat{C}$ is some constant larger than C^* in part (ii) of Theorem 3.4.

Remark 3.2. Note that $d^{(l)}, l = 1, 2, ..$ are independent of d_1, d_2 and r_{12}. If we adjust d_2 so that $d_2 \in (d^{(k+1)}, d^{(k)})$ as described above, then Theorem 3.5 (i) and (ii) claim that there must be non-constant solution provided r_{12} is sufficiently large.

Theorem 3.5 will be proved by using the following Lemmas 3.1 to 3.3. In order to present these lemmas, we recall that the system (3.36) is transformed in (3.39) by the mapping $G(u)$ of (3.37) and its inverse $H(\psi)$. Here, we write $G = G(u; r_{ij})$ and $H = H(u; r_{ij})$ in order to indicate their dependence on r_{ij}. To study the solutions of (3.39), we will deform the problem by homotopy to the situation when $r_{ij} = 0$. For each $s \in [0, 1]$, we consider the following system:

$$\begin{cases} d_1 \Delta \psi_1 + h_1(\psi; sr_{ij}) f_1(H(\psi; sr_{ij})) = 0 & \text{in } \Omega, \\ d_2 \Delta \psi_2 + h_2(\psi; sr_{ij}) f_2(H(\psi; sr_{ij})) = 0 & \text{in } \Omega, \\ \frac{\partial \psi_1}{\partial \nu} = \frac{\partial \psi_2}{\partial \nu} = 0 \quad \text{on } \partial\Omega, \ \psi_1 > 0, \ \psi_2 > 0 & \text{in } \Omega. \end{cases} \tag{3.52}$$

Note that for $s = 1$, (3.52) is the same as (3.39); and for $s = 0$, it reduces to the steady-state of (3.3) with $s_{i,j} \equiv 0, i, j = 1, 2$. For convenience, define

$$E = C(\bar{\Omega}) \times C(\bar{\Omega}), \ \ P = \{u = (u_1, u_2) \in E |\, u_1, u_2 \geq 0\}.$$

For each $s \in [0, 1]$, let $T_s : P \to E$ be the operator defined by

$$T_s(\psi) = \begin{bmatrix} (-d_1 \Delta + I)^{-1} [\psi_1 + h_1(\psi; sr_{ij}) f_1(H(\psi; sr_{ij})] \\ (-d_2 \Delta + I)^{-1} [\psi_2 + h_2(\psi; sr_{ij}) f_2(H(\psi; sr_{ij})] \end{bmatrix}, \tag{3.53}$$

where $(-d_i \Delta + I)^{-1}, i = 1, 2$, is the inverse of $-d_i \Delta + I$ subject to homogeneous Neumann boundary condition, and I is the identity map on $C(\bar{\Omega})$. By L^p estimates and Sobolev embedding, we know that T_s is a continuous and compact operator for each $s \in [0, 1]$. Moreover, ψ is a fixed point of T_s if and only if ψ is a solution of (3.52). In particular, u^* is a fixed point of T_0.

In order to use homotopy invariance, we need uniform bound for solutions of (3.52) for all $s \in [0, 1]$. For this purpose, we will use the following lemma.

Lemma 3.1. *Assume the hypotheses of Theorem 3.5 for problem (3.36). For any $\eta > 0$, if*

$$min\{d_1, d_2\} \geq \eta, \ \ and \ \ max\{r_{12}, r_{21}\} \leq \frac{1}{\eta},$$

then there exist two positive constants $\underline{C}(\eta) < \bar{C}(\eta)$, which are independent of d_i and $r_{ij}, (i, j = 1, 2)$, such that any solution of (3.36) must satisfy

$$\underline{C}(\eta) \leq u_i(x) \leq \bar{C}(\eta)$$

for all $x \in \bar{\Omega}$ and $i = 1, 2$.

In order not to distract the main idea of the proof of Theorem 3.5, we postpone the proof of the above lemma, and first consider the next two lemmas concerning the indices of the maps T_0 and T_1 at the fixed points u^* and $\psi^* = G(u^*; r_{ij})$ respectively. They are denoted as $index(T_0, u^*)$ and $index(T_1, \psi^*)$.

Lemma 3.2. *Let $f(u)$ as described in (3.2), and suppose that the equation $f(u) = 0$ has a solution in the first open quadrant. Let the mapping T_0 be as defined in (3.53) with fixed point u^*.*

(i) If $b_1/b_2 > a_1/a_2 > c_1/c_2$, then $index(T_0, u^*) = 1$.

(ii) If $b_1/b_2 < a_1/a_2 < c_1/c_2$ then there exists a positive constant C independent of d_1, d_2 such that for $max\{d_1, d_2\} \geq C$, $index(T_0, u^*) = -1$.

Proof. Since $H(\psi, 0) \equiv \psi$, from (3.53) we have

$$T_0(\psi) = \begin{bmatrix} (-d_1\Delta + I)^{-1}[\psi_1 + \psi_1 f_1(\psi)] \\ (-d_2\Delta + I)^{-1}[\psi_2 + \psi_2 f_2(\psi)] \end{bmatrix}.$$

Direct calculations gives

$$DT_0(u^*)(\psi) = \begin{bmatrix} (-d_1\Delta + I)^{-1}[(1 - b_1 u_1^*)\psi_1 - c_1 u_1^* \psi_2] \\ (-d_2\Delta + I)^{-1}[-b_2 u_2^* \psi_1 + (1 - c_2 u_2^*)\psi_2] \end{bmatrix}.$$

If u^* is an isolated fixed point of T_0, by Leray-Schauder degree theory (cf. Theorem A2-3) in Chapter 6, we have

$$index(T_0, u^*) = (-1)^\sigma, \tag{3.54}$$

where σ is the number of negative eigenvalues of $I - DT_0(u^*)$ (counting algebraic multiplicity). If $-\rho \leq 0$ is an eigenvalue of $I - DT_0(u^*)$, then there exists a nontrivial $\psi = (\psi_1, \psi_2)$ such that

$$\begin{cases} -d_1(1+\rho)\Delta\psi_1 + \rho\psi_1 = -b_1 u_1^* \psi_1 - c_1 u_1^* \psi_2 & \text{in } \Omega, \\ -d_2(1+\rho)\Delta\psi_2 + \rho\psi_1 = -b_2 u_2^* \psi_1 - c_2 u_2^* \psi_2 & \text{in } \Omega, \\ \frac{\partial\psi_1}{\partial\nu} = \frac{\partial\psi_2}{\partial\nu} = 0 & \text{on } \partial\Omega. \end{cases} \tag{3.55}$$

For each eigenvalue $-\mu_l$ of the Laplacian as described above, we define the matrix

$$M_l(\rho) = \begin{bmatrix} d_1(1+\rho)\mu_l + \rho + b_1 u_1^* & c_1 u_1^* \\ b_2 u_2^* & d_2(1+\rho)\mu_l + \rho + c_2 u_2^* \end{bmatrix}.$$

Problem (3.55) has a non-trivial solution if and only if the determinant satisfies $det\, M_l(\rho) = 0$ for some $\rho \geq 0$ and $l \geq 0$. We readily obtain the determinant

$$
\begin{aligned}
det\ M_l(\rho) &= [d_1(1+\rho)\mu_l + \rho][d_2(1+\rho)\mu_l + \rho] + b_1 u_1^*[d_2(1+\rho)\mu_l + \rho] \\
&\quad + c_2 u_2^*[d_1(1+\rho)\mu_l + \rho] + u_1^* u_2^*(b_1 c_2 - b_2 c_1).
\end{aligned} \tag{3.56}
$$

Thus, for all $\rho \geq 0$ and $l \geq 0$, we must have $det\, M_l(\rho) > 0$ if $b_1/b_2 > c_1/c_2$. This implies that $I - DT_0(u^*)$ has no non-positive eigenvalues. By Theorem A2-3 in Chapter 6, we find that u^* is an isolated fixed point, $\sigma = 0$ and $index(T_0, u^*) = 1$ for case (i).

For case (ii), $b_1/b_2 < c_1/c_2$, we readily see that $det\, M_0(0) \neq 0$ and the equation $det\, M_0(\rho) = 0$ has a unique positive root. If $l \geq 1$ and $\rho \geq 0$, we obtain from (3.56)

$$det\, M_l(\rho) > b_1 u_1^* \mu_1 d_2 + c_2 u_2^* \mu_1 d_1 + u_1^* u_2^*(b_1 c_2 - b_2 c_1).$$

Thus, we find $det\, M_l(\rho) > 0$ for all $\rho \geq 0$ and $l \geq 1$ provided that:

$$max\,\{d_1, d_2\} \geq C := max\{\frac{u_2^*(b_2 c_1 - b_1 c_2)}{b_1 \mu_1}, \frac{u_1^*(b_2 c_1 - b_1 c_2)}{c_2 \mu_1}\} > 0.$$

Consequently, u^* is isolated and $I - DT_0(u^*)$ has a unique negative eigenvalue which is simple, i.e. $\sigma = 1$. We then conclude from (3.54) that $index(T_0, u^*) = -1$.

The following lemma concerning the indices of the mapping T_1 in (3.53) will also be needed.

Lemma 3.3. *Assume $f(u)$ satisfies the condition described in Lemma 3.2 and hypothesis (3.50). Let $d_1 > 0, r_{11} \geq 0, r_{21} \geq 0, r_{22} \geq 0$ be arbitrary given constants and $d_2 \in (d^{(k+1)}, d^{(k)})$ for some $k \geq 1$. Then there exists a positive constant $\Lambda = \Lambda(d_1, d_2, r_{11}, r_{21}, r_{22})$ such that if $r_{12} \geq \Lambda$,*

$$index(T_1, \psi^*) = \begin{cases} (-1)^{\sum_{l=1}^{k} m_l} & \text{if } b_1 c_2 - b_2 c_1 > 0; \\ (-1)^{\sum_{l=1}^{k} m_l + 1} & \text{if } b_1 c_2 - b_2 c_1 < 0, \end{cases}$$

where T_1 is defined in (5.53).

We first prove Theorem 3.5 before we complete the proof of Lemma 3.1 and Lemma 3.3 below.

Proof (of Theorem 3.5). Let $d_1 > 0, r_{11} \geq 0, r_{21} \geq 0, r_{22} \geq 0$ and $d_2 \in (d^{(k+1)}, d^{(k)})$. Assume $r_{12} \geq \Lambda$, where Λ is described in Lemma 3.3. Choose $\eta > 0$ small enough so that

$$min\,\{d_1, d_2\} \geq \eta, \quad max\,\{r_{12}, r_{21}\} \leq 1/\eta.$$

By Lemma 3.1 there exist two positive constants $\underline{P}(\eta) < \bar{P}(\eta)$ such that

$$\underline{P}(\eta) \leq \psi_i(x) \leq \bar{P}(\eta), \quad \text{for all } x \in \Omega, i = 1, 2, \tag{3.57}$$

for any positive solution $\psi = (\psi_1, \psi_2)$ of (3.52). Let

$$S = \{\psi = (\psi_1, \psi_2) \in E \,|\, \frac{\underline{P}(\eta)}{2} \leq \psi_i(x) \leq 2\bar{P}(\eta), i = 1, 2, \text{ for all } x \in \bar{\Omega}\}.$$

Since T_s has no fixed point on the boundary of S for all $s \in [0, 1]$, by homotopy invariance, we have

$$deg(I - T_1, S, 0) = deg(I - T_0, S, 0). \tag{3.58}$$

Suppose that (3.39) has no non-constant solution, i.e. T_1 has a unique fixed point ψ^* in S. Then by Lemma 3.3 and the assumption that $\sum_{l=1}^{k} m_l$ is odd, we obtain

$$\begin{aligned} deg(I - T_1, S, 0) &= index(T_1, \psi^*) \\ &= \begin{cases} (-1)^{\sum_{l=1}^{k} m_l} = -1 & \text{if } \ b_1c_2 - b_2c_1 > 0; \\ (-1)^{\sum_{l=1}^{k} m_l + 1} = 1 & \text{if } \ b_1c_2 - b_2c_1 < 0. \end{cases} \end{aligned} \tag{3.59}$$

Under the assumptions of this theorem, Theorem 3.4 implies T_0 has a unique fixed point in S. From Lemma 3.2, we obtain

$$\begin{aligned} deg(I - T_0, S, 0) &= index(T_0, u^*) \\ &= \begin{cases} 1 & \text{if } \ b_1c_2 - b_2c_1 > 0, \\ -1 & \text{if } \ b_1c_2 - b_2c_1 < 0 \text{ and } d_1 \text{ is large enough.} \end{cases} \end{aligned}$$

This contradicts (3.58) and (3.59), and completes the proof of Theorem 3.5.

Proof (of Lemma 3.3). To calculate $index(T_1, \psi^*)$, we need to consider the Fréchet derivative

$$\begin{aligned} DT_1(\psi^*) \begin{bmatrix} \psi_1 \\ \psi_2 \end{bmatrix} &= \begin{bmatrix} (-d_1\Delta + I)^{-1}[(1 + u_1^*(-b_1\frac{\partial h_1}{\partial \psi_1}(\psi^*) - c_1\frac{\partial h_2}{\partial \psi_1}(\psi^*)))\psi_1] \\ (-d_2\Delta + I)^{-1}[(1 + u_2^*(-b_2\frac{\partial h_1}{\partial \psi_2}(\psi^*) - c_2\frac{\partial h_2}{\partial \psi_2}(\psi^*)))\psi_2] \end{bmatrix} \\ &+ \begin{bmatrix} (-d_1\Delta + I)^{-1}[u_1^*(-b_1\frac{\partial h_1}{\partial \psi_2}(\psi^*) - c_1\frac{\partial h_2}{\partial \psi_2}(\psi^*))\psi_2] \\ (-d_2\Delta + I)^{-1}[u_2^*(-b_2\frac{\partial h_1}{\partial \psi_1}(\psi^*) - c_2\frac{\partial h_2}{\partial \psi_1}(\psi^*))\psi_1] \end{bmatrix}. \end{aligned}$$

As in the proof of Lemma 3.2, $-\rho \leq 0$ is an eigenvalue of $I - DT_1(\psi^*)$ if and only if the matrix

$$\begin{aligned} N_l(\rho) &= \begin{bmatrix} d_1(1 + \rho)\mu_l + \rho & 0 \\ 0 & d_2(1 + \rho)\mu_l + \rho \end{bmatrix} \\ &+ \begin{bmatrix} u_1b_1 & u_1c_1 \\ u_2b_2 & u_2c_2 \end{bmatrix}_{u=u^*} \begin{bmatrix} \frac{\partial h_1}{\partial \psi_1} & \frac{\partial h_1}{\partial \psi_2} \\ \frac{\partial h_2}{\partial \psi_1} & \frac{\partial h_2}{\partial \psi_2} \end{bmatrix}_{\psi=\psi^*} \end{aligned}$$

is singular for some $\rho \geq 0$ and some $l \geq 0$. We are thus led to the quadratic equations in ρ for each $l \geq 0$:

$$\det N_l(\rho) = 0. \tag{3.60}$$

To analyze this equation, we note that for large r_{12}, we have

$$\begin{bmatrix} \frac{\partial h_1}{\partial \psi_1} & \frac{\partial h_1}{\partial \psi_2} \\ \frac{\partial h_2}{\partial \psi_1} & \frac{\partial h_2}{\partial \psi_2} \end{bmatrix}_{\psi=\psi^*} = \begin{bmatrix} o(1) & \frac{-u_1^*}{u_2^*(1+2r_{22}u_2^*)} + o(1) \\ o(1) & \frac{1}{1+2r_{22}u_2^*} + o(1) \end{bmatrix}, \tag{3.61}$$

and

$$\det \begin{bmatrix} \frac{\partial h_1}{\partial \psi_1} & \frac{\partial h_1}{\partial \psi_2} \\ \frac{\partial h_2}{\partial \psi_1} & \frac{\partial h_2}{\partial \psi_2} \end{bmatrix}_{\psi=\psi^*} = o(1). \tag{3.62}$$

Thus for $l = 0$ and large r_{12}, (3.60) can be simplified into
(3.63)

$$\rho^2 + \rho[\frac{-u_1^* b_2 + u_2^* c_2}{1 + 2r_{22}u_2^*} + o(1)] + u_1^* u_2^* (b_1 c_2 - b_2 c_1) \cdot (\det \begin{bmatrix} \frac{\partial h_1}{\partial \psi_1} & \frac{\partial h_1}{\partial \psi_2} \\ \frac{\partial h_2}{\partial \psi_1} & \frac{\partial h_2}{\partial \psi_2} \end{bmatrix}_{\psi=\psi^*}) = 0.$$

If $(b_1 c_2 - b_2 c_1) > 0$, then from hypothesis (3.50) and the fact that

$$0 < \det \begin{bmatrix} \frac{\partial h_1}{\partial \psi_1} & \frac{\partial h_1}{\partial \psi_2} \\ \frac{\partial h_2}{\partial \psi_1} & \frac{\partial h_2}{\partial \psi_2} \end{bmatrix}_{\psi=\psi^*} = o(1),$$

we can readily obtain from the quadratic formula that equation (3.63) has exactly two positive roots, denoted by $\rho_{0,1}, \rho_{0,2}$. If $(b_1 c_2 - b_2 c_1) < 0$, then equation (3.63) has exactly one positive root.

For $l \geq 1$, solving (3.60) is equivalent to solving the equation

$$g_l(\rho) = d_2, \tag{3.64}$$

where we define
(3.65)

$$g_l(\rho) := -\frac{\rho}{\mu_l(1+\rho)} + \frac{[d_1(1+\rho)\mu_l+\rho]u_2^* \cdot \det \begin{bmatrix} -b_2 & \frac{\partial h_2}{\partial \psi_2} \\ c_2 & \frac{\partial h_1}{\partial \psi_2} \end{bmatrix}_{\psi=\psi^*} - u_1^* u_2^* (b_1 c_2 - b_2 c_1) \cdot \det \begin{bmatrix} \frac{\partial h_1}{\partial \psi_1} & \frac{\partial h_1}{\partial \psi_2} \\ \frac{\partial h_2}{\partial \psi_1} & \frac{\partial h_2}{\partial \psi_2} \end{bmatrix}_{\psi=\psi^*}}{\mu_l(1+\rho) \left[d_1(1+\rho)\mu_l + \rho - u_1^* \cdot \det \begin{bmatrix} -b_1 & \frac{\partial h_2}{\partial \psi_1} \\ c_1 & \frac{\partial h_1}{\partial \psi_1} \end{bmatrix}_{\psi=\psi^*} \right]}.$$

From (3.61), (3.62) and (3.65), we obtain for large r_{12}

$$g_l(\rho) = \frac{-\rho}{\mu_l(1+\rho)} + \frac{[d_1(1+\rho)\mu_l + \rho][\mu_l d^{(l)} + o(1)] + o(1)}{\mu_l(1+\rho)[d_1(1+\rho)\mu_l + \rho + o(1)]}, \tag{3.66}$$

where the $o(1)$ terms are independent of ρ. From (3.66) we find

$$\frac{dg_l}{d\rho}(\rho) = \frac{-1 - \mu_l d^{(l)} + o(1)}{\mu_l(1+\rho)^2} < 0. \tag{3.67}$$

From (3.66) and (3.67), we readily obtain the following properties for the functions $g_l(\rho)$ for all $l \geq 1$: There exists $\Lambda_1 = \Lambda_1(d_1, r_{11}, r_{21}, r_{22})$ such that if $r_{12} \geq \Lambda_1$, then

$$\begin{aligned} &|g_l(0) - d^{(l)}| \leq \tfrac{C}{r_{12}}, \quad \tfrac{dg_l}{d\rho}(\rho) < 0 \text{ for all } \rho \geq 0; \text{ and} \\ &lim_{\rho \to +\infty}\, g_l(\rho) = -\tfrac{1}{\mu_l} \end{aligned} \tag{3.68}$$

where C is a positive constant independent of d_2 and r_{12}.

As in the proof of Lemma 3.2, we see that if 0 is an eigenvalue of $I - DT_1(\psi^*)$, then $det\, N_l(0) = 0$ for some $l \geq 0$. For the case $l = 0$, we note that (3.63) is not valid for $\rho = 0$, implying $det\, N_0(0) \neq 0$. For the case $l \geq 1$, we see from (3.60) and (3.64) that 0 is an eigenvalue of $I - DT_1(\psi^*)$ if and only if $g_l(0) = d_2$ for some l. Thus from the first property of (3.68), we must have for $r_{12} \geq \Lambda_1$, the inequality

$$|d_2 - d^{(l)}| \leq \frac{C}{r_{12}},$$

for some $l \geq 1$, where C is some positive constant independent of d_2 and r_{12}. Since $d_2 \in (d^{(k+1)}, d^{(k)})$ by assumption, we must have

$$min\,\{|d_2 - d^{(k)}|,\, |d_2 - d^{(k+1)}|\} \leq \frac{C}{r_{12}}$$

in order that $g_l(0) = d_2$ for some $l \geq 1$. However, the above inequality is impossible if we choose $r_{12} \geq \Lambda_2$, where

$$\Lambda_2 := \frac{C}{min\,\{|d_2 - d^{(k)}|,\, |d_2 - d^{(k+1)}|\}} + \Lambda_1.$$

Thus we conclude that if $r_{12} \geq \Lambda_2$, ψ^* is an isolated fixed point of T_1 and $index(T_1, \psi^*)$ is well-defined.

We next proceed to show that

$$\sigma = \begin{cases} \sum_{l=1}^{k} m_l + 2 & \text{if } b_1c_2 - b_2c_1 > 0, \\ \sum_{l=1}^{k} m_l + 1 & \text{if } b_1c_2 - b_2c_1 < 0, \end{cases} \tag{3.69}$$

where σ is the number negative eigenvalues of $I - DT_1(\psi^*)$ (counting algebraic multiplicity).

First consider the case $b_1c_2 - b_2c_1 > 0$. Recall that from (3.63), the equation $det\, N_0 = 0$ has exactly two positive roots $\rho_{0,1}, \rho_{0,2}$. That is, $-\rho_{0,1}$ and $-\rho_{0,2}$

account for two negative eigenvalues of $I - DT_1(\psi^*)$. We now claim that for each $l \in \{1, 2, ..., k\}$, the equation $g_l(\rho) = d_2$ has at least one positive root, which we shall denote by ρ_l. To see this, we note that by the first property in (3.68), the monotonicity of $\{d_l\}_{l=1}^{\infty}$, and the choice of Λ_2 described above, we have

$$g_l(0) \geq d^{(l)} - \frac{C}{r_{12}} \geq d^{(k)} - \frac{C}{r_{12}} > d_2$$

if $1 \leq l \leq k$ and $r_{12} \geq \Lambda_2$. On the other hand, from the third property in (3.68), we find

$$g_l(\rho) < 0 < d_2$$

for large ρ and each $l \geq 1$. By the Intermediate Value Theorem, there exists at least one positive root, ρ_l to the equation $g_l(\rho) = d_2$ for each $l \in \{1, 2, ..., k\}$ as claimed above. Since the multiplicity of μ_l is m_l for each $l \in \{1, 2.., k\}$, the multiplicity of $-\rho_i$ as an eigenvalue of $I - DT_1(\psi^*)$ is at least m_i. Consequently, we have

$$\sigma \geq 2 + \sum_{l=1}^{k} m_l \ \text{ if } b_1c_2 - b_2c_1 > 0. \tag{3.70}$$

We next show that the inequality for σ in (3.70) is also valid in reverse. We will see that if $-\tilde{\rho} < 0$ is an eigenvalue of $I - DT_1(\psi^*)$, then $\tilde{\rho}$ must be equal to one of $\rho_{0,1}, \rho_{0,2}, \rho_1, ..., \rho_k$. We know that the equation $det\, N_0(\rho) = 0$ has exactly two positive roots $\rho_{0,1}, \rho_{0,2}$. We next claim that if $g_l(\rho) = d_2$ for some $l \geq 1$, then we must have $l \in \{1, 2, ..., k\}$. Since $g_l(\rho)$ is strictly decreasing for positive ρ, from the second property of (3.68), it suffices to show that if $r_{12} \geq \Lambda_2$, then $d_2 > g_l(0)$ for all $l \geq k + 1$. From (3.68) and the definition of Λ_2, we have

$$d_2 \geq d^{(k+1)} + \frac{C}{r_{12}} \geq g_l(0) + (d^{(l)} - g_l(0) + \frac{C}{r_{12}}) > g_l(0)$$

provided that $l \geq k + 1$. This shows that the claim is valid.

Again, since g_l is strictly decreasing for $l \geq 1$, the equation $g_l(\rho) = d_2$ can have at most one positive root $\rho = \rho_l$ for some $l \in \{1, 2, ..., k\}$. Moreover, from the equation for N_l and (3.65), we find

$$\frac{d}{d\rho}(det\, N_l(\rho))|_{\rho=\rho_l} = -\frac{dg_l}{d\rho}(\rho_l)\mu_l(1 + \rho_l)[d_1(1 + \rho_l)\mu_l + \rho_l + o(1)] \neq 0.$$

Thus ρ_l is a simple eigenvalue of the matrix $N_l(\rho)$. From the discussion above, we see that the multiplicity for $-\rho_l$ as an eigenvalue of $I - DT_1(\psi_*)$ is exactly m_l. Therefore, we obtain

$$\sigma \leq 2 + \sum_{l=1}^{k} m_l \ \text{ if } b_1c_2 - b_2c_1 > 0. \tag{3.71}$$

We thus obtain the first half of the formula (3.69) from (3.70) and (3.71), and we find from Leray-Schauder degree theory that

$$index(T_1, \psi_*) = (-1)^\sigma = (-1)^{\sum_{l=1}^k m_l}, \quad \text{if } b_1c_2 - b_2c_1 > 0,$$

under the assumptions of the lemma, for $r_{12} \geq \Lambda_2$.

In case $b_1c_2 - b_2c_1 < 0$, we can calculate σ in a similar manner. The only difference is that we have now exactly one positive root rather than two for the equation $det\, N_0(\rho) = 0$. We thus show that there exists a positive constant $\Lambda_3 = \Lambda(d_1, r_{11}, r_{21}, r_{22})$ such that if $r_{12} \geq \Lambda_3$, then

$$\sigma = \sum_{l=1}^{k} m_l + 1, \quad \text{if } b_1c_2 - b_2c_1 < 0.$$

Finally, the proof of the Lemma 3.3 is complete by choosing $\Lambda = max\,\{\Lambda_2, \Lambda_3\}$.

Proof (of Lemma 3.1). From (3.15) and (3.16) in the proof of Theorem 3.1, we have

$$max_{\bar{\Omega}}\, u_1 \leq C(1 + \frac{\alpha_{12}}{d_1}) \leq C(1 + \frac{1}{\eta^2}), \quad max_{\bar{\Omega}}\, u_2 \leq C(1 + \frac{\alpha_{21}}{d_2}) \leq C(1 + \frac{1}{\eta^2}),$$

for any solution (u_1, u_2) of (3.36), where $C = C(a_i, b_i, c_i), i = 1, 2$. Choosing

$$\bar{C}(\eta) := C(1 + \frac{1}{\eta^2}),$$

we obtain the upper bound in the statement of Lemma 3.1.

To deduce a positive lower bound, we let $\psi_1 = u_1(1 + r_{11}u_1 + r_{12}u_2)$, and note that ψ_1 satisfies

$$\Delta\psi_1 + c(x)\psi_1 = 0 \ \text{ in } \Omega, \quad \frac{\partial\psi_1}{\partial\nu} = 0 \text{ on } \partial\Omega,$$

where

$$c(x) := \frac{f_1(u(x))}{d_1(1 + r_{11}u_1(x) + r_{12}u_2(x))}.$$

Using the upper bound for u and the assumption on d_i, we have

$$||c||_\infty \leq \frac{1}{\eta} max_{0 \leq u_1, u_2 \leq \bar{C}(\eta)} |f_1(u)|.$$

Thus by the Harnack type inequality in Theorem A5-5 of Chapter 6, there exists a positive constant $C_1(\eta)$ such that

$$min_{\bar{\Omega}}\, \psi_1 \geq C_1(\eta) max_{\bar{\Omega}}\, \psi_1. \tag{3.72}$$

From the definition of ψ_1, we have

$$max_{\bar{\Omega}}\,\psi_1 \geq (1 + r_{11}\,max_{\bar{\Omega}}\,u_1)max_{\bar{\Omega}}\,u_1, \quad \text{and}$$

$$min_{\bar{\Omega}}\,\psi_1 \leq (1 + r_{11}\,min_{\bar{\Omega}}\,u_1 + r_{12}\,max_{\bar{\Omega}}\,u_2)\,min_{\bar{\Omega}}\,u_1.$$

Thus from the (3.72) and the upper bound of u_2, we obtain

$$(1 + r_{11}\,max_{\bar{\Omega}}\,u_1)max_{\bar{\Omega}}\,u_1 \leq C_2(\eta)(1 + r_{11}\,min_{\bar{\Omega}}\,u_1)min_{\bar{\Omega}}\,u_1 \tag{3.73}$$

for some positive constant $C_2(\eta)$. If $r_{11} \leq 1$, then

$$max_{\bar{\Omega}}\,u_1 \leq C_2(\eta)(1 + \bar{C}(\eta))min_{\bar{\Omega}}\,u_1. \tag{3.74}$$

In case $r_{11} \geq 1$, then by (3.73) again we have

$$\begin{aligned}(max_{\bar{\Omega}}\,u_1)^2 &\leq C_2(\eta)(\tfrac{1}{r_{11}} + min_{\bar{\Omega}}\,u_1)min_{\bar{\Omega}}\,u_1 \\ &\leq C_2(\eta)(1 + \bar{C}(\eta))min_{\bar{\Omega}}\,u_1.\end{aligned} \tag{3.75}$$

From (3.74) and (3.75), we find

$$min_{\bar{\Omega}}\,u_1 \geq \tilde{C}(\eta)\,min\,\{max_{\bar{\Omega}}\,u_1, 1\}\,max_{\bar{\Omega}}\,u_1 \tag{3.76}$$

for some positive constant $\tilde{C}(\eta)$. Similarly, we show

$$min_{\bar{\Omega}}\,u_2 \geq \tilde{C}(\eta)\,min\,\{max_{\bar{\Omega}}\,u_2, 1\}\,max_{\bar{\Omega}}\,u_2. \tag{3.77}$$

In view of (3.76) and (3.77), in order to establish Lemma 3.1, it suffices to show

$$max_{\bar{\Omega}}\,u_i \geq C_3(\eta), \quad i = 1, 2 \tag{3.78}$$

for some positive constant $C_3(\eta)$. Suppose (3.78) is not true for some constant η_0. Then there exist sequences $\{d_{i,k}\}_{k=1}^{\infty}$, $\{r_{ij,k}\}_{k=1}^{\infty}$ and $\{u_{i,k}\}_{k=1}^{\infty}$ with $i, j = 1, 2$ and

$$d_{i,k} \geq \eta_0, \; r_{12,k} \leq \frac{1}{\eta_0}, \quad r_{21,k} \leq \frac{1}{\eta_0}$$

such that either $max_{\bar{\Omega}}\,u_{1,k} \to 0$ or $max_{\bar{\Omega}}\,u_{2,k} \to 0$, where $u_k = (u_{1,k}, u_{2,k})$ satisfies

$$\begin{cases} d_{1,k}\Delta[(1 + r_{11,k}u_{1,k} + r_{12,k}u_{2,k})u_{1,k}] + u_{1,k}f_1(u_k) = 0 \;\; \text{in } \Omega, \\ d_{2,k}\Delta[(1 + r_{21,k}u_{1,k} + r_{22,k}u_{2,k})u_{2,k}] + u_{2,k}f_2(u_k) = 0 \;\; \text{in } \Omega, \\ \frac{\partial u_{1,k}}{\partial \nu} = \frac{\partial u_{2,k}}{\partial \nu} = 0 \;\; \text{on } \partial\Omega, \; u_{1,k} > 0, \; u_{2,k} > 0 \;\text{in } \Omega. \end{cases} \tag{3.79}$$

Integrating the first equation of (3.79) in Ω, we find

$$\int_\Omega u_{1,k} f_1(u_k)\,dx = 0.$$

Since $f_1(0,0) > 0$, the above equation implies that we cannot have both $max_{\bar{\Omega}}\, u_{1,k} \to 0$ and $max_{\bar{\Omega}}\, u_{2,k} \to 0$ simultaneously. We may thus without loss of generality assume that $max_{\bar{\Omega}}\, u_{1,k} \to 0$ and $max_{\bar{\Omega}}\, u_{2,k} \to \bar{u}_2$, where $\bar{u}_2$ is a positive constant. Thus, we obtain from (3.77),

$$min_{\bar{\Omega}}\, u_{2,k} \geq \tilde{C}(\eta_0) min\{\frac{\bar{u}_2}{2}, 1\}\frac{\bar{u}_2}{2} > 0 \tag{3.80}$$

for sufficiently large k.

We now consider the following two cases:

Case 1. $\{r_{22,k}\}_{k=1}^\infty$ is bounded. Define

$$\psi_{2,k} = u_{2,k}(1 + r_{21,k}u_{1,k} + r_{22,k}u_{2,k}). \tag{3.81}$$

From the uniform upper bound for u_k, we have $||\psi_{2,k}||_\infty \leq C$ for all $k \geq 1$. Since $\psi_{2,k}$ satisfies

$$\Delta\psi_{2,k} + \frac{f_2(u_k)}{d_{2,k}(1 + r_{21,k}u_{1,k} + r_{22,k}u_{2,k})}\psi_{2,k} = 0 \text{ in } \Omega, \quad \frac{\partial\psi_{2,k}}{\partial\nu} = 0 \text{ on } \partial\Omega,$$

we obtain by L^p estimates and Sobolev embedding that

$$||\psi_{2,k}||_{C^{1,\alpha}(\bar{\Omega})} \leq C||\psi_{2,k}||_{W^{2,p}(\Omega)} \leq C$$

for some $\alpha \in (0,1)$. Choosing a subsequence if necessary, we may thus assume that $\psi_{2,k} \to \hat{\psi}_2$ in $C^1(\bar{\Omega}), r_{22,k} \to \hat{r}_{22} \in [0, +\infty)$ and $d_{2,k} \to \hat{d}_2 \in [\eta_0, +\infty]$.

We next show that $u_{2,k}$ converges uniformly to a positive constant. For this purpose, we consider the following two situations:(a) $\hat{r}_{22} \in (0,\infty)$, or (b) $\hat{r}_{22} = 0$.

Consider situation (a), when $\hat{r}_{22} \in (0,\infty)$. From definition (3.81), we obtain by means of the quadratic formula and the fact that $u_{1,k} \to 0$ that

$$u_{2,k} \to \hat{u}_2 := \frac{-1 + \sqrt{1 + 4\hat{r}_{22}\hat{\psi}_2}}{2\hat{r}_{22}} \quad \text{in } C(\bar{\Omega}).$$

Thus $\hat{\psi}_2$ satisfies the following equation weakly

$$\Delta\hat{\psi}_2 + \frac{f_2(0,\hat{u}_2)}{\hat{d}_2(1 + \hat{r}_{22}\hat{u}_2)}\hat{\psi}_2 = 0 \text{ in } \Omega, \quad \frac{\partial\hat{\psi}_2}{\partial\nu} = 0 \text{ on } \partial\Omega.$$

By standard elliptic regularity theory, we conclude that $\hat{\psi}_2 \in C^2(\bar{\Omega})$, and is a classical solution of the above problem. If $\hat{d}_2 = +\infty$, we readily see that

$\hat{\psi}_2 \equiv constant$, and consequently we have $\hat{u}_2 \equiv constant$. On the other hand, if $\hat{d}_2 \in [\eta_0, \infty)$, then $\hat{u}_2$ is a non-negative solution of

$$\hat{d}_2 \Delta[(1 + \hat{r}_{22}\hat{u}_2)\hat{u}_2] + f_2(0, \hat{u}_2)\hat{u}_2 = 0 \text{ in } \Omega, \ \frac{\partial \hat{u}_2}{\partial \nu} = 0 \text{ on } \partial\Omega.$$

We can next use the arguments following equation (3.22) in the proof of Theorem 3.1 to show that $\hat{u}_2 \equiv \bar{u}_2$, where $\bar{u}_2$ is a non-negative constant, which must be positive by (3.80).

For situation (b) $\hat{r}_{22} = 0$, observe that

$$u_{2,k} - \hat{\psi}_2 = -u_{2,k}(r_{21,k}u_{1,k} + r_{22,k}u_{2,k}) + (\psi_{2,k} - \hat{\psi}_2) \to 0.$$

We can thus follow the same arguments as in situation (a) above to show that $u_{2,k}$ converges to a positive constant $\bar{u}_2$.

Consequently, in Case 1 when $\{r_{22,k}\}_{k=1}^{\infty}$ is bounded, we have shown that $(u_{1,k}, u_{2,k}) \to (0, \bar{u}_2)$ uniformly as $k \to \infty$. We can then use the same arguments as in the proof of Theorem 3.1 to show that $f(0, \bar{u}_2) = 0$. However, we have $\bar{u}_2 > 0$, which contradicts the assumption on the functions $f(u_1, u_2)$.

Case 2. $\{r_{22,k}\}_{k=1}^{\infty}$ is unbounded. Define

$$\tilde{\psi}_{2,k} = u_{2,k}\left(\frac{1}{r_{22,k}} + \frac{r_{21,k}}{r_{22,k}}u_{1,k} + u_{2,k}\right). \tag{3.82}$$

By choosing subsequence if necessary, we may assume $r_{22,k} \to \infty$ as $k \to \infty$. Using arguments similar to that in Case 1, we deduce that by passing to a subsequence, $\tilde{\psi}_{2,k} \to \tilde{\psi}_2$ in $C^1(\bar{\Omega})$, where $\tilde{\psi}_2$ satisfies

$$\Delta\tilde{\psi}_2 = 0 \text{ in } \Omega, \ \frac{\partial \tilde{\psi}_2}{\partial \nu} = 0 \text{ on } \partial\Omega,$$

which implies that $\tilde{\psi}_2 \equiv \bar{\psi}_2$, a non-negative constant. From (3.82), we have

$$u_{2,k}^2 - \bar{\psi}_2 = (\tilde{\psi}_{2,k} - \tilde{\psi}_2) - u_{2,k}\left(\frac{1}{r_{22,k}} + \frac{r_{21,k}}{r_{22,k}}u_{1,k}\right) \to 0.$$

It follows that $u_{2,k} \to \bar{u}_2 := \sqrt{\bar{\psi}_2}$, and we then obtain a contradiction as in Case 1.

We have thus established inequality (3.78), and the proof of Lemma 3.1 is complete. This also finish the proof of Theorem 3.5.

Recall that $0 = \mu_0 < \mu_1 < \cdots < \mu_k < \cdots$ are the eigenvalues of the negative Laplace operator in Ω under homogeneous Neumann boundary condition on $\partial\Omega$. Note that any of the two conditions $b_1/b_2 > a_1/a_2 > c_1/c_2$ or $b_1/b_2 < a_1/a_2 < c_1/c_2$ implies that the equation $f(u) = 0$ has a unique root in the first open quadrant as described in the hypothesis in Theorem 3.5. Theorem 3.5 can be rephrased more conveniently from a different viewpoint for application in the following two corollaries. We note that there will be non-constant solution in case (i) when the cross-diffusion pressure α_{12} is large, and in case (ii) when the cross diffusion pressure α_{21} is large.

Corollary 3.6 (Non-Constant Solution in Weak Competition Case). *Consider problem (3.1) with $b_1/b_2 > a_1/a_2 > c_1/c_2$. Suppose that for some $\tilde{k} \geq 1$, the eigenvalue $\mu_{\tilde{k}}$ of the operator $-\Delta$ on Ω has odd multiplicity.*

(i) If $a_1/a_2 > \frac{1}{2}[(b_1/b_2) + (c_1/c_2)]$, then there exist positive constants $K_1 = K_1(a_i, b_i, c_i) < K_2 = K_2(a_i, b_i, c_i)$ and $\Lambda_1 = \Lambda_1(d_i, a_i, b_i, c_i, \alpha_{11}, \alpha_{21}, \alpha_{22})$ such that for any $d_1 > 0$, $\alpha_{11} \geq 0$, $\alpha_{21} \geq 0$, the problem (3.1) has at least one non-constant solution provided that $\alpha_{12} \geq \Lambda_1$ and $d_2 + 2u_2^*\alpha_{22} \in (K_1, K_2)$.

(ii) If $a_2/a_1 > \frac{1}{2}[(b_2/b_1) + (c_2/c_1)]$, then there exist positive constants $\hat{K}_1 = \hat{K}_1(a_i, b_i, c_i) < \hat{K}_2 = \hat{K}_2(a_i, b_i, c_i)$ and $\hat{\Lambda}_1 = \hat{\Lambda}_1(d_i, a_i, b_i, c_i, \alpha_{11}, \alpha_{12}, \alpha_{22})$ such that for any $d_2 > 0$, $\alpha_{12} \geq 0$, $\alpha_{22} \geq 0$, the problem (3.1) has at least one non-constant solution provided that $\alpha_{21} \geq \hat{\Lambda}_1$ and $d_1 + 2u_1^*\alpha_{11} \in (\hat{K}_1, \hat{K}_2)$.

Proof. As noted above, the hypothesis concerning the location of the root of $f(u) = 0$ for Theorem 3.5 is satisfied. We now consider case (i). Observe that here we are also assuming (3.50) stated for Theorem 3.5. Furthermore, the assumption that $\mu_{\tilde{k}}$ has odd multiplicity for some $\tilde{k} \geq 1$ implies that there exists some $k \geq 1$ such that $\sum_{l=1}^{k} m_l$ is odd. Hence in order to apply Theorem 3.5(i), it suffices to have d_2 satisfying $d_2 \in (d^{(k+1)}, d^{(k)})$. This condition is satisfied if

$$\frac{b_2 u_1^* - c_2 u_2^*}{\mu_{k+1}} < d_2 + 2u_2^*\alpha_{22} < \frac{b_2 u_1^* - c_2 u_2^*}{\mu_k}.$$

Thus we obtain the conclusion in part (i) by applying Theorem 3.5(i), and choosing $K_1 = (b_2 u_1^* - c_2 u_2^*)/\mu_{k+1}$ and $K_2 = (b_2 u_1^* - c_2 u_2^*)/\mu_k$.

Part (ii) can be proved in the same way as part (i) by applying an analog of Theorem 3.5(i), interchanging the role of the first and second equation of (3.1).

Corollary 3.7 (Non-Constant Solution in Strong Competition Case). *Consider problem (3.1) with $b_1/b_2 < a_1/a_2 < c_1/c_2$. Suppose that for some $\tilde{k} \geq 1$, the eigenvalue $\mu_{\tilde{k}}$ of the operator $-\Delta$ on Ω has odd multiplicity.*

(i) If $a_1/a_2 < \frac{1}{2}[(b_1/b_2) + (c_1/c_2)]$, then there exist positive constants $K_3 = K_3(a_i, b_i, c_i) < K_4 = K_4(a_i, b_i, c_i), K_5 = K_5(a_i, b_i, c_i)$ and $\Lambda_2 = \Lambda_2(d_i, a_i, b_i, c_i, \alpha_{11}, \alpha_{21}, \alpha_{22})$ such that for any $d_1 \geq K_5$, $\alpha_{11} \geq 0$, $\alpha_{21} \geq 0$, the problem (3.1) has at least one non-constant solution provided that $\alpha_{12} \geq \Lambda_2$ and $d_2 + 2u_2^*\alpha_{22} \in (K_3, K_4)$.

(ii) If $a_2/a_1 < \frac{1}{2}[(b_2/b_1) + (c_2/c_1)]$, then there exist positive constants $\hat{K}_3 = \hat{K}_3(a_i, b_i, c_i) < \hat{K}_4 = \hat{K}_4(a_i, b_i, c_i), \hat{K}_5 = \hat{K}_5(a_i, b_i, c_i)$ and $\hat{\Lambda}_2 = \hat{\Lambda}_2(d_i, a_i, b_i, c_i,$

$\alpha_{11}, \alpha_{12}, \alpha_{22}$) such that for any $d_2 \geq \hat{K}_5$, $\alpha_{12} \geq 0$, $\alpha_{22} \geq 0$, the problem (3.1) has at least one non-constant solution provided that $\alpha_{21} \geq \hat{\Lambda}_2$ and $d_1 + 2u_1^* \alpha_{11} \in (\hat{K}_3, \hat{K}_4)$.

Proof. Part (i) is proved by applying Theorem 3.5(ii). Part (ii) is proved by using an analog of Theorem 3.5(ii), interchanging the role of the first and second equation in (3.1).

The study of cross-diffusion has been made rigorously for larger systems. For example, Wang [226] considers the following 3-species problem of two preys with one predator. The cross-diffusions are included in such a way that the predator chases the prey and the prey runs away from the predator.

$$(3.83) \quad \begin{cases} -(K_{11}(u)u_{1x} + K_{13}(u)u_{3x})_x = u_1(1 - u_1 - cu_2 - u_3), \\ -(K_{22}(u)u_{2x} + K_{23}(u)u_{3x})_x = u_2(a - bu_1 - u_2 - ku_3), \quad 0 < x < l, \\ -(-\sum_{i=1}^2 K_{3i}(u)u_{ix} + K_{33}(u)u_{3x})_x = qu_3(u_1 + \rho k u_2 - r), \\ u_{1x} = u_{2x} = u_{3x} = 0, \quad x = 0, l. \end{cases}$$

$$(3.84) \quad K_{ii}(u) = d_i + b_i u_i + u_i \sum_{j=1}^3 \alpha_{ij} u_j, \quad i = 1, 2, 3,$$

$$K_{13}(u) = \beta_{13}u_1u_2, \; K_{23}(u) = \beta_{23}u_2u_1, \; K_{31}(u) = \beta_{31}u_3u_2, \; K_{32}(u) = \beta_{32}u_3u_1,$$

where d_i, b_i, α_{ij} and $\beta_{ij}, i, j = 1, 2, 3$ are positive constants, $u = (u_1, u_2, u_3)$. The parameters a, b, c, k, q, ρ and r are also positive constants. Here, $u_i, i = 1, 2$ are the concentrations of the two preys, and u_3 is the concentration of the predator. (We can interpret $K_{12}(u) = K_{21}(u) \equiv 0$). In this model, $J_1 = -K_{11}(u)u_{1x} - K_{13}(u)u_{3x}$, $J_2 = -K_{22}(u)u_{2x} - K_{23}(u)u_{3x}$ and $J_3 = \sum_{i=1}^2 K_{3i}(u)u_{ix} - K_{33}(u)u_{3x}$ indicate the population fluxes of u_1, u_2 and u_3 respectively. The terms $K_{11}(u)$, $K_{22}(u)$ and $K_{33}(u)$ represent the self-diffusion, and they are positive for $u \geq 0$. The terms $K_{ij}(u), i \neq j$ represent cross-diffusion. The fact that $-K_{i3}(u) \leq 0, i = 1, 2$ in J_i means that the prey $u_i, i = 1, 2$ is directed toward decreasing population of predator u_3. On the other hand, $K_{3i}(u) \geq 0, i = 1, 2$ in J_3 means that the predator u_3 is directed toward increasing population of prey u_i. That is, we have the prey running away from the predator, and the predator chasing the prey.

We assume that problem (3.83) has a unique positive constant solution $u^* = (u_1^*, u_2^*, u_3^*)$, and define constants $k_{ij} = K_{ij}(u^*), i, j = 1, 2, 3$. (Note that this assumption is valid if we assume, for example, the parameters in the right hand side of the system (3.83) satisfy:

$$k < a < \frac{1}{c} < b, \; ck + b\rho k < 1 + \rho k^2, \; \frac{\rho k(a-k)}{1-ck} < r < \frac{a-k}{b-k}$$

as in biological studies by Kan-on [98] and Kan-on and Mimura [99], without cross-diffusion.) From (3.84), we have $k_{ij} \geq 0$. Assume $b\rho k - 1 > 0$, and set

$$b_{13} := q(b\rho k - 1)u_2^* u_3^* > 0, \;\; a_{13} := (k_{31} - bk_{32})u_2^*,$$

$$\mu(k_{22}) := \frac{\sqrt{a_{13}^2 + 4k_{22}k_{31}b_{13}} - a_{13}}{2k_{22}b_{31}} > 0.$$

It is shown in Wang [226], by means of above methods in this section and bifurcation techniques as in earlier chapters, that non-constant equilibrium can exist under large enough cross-diffusion effect. More precisely, the following theorem is proved.

Theorem 3.8. *Assume that $\mu(k_{22}) \in ((n\pi/l)^2, ((n+1)\pi/l)^2)$ for some $k_{22} > 0$, and $n \geq 1$. If n is odd, then there exists a positive constant C such that problem (3.83) has at least one positive non-constant solution provided that $k_{13} > C$.*

Note the similarity between Theorem 3.8 and Theorem 3.5. Many more general results are found in [226]. The details are too lengthy to be presented here. More developments on the problem (3.1) when cross-diffusion pressures are extremely large can be found in Lou and Ni [165]. Other recent study on cross-diffusion with more general coupling terms and boundary conditions using index theory are found in Ryu and Ahn [196].

4.4 Degenerate and Density-Dependent Diffusions, Non-Extinction in Highly Spatially Heterogenous Environments

Part A: Weak Upper and Lower Solutions for Degenerate or Non-Degenerate Elliptic Systems.

We first consider positive solutions for the following degenerate elliptic systems with homogeneous Dirichlet boundary conditions.

$$\Delta\psi(w_i) + f_i(x, w_1, w_2) = 0 \text{ in } \Omega, \;\; w_i = 0 \text{ on } \partial\Omega, \;\; i = 1, 2. \tag{4.1}$$

Here, the function $\psi(s)$ satisfies the conditions $\psi \in C^1[0, \infty), \psi(0) = 0$ and $\psi'(s) > 0$ for $s > 0$. The equations become degenerate in the sense that we may also allow $\psi'(0) = 0$. Problems of this nature are of interest in reaction-diffusion processes in biology and chemistry. For example, the case for $\psi(u) = u^m, m > 1$ or $m \in (0, 1)$ for single parabolic equations (i.e. $u_t = \Delta u^m + f(x, u)$) has been studied for porous medium analysis and population dynamics (cf. Peletier [185] and Pozio and Tesei [186]). As $t \to \infty$ these solutions may tend to a solution of the corresponding elliptic scalar equation. They can also lead to the study of free boundary as in Diaz and Hernandez [47]. In various other problems (see e.g. early works in Aronson, Crandall and Peletier [5], DeMottoni, Schiaffino and Tesei [46], Schatzman [204] and Pozio and Tesei [187]), the function u^m is replaced with $\psi(u)$ satisfying conditions described above. For more general applications, we will allow the functions f_i in (4.1) to be discontinuous in x, and not necessarily Lipschitz in w_i.

We first discuss some existence and uniqueness theorems for the scalar equation:

$$\Delta\psi(w) + f(x,w) = 0 \text{ in } \Omega, \quad w = 0 \text{ on } \partial\Omega. \tag{4.2}$$

Monotone iteration is used to obtain a sequence which converges in $W^{2,p}(\Omega) \cap W_0^{1,p}(\Omega)$ to a maximal solution. We then deduce an existence theorem for systems of the form (4.1). We use Schauder's fixed point theorem to find a positive solution in $W^{2,p}(\Omega) \cap W_0^{1,p}(\Omega)$ for the system between appropriate weak upper and lower solutions. We will apply the results to simple ecological prey-predator models. Comparing the results here with those for nondegenerate case (m=1) in earlier chapters, we find here a much less stringent sufficient condition for coexistence. For example, we do not assume that the intrinsic growth rates of the species are larger than the principal eigenvalue of the domain. We allow the intrinsic growth rate $a(x)$ to be discontinuous and to have negative values somewhere. Again, note that when $\psi(u) = u$, the results in this section include the case of nondegenerate diffusion. In part A of this section, we essentially follow the results in Leung and Fan [135].

More precisely, in this section Ω is a bounded domain in $R^N (N \geq 2)$ with boundary $\partial\Omega \in C^2$. The functions $\psi : [0,\infty) \to [0,\infty), f : \Omega \times [0,\infty) \to R^1$ are assumed to satisfy some of the following hypotheses:

[H1] $\psi \in C^1[0,\infty), \psi(0) = 0$ and $\psi'(s) > 0$ for $s > 0$.

[H2] There is a bounded interval $[0,b]$ such that

(i) $f \in L^\infty(\Omega \times [0,b])$;

(ii) for any fixed $x \in \Omega$ *a.e.* the function $f(x,y)$ is continuous in y for all $y \in [0,b]$;

(iii) there is a constant $M > 0$ such that $f(x,y_2) - f(x,y_1) \geq -M(\psi(y_2) - \psi(y_1))$ for $x \in \Omega$ *a.e.*, $0 \leq y_1 \leq y_2 \leq b$.

[H3] For each fixed $x \in \Omega$ *a.e*, the function $f(x,y)/\psi(y)$ is a strictly monotonic increasing or decreasing function in y for $y \in [0,b]$.

Definition 4.1. A function $w \in C(\bar{\Omega})$ is called a non-negative solution of (4.2) if $w(x) \geq 0$ in Ω and $u = \psi(w) \in W^{2,p}(\Omega) \cap W_0^{1,p}(\Omega), (p > N)$, satisfies

$$\Delta u + f(x, \psi^{-1}(u)) = 0 \text{ a.e. in } \Omega, \quad u = 0 \text{ on } \partial\Omega, \tag{4.3}$$

where the derivatives of u are taken in the weak sense. A function w is called a positive solution of (4.2) if, in addition $w(x) > 0$ in Ω.

We first prove an existence result for a non-negative solution as defined above between the "upper" and "lower" solutions in the sense of (4.4) below.

Lemma 4.1. *Suppose that [H1], [H2,i] to [H2,iii] are satisfied. Assume that there are functions $\underline{w}, \bar{w}$ in $C(\bar{\Omega})$ with $0 \le \underline{w} \le \bar{w} \le b$ in $\bar{\Omega}$ and that $\psi(\underline{w}), \psi(\bar{w})$ are in $W^{1,p}(\Omega), (p > N)$, satisfying the inequalities*

$$(4.4)\qquad \begin{aligned} &-\int_\Omega \nabla\psi(\underline{w})\nabla\phi\, dx + \int_\Omega f(x,\underline{w})\phi\, dx \ge 0, \quad \underline{w} = 0 \text{ on } \partial\Omega, \\ &-\int_\Omega \nabla\psi(\bar{w})\nabla\phi\, dx + \int_\Omega f(x,\bar{w})\phi\, dx \le 0 \end{aligned}$$

for all $\phi \in C_0^1(\Omega), \phi \ge 0$. Then there exists at least one non-negative solution w of (4.2) satisfying $\underline{w} \le w \le \bar{w}$ in $\bar{\Omega}$.

Proof. For any given $u \in C(\bar{\Omega})$ with $0 \le u \le \psi(b)$, by hypothesis [H1], we have $0 \le \psi^{-1}(u) \le b, \psi^{-1}(u) \in C(\bar{\Omega})$; and $f(x, \psi^{-1}(u)) \in L^\infty(\Omega)$ by hypothesis [H2,i]. Since Ω is a bounded domain, we obtain $L^\infty(\Omega) \subset L^p(\Omega)$ for all $1 \le p < \infty$, and $Mu + f(x, \psi^{-1}(u)) \in L^p(\Omega)$. Here M is given in hypothesis [H2,iii]. It follows from linear elliptic L^p-theory that the problem

$$(4.5)\qquad \Delta v - Mv + Mu + f(x, \psi^{-1}(u)) = 0 \text{ a.e. in } \Omega, \quad v = 0 \text{ on } \partial\Omega$$

has a unique solution v, say $S(u)$, in $W^{2,p}(\Omega) \cap W_0^{1,p}(\Omega) \subset C(\bar{\Omega})$ satisfying

$$(4.6)\qquad ||S(u)||_{2,p} \le \bar{C}||Mu + f(x, \psi^{-1}(u))||_p,$$

where $\bar{C}$ is a positive constant which depends only on Ω and p.

Letting $\underline{u} = \psi(\underline{w}), \bar{u} = \psi(\bar{w})$, we have by hypothesis [H1] that $0 \le \underline{u} \le \bar{u} \le \psi(b)$ in $\bar{\Omega}$, and $\underline{u}, \bar{u} \in C(\bar{\Omega})$. Hence, as above, we obtain $S(\underline{u}), S(\bar{u})$ in $W^{2,p}(\Omega) \cap W_0^{1,p}(\Omega)$ as the unique solution of (4.5) corresponding to $\underline{u}$ and $\bar{u}$ respectively

We now construct a monotone sequence $\{u_i\}$ which will converge in $W^{2,p}(\Omega)$ to a solution of (4.3). First, define $u_0 = \bar{u}$ in $\bar{\Omega}$. From the arguments above, we can define $u_{i+1}, i = 1, 2, \ldots$ iteratively as the solution of

$$(4.7)\qquad \Delta v - Mv + Mu_i + f(x, \psi^{-1}(u_i)) = 0 \text{ a.e. in } \Omega, \quad v = 0 \text{ on } \partial\Omega$$

provided that each successive $u_i \ge 0$ in Ω so that $f(x, \psi^{-1}(u_i))$ is defined. We then have $u_{i+1} = S(u_i) \in W^{2,p}(\Omega) \cap W_0^{1,p}(\Omega) \subset C(\bar{\Omega})$ for $i = 1, 2, \ldots$.

We first show that these u_i are properly defined and that

$$(4.8)\qquad 0 \le \underline{u} \le \cdots \le u_2 \le u_1 \le u_0 = \bar{u} \quad \text{in } \bar{\Omega}.$$

Since $u_0 \geq 0$, equation (4.7) is meaningful for $i = 0$. Multiplying (4.7) by $\phi \in C_0^1(\Omega)$ and integrating on Ω, we obtain for $i = 0$ that

$$\int_\Omega (\Delta u_{i+1} - M u_{i+1})\phi \, dx + \int_\Omega [M u_i + f(x, \psi^{-1}(u_i))]\phi \, dx = 0 \tag{4.9}$$

for all $\phi \in C_0^1(\Omega)$. Since

$$\int_\Omega \Delta u_{i+1} \phi \, dx = -\int_\Omega \nabla u_{i+1} \nabla \phi \, dx \quad \text{for all } \phi \in C_0^1(\Omega),$$

(4.9) yields

$$-\int_\Omega \nabla u_{i+1} \nabla \phi \, dx + \int_\Omega [-M u_{i+1} + M u_i + f(x, \psi^{-1}(u_i))]\phi \, dx = 0 \tag{4.10}$$

for all $\phi \in C_0^1(\Omega)$. From the definition of $u_0 = \bar{u} = \psi(\bar{w})$ and the second hypothesis in (4.4), we have

$$-\int_\Omega \nabla u_0 \nabla \phi \, dx + \int_\Omega f(x, \psi^{-1}(u_0))\phi \, dx \leq 0 \tag{4.11}$$

for all $\phi \in C_0^1(\Omega)$ with $\phi \geq 0$. Setting $i = 0$ in (4.10) and subtracting (4.11), we obtain

$$-\int_\Omega \nabla(u_1 - u_0)\nabla \phi \, dx - M \int_\Omega (u_1 - u_0)\phi \, dx \geq 0 \tag{4.12}$$

for all $\phi \in C_0^1(\Omega)$ with $\phi \geq 0$. It follows from the weak maximum principle for weak solutions (see e.g [71], p. 179) that

$$sup_\Omega \, (u_1 - u_0) \leq sup_{\partial\Omega} \, (u_1 - u_0)^+ = 0,$$

hence, $u_1 \leq u_0 = \bar{u}$ in Ω. Similarly, using (4.10) with $i = 0$ and the first hypothesis in (4.4), we deduce that $\underline{u} \leq u_1$. We next inductively assume that

$$\underline{u} \leq u_j \leq u_{j-1} \leq \bar{u} \quad \text{in } \bar{\Omega} \tag{4.13}$$

for $j \geq 1$. Thus equations (4.7), (4.9) and (4.10) are meaningful for $i = j$ and $j - 1$. Letting $i = j$ and $j - 1$ in (4.7), we subtract to obtain

$$\begin{cases} \Delta(u_{j+1} - u_j) - M(u_{j+1} - u_j) + M(u_j - u_{j-1}) + f(x, \psi^{-1}(u_j)) \\ \quad -f(x, \psi^{-1}(u_{j-1})) = 0 \qquad \text{a.e. in } \Omega, \\ u_{j+1} - u_j = 0 \qquad \text{on } \partial\Omega. \end{cases} \tag{4.14}$$

Since $0 \le \psi^{-1}(u_j) \le \psi^{-1}(u_{j-1}) \le b$, we obtain from [H2,iii] that

$$M(u_{j-1} - u_j) + f(x, \psi^{-1}(u_{j-1})) - f(x, \psi^{-1}(u_j)) \ge 0 \text{ a.e. in } \Omega,$$

and (4.14) yields

$$(4.15) \quad \Delta(u_{j+1} - u_j) - M(u_{j+1} - u_j) \ge 0 \text{ a.e. in } \Omega, \quad u_{j+1} - u_j = 0 \text{ on } \partial\Omega.$$

It follows from maximum principle (see [71], p. 225 or Theorem A3-1 in Chapter 6) that $u_{j+1} \le u_j$ in $\bar{\Omega}$. Analogously, using (4.10) for $i = j$ and the first hypothesis of (4.4) as before, we obtain by the maximum principle that $\underline{u} \le u_{j+1}$ in $\bar{\Omega}$. By induction, we have

$$\underline{u} \le \cdots \le u_{i+1} \le u_i \le \cdots \le u_2 \le u_1 \le u_0 \text{ in } \bar{\Omega}.$$

We can therefore define by pointwise convergence in $\bar{\Omega}$

$$u(x) = \lim_{i \to \infty} u_i(x) \text{ in } \bar{\Omega}.$$

By the Lebesgue Convergence Theorem, $\{Mu_i + f(x, \psi^{-1}(u_i))\}$ must be a Cauchy sequence in $L^p(\Omega)$. From the equations satisfied by $u_{i+1} - u_{j+1}$, we obtain the estimate as in (4.6) that

$$(4.16) \quad ||u_{i+1} - u_{j+1}||_{2,p} \le \bar{C}||M(u_j - u_i) + f(x, \psi^{-1}(u_j)) - f(x, \psi^{-1}(u_i))||_p.$$

Consequently, $\{u_i\}$ is a Cauchy sequence in $W^{2,p}(\Omega)$, and $u_i \to u$ in $W^{2,p}(\Omega)$ as $i \to \infty$. Passing to the limit in (4.7), we have

$$\Delta u + f(x, \psi^{-1}(u)) = 0 \text{ a.e. in } \Omega, \quad u = 0 \text{ on } \partial\Omega,$$

where the derivatives are taken in the weak sense and $u \in W^{2,p}(\Omega)$ (Note that $v = u_{i+1}$ in (4.7)). Furthermore, since the u_i are in $W_0^{1,p}(\Omega)$, which is a closed subspace of $W^{1,p}(\Omega)$, and $u_i \to u$ in $W^{1,p}(\Omega)$, we must also have $u \in W_0^{1,p}(\Omega)$. Letting $w = \psi^{-1}(u)$, we obtain w as a non-negative solution of (4.2) with $\underline{w} \le w \le \bar{w}$.

With the addition of hypothesis [H3] and the assumption that the lower solution $\underline{w}$ is positive in Ω we now deduce a uniqueness result.

Lemma 4.2. *Assume all the hypotheses of Lemma 4.1. In addition, suppose that [H3] is valid and that $\underline{w} > 0$ in Ω. Then there exists a unique positive solution w^* of (4.2) with the property:*

$$0 < \underline{w} \le w^* \le \bar{w} \text{ in } \Omega.$$

Proof. Let w be the solution of (4.2) obtained from the monotonic sequence in Lemma 4.1. Now $w > 0$ in Ω, since $w \ge \underline{w} > 0$ in Ω. Let z be any positive solution of (4.2) with $\underline{w} \leq z \leq \bar{w}$ in Ω. Then $u = \psi(w), v = \psi(z)$ are two positive solutions of (4.3) in $W^{2,p}(\Omega) \cap W_0^{1,p}(\Omega)$ with $0 < \underline{u} \leq u \leq \bar{u}, 0 < \underline{u} \leq v \leq \bar{u}$ in Ω. By applying the same argument as that used in the proof of Lemma 4.1, we obtain $\underline{u} \leq v \leq u_i \leq \bar{u}$ in Ω for each $i = 0, \ldots$. Hence, we have the inequality

$$0 \leq \underline{u} \leq v \leq u \leq \bar{u} \text{ in } \Omega. \tag{4.17}$$

It remains to show that $v = u$ in $\bar{\Omega}$. Since both u, v are in $W_0^{1,p}(\Omega)$, there are two sequences $\{u_n\}, \{v_n\}$ in $C_0^\infty(\Omega)$ which converge to u, v, respectively, in $W^{1,p}(\Omega)$. Since u, v are solutions of (4.3) in $W^{2,p}(\Omega)$, and $\{u_n\}, \{v_n\}$ have compact support in Ω, we use the definition of weak derivative to obtain

$$\begin{aligned} &\textstyle\int_\Omega u \Delta v_n \, dx + \int_\Omega f(x, \psi^{-1}(u)) v_n \, dx = 0, \\ &\textstyle\int_\Omega v \Delta u_n \, dx + \int_\Omega f(x, \psi^{-1}(v)) u_n \, dx = 0 \end{aligned} \tag{4.18}$$

for $n = 1, 2, \ldots$. Subtracting the two previous equations, we obtain

$$\int_\Omega [v \Delta u_n - u \Delta v_n] \, dx = \int_\Omega [f(x, \psi^{-1}(u)) v_n - f(x, \psi^{-1}(v)) u_n] \, dx. \tag{4.19}$$

It follows from the definition of the weak derivative and $u \in W^{1,p}(\Omega), u_n, v_n \in C_0^\infty(\Omega)$ that we have

$$\begin{aligned} &\textstyle\int_\Omega u \Delta v_n \, dx = -\int_\Omega \nabla u \nabla v_n \, dx, \\ &\textstyle\int_\Omega v \Delta u_n \, dx = -\int_\Omega \nabla v \nabla u_n \, dx. \end{aligned}$$

Hence, the left hand side of (4.19) becomes

$$\begin{aligned} \textstyle\int_\Omega [v \Delta u_n - u \Delta v_n] \, dx &= \textstyle\int_\Omega [\nabla u \nabla v_n - \nabla v \nabla u_n] \, dx \\ &= \textstyle\int_\Omega \nabla u (\nabla v_n - \nabla v) \, dx - \int_\Omega \nabla v (\nabla u_n - \nabla u) \, dx. \end{aligned} \tag{4.20}$$

From the Schwarz inequality, we have

$$\int_\Omega |\nabla u (\nabla v_n - \nabla v)| \, dx \leq ||\nabla u||_{L^2(\Omega)} ||\nabla (v_n - v)||_{L^2(\Omega)}.$$

Since $v_n \to v$ in $W^{1,p}(\Omega), p > N \geq 2$, it follows that

$$\int_\Omega \nabla u (\nabla v_n - \nabla v) \, dx \to 0, \text{ as } n \to \infty.$$

Similarly, one also has

$$\int_\Omega \nabla v (\nabla u_n - \nabla u) \, dx \to 0, \text{ as } n \to \infty.$$

Consequently, it follows from (4.20) that

$$\int_\Omega [v\Delta u_n - u\Delta v_n]\,dx \to 0, \quad \text{as } n \to \infty. \tag{4.21}$$

Equation (4.21) and (4.19) lead to the property that

$$\int_\Omega [f(x,\psi^{-1}(u))v_n - f(x,\psi^{-1}(v))u_n]\,dx \to 0, \quad \text{as } n \to \infty. \tag{4.22}$$

On the other hand, from Sobolev's Imbedding Theorem, u_n and v_n are uniformly bounded in $\bar{\Omega}$, so the Dominated Convergence Theorem leads to

$$\int_\Omega [f(x,\psi^{-1}(u))v_n - f(x,\psi^{-1}(v))u_n]\,dx \to \int_\Omega [f(x,\psi^{-1}(u))v - f(x,\psi^{-1}(v))u]\,dx, \tag{4.23}$$

as $n \to \infty$. From (4.22) and (4.23), we deduce that

$$\int_\Omega [f(x,\psi^{-1}(u))v - f(x,\psi^{-1}(v))u]\,dx = 0. \tag{4.24}$$

Suppose that v is not identically equal to u in $\bar{\Omega}$. The set

$$\Omega_1 := \{x \in \Omega |\, v(x) < u(x)\}$$

then has measure greater than zero. From assumption [H3], we have

$$\begin{aligned} & f(x,\psi^{-1}(u))v - f(x,\psi^{-1}(v))u \\ & = uv[\tfrac{f(x,\psi^{-1}(u))}{u} - \tfrac{f(x,\psi^{-1}(v))}{v}] > 0 \text{ (or } < 0) \text{ a.e. in } \Omega_1 \end{aligned} \tag{4.25}$$

(recall that $v(x) \geq \psi(\underline{w}(x)) > 0$ for all x in Ω). This lead to

$$\int_\Omega [f(x,\psi^{-1}(u))v - f(x,\psi^{-1}(v))u]\,dx > 0 \text{ (or } < 0), \tag{4.26}$$

which contradicts (4.24), since

$$\begin{aligned} 0 & = \textstyle\int_\Omega [f(x,\psi^{-1}(u))v - f(x,\psi^{-1}(v))u]\,dx \\ & = \textstyle\int_{\Omega_1} [f(x,\psi^{-1}(u))v - f(x,\psi^{-1}(v))u]\,dx \neq 0. \end{aligned} \tag{4.27}$$

This completes the proof of the Lemma.

Remark 4.1. As mentioned above, Definition 4.1, Lemmas 4.1 and 4.2 apply also for the usual non-degenerate case $\psi(u) = u$.

From the above remark, we can apply Lemmas 4.1 and 4.2 to the problem

$$\Delta w + w(a(x) - bw) = 0 \text{ a.e. in } \Omega, \quad w = 0 \text{ on } \partial\Omega. \tag{4.28}$$

Here we will not make the strong assumption that the growth rate satisfies $a(x) > \lambda_1$ for all $x \in \Omega$, where λ_1 is the principal eigenvalue for the operator $-\Delta$ on the domain Ω under homogeneous Dirichlet boundary condition. In a highly spatially heterogeneous habitat in ecological problems, $a(x)$ is relatively large in part of Ω, and may be small or even negative in other parts of Ω. Furthermore, it may be discontinuous. This motivates the following example.

Example 4.1 (Highly Spatially Heterogeneous Environment). Let Ω_s be a subdomain of Ω, with boundary $\partial\Omega_s \in C^2$; and let λ_s be the principal eigenvalue for Ω_s, i.e. the first eigenvalue of $\Delta u + \lambda u = 0$ in $\Omega_s, u = 0$ on $\partial\Omega_s$. Consider (4.28), where b is a positive constant, $a(x) \in L^\infty(\Omega)$ and

$$a(x) = \begin{cases} a_1(x) & \text{in } \Omega_s, \\ a_2(x) & \text{in } \Omega\backslash\Omega_s, \end{cases} \tag{4.29}$$

with

$$a_1(x) > \lambda_s, \quad \text{a.e. for } x \in \Omega_s. \tag{4.30}$$

Then the Dirichlet problem (4.28) has one and only one non-negative nontrivial solution w (in the sense of Definition 4.1 with $\psi(w) = w$) satisfying $0 \leq w \leq \bar{a}/b$. Here $\bar{a} := ess\,sup_{x\in\Omega}\, a(x)$. Moreover, $w > 0$ in Ω. (Note that $a(x)$ can possibly have negative values somewhere in Ω.)

(Proof of assertion of Example 4.1.) Let $\psi(s) = s$ for $s \geq 0$ and $f(x, w) = w(a(x) - bw)$. One readily verifies that $[H1], [H2, i - H2, iii]$ are satisfied. We will apply Lemma 4.1 to prove the existence of solution. Let $\bar{w}, \underline{w}$ be defined as

$$\bar{w} := \frac{\bar{a}}{b} \text{ in } \bar{\Omega}, \qquad \underline{w} := \begin{cases} \delta\theta(x) & x \in \Omega_s, \\ 0 & x \in \bar{\Omega}\backslash\Omega_s; \end{cases} \tag{4.31}$$

where $\theta(x)$ is a positive principal eigenfunction associated with the principal eigenvalue λ_s of the domain Ω_s, and $\delta > 0$ is to be determined. For $\delta > 0$ sufficiently small, we clearly have $0 \leq \underline{w} \leq \bar{w}$ in Ω. The constant function $\bar{w}$ is in $W^{1,p}(\Omega)$, and we now verify that $\underline{w} \in W^{1,p}(\Omega)$. By the definition of $\underline{w}$, we have, for $|\alpha| = 1, \phi \in C_0^1(\Omega)$,

$$-\int_\Omega \underline{w} D^\alpha \phi\, dx = -\int_{\Omega_s} \delta\theta(x) D^\alpha \phi\, dx. \tag{4.32}$$

Integrating by parts, we obtain

$$-\int_\Omega \underline{w} D^\alpha \phi\, dx = \int_{\Omega_s} \delta D^\alpha \theta(x) \phi\, dx, \tag{4.33}$$

since $\theta(x) \equiv 0$ on $\partial\Omega_s$. Hence the α-th weak derivative of $\underline{w}$ is

$$D^\alpha \underline{w}(x) = \begin{cases} \delta D^\alpha \theta(x), & x \in \Omega_s, \\ 0, & x \in \bar{\Omega}\backslash\Omega_s. \end{cases} \tag{4.34}$$

Since $D^\alpha \underline{w} \in L^p(\Omega)$, we obtain $\underline{w} \in W^{1,p}(\Omega)$. To see whether the second inequality in (4.4) holds, we calculate

$$-\int_\Omega \nabla\bar{w}\nabla\phi\, dx + \int_\Omega \bar{w}(a(x) - b\bar{w})\phi\, dx = \int_\Omega \frac{\bar{a}}{b}(a(x) - b\frac{\bar{a}}{b})\phi\, dx \le 0 \tag{4.35}$$

for all $\phi \in C_0^1(\Omega), \phi \ge 0$. To verify the first inequality in (4.4), one has

$$\begin{aligned} &- \textstyle\int_\Omega \nabla\underline{w}\nabla\phi\, dx + \int_\Omega \underline{w}(a(x) - b\underline{w})\phi\, dx \\ &= - \textstyle\int_{\Omega_s} \delta\nabla\theta\nabla\phi\, dx + \int_{\Omega_s} \delta\theta(x)(a_1 - b\delta\theta(x))\phi\, dx \\ &= \textstyle\int_{\Omega_s} \delta\Delta\theta\, \phi\, dx - \int_{\partial\Omega_s} \delta\frac{\partial\theta}{\partial\nu}\phi\, ds + \int_{\Omega_s} \delta\theta(x)(a_1 - b\delta\theta(x))\phi\, dx \\ &= \textstyle\int_{\Omega_s} [-\lambda_s + a_1 - b\delta\theta(x)]\delta\theta\phi\, dx - \int_{\partial\Omega_s} \delta\frac{\partial\theta}{\partial\nu}\phi\, ds \end{aligned} \tag{4.36}$$

which is positive for $\delta > 0$ sufficiently small, by hypothesis (4.30) and the fact that $\partial\theta/\partial\nu \le 0$ on $\partial\Omega_s$. Applying Lemma 4.1, we conclude that (4.28) has a non-negative solution w in $W^{2,p}(\Omega) \cap W_0^{1,p}(\Omega)$ with $0 \le w \le \bar{a}/b$ in Ω and $w > 0$ in Ω_s.

To prove that $w > 0$ in Ω, let $u(t, x) = e^{ct}w(x)$ for $(t, x) \in [0, +\infty) \times \bar{\Omega}$, where c is a positive constant such that $c \ge ess\, sup_{x\in\Omega}(bw(x) - a(x))$. Thus, since w satisfies (4.28) a.e., we also have $\Delta u + u(a(x) - bw(x)) = 0$ a.e. in $[0, +\infty) \times \Omega$; hence

$$\begin{aligned} u_t &= cu = cu + \Delta u + u(a(x) - bw(x)) \\ &= \Delta u + u(c + a(x) - bw(x)) \ge \Delta u \quad \text{a.e. in } (0, +\infty) \times \Omega, \end{aligned} \tag{4.37}$$

by the choice of c. Thus u is an upper solution (in the weak sense as described in Definition A5-1 for Theorem A5-6 in Chapter 6) to the problem

$$\begin{cases} v_t = \Delta v & \text{in } (0, +\infty) \times \Omega, \\ v = 0 & \text{on } (0, +\infty) \times \partial\Omega, \\ v(0, x) = w(x) & \text{in } \Omega. \end{cases} \tag{4.38}$$

By means of semigroup and Schauder's theory, we know that problem (4.38) has a classical solution. Thus, if v is the classical solution of (4.38), we have

$u(t,x) \geq v(t,x) > 0$ in $(0,+\infty) \times \Omega$ (the fact that $v(x,t) > 0$ in $(0,+\infty) \times \Omega$ follows from the strong maximum principle). Thus $w(x) > 0$ in Ω, by the definition of u. (Here, we have used the comparison Theorem A5-6 for upper solutions in the weak sense for problem (5.3), with $f \equiv 0$ and $\eta(u) = u$ in Chapter 6.)

Finally, we prove that such a w is unique. Let w^* be the solution of (4.28) obtained from the monotonic convergence sequence as in Lemma 4.1, using $w_0 = \bar{a}/b$ as the first iterate and defining $w_{j+1} = S(w_j)$, (recall $\psi(s) = s$). Using the fact that $w \leq w_0$; we can prove that $w \leq w_j$ in Ω by using the maximum principle as in (4.15) with u_{j+1} and u_j respectively replaced by w and w_j. This leads to the fact that $0 < w \leq w^*$ in Ω. Let z be any non-negative (nontrivial) solution of (4.28) with $z \leq \bar{a}/b$ in Ω. As above, we have $z \leq w^*$ in Ω. Let $f(x,w) = w(a(x) - bw)$. We follow the proof of Lemma 4.2, with the role of u, v respectively replaced by w^*, z until (4.24). Then (4.24) implies that

$$
\begin{aligned}
0 &= \textstyle\int_\Omega [f(x,w^*)z - f(x,z)w^*]\,dx = \int_{D_1} [f(x,w^*)z - f(x,z)w^*]\,dx \\
&= \textstyle\int_{D_2} [f(x,w^*)z - f(x,z)w^*]\,dx,
\end{aligned}
\tag{4.39}
$$

where $D_1 = \{x \in \Omega \,|\, z(x) < w^*(x)\}$ and $D_2 = \{x \in \Omega \,|\, 0 < z(x) < w^*(x)\}$. The last equality follows from the fact that $f(x,0) = 0$ for x a.e. in Ω. However,

$$
f(x,w^*)z - f(x,z)w^* < 0 \quad \text{in } D_2. \tag{4.40}
$$

We therefore conclude that $w^* = z$ in the set $D = \{x \in \Omega \,|\, 0 < z(x)\}$. We observe that the set D is open in Ω. Moreover, the set D is also closed in Ω for the following reason: Let $x_n \in D, x_n \to x \in \Omega$. Then

$$
z(x) = \lim_{n\to\infty} z(x_n) = \lim_{n\to\infty} w^*(x_n) = w^*(x); \tag{4.41}
$$

however, $w^*(x) > 0$ in Ω, therefore $z(x) > 0$, and $x \in D$. Consequently, we must have $D = \Omega$. In D, we have concluded that $w^* = z$. Thus every non-negative nontrivial solution bounded above by $\bar{a}/b$ must be identically equal to the same w^* in Ω. This completes the proof for the assertions of the example.

So far in Example 4.1, we have assumed Ω to be connected. However, if Ω is not connected, we do not have $\phi > 0$ in Ω but rather $\phi \geq 0$ and ϕ not identically zero in Ω. Thus we have the following conclusion.

Theorem 4.1. *In Example 4.1, suppose that Ω is not connected and that Ω_s is connected. Assume that $a(x) \in L^\infty(\Omega)$, and that hypotheses (4.29) and (4.30) hold. Then the Dirichlet problem (4.28) has at least one non-negative solution w (in the sense of Definition 4.1 with $\psi(w) = w$) satisfying $0 \leq w \leq \bar{a}/b$ in Ω and $w > 0$ in the component of Ω which contains Ω_s. Furthermore, if there* *is a positive solution* v *of (4.28), with* $0 < v \leq \bar{a}/b$ *in* Ω*, then it is the unique non-negative (nontrivial) solution of (4.28) satisfying* $0 < w \leq \bar{a}/b$ *in* Ω.

The proof of the part of the above theorem concerning uniqueness follows the method of proving Lemma 4.2.

Remark 4.2. If the function $a(x)$ is in $C^{\alpha}(\Omega), 0 < \alpha < 1$, and we have $a(x) < \lambda_1$ in Ω, then we know from earlier chapters that the only non-negative solution of (4.28) is the trivial solution. (Here, λ_1 is the principal eigenvalue of the operator $-\Delta$ on the domain Ω). More general results for the degenerate case corresponding to (4.28) with Δw replaced by $\Delta\psi(w)$ can be found in e.g. [187].

With the help of the Lemmas 4.1 and 4.2 above concerning scalar equations, we are now ready to study the existence of positive solutions for elliptic systems of the form:

$$\begin{cases} \Delta\psi(w_1) + f_1(x, w_1, w_2) = 0 & \text{a.e. in } \Omega, \\ \Delta\psi(w_2) + f_2(x, w_1, w_2) = 0 & \text{a.e. in } \Omega, \\ w_1 = w_2 = 0 & \text{on } \partial\Omega, \end{cases} \tag{4.42}$$

where the derivatives are taken in the weak sense; The functions $\psi : [0, \infty) \to [0, \infty), f_i : \Omega \times [0, \infty) \times [0, \infty) \to R^1$ are assumed to satisfy the following hypotheses:

$[\tilde{H}1]$ $\psi \in C^1[0, \infty), \psi(0) = 0$ and $\psi'(s) > 0$ for $s > 0$.

$[\tilde{H}2]$ There are two positive constants b_1, b_2 such that

(i) $f_i \in L^{\infty}(\Omega \times [0, b_1] \times [0, b_2])$ for $i = 1, 2$;

(ii) for any fixed $x \in \Omega$ *a.e.* the functions $f_i(x, y_1, y_2)$ are continuous in (y_1, y_2) for all $(y_1, y_2) \in [0, b_1] \times [0, b_2], i = 1, 2$;

(iii) there is a constant $M > 0$ such that

$$f_1(x, \xi, y_2) - f_1(x, \eta, y_2) \geq -M(\psi(\xi)) - \psi(\eta))$$
$$\text{for } x \in \Omega \text{ a.e.}, \; y_2 \in [0, b_2], 0 \leq \eta \leq \xi \leq b_1;$$

$$f_2(x, y_1, \xi) - f_2(x, y_1, \eta) \geq -M(\psi(\xi)) - \psi(\eta))$$
$$\text{for } x \in \Omega \text{ a.e.}, \; y_1 \in [0, b_1], 0 \leq \eta \leq \xi \leq b_2.$$

Definition 4.2. A pair of continuous functions (w_1, w_2) in $C(\bar{\Omega}) \times C(\bar{\Omega})$, with $w_i(x) > 0$ in Ω, $i = 1, 2$, is called a positive solution of (4.42) if $\psi(w_i) \in W^{2,p}(\Omega) \cap W_0^{1,p}(\Omega)$ $(p > N)$, and (4.42) holds.

Theorem 4.2 (Main Existence Theorem). *Assume hypotheses* $[\tilde{H}1]$ *and* $[\tilde{H}2]$. *Suppose that there are functions* $\underline{w}_i(x), \bar{w}_i(x)$ $(i = 1, 2)$ *in* $C(\bar{\Omega})$ *with* $\psi(\underline{w}_i), \psi(\bar{w}_i)$ $(i = 1, 2)$ *in* $W^{1,p}(\Omega)$ $(p > N)$ *satisfying the inequalities:*

(4.43)

$$-\int_\Omega \nabla\psi(\underline{w}_1)\nabla\phi\, dx + \int_\Omega f_1(x, \underline{w}_1, w_2)\phi\, dx \geq 0 \quad \text{for } \underline{w}_2 \leq w_2 \leq \bar{w}_2,$$

$$-\int_\Omega \nabla\psi(\bar{w}_1)\nabla\phi\, dx + \int_\Omega f_1(x, \bar{w}_1, w_2)\phi\, dx \geq 0 \quad \text{for } \underline{w}_2 \leq w_2 \leq \bar{w}_2,$$

$$-\int_\Omega \nabla\psi(\underline{w}_2)\nabla\phi\, dx + \int_\Omega f_2(x, w_1, \underline{w}_2)\phi\, dx \geq 0 \quad \text{for } \underline{w}_1 \leq w_1 \leq \bar{w}_1,$$

$$-\int_\Omega \nabla\psi(\bar{w}_2)\nabla\phi\, dx + \int_\Omega f_2(x, w_1, \bar{w}_2)\phi\, dx \geq 0 \quad \text{for } \underline{w}_1 \leq w_1 \leq \bar{w}_1$$

for all $\phi \in C_0^1(\Omega), \phi \geq 0$. *Here* $w_i = w_i(x)$ *are assumed to be continuous in* $\bar{\Omega}$, *and* $0 \leq \underline{w}_i \leq w_i \leq \bar{w}_i \leq b_i$ *in* $\bar{\Omega}$, $\underline{w}_i > 0$ *in* Ω, *and* $\underline{w}_i = 0$ *on* $\partial\Omega$. *Then there exists at least one positive solution* (w_1^*, w_2^*) *of (4.42) satisfying* $\underline{w}_i \leq w_i^* \leq \bar{w}_i$ *in* $\bar{\Omega}$.

Proof. Let $\underline{u}_i = \psi(\underline{w}_i), \bar{u}_i = \psi(\bar{w}_i), X_i = \{u \in C(\bar{\Omega}), \underline{u}_i \leq u_i \leq \bar{u}_i \text{ in } \bar{\Omega}\}, i = 1, 2$, and let M be described as in $[\tilde{H}2, iii]$. The set $X_1 \times X_2$ is a bounded closed convex set in $C(\bar{\Omega}) \times C(\bar{\Omega})$. We define the map $T : X_1 \times X_2 \to X_1 \times X_2$ as

$$T(u_1, u_2) = (v_1, v_2) \quad \text{for} \quad (u_1, u_2) \in X_1 \times X_2,$$

where $v_1, v_2 \in W^{2,p}(\Omega) \cap W_0^{1,p}(\Omega) \subset C(\bar{\Omega})$ $(p > N)$, and (v_1, v_2) is uniquely determined as the solution of the (decoupled) system

$$\Delta v_i - Mv_i + f_i(x, \psi^{-1}(u_1), \psi^{-1}(u_2)) + Mu_i = 0 \quad \text{in } \Omega, \ i = 1, 2. \tag{4.44}$$

(Here the derivatives are meant in the weak sense.)

We first show that $(v_1, v_2) \in X_1 \times X_2$. From equation (4.44), and hypotheses $[\tilde{H}1], [\tilde{H}2, iii]$ and the second line in (4.43), we have, for any $\phi \in C_0^1(\Omega), \phi \geq 0$,

(4.45)

$$-\int_\Omega \nabla(\bar{u}_1 - v_1)\nabla\phi\, dx - M\int_\Omega(\bar{u}_1 - v_1)\phi\, dx \leq -\int_\Omega[f_1(x, \psi^{-1}(\bar{u}_1), \psi^{-1}(u_2))$$

$$-f_1(x, \psi^{-1}(u_1), \psi^{-1}(u_2))]\phi\, dx - M\int_\Omega(\bar{u}_1 - u_1)\phi\, dx \ \leq 0.$$

Hence the weak maximum principle implies that $\bar{u}_1 \geq v_1$. Analogously, since

(4.46)

$$-\int_\Omega \nabla(\underline{u}_1 - v_1)\nabla\phi\, dx - M\int_\Omega(\underline{u}_1 - v_1)\phi\, dx \geq -\int_\Omega[f_1(x, \psi^{-1}(\underline{u}_1), \psi^{-1}(u_2))$$

$$-f_1(x, \psi^{-1}(u_1), \psi^{-1}(u_2))]\phi\, dx - M\int_\Omega(\underline{u}_1 - u_1)\phi\, dx \ \geq 0$$

for any $\phi \in C_0^1(\Omega), \phi \geq 0$, we deduce that $\underline{u}_1 \leq v_1$. We apply the same procedure to prove that $v_2 \in X_2$.

We next show that T is a continuous operator from $X_1 \times X_2$ into itself. Let $(u_1^{(n)}, u_2^{(n)})$ be a sequence in $X_1 \times X_2$, which converges to (u_1, u_2) in $X_1 \times X_2$. Define $(v_1^{(n)}, v_2^{(n)}) = T(u_1^{(n)}, u_2^{(n)})$ and $(v_1, v_2) = T(u_1, u_2)$ as in (4.44). By the classical L^p-estimate for the linear problem (4.44), we have

$$(4.47) \qquad ||v_i^{(n)}||_{2,p} \leq \bar{C}_i ||f_i(x, \psi^{-1}(u_1^{(n)}), \psi^{-1}(u_2^{(n)})) + M u_i^{(n)}||_p$$

with $\underline{u}_i \leq v_i^{(n)} \leq \bar{u}_i$ for $n = 1, 2, \ldots$, where $\bar{C}_i$ are positive constants. By $[\tilde{H}2, i]$, there exist constants $M_i > 0$ such that

$$(4.48) \qquad |f_i(x, y_1, y_2)| \leq M_i \text{ for almost all } (x, y_1, y_2) \in \Omega \times [0, b_1] \times [0, b_2].$$

Since Ω is a bounded domain, (4.48) implies that $\{f_i(x, \psi^{-1}(u_1^{(n)}), \psi^{-1}(u_2^{(n)}))\}$, $i = 1, 2$, are bounded sequences in $L^p(\Omega)$. It follows from (4.47) that $\{v_i^{(n)}\}$ is a bounded sequence in $W^{2,p}(\Omega) \cap W_0^{1,p}(\Omega)(\Omega)$ $(p > N)$. Applying Sobolev's Theorem, we can select a subsequence $\{v_i^{(n_k)}\}$ from $\{v_i^{(n)}\}$ such that $\{v_i^{(n_k)}\}$ converges in $C(\bar{\Omega})$ to say, v_i^*. To see whether $\{v_i^{(n_k)}\}$ actually converges to v_i^* in $W^{2,p}(\Omega)$, we first deduce from $[\tilde{H}2, ii]$ that

$$(4.49) \qquad f_i(x, \psi^{-1}(u_1^{(n_k)}), \psi^{-1}(u_2^{(n_k)})) \to f_i(x, \psi^{-1}(u_1), \psi^{-1}(u_2))$$

pointwise in Ω. Since Ω is bounded, the Lebesgue Convergence Theorem implies that the convergence in (4.49) is true in the $L^p(\Omega)$ norm $(N < p < \infty)$. The estimate (4.47) hence implies that $\{v^{(n_k)}\}$ converges to v_i^* in $W^{2,p}(\Omega)$. By the definition of $\{v^{(n_k)}\}$, we have, for $k = 1, 2, \ldots$ that

$$(4.50) \qquad \Delta v_i^{(n_k)} - M v_i^{(n_k)} + f_i(x, \psi^{-1}(u_1^{(n_k)}), \psi^{-1}(u_2^{(n_k)})) + M u_i^{(n_k)} = 0 \text{ in } \Omega, \ i = 1, 2.$$

Passing to the limit in (4.50), we obtain

$$(4.51) \qquad \Delta v_i^* - M v_i^* + f_i(x, \psi^{-1}(u_1), \psi^{-1}(u_2)) + M u_i = 0 \text{ in } \Omega, \ i = 1, 2.$$

From (4.44) and (4.51), we see that both (v_1^*, v_2^*) and (v_1, v_2) are positive solutions of the same linear problem. We conclude by uniqueness of the positive solution of the linear problem (4.44) that $(v_1^*, v_2^*) = (v_1, v_2)$. Hence we have $\{v_i^{(n_k)}\} \to v_i$ in $C(\bar{\Omega})$. Finally, we claim that the full sequence $\{v_i^{(n)}\} \to v_i$ in $C(\bar{\Omega})$ as $i \to \infty$. Suppose not; then there exist a subsequence $\{v_i^{(n_j)}\}$ and a constant $\epsilon_0 > 0$ such that

$$(4.52) \qquad ||v_i^{(n_j)} - v_i|| \geq \epsilon_0 \text{ for } j = 1, 2, \ldots .$$

Here the norm is taken in $C(\bar{\Omega})$. Using the same argument as that used above, by replacing $\{v_i^{(n)}\}$ with $\{v_i^{(n_j)}\}$, we can select a subsequence of $\{v_i^{(n_j)}\}$ which converges to v_i in $C(\bar{\Omega})$. This contradicts the inequality (4.52). Consequently, $\{v_i^{(n)}\}$ converges to v_i in $C(\bar{\Omega})$ as $i \to \infty$. This leads to the conclusion that T is a continuous operator from $X_1 \times X_2$ into itself.

We finally show that T is a compact operator. From (4.47), T maps a bounded set in $X_1 \times X_2$ into a bounded set in $W_0^{1,p}(\Omega) \times W_0^{1,p}(\Omega)$. By Sobolev Compact Imbedding Theorem, the identity map from $W_0^{1,p}(\Omega)$ to $C(\bar{\Omega})$ is compact. Hence, we can view T as a composition of a bounded map form $X_1 \times X_2$ to $W_0^{1,p}(\Omega) \times W_0^{1,p}(\Omega)$ followed by a compact identity map from $W_0^{1,p}(\Omega) \times W_0^{1,p}(\Omega)$ to $X_1 \times X_2$; and we conclude that T is a compact operator form $X_1 \times X_2$ into itself. Schauder's fixed point theorem asserts that T has a fixed point (u_1^*, u_2^*) in $X_1 \times X_2$. It follows from (4.44) that

$$
(4.53) \qquad \begin{cases} \Delta u_1^* + f_1(x, \psi^{-1}(u_1^*), \psi^{-1}(u_2^*)) = 0 & \text{a.e. in } \Omega, \\ \Delta u_2^* + f_2(x, \psi^{-1}(u_1^*), \psi^{-1}(u_2^*)) = 0 & \text{a.e. in } \Omega, \\ u^* = u_2^* = 0 & \text{on } \partial\Omega. \end{cases}
$$

The fact that (u_1^*, u_2^*) is in $X_1 \times X_2$ implies that (u_1^*, u_2^*) is in $W^{2.p}(\Omega) \cap W_0^{1,p}(\Omega)$ and that $\underline{\mathrm{u}}_1 \le u_i^* \le \bar{u}_i$ is in $\bar{\Omega}$ for $i = 1, 2$. Consequently, $(w_1^*, w_2^*) = (\psi^{-1}(u_1^*), \psi^{-1}(u_2^*))$ is a positive solution of (4.42) with $\underline{\mathrm{w}}_i \le w_i^* \le \bar{w}_i$ in $\bar{\Omega}$.

The following corollary is sometimes more readily applicable than Theorem 4.2.

Corollary 4.3. *Assume hypothesis $[\tilde{H}1]$ and $[\tilde{H}2]$. Suppose that there are functions $\underline{w}_i(x), \bar{w}_i(x)$ $(i = 1, 2)$ in $C(\bar{\Omega})$ with $\psi(\underline{w}_i), \psi(\bar{w}_i)$ $(i = 1, 2)$ in $W^{2,p}(\Omega)$ $(p > N)$ satisfying the inequalities*

$$
(4.54) \qquad \begin{cases} \Delta\psi(\underline{w}_1) + f_i(x, \underline{w}_1, w_2) \ge 0 \;\; a.e \; in \; \Omega, \;\; for \; \underline{w}_2 \le w_2 \le \bar{w}_2, \\ \Delta\psi(\bar{w}_1) + f_i(x, \bar{w}_1, w_2) \le 0 \;\; a.e \; in \; \Omega, \;\; for \; \underline{w}_2 \le w_2 \le \bar{w}_2, \\ \Delta\psi(\underline{w}_2) + f_2(x, w_1, \underline{w}_2) \ge 0 \;\; a.e \; in \; \Omega, \;\; for \; \underline{w}_1 \le w_1 \le \bar{w}_1, \\ \Delta\psi(\bar{w}_2) + f_2(x, w_1, \bar{w}_2) \le 0 \;\; a.e \; in \; \Omega, \;\; for \; \underline{w}_1 \le w_1 \le \bar{w}_1, \end{cases}
$$

where the derivatives are taken in the weak sense. Here $w_i = w_i(x)$ are assumed to be continuous in $\bar{\Omega}$, and $0 \le \underline{w}_i \le w_i \le \bar{w}_i \le b_i$ in $\bar{\Omega}$, $\underline{w}_i > 0$ in Ω and $\underline{w}_i = 0$ on $\partial\Omega$. Then there exists at least one positive solution (w_1^*, w_2^*) of (4.42) satisfying $\underline{w}_i \le w_i^* \le \bar{w}_i$ in $\bar{\Omega}$.

Proof. This is an immediate result of Theorem 4.2 since (4.54) implies (4.43). To see this, we let $\phi \in C_0^1(\Omega), \phi \ge 0$, and multiply the first equation in (4.54) by ϕ. We integrate over Ω to find

$$\int_\Omega \Delta\psi(\underline{w}_1)\phi\, dx + \int_\Omega f_1(x, \underline{w}_1, w_2)\phi\, dx \geq 0 \quad \text{for} \quad \underline{w}_2 \leq w_2 \leq \bar{w}_2. \tag{4.55}$$

It follows from the definition of the weak derivative that

$$-\int_\Omega \nabla\psi(\underline{w}_1)\nabla\phi\, dx + \int_\Omega f_1(x, \underline{w}_1, w_2)\phi\, dx \geq 0 \quad \text{for} \quad \underline{w}_2 \leq w_2 \leq \bar{w}_2. \tag{4.56}$$

Similarly, we can verify the rest of the inequalities in (4.43). By application of Theorem 4.2, the proof is completed.

We now apply Corollary 4.3 to the following prey-predator ecological model with degenerate density-dependent diffusion.

Example 4.2 (Degenerate Prey-Predator Model).

$$\begin{cases} \Delta u^m + u(a(x) - bu^k - cv) = 0 & \text{in } \Omega, \\ \Delta v^m + v(e(x) + fu - gv^k) = 0 & \text{in } \Omega, \\ u = v = 0 & \text{on } \partial\Omega. \end{cases} \tag{4.57}$$

Here Ω is a bounded domain in $R^N, N \geq 2$ with boundary $\partial\Omega \in C^2$, and m, k, b, c, f, g are positive constants with $1 + k > m > 1$. We assume that $a(x), e(x)$ are two positive functions in $L^\infty(\Omega)$ with

$$\underline{a} := ess\, inf_{x\in\Omega}\, a(x) > 0, \quad \text{and} \quad \underline{e}(x) := ess\, inf_{x\in\Omega}\, e(x) > 0. \tag{4.58}$$

For convenience, we denote $\bar{a} := ess\ sup_{x\in\Omega}\, a(x) > 0$ and $\bar{e}(x) := ess\ sup_{x\in\Omega}\, e(x) > 0$. The following theorem gives sufficient conditions for the coexistence of two species. If one compares them with the results in Chapter 2 concerning nondegenerate case, we see that the conditions here are much more readily satisfied. For example. there is no need for the intrinsic growth rates $a(x)$ and $e(x)$ to be larger than the principal eigenvalue for the domain Ω.

Theorem 4.4. *Assume $1 + k > m > 1$, hypothesis (4.58) and*

$$g(\frac{\underline{a}}{c})^k > \bar{e} + f(\frac{\bar{a}}{b})^{1/k}. \tag{4.59}$$

Then there exists a positive solution (u, v) of (4.57) with $u, v \in C(\bar{\Omega})$ and $u, v \in W^{2,p}(\Omega) \cap W_0^{1,p}(\Omega)$ $(p > N)$. Moreover, the solution satisfies

$$0 < u \leq (\frac{\bar{a}}{b})^{1/k}, \quad 0 < v \leq [g^{-1}(\bar{e} + f(\frac{\bar{a}}{b}))^{1/k}]^{1/k} \ \text{ in } \Omega. \tag{4.60}$$

Proof. We will apply Corollary 4.3. Let $b_1 = (\bar{a}/b)^{1/k}$, $b_2 = [g^{-1}(\bar{e} + f(\bar{a}/b))^{1/k}]^{1/k}$. Define

$$\psi(s) = s^m \quad \text{for } s \geq 0$$

and

$$f_1(x, y_1, y_2) = y_1(a(x) - by_1^k - cy_2),$$

$$f_2(x, y_1, y_2) = y_2(e(x) + fy_1 - gy_2^k) \quad \text{for } (x, y_1, y_2) \in \Omega \times [0, \infty) \times [0, \infty).$$

Then one can immediately verify that $[\tilde{H}1]$ and $[\tilde{H}2, i - \tilde{H}2, ii]$ are satisfied. Since

$$\begin{aligned} f_1(x, \xi, y_2) - f_1(x, \eta, y_2) &= \xi(a(x) - b\xi^k - cy_2) - \eta(a(x) - b\eta^k - cy_2) \\ &= (a(x) - cy_2)(\xi - \eta) - b(\xi^{k+1} - \eta^{k+1}) \\ &\geq (\underline{a} - cb_2)(\xi - \eta) - b(\xi^{k+1} - \eta^{k+1}) \\ &\qquad \text{for } x \in \Omega \text{ a.e.}, \quad y_2 \in [0, b_2], \ 0 \leq \eta \leq \xi \leq b_1, \end{aligned}$$

we can verify the first part of $[\tilde{H}2, iii]$ by showing that there is a constant $M > 0$ such that

$$(\underline{a} - cb_2)(\xi - \eta) - b(\xi^{k+1} - \eta^{k+1}) > -M(\xi^m - \eta^m), \quad 0 \leq \eta \leq \xi \leq b_1. \tag{4.61}$$

From hypothesis (4.59), we have $\underline{a} - cb_2 > 0$; thus (4.61) is satisfied if

$$M(\xi^m - \eta^m) \geq b(\xi^{k+1} - \eta^{k+1}) \quad \text{for } 0 \leq \eta \leq \xi \leq b_1. \tag{4.62}$$

However, (4.62) can be readily verified if we note that the function $h(\xi) = M\xi^m - b\xi^{k+1}$ is increasing in $[0, b_1]$ by choosing $M > (b/m)(k+1)b_1^{k+1-m}$. Similarly, we verify the second part of $[\tilde{H}2, iii]$,

$$\begin{aligned} f_2(x, y_1, \xi) - f_2(x, y_1, \eta) &= \xi(e(x) + fy_1 - g\xi^k) - \eta(e(x) + fy_1 - g\eta^k) \\ &\geq \underline{e}(\xi - \eta) - g(\xi^{k+1} - \eta^{k+1}) \\ &\geq -M(\psi(\xi) - \psi(\eta)), \end{aligned}$$

for $x \in \Omega$ a.e., $y_1 \in [0, b_1], 0 \leq \eta \leq \xi \leq b_2$ if $M > (g/m)(k+1)b_2^{k+1-m}$.

To construct upper and lower solutions $(\underline{u}, \underline{v})$ and $(\bar{u}, \bar{v})$, we let $\lambda_1 > 0$ be the principal eigenvalue for the problem: $\Delta w + \lambda w = 0$ in Ω, $w = 0$ on $\partial\Omega$, and $\phi(x)$ be a positive principal eigenfunction. We define $\underline{u} = \underline{v} = (\delta\phi)^{1/m}$ in Ω for a small $\delta > 0$ to be determined. Thus they satisfy $\underline{u} = \underline{v} = 0$ on $\partial\Omega$, $\underline{u} = \underline{v} > 0$ in Ω. Also, we define $\bar{u} = b_1, \bar{v} = b_2$. We verify that

$$\Delta \bar{u}^m + \bar{u}(a(x) - b\bar{u}^k - cv) \leq \bar{u}(a(x) - b\bar{u}^k)$$

$$= (\bar{a}/b)^{1/k}(a(x) - b(\bar{a}/b)) \leq 0 \qquad \text{a.e. in } \Omega, \quad \text{for all } \underline{v} \leq v \leq \bar{v},$$

$$\Delta \bar{v}^m + \bar{v}(e(x) + fu - g\bar{v}^k) \leq \bar{v}(\bar{e} + f\bar{u} - g\bar{v}^k) = 0 \;\; \text{for all } \underline{u} \leq u \leq \bar{u}.$$

Moreover, we have

$$\Delta \underline{u}^m + \underline{u}(a(x) - b\underline{u}^k - cv) = -\lambda_1(\delta\phi) + (\delta\phi)^{1/m}(a(x) - b(\delta\phi)^{k/m} - cv)$$

$$\geq (\delta\phi)^{1/m}(-\lambda_1(\delta\phi)^{1-1/m} + \underline{a} - b(\delta\phi)^{k/m} - c\bar{v}) \geq 0$$

for all x *a.e.* in Ω and all $\underline{v} \leq v \leq \bar{v}$, when δ is sufficiently small, since $\underline{a} - c\bar{v} > 0$ by assumption (4.59). Finally

$$\Delta \underline{v}^m + \underline{v}(e(x) + fu - g\underline{v}^k) = -\lambda_1(\delta\phi) + (\delta\phi)^{1/m}(e(x) + fu - g(\delta\phi)^{k/m})$$

$$\geq (\delta\phi)^{1/m}(-\lambda_1(\delta\phi)^{1-1/m} + \underline{e} + f\underline{u} - g(\delta\phi)^{k/m}) \geq 0$$

for x *a.e.* in Ω all $\underline{u} \leq u \leq \bar{u}$, when δ is sufficiently small, since $\underline{e} + f\underline{u} > 0$. From Corollary 4.3, the four inequalities above imply that there is a positive solution (u, v) of (4.57) with u, v in $C(\bar{\Omega}); u^m, v^m \in W^{2,p}(\Omega) \cap W_0^{1,p}(\Omega)$; and (4.60) is satisfied.

Many other results are found recently for systems with degenerate diffusion. Suppose that there are smoother density-dependence on diffusion, and smoother dependence on reaction. More specifically, assume

[P1] For $i = 1, 2$, $\varphi_i(s)$ is C^2 in $[0, \infty)$, with $\varphi_i(0) = 0$, and $\varphi_i'(s) > 0$ for all $s > 0$.

[P2] $\tilde{f}(u, v), \tilde{g}(u, v)$ are in $C^1([0, \infty) \times [0, \infty))$, and $\tilde{f}_u < 0, \tilde{g}_v < 0$ for all $(u, v) \in [0, \infty) \times [0, \infty)$.

In Ryu and Ahn [196], classical positive solutions are found for the following degenerate system with homogeneous Dirichlet boundary condition.

$$\text{(4.63)} \qquad \begin{cases} \Delta[\varphi_1(u)u] + u\tilde{f}(u, v) = 0 & \text{in } \Omega, \\ \Delta[\varphi_2(v)v] + v\tilde{g}(u, v) = 0 & \text{in } \Omega, \\ u = v = 0 & \text{on } \partial\Omega, \end{cases}$$

where Ω is a bounded domain in R^N with smooth boundary.

Theorem 4.5. *Consider problem (4.63) under hypotheses [P1] and [P2].*

(I) (Prey-Predator case) Suppose: (a) $\tilde{f}_v < 0$ *and* $\tilde{g}_u > 0$ *for all* $(u,v) \in [0,\infty) \times [0,\infty)$*; (b) there exist positive constants* C_1, C_2 *such that* $\tilde{f}(C_1, 0) \leq 0, \tilde{g}(C_1, C_2) \leq 0$*; and (c) if* $min\{\tilde{f}(0, C_2), \tilde{g}(0,0)\} > 0$*, then problem (4.63) has a positive coexistence solution.*

(II) (Competing case) Suppose: (a) $\tilde{f}_v, \tilde{g}_u < 0$ *for all* $(u,v) \in [0,\infty) \times [0,\infty)$*; (b) there exist positive constants* C_3, C_4 *such that* $\tilde{f}(C_3, 0) \leq 0, \tilde{g}(0, C_4) \leq 0$*; and (c) if* $min\{\tilde{f}(0, C_4), \tilde{g}(C_3, 0)\} > 0$*, then problem (4.63) has a positive coexistence solution.*

(III) (Cooperating case) Suppose: (a) $\tilde{f}_v, \tilde{g}_u > 0$ *for all* $(u,v) \in [0,\infty) \times [0,\infty)$*; (b) there exist positive constants* C_5, C_6 *such that* $\tilde{f}(C_5, C_6) = \tilde{g}(C_5, C_6) = 0$*; and (c) if* $min\{\tilde{f}(0,0), \tilde{g}(0,0)\} > 0$*, then problem (4.63) has a positive coexistence solution.*

Here, the solutions (u,v) *are in* $C^2(\bar{\Omega}) \times C^2(\bar{\Omega})$*, and* $u, v > 0$ *in* Ω*.*

The proof of the above theorem is based on a combination of bifurcation and index method, as explained in Chapter 1. We obtain classical solutions because the system is smoother by assumption. The details of the proof are too lengthy to be presented in this section. Note that, as in Theorem 4.4, we need much smaller intrinsic growth rates for coexistence compared with problems with nondegenerate diffusion.

Part B: Lower Bounds for Density-Dependent Diffusive Systems with Regionally Large Growth Rates.

For the remaining part of this section, we consider some time dependent behavior of reaction-diffusive systems involving nonlinear density-dependent diffusion. We only restrict discussion to the non-degenerate case. Moreover, in order to avoid excessive technicalities, we assume more smoothness in the system than the beginning part of this section. More precisely, we assume for each $i = 1, ..., n$

[C1] $\sigma_i(0) > 0, \sigma_i'(s) \geq 0$ in $[0,\infty), \sigma_i''(s)$ is continuous in $(-\infty, \infty)$.

[C2] The functions $h_i : R^n \to R$ have Hölder continuous partial derivatives up to second order in compact sets.

[C3] $\tilde{a}_i(x) \in C^{1+\alpha}(\bar{\Omega})$ for some $\alpha \in (0,1)$, and $\tilde{a}_i(x) \geq 0$ in $\bar{\Omega}$.

We consider the following initial Dirichlet boundary value problem for $i = 1, ..., n$, $(x,t) = (x_1, ..., x_N, t)$:

(4.64)
$$\begin{cases} \frac{\partial u_i}{\partial t} = div(\sigma_i(u_i)\nabla u_i) + u_i[\tilde{a}_i(x) + h_i(u_1, ..., u_n)], & (x,t) \in \Omega \times (0,T],\ T > 0; \\ u_i(x,0) = \phi_i(x),\ x \in \bar{\Omega}; \quad u_i(x,t) = \Phi_i(x),\ (x,t) \in \partial\Omega \times [0,T]. \end{cases}$$

Here, Ω is a bounded domain in R^N with boundary $\partial\Omega$ belonging to $C^{2+\alpha}$ for the remaining part of this section. For simplicity, we limit ourselves to application of system (4.64) to competing interacting species. That is, we assume for each $i, j = 1, ..., n$

$$\text{(4.65)} \quad \begin{cases} \frac{\partial h_i}{\partial u_j} < 0 \text{ in } \{(u_1, ..., u_n) \,|\, u_i \geq 0, i = 1, ..., n\}, \;\; h_i(0,,,,0) = 0, \text{ and} \\ \\ sup_{s\geq 0} \frac{\partial h_i}{\partial u_i}(0, ..0, s, 0, ..0) := r_i < 0, \end{cases}$$

where $s \geq 0$ above occurs at the i-th component. The last assumption is the usual one concerning the crowding effect of each species on its own growth rate. In the earlier chapters, conditions for existence of positive coexistence steady-states were usually of the nature that growth rates of the species are uniformly larger than certain positive constants related to the first eigenvalue. Here, we will assume that the intrinsic growth rate, $\tilde{a}_k(x)$, of a particular k-th species is locally high in a subdomain Ω_s of Ω. We will obtain a criteria on $\tilde{a}_k$ which ensures that the population $u_k(x, t)$ will be bounded below by a positive constant in compact subsets of Ω_s for all t. Such criteria can thus be interpreted as a non-extinction condition for the k-th species. Such condition is much more realistic, because the growth rate does not have to be large on the entire domain. One only needs regionally large growth rates to sustain survival. Before we formulate such conditions, we first consider the existence problem of non-negative classical solution in arbitrary finite interval $[0, T]$ for the initial-boundary value problem. For convenience, we let $\Omega_T = \Omega \times (0, T)$.

Theorem 4.6 (Existence of Solution for Initial-Boundary Value Problem). *Consider problem (4.64) under smoothness conditions [C1]-[C3], and reaction assumptions (4.65). Let the initial boundary functions ϕ_i, Φ_i satisfy: $\phi_i(x) = \Phi_i(x)$ for $x \in \partial\Omega, \phi_i(x) \geq 0$ in $\bar{\Omega}$, ϕ_i has all third derivatives continuous in $\bar{\Omega}$, and*

$$\text{(4.66)} \qquad \{div(\sigma_i(\phi_i)\nabla\phi_i) + \phi_i[\tilde{a}_i(x) + h_i(\phi_1(x), ..., \phi_n(x)]\}|_{x\in\partial\Omega} = 0$$

for $i = 1, ..., n$. Then for any $T > 0$, in the class of functions in $C^{2+\alpha, 1+\alpha/2}(\bar{\Omega}_T)$, there exists a unique solution for the initial-boundary value problem (4.64).

Proof. Let d_i be positive numbers satisfying

$$d_i \geq |r_i|^{-1} max\{\tilde{a}(x) \,|\, x \in \bar{\Omega}\}, \quad \text{and } 0 \leq \phi_i(x) \leq d_i, \;\; x \in \bar{\Omega}$$

for $i = 1, ..., n$. Define $c_i(x, u_1, ..., u_n), i = 1, ..., n, (x, u_1, ..., u_n) \in \bar{\Omega} \times R^n$ by

$$c_i(x, u_1, ..., u_n) = k_i(u_i)[\tilde{a}_i(x) + h_i(k_1(u_1), ..., k_n(u_n)],$$

$$\text{where } k_i(s) = \begin{cases} s & \text{if } |s| \leq d_i \\ \rho_i(s) & \text{if } |s| > d_i \end{cases}$$

with $\rho_i(s)$ a twice continuously differentiable function for $|s| \geq d_i$, and $|\rho_i(s)| \leq 2d_i, \rho_i(\pm d_i) = \pm d_i, \rho_i'(\pm d_i) = 1$ and $\rho''(\pm d_i) = 0$. Extend $\tilde{\sigma}_i(s)$ positively to $(-\infty, 0)$ by letting $\tilde{\sigma}_i(s) = \sigma_i(s)$ for $s \in [0, \infty)$, with $\tilde{\sigma}_i(s)$ twice continuously differentiable for $s \in (-\infty, \infty)$, and $\tilde{\sigma}_i(s) \geq (\sigma_i(0)/2) > 0$ for $s \in (-\infty, 0), i = 1, ..., n$.

We consider the initial boundary value problem:

(4.67)

$$\frac{\partial z_i}{\partial t}(x,t) = \tilde{\sigma}_i(k_i(z_i + \phi_i(x))\Delta z_i + \tilde{\sigma}_i'(k_i(z_i + \phi_i)) \sum_{j=1}^n [(z_i)_{x_j} + 2(\phi_i)_{x_j}] \cdot (z_i)_{x_j}$$

$$+ \tilde{\sigma_i}'(k_i(z_i + \phi_i)) \sum_{j=1}^n (\phi_i)_{x_j}^2 + \tilde{\sigma}_i(k_i(z_i + \phi_i))\Delta\phi_i + c_i(x, z_1 + \phi_1, ..., z_n + \phi_n)$$

for $(x,t) \in \Omega \times (0,T], i = 1, ..., n$;

$$z_i(x,0) = 0 \text{ in } \bar{\Omega} \text{ and } z_i(x,t) = 0 \text{ for } (x,t) \in \partial\Omega \times [0,T]. \tag{4.68}$$

(Note that if we let $u_i(x,t) = z_i(x,t) + \phi_i(x)$ and if $0 \leq u_i(x,t) \leq d_i, i = 1, ..., n$ then $u_i(x,t)$ satisfies:

$$\frac{\partial u_i}{\partial t} = \sigma_i(u_i)\Delta u_i + \sigma_i'(u_i)|\nabla u_i|^2 + u_i[\tilde{a}_i + h_i(u_1, ..., u_n)]. \tag{4.69}$$

Moreover, u_i satisfies the initial boundary conditions of (4.64)). Apply Theorem 7.1 on p. 596 of [113]. The positivity of $\tilde{\sigma}_i$ and the boundedness of the last three terms of (4.67) imply that the condition (a) in Theorem 7.1 is satisfied. (6.3) of (b) in [113] is satisfied by letting $P(|p|, |u|) = C(1 + |p|)^{-2}$ for some large constant C and $\epsilon(|u|) = 0$. The smoothness of $\phi_i, \tilde{\sigma}_i$ and k_i ensure that (c) is satisfied. Compatibility condition (4.66) gives (d). Consequently, Theorem 7.1 on p. 586 of [113] gives a unique solution $z = (z_1(x,t), ..., z_n(x,t))$ to (4.67), (4.68) for $(x,t) \in \bar{\Omega} \times [0,T]$ in the class $C^{2+\alpha, 1+\alpha/2}(\bar{\Omega}_T)$.

We next show that $0 \leq z_i(x,t) + \phi_i(x) \leq d_i, i = 1, ..., n$. Let $\hat{\alpha}_i(x,t) \equiv 0$ and $\hat{\beta}_i(x,t) \equiv d_i, i = 1, ..., n$. We clearly have in $\Omega \times (0,T]$,

$$div(\sigma_i(\hat{\alpha}_i)\nabla\hat{\alpha}_i) + \hat{\alpha}_i[\tilde{a}_i(x) + h_i(\hat{\beta}_1, ..., \hat{\beta}_{i-1}, \hat{\alpha}_i, \hat{\beta}_{i+1}, ..., \hat{\beta}_n)] - \frac{\partial\hat{\alpha}_i}{\partial t} \geq 0, \tag{4.70}$$

for each i. Each function $\hat{\beta}_i$ satisfies

(4.71)

$$div(\sigma_i(\hat{\beta}_i)\nabla\hat{\beta}_i) + \hat{\beta}_i[\tilde{a}_i(x) + h_i(\hat{\alpha}_1, ..., \hat{\alpha}_{i-1}, \hat{\beta}_i, \hat{\alpha}_{i+1}, ..., \hat{\alpha}_n)] - \frac{\partial\hat{\beta}_i}{\partial t}$$

$$= d_i[\tilde{a}_i(x) + h_i(0, ..., 0, d_i, 0, ..., 0)]$$

$$= d_i[\tilde{a}_i(x) + \int_0^{d_i} \frac{\partial h_i}{\partial s_i}(0, ..., 0, s_i, 0, ..., 0)\, ds_i]$$

$$\leq d_i[\tilde{a}_i(x) + r_i d_i] \leq d_i[\tilde{a}_i(x) - max\{\tilde{a}_i(x) | x \in \bar{\Omega}\}] \leq 0$$

in $\Omega \times (0,T]$. For $i = 1,..,n, (x,t) \in \bar{\Omega} \times [0,T]$, the function

$$u_i(x,t) := z_i(x,t) + \phi_i(x)$$

satisfies (4.69) for $x \in \Omega, 0 < t \leq t_1 \leq T$ as long as $\hat{\alpha}_i \leq u_i(x,t) \leq \hat{\beta}_i$ for $(x,t) \in \bar{\Omega} \times [0,t_1]$. From (4.70) and (4.71), we obtain from a generalization of the comparison Theorem 1.2-6 in [125] to density-dependent diffusion case, that $\hat{\alpha}_i \leq u_i \leq \hat{\beta}_i$ for all $(x,t) \in \bar{\Omega} \times [0,T], i = 1,...,n$. (The same arguments for the generalization will be used in proving the next comparison Theorem 4.7 for a even more difficult case involving non-smooth lower solution. Note that our present situation is simpler because $\hat{\alpha}_i \equiv 0$, and less argument is needed to justify $u_j \geq \hat{\alpha}_j$.) The a-priori bound, $\hat{\alpha}_i \leq u_i \leq \hat{\beta}_i$ in $\bar{\Omega} \times [0,T]$, consequently implies that $u(x,t)$ is the unique solution of the initial-boundary value problem (4.64), in $C^{2+\alpha,1+\alpha/2}(\bar{\Omega}_T)$.

When the competing reaction relation (4.65) is replaced by prey-predator, cooperating or the general food-pyramid condition, the existence Theorem 4.6 can be generalized and proved similarly by changing the upper solution to be time-dependent. The procedure is similar to the case when diffusion does not depend on density as described in Theorem 2.1-1 in [125] or [231].

As in Example 4.1 and Theorem 4.1 above, when the intrinsic growth rate is highly heterogeneous, one is led to the use of non-smooth lower solution to show the existence of positive solutions. We are thus motivated to prove the following comparison Theorem, using nonsmooth lower bound for the smooth time-dependent problem (4.64) as well.

Theorem 4.7 (Comparison). *Let $\Omega_s \subseteq \Omega$ be a subdomain, $\partial\Omega_s \in C^2$, with λ_s as its principal eigenvalue for the operator $-\Delta$ on Ω_s with Dirichlet boundary condition, and $\theta(x)$ is a positive eigenfunction in Ω_s. Let j be an integer, $1 \leq j \leq n; \alpha_i(x,t) \equiv 0$ in Ω_T if $i \neq j, 1 \leq i \leq n$, and*

$$\alpha_j(x,t) = \begin{cases} \delta\theta(x) & \text{if } (x,t) \in \Omega_s \times [0,T] \\ 0 & \text{if } (x,t) \in (\bar{\Omega} \backslash \Omega_s) \times [0,T] \end{cases}$$

where $\delta > 0$ is a constant. For $i = 1,...,n$, let $\beta_i(x,t)$ be a non-negative function in $C^{2+\alpha,1+\alpha/2}(\bar{\Omega}_T)$. Suppose that α_i, β_i satisfy:

$$\alpha_i(x,t) \leq \beta_i(x,t) \quad \text{for } (x,t) \in \bar{\Omega}_T;$$

$$(4.72) \quad div(\sigma_i(\alpha_i)\nabla\alpha_i) + \alpha_i[\tilde{a}_i(x) + h_i(\beta_1,...,\beta_{i-1},\alpha_i,\beta_{i+1},...,\beta_n)] - \frac{\partial\alpha_i}{\partial t} \geq 0,$$

$$(4.73) \quad div(\sigma_i(\beta_i)\nabla\beta_i) + \beta_i[\tilde{a}_i(x) + h_i(\alpha_1,...,\alpha_{i-1},\beta_i,\alpha_{i+1},...,\alpha_n)] - \frac{\partial\beta_i}{\partial t} \leq 0$$

for $(x,t) \in \Omega \times (0,T], i = 1,...,n$, except for $i = j$ in (4.72) valid only for $(x,t) \in (\Omega \backslash \partial\Omega_s) \times (0,T]$. (Here, we assume conditions [C1-C3] and (4.65) for the equations). Let $(u_1,...,u_n), u_i \in C^{2+\alpha,1+\alpha/2}(\bar{\Omega}_T)$, be a solution of the differential equation in (4.64) with initial boundary conditions satisfying

$$\begin{aligned} &\alpha_i(x,0) \leq u_i(x,0) \leq \beta_i(x,0), \quad x \in \bar{\Omega}, \\ &\alpha_i(x,t) \leq u_i(x,t) \leq \beta_i(x,t), \quad (x,t) \in \partial\Omega \times [0,T] \end{aligned} \tag{4.74}$$

for $i = 1,...,n$. Then we have

$$\alpha_i(x,0) = \alpha_i(x,t) \leq u_i(x,t) \leq \beta_i(x,t) \tag{4.75}$$

for $(x,t) \in \bar{\Omega} \times [0,T], i = 1,...,n$.

Proof. Since u_i, α_i and $\beta_i \in C^{2+\alpha,1+\alpha/2}(\bar{\Omega}_T)$, there are constants K and M such that $|\alpha_i| \leq K, |\beta_i| \leq K, |u_i| \leq K, |\Delta u_i| \leq M, |\nabla u_i|^2 \leq M$ for all $(x,t) \in \bar{\Omega}_T, i = 1,...,n$. The assumption on $h_i, \tilde{a}_i$ and σ_i imply that there are constants R and B so that for each $i = 1,...,n$, we have $|\sigma_i'(s)| \leq R, |\sigma_i''(s)| \leq R$ for $-2K \leq s \leq 2K$, and $|\tilde{a}_i(x) + h_i(s_1,...,s_n)| \leq B$ for $x \in \bar{\Omega}, -2K \leq s \leq 2K, i = 1,...,n$.

Let $0 < \epsilon < K[1+3(B+2MR+KLn)T]^{-1}$, where $(1/2)L$ is a bound for the absolute values of all first partial derivatives of $h_i(s_1,...,s_n), 0 \leq s_i \leq 2K, i = 1,...,n$. Define, for $(x,t) \in \bar{\Omega}_T, i = 1,...,n$,

$$\begin{aligned} &u_i^+(x,t) = u_i(x,t) + \epsilon[1 + 3(B + 2MR + KLn)t], \\ &u_i^-(x,t) = u_i(x,t) - \epsilon[1 + 3(B + 2MR + KLn)t]. \end{aligned} \tag{4.76}$$

By hypothesis (4.74), we have

$$\alpha_i(x,t) < u_i^+(x,t) \text{ and } u_i^-(x,t) < \beta_i(x,t) \tag{4.77}$$

for $x \in \bar{\Omega}, t = 0, i = 1,...,n$. Suppose one of these inequalities fails at some point in $\bar{\Omega} \times (0,\tau_1)$, where $\tau_1 = min\{T, 1/(3(B + 2MR + KLn))\}$; and (x_1,t_1) is a point in $\bar{\Omega} \times (0,\tau_1)$ with minimal t_1 where (4.77) fails. At $(x_1,t_1), \alpha_i = u_i^+$ or $u_i^- = \beta_i$ for some i. Assume the former is the case; a similar proof holds for the latter case.

Suppose further that at $(x_1,t_1), \alpha_j = u_j^+$ (a similar proof will work if $\alpha_i = u_i^+$ at (x_1,t_1) for $i \neq j$), we consider separately the situations for $x_1 \in (\Omega \backslash \Omega_s)$ or $x_1 \in \Omega_s$. If $x_1 \in \Omega \backslash \Omega_s$, we have $u_j^+(x,t) > 0$ for $t < t_1, x \in \bar{\Omega}$ and $u_j^+(x_1,t_1) = 0$. Observe that $x_1 \notin \partial\Omega$ because $u_j^+(x,t_1) > u_j(x,t_1) \geq 0$ for $x \in \partial\Omega$, by (4.74). However, for $(x,t) \in \Omega \times (0,T]$:

$$
\begin{aligned}
\tfrac{\partial}{\partial t}(-u_j^+) &= -\tfrac{\partial}{\partial t}(u_j) - \epsilon 3(B+2MR+KLn) \\
&= -\sigma_j(u_j)\Delta u_j - \sigma_j'(u_j)|\nabla u_j|^2 - u_j[\tilde{a}_j(x) + h_j(u_1,...,u_n)] \\
&\quad - \epsilon 3(B+2MR+KLn) \\
&= -\sigma_j(u_j^+)\Delta u_j + [\sigma_j(u_j^+) - \sigma_j(u_j)]\Delta u_j - \sigma_j'(u_j)|\nabla u_j|^2 \\
&\quad + [u_j^+ - u_j][\tilde{a}_j + h_j(u_1,...,u_n)] - u_j^+[\tilde{a}_j + h_j(u_1,...,u_n)] \\
&\quad - \epsilon 3(B+2MR+KLn).
\end{aligned} \tag{4.78}
$$

Recalling that $\sigma_j(u_j^+) > 0$, and at (x_1, t_1) we have $\nabla u_j = \nabla u_j^+ = 0$, $\Delta u_j = \Delta u_j^+ \geq 0$, (4.78) implies

$$
\begin{aligned}
\tfrac{\partial}{\partial t}(-u_j^+)|_{(x_1,t_1)} &\leq R\epsilon[1 + 3(B+2MR+KLn)t_1]M \\
&\quad + \epsilon[1+3(B+2MR+KLn)t_1]B - \epsilon 3(B+2MR+KLn) \\
&\leq MR\epsilon 2 + B\epsilon 2 - \epsilon 3[B+2MR+KLn] < 0,
\end{aligned} \tag{4.79}
$$

contradicting the definition of (x_1, t_1).

If $x_1 \in \Omega_s$, we have $u_j^+(x,t) > \alpha_j(x,t)$ for $t < t_1, x \in \bar{\Omega}$; and $u_j^+(x_1,t_1) = \alpha_j(x_1,t_1) = \delta\theta(x_1)$. But for $(x,t) \in \Omega_s \times (0,T]$

$$
\begin{aligned}
\tfrac{\partial}{\partial t}(\alpha_j - u_j^+) &\leq div(\sigma_j(\alpha_j)\nabla\alpha_j) + \alpha_j[\tilde{a}_j + h_j(\beta_1,...,\beta_{j-1},\alpha_j,\beta_{j+1},...,\beta_n)] \\
&\quad - div(\sigma_j(u_j)\nabla u_j) - u_j[\tilde{a}_j + h_j(u_1,...,u_n)] - \epsilon 3(B+2MR+KLn) \\
&= \sigma_j(\alpha_j)\Delta\alpha_j + \sigma_j'(\alpha_j)|\nabla\alpha_j|^2 + \alpha_j[\tilde{a}_j + h_j(\beta_1,...,\beta_{j-1},\alpha_j,\beta_{j+1},...,\beta_n)] \\
&\quad - \sigma_j(u_j^+)\Delta u_j + [\sigma_j(u_j^+) - \sigma_j(u_j)]\Delta u_j + [\sigma_j'(u_j^+) - \sigma_j'(u_j)]|\nabla u_j|^2 \\
&\quad - \sigma_j'(u_j^+)|\nabla u_j|^2 + (u_j^+ - u_j)[\tilde{a}_j + h_j(u_1,...,u_n)] - u_j^+[\tilde{a}_j + h_j(u_1,...,u_n)] \\
&\quad - \epsilon 3(B+2MR+KLn).
\end{aligned} \tag{4.80}
$$

At (x_1, t_1). we have $u_j^+ = \alpha_j, \nabla u_j = \nabla u_j^+ = \nabla\alpha_j, \Delta(\alpha_j - u_j) = \Delta(\alpha_j - u_j^+) \leq 0$, thus (4.80) gives

(4.81)
$$\frac{\partial}{\partial t}(\alpha_j - u_j^+)|_{(x_1,t_1)} = \sigma_j(\alpha_j)\Delta(\alpha_j - u_j) + \alpha_j[h_j(\beta_1, ..., \beta_{j-1}, \alpha_j, \beta_{j+1}, ..., \beta_n)$$
$$- h_j(u_1, ..., u_n)] + [\sigma_j(u_j^+) - \sigma_j(u_j)]\Delta u_j + [\sigma_j'(u_j^+) - \sigma_j'(u_j)]|\nabla u_j|^2$$
$$+ (u_j^+ - u_j)[\tilde{a}_j + h_j(u_1, ..., u_n)] - \epsilon 3(B + 2MR + KLn).$$

Moreover, at (x_1, t_1) we have

$$h_j(\beta_1, ..., \beta_{j-1}, \alpha_j, \beta_{j+1}, ..., \beta_n) - h_j(u_1, ..., u_n)$$
$$\le h_j(\tilde{u}_1^-, ..., \tilde{u}_{j-1}^-, u_j^+, \tilde{u}_{j+1}^-, ..., \tilde{u}_n^-) - h_j(u_1, ..., u_n) \tag{4.82}$$
$$\le L\epsilon[1 + 3(B + 2MR + KLn)t_1]n$$

where $\tilde{u}_i^-(x_1, t_1) = max\{u_i^-(x_1, t_1), \alpha_i(x_1, t_1)\}$, because $|u_i - \tilde{u}_i^-| \le |u_i^+ - u_i^-|$. Consequently (4.81) gives

(4.83)
$$\frac{\partial}{\partial t}(\alpha_j - u_j^+)|_{(x_1,t_1)} \le KLn\epsilon[1 + 3(B + 2MR + KLn)t_1]$$
$$+(2MR\epsilon + B\epsilon)[1 + 3(B + 2MR + KLn)t_1] - \epsilon 3(B + 2MR + KLn)$$
$$\le KLn\epsilon 2 + 4MR\epsilon + 2B\epsilon - \epsilon 3(B + 2MR + KLn) < 0,$$

contradicting the definition of (x_1, t_1). From these contradictions we conclude that $u_j^+(x, t) > \alpha_j(x.t)$ for $(x, t) \in \bar{\Omega} \times [0, \tau_1)$. Passing to the limit as $\epsilon \to 0^+$, we obtain $u_j(x, t) \ge \alpha_j(x, t)$ in $\bar{\Omega} \times [0, \tau_1]$.

If at (x_1, t_1), we have $\alpha_m = u_m^+$ for $m \ne j$, then $u_m^+(x, t) > 0$ for $t < t_1, x \in \bar{\Omega}$ and $u_m^+(x_1, t_1) = 0$, with $x_1 \notin \partial\Omega$. For $x \in \Omega$, we repeat the arguments in (4.78) to (4.79), with j replaced by m. (There is no need for arguments analogous to (4.80) to (4.83)). We obtain $u_m^+ > \alpha_m = 0$ for $(x, t) \in \bar{\Omega} \times [0, \tau_1)$, and consequently $u_m \ge \alpha_m = 0$ for $(x, t) \in \bar{\Omega} \times [0, \tau_1]$.

If at $(x_1, t_1), u_i^- = \beta_i$ for some i, we show that

$$\frac{\partial}{\partial t}(\beta_i - u_i^-)|_{(x_1,t_1)} > 0$$

by means of (4.73), in a way similar to the arguments that led to (4.80) to (4.83), but with inequalities reversed. Passing to the limit as $\epsilon \to 0^+$, we obtain $u_i \le \beta_i$ for $(x, t) \in \bar{\Omega} \times [0, \tau_1]$.

If $\tau_1 < T$, we repeat the above arguments by starting to define u_i^+, u_i^- with (4.76), with t in the square brackets on the right side of the formulas replaced by $(t - \tau_1)$. This leads to $\alpha_i \le u_i \le \beta_i$ for $x \in \bar{\Omega}, \tau_1 \le t \le min\{T, (2/3)(B + 2MR + KLn)^{-1}\}$ etc. Eventually, we obtain (4.75) in $\bar{\Omega} \times [0, T]$.

Remark 4.3. Note that the assumption $\sigma_i'(s) \geq 0$, in $[0, \infty), i = 1, ..., n$ has never been used in the proof of Theorem 4.7. However, $\sigma_k' \geq 0$ is essential for establishing the positivity of expression (4.91), in the proof of Theorem 4.8 below.

We next assume that the intrinsic growth rate $\tilde{a}_k$ of a particular k-th species is locally high in a subdomain Ω_s of Ω. We will obtain a criteria on $\tilde{a}_k(x)$ which ensures that the population $u_k(x,t)$ will be bounded below by a positive constant in compact subsets of Ω_s for all t. Such criteria can thus be interpreted as a non-extinction condition for the k-th species. Such result is more realistic than those in earlier chapters because the growth rate can be very small in parts of the entire domain.

Theorem 4.8 (Non-Extinction of Certain Species with Regionally High Growth Rate). *Let k be an integer, $1 \leq k \leq n$, Let $u = (u_1, ..., u_n)$ be a solution of (4.64), under assumptions [C1-C3] and (4.65), in the class $C^{2+\alpha,1+\alpha/2}(\bar{\Omega}_T)$, $T > 0$ satisfying*

$$0 \leq u_i(x,0) \leq b_i, \quad x \in \bar{\Omega}, \ i = 1, ..., n \tag{4.84}$$

where b_i are positive numbers satisfying $b_i \geq |r_i|^{-1} max\{\tilde{a}_i(x) \ x \in \bar{\Omega}\}$. Suppose that there exists a subdomain $\Omega_s \subseteq \Omega$ (with smooth boundary and principal eigenvalue $\lambda_s > 0$ for the operator $-\Delta$ with homogeneous Dirichlet condition on $\partial\Omega_s$) with the properties

$$\begin{aligned} &0 < u_k(x,0), \quad x \in \bar{\Omega}_s, \\ &\tilde{a}_k(x) - \sigma_k(0)\lambda_s + h_k(b_1, ..., b_{k-1}, 0, b_{k+1}, ..., b_n) > 0 \end{aligned} \tag{4.85}$$

for all $x \in \bar{\Omega}_s$. Then the solution satisfies:

$$0 < u_k(x,t) \ \ for \ (x,t) \in \Omega_s \times [0,T]. \tag{4.86}$$

Moreover, $u_k(x,t) \geq \delta > 0$ for all x in any compact set contained in $\Omega_s, 0 \leq t \leq T$ (where δ is some constant which depends on the compact set, independent of T); and

$$0 \leq u_i(x,t) \leq b_i \qquad \text{for } \ (x,t) \in \bar{\Omega}_T, i = 1, ..., n. \tag{4.87}$$

Proof. We shall construct lower and upper solutions v_i, w_i satisfying differential inequalities (4.72), (4.73), with v_i, w_i replacing α_i, β_i respectively. Then, we apply comparison Theorem 4.7 above to conclude $u_k(x,t) \geq v_k(x,t)$ in $\bar{\Omega}_T$. The function v_k will be positive for x in the interior of Ω_s, thus implying the survival of the k-th species. Let $\theta(x)$ be a positive eigenfuction in Ω_s associated with the principal eigenvalue λ_s. Define $v_i(x,t) \equiv 0$ in $\bar{\Omega}_T$ for $i \neq k, 1 \leq i \leq n$; and

$$v_k(x,t) = \begin{cases} \epsilon\theta(x) & \text{if } x \in \Omega_s \\ 0 & \text{if } x \in \bar{\Omega}\backslash\Omega_s \end{cases} \tag{4.88}$$

in $\bar{\Omega}_T$. Here ϵ is a sufficiently small positive constant to be determined later. For $i = 1, ..., n$, define $w_i(x,t) \equiv b_i$ in $\bar{\Omega}_T$. By the relation (4.65) and the size of b_i, we can readily obtain

$$div(\sigma_i(w_i)\nabla w_i) + w_i[\tilde{a}_i + h_i(v_1, ..., v_{i-1}, w_i, v_{i+1}, ..., v_n)] - \frac{\partial w_i}{\partial t} \leq 0 \tag{4.89}$$

for $(x,t) \in \Omega \times [0,T]$. For $i \neq k$, clearly we have

$$div(\sigma_i(v_i)\nabla v_i) + v_i[\tilde{a}_i + h_i(w_1, ..., w_{i-1}, v_i, w_{i+1}, ..., w_n)] - \frac{\partial v_i}{\partial t} = 0 \tag{4.90}$$

for $(x,t) \in \Omega \times [0,T]$. For $i = k$, (4.90) is clearly valid for $(x,t) \in (\Omega\backslash\bar{\Omega}_s) \times [0,T]$. If $(x,t) \in \Omega_s \times [0,T]$, we have

(4.91)

$$div(\sigma_k(v_k)\nabla v_k) + v_k[\tilde{a}_k + h_k(w_1, ..., w_{k-1}, v_k, w_{k+1}, ..., w_n)] - \tfrac{\partial v_k}{\partial t}$$

$$= \sigma_k(v_k)\Delta v_k + \sigma_k'(v_k)|\nabla v_k|^2 + \epsilon\theta(x)[\tilde{a}_k(x) + h_k(w_1, ..., v_k, ..., w_n)]$$

$$= \epsilon\theta(x)[\tilde{a}_k(x) - \sigma_k(\epsilon\theta(x))\lambda_s + h_k(b_1, ..., b_{k-1}, \epsilon\theta, b_{k+1}, ..., b_n)] + \sigma_k'(\epsilon\theta)|\nabla v_k|^2.$$

Now, choose $\epsilon > 0$ sufficiently small so that the expression in (4.91) is positive in $\Omega_s \times [0,T]$. (This is possible due to hypotheses [C1] and (4.85)). Let $(u_1, ..., u_n)$ be a solution of (4.64) satisfying (4.84) and (4.85) as stated. Reduce the choice of $\epsilon > 0$, if necessary, so that $u_k(x,0) > v_k(x,0) = \epsilon\theta(x)$ for $x \in \bar{\Omega}_s$ (note that this will not affect the sign of the expression in (4.91)). Utilizing inequalities (4.89) to (4.91) and Theorem 4.7 above, we conclude that

$$0 \leq v_i \leq u_i(x,t) \leq w_i = b_i, \quad i = 1, ..., n$$

for $(x,t) \in \bar{\Omega} \times [0,T]$. From the definition of v_k in (4.88), we have (4.86) and the strict positivity of u_k in compact subsets of Ω_s as stated in the theorem.

The following is an immediate consequence of Theorem 4.8. It gives a sufficient condition for the long term survival of r species, $0 < r \leq n$.

Corollary 4.9 (Different Species Surviving in Different Regions). *Let $b_i \geq |r_i|^{-1} max\{\tilde{a}_i(x) \,|\, x \in \bar{\Omega}\}, i = 1, ..., n$. Suppose there are subdomains $\Omega_{k_1}, ..., \Omega_{k_r}, (0 < r \leq n,\ k_1, .., k_r$ are distinct positive integers $\leq n)$ in Ω, with the property*

$$\tilde{a}_{k_i}(x) - \sigma_{k_i}(0)\lambda_{k_i} + h_{k_i}(b_1, ..., b_{k_{i-1}}, 0, b_{k_{i+1}}, ..., b_n) > 0 \tag{4.92}$$

for $x \in \Omega_{k_i}, i = 1, ..., r$. *(Here,* $\lambda = \lambda_{k_i} > 0$ *is the first eigenvalue for the problem* $\Delta\phi + \lambda\phi = 0$ *in* $\Omega_{k_i}, \phi = 0$ *on* $\partial\Omega_{k_i}$*). Let* $(u_1, ..., u_n)$ *be a solution of (4.64), under assumptions [C1-C3] and (4.65), with each component in* $C^{2+\alpha,1+\alpha/2}(\bar{\Omega}_T), T > 0$*, and assume initially that*

$$(4.93)\qquad \begin{aligned} &0 \le u_i(x,0) \le b_i, && x \in \bar{\Omega},\ i = 1, ..., n,\\ &0 < u_{k_i}(x,0), && x \in \bar{\Omega}_{k_i},\ i = 1, ..., r. \end{aligned}$$

Then the solution satisfies

$$(4.94)\qquad 0 < u_{k_i}(x,t), \quad (x,t) \in \bar{\Omega}_{k_i} \times [0,T], \ i = 1, ..., r.$$

Moreover $u_{k_i}(x,t) \ge \delta > 0$ *for all* x *in any compact set contained in* $\Omega_{k_i}, 0 \le t \le T$ *(where* δ *is some constant which depends on the compact set, independent of* T*); and*

$$(4.95)\qquad 0 \le u_i(x,t) \le b_i, \quad (x,t) \in \bar{\Omega}_T.$$

Note that the k_i-th species will have, for all time under consideration, its concentration bounded below by positive constants in compact subset od Ω_{k_i}. The simplest situation happens when $\Omega_{k_1} = \Omega_{k_2} \cdots = \Omega_{k_r}$; otherwise, the different species will primarily survive at different subregions in Ω.

We finally discuss some criteria for the extinction of a certain k-th species (that is, the situation when u_k tends to zero as $t \to \infty$).

Theorem 4.10 (Exitinction of Certain Species). *Consider problem (4.64) under all the hypotheses of Theorem 4.6. Suppose further that* $\Phi_k(x) \equiv 0, x \in \partial\Omega$ *for a particular* k*-th component, and*

$$(4.96)\qquad \tilde{a}_k(x) < \sigma_k(0)\lambda^1, \quad \textit{for all } x \in \bar{\Omega}$$

(where $\lambda = \lambda^1$ *is the first eigenvalue of* $\Delta w + \lambda w = 0$ *in* $\Omega, w = 0$ *on* $\partial\Omega$*). Let* $C_i > 0, i = 1, ..., n$ *be such that for* $x \in \bar{\Omega}$

$$(4.97)\qquad \tilde{a}_i(x) + h_i(0, ..., 0, C_i, 0, ..., 0) \le 0$$

(here C_i *appears in the* i*-th component), and let* $\bar{C}_i := max\{C_i, sup_{x\in\Omega}\phi_i(x)\}$*. Then there exists a (small) constant* $q > 0$ *so that, provided*

$$(4.98)\qquad \sigma_k'(s) \le q \ \textit{for all } 0 \le s \le \bar{C}_k,$$

any solution $(u_1, ..., u_n)$ *of (4.64), under the stated hypotheses, with each component in* $C^{2+\alpha,1+\alpha/2}(\bar{\Omega}_T), T > 0$ *must satisfy*

$$(4.99)\qquad 0 \le u_k(x,t) \le Ke^{-\epsilon t} \quad \textit{in } \bar{\Omega}_T,$$

where K, ϵ are positive constants dependent on q and independent of T. Moreover, we have

$$0 \le u_i(x,t) \le \bar{C}_i, \;\; i = 1, ..., n \text{ in } \bar{\Omega}_T.$$

Note that by assumptions (4.65), the constants C_i defined above must exist. Theorem 4.10 is proved by means of comparison with appropriately constructed upper and lower solutions as in the earlier theorems. In order that the k-th component of the upper solution decays as described in right of inequality (4.99), the conditon on $\sigma_k'(s)$ described in (4.98) is used. The details of the proof can be found in Leung [124] and is too lengthy to be included here.

Notes.

Theorem 1.1 is proved in Hale and Waltman [81]. Theorems 1.2 to 1.5 and Theorem 1.7 are obtained from Cantrell, Cosner and Hutson [22]. Theorem 1.6 is found in Hutson [92]. Theorem 1.8 is given in Hale and Waltman [81]. Theorem 2.1 to Theorem 2.4 are adopted from Leung [125] and Pao [183]. Theorems 2.5 and 2.6 can be found in Pao [183]. Theorem 2.7 and Corollary 2.8 are obtained from Cantrell and Cosner [19]. Theorems 2.9 to 2.11 and Corollary 2.12 are obtained from Leung and Zhang [143]. Theorems 3.1 to 3.5 and Corollaries 3.6 to 3.7 are proved in Lou and Ni [164]. Theorem 3.8 is found in Wang [226]. Theorems 4.1 to 4.4 are presented in Leung and Fan [135]. Theorem 4.5 is due in Ryu and Ahn [198]. Theorems 4.6 to 4.10 are obtained from Leung [124].

Chapter 5

Traveling Waves, Systems of Waves, Invariant Manifolds, Fluids and Plasma

5.1 Traveling Wave Solutions for Competitive and Monotone Systems

Many results concerning traveling wave solutions for ecological reaction-diffusion systems are published in the last seventies and eighties. Early results are obtained by Tang and Fife [215], Gardner [68], Conley and Gardner [31], Dunbar [55] [56], Mischaikow and Hutson [174]. These results were deduced by dynamical systems and ordinary differential equations methods, using theories concerning stable and unstable manifolds. On the other hand, Volpert et al. [224] use comparison method of parabolic systems and upper-lower solutions to find traveling wave solutions. One can find many interesting results related to competing species, prey-predator and interacting populations in the review article by Gardner [69]. This section describes some more recent results concerning competing populations and monotone systems. We use the method of upper-lower solutions for elliptic systems to prove the existence of traveling wave solutions. Such methods are used by Volpert et al. [224], Ye and Wang [238], and Zou and Wu [243]. The techniques have been extended to finding traveling wave solutions for delayed equations (see e.g. Wu and Zou [234] and Boumenir and Nguyen [14]). More specifically, we first consider traveling wave solutions of the system:

$$(1.1)\qquad \begin{cases} u_t = du_{xx} + u(a_1 - b_1 u - c_1 v), \\ v_t = dv_{xx} + v(a_2 - b_2 u - c_2 v) \end{cases} \qquad \text{(x,t)} \in R \times [0, \infty),$$

where $u = u(x,t), v = v(x,t)$; and d, a_i, b_i, c_i are positive constants. The system describes two interacting species with diffusive effects. The quantities u and v are the concentrations of two competing populations. The parameters a_i, $i = 1, 2$ are the intrinsic growth rates of the species; b_1, c_2 are the crowding-effect coefficients. The parameters c_1 and b_2 are the coefficients describing competition between the species; and d is the diffusion constant.

In the Part A of this section, we will always assume the following hypotheses:

[P1] $\dfrac{a_1}{b_1} < \dfrac{a_2}{b_2}$; and

[P2] $\dfrac{a_2}{c_2} < \dfrac{a_1}{c_1}$.

Under conditions [P1] and [P2], Tang and Fife [215] proved the existence of traveling wave solutions moving from $(0,0)$ to the positive coexistence equilibrium. In each of [215], [31] and [68], dynamical system and ordinary differential equation methods are used to prove the existence of various traveling wave solutions for competing species models. More recently, Li, Weinberger and Lewis [147] considers a special example of (1.1) satisfying [P1],[P2], and the existence of traveling wave connecting a mono-culture (i.e semi-trivial) state to co-existence state is proved by taking limit for recursion monotone maps as in Lui [166]. The presentation and method in this section is different and more readily applicable.

It is a general belief among biologists that if the two competing species have different preferences in resource usage, then the intra-specific competition between the two species is stronger than the inter-specific competition, therefore, the co-existence of the two species is possible. We will show rigorously that under some mild conditions, there can be traveling waves moving from monoculture (i.e. semi-trivial) steady state to coexistence steady-state. Moreover, the traveling wave solution has certain speed and a particular shape. More precisely, The traveling wave solutions have the form $(u(\sqrt{\frac{a_1}{d}}x + ca_1t), v(\sqrt{\frac{a_1}{d}}x + ca_1t))$, $c \geq 2\sqrt{\frac{b_1(a_1c_2 - a_2c_1)}{a_1(b_1c_2 - b_2c_1)}}$ and join the equilibria $(0, \frac{a_2}{c_2})$ and $(\frac{a_1c_2 - a_2c_1}{b_1c_2 - b_2c_1}, \frac{a_2b_1 - a_1b_2}{b_1c_2 - b_2c_1})$ as $\sqrt{\frac{a_1}{d}}x + ca_1t$ moves from $-\infty$ to $+\infty$. This means that the first species move from extinction to the positive co-existence state, while the second species move from carrying capacity state to co-existence state.

In Theorem 1.1 and Corollary 1.2, we use the method of upper-lower solutions to prove the existence of traveling wave solutions. In order to simplify the proof, we make a change of variable in the second equation of (1.1) by reversing order so that the resulting system becomes monotone (with respect to the other component). For such system, we can utilize available theorems in [224] to simplify the proof of our present theorem. In order to apply the theorem, we simply need to construct upper solution with appropriate limiting values for the system. The construction of upper solution for the system is made by using traveling wave solutions with appropriate dichotomy properties for the scalar KPP (Kolmogorov, Petrovskii and Piskunov [106]) equations. The use of such method avoids the difficulty of analyzing the stable and unstable manifolds used in dynamical system method in [215] and [31]. The proof of Volpert's theorem obtains the traveling wave solution from the limit of a subsequence chosen from shifts of solutions of an appropriate elliptic system of one independent variable. Further, the solutions of the elliptic system are obtained from limits of solutions of corresponding parabolic system as time tends to infinity. There is no need to explicitly construct a non-trivial lower solution. The presentation of Part A of this section is obtained from the beginning part of Hou and Leung [90].

In Part B of this section, the traveling wave solution is constructed by an iterative scheme of solutions of elliptic systems, without using any theory from parabolic equations. The results of this part is adopted from Wu and Zou [234], [235] and Boumenir and Nguyen [14]. The results of this method are applied to study system (1.1) for the case:

[P2] $\dfrac{a_2}{c_2} < \dfrac{a_1}{c_1}$; and

[P3] $\dfrac{a_2}{b_2} < \dfrac{a_1}{b_1}$.

Under some more mild conditions in this case, the traveling waves $(u(\sqrt{\frac{a_1}{d}}x + ca_1 t),\ v(\sqrt{\frac{a_1}{d}}x + ca_1 t))$ connects the steady state $(0, a_2/c_2)$ and $(a_1/b_1, 0)$ as $\sqrt{\frac{a_1}{d}}x + ca_1 t$ moves from $-\infty$ to $+\infty$. That is, the first species move from extinction to carrying capacity, while the second species move from carrying capacity to extinction. Although both upper and lower solutions have to be constructed, the lower solution is not required to satisfy the stringent limiting conditions at $\pm\infty$. Moreover, one obtains more precise estimate of the asymptotic behavior of the traveling waves at $\pm\infty$. The two methods in Part A and B serve to supplement each other when one of them does not apply because of various additional constraint conditions.

Part A: Existence of Traveling Wave Connecting a Semi-Trivial Steady-State to a Coexistence Steady-State.

Letting:

$$\tau = a_1 t \quad \text{and} \quad x = \sqrt{a_1^{-1} d}\ \tilde{x}, \tag{1.2}$$

equation (1.1) can be written as:

$$(1.3) \qquad \begin{cases} u_\tau = u_{\tilde{x}\tilde{x}} + u(1 - a_1^{-1} b_1 u - a_1^{-1} c_1 v), \\ v_\tau = v_{\tilde{x}\tilde{x}} + v(a_1^{-1} a_2 - a_1^{-1} b_2 u - a_1^{-1} c_2 v). \end{cases} \qquad (\tilde{x}, \tau) \in R \times R^+$$

Let

$$(1.4) \qquad u = kw, \; v = qz,$$

where $k = a_1 b_1^{-1}$ and q is a constant satisfying:

$$(1.5) \qquad a_2 c_2^{-1} < q < a_1 c_1^{-1}.$$

System (1.3) becomes

$$(1.6) \qquad \begin{cases} w_\tau = w_{\tilde{x}\tilde{x}} + w(1 - w - rz), \\ z_\tau = z_{\tilde{x}\tilde{x}} + z(\epsilon_1 - bw - \epsilon_1(1 + \epsilon_2) z) \end{cases}$$

where

$$(1.7) \qquad \begin{aligned} r &= a_1^{-1} c_1 q \;, \; \epsilon_1 = a_1^{-1} a_2, \\ b &= b_2 b_1^{-1} \;, \; \epsilon_2 = a_2^{-1} c_2 q - 1. \end{aligned}$$

We will study system (1.6) which is related to (1.1) by the change of variables (1.2) and (1.4).

Recall that in Part A of this section we will always assume the following hypotheses:

[P1] $\dfrac{a_1}{b_1} < \dfrac{a_2}{b_2}$; and

[P2] $\dfrac{a_2}{c_2} < \dfrac{a_1}{c_1}$.

We first consider the system of ordinary differential equations corresponding to (1.1), i.e. $d = 0$. Under conditions [P1] and [P2], The points $(0, \frac{a_2}{c_2})$ and $(\frac{a_1 c_2 - a_2 c_1}{b_1 c_2 - b_2 c_1}, \frac{a_2 b_1 - a_1 b_2}{b_1 c_2 - b_2 c_1})$ are equilibria, with the first equilibrium being unstable and the second stable. (See Figure 5.1.1.)

We will prove in this section the existence of traveling wave solution of system (1.6), with $d > 0$, of the form $(u(\sqrt{\frac{a_1}{d}} x + c a_1 t), v(\sqrt{\frac{a_1}{d}} x + c a_1 t))$ joining the equilibria $(0, \frac{a_2}{c_2})$ and $(\frac{a_1 c_2 - a_2 c_1}{b_1 c_2 - b_2 c_1}, \frac{a_2 b_1 - a_1 b_2}{b_1 c_2 - b_2 c_1})$ as $\frac{\sqrt{a_1}}{d} x + c a_1 t$ moves from $-\infty$ to $+\infty$.

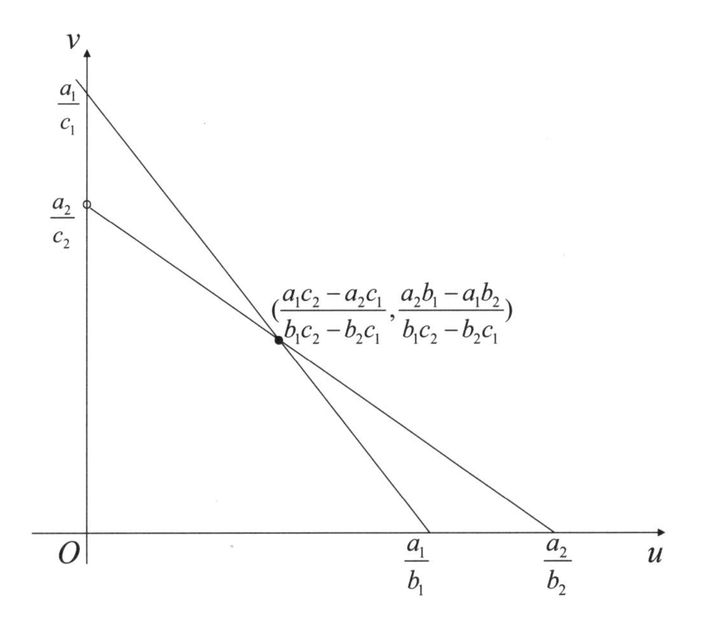

Figure 5.1.1: Nullclines for ODE system corresponding to (1.1) under hypotheses [P1] and [P2]; the symbols · and ∘ represent stable or unstable equilibrium respectively.

As mentioned above, the changes of variables described from (1.2) to (1.5) transform system (1.1) into system (1.6) with parameters given in (1.7). Note that hypotheses [P1] and (1.5), lead to the relation

$$0 < b < \epsilon_1, 0 < r < 1, \epsilon_2 > 0. \tag{1.8}$$

Observe that by choosing q close to $a_2c_2^{-1}$ in (1.5), ϵ_2 can be made arbitrarily small. We now summarize the transformations in the following lemma.

Lemma 1.1. *Consider system (1.1) under hypotheses [P1] and [P2]. The change of variables (1.2), (1.4) with q satisfying (1.5) transforms (1.1) into system (1.6). The parameters in (1.6) are related to those in (1.1) by (1.7). The parameters r, ϵ_1, ϵ_2 and b satisfy the inequalities in (1.8).*

One can readily verify that if $(w(\tilde{x},\tau), z(\tilde{x},\tau))$ is a solution of (1.6), then

$$(u(x,t), v(x,t)) = (u(\sqrt{\frac{d}{a_1}}\,\tilde{x}, a_1^{-1}\tau), v(\sqrt{\frac{d}{a_1}}\,\tilde{x}, a_1^{-1}\tau)) = (kw(\tilde{x},\tau), qz(\tilde{x},\tau)) \tag{1.9}$$

is a solution of (1.1), where k, q are described in (1.4), (1.5). We will look for solution of (1.6) with the form $(w(\tilde{x},\tau), z(\tilde{x},\tau)) = (w(\xi), z(\xi))$, $\xi = \tilde{x}+c\tau$, where

$c > 0$ satisfying:

$$
(1.10) \qquad \begin{cases} \lim_{\xi\to-\infty}(w(\xi), z(\xi)) = (0, \dfrac{1}{1+\epsilon_2}), \\ \\ \lim_{\xi\to\infty}(w(\xi), z(\xi)) = (\dfrac{\epsilon_1(1+\epsilon_2-r)}{\epsilon_1(1+\epsilon_2)-br}, \dfrac{\epsilon_1 - b}{\epsilon_1(1+\epsilon_2)-br}). \end{cases}
$$

Relating back to (1.6), we are thus looking for solutions of

$$
(1.11) \qquad \begin{aligned} -w_{\xi\xi} + cw_\xi &= w(1-w-rz), \\ -z_{\xi\xi} + cz_\xi &= z(\epsilon_1 - bw - \epsilon_1(1+\epsilon_2)z), \end{aligned} \qquad -\infty < \xi < \infty
$$

with boundary conditions (1.10).

Theorem 1.1 (Traveling Waves Connecting Semi-Trivial to Coexistence State). *Under hypotheses [P1] and [P2], system (1.1) has a traveling wave solution of the form:*

$$
(1.12) \qquad (u(x,t), v(x,t)) = (kw(\sqrt{\frac{a_1}{d}}x + ca_1t), qz(\sqrt{\frac{a_1}{d}}x + ca_1t))
$$

for any $c \geq 2\sqrt{\frac{b_1}{a_1}(\frac{a_1c_2-a_2c_1}{b_1c_2-b_2c_1})}$. *Here* (w, z) *is a function of one single variable* ξ *satisfying (1.11) for* $-\infty < \xi < \infty$, *and (1.10) as* $\xi \to \pm\infty$. *Moreover,* $w(\xi)$ *and* $z(\xi)$ *are positive monotonic functions for* $-\infty < \xi < \infty$.

Remark 1.1. The function (w, z) in (1.12) now satisfies (1.11) with parameters satisfying (1.7), (1.8) as described above. Moreover, we have

$$
(1.13) \qquad \begin{aligned} &\lim_{t\to-\infty}(u(x,t), v(x,t)) = (0, a_2/c_2), \\ &\lim_{t\to+\infty}(u(x,t), v(x,t)) = (\frac{a_1c_2 - a_2c_1}{b_1c_2 - b_2c_1}, \frac{a_2b_1 - a_1b_2}{b_1c_2 - b_2c_1}). \end{aligned}
$$

Proof. Note that by (1.7), we always have relation

$$
(1.14) \qquad \frac{(1+\epsilon_2 - r)\epsilon_1}{\epsilon_1(1+\epsilon_2) - rb} = \frac{b_1}{a_1}\frac{(a_1c_2 - a_2c_1)}{(b_1c_2 - b_2c_1)} > 0.
$$

The last inequality above is due to [P1] and [P2]. After transforming (1.1) into (1.6) and (1.11) as described above, we next introduce the transformation

$$
(1.15) \qquad \begin{cases} u_1(\xi) = w(\xi), \\ \\ u_2(\xi) = \dfrac{1}{1+\epsilon_2} - z(\xi) \end{cases}
$$

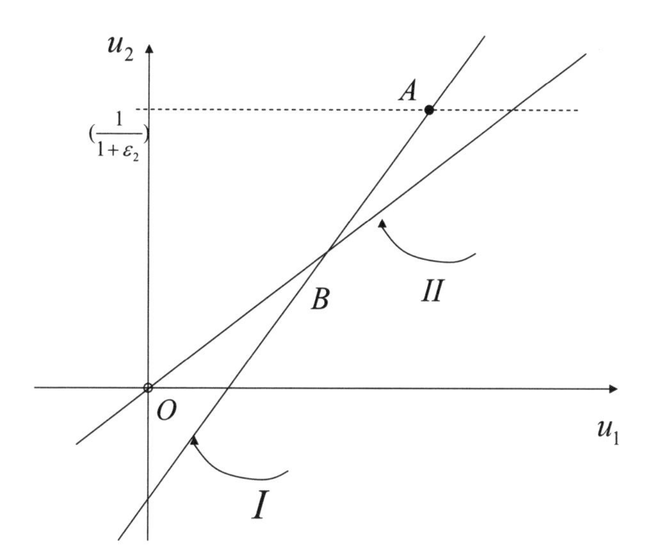

Figure 5.1.2: Nullclines for system (1.16); $A = (1, \frac{1}{1+\epsilon_2})$, $B = (\frac{(1+\epsilon_2-r)\epsilon_1}{\epsilon_1(1+\epsilon_2)-rb}, \frac{b(1+\epsilon_2-r)}{(1+\epsilon_2)(\epsilon_1(1+\epsilon_2-rb)})$. I and II represent the nullclines $1 - \frac{r}{1+\epsilon_2} - u_1 + ru_2 = 0$ and $bu_1 - \epsilon_1(1+\epsilon_2)u_2 = 0$ respectively.

for $-\infty < \xi < \infty$. This transforms (1.11) into

$$\begin{cases} -(u_1)_{\xi\xi} + c(u_1)_\xi = u_1(\dfrac{1+\epsilon_2-r}{1+\epsilon_2} - u_1 + ru_2), \\ \\ -(u_2)_{\xi\xi} + c(u_2)_\xi = (\dfrac{1}{1+\epsilon_2} - u_2)(bu_1 - \epsilon_1(1+\epsilon_2)u_2) \end{cases} \tag{1.16}$$

for $-\infty < \xi < \infty$, which is monotone (with respect to the other component) for $0 \le u_1$, $0 \le u_2 \le \frac{1}{1+\epsilon_2}$. We illustrate the transformed system (1.16) by the above Fig. 5.1.2.

Note that the two expressions on the right of (1.16) now vanish simultaneously at $(u_1, u_2) = (\frac{(1+\epsilon_2-r)\epsilon_1}{\epsilon_1(1+\epsilon_2)-rb}, \frac{b(1+\epsilon_2-r)}{(1+\epsilon_2)(\epsilon_1(1+\epsilon_2)-rb)})$ in the first quadrant. The second component satisfies

$$\frac{b(1+\epsilon_2-r)}{(1+\epsilon_2)(\epsilon_1(1+\epsilon_2)-rb)} < \frac{1}{1+\epsilon_2} \tag{1.17}$$

because $0 < b < \epsilon_1$. We now construct a pair of coupled upper solutions for system (1.16). Let $u_\rho(\xi), -\infty < \xi < \infty$, be the monotone solution of the following KPP equation:

$$u_\rho'' - cu_\rho' + u_\rho(\frac{(1+\epsilon_2-r)\epsilon_1}{\epsilon_1(1+\epsilon_2)-rb} - u_\rho) = 0, \tag{1.18}$$

with $\lim_{\xi\to-\infty} u_\rho(\xi) = 0$, and $\lim_{\xi\to\infty} u_\rho(\xi) = \frac{(1+\epsilon_2-r)\epsilon_1}{\epsilon_1(1+\epsilon_2)-rb}$. From [106], such solution $u_\rho(\xi)$ exists for $c \geq 2\sqrt{\frac{(1+\epsilon_2-r)\epsilon_1}{\epsilon_1(1+\epsilon_2)-rb}} = 2\sqrt{\frac{b_1}{a_1}\frac{(a_1c_2-a_2c_1)}{(b_1c_2-b_2c_1)}}$.

Define

$$\bar{u}_1(\xi) = u_\rho(\xi),\ \bar{u}_2(\xi) = \frac{b}{\epsilon_1(1+\epsilon_2)}u_\rho(\xi).$$

Observe that the system (1.16) is monotone in the region:

$$\begin{aligned} &0 < u_1 < \frac{(1+\epsilon_2-r)\epsilon_1}{\epsilon_1(1+\epsilon_2)-rb} := u_1^* = \lim_{\xi\to\infty}\bar{u}_1(\xi), \\ &0 < u_2 < \frac{b(1+\epsilon_2-r)}{(1+\epsilon_2)(\epsilon_1(1+\epsilon_2)-rb)} := u_2^* = \lim_{\xi\to\infty}\bar{u}_2(\xi). \end{aligned} \tag{1.19}$$

(That is, the expressions on the right of (1.16) are nondecreasing in the off diagonal variables.)

For $0 \leq u_2 \leq \bar{u}_2(\xi)$, one readily verifies that

$$\begin{aligned} &-(\bar{u}_1)_{\xi\xi} + c(\bar{u}_1)_\xi - \bar{u}_1(\frac{1+\epsilon_2-r}{1+\epsilon_2} - \bar{u}_1 + ru_2) \\ &\geq \bar{u}_1(\frac{(1+\epsilon_2-r)\epsilon_1}{\epsilon_1(1+\epsilon_2)-rb} - \bar{u}_1 - \frac{1+\epsilon_2-r}{1+\epsilon_2} + \bar{u}_1 - r\bar{u}_2) \\ &\geq \bar{u}_1(\frac{(1+\epsilon_2-r)(1+\epsilon_2)\epsilon_1 - \epsilon_1(1+\epsilon_2-r)(1+\epsilon_2) + rb(1+\epsilon_2-r)}{(\epsilon_1(1+\epsilon_2)-rb)(1+\epsilon_2)} \\ &-\frac{rb}{\epsilon_1(1+\epsilon_2)}u_1^*) \\ &= \bar{u}_1\frac{(1+\epsilon_2-r)(1+\epsilon_2)(\epsilon_1-\epsilon_1)}{(\epsilon_1(1+\epsilon_2)-rb)(1+\epsilon_2)} = 0. \end{aligned} \tag{1.20}$$

For $0 \leq u_1 \leq \bar{u}_1(\xi)$, one verifies

$$\begin{aligned} &-(\bar{u}_2)_{\xi\xi} + c(\bar{u}_2)_\xi - (\frac{1}{1+\epsilon_2} - \bar{u}_2)(bu_1 - \epsilon_1(1+\epsilon_2)\bar{u}_2) \\ &\geq \frac{b}{\epsilon_1(1+\epsilon_2)}u_\rho(\frac{(1+\epsilon_2-r)\epsilon_1}{\epsilon_1(1+\epsilon_2)-rb} - u_\rho) \\ &-(\frac{1}{1+\epsilon_2} - \frac{b}{\epsilon_1(1+\epsilon_2)}u_\rho)(bu_\rho - \epsilon_1(1+\epsilon_2)\bar{u}_2) \\ &= \frac{b}{\epsilon_1(1+\epsilon_2)}u_\rho(\frac{(1+\epsilon_2-r)\epsilon_1}{\epsilon_1(1+\epsilon_2)-rb} - u_\rho) > 0. \end{aligned} \tag{1.21}$$

The pair of functions defined by:

$$\bar{\rho}_1(\xi) = \bar{u}_1(-\xi)\ ,\ \bar{\rho}_2(\xi) = \bar{u}_2(-\xi) \tag{1.22}$$

form a pair of upper solutions for the monotone system

$$\begin{cases} (\rho_1)_{\xi\xi} + c(\rho_1)_\xi + \rho_1(\dfrac{1+\epsilon_2-r}{1+\epsilon_2} - \rho_1 + r\rho_2) = 0, \\ \\ (\rho_2)_{\xi\xi} + c(\rho_2)_\xi + (\dfrac{1}{1+\epsilon_2} - \rho_2)(b\rho_1 - \epsilon_1(1+\epsilon_2)\rho_2) = 0 \end{cases} \tag{1.23}$$

for $-\infty < \xi < \infty$, in the sense that

$$\begin{cases} (\bar{\rho}_1)_{\xi\xi} + c(\bar{\rho}_1)_\xi + \bar{\rho}_1(\dfrac{1+\epsilon_2-r}{1+\epsilon_2} - \bar{\rho}_1 + r\rho_2) \leq 0, \\ \\ (\bar{\rho}_2)_{\xi\xi} + c(\bar{\rho}_2)_\xi + (\dfrac{1}{1+\epsilon_2} - \bar{\rho}_2)(b\rho_1 - \epsilon_1(1+\epsilon_2)\bar{\rho}_2) \leq 0 \end{cases} \tag{1.24}$$

for $-\infty < \xi < \infty$, all $0 \leq \rho_2 \leq \bar{\rho}_2(\xi)$, $0 \leq \rho_1 \leq \bar{\rho}_1(\xi)$. (Note that the system (1.23) is monotone (with respect to the other component) in the region: $0 \leq \rho_1 \leq u_1^*$, $0 \leq \rho_2 \leq u_2^*$, where u_1^*, u_2^* are defined in (1.19). In particular, (1.24) is true when $\rho_1 = \bar{\rho}_1(\xi)$ in the first equation, $\rho_2 = \bar{\rho}_2(\xi)$ in the second equation for all $-\infty < \xi < \infty$. Let

$$F_1(\rho_1, \rho_2) = \rho_1(\frac{1+\epsilon_2-r}{1+\epsilon_2} - \rho_1 + r\rho_2),$$

$$F_2(\rho_1, \rho_2) = (\frac{1}{1+\epsilon_2} - \rho_2)(b\rho_1 - \epsilon_1(1+\epsilon_2)\rho_2).$$

We clearly have for $i = 1, 2$ that

$$F_i(s, \frac{sb}{2\epsilon_1(1+\epsilon_2)}) > 0$$

for $s > 0$ sufficiently small. We now apply Theorem 4.2 in Volpert [224], p. 176, (or Theorem A5-7 in Chapter 6), with $\overrightarrow{F} = (F_1, F_2)$, $\overrightarrow{w_+} = (0, 0)$, $\overrightarrow{w_-} = (\frac{(1+\epsilon_2-r)\epsilon_1}{\epsilon_1(1+\epsilon_2)-rb}, \frac{b(1+\epsilon_2-r)}{(1+\epsilon_2)(\epsilon_1(1+\epsilon_2)-rb)})$ and K the class of vector-valued functions $\overrightarrow{\rho}(\xi) = (\rho_1(\xi), \rho_2(\xi))$, $\rho_i(\xi) \in C^2(-\infty, \infty), i = 1, 2$, decreasing monotonically and satisfying $\lim_{\xi \to \pm\infty} \overrightarrow{\rho}(\xi) = \overrightarrow{w_\pm}$. The existence of the function $\overrightarrow{\rho}(\xi) = (\bar{\rho}_1(\xi), \bar{\rho}_2(\xi))$ satisfying (1.24) implies that

$$w^* := \inf_{\overrightarrow{\rho} \in K} \{\sup_{\xi, k} \frac{\rho_k'' + F_k(\overrightarrow{\rho}(\xi))}{-\rho_k'(\xi\}}\}$$

is finite, and we have $c \geq w^*$. We can thus apply Theorem 4.2 in [224] described above to assert that the system (1.23) has a solution $(\hat{\rho}_1(\xi), \hat{\rho}_2(\xi))$ with each component monotonically decreasing for $-\infty < \xi < \infty$ and $\lim_{\xi \to \pm\infty}(\hat{\rho}_1(\xi), \hat{\rho}_2(\xi)) = \overrightarrow{w_{\pm}}$. (See Fig. 5.1.3 below.)

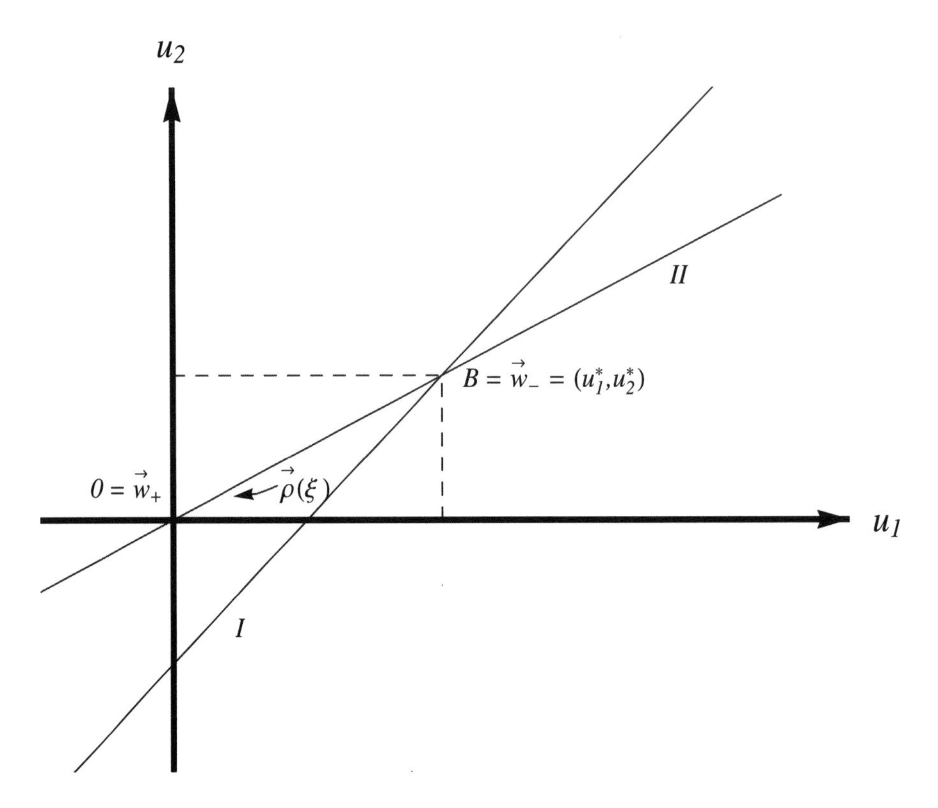

Figure 5.1.3: Traveling wave connecting $w_- = (u_1^*, u_2^*)$ with $w_+ = (0, 0)$.

The function

$$(\hat{u}_1(\xi), \hat{u}_2(\xi)) := (\hat{\rho}_1(-\xi), \hat{\rho}_2(-\xi)) \tag{1.25}$$

is then a solution of the system (1.16). Finally, setting $w(\xi) = \hat{u}_1(\xi)$, $z(\xi) = \frac{1}{1+\epsilon_2} - \hat{u}_2(\xi)$ for $-\infty < \xi < \infty$ as in (1.15), then $(u(x,t), v(x,t))$ as defined in (1.12) is a traveling wave solution of system (1.1) for $-\infty < x < \infty$, $-\infty < t < \infty$, satisfying (1.13) as described in Remark 1.1 following the statement of Theorem 1.1.

Remark 1.2. The wave speed of the solution $(u(x,t), v(x,t))$ for system (1.1) described in (1.12) in the original space variable is $c\sqrt{a_1 d}$.

Interchanging the roles of a_1, b_1 and c_1 respectively with a_2, c_2 and b_2, we obtain readily the following corollary from Theorem 1.1.

Corollary 1.2. *Assume hypotheses [P1] and [P2]. Then for any* $c \geq 2\sqrt{\dfrac{c_2\,(a_2b_1 - a_1b_2)}{a_2\,(c_2b_1 - c_1b_2)}}$, *system (1.1) has a traveling wave solution of the form:*

$$(u(x,t), v(x,t)) = (\tilde{u}(\sqrt{\frac{a_2}{d}}x + ca_2t), \tilde{v}(\sqrt{\frac{a_2}{d}}x + ca_2t)). \tag{1.26}$$

Here $\tilde{u}, \tilde{v} \in C^2(-\infty, \infty)$ *are monotonic functions of one single variable, and*

$$\begin{cases} \lim_{t \to -\infty}(u(x,t), v(x,t)) = (a_1/b_1, 0), \\ \\ \lim_{t \to +\infty}(u(x,t), v(x,t)) = (\dfrac{a_1c_2 - a_2c_1}{b_1c_2 - b_2c_1}, \dfrac{a_2b_1 - a_1b_2}{b_1c_2 - b_2c_1}). \end{cases} \tag{1.27}$$

Remark 1.3. The wave speed of the solution $(u(x,t), v(x,t))$ for system (1.1) described in (1.26) in the original space variable is $c\sqrt{a_2 d}$.

Part B: Iterative Method for obtaining Traveling Wave for General Monotone Systems.

As we see in part A, the existence of traveling wave solution of parabolic systems can be reduced to the study of a system of ordinary differential equations of the form:

$$D\rho''(t) - c\rho'(t) + \tilde{F}(\rho(t)) = 0, \qquad t \in R \tag{1.28}$$

where $\rho \in R^n, D = diag.(d_1, ..., d_n)$ with $d_i > 0, i = 1, ..., n$. Here, we assume that $\tilde{F} : R^n \to R^n$ is continuous, and $c > 0$ is a positive constant corresponding to the wave speed. We will discuss the existence of solutions of (1.28) with assigned limits at $\pm\infty$, under quasimonotone condition on $\tilde{F} = (\tilde{F}_1, ..., \tilde{F}_n)^T$, using the method of upper and lower solution. It is sometimes difficult to construct $C^2(R, R^n)$ upper and lower solutions of (1.28) with appropriate limits at $\pm\infty$. We are thus led to the following concept.

Definition 1.1. A function $\rho \in C^1(R, R^n)$ is called a quasi-upper solution of (1.28) if

(1) $sup_{t \in R}||\rho(t)|| < \infty, \quad sup_{t \in R}||\rho'(t)|| < \infty,$

(2) $\rho''(t)$ exists and is continuous on $R\backslash\{0\}$, and

$$sup_{t \in R\backslash\{0\}}||\rho''(t)|| < \infty,$$

(3) $\lim_{t \to 0^-} \rho''(t)$ and $\lim_{t \to 0^+} \rho''(t)$ exist,

(4) $\rho(t)$ satisfies

$$D\rho''(t) - c\rho'(t) + \tilde{F}(\rho(t)) \leq 0, \text{ for all } t \in R\backslash\{0\}.$$

A quasi-lower solution of (1.28) is defined in the same way by reversing the inequality in the line immediately above.

We will look for solutions of (1.28) in the family of functions Γ as follows:

$$\Gamma := \{\varphi \in C(R, R^n) : \varphi \text{ is non-decreasing, and } lim_{t\to-\infty}\varphi(t) = \mathbf{0},\ lim_{t\to+\infty}\varphi(t) = \mathbf{K}\},$$

where $\mathbf{K} = (K_1, ..., K_n)^T$ is given fixed, with $K_i > 0$ for $i = 1, ..., n$. Here, φ is non-decreasing is interpreted componentwise. For comparing two vectors $y = (y_1, ..., y_n)^T, z = (z_1, ..., z_n)^T \in R^n$, we write $y \geq z$ if and only if $y_i \geq z_i$ for $i = 1, ..., n$; and if there is at least one i such that $y_i > z_i$, we write $y > z$. We will assume that $\tilde{F}$ has the following properties concerning its zeros and quasi-monotonicity:

[A1] $\tilde{F}(\mathbf{0}) = \tilde{F}(\mathbf{K}) = \mathbf{0}$ and $\tilde{F}(u) \neq \mathbf{0}$ for $u \in R^n$ with $\mathbf{0} < u < \mathbf{K}$, where $\mathbf{K} = (K_1, K_2, ..., K_n)^T$, $K_i > 0$, for $i = 1, ..., n$.

[A2] There exists a matrix $\beta = diag(\beta_1, ..., \beta_n)$ with $\beta_i \geq 0$ such that

$$H(\phi) - H(\psi) \geq 0, \text{ for all } \phi, \psi \in R^n, \text{ satisfying } \mathbf{0} \leq \psi \leq \phi \leq \mathbf{K}, \text{ where}$$

$$H(v) = \tilde{F}(v) + \beta v \quad \text{for } v \in R^n. \tag{1.29}$$

We will be iterating from a quasi-upper solution of (1.28), and will be led to solve the linear problem

$$Dx''(t) - cx'(t) - \beta x(t) + H(\phi(t)) = 0, \ \ t \in R$$

for given $\phi(t)$. Let

$$\lambda_{1i} := \frac{c - \sqrt{c^2 + 4\beta_i d_i}}{2d_i} < 0, \ \ \lambda_{2i} := \frac{c + \sqrt{c^2 + 4\beta_i d_i}}{2d_i} > 0 \tag{1.30}$$

which are the real roots of the equation

$$d_i\lambda^2 - c\lambda - \beta_i = 0, \quad i = 1, 2, ..., n, \text{ with } \beta_i > 0.$$

The solution of the above linear problem leads to the following operator:

$$G(H(\phi))(t) = (G_1(H_1(\phi))(t), ..., G_n(H_n(\phi))(t))^T, \tag{1.31}$$

where $H = (H_1, ..., H_n)^T$,

$$\begin{aligned} G_i(H_i(\phi))(t) := \tfrac{1}{d_i(\lambda_{2i} - \lambda_{1i})}\Big(&\textstyle\int_{-\infty}^{t} e^{\lambda_{1i}(t-s)} H_i(\phi(s))ds \\ &+ \textstyle\int_t^{+\infty} e^{\lambda_{2i}(t-s)} H_i(\phi(s))ds\Big). \end{aligned} \tag{1.32}$$

Before solving (1.28), we first review some basic definition and theorem concerning scalar linear nonhomogeneous second order ordinary differential equation. Consider

$$u''(t) + \alpha u'(t) - \hat{\beta} u(t) + f(t) = 0, \quad t \in R, \ u(t) \in R \tag{1.33}$$

where $f(t)$ is a continuous and bounded function on $R\backslash\{0\}$ with left and right limits at $t = 0$, $f(0^-)$ and $f(0^+)$. The parameters α and $\hat{\beta}$ are real with $\hat{\beta} > 0$. The characteristic equation

$$\lambda^2 + \alpha\lambda - \hat{\beta} = 0$$

has two distinct roots of opposite signs, $\lambda_1 < 0$ and $\lambda_2 > 0$. Since we will allow f to have jump discontinuity at $t = 0$ (cf. (1.39) below), we need to clarify our terminology.

Definition 1.2. Suppose that f is bounded and continuous on $R\backslash\{0\}$, and both $f(0^-)$ and $f(0^+)$ exist. A function u defined on R is said to be a generalized C^1 bounded solution of (1.33) if

(1) u satisfies (1.33) on $R\backslash\{0\}$,

(2) u and u' are bounded and continuous on R,

(3) u'' exists and is continuous on $R\backslash\{0\}$, and both $u''(0^-)$ and $u''(0^+)$ exist.

Lemma 1.2. *Consider equation (1.33) with $\hat{\beta} > 0$. Suppose that f is bounded and continuous on $R\backslash\{0\}$, and both $f(0^-)$ and $f(0^+)$ exist. Then (1.33) has a unique generalized C^1 bounded solution (in the sense of Definition 1.2) which is given by*

$$\hat{u}(t) = \frac{1}{\lambda_2 - \lambda_1}\left(\int_{-\infty}^{t} e^{\lambda_1(t-s)} f(s)ds + \int_{t}^{+\infty} e^{\lambda_2(t-s)} f(s)ds\right). \tag{1.34}$$

Proof. The function $\hat{u}(t)$ define by formula (1.34) is well-defined and is bounded on R. Direct calculation shows that it is continuously differentiable on R and $\hat{u}'(t)$ is bounded on R. Moreover, $\hat{u}''(t)$ exists on the whole R except possibly at $t = 0$, while $\hat{u}''(0^-)$ and $\hat{u}''(0^+)$ exist. In fact $\hat{u}(t)$ is a classical solution of (1.33) on each of the two intervals $(-\infty, 0)$ and $(0, +\infty)$. It remains to show uniqueness, that is $u(t) = \hat{u}(t)$ for any generalized C^1 bounded solution $u(t)$ of (1.33). Since the function $u(t) - \hat{u}(t)$ is a classical solution of (1.33) on $(0, \infty)$, we have

$$u(t) - \hat{u}(t) = c_1 e^{\lambda_1 t} + c_2 e^{\lambda_2 t}, \quad \text{for } t > 0,$$

where c_1, c_2 are constants. From the boundedness of u and $\hat{u}$, we deduce $c_2 = 0$, i.e.

$$u(t) - \hat{u}(t) = c_1 e^{\lambda_1 t}, \quad \text{for } t > 0.$$

Similarly, since $u - \hat{u}$ is a bounded classical solution of (1.33) on $(-\infty, 0)$, we find

$$u(t) - \hat{u}(t) = c_2 e^{\lambda_2 t}, \quad \text{for } t < 0.$$

Since $u - \hat{u}$ is continuous at $t = 0$, the last two formulas yield

$$c_1 = c_2. \tag{1.35}$$

Similarly, since $u - \hat{u}$ has continuous derivative at $t = 0$, we obtain

$$\lambda_1 c_1 = \lambda_2 c_2. \tag{1.36}$$

Since $\lambda_1 < 0$ and $\lambda_2 > 0$, (1.35) and (1.36) imply that $c_1 = c_2 = 0$.

We now begin to find solution of (1.28) in the family of non-decreasing functions Γ with appropriate limits at $\pm\infty$.

Lemma 1.3. *Suppose [A1] and [A2] hold. Then for any $\phi \in \Gamma$, we have*
(i) $H(\phi(t)) \geq 0$, for $t \in R$,
(ii) $H(\phi(t))$ is non-decreasing in $t \in R$,
(iii) $H(\psi(t)) \leq H(\phi(t))$ for all $t \in R$, if $\psi \in C(R, R^n)$ is given so that $\mathbf{0} \leq \psi(t) \leq \phi(t) \leq \mathbf{K}$ for $t \in R$.

Proof. Assertion (i) follows readily from assumptions [A1] and [A2] since $\tilde{F}(\mathbf{0}) = \mathbf{0}$. Assertion (iii) is a direct consequence of [A2]. To prove (ii), let $t \in R$ and $s > 0$ be given. Then, since $\phi \in \Gamma$, we have

$$\mathbf{0} \leq \phi(t) \leq \phi(t+s) \leq \mathbf{K}.$$

Consequently [A2] implies that $H(\phi(t+s)) - H(\phi(t)) \geq 0$. This completes the proof.

Lemma 1.4. *Assume [A1] and [A2]. Let $\tilde{\rho}(t)$ and $\bar{\rho}(t)$ be quasi-lower and quasi-upper solutions of (1.28), with $\bar{\rho} \in \Gamma$ and additional properties*
[H1] $\mathbf{0} \leq \tilde{\rho}(t) \leq \bar{\rho}(t) \leq \mathbf{K}$, $t \in R$,
[H2] $\tilde{\rho}(t) \not\equiv \mathbf{0}, t \in R$.
Suppose that λ_{i1} and λ_{i2}, $i = 1, ..., n$ are given by (1.30) with $\beta_i > 0$, and define

$$\phi_1 := G(H(\bar{\rho}))(t), \text{ for } t \in R, \tag{1.37}$$

where $G(H)$ is given in (1.31) and (1.32). Then
(i) $\phi_1 \in C^2(R, R^n)$, and satisfies

$$D\phi_1''(t) - c\phi_1'(t) - \beta\phi_1(t) + H(\bar{\rho}(t)) = 0 \text{ for } t \in R, \tag{1.38}$$

(ii) $\phi_1 \in \Gamma$,

(iii) $\tilde{\rho}(t) \leq \phi_1(t) \leq \bar{\rho}(t)$, $t \in R$; *and*

(iv) ϕ_1 *is an upper solution of (1.28), i.e. (1.28) is satisfied for all* $t \in R$ *with "=" replaced by "≤".*

Proof. Since $\tilde{F}$ is continuous in R^n and $\bar{\rho}$ is in $C^1(R, R^n)$, direct computation shows that $\phi_1 \in C^2(R, R^n)$ and satisfies (1.38) for all $t \in R$. To show (ii), we first consider the limits of $\phi_1(t) := (\phi_{11}(t), ..., \phi_{1n}(t))^T$ as $t \to \pm\infty$. Applying L'Hospital's rule to the formula (1.37) and (1.32), and using [A1] and $\bar{\rho} \in \Gamma$, we obtain

$$\lim_{t\to-\infty} \phi_{1i}(t) = \frac{1}{d_i(\lambda_{2i}-\lambda_{1i})} \lim_{t\to-\infty}[\frac{H_i(\bar{\rho}(t))}{-\lambda_{1i}} - \frac{H_i(\bar{\rho}(t))}{-\lambda_{2i}}]$$

$$= \frac{1}{d_i(\lambda_{2i}-\lambda_{1i})}[\frac{0}{-\lambda_{1i}} + \frac{0}{\lambda_{2i}}] = 0, \quad i = 1, .., n,$$

and

$$\lim_{t\to+\infty} \phi_{1i}(t) = \frac{1}{d_i(\lambda_{2i}-\lambda_{1i})} \lim_{t\to+\infty}[\frac{H_i(\bar{\rho}(t))}{-\lambda_{1i}} - \frac{H_i(\bar{\rho}(t))}{-\lambda_{2i}}]$$

$$= \frac{1}{d_i(\lambda_{2i}-\lambda_{1i})}[\frac{\beta_i K_i}{-\lambda_{1i}} + \frac{\beta_i K_i}{\lambda_{2i}}]$$

$$= \frac{\beta_i K_i}{-\lambda_{1i}\lambda_{2i}d_i} = \frac{\beta_i K_i}{\beta_i} = K_i, \quad i = 1, .., n.$$

Thus, we obtain $\lim_{t\to-\infty} \phi_1(t) = (0, ..., 0)^T = \mathbf{0}$ and $\lim_{t\to+\infty} = (K_1, ..., K_n)^T = \mathbf{K}$.

We next show that each component of ϕ_1 is non-decreasing in R. Let $t \in R$ and $s > 0$ be given. Then

$$\begin{aligned}
&\phi_{1i}(t+s) - \phi_{1i}(t) \\
&= \frac{1}{d_i(\lambda_{2i}-\lambda_{1i})}[\int_{-\infty}^{t+s} e^{\lambda_{1i}(t+s-\theta)} H_i(\bar{\rho}(\theta))d\theta + \int_{t+s}^{\infty} e^{\lambda_{2i}(t+s-\theta)} H_i(\bar{\rho}(\theta))d\theta] \\
&\quad - \frac{1}{d_i(\lambda_{2i}-\lambda_{1i})}[\int_{-\infty}^{t} e^{\lambda_{1i}(t-\theta)} H_i(\bar{\rho}(\theta))d\theta + \int_{t}^{\infty} e^{\lambda_{2i}(t-\theta)} H_i(\bar{\rho}(\theta))d\theta] \\
&= \frac{1}{d_i(\lambda_{2i}-\lambda_{1i})}[\int_{-\infty}^{t} e^{\lambda_{1i}(t-\theta)}[H_i(\bar{\rho}(\theta+s)) - H_i(\bar{\rho}(\theta))]d\theta \\
&\quad + \frac{1}{d_i(\lambda_{2i}-\lambda_{1i})}[\int_{t}^{\infty} e^{\lambda_{2i}(t-\theta)})[H_i(\bar{\rho}(\theta+s)) - H_i(\bar{\rho}(\theta))]d\theta \\
&\geq 0.
\end{aligned}$$

The last inequality is due to Lemma 1.3(ii). This completes the proof of part (ii) of this lemma.

To show part (iii), we denote

$$w_i(t) := \phi_{1i}(t) - \bar{\rho}_i(t) \quad \text{for } t \in R, \ i = 1, ..., n.$$

From (1.38), we find for each $i = 1, ..., n$,

$$\begin{aligned} d_i w_i''(t) - c w_i'(t) - \beta_i w_i(t) &= -d_i \bar{\rho}_i''(t) + c\bar{\rho}_i'(t) + \beta_i \bar{\rho}_i(t) - H_i(\bar{\rho}(t)) \\ &= -d_i \bar{\rho}_i''(t) + c\bar{\rho}_i'(t) - \tilde{F}_i(\bar{\rho}(t)) \\ &= -r_i(t), \end{aligned} \tag{1.39}$$

where $r_i(t) := d_i \bar{\rho}_i''(t) - c\bar{\rho}_i'(t) + \tilde{F}_i(\bar{\rho}(t)) \leq 0$ for all $t \in R\backslash\{0\}$, since $\bar{\rho}$ is a quasi-upper solution of (1.28). Thus $w_i(t)$ is a generalized C^1 bounded solution of (1.39) in the sense of Definition 1.2. By Lemma 1.2, we have

$$w_i(t) = \frac{1}{d_i(\lambda_{2i} - \lambda_{1i})}\Big[\int_{-\infty}^{t} e^{\lambda_{1i}(t-s)} r_i(s) ds + \int_{t}^{+\infty} e^{\lambda_{2i}(t-s)} r_i(s) ds\Big]. \tag{1.40}$$

Since $r_i \leq 0$ in $R\backslash\{0\}$, we find from (1.40) that $\phi_{1i}(t) - \bar{\rho}_i(t) = w_i(t) \leq 0$ for $t \in R$. This proves $\phi_{1i}(t) \leq \bar{\rho}_i(t)$, for $t \in R$. Similarly, we show that $\tilde{\rho}(t) \leq \phi_1(t)$ for $t \in R$, since $\tilde{\rho}$ is a quasi-lower solution. This proves assertion (iii).

To show that ϕ_1 is an upper solution of (1.28), we verify for all $t \in R$,

$$\begin{aligned} D\phi_1''(t) - c\phi_1'(t) + \tilde{F}(\phi_1(t)) &= D\phi_1''(t) - c\phi_1'(t) - \beta\phi_1(t) + H(\phi_1(t)) \\ &= D\phi_1''(t) - c\phi_1'(t) - \beta\phi_1(t) + H(\bar{\rho}(t)) + [H(\phi_1(t)) - H(\bar{\rho}(t))] \\ &= H(\phi_1(t)) - H(\bar{\rho}(t)) \leq \mathbf{0}. \end{aligned}$$

The last inequality is true by means of Lemma 1.3(iii) and $\mathbf{0} \leq \phi_1(t) \leq \bar{\rho}(t) \leq \mathbf{K}$ for $t \in R$. This proves Lemma 1.4.

We next inductively define for $t \in R$

$$\phi_m(t) := G(H(\phi_{m-1}(t)), \quad m = 2, 3, ... \tag{1.41}$$

We can inductively prove as in the last lemma the following.

Lemma 1.5. *Assume all the hypotheses of Lemma 1.4. The function $\phi_m(t), m = 2, 3, \ldots$ defined by (1.41) satisfies:*
(i) ϕ_m is the classical $C^2(R, R^n)$ solution of

$$D\phi_m''(t) - c\phi_m'(t) - \beta\phi_m(t) + H(\phi_{m-1}(t)) = 0, \quad t \in R,\ m = 2, 3, ..., ; \tag{1.42}$$

(ii) $\phi_m \in \Gamma$;
(iii) $\tilde{\rho}(t) \leq \phi_m(t) \leq \phi_{m-1}(t) \leq \bar{\rho}(t), \ t \in R$; and
(iv) each ϕ_m is an upper solution of (1.28).

Theorem 1.3 (Main Existence Theorem). *Assume [A1] and [A2]. Let $\tilde{\rho}(t)$ and $\bar{\rho}(t)$ be quasi-lower and quasi-upper solutions of (1.28), with $\bar{\rho} \in \Gamma$ and additional properties*

[H1] $\mathbf{0} \leq \tilde{\rho}(t) \leq \bar{\rho}(t) \leq \mathbf{K}$, $t \in R$,

[H2] $\tilde{\rho}(t) \not\equiv \mathbf{0}, t \in R$.

Suppose that λ_{i1} and λ_{i2}, $i = 1, ..., n$ are given by (1.30) with $\beta_i > 0$; let $\phi_1(t)$ be defined by (1.37), and $\phi_m(t), m = 2, 3, \ldots$ be defined by (1.41). Then the function $\phi(t) := lim_{m\to\infty}\phi_m(t)$ exists and satisfies

$$\tilde{\rho}(t) \leq \phi(t) \leq \bar{\rho}(t), \quad t \in R. \tag{1.43}$$

Further, the function $\phi(t)$ is non-decreasing for $t \in R$ with the property

$$\lim_{t\to-\infty} \phi(t) = \mathbf{0}, \quad \lim_{t\to+\infty} \phi(t) = \mathbf{K}, \tag{1.44}$$

and is a classical solution of (1.28) for $t \in R$. (Note that the quasi-lower solution $\tilde{\rho}(t)$ may not belong to the family Γ of functions.)

Proof. Since $\phi_m(t)$ is non-decreasing by Lemma 1.5(ii), we have the limit function $\phi(t)$ non-decreasing for $t \in R$. Similarly, (1.43) follows from Lemma 1.5(iii). Denote the i-th component of $\phi(t)$ and $\phi_m(t)$ by $(\phi(t))_i$ and $\phi_{mi}(t)$ respectively for $i = 1, .., n$. From (1.41) and Lebesgue's dominated convergence theorem, we have

$$\begin{aligned}
(\phi(t))_i &= \lim_{m\to\infty} \phi_{mi}(t) \\
&= \tfrac{1}{d_i(\lambda_{2i}-\lambda_{1i})} \lim_{m\to\infty}[\textstyle\int_{-\infty}^{t} e^{\lambda_{1i}(t-s)} H_i(\phi_{m-1}(s))ds \\
&\qquad + \textstyle\int_{t}^{+\infty} e^{\lambda_{2i}(t-s)} H_i(\phi_{m-1}(s))ds] \\
&= \tfrac{1}{d_i(\lambda_{2i}-\lambda_{1i})}[e^{\lambda_{1i}t} \textstyle\int_{-\infty}^{t} e^{-\lambda_{1i}s} H_i(\phi(s))ds + e^{\lambda_{2i}t} \textstyle\int_{t}^{+\infty} e^{-\lambda_{2i}s} H_i(\phi(s))ds]
\end{aligned} \tag{1.45}$$

for $t \in R, i = 1, 2, ..., n$. Direct calculation gives

$$\begin{aligned}
(\phi'(t))_i &= \lambda_{1i} e^{\lambda_{1i}t} \textstyle\int_{-\infty}^{t} \tfrac{e^{-\lambda_{1i}s}}{d_i(\lambda_{2i}-\lambda_{1i})} H_i(\phi(s))ds \\
&\quad + \lambda_{2i} e^{\lambda_{2i}t} \textstyle\int_{t}^{+\infty} \tfrac{e^{-\lambda_{2i}s}}{d_i(\lambda_{2i}-\lambda_{1i})} H_i(\phi(s))ds,
\end{aligned} \tag{1.46}$$

and

$$\begin{aligned}
(\phi''(t))_i &= \lambda_{1i}^2 e^{\lambda_{1i}t} \textstyle\int_{-\infty}^{t} \tfrac{e^{-\lambda_{1i}s}}{d_i(\lambda_{2i}-\lambda_{1i})} H_i(\phi(s))ds \\
&\quad + \lambda_{2i}^2 e^{\lambda_{2i}t} \textstyle\int_{t}^{+\infty} \tfrac{e^{-\lambda_{2i}s}}{d_i(\lambda_{2i}-\lambda_{1i})} H_i(\phi(s))ds + \tfrac{H_i(\phi(t))}{d_i(\lambda_{2i}-\lambda_{1i})}[\lambda_{1i} - \lambda_{2i}].
\end{aligned} \tag{1.47}$$

Thus, from (1.46) and (1.47), we obtain

$$
\begin{aligned}
&d_i(\phi''(t))_i - c(\phi'(t))_i - \beta_i(\phi(t))_i \\
&\quad = (d_i\lambda_{1i}^2 - c\lambda_{1i} - \beta_i)e^{\lambda_{1i}t}\int_{-\infty}^{t} \frac{e^{-\lambda_{1i}s}}{d_i(\lambda_{2i}-\lambda_{1i})}H_i(\phi(s))ds \\
&\quad\quad + (d_i\lambda_{2i}^2 - c\lambda_{2i} - \beta_i)e^{\lambda_{2i}t}\int_{t}^{+\infty} \frac{e^{-\lambda_{2i}s}}{d_i(\lambda_{2i}-\lambda_{1i})}H_i(\phi(s))ds - H_i(\phi(t)) \\
&\quad = -H_i(\phi(t)), \qquad t \in R,\ \ i = 1, ..., n.
\end{aligned}
$$

From the above equation, we find

$$
D\phi''(t) - c\phi'(t) - \beta\phi(t) = -H(\phi(t)) = -\tilde{F}(\phi(t)) - \beta\phi(t), \ \ t \in R.
$$

That is $\phi(t)$ is a solution of (1.28). From [H1], (1.43) and the fact that $\lim_{t\to-\infty}\bar{\rho}(t) = \mathbf{0}$, we obtain $\lim_{t\to-\infty}\phi(t) = \mathbf{0}$. On the other hand, $\phi(t)$ is non-decreasing and bounded from above by $\mathbf{K}$. We must have $\lim_{t\to+\infty}\phi(t) = \mathbf{Q} := (Q_1, ..., Q_n)$ exists and $Q_i \leq K_i, i = 1, ..., n$. From [H2] and (1.43), we obtain $\mathbf{Q} \neq \mathbf{0}$. Applying l'Hospital's rule to (1.45), we obtain

$$
\begin{aligned}
Q_i = \lim_{t\to+\infty}(\phi(t))_i &= \lim_{t\to+\infty}\frac{1}{d_i(\lambda_{2i}-\lambda_{1i})}\left[\frac{H_i(\phi(t))}{-\lambda_{1i}} + \frac{H_i(\phi(t))}{\lambda_{2i}}\right] \\
&= \frac{\tilde{F}_i(\mathbf{Q})+\beta_i Q_i}{-d_i\lambda_{2i}\lambda_{1i}} \\
&= \frac{\tilde{F}_i(\mathbf{Q})+\beta_i Q_i}{\beta_i} \\
&= \frac{\tilde{F}_i(\mathbf{Q})}{\beta_i} + Q_i, \quad i = 1, ..., n.
\end{aligned} \tag{1.48}
$$

This leads to $\tilde{F}(\mathbf{Q}) = \mathbf{0}$. By assumption [A1], we must have $\mathbf{Q} = \mathbf{K}$. That is $\lim_{t\to+\infty}\phi(t) = \mathbf{K}$. This completes the proof of the theorem.

We now use a variant of Theorem 1.3 to investigate the competitive system (1.1), under different assumptions on the parameters as in Part A. We will assume:

[P2] $\dfrac{a_2}{c_2} < \dfrac{a_1}{c_1}$,

[P3] $\dfrac{a_2}{b_2} < \dfrac{a_1}{b_1}$.

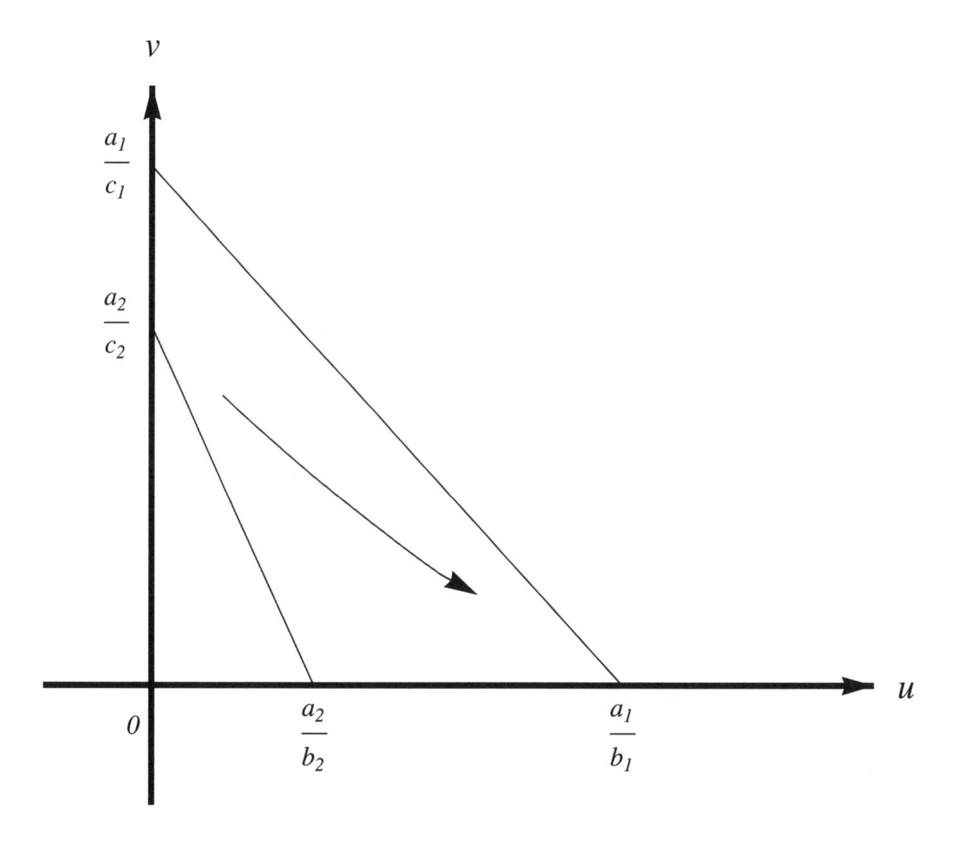

Figure 5.1.4: Traveling wave connecting $(0, \frac{a_2}{c_2})$ and $(\frac{a_1}{b_1}, 0)$.

Recall [P2] was introduced in Part A. Under conditions [P2] and [P3], the two straight lines $a_i - b_i u - c_i v = 0, i = 1, 2$ do not intersect in the first quadrant. Under another additional assumption, we will use a variant of Theorem 1.3 to prove the existence of traveling wave solution of the form $(u(\sqrt{\frac{a_1}{d}}x + ca_1 t), v(\sqrt{\frac{a_1}{d}}x + ca_1 t))$ connecting the steady-states $(0, \frac{a_2}{c_2})$ and $(\frac{a_1}{b_1}, 0)$ as $\sqrt{\frac{a_1}{d}}x + ca_1 t$ moves from $-\infty$ to ∞. This means that the first species move from extinction to carrying capacity, while the second species move from carrying capacity to extinction. (See Fig. 5.1.4.) We call such traveling waves exclusive in Theorem 1.4 below.

Lemma 1.6. *Consider system (1.1) under hypotheses [P2] and [P3]. The change of variables (1.2), (1.4) with q satisfying (1.5) transforms (1.1) into system (1.6). The parameters in (1.6) are related to those in (1.1) by (1.7). The parameters r, ϵ_1, ϵ_2 and b satisfy the inequalities in (1.49) below:*

$$0 < \epsilon_1 < b, \; 0 < r < 1, \; \epsilon_2 > 0. \tag{1.49}$$

Note that the relation between ϵ_1 and b is now different from (1.8) because of the different hypotheses. One can readily verify that if $(w(\tilde{x}, \tau), z(\tilde{x}, \tau))$ is a solution of (1.6), then

(1.50)

$$(u(x,t),v(x,t)) = (u(\sqrt{\frac{d}{a_1}}\,\tilde{x}, a_1^{-1}\tau), v(\sqrt{\frac{d}{a_1}}\,\tilde{x}, a_1^{-1}\tau)) = (kw(\tilde{x},\tau), qz(\tilde{x},\tau))$$

is a solution of (1.1), where k, q are described in (1.4), (1.5). We will look for solution of (1.6) with the form $(w(\tilde{x},\tau), z(\tilde{x},\tau)) = (w(\xi), z(\xi))$, $\xi = \tilde{x}+c\tau$, where $c > 0$, satisfying

$$\text{(1.51)} \qquad \begin{cases} \lim_{\xi\to-\infty}(w(\xi), z(\xi)) = (0, \dfrac{1}{1+\epsilon_2}), \\ \lim_{\xi\to\infty}(w(\xi), z(\xi)) \ = (1,0). \end{cases}$$

Relating back to (1.6), we are thus looking for solutions of

$$\text{(1.52)} \qquad \begin{aligned} -w_{\xi\xi} + cw_\xi &= w(1-w-rz), \\ -z_{\xi\xi} + cz_\xi &= z(\epsilon_1 - bw - \epsilon_1(1+\epsilon_2)z), \end{aligned} \qquad -\infty < \xi < \infty$$

with boundary conditions (1.51).

Theorem 1.4 (Exclusive Traveling Waves for Competing Species). *Under hypotheses [P2], [P3] and*

[P4] $\qquad \frac{b_2}{b_1} + \frac{c_1 a_2}{c_2 a_1} \le 1 + \frac{a_2}{a_1},$

system (1.1) has a traveling wave solution of the form:

$$\text{(1.53)} \qquad (u(x,t),v(x,t)) = (kw(\sqrt{\frac{a_1}{d}}x + ca_1 t), qz(\sqrt{\frac{a_1}{d}}x + ca_1 t))$$

for any $c \ge 2\sqrt{1-\frac{c_1a_2}{c_2a_1}}$. *Here* (w,z) *is a function of one single variable* ξ *satisfying (1.52) for* $-\infty < \xi < \infty$, *and (1.51) as* $\xi \to \pm\infty$. *Moreover,* $w(\xi)$ *and* $z(\xi)$ *are positive monotonic functions for* $-\infty < \xi < \infty$.

Remark 1.4. Note that in Theorem 1.4, we have

$$\text{(1.54)} \qquad \begin{aligned} \lim_{t\to-\infty}(u(x,t),v(x,t)) &= (0, \tfrac{a_2}{c_2}), \\ \lim_{t\to+\infty}(u(x,t),v(x,t)) &= (\tfrac{a_1}{b_1}, 0). \end{aligned}$$

Proof. First, we note that from (1.7), no matter how q is chosen in the interval described in (1.5), we always have the relation:

$$1 - \frac{r}{1+\epsilon_2} = 1 - \frac{c_1 a_2}{a_1 c_2}.$$

We begin by transforming (1.11) or (1.52) into (1.16) by the change of variable (1.15) as in the proof of Theorem 1.1. The system (1.16) is now monotone for $0 \le u_1, 0 \le u_2 \le \frac{1}{1+\epsilon_2}$. We now modify the proof of Theorem 1.1 by defining $Y(\xi)$ to be the increasing function satisfying:

$$-Y_{\xi\xi} + cY_\xi = (1 - \frac{r}{1+\epsilon_2})Y(1-Y) \tag{1.55}$$

for $-\infty < \xi < \infty$ with $\lim_{\xi\to-\infty} Y(\xi) = 0$, $\lim_{\xi\to\infty} Y(\xi) = 1$. Such solution $Y(\xi)$ exists for $c \ge 2\sqrt{1 - \frac{c_1 a_2}{a_1 c_2}}$, by the choice of $1 - \frac{r}{1+\epsilon_2}$ as explained above and the results in [106]. We now construct a new pair of coupled upper solutions for the system (1.16). Define

$$\bar{u}_1(\xi) = Y(\xi)\ ,\ \bar{u}_2(\xi) = \frac{1}{1+\epsilon_2}Y(\xi). \tag{1.56}$$

For $0 \le u_2 \le \bar{u}_2(\xi)$, we readily verify that

$$\begin{aligned}
&-\bar{u}_1''(\xi) + c\bar{u}_1'(\xi) - \bar{u}_1(\frac{1+\epsilon_2 - r}{1+\epsilon_2} - \bar{u}_1 + ru_2)\\
&= (1 - \frac{r}{1+\epsilon_2})\bar{u}_1(1-\bar{u}_1) - \bar{u}_1(1 - \frac{r}{1+\epsilon_2} - \bar{u}_1 + ru_2)\\
&= \bar{u}_1\left\{(1 - \frac{r}{1+\epsilon_2})(1-\bar{u}_1) - 1 + \frac{r}{1+\epsilon_2} + \bar{u}_1 - ru_2\right\}\\
&= \bar{u}_1\left\{\frac{r}{1+\epsilon_2}\bar{u}_1 - ru_2\right\}\\
&\ge \bar{u}_1\left\{\frac{r}{1+\epsilon_2}\bar{u}_1 - r\bar{u}_2\right\} \equiv 0
\end{aligned} \tag{1.57}$$

for all $-\infty < \xi < \infty$. For $0 \le u_1 \le \bar{u}_1(\xi)$, one verifies

$$\begin{aligned}
&-\bar{u}_2''(\xi) + c\bar{u}_2'(\xi) - (\frac{1}{1+\epsilon_2} - \bar{u}_2)(bu_1 - \epsilon_1(1+\epsilon_2)\bar{u}_2)\\
&= \frac{1}{1+\epsilon_2}\{-Y'' + cY' - (1-Y)(bu_1 - \epsilon_1 Y)\}\\
&= \frac{1}{1+\epsilon_2}\left\{(1 - \frac{r}{1+\epsilon_2})Y(1-Y) + (1-Y)(\epsilon_1 Y - bu_1)\right\}\\
&\ge \frac{1}{1+\epsilon_2}(1-Y)\left\{(1 - \frac{r}{1+\epsilon_2})Y + \epsilon_1 Y - bY\right\}\\
&= \frac{1}{1+\epsilon_2}(1-Y)Y\left\{1 - \frac{r}{1+\epsilon_2} + \epsilon_1 - b\right\} \ge 0.
\end{aligned} \tag{1.58}$$

The last inequality is true provided $b < 1 - \frac{r}{1+\epsilon_2} + \epsilon_1$, which is valid due to hypothesis [P4].

We next construct quasi-lower solution for the system (1.16) in the sense described in Definition 1.1. Actually the quasi-lower solution will even be in $C^2(R)$, and is really a lower solution. We define $Z(\xi)$ to be the increasing function satisfying:

$$-Z_{\xi\xi} + cZ_\xi = (1 - \frac{r}{1+\epsilon_2})Z(1 - \frac{1 - \frac{lr}{1+\epsilon_2}}{1 - \frac{r}{1+\epsilon_2}}Z) \tag{1.59}$$

for $-\infty < \xi < \infty$ with $\lim_{\xi\to-\infty} Z(\xi) = 0$, $\lim_{\xi\to\infty} Z(\xi) = \frac{1-\frac{r}{1+\epsilon_2}}{1-\frac{lr}{1+\epsilon_2}}$. Here l is some number in the interval $(0,1)$ to be determined. One can readily verify that the solutions of (1.55) and (1.59) are related by the following

$$Z(\xi) = \frac{1 - \frac{r}{1+\epsilon_2}}{1 - \frac{lr}{1+\epsilon_2}}Y(\xi), \quad \xi \in R. \tag{1.60}$$

Since $0 < l < 1$, we have

$$Z(\xi) < Y(\xi), \quad \xi \in R. \tag{1.61}$$

We define the quasi-lower solutions of (1.16) by setting

$$\tilde{u}_1 = Z, \quad \tilde{u}_2 = \frac{l}{1+\epsilon_2}Z, \tag{1.62}$$

where $l \in (0,1)$ is to be determined. We readily verify that they satisfy

$$\begin{aligned} &-\tilde{u}_1''(\xi) + c\tilde{u}_1'(\xi) - \tilde{u}_1(\frac{1+\epsilon_2 - r}{1+\epsilon_2} - \tilde{u}_1 + r\tilde{u}_2) \\ &= Z\left\{(1 - \frac{r}{1+\epsilon_2}) - (1 - \frac{lr}{1+\epsilon_2})Z - (\frac{1+\epsilon_2 - r}{1+\epsilon_2}) + Z - \frac{rl}{1+\epsilon_2}Z\right\} \\ &= 0. \end{aligned} \tag{1.63}$$

Moreover, we have

(1.64)

$$
\begin{aligned}
&-\tilde{u}_2''(\xi) + c\tilde{u}_2'(\xi) - (\frac{1}{1+\epsilon_2} - \tilde{u}_2)(b\tilde{u}_1 - \epsilon_1(1+\epsilon_2)\tilde{u}_2) \\
&= \frac{l}{1+\epsilon_2} Z \left\{ 1 - \frac{r}{1+\epsilon_2} - (1 - \frac{lr}{1+\epsilon_2}) Z \right\} - (\frac{1}{1+\epsilon_2} - \frac{l}{1+\epsilon_2} Z) \{bZ - \epsilon_1 l Z\} \\
&= \frac{l}{1+\epsilon_2} Z \left\{ (1-Z) - r(\frac{1}{1+\epsilon_2} - \frac{l}{1+\epsilon_2} Z) \right\} \\
&\quad -(\frac{1}{1+\epsilon_2} - \frac{l}{1+\epsilon_2} Z) \{bZ - \epsilon_1 l Z\} \\
&\leq (\frac{1}{1+\epsilon_2} - \frac{l}{1+\epsilon_2} Z) \left\{ -\frac{rl}{1+\epsilon_2} Z - bZ + \epsilon_1 l Z \right\} + (1 - lZ)\frac{lZ}{1+\epsilon_2} \\
&= (\frac{1}{1+\epsilon_2} - \frac{l}{1+\epsilon_2} Z) \left\{ -\frac{rl}{1+\epsilon_2} - b + \epsilon_1 l + l \right\} Z \\
&\leq 0.
\end{aligned}
$$

The last inequality is valid provided that

$$
0 < l \leq min.\{1, \frac{b}{1+\epsilon_1 - \frac{r}{1+\epsilon_2}}\}. \tag{1.65}
$$

Now we apply a slight variant of Theorem 1.3, with

$$
\begin{cases}
\tilde{F}_1(\rho_1, \rho_2) = \rho_1(\frac{1+\epsilon_2 - r}{1+\epsilon_2} - \rho_1 + r\rho_2), \\
\tilde{F}_2(\rho_1, \rho_2) = (\frac{1}{1+\epsilon_2} - \rho_2)(b\rho_1 - \epsilon_1(1+\epsilon_2)\rho_2)
\end{cases} \tag{1.66}
$$

for equation (1.28), $\beta = diag.(\delta, \delta)$, δ any positive constant for (1.29), $\mathbf{K} = (1, \frac{1}{1+\epsilon_2})$, $\tilde{\rho}(t) = (Z(t), \frac{l}{1+\epsilon_2} Z(t))$, and $\bar{\rho}(t) = (Y(t), \frac{1}{1+\epsilon_2} Y(t)), t \in R$, as described above. A slight variant is needed because [A1] is not true, with the additional zero of $\tilde{F}$ at $(0, \frac{1}{1+\epsilon_2})$ between $\mathbf{0}$ and $\mathbf{K}$. By means of the iterative procedure in the proof of Theorem 1.3 we obtain a solution of (1.16) $\phi(t) := (u_1(t), u_2(t))$, satisfying the inequality (1.43). From the comparison argument with $\bar{\rho}(t)$ in the proof of Theorem 1.3, we have

$$
\lim_{t \to -\infty} u_1(t) = \lim_{t \to -\infty} u_2(t) = 0.
$$

Again, by the limit and comparison argument in the proof of Theorem 1.3, we obtain

$$
\lim_{t \to \infty} u_i(t) = Q_i, \quad i = 1, 2, \quad \text{and} \quad \tilde{F}(Q_1, Q_2) = 0, \tag{1.67}
$$

where

$$0 < \lim_{t\to\infty} Z(t) \le Q_1 \le K_1 = 1, \quad 0 < \lim_{t\to\infty} \frac{l}{1+\epsilon_2} Z(t) \le Q_2 \le K_2 = \frac{1}{1+\epsilon_2}. \tag{1.68}$$

From (1.66), (1.67) and (1.68), we conclude that we must have

$$Q_1 = K_1 = 1, \quad Q_2 = K_2 = \frac{1}{1+\epsilon_2}. \tag{1.69}$$

Relating the solutions of (1.16), (1.11) and (1.1), we obtain the assertions of the theorem.

Example 1.1. Note that by hypotheses [P2] and [P3], we have $1 - \frac{c_1 a_2}{a_1 c_2} > 0$. In case [P2] and [P3] are satisfied for the system (1.1), the additional hypothesis [P4] will always be satisfied by reducing all the coefficients (except d) of the second equation by the same factor, while holding the first equation unchanged. As an indication, the system

$$u_t = du_{xx} + u(5 - 3u - 3v), \quad v_t = dv_{xx} + v(0.15 - 0.1u - 0.1v)$$

satisfies all the hypotheses [P2],[P3] and [$P4$]. From Theorem 1.4, we obtain traveling waves with wave speed $c \ge 2/\sqrt{10}$.

From the comparison relation (1.43) in Theorem 1.3, used in proving Theorem 1.4, we actually obtain the following result with more accurate description of the asymptotic behavior of the traveling wave solution for system (1.1) at $\pm\infty$.

Corollary 1.5 (Asymptotic Rates of the Traveling Waves). *Under hypotheses [P2], [P3] and [P4], system (1.1) has a traveling wave solution of the form (1.53) for any* $c \ge 2\sqrt{1 - \frac{c_1 a_2}{c_2 a_1}}$. *Here* (w, z) *is a function of one single variable* ξ *satisfying (1.11) or (1.52) for* $-\infty < \xi < \infty$, *and (1.51) as* $\xi \to \pm\infty$. *Moreover,* $w(\xi)$ *and* $z(\xi)$ *are positive monotonic functions for* $-\infty < \xi < \infty$, *satisfying*

$$\begin{aligned} &0 < Z(\xi) \le w(\xi) \le Y(\xi), \\ &0 < \tfrac{l}{1+\epsilon_2} Z(\xi) \le \tfrac{1}{1+\epsilon_2} - z(\xi) \le \tfrac{1}{1+\epsilon_2} Y(\xi). \end{aligned} \tag{1.70}$$

Here $Y(\xi)$ *is the solution of (1.55) with* $\lim_{\xi\to-\infty} Y(\xi) = 0, \lim_{\xi\to\infty} Y(\xi) = 1$; *and* $Z(\xi)$ *is a solution of (1.59) with* $\lim_{\xi\to-\infty} Z(\xi) = 0$, $\lim_{\xi\to\infty} Z(\xi) = \frac{1-\frac{r}{1+\epsilon_2}}{1-\frac{lr}{1+\epsilon_2}}$, *with* l *satisfying (1.65).*

Remark 1.5. Recently, correction for [234] is published in [235]. It is found that the "lower" solution require more stringent smoothness than just continuity (i.e. C^0), as initially described in [234]. Such correction has been included in the presentation for Theorem 1.3 in Part B above. Note that we assume in Theorem 1.3 that the "quasi-lower" solution is in C^1. Unfortunately, the original result in [234] was used to obtain upper and lower bound of the traveling wave solutions in [136] and [90]. Such bounds lead to asymptotic estimate of the traveling wave at $\pm\infty$, and eventually to the stability (with a shift) result in an appropriate weighted Banach space in [136]. The needed correction for [234] would necessitate modification of proof in [136] in order to obtain stability (with a shift) result. In order to retain the upper and lower estimate and the stability result just mentioned for [136], it suffices to re-construct a C^1 lower-solution or one with some additional condition of its derivative at $t = 0$ as described in [235]. So far this problem remains unsolved. Note that although we have C^2 upper-lower solutions above in Theorem 1.4 and Corollary 1.5, Part B, the hypothesis [P4] in Theorem 1.4 is different from the additional hypothesis called [H3] in Corollary 2.3 of [136], used for stability analysis of traveling wave in [136]. Theorem 1.4 is thus an adjustment of the result in [136]. On the other hand, the theory presented in Part A does not use any nonsmooth lower solution, and no correction from the original is necessary.

5.2 Positive Solutions for Systems of Wave Equations and Their Stabilities

Nonlinear hyperbolic systems of partial differential equations arise in the study of physics and mechanics as the coupled sine-Gordon equations (see e.g. Temam [217]). In Salvatori and Vitillaro [199], recent study is made by energy method concerning the decay of the solutions for similar damped systems. This section considers a system of nonlinear damped wave equations with symmetric linear part. This can be called a coupled system of many vibrating strings. We will find that a positive steady-state bifurcates from the trivial solution as a parameter changes. The spectrum of the linearized operator is studied. Then the stability of the positive steady-state is considered as a solution of the nonlinear hyperbolic system. Asymptotic stability results are found for solutions in R^N, for some $N \geq 1$. Bifurcation methods are used to find the steady-states; semigroup methods and local bifurcation analysis are used to study stability. Stability results are obtained although the semigroup is not analytic.

More precisely, we consider the positive steady-state solutions and their stabilities for the following nonlinear system of damped wave equations:

$$
(2.1) \qquad \begin{cases} u_{tt} + \beta u_t = \Delta u + \lambda B(x)u + \lambda G(u)u & \text{for (x,t)} \in \Omega \times [0,\infty), \\ u = 0 & \text{for (x,t)} \in \partial\Omega \times [0,\infty). \end{cases}
$$

Here, $u = \text{col.}(u_1, \ldots, u_n), \beta > 0, \lambda$ is a real parameter, $G = [g_{ij}]$ is an $n \times n$ matrix function with each entry in $C^2(R^n)$ and $g_{ij}(0, \ldots, 0) = 0$ for $i, j = 1, \ldots, n$. Ω is a bounded domain in R^N, with boundary $\partial\Omega$ of class $C^{2+\mu}, 0 < \mu < 1$. The $n \times n$ matrix $B(x) = [b_{ij}(x)]$ is assumed to satisfy

[$H1$] $B(x)$ is a real symmetric matrix, with each entry $b_{ij}(x)$ in $C^\mu(\bar{\Omega})$, $0 < \mu < 1$, and non-negative in $\bar{\Omega}$.

[$H2$] There is a permutation $\{r_1, r_2, \ldots, r_n\}$ of $\{1, 2, \ldots, n\}$ such that $b_{r_1 r_2} \not\equiv 0, b_{r_2 r_3} \not\equiv 0, ..., b_{r_{n-1} r_n} \not\equiv 0$, and $b_{r_n r_1} \not\equiv 0$ in $\bar{\Omega}$.

(The hypothesis [$H1$] will be modified to [$H1^*$] below such that the diagonal entries of $B(x)$ may change sign). Nonlinear hyperbolic systems of similar nature arise in the study of physics and mechanics as the coupled sine-Gordon equations. This section follows the presentation in Leung [127]. The development can be considered as an extension of the scalar results in Webb [230], where both $n = N = 1$, to systems with symmetric linear part.

We first show that a positive steady-state bifurcates from the trivial solution as the parameter λ changes. Moreover, the corresponding elliptic system is linearized at the positive steady-state and the spectrum of the linear operator is investigated under appropriate conditions on the nonlinear terms. Next, the stability of the positive steady-state is considered as a solution of hyperbolic problem (2.1). Note that the corresponding semigroup here is not analytic, and the stability theorem in Henry [84] (i.e. Theorem A4-11 in Chapter 6) does not apply. The stability results here will be shown to be applicable to solutions in R^1 by means of Morrey's inequality, and to solutions in R^3 or R^4 by means of Gagliardo-Nirenberg-Sobolev inequality (cf. Evans [57]). Finally, positive steady-state is found for $\lambda = 1$ by means of global bifurcation, under further assumptions.

For convenience, we will adopt the following conventions. Let $E := \{w = \text{col.}(w_1, \ldots, w_n) : w_i \in C^1(\bar{\Omega}), w_i = 0 \text{ on } \partial\Omega, i = 1, \ldots, n\}$, $P := \{w = \text{col.}(w_1, \ldots, w_n) \in E : w_i \geq 0 \text{ in } \bar{\Omega}, i = 1, \ldots, n\}$, and $Y := \{w = \text{col.}(w_1, \ldots, w_n) : w_i \in C^{2+\alpha}(\bar{\Omega}), w_i = 0 \text{ on } \partial\Omega, i = 1, \ldots, n\}$.

We first proceed to show that as the parameter λ passes a certain eigenvalue of the linearized system, a positive steady-state bifurcates from the trivial solution, using the theory of Crandall and Rabinowitz [33] Then, we will analyze the spectrum of the elliptic part of the operator equation linearized at the positive steady-state. The spectral analysis is quite different from those in Chapter 2, because the system (2.1) here involves two time derivatives.

For convenience, we define the operator L_q with n components as: $L_q \equiv (-\Delta + q_1(x), \ldots, -\Delta + q_n(x))$, where $q_i(x), i = 1, \ldots, n$ are any non-negative functions in $C^\mu(\bar{\Omega})$.

Theorem 2.1. *Under hypotheses [H1] and [H2], there exists* $(\lambda_0, u^0) \in R \times Y$, *such that*

$$L_q[u^0] = \lambda_0 B u^0 \text{ in } \Omega, \;\; u^0 = 0 \text{ on } \partial\Omega, \tag{2.2}$$

with $\lambda_0 > 0$, *and each component* $u_i^0 > 0$ *in* $\Omega, \partial u_i^0/\partial\nu < 0$ *on* $\partial\Omega$ *for* $i =$ *1,...,n. Furthermore, the eigenfunction corresponding to the eigenvalue* $1/\lambda_0$ *for the operator* $L_q^{-1}B : [C^1(\bar{\Omega})]^n \to [C^1(\bar{\Omega})]^n$ *is unique up to a multiple. Also, the number* $\lambda = \lambda_0$ *is the unique positive number so that the problem* $u = \lambda L_q^{-1}Bu$ *has a nontrivial non-negative solution for* $u \in P$.

Proof. The operator $L_q^{-1}B$ is completely continuous and positive with respect to the cone P. By means of Theorem 2.5 in Krasnosel'skii [108] (i.e. Theorem A3-10 in Chapter 6) we can obtain a nontrivial $u^0 \in P$ such that $L_q^{-1}Bu^0 = \rho_0 u^0$ for some $\rho_0 > 0$ (i.e. (2.2), with $\lambda_0 = 1/\rho_0$). By the nonnegativity of all entries of B and hypothesis $[H2]$, we show with maximum principle that $u_i^0 > 0$ in Ω and $\partial u_i^0/\partial\nu < 0$ on $\partial\Omega$ for each $i = 1, \ldots, n$. By using a comparison principle for systems (Lemma 6.1 in Section 2.6), we can show in the same way as Theorem 6.1 in Section 2.6 that eigenfunctions corresponding to the eigenvalue $1/\lambda_0$ for the operator $L_q^{-1}B$ is unique up to a multiple, and λ_0 is the unique positive number with a non-negative eigenfunction as described above.

For a more general situation, we will allow the diagonal entries of $B(x)$ to change sign. For $i = 1, \ldots, n$, let $b_{ii}(x) = b_{ii}^+(x) + b_{ii}^-(x)$, where $b_{ii}^+(x) = \max\{b_{ii}(x), 0\}$ and $b_{ii}^-(x) = \min\{b_{ii}(x), 0\}$. We introduce following hypothesis:

[$H1^*$] $B(x)$ is a real symmetric matrix with each entry $b_{ij}(x)$ in $C^\mu(\bar{\Omega})$. For all $i \neq j$, b_{ij} is non-negative in $\bar{\Omega}$, and there exists an integer $k, 1 \leq k \leq n$, such that $b_{kk}^+(x) \not\equiv 0$ in $\bar{\Omega}$.

The following is an extension for Theorem 2.1.

Theorem 2.2. *Under hypotheses* [$H1^*$] *and* [$H2$], *there exists* $(\hat{\lambda}_0, v^0) \in R \times Y$ *such that*

$$-\Delta v^0 = \hat{\lambda}_0 B(x) v^0 \text{ in } \Omega, \;\; v^0 = 0 \text{ on } \partial\Omega, \tag{2.3}$$

with $\hat{\lambda}_0 > 0$, *each component* $v_i^0 > 0$ *in* Ω *and* $\partial v_i^0/\partial\nu < 0$ *on* $\partial\Omega$ *for* $i =$ $1, ..., n$. *Furthermore, the eigenfunction corresponding to the eigenvalue* $1/\hat{\lambda}_0$ *for the operator* $(-\Delta)^{-1}B : [C^1(\bar{\Omega})]^n \to [C^1(\bar{\Omega})]^n$ *is unique up to a multiple..*

Proof. We will use Theorem 2.1 to prove this theorem. For convenience, define $\tilde{B} = \{\tilde{b}_{ij}(x)\}$ to be the $n \times n$ matrix function on $\bar{\Omega}$ as follows:

$$\tilde{b}_{ij}(x) = b_{ij}(x) \text{ if } i \neq j, \quad \tilde{b}_{ii}(x) = b_{ii}^{+}(x)$$

for $i, j = 1, \ldots, n, x \in \bar{\Omega}$. For each $\lambda \geq 0$, define the n component vector operator

$$\tilde{L}_{\lambda} \equiv (-\Delta - \lambda b_{11}^{-}(x), -\Delta - \lambda b_{22}^{-}(x), \ldots, -\Delta - \lambda b_{nn}^{-}(x)),$$

and consider the eigenvalue problem

$$\tilde{L}_{\lambda} u = \rho \tilde{B} u \text{ in } \Omega, \quad u|_{\partial\Omega} = 0, \tag{2.4}$$

with eigenvalue ρ. Since $\tilde{B}$ satisfies the conditions in Theorem 2.1, the problem (2.4) has a unique positive solution $\rho = \hat{\rho}(\lambda)$, with corresponding eigenfunction u_{λ} whose components are all positive in Ω.

To proceed with the proof, we need the following two lemmas.

Lemma 2.1. *Under the hypotheses of Theorem 2.2, the function $\hat{\rho}(\lambda)$ is bounded for all $\lambda \in [0, \infty)$.*

Proof. By hypothesis $[H1^*]$, there is an open set D in Ω with its closure in Ω, such that $b_{kk} = b_{kk}^{+}$ in D for some k. Let Φ be a nontrivial C^{∞} function with compact support contained in D. We clearly have $\int_D b_{kk} \Phi^2 dx > 0$. Let u_{λ} be as described above, and set $w_{\lambda}(x) = \ln(u_{\lambda})_k(x)$ for $x \in \Omega$. Thus we have in D that

$$-\Delta w_{\lambda} - \sum_{i=1}^{N} (\partial(w_{\lambda})/\partial x_i)^2 = [1/(u_{\lambda})_k][-\Delta(u_{\lambda})_k - b_{kk}^{-}(x)(u_{\lambda})_k]$$

$$= [\hat{\rho}(\lambda)/(u_{\lambda})_k] \sum_{j=1}^{n} \tilde{b}_{kj}(x)(u_{\lambda})_j \geq b_{kk}(x)\hat{\rho}(\lambda).$$

Multiplying by Φ^2 and integrating over D by parts on the left, we obtain

$$\int_D < \Phi \nabla w_{\lambda}, 2\nabla\Phi - \Phi \nabla w_{\lambda} > dx \;\geq\; \hat{\rho}(\lambda) \int_D b_{kk} \Phi^2 \text{dx}.$$

From this we deduce

$$\frac{\int_D < \nabla\Phi, \nabla\Phi > dx}{\int_D b_{kk} \Phi^2 \, dx} \;\geq\; \hat{\rho}(\lambda) > 0$$

for all $\lambda \in [0, \infty)$.

Lemma 2.2. *Under the hypotheses of Theorem 2.2, the function $\hat{\rho}(\lambda)$ defined for problem (2.4) is continuous for $\lambda \in [0, \infty)$.*

The proof of the two lemmas above are similar to that for Lemmas 6.2 and 6.3 in Section 2.6. Further details will thus be omitted.

To complete the proof of Theorem 2.2, we solve the equation $\hat{\rho}(\lambda) - \lambda = 0$ for $\lambda = \hat{\lambda}_0$ as described in (2.3). The solution of the equation must exist due to the properties of $\hat{\rho}(\lambda)$ described in Lemmas 2.1 and 2.2. The uniqueness of eigenfunction up to a multiple for the eigenvalue $1/\hat{\lambda}_0$ follows from (2.4) and Theorem 2.1.

Note that the pair $(\hat{\lambda}_0, v^0)$ in Theorem 2.2 also satisfies

$$-\Delta v^0 = \hat{\lambda}_0 B^T(x) v^0, \tag{2.5}$$

since $B = B^T$. For convenience, we define an operator $F : R^+ \times E \to E$ by

$$F(\lambda, u) := u - \lambda(-\Delta)^{-1}[B + G(u)]u \quad \text{for } (\lambda, u) \in R^+ \times E. \tag{2.6}$$

The steady-state solution of (2.1) can be written as

$$F(\lambda, u) = 0. \tag{2.7}$$

Defining operators

$$\begin{aligned} L_0 : E \to E \quad & \text{by} \quad L_0 := I - \hat{\lambda}_0(-\Delta)^{-1}B, \\ L_1 : E \to E \quad & \text{by} \quad L_1 := \Delta^{-1}B, \ \text{ and} \\ \tilde{G} : R^+ \times E \to E \quad & \text{by} \quad \tilde{G}(\lambda, u) := \lambda\Delta^{-1}[G(u)u], \end{aligned}$$

equation (2.7) becomes

$$L_0 u + (\lambda - \hat{\lambda}_0) L_1 u + \tilde{G}(\lambda, u) = 0, \ \text{for } (\lambda, u) \in R^+ \times E. \tag{2.8}$$

As in Section 2.6, we readily obtain the following.

Lemma 2.3. *Under the hypotheses of Theorem 2.2, the null space and range of* L_0*, denoted respectively by* $N(L_0)$ *and* $R(L_0)$*, satisfy:*
(i) $N(L_0)$ *is one-dimensional, spanned by* v^0*;*
(ii) $\dim[E/R(L_0)] = 1$;
(iii) $L_1 v^0 \notin R(L_0)$.

Applying the bifurcation theorem in Crandall and Rabinowitz [33] or Theorem A1-3 in Chapter 6, we obtain the following.

Theorem 2.3 (Bifurcation of Steady-States). *Assume hypotheses [H1*], [H2] and each entry of G is in* $C^2(R^n)$. *Then the point* $(\hat{\lambda}_0, 0)$ *is a bifurcation point for the problem (2.7). Moreover, there exists a* $\delta > 0$ *and a* C^1*-curve* $(\hat{\lambda}(s), \hat{\phi}(s)) : (-\delta, \delta) \to R \times E$ *with* $\hat{\lambda}(0) = \hat{\lambda}_0, \hat{\phi}(0) = 0$, *such that in a neighborhood of* $(\hat{\lambda}_0, 0)$, *any solution of (2.7) is either of the form* $(\lambda, 0)$ *or on the curve* $(\hat{\lambda}(s), s[v^0 + \hat{\phi}(s)])$ *for* $|s| < \delta$, *where* $s[v^0 + \hat{\phi}(s)] > 0$ *in* Ω. *(Here,* $(\hat{\lambda}_0, v^0)$ *is described in Theorem 2.2).*

We next investigate the spectrum of the linearized equation at the solution on the bifurcation curve described in Theorem 2.3. We observe that by further application of the theory in Crandall and Rabinowitz [33], we can assert that there exists $\delta_1 \in (0, \delta)$ and two functions $(\sigma(.), z(.)) : (\hat{\lambda}_0 - \delta_1, \hat{\lambda}_0 + \delta_1) \to R \times E$, and $(\eta(.), h(.)) : [0, \delta_1) \to R \times E$ with $(\sigma(\hat{\lambda}_0), z(\hat{\lambda}_0)) = (\eta(0), h(0)) = (0, v^0)$ such that

$$D_2F(\lambda, 0)z(\lambda) = \sigma(\lambda)\Delta^{-1}(z(\lambda)), \text{ and}$$

$$D_2F(\hat{\lambda}(s), s[v^0 + \hat{\phi}(s)])h(s) = \eta(s)\Delta^{-1}(h(s)).$$

Here, $\sigma(\lambda)$ and $\eta(s)$ are respectively Δ^{-1}-simple eigenvalues with eigenfunctions $z(\lambda)$ and $h(s)$. The theory in [33] leads to the following.

Lemma 2.4. *Assume all the hypotheses in Theorem 2.3. There exists* $\rho > 0$ *such that for each* $s \in [0, \delta_1)$, *there is a unique (real) eigenvalue* $\eta(s)$ *for the linear operator*

$$F_s^* := \Delta D_2F(\hat{\lambda}(s), s[v^0 + \hat{\phi}(s)]) : Y \to [C^\mu(\bar{\Omega})]^n \tag{2.9}$$

satisfying $|\eta(s)| < \rho$ *with eigenfunction* $h(s) \in$ Y. *That is,*

$$F_s^* h(s) \equiv \Delta[h(s)] + \hat{\lambda}(s)[B + G(u_s) + G_u(u_s)u_s]h(s) = \eta(s)h(s), \tag{2.10}$$

where $u_s := s[v^0 + \hat{\phi}(s)]$. *Here* $G_u(u_s)u_s$ *denotes the* $n \times n$ *matrix whose i-th column is* $\{\partial G/\partial u_i\}(u_s)u_s$.

In order to determine the signs of the real part of the eigenvalues we need additional assumptions.

Lemma 2.5. *Assume all the hypotheses of Theorem 2.3. Suppose further that*

[H3] $\{\partial g_{ij}/\partial u_k\}(0) \le 0$ *for all* $i, j, k = 1, ..., n$ *with at least one inequality being strict.*

Then the function $\hat{\lambda}(s)$ *satisfies* $\hat{\lambda}'(0) > 0$.

Proof. Theorem 2.3 asserts that $\hat{\lambda}'(0)$ exists; moreover, for $s \in [0, \delta)$, we have

$$\Delta s[v^0 + \hat{\phi}(s)] + \hat{\lambda}(s)B(x)s[v^0 + \hat{\phi}(s)] + \hat{\lambda}(s)G(s[v^0 + \hat{\phi}(s])s[v^0 + \hat{\phi}(s)] = 0.$$

Dividing by s, then differentiating with respect to s and setting $s = 0$, we obtain

$$\Delta\hat{\phi}'(0) + \hat{\lambda}'(0)B(x)v^0 + \hat{\lambda}_0 B(x)\hat{\phi}'(0) + \hat{\lambda}_0 \frac{d}{ds}G(s[v^0 + \hat{\phi}(s)])|_{s=0}v^0 = 0.$$

Multiplying by $(v^0)^T$ and integrating by parts over Ω, we find

$$\hat{\lambda}'(0) = \frac{-\hat{\lambda}_0 \int_\Omega (v^0)^T \frac{d}{ds}G(s[v^0 + \hat{\phi}(s)])|_{s=0}v^0 dx}{\int_\Omega (v^0)^T B v^0 dx}$$

$$= \frac{-\hat{\lambda}_0 \int_\Omega (v^0)^T [\sum_{i=1}^n v_i^0 \frac{\partial G}{\partial u_i}(0)]v^0 dx}{(1/\hat{\lambda}_0)\int_\Omega |\nabla v^0|^2 dx} > 0.$$

The last inequality is due to hypothesis $[H3]$.

Lemma 2.6. *Under all the hypotheses of Theorem 2.3. The function $\sigma(\lambda)$ satisfies $\sigma'(\hat{\lambda}_0) > 0$.*

The proof of the last lemma is similar to that for Lemma 6.3 in Section 2.6, while using the additional fact that $B = B^T$ here. The details will be omitted.

Lemma 2.7. *Under all the hypotheses of Theorem 2.3, and $[H3]$, there exists $\delta_2 \in (0, \delta_1)$ such that $\eta(s) < 0$ for all $s \in (0, \delta_2)$.*

Proof. From Theorem 1.16 in [33], we find $-s\hat{\lambda}'(s)\sigma'(\hat{\lambda}_0)$ and $\eta(s)$ have the same sign for $s > 0$ near 0. The conclusion follows from Lemmas 2.5 and 2.6 above.

The linearized eigenvalue problem for (2.7) at the bifurcating solution $u = s[v^0 + \hat{\phi}(s)]$ is precisely (2.10). When $s = 0, \lambda = \hat{\lambda}(0) = \hat{\lambda}_0$, the eigenvalue problem corresponding to (2.10) becomes

$$\Delta[h] + \hat{\lambda}_0 Bh = \eta h, \qquad h \in Y, \tag{2.11}$$

where η is the eigenvalue. Under hypotheses $[H1^*]$ and $[H2]$, Theorem 2.2 asserts that $\eta = 0$ is an eigenvalue for (2.11) with positive eigenfunction. Using this property and the fact that the off-diagonal terms of B are all non-negative, we can show the following, as in Leung and Ortega [138].

Lemma 2.8. *Under hypotheses $[H1^*]$ and $[H2]$, all eigenvalues in equation (2.11) except $\eta = 0$ satisfies $Re(\eta) < -r$ for some positive number r.*

Theorem 2.4. *Under the hypotheses of Theorem 2.3 and $[H3]$, there exists a number $\delta^* \in (0, \delta)$ and a positive function $\hat{\eta}(s)$ for $s \in (0, \delta^*)$ such that the* *real parts of all the numbers in the point spectrum of the linear operator* F_s^* *are contained in the interval* $(-\infty, -\hat{\eta}(s))$, *for* $s \in (0, \delta^*)$. *(Here,* δ *is described in Theorem 2.3 and* F_s^* *is described in (2.9) in Lemma 2.4).*

The proof of Lemma 2.8 is the same as the proof of Lemma 2.8 in [138]; and the proof of Theorem 2.4, using the assertions in Lemmas 2.4 to 2.8 and perturbation arguments, is the same as Theorem 2.2 in [138]. The details are thus omitted. More details of proof of lemmas and theorems similar to those in this section can also be found in Section 2.6.

We are now ready to discuss the stability of the positive steady-states found in Theorem 2.3 with respect to the original system (2.1) of damped wave equations. For each $s \in (0, \delta^*)$, the function $u_s := s[v^0 + \hat{\phi}(s)]$ described in Theorem 2.3 and Lemma 2.4 can be considered as a steady-state solution of (2.1) with $\lambda = \hat{\lambda}(s)$. We now consider the time asymptotic stability of this steady state as a solution of the system of hyperbolic equations (2.1). We convert (2.1) into a first order system by letting

$$(2.12) \qquad \begin{cases} \partial u_i/\partial t = v_i \qquad i = 1, \dots, n, \\ \partial v_i/\partial t = \Delta u_i - \beta v_i + \lambda \sum_{j=1}^n [b_{ij}(x) + g_{ij}(u_1, \dots, u_n)] u_j. \end{cases}$$

Let $J(u_s(x))$ be the linearization of $G(u)u$ (i.e. $\partial[G(u)u]/\partial u$) evaluated at $u = u_s$. For convenience, define $\bar{B}(u_s(x)) := B(x) + J(u_s(x))$ and let $\bar{b}_{ij}$ denotes the ij-th entry of $\bar{B}(u_s(x))$. The system (2.12) linearized at u_s and $\lambda = \hat{\lambda}(s)$ can be written in the form

$$(2.13) \qquad \partial \xi/\partial t = A_s \xi$$

where $\xi = \text{col.}(\hat{u}_1, \hat{v}_1, \dots, \hat{u}_n, \hat{v}_n)$, and A_s is the differential linear operator on the $2n$ components of ξ as follows:

$$A_s = \begin{bmatrix} A_{11}^s \cdots\cdots A_{1n}^s \\ \cdots\cdots\cdots\cdots \\ \cdots\cdots\cdots\cdots \\ \cdots\cdots\cdots\cdots \\ A_{n1}^s \cdots\cdots A_{nn}^s \end{bmatrix}, \text{ where } A_{ij}^s \text{ are } 2 \times 2 \text{ blocks,}$$

$$A_{ii}^s = \begin{bmatrix} 0 & 1 \\ \Delta + \hat{\lambda}(s)\bar{b}_{ii} & -\beta \end{bmatrix}, \quad A_{ij}^s = \begin{bmatrix} 0 & 0 \\ \hat{\lambda}(s)\bar{b}_{ij} & 0 \end{bmatrix}$$

for $i \neq j$, $i, j = 1, ..., n$. Here, each A_{ij}^s can be considered as an operator from

$[H^2(\Omega)\bigcap H_0^1(\Omega)]\times H_0^1(\Omega)$ into $H_0^1(\Omega)\times L^2(\Omega)$. Direct calculation shows that if θ is an eigenvalue of A_s, then $\theta^2+\beta\theta$ is an eigenvalue of F_s^*. That is, if

$$A_s\bar{\xi}=\theta\bar{\xi} \qquad \text{for some } \bar{\xi}\in([H^2(\Omega)\bigcap H_0^1(\Omega)]\times H_0^1(\Omega))^n, \text{ then}$$

$$\Delta\bar{u}+\hat{\lambda}(s)\bar{B}(u_s)\bar{u}=(\theta^2+\beta\theta)\bar{u}$$

where $\bar{\xi}=\mathrm{col}(\bar{u}_1,\bar{v}_1,\ldots,\bar{u}_n,\bar{v}_n)$ and $\bar{u}=\mathrm{col}(\bar{u}_1,\ldots,\bar{u}_n)$. In order to obtain the stability of the steady-state u_s, we will impose an additional assumption:

$[H4]$ $\bar{B}(u_s(x))$ is symmetric in $\bar{\Omega}$ for each $s\in(0,\delta^*)$.

By Theorem 2.4 and hypotheses $[H4]$, as an operator on Y, the eigenvalues of $F_s^*=\Delta+\hat{\lambda}(s)\bar{B}(u_s)$ are all strictly negative for $s\in(0,\delta^*)$. However, the eigenvalues of $\Delta+\hat{\lambda}(s)\bar{B}(u_s)$ are the same, as an operator on Y or $[H^2(\Omega)\bigcap H_0^1(\Omega)]^n$. Consequently, the eigenvalues θ of $A_s, s\in(0,\delta^*)$, satisfies

$$\theta^2+\beta\theta-\eta=0,$$

where η are negative real numbers. Thus, the real parts of θ are all strictly negative. The operator A_s generates a strongly continuous semigroup $T_s(t)$ on

$$X:=[H_0^1(\Omega)\times L^2(\Omega)]^n,$$

with domain $D(A_s):=([H^2(\Omega)\bigcap H_0^1(\Omega)]\times H_0^1(\Omega))^n$, (see e.g. [184]). Using the fact that the real parts of θ are negative, we now show the following stability result for the linearized system (2.13).

Theorem 2.5 (Stability of the Linearized Equations). *Assume hypotheses $[H1^*],[H2],[H3]$ and each entry of G is in $C^2(R^n)$. Suppose further that $\bar{B}(u_s(x))$ is symmetric as described in $[H4]$, then the semigroup of bounded linear operator $T_s(t), 0\le t<\infty$, on X (generated by A_s) satisfies*

$$\| T_s(t)\|\le Me^{-kt}$$

for some positive constants M and k, which may depend on s.

Proof. Let p be a large enough positive constant such that

$$y^T[\hat{\lambda}(s)\bar{B}(u_s(x))-pI]y\le-\omega(y^T.y)$$

for all $s\in(0,\delta^*), x\in\bar{\Omega}$, where ω is a positive constant. Define the operator $S:=-(\Delta+\hat{\lambda}\bar{B}-pI)^{-1}$ from $[L^2(\Omega)]^n$ into $[L^2(\Omega)]^n$ as follows. For each $f\in[L^2(\Omega)]^n$, let $g:=S[f]$ where $g\in[H_0^1(\Omega)]^n\cap[H^2(\Omega)]^n$, and $<<\nabla g,\nabla h>>$ $+\int_\Omega h^T(-\hat{\lambda}\bar{B}+pI)g\,dx=\int_\Omega h^T f\,dx$ for all $h\in[H_0^1(\Omega)]^n$. Here $<<\ >>$ is the real inner product in $[L^2(\Omega)]^{nN}$. The operator $S : [L^2(\Omega)]^n \to [L^2(\Omega)]^n$ is well-defined and compact by means of the Lax-Milgram Theorem and Sobolev imbedding (cf. [57]). From the symmetric property of $\bar{B}$, we can readily verify that the operator S is symmetric as follows. Let $f, h \in [L^2(\Omega)]^n$, $g = S[f], q = S[h]$, and $< \ >$ denotes the real inner product in $[L^2(\Omega)]^n$. Then we have

$$< Sf, h > = < g, -(\Delta + \hat{\lambda}\bar{B} - pI)q > \; = \; < -\Delta g, q > + < g, -(\hat{\lambda}\bar{B} - pI)q >$$

$$= < -\Delta g, q > + < -(\hat{\lambda}\bar{B} - pI)g, q > \; = \; < f, q > \; = \; < f, Sh > .$$

From the theory of compact symmetric operators, we assert that there exists a countable orthonormal basis of $[L^2(\Omega)]^n$ consisting of eigenfunctions of S. Direct computation shows that if η is an eigenvalue of $(\Delta + \hat{\lambda}\bar{B})$, then $1/(p - \eta)$ is an eigenvalue of S. Theorem 2.4 and the additional assumption that $\bar{B}$ is symmetric thus imply that all the eigenvalues of S are of the form $1/(p+\alpha_i)$, with $0 < \alpha_1 < \alpha_2 < \ldots$,(note that α_i depends on s). There exist corresponding eigenfunctions $\{\phi_m\}_{m=1}^{\infty}$, which form an orthonormal basis in $[L^2(\Omega)]^n$.

Let $[\tilde{H}_0^1(\Omega)]^n$ denotes the real Hilbert space of functions in $[H_0^1(\Omega)]^n$ with inner product $\tilde{J}[g, q] := << \nabla g, \nabla q >> \; + \int_\Omega q^T[-\hat{\lambda}\bar{B} + pI]g \, dx$. We verify that

$$\tilde{J}[\phi_j, \phi_k] \; = < -\Delta\phi_j, \phi_k > + \int_\Omega \phi_k^T[-\hat{\lambda}\bar{B} + pI]\phi_j \, dx \; = < (\alpha_j + p)\phi_j, \phi_k > .$$

Thus $\{\phi_m/(\alpha_m + p)^{1/2}\}_{m=1}^{\infty}$ form an orthonormal set in $[\tilde{H}_0^1(\Omega)]^n$. Further, the identity

$$\tilde{J}[\phi_j, g] = \int_\Omega g^T[-\Delta\phi_j - \hat{\lambda}\bar{B} + pI]\phi_j \, dx \; = < (\alpha_j + p)\phi_j, g >$$

implies that if $\tilde{J}[\phi_j, g] = 0$ for each $j = 1, 2 \ldots$, then we also have $< \phi_j, g >= 0$. This implies that $\{\phi_m/(\alpha_m + p)^{1/2}\}_{m=1}^{\infty}$ form an orthonormal basis in $[\tilde{H}_0^1(\Omega)]^n$. If $\hat{u} \in [\tilde{H}_0^1(\Omega)]^n$, we can assert that the series

$$\hat{u} = \sum_{m=1}^{\infty} < \hat{u}, \phi_m > \phi_m \qquad \text{converges in } [L^2(\Omega)]^n,$$

and further

$$\hat{u} = \textstyle\sum_{m=1}^{\infty} \tilde{J}[\hat{u}, \phi_m/(\alpha_m + p)^{1/2}]\{\phi_m/(\alpha_m + p)^{1/2}\}$$

$$= \textstyle\sum_{m=1}^{\infty}(\alpha_m + p)^{1/2} < \hat{u}, \phi_m > \{\phi_m/(\alpha_m + p)^{1/2}\} = \sum_{m=1}^{\infty} < \hat{u}, \phi_m > \phi_m$$

actually converges in $[\tilde{H}_0^1(\Omega)]^n$.

For $\phi \in D(A_s)$, let $\hat{w} = \hat{w}(t;\phi) := T_s(t)\phi \in D(A_s)$, we know from semigroup theory that $d\hat{w}/dt \in X$ for $t > 0$. That is, if we let $\hat{w} = (\hat{u}_1, \hat{v}_1, \ldots, \hat{u}_n, \hat{v}_n)$, and consider $\hat{u}_i$ as functions from $[0,\infty)$ into X, then $d\hat{u}_i/dt \in H_1^0(\Omega)$, and $d^2\hat{u}_i/dt^2 \in L^2$. This implies that if we let $\hat{u} = \text{col.}\ (\hat{u}_1, \ldots, \hat{u}_n)$, $\hat{v} = \text{col}\ (\hat{v}_1, \ldots, \hat{v}_n)$ and $r_m(t) := \ <\hat{u}(t), \phi_m>$, then the series

$$d\hat{u}/dt = \sum_{m=1}^{\infty} r'_m(t)\phi_m \qquad \text{converges in} \quad [H_1^0(\Omega)]^n, \text{ and}$$

$$d^2\hat{u}/dt^2 = \sum_{m=1}^{\infty} r''_m(t)\phi_m \qquad \text{converges in } [L^2(\Omega)]^n,$$

for $t > 0$, where $r'_m(t)$ and $r''_m(t)$ are the first and second derivatives of $r_m(t)$. From semigroup theory and the structure of the operator A_s, we find that $\hat{u}(t)$ satisfies

$$d^2\hat{u}/dt^2 + \beta d\hat{u}/dt = [\Delta + \hat{\lambda}(s)\bar{B}(u_s)]\hat{u}, \quad \text{for } t > 0.$$

Taking inner product with ϕ_m, we obtain

$$r''_m(t) + \beta r'_m(t) = -\alpha_m r_m(t) \quad \text{for } t > 0,\ m = 1, 2, \ldots \tag{2.14}$$

Denote $\phi := \text{col.}(g_1, h_1, \ldots, g_n, h_n) \in D(A_s) = [(H^2 \cap H_0^1) \times H_0^1]^n$, with

$$\begin{aligned} g &= \textstyle\sum_{m=1}^{\infty} < g, \phi_m > \phi_m \\ &= \textstyle\sum_{m=1}^{\infty} (\alpha_m + p)^{1/2} < g, \phi_m > [\phi_m/(\alpha_m + p)^{1/2}] \quad \text{in } [H_0^1(\Omega)]^n, \\ h &= \textstyle\sum_{m=1}^{\infty} < h, \phi_m > \phi_m \qquad\qquad \text{in } [L^2(\Omega)]^n. \end{aligned}$$

Let $T_s(t)\phi := \text{col}\ (\hat{u}_1, \hat{v}_1, \ldots, \hat{u}_n, \hat{v}_n)$. From (2.14), we find

$$\begin{aligned} r_m(t) &= < g, \phi_m > y_m(t) + < h, \phi_m > z_m(t), \\ r'_m(t) &= < g, \phi_m > y'_m(t) + < h, \phi_m > z'_m(t) \end{aligned}$$

where $y_m(t)$ and $z_m(t)$ satisfy the same equations for $r_m(t)$ in (2.14), with initial conditions:

$$y_m(0) = 1,\ y'_m(0) = 0 \quad \text{and} \quad z_m(0) = 0,\ z'_m(0) = 1.$$

For those m with $4\alpha_m > \beta^2$, we have

$$\begin{cases} y_m(t) = e^{-\beta t/2}\cos(4\alpha_m - \beta^2)^{1/2}t/2, \\ z_m(t) = 2(4\alpha_m - \beta^2)^{-1/2}e^{-\beta t/2}\sin(4\alpha_m - \beta^2)^{1/2}t/2. \end{cases} \tag{2.15}$$

Expanding $\hat{u}$ in $[\tilde{H}_0^1(\Omega)]^n$, and using (2.15) for all large m, we find

$$\| \hat{u} \|_{\tilde{H}^1} = \sum_{m=1}^{\infty} [< g, \phi_m > y_m(t) + < h, \phi_m > z_m(t)]^2 (\alpha_m + p)$$

$$\leq \hat{K} e^{-\epsilon t} \{ \sum_{m=1}^{\infty} < g, \phi_m >^2 (\alpha_m + p) + \sum_{m=1}^{\infty} < h, \phi_m >^2 \},$$

$$\| \hat{v} \|_{L^2} = \textstyle\sum_{m=1}^{\infty} [< g, \phi_m > y'_m(t) + < h, \phi_m > z'_m(t)]^2$$

$$\leq \hat{K} e^{-\epsilon t} \{ \sum_{m=1}^{\infty} < g, \phi_m >^2 (\alpha_m + p) + \sum_{m=1}^{\infty} < h, \phi_m >^2 \},$$

for some positive constants $\hat{K}$ and ϵ. Since $[\tilde{H}_0^1(\Omega)]^n$ and $[H_0^1(\Omega)]^n$ are equivalent, we obtain

$$\| T_s(t)\phi \|_X \leq M e^{-\epsilon t} \| \phi \|_X$$

for all ϕ in $D(A_s)$. Since $D(A_s)$ is dense in X, the proof of Theorem 2.5 is complete.

If $(u_1(t), v_1(t), \ldots, u_n(t), v_n(t))$ is a solution of (2.12), then its difference with the steady-state, i.e. $\text{col.}(w_1(t), w_2(t), \ldots, w_{2n}(t)) := \text{col.}(u_1(t), v_1(t), \ldots, u_n(t), v_n(t)) - \text{col.}((u_s)_1, 0, \ldots, (u_s)_n, 0)$, can be interpreted as a solution of:

$$\frac{dw}{dt} = A_s w + \hat{\lambda}(s)\gamma(w), \quad w = \text{col.}(w_1, w_2, \ldots, w_{2n}), \tag{2.16}$$

where $\gamma(w) := \text{col.}(0, z_1(\tilde{w}), 0, z_2(\tilde{w}), \ldots, 0, z_n(\tilde{w}))$, $\tilde{w} = \text{col.}(\tilde{w}_1, \tilde{w}_2, \ldots, \tilde{w}_n) := \text{col.}(w_1, w_3, \ldots, w_{2n-1})$, $z = (z_1(\tilde{w}), z_2(\tilde{w}), \ldots, z_n(\tilde{w})) := [G(u_s + \tilde{w})(u_s + \tilde{w}) - G(u_s)u_s - J(u_s)\tilde{w}]$.

We now clarify some terminologies and hypotheses which we will be using. A function $h : X \to X$ is said to satisfy a local Lipschitz condition if for every constant $c \geq 0$, there is a constant $L(c)$ such that

$$\| h(w^*) - h(w^{**}) \|_X \leq L(c) \| w^* - w^{**} \|_X$$

holds for all $w^*, w^{**} \in X$ with $\| w^* \|_X \leq c$ and $\| w^{**} \|_X \leq c$. The function $\gamma(w)$ described above is said to satisfy property $[P]$ if

$[P]$ $\quad \| z(\tilde{w}) \|_{L^2(\Omega)]^n} = o(\| \tilde{w} \|_{H^1(\Omega)]^n}), \quad$ as $\| \tilde{w} \|_{H^1(\Omega)]^n}$ tends to 0.

Remark 2.1. Since the norm in X uses the L^2-norm of the even components and the entries of $G(u)$ is in $C^2(R^n)$, it can be readily verified by using Sobolev embedding or Morrey's inequality that the function $\gamma : X \to X$ satisfies a local Lipschitz condition, in the case $N = 1$. By Theorem 1.4 of Chapter 6 in Pazy [184], equation (2.16) has a unique mild solution w on $[0, t_{\max})$, for every given initial condition: $w(0) = w_0 \in$ X. Moreover, if $t_{\max} < \infty$ then $\| w(t) \|_X \to \infty$, as $t \to \infty$.

Remark 2.2. Using the assumption that the entries of G are in C^2 and Sobolev embedding again, we can readily verify that in the case $N = 1$, the function $\gamma(w)$ described above satisfies property $[P]$.

Remark 2.3. The even components of $\gamma(w)$ are expressed by $z(\tilde{w})$, where $\tilde{w}$ is an n-vector function consisting of the odd components of w. Careful calculations shows that the j-th component, $1 \leq j \leq n$, of $z(\tilde{w})$ can be written as:

$$\begin{aligned} & \textstyle\sum_{k=1}^{n} (u_s)_k < \nabla g_{jk}(u_s + \tau_{jk}^{*}\tilde{w}) - \nabla g_{jk}(u_s), \tilde{w} > \\ & \qquad + \textstyle\sum_{k=1}^{n} \tilde{w}_k (g_{jk}(u_s + \tilde{w}) - g_{jk}(u_s)) \\ = \ & \textstyle\sum_{k=1}^{n} (u_s)_k \sum_{i=1}^{n} < (\nabla(\partial g_{jk}/\partial u_i)(u_s + \tau_{jk}^{**}\tilde{w}), \tau_{jk}^{*}\tilde{w} > \tilde{w}_i \\ & \qquad + \textstyle\sum_{k=1}^{n} \tilde{w}_k < (\nabla g_{jk})(u_s + \hat{\tau}_{jk}\tilde{w}), \tilde{w} >, \end{aligned}$$

where $0 < \tau_{jk}^{**} < \tau_{jk}^{*} < 1, 0 < \hat{\tau}_{jk} < 1$. When the space dimension N is 3 or 4, Gagliardo-Nirenberg-Sobolev inequality asserts that the L^4 norm of a function in $\Omega \subset R^N$ is bounded by a constant multiple of its $H^1(\Omega)$ norm. If we assume all the first and second partial derivatives of $g_{jk}, 1 \leq j, k \leq n$, are bounded in R^n, then from the above formula, we can readily obtain

$$\| \gamma(w) \|_X = \| z(\tilde{w}) \|_{[L^2(\Omega)]^n} \leq K(\| \tilde{w} \|_{[L^4(\Omega)]^n})^2$$

for all $w \in X = [H_0^1(\Omega) \times L^2(\Omega)]^n$. Thus property $[P]$ is satisfied. The local Lipschitz condition of $\gamma : X \to X$ can be verified similarly. In short, if all the entries of $G(u)$ are in $C^2(R^n)$ and has bounded first and second partial derivatives in R^n, then the function γ satisfies the local Lipschitz condition and property $[P]$ described above, provided the space dimension N of Ω is 3 or 4.

Using Gronwall inequality type argument, we then use semigroup theory as in [84] or [184] to prove the following asymptotic stability theorem by means of Theorem 2.5.

Theorem 2.6 (Asymptotic Stability in Semigroup Formulation). *Assume hypotheses $[H1^*], [H2], [H3], G$ is in $C^2(R^n)$, and $\bar{B}(u_s)$is symmetric for each $s \in (0, \delta^*)$ as described in $[H4]$. Suppose that in equation (2.16), the function γ satisfies a local Lipschitz condition and property $[P]$, then the trivial steady-state solution $w = 0$ of equation (2.16) is locally asymptotically stable.*

That is, there exist positive constants $\epsilon, \hat{M}, \hat{k}$ *such that if the initial condition* $w(0) := (w_1(0), w_2(0), \ldots, w_{2n}(0))$ *satisfies* $\| w(0) \|_X \leq \epsilon$*, then the unique mild solution of (2.16) satisfies*

$$\| w(t) \|_X \leq \hat{M} e^{-\hat{k}t} \| w(0) \|_X . \tag{2.17}$$

Proof. The local Lipschitz condition of γ insures the existence of a unique mild solution $w(t) \in X$ of (2.16) as described in Theorem 1.4 in Chapter 6 in [184]. The mild solution $w(t)$ can be expressed by:

$$w(t) = T_s(t)w(0) + \int_0^t T_s(t-\tau)\hat{\lambda}(s)\gamma(w(\tau))d\tau, \quad w(0) \in X, \tag{2.18}$$

where T_s is the strongly continuous semigroup generated by A_s. Let M and k be the positive constants given by Theorem 2.5 for T_s, we obtain for $t > 0$ inside the interval of existence of $w(t)$:

$$\| w(t) \|_X \leq Me^{-kt} \| w(0) \|_X + \int_0^t Me^{-k(t-\tau)} \| \hat{\lambda}(s)\gamma(w(\tau)) \|_X \, d\tau. \tag{2.19}$$

Without loss of generality, we may assume $M > 1$. By property $[P]$ for γ, there exist $\rho > 0$ so small such that

$$\| \hat{\lambda}(s)\gamma(w) \|_X \leq [k/(2M)] \| \tilde{w} \|_{[H^1]^n} \leq [k/(2M)] \| w \|_X \tag{2.20}$$

as long as $\| w \|_X \leq \rho$. Let $\| w(0) \|_X \leq \rho/M$, the solution will satisfy $\| w(t) \|_X \leq \rho$ for sufficiently small $t > 0$. For such small $t > 0$, (2.19) and (2.20) lead to

$$e^{kt} \| w(t) \|_X \leq M \| w(0) \|_X + (k/2) \int_0^t e^{k\tau} \| w(\tau) \|_X \, d\tau. \tag{2.21}$$

Thus Gronwall's inequality gives

$$\| w(t) \|_X \leq M \| w(0) \|_X \, e^{-(k/2)t} \leq \rho e^{-(k/2)t} \tag{2.22}$$

as long as $\| w(t) \|_X \leq \rho$. Thus $w(t)$ exists for all $t > 0$, if we choose $w(0)$ to satisfy $\| w(0) \|_X \leq \rho/M$; and (2.20) is valid for all $t > 0$ in such cases. Consequently, Theorem 2.6 is valid by choosing $\epsilon = \rho/M, \hat{M} = M$, and $\hat{k} = k/2$.

We now return to equation (2.12), which is the first order system form of (2.1). Equation (2.12) can be written as:

$$\frac{\partial y}{\partial t} = \tilde{A}_s y + \hat{\lambda}(s)q(y) \tag{2.23}$$

where $y = \text{col. } (y_1, y_2, \ldots, y_{2n}) = \text{col.}(u_1, v_1, \ldots, u_n, v_n)$, $\tilde{A}_s$ is the same as A_s with all $\bar{b}_{ij}$ replaced by the corresponding b_{ij} , and $q(y) := \text{col.}(0, \hat{q}_1(\hat{y}), \ldots, 0, \hat{q}_n(\hat{y}))$, $\hat{y} = \text{col.}(y_1, y_3, \ldots, y_{2n-1})$, $\text{col.}(\hat{q}_1(\hat{y}), \ldots, \hat{q}_n(\hat{y})) := G(\hat{y})\hat{y}$. We will express the results in Theorem 2.6 in terms of the mild solutions in X of the semigroup form of (2.23), that is:

$$\frac{dy}{dt} = \tilde{A}_s y + \hat{\lambda}(s)q(y), \quad y(t) \in X,\, t \geq 0. \tag{2.24}$$

The constant function $\eta(t) := \text{col.}((u_s)_1, 0, \ldots, (u_s)_n, 0)$ is a strong solution of

$$\frac{d\eta}{dt} = A_s\eta + \hat{\lambda}(s)h(u_s) \tag{2.25}$$

where $h(u_s) = \text{col.}(0, \hat{h}_1(u_s), \ldots, 0, \hat{h}_n(u_s))$, and $\text{col.}(\hat{h}_1(u_s), \ldots, \hat{h}_n(u_s)) = G(u_s)u_s - J(u_s)u_s$. Thus $\eta(t)$ satisfies

$$\eta(t) = T_s(t)\eta(0) + \int_0^t T_s(t-\tau)\hat{\lambda}(s)h(u_s)d\tau.$$

If $w(t)$ is a mild solution of (2.16), then $w(t) + \eta$ satisfies

$$w(t) + \eta = T_s(t)[w(0) + \eta] + \int_0^t T_s(t-\tau)\hat{\lambda}(s)\{\gamma(w(\tau)) + h(u_s)\}d\tau. \tag{2.26}$$

If $w(0) \in D(A_s)$, then $y = w(t) + \eta$ is a strong solution of

$$\begin{aligned}\frac{dy}{dt} &= A_s y + \hat{\lambda}(s)[\gamma(w) + h(u_s)] \\ &= \tilde{A}_s y + \hat{\lambda}(s)[\gamma(w) + h(u_s) + r(y)]\end{aligned}$$

where $r(y) = \text{col.}(0, \hat{r}_1(\hat{y}), \ldots, 0, \hat{r}_n(\hat{y}))$, $\text{col.}(\hat{r}_1(\hat{y}), \ldots, \hat{r}_n(\hat{y})) = J(u_s)\hat{y}$. Thus, if $w(0) \in D(A_s)$, we also has

$$\begin{aligned}w(t) + \eta &= \tilde{T}_s(t)[w(0) + \eta] \\ &+ \textstyle\int_0^t \tilde{T}_s(t-\tau)\hat{\lambda}(s)\{\gamma(w(\tau)) + h(u_s) + r(w(\tau) + \eta)\}d\tau\end{aligned} \tag{2.27}$$

where $\tilde{T}_s$ is the continuous semigroup generated by $\tilde{A}_s$. By the density of $D(A_s)$ in X and the strong continuity of T_s and $\tilde{T}_s$, the right-hand sides of both (2.26) and (2.27) are still equal for all cases with $w(0) \in$ X. Since $\gamma(w) + h(u_s) + r(w + \eta) = q(w + \eta)$, we obtain from (2.27) that $w(t) + \eta$ is a mild solution of equation (2.24).

Theorem 2.6 and the above arguments lead to the following conclusion concerning solutions of (2.1) or (2.12).

Theorem 2.7 (Asymptotic Stability of Bifurcating Steady-States). *Assume all the hypotheses of Theorem 2.6 (including those concerning γ). Then if the initial condition $(y_1(0), y_2(0), \ldots, y_{2n}(0)) := (u_1(0), v_1(0), \ldots, u_n(0), v_n(0))$ is given sufficiently close to $((u_s)_1, 0, \ldots, (u_s)_n, 0)$ in X, the unique mild solution $y(t) = (u_1(t), v_1(t), \ldots, u_n(t), v_n(t))$ of the initial value problem corresponding to (2.24) exists for all $t > 0$. Moreover, there exist positive constants $\epsilon, \hat{M}, \hat{k}$ such that if*

$$\| (u_1(0), v_1(0), \ldots, u_n(0), v_n(0)) - ((u_s)_1, 0, \ldots, (u_s)_n, 0) \|_X \leq \epsilon,$$

then the mild solution of (2.24) satisfies

$$\| (u_1(t), v_1(t), \ldots, u_n(t), v_n(t)) - ((u_s)_1, 0, \ldots, (u_s)_n, 0) \|_X \leq$$

$$\hat{M} e^{-\hat{k}t} \| (u_1(0), v_1(0), \ldots, u_n(0), v_n(0)) - ((u_s)_1, 0, \ldots, (u_s)_n, 0) \|_X$$

for all $t > 0$. Recall that (2.24) is the semigroup form of (2.1) or (2.12)).

For applications of the theorems above, consider the following examples.

Example 2.1. Let $\Omega = \{x = (x_1, x_2, x_3) : x_1^2 + x_2^2 + x_3^2 < 1\}$. Consider (2.1), with $\beta = 1, u = \text{col.}(u_1, u_2, u_3)$,

$$B(x) = \begin{bmatrix} \sin 2\pi(x_1^2 + x_2^2 + x_3^2) & 1 & 2 \\ 1 & \cos 2\pi(x_1^2 + x_2^2 + x_3^2) & 3 \\ 2 & 3 & -(x_1^2 + x_2^2 + x_3^2) \end{bmatrix},$$

$$G(u)u = \begin{bmatrix} -u_1 & -u_1 & 0 \\ -u_1/2 & -\sin u_2 & 0 \\ 0 & 0 & 0 \end{bmatrix} \begin{bmatrix} u_1 \\ u_2 \\ u_3 \end{bmatrix} = \begin{bmatrix} -u_1^2 - u_1 u_2 \\ -(1/2)u_1^2 - u_2 \sin u_2 \\ 0 \end{bmatrix}.$$

It is clear that $B(x)$ satisfies $[H1^*]$ and $[H2], G(u)$ satisfies $[H3]$ with every entry in C^2, and

$$\bar{B}(u_s(x)) = B(x) + \begin{bmatrix} -2(u_s)_1 - (u_s)_2 & -(u_s)_1 & 0 \\ -(u_s)_1 & -(u_s)_2 \cos(u_s)_2 - \sin(u_s)_2 & 0 \\ 0 & 0 & 0 \end{bmatrix}$$

is symmetric as specified in $[H4]$. Since the first and second partial derivatives of $G(u)$ are bounded, by Remark 2.3 the function γ satisfies the local Lipschitz condition and property $[P]$. Thus all assumptions of Theorem 2.7 are satisfied, and the steady-state $((u_s)_1, 0, (u_s)_2, 0, (u_s)_3, 0)$ is locally asymptotically stable in the sense described by Theorem 2.7 for the equation (2.24), which is the semigroup form for (2.1). Here, we are considering those $\lambda = \hat{\lambda}(s)$ close to the right of $\hat{\lambda}_0, s \in (0, \delta^*)$, as described in Theorem 2.4.

Example 2.2. Let $\Omega = [0, \pi]$, consider (2.1) with $\beta = 1, u = \text{col.}(u_1, u_2, u_3)$,

$$B(x) = \begin{bmatrix} 5 & 6 & 7 \\ 6 & 10 & 4 \\ 7 & 4 & -1 \end{bmatrix},$$

$$G(u)u = \begin{bmatrix} -u_1 & -\sin(u_1u_2) & 0 \\ -\sin(u_1u_2) & -u_2^2 & 0 \\ 0 & 0 & 0 \end{bmatrix} \begin{bmatrix} u_1 \\ u_2 \\ u_3 \end{bmatrix} = \begin{bmatrix} -u_1^2 - u_2\sin(u_1u_2) \\ -u_1\sin(u_1u_2) - u_2^3 \\ 0 \end{bmatrix}.$$

It can be verified readily that $\bar{B}(u_s(x))$ is symmetric and all the hypotheses of Theorem 2.7 are satisfied. Here, $N = 1$, by Remarks 2.1 and 2.2 we find γ satisfies the local Lipschitz condition and $[P]$, without requiring the boundedness of the first and second partial derivatives of G.

We finally discuss some results and examples involving global (rather than local) bifurcation. We consider the global behavior of the bifurcation curve of positive solution of equation (2.7). We find conditions so that equation (2.1) has a positive steady-state solution with $\lambda = 1$. We now introduce a few notation and hypotheses. Let $\mathcal{G} := \{(\lambda, w) \in R^+ \times P : F(\lambda, w) = 0, \lambda > 0 \text{ and } w \in P\backslash\{0\}\}, \bar{\mathcal{G}}$ denotes the closure of $\mathcal{G}$; and define the following conditions:

[A1] There exists an integer $m, 1 \leq m \leq n$ such that $b_{mj}(x) > 0, b_{jm}(x) > 0$ in $\bar{\Omega}$ for all $j \neq m$, and the m-th component of $G(u)u$ is $\equiv 0$ in $\bar{\Omega}$.

[A2] For each $j \neq m$, the j-th component of $G(u)u$ is expressible as $k_j(u_1, \ldots, u_n)u_j$ with $k_j(u_1, \ldots, u_n) \in C^0$ for $u_i \geq 0, i = 1, \ldots, n$.

Theorem 2.8 (Global Bifurcation). *Under hypotheses* $[H1], [H1^*], [H2], [A1]$, *and* $[A2]$, *the component of* $\bar{\mathcal{G}}$ containing the point $(\hat{\lambda}_0, 0)$ *is unbounded, and* $\bar{\mathcal{G}} \bigcap (R \times \partial P) = (\hat{\lambda}_0, 0)$. *(Recall that* $(\hat{\lambda}_0, 0)$ *is defined in Theorems 2.2 and 2.3.)*

Proof. Theorems 2.1, 2.2 and (2.6) imply by means of Theorem 29.2 in Diemling [49] that the component of $\bar{\mathcal{G}}$ containing the point $(\hat{\lambda}_0, 0)$ is unbounded. Let $(\lambda_i, w_i) \in \mathcal{G}$, $i = 1, 2, ..$ be a sequence tending to a limit point $(\bar{\lambda}, \bar{w})$ in $R \times \partial P$, and $(\bar{\lambda}, \bar{w}) \neq (\hat{\lambda}_0, 0)$. We now show $\bar{w} = \text{col.}(\bar{w}_1, \ldots, \bar{w}_n)$ must satisfy $\bar{w}_i \equiv 0$, for each i, if $\bar{\lambda} > 0$. Consider the first case when there exists some $x_0 \in \Omega$ where $\bar{w}_m(x_0) = 0$. The equation $-\Delta\bar{w}_m = \bar{\lambda}\sum_{j=1}^n b_{mj}\bar{w}_j \geq 0$ implies that $\bar{w}_m \equiv 0$ in $\bar{\Omega}$; and subsequently, the right hand side of this equation and [A1] imply that $\bar{w}_j \equiv 0$ for each $j \neq m$ too. Hence, $\bar{w} \equiv 0$ in this case.

Consider the second case when $\bar{w}_m(x) > 0$ for all $x \in \Omega$. For each $j \neq m$, consider the problem:

$$(2.28) \qquad \begin{cases} -\Delta z(x) = \bar{\lambda} b_{jm}\bar{w}_m + \bar{\lambda}\sum_{k\neq m,j} b_{jk}\bar{w}_k + \bar{\lambda} b_{jj} z(x) \\ \qquad\qquad + \bar{\lambda} k_j(\bar{w}_1, .., \bar{w}_{j-1}, z(x), \bar{w}_{j+1}, .., \bar{w}_n) z \qquad \text{in } \Omega, \\ z = 0 \qquad \text{on } \partial\Omega. \end{cases}$$

The function $z \equiv 0$ is a lower solution. From [A1], [A2] and $\partial \bar{w}_m/\partial\nu < 0$ on $\partial\Omega$, we have

$$\lambda_1 \delta\phi(x) < \bar{\lambda} b_{jm}(x)\bar{w}_m + \bar{\lambda}\sum_{k\neq m,j} b_{jk}\bar{w}_k + \bar{\lambda} b_{jj}\delta\phi$$

$$+ \bar{\lambda} k_j(\bar{w}_1, \ldots, \bar{w}_{j-1}, \delta\phi, \bar{w}_{j+1}, \ldots, \bar{w}_n)\delta\phi \qquad \text{in} \quad \Omega$$

for $\delta > 0$ sufficiently small. Thus $z = \delta\phi$ is a family of lower solution for problem (2.28) for such δ. Hence the solution of (2.28) satisfies $\bar{w}_j(x) > \delta\phi(x)$ in Ω. This contradicts the fact that $\bar{w} \in \partial P$. We must thus have $\bar{w} \equiv 0$ in Ω in any case, for $\bar{\lambda} \geq 0$.

Next, define $z_i := w_i / \parallel w_i \parallel_E, i = 1, 2, \ldots$; they satisfy

$$(2.29) \qquad z_i + \lambda_i \Delta^{-1}[B z_i] + \lambda_i \Delta^{-1}[G(w_i)w_i] / \parallel w_i \parallel_E = 0.$$

Since $\Delta^{-1}B$ is compact, there exists a subsequence (again denoted by $\{z_i\}$ for convenience) such that $\Delta^{-1}[Bz_i]$ converges in E. Since $\Delta^{-1}[G(w_i)w_i] / \parallel w_i \parallel_E$ tends to zero in E, as $\parallel w_i \parallel_E \rightarrow \parallel \bar{w} \parallel_E = 0$, equation (2.29) implies that $\{z_i\}$ converges in E to a function z_0 say, and

$$z_0 = -\bar{\lambda}\Delta^{-1}[Bz_0].$$

Moreover, we have $z_0 \in P$, since $w_i \in P$; and $z_0 \not\equiv 0$ since $\parallel z_i \parallel_E = 1$. Hence, we must have $\bar{\lambda} > 0$. The uniqueness part Theorem 2.1 implies that $\bar{\lambda} = \hat{\lambda}_0$. Consequently, we must have $(\bar{\lambda}, \bar{w}) = (\hat{\lambda}_0, 0)$. This completes the proof of Theorem 2.8.

Let $-\lambda_1$ be the principal eigenvalue of the Laplace operator on Ω with zero Dirichlet boundary data, and $\phi > 0$ be a positive eigenfunction.

Theorem 2.9 (Steady-State for Given $\lambda = 1$). *Assume all the hypotheses in Theorem 2.8 and $b_{mm}(x) \equiv 0$ in assumption [A1]. Suppose further that:*

[A3] *There exists constant vector $\vec{g} = col.(g_1, ..., g_n)$, $g_i > 0$, such that $B(x)\vec{g} > \lambda_1 \vec{g}$ for all $x \in \Omega$.*

[A4] *For each $j \neq m$, the function k_j described in [A2] has the property*

$\limsup_{N\to\infty} N^{-1} \max\{k_j(u_1, .., u_{j-1}, N, u_{j+1}, ..., u_n) : 0 \le u_\tau \le N,$
$\tau \ne j\} < 0.$

Then (2.1) has a steady-state solution in P for $\lambda = 1$.

Proof. Let $\hat{z} = \phi\vec{g}$, [A3] implies that there exists $\mu > 1$ such that $B\hat{z} \ge \mu\lambda_1\hat{z}$ in $\bar{\Omega}$. Hence, $(-\Delta)^{-1}(B\hat{z}) \ge \mu\lambda_1(-\Delta)^{-1}(\phi\vec{g}) = \mu\phi\vec{g} = \mu\hat{z}$. By Theorem 2.5 in [108] or Theorem A3-10 in Chapter 6, we have $1/\hat{\lambda}_0 \ge \mu > 1$, that is $\hat{\lambda}_0 < 1$. (Here, $\hat{\lambda}_0$ is defined in Theorem 2.8.)

The equation for u_m and $b_{mm} \equiv 0$ imply that if $\lambda \in [0, C]$ for some C, and if all the components $u_j, j \ne m$ of a steady state solution of (2.1) have the property $|u_j| \le M$, then u_m must satisfy $|u_m| \le KM$ for some constant K independent of M. We now show that there must indeed exist some constant M, such that if $\lambda \in (0, C]$ and $(\lambda, u) \in \mathcal{G}$, then $0 \le u_j \le M$ for all $j \ne$ m. Otherwise, there exists $u_r, r \ne m$, and a point $x^* \in \Omega$ where $u_r(x^*) = \max_{j\ne m}\{\sup_{x\in\bar{\Omega}} u_j(x)\} > M$. Let $p = u_r(x^*)/M > 1$, then we have the property: $0 \le u_j(x) \le pM$ for all $x \in \bar{\Omega}, j \ne m$ Consider the equation satisfied by u_r at the point $x^* \in \Omega$.

$$\begin{aligned}
-\Delta u_r(x^*) &= \textstyle\sum_{j\ne m,r} \lambda b_{rj}u_j + \lambda b_{rm}u_m \\
&\qquad + \lambda k_r(u_1, .., u_{r-1}(x^*), pM, u_{r+1}(x^*), .., u_n)\, u_r(x^*) \\
&\le \lambda\,[\textstyle\sum_{j\ne m,r} b_{rj}pM + b_{rm}KpM + \\
&\quad \max\{(pM)^{-1}k_r(u_1, .., u_{r-1}, pM, u_{r+1}, .., u_n) : 0 \le u_\tau \le pM, \tau \ne r\}(pM)^2] \\
&< 0.
\end{aligned}$$

The last inequality is satisfied for M sufficiently large due to hypothesis [A4]. This contradicts the definition of x^*. We thus assert that if $\lambda \in [0, C]$, then $|u_i|$ are bounded for each $i = 1, \ldots, n$, if $(\lambda, u) \in \mathcal{G}$. Finally, using gradient estimates by means of (2.7), we conclude that $\mathcal{G}$ cannot be unbounded if $\lambda \in [0, C]$, for some C. Thus a solution of (2.7) must exist for $\lambda = 1$.

Example 2.3. We consider Example 2.2 with only a single modification. We change the entry b_{33} of B from -1 to 0. Clearly, condition [A1] is satisfied with $m = 3$. For [A2], we define

$$k_1(u_1, u_2) = \begin{cases} -u_1 - (u_2/u_1)\sin(u_1u_2) & \text{if} \quad u_1 \ne 0 \\ -u_2^2 & \text{if} \quad u_1 = 0, \end{cases}$$

$$k_2(u_1,u_2) = \begin{cases} -(u_1/u_2)\sin(u_1u_2) - u_2^2 & \text{if} \quad u_2 \neq 0 \\ \\ -u_1^2 & \text{if} \quad u_2 = 0, \end{cases}$$

then k_1 and k_2 are continuous for $u_i \geq 0$. Condition [A3] is satisfied by choosing $\vec{g} =$ col.(1,1,1). Using the formulas for k_1 and k_2, we can verify [A4] is satisfied. Thus Theorem 2.9 can be applied, and we conclude that (2.1) has a steady-state solution for $\lambda = 1$ in this case. As for Example 2.2, we also find the bifurcating positive steady-state is asymptotically stable for λ close to $\hat{\lambda}_0$ for this example. The stability of the steady-state when $\lambda = 1$ remains to be considered.

5.3 Invariant Manifolds for Coupled Navier-Stokes and Second Order Wave Equations

This section considers the dynamics of a coupled system of incompressible Navier-Stokes equations with second order wave equations. The system may be used to approximate the interaction of ionized plasma particles with an electromagnetic field. Under appropriate assumptions and provided that the viscosity of the fluid is sufficiently large, we prove the existence of an invariant manifold as described in Definition 3.1 below. Moreover, the manifold is attractive as $t \to +\infty$ for all close neighboring solutions.

More precisely, we consider the coupled parabolic-hyperbolic system:

$$\text{(3.1)} \qquad \begin{cases} \mathbf{u}_t = \epsilon^{-1}\Delta\mathbf{u} - (\mathbf{u}\cdot\nabla)\mathbf{u} - \nabla p + \mathbf{f}(\mathbf{u},\mathbf{v}) & \text{in } \Omega\times(0,\infty), \\ \\ \text{div } \mathbf{u}(x,t) = 0 & \text{in } \Omega\times(0,\infty), \\ \\ \mathbf{v}_{tt} + \mu\mathbf{v}_t = \Delta\mathbf{v} + B(x)\mathbf{v} + \mathbf{g}(\mathbf{u},\mathbf{v}) & \text{in } \Omega\times(0,\infty), \\ \\ \mathbf{u} = \mathbf{h}, \quad \mathbf{v} = \mathbf{q} & \text{on } \partial\Omega\times[0,\infty). \end{cases}$$

The system applies to the study of physical problems involving the interactions of fluid motion with other forms of waves. As an example, the vector $\mathbf{u}$ may describe the three components of the velocity of ionized plasma particles in position x and time t. The vector $\mathbf{v}$ may describe the three components of a magnetic field in position x and time t interacting with the ionized particles there. The scalar p denotes the pressure function of x and t. In the dynamic theory of plasma physics, one studies the motion of a conducting fluid interacting with electric and magnetic fields. This leads to the coupling of the Navier-Stokes equation for the ionized particles with the Maxwell's equations for the electromagnetic force fields. In many investigations, the term $\epsilon^{-1}\Delta\mathbf{u}$

is omitted. However, in order to include the viscosity effect of the fluid, it is sometimes introduced to obtain the Navier-Stokes equation (cf. p. 131-134 or p.89 in Thompson [219]). In Chapter 4 of [219], we assume that the current $\mathbf{j}$ satisfies $\mathbf{j} = \sigma(\mathbf{E}+c^{-1}\mathbf{u}\times\mathbf{v})$, where $\mathbf{E}$ is the electric field; while the magnetic and electric fields satisfy the Maxwell equations $\rho^{-1}\mathbf{curl}\ \mathbf{v} = 4\pi c^{-1}\mathbf{j} + \delta c^{-1}\partial\mathbf{E}/\partial t$ and $\mathbf{curl}\ \mathbf{E} = -c^{-1}\partial\mathbf{v}/\partial t$. Here, σ, c, ρ, and δ are physical constants. It is shown that explicit dependence on the electric field $\mathbf{E}$ can be eliminated from the Maxwell's equations. Hence, we obtain first-order time and second-order space differential equations in the magnetic field $\mathbf{v}$, coupling with the effects of the velocity field $\mathbf{u}$. Observe at p. 44 or 47-48 of [219] the author makes the simplifying assumption that for slow motions, the displacement current involving $\partial\mathbf{E}/\partial t$ can be neglected. Thus, even after eliminating the dependence on the electric field, the first order time derivative of $\mathbf{v}$ only will appear in the equations of magnetohydrodynamics. However, in the more general situations, one would obtain three equations for the three components of the particle motion, coupled with three second order wave equations for the three components of the magnetic field as described in (3.1) above. For simplicity, we restrict ourselves to the case of incompressible fluid, which leads to the divergence-free condition on the second equation in (3.1). In the study of plasma physics, one is interested in the large-time asymptotic behavior of the system as well as the existence and stabilities of steady-states and invariant manifolds. The search for a stable confining configuration is a major challenge. For further investigations, one may impose also div $\mathbf{v} = 0$, or a more precise description of the nonlinear terms $\mathbf{f}$ and $\mathbf{g}$. Conceivably, the system (3.1) may be adapted to analyze other physical problems as well.

We let Ω be a bounded domain in R^n with smooth boundaries. We will denote $\mathbf{L}^2 := L^2(\Omega, R^3), \mathbf{H}^{1,2} := H^{1,2}(\Omega, R^3)$ and $\mathbf{H}_0^{1,2} := H_0^{1,2}(\Omega, R^3)$. For $\boldsymbol{\phi} = (\phi_1, \phi_2, \phi_3)^T \in \mathbf{L}^2$, define the norm $\|\boldsymbol{\phi}\|_{\mathbf{L}^2} := (\sum_{i=1}^3 \|\phi_i\|_{L^2}^2)^{1/2}$; and for $\boldsymbol{\psi} = (\psi_1, \psi_2, \psi_3)^T \in \mathbf{H}^{1,2}$, let $\|\boldsymbol{\psi}\|_{\mathbf{H}^{1,2}} := (\sum_{i=1}^3 \|\psi_i\|_{H^{1,2}}^2)^{1/2}$. (Note that $H^{1,2}(\Omega)$ is denoted as $W^{1,2}(\Omega)$ in some other sections). The parameter ϵ is a positive constant with ϵ^{-1} representing the viscosity of the fluid. The parameter μ for the wave equations is a constant which can have any sign. We do not make any restriction on the sign of μ until the last Theorem 3.3, even though it should be positive in application to magnetohydrodynamics. The symbols Δ and ∇ denote the Laplacian and gradient operators respectively. $B(x)$ is a 3 by 3 matrix with each entry in $L^\infty(\Omega)$. The following symbols and notations will also be used throughout this section for the three fluid components:

$$\tilde{\mathbf{H}}_0^{1,2} := \tilde{H}_0^{1,2}(\Omega, R^3) := \{\mathbf{u} \in H_0^{1,2}(\Omega, R^3) : \text{ div } \mathbf{u} = 0 \text{ in } \Omega\},$$

$$\mathbf{X} := \tilde{\mathbf{L}}^2 := \tilde{L}^2(\Omega, R^3) \text{ is the closure of } \tilde{\mathbf{H}}_0^{1,2} \text{ in } \mathbf{L}^2 = L^2(\Omega, R^3),$$

$$\mathbf{X}^\perp = \{\mathbf{w} \in \mathbf{L}^2 : \mathbf{w} = \nabla p \text{ for some } p \in H^{1,2}(\Omega)\}.$$

We will use the Helmholtz-Weyl orthogonal decomposition of vector fields (cf. Galdi [67] or McOwen [169]), which may be expressed as:

$$\mathbf{L}^2 = \mathbf{X} \oplus \mathbf{X}^\perp = \tilde{\mathbf{L}}^2 \oplus \mathbf{X}^\perp.$$

We will also use the following Banach spaces for the magnetic field components:

$$\mathbf{Y} = [H_0^{1,2}(\Omega) \times L^2(\Omega)]^3, \quad \tilde{\mathbf{Y}} = \mathbf{H}_0^{1,2} \times \mathbf{L}^2.$$

Here, the norm for $\mathbf{y} = (y_1, .., y_6)^T \in \mathbf{Y}$ is defined as $\|\mathbf{y}\|_{\mathbf{Y}} = (\|(y_1, y_3, y_5)^T\|^2_{\mathbf{H}^{1,2}} + \|(y_2, y_4, y_6)^T\|^2_{\mathbf{L}^2})^{1/2}$, and the norm for $\mathbf{y} = (y_1, .., y_6)^T \in \tilde{\mathbf{Y}}$ is defined as $\|\mathbf{y}\|_{\tilde{\mathbf{Y}}} = (\|(y_1, y_2, y_3)^T\|^2_{\mathbf{H}^{1,2}} + \|(y_4, y_5, y_6)^T\|^2_{\mathbf{L}^2})^{1/2}$. For any $\mathbf{y} = (y_1, y_2, y_3, y_4, y_5, y_6)^T \in \mathbf{Y}$, for convenience we will denote by $\mathbf{y}_O = (y_1, y_3, y_5)^T$ the triplet of the odd components of $\mathbf{y}$.

For any $\mathbf{u} \in \mathbf{L}^2$, we denote by $P\mathbf{u}$ the projection of $\mathbf{u}$ into $\mathbf{X}$. The operator $A := -\epsilon^{-1}\Delta$ is an unbounded operator on $\mathbf{X}$ with domain $D(A) = \tilde{\mathbf{H}}_0^{1,2} \bigcap H^{2,2}(\Omega, R^3)$. In Lemma 1 at p. 328 of [169], it is shown that $A : D(A) \to \mathbf{X}$ is a densely defined, closed linear operator on $\mathbf{X}$ with compact inverse. Moreover, A is the infinitesimal generator of a linear semigroup on $\mathbf{X}$, and A admits a complete orthonormal set of eigenfunctions on $\mathbf{X}$. Thus, for $1 \geq \alpha \geq 0$, we can use the eigenfunction expansion to define an operator A^α on $\mathbf{X}$ with domain $\mathbf{X}_\alpha$ in the usual manner. (See e.g. Section 11.2 in [169], or more details in Part A below, or Section 6.4 for X^α in Chapter 6.) It is well known that $\mathbf{X}_\alpha$ is a Banach space equipped with the graph norm $\|\cdot\|_\alpha$. Note that we have $D(A) \subset \mathbf{X}_\alpha \subset \mathbf{X}$.

In applications, the nonlinear terms $\mathbf{f}$ and $\mathbf{g}$ in (3.1) should include first spatial derivatives of the $\mathbf{u}$ and $\mathbf{v}$. Consequently, throughout this section we restrict α to the interval:

$$3/4 < \alpha < 1.$$

More precisely, the function $\mathbf{f} : (\mathbf{X}_\alpha, \mathbf{H}^{1,2}) \to \mathbf{L}^2$ in (3.1) will satisfy the hypothesis:

$$\|\mathbf{f}(\mathbf{u}, \mathbf{v}) - \mathbf{f}(\mathbf{y}, \mathbf{z})\|_{\mathbf{L}^2} \leq \hat{C}(\rho)(\|\mathbf{u} - \mathbf{y}\|_\alpha + \|\mathbf{v} - \mathbf{z}\|_{\mathbf{H}^{1,2}}) \tag{3.2}$$

for all $\mathbf{u}, \mathbf{y}$ in $\mathbf{X}_\alpha$ with $\|\mathbf{u}\|_\alpha, \|\mathbf{y}\|_\alpha \leq \rho$, and all $\mathbf{v}, \mathbf{z}$ in $\mathbf{H}^{1,2}$, where $\hat{C}(\rho)$ is a constant which depends on ρ. The function $\mathbf{g} : (\mathbf{X}_\alpha, \mathbf{H}^{1,2}) \to \mathbf{L}^2$ in (3.1) will satisfy:

$$\|\mathbf{g}(\mathbf{u}, \mathbf{v}) - \mathbf{g}(\mathbf{y}, \mathbf{z})\|_{\mathbf{L}^2} \leq K(\|\mathbf{u} - \mathbf{y}\|_\alpha + \|\mathbf{v} - \mathbf{z}\|_{\mathbf{H}^{1,2}}) \tag{3.3}$$

for all $\mathbf{u}, \mathbf{y}$ in $\mathbf{X}_\alpha$, and all $\mathbf{v}, \mathbf{z}$ in $\mathbf{H}^{1,2}$, where K is a positive constant. For each constant $R > 0$, there exists $N(R) > 0$, such that:

$$\|\mathbf{f}(\mathbf{u},\mathbf{v})\|_{\mathbf{L}^2} \leq N(R) \tag{3.4}$$

for all $\mathbf{u}$ in $\mathbf{X}_\alpha$ with $\|\mathbf{u}\|_\alpha \leq R$ and all $\mathbf{v}$ in $\mathbf{H}^{1,2}$. Note that in (3.2)–(3.4) we consider $\mathbf{v}, \mathbf{z}$ in $\mathbf{H}^{1,2}$. Thus, one spatial derivative for $\mathbf{v}$ is allowed in the nonlinear terms $\mathbf{f}$ and $\mathbf{g}$ in (3.1).

In Part A, we will prove that for ϵ sufficiently small, there exists an invariant manifold for system (3.1). Roughly speaking, there exists a relationship between the vector fields $\mathbf{u}$, $\mathbf{v}$ and $\mathbf{v}_t$, such that if the relationship is satisfied by a solution $(\mathbf{u}(\cdot,t), \mathbf{v}(\cdot,t))$ of (3.1) at some time t, then it will be satisfied for all $t \in (-\infty, +\infty)$. In Part B, it will be shown that the invariant manifold is asymptotically stable, i.e. an attractor for neighboring solutions. Solutions of (3.1) which start close to the invariant manifold will tend to the manifold as $t \to +\infty$. Finally as an application to a special situation, we show that if the trivial solution is an asymptotically stable equilibrium on the invariant manifold, it will be asymptotically stable for the full system (3.1). That is, if the origin attracts neighboring solutions on the manifold, it will attract all neighboring solutions of the full system. The study of the relationship of the full initial-boundary value problem with that of the reduced problem on the manifold is analogous to the study of an invariant manifold for coupled parabolic partial differential equations as in Hale [79] and Henry [84]. It is also analogous to the study of reduced problem for singularly perturbed ordinary differential equations as in Hoppensteadt [89], Tihonov [220] and Wasow [229]. Here, we extend the method for analyzing invariant manifolds for prototype coupled parabolic-hyperbolic equations in Leung [128] to the situation of coupled Navier-Stokes and other nonlinear wave systems. Many theories on Navier-Stokes and nonlinear wave equations can be found in Galdi [67], Strauss [214] and Temam [216]. Hopefully, more satisfactory results for system (3.1) will be obtained in the future under less restrictive assumptions than (3.2)–(3.4). In remark 3.7, we illustrate how the results here may be adapted to the case when condition (3.4) on $\mathbf{f}$ is removed. The presentation of this section follows the results found in Leung [130].

Part A: Main Theorem for the Existence of Invariant Manifold.

We now proceed to prove the Main Theorem for the existence of invariant manifold. Problem (3.1) will be converted by semigroup method into a system of equations in $\mathbf{X}_\alpha \times \mathbf{H}_0^{1,2}$, or $(\mathbf{u}, \mathbf{v}, \mathbf{v}_t)$ in $\mathbf{X}_\alpha \times \mathbf{H}_0^{1,2} \times \mathbf{L}^2$. This will be described in system (3.9) below and clarified in remark 3.3. A solution of (3.1) will thus be interpreted in semigroup setting as described in remark 3.3 below.

Definition 3.1. A set $\mathbf{S} \subset R \times \mathbf{X}_\alpha \times \tilde{\mathbf{Y}} = R \times \mathbf{X}_\alpha \times \mathbf{H}_0^{1,2} \times \mathbf{L}^2$ is a local invariant manifold for problem (3.1) provided that, for any $(t_0, \mathbf{u}_0, (\boldsymbol{\eta}_{01}, \boldsymbol{\eta}_{02})) \in \mathbf{S}$, there exists a solution $(\mathbf{u}(\cdot,t),\mathbf{v}(\cdot,t))$ of (3.1) on an open interval (t_1,t_2) containing t_0 with $\mathbf{u}(\cdot,t_0)=\mathbf{u}_0,\mathbf{v}(\cdot,t_0)=\boldsymbol{\eta}_{01},\mathbf{v}_t(\cdot,t_0)=\boldsymbol{\eta}_{02}$ and $(t,\mathbf{u}(t),(\mathbf{v}(t).\mathbf{v}_t(t)))\in\mathbf{S}$ for $t_1<t<t_2$. $\mathbf{S}$ is an invariant manifold if we can always choose $(t_1,t_2)=(-\infty,\infty)$. Since the problem (3.1) is autonomous, we may call $\mathbf{S}_1$ an invariant manifold when $\mathbf{S}=R\times\mathbf{S}_1$.

We will show the existence of an invariant manifold for problem (3.1) under appropriate assumptions. We first impose some smoothness conditions on the boundary functions $\mathbf{h}$ and $\mathbf{q}$, and then transform problem (3.1) into a convenient formulation. We assume that $\mathbf{h}$ is the trace of some $\hat{\mathbf{h}}\in\mathbf{H}^{1,2}$ with div $\hat{\mathbf{h}}=0$ in Ω. The problem:

$$(3.5)\qquad \begin{cases}\epsilon^{-1}\Delta\mathbf{w}+\epsilon^{-1}\Delta\hat{\mathbf{h}}=\nabla\hat{p} & \text{in } \Omega,\\ \operatorname{div}\mathbf{w}=0 & \text{in } \Omega,\\ \mathbf{w}=0 & \text{on } \partial\Omega,\end{cases}$$

uniquely defines $\mathbf{w}\in\mathbf{X}$, while $\hat{p}=\hat{p}(x)$ is only unique up to an additive constant. (See e.g. [67] or [169].) The function $\mathbf{u}^0:=\mathbf{w}+\hat{\mathbf{h}}$ in Ω is a function in $\mathbf{H}^{1,2}$, with div $\mathbf{u}^0=0$ in Ω, and $\mathbf{u}^0=\mathbf{h}$ on $\partial\Omega$. If $\mathbf{u}^0_i=\mathbf{w}_i+\hat{\mathbf{h}}_i$, $i=1,2$, are the two functions with $\mathbf{w}_i$ and $\hat{\mathbf{h}}_i$ having the corresponding properties for $\mathbf{w}$ and $\hat{\mathbf{h}}$ above, then $\mathbf{z}:=\mathbf{u}^0_1-\mathbf{u}^0_2$ would satisfy: $\epsilon^{-1}\Delta\mathbf{z}=\nabla\tilde{p}$ in Ω for some $\tilde{p}$, with div $\mathbf{z}=0$ in Ω, and $\mathbf{z}=0$ on $\partial\Omega$. By the same uniqueness reason as above, we obtain $\mathbf{z}=0$ in Ω. Thus the function $\mathbf{u}^0=\mathbf{w}+\hat{\mathbf{h}}$ is uniquely defined in Ω by using (3.5) if there is any divergence-free $\hat{\mathbf{h}}$ in $\mathbf{H}^{1,2}$ with the same trace $\mathbf{h}$ on $\partial\Omega$.

We next consider the problem:

$$(3.6)\qquad \Delta\mathbf{z}+B(x)\mathbf{z}+\mathbf{g}(\mathbf{u}^0,\mathbf{z})=0 \text{ in } \Omega,\qquad \mathbf{z}=\mathbf{q}\quad \text{on } \partial\Omega.$$

We assume that problem (3.6) has a solution $\mathbf{z}=\mathbf{v}^0\in\mathbf{H}^{1,2}$. (Here, there may exist more than one $\mathbf{v}^0$.)

We will look for solution of problem (3.1) with the $\mathbf{u}$ components in a smoother subspace of $\mathbf{X}$. The operator $A=A(\epsilon):=-\epsilon^{-1}\Delta$ on $\mathbf{X}$ admits a complete orthonormal set of eigenfunctions $\boldsymbol{\psi}_1,\boldsymbol{\psi}_2,\ldots$, with their corresponding eigenvalues $0<\lambda_1\le\lambda_2\le\ldots$. We define $A^\alpha\boldsymbol{\psi}:=A^\alpha(\sum_{j=1}^\infty a_j\boldsymbol{\psi}_j)=\sum_{j=1}^\infty\lambda_j^\alpha a_j\boldsymbol{\psi}_j$ whenever $\boldsymbol{\psi}=\sum_{j=1}^\infty a_j\boldsymbol{\psi}_j$ is in $D(A^\alpha)$. Here, we have $D(A^\alpha)=\mathbf{X}_\alpha:=\{\boldsymbol{\psi}=\sum_{j=1}^\infty a_j\boldsymbol{\psi}_j\in\mathbf{X}:A^\alpha\boldsymbol{\psi}\in\mathbf{X}\}$. We thus consider $\mathbf{X}_\alpha$ as a Banach space with the graph norm $\|\boldsymbol{\psi}\|_\alpha=\|\boldsymbol{\psi}\|_{\mathbf{X}}+\|A^\alpha\boldsymbol{\psi}\|_{\mathbf{X}}$. We have $D(A)\subset\mathbf{X}_\alpha\subset\mathbf{X}$. (See [169] or [84]).

Let the semigroup of bounded linear operators on $\mathbf{X}$ generated by $-A(\epsilon)$ be denoted by $\{e^{-A(\epsilon)t}\}$. We can express

$$(A(\epsilon))^\alpha e^{-A(\epsilon)t}\psi = \sum_{j=1}^{\infty} \lambda_j^\alpha e^{-\lambda_j t} a_j \psi_j.$$

For a fixed $r \in (0,1)$, and each $j \geq 1$, $t > 0$, consider the expression:

$$\begin{aligned}\lambda_j^\alpha e^{-\lambda_j t} &= t^{-\alpha}[(\lambda_j t)^\alpha e^{-r\lambda_1 t} e^{(-\lambda_j t + r\lambda_1 t)}] \\ &\leq t^{-\alpha} e^{-r\lambda_1 t}[(\lambda_j t)^\alpha e^{-(1-r)\lambda_j t}].\end{aligned}$$

We thus obtain

$$\|(A(\epsilon))^\alpha e^{-A(\epsilon)t}\psi\|_{\mathbf{L}^2} \leq \hat{M} t^{-\alpha} e^{-r\lambda_1 t}\|\psi\|_{\mathbf{L}^2} \qquad \text{for } t > 0,$$

for some constant $\hat{M}$. Since $\lambda_1 = \lambda_1(\epsilon)$ depends on ϵ, and tends to $+\infty$ as $\epsilon \to 0^+$, we find the operator norms of $\{e^{-A(\epsilon)t}\}$ and $\{(A(\epsilon))^\alpha e^{-A(\epsilon)t}\}$ on $\mathbf{X}$ satisfy:

$$(3.7) \qquad \|e^{-A(\epsilon)t}\| \leq \hat{M} e^{-\beta(\epsilon)t}, \quad \|(A(\epsilon))^\alpha e^{-A(\epsilon)t}\| \leq \hat{M} t^{-\alpha} e^{-\beta(\epsilon)t} \quad \text{for } t > 0.$$

Here, $\beta(\epsilon)$ is a positive constant with $\beta(\epsilon) \to +\infty$ as $\epsilon \to 0^+$, and $\hat{M}$ is independent of ϵ. (See also e.g. Section 2.6 or 8.4 in [184], or Remark 2.1 in [128].) By slightly increasing $\hat{M}$ and reducing $\beta(\epsilon)$, we may assume without loss of generality that:

$$(3.8) \qquad \|e^{-A(\epsilon)t}\theta\|_\alpha \leq \hat{M} t^{-\alpha} e^{-\beta(\epsilon)t}\|\theta\|_{\mathbf{X}} \qquad \text{for all } t > 0,\ \theta \in \mathbf{X},$$

where $\beta(\epsilon)$ and $\hat{M}$ still have all the properties described above. We are now ready for the following existence theorem.

Theorem 3.1 (Existence of Invariant Manifold). *Consider system* (3.1) *under hypotheses (3.2)-(3.4). Let* $\mathbf{u}^0, \mathbf{v}^0$ *be determined by the boundary functions* $\mathbf{h}$,$\mathbf{q}$ *by means of problems (3.5), (3.6) as described above. For* $\epsilon > 0$ *sufficiently small, there exists an invariant manifold for system (3.1) of the form:*

$$\mathbf{S} = \{(t, \mathbf{u}, \mathbf{v}, \mathbf{v}_t) : \mathbf{u} = \mathbf{u}^0 + \sigma^*(\mathbf{v} - \mathbf{v}^0, \mathbf{v}_t), -\infty < t < \infty, (\mathbf{v} - \mathbf{v}^0, \mathbf{v}_t)^T \in \tilde{\mathbf{Y}}\},$$

with $\sigma^* : \tilde{\mathbf{Y}} \to \mathbf{X}_\alpha$ *satisfying* $\|\sigma^*(\mathbf{y})\|_\alpha \leq K_1$, *and* $\|\sigma^*(\mathbf{y}) - \sigma^*(\hat{\mathbf{y}})\|_\alpha \leq K_2\|\mathbf{y} - \hat{\mathbf{y}}\|_{\tilde{\mathbf{Y}}}$. *(Here,* K_1 *and* K_2 *are arbitrary given positive constants which determine the size of* ϵ *such that* $\mathbf{S}$ *exists.)*

Remark 3.1. Roughly speaking, there exists a relationship σ^* between the vector fields $\mathbf{u}$, $\mathbf{v}$ and $\mathbf{v}_t$, such that if the relationship is satisfied by a solution $(\mathbf{u}(\cdot,t), \mathbf{v}(\cdot,t))$ of (3.1) at some time t, then it will be satisfied for all $t \in (-\infty, +\infty)$.

Remark 3.2. In Remark 3.7, we will illustrate situations when the theorem can still be applied when the boundedness condition (3.4) on $\mathbf{f}$ is removed.

Before proving Theorem 3.1, we will first show three preliminary lemmas. For convenience, let $\tilde{\mathbf{u}} = \mathbf{u} - \mathbf{u}^0, \tilde{\mathbf{v}} = \mathbf{v} - \mathbf{v}^0$, and define functions $\mathbf{F} : \mathbf{X}_\alpha \times \mathbf{H}_0^{1,2} \to \mathbf{L}^2$, $\mathbf{G} : \mathbf{X}_\alpha \times \mathbf{H}_0^{1,2} \to \mathbf{L}^2$ by:

$$\mathbf{F}(\tilde{\mathbf{w}}, \tilde{\mathbf{z}}) := \mathbf{f}(\tilde{\mathbf{w}} + \mathbf{u}^0, \tilde{\mathbf{z}} + \mathbf{v}^0) - ((\tilde{\mathbf{w}} + \mathbf{u}^0) \cdot \nabla)(\tilde{\mathbf{w}} + \mathbf{u}^0),$$

$$\mathbf{G}(\tilde{\mathbf{w}}, \tilde{\mathbf{z}}) := \mathbf{g}(\tilde{\mathbf{w}} + \mathbf{u}^0, \tilde{\mathbf{z}} + \mathbf{v}^0) - \mathbf{g}(\mathbf{u}^0, \mathbf{v}^0)$$

for any $(\tilde{\mathbf{w}}, \tilde{\mathbf{z}}) \in \mathbf{X}_\alpha \times \mathbf{H}_0^{1,2}$.

Problem (3.1) can be converted into a system of equations for $(\tilde{\mathbf{u}}, \tilde{\mathbf{v}}) \in \mathbf{X}_\alpha \times \mathbf{H}_0^{1,2}$ as follows:

$$(3.9) \qquad \begin{cases} \tilde{\mathbf{u}}_t = -A\tilde{\mathbf{u}} + P[\mathbf{F}(\tilde{\mathbf{u}}, \tilde{\mathbf{v}})] & \text{for } \tilde{\mathbf{u}}(t) \in \mathbf{X}_\alpha, \\ \tilde{\mathbf{v}}_{tt} + \mu\tilde{\mathbf{v}}_t = \Delta\tilde{\mathbf{v}} + B(x)\tilde{\mathbf{v}} + \mathbf{G}(\tilde{\mathbf{u}}, \tilde{\mathbf{v}}) & \text{for } \tilde{\mathbf{v}}(t) \in \mathbf{H}_0^{1,2}. \end{cases}$$

We next deduce a Lipschitz condition for the nonlinear terms in (3.9) corresponding to (3.2) and (3.3).

Lemma 3.1. *The functions* $\mathbf{F}$ *and* $\mathbf{G}$ *defined above satisfy:*

$$(3.10) \qquad \|P\mathbf{F}(\tilde{\mathbf{u}}, \tilde{\mathbf{v}}) - P\mathbf{F}(\tilde{\mathbf{y}}, \tilde{\mathbf{z}})\|_{\mathbf{X}} \le C(\rho)(\|\tilde{\mathbf{u}} - \tilde{\mathbf{y}}\|_\alpha + \|\tilde{\mathbf{v}} - \tilde{\mathbf{z}}\|_{\mathbf{H}^{1,2}})$$

for all $\tilde{\mathbf{u}}, \tilde{\mathbf{y}}$ *in* $\mathbf{X}_\alpha$ *with* $\|\tilde{\mathbf{u}}\|_\alpha, \|\tilde{\mathbf{y}}\|_\alpha \le \rho$ *and all* $\tilde{\mathbf{v}}, \tilde{\mathbf{z}}$ *in* $\mathbf{H}_0^{1,2}$;

$$(3.11) \qquad \|\mathbf{G}(\tilde{\mathbf{u}}, \tilde{\mathbf{v}}) - \mathbf{G}(\tilde{\mathbf{y}}, \tilde{\mathbf{z}})\|_{\mathbf{L}^2} \le L(\|\tilde{\mathbf{u}} - \tilde{\mathbf{y}}\|_\alpha + \|\tilde{\mathbf{v}} - \tilde{\mathbf{z}}\|_{\mathbf{H}^{1,2}})$$

for all $\tilde{\mathbf{u}}, \tilde{\mathbf{y}}$ *in* $\mathbf{X}_\alpha$ *and all* $\tilde{\mathbf{v}}, \tilde{\mathbf{z}}$ *in* $\mathbf{H}_0^{1,2}$. *Here,* $C(\rho)$ *is a positive constant which depends on* ρ, *and* L *is another positive constant.*

Proof. Let $\tilde{\mathbf{u}} = (\tilde{u}_1, \tilde{u}_2, \tilde{u}_3)^T$, $\tilde{\mathbf{y}} = (\tilde{y}_1, \tilde{y}_2, \tilde{y}_3)^T$ and $\mathbf{u}^0 = (u_1^0, u_2^0, u_3^0)^T$. In addition to terms involving $\mathbf{f}$, the components of $\mathbf{F}(\tilde{\mathbf{u}}, \tilde{\mathbf{v}}) - \mathbf{F}(\tilde{\mathbf{y}}, \tilde{\mathbf{z}})$ consist of sums of terms of the forms: $\tilde{u}_i(\tilde{u}_j)_{x_k} - \tilde{y}_i(\tilde{y}_j)_{x_k}$, $(\tilde{u}_i - \tilde{y}_i)(u_j^0)_{x_k}$, and $u_j^0((\tilde{u}_i)_{x_k} - (\tilde{y}_i)_{x_k})$, for $i, j = 1, 2, 3, k = 1, 2, ..., n$. The terms of the first type are estimated as follows:

$$\begin{aligned} \|\tilde{u}_i(\tilde{u}_j)_{x_k} - \tilde{y}_i(\tilde{y}_j)_{x_k}\|_{L^2} &= \|\tilde{u}_i((\tilde{u}_j)_{x_k} - (\tilde{y}_j)_{x_k}) + (\tilde{u}_i - \tilde{y}_i)(\tilde{y}_j)_{x_k}\|_{L^2} \\ &\le \|\tilde{u}_i\|_{L^\infty}\|\tilde{u} - \tilde{y}\|_{\mathbf{H}^{1,2}} + \|\tilde{u}_i - \tilde{y}_i\|_{\mathbf{L}^\infty}\|\tilde{y}\|_{\mathbf{H}^{1,2}}. \end{aligned}$$

Since $3/4 < \alpha < 1$, the embedding theorems imply that $\mathbf{X}_\alpha \subset L^\infty(\Omega, R^3)$ and $\mathbf{X}_\alpha \subset \mathbf{H}^{1,2}$, with

$$\|\cdot\|_{\mathbf{L}^\infty} \leq C\|\cdot\|_\alpha, \qquad \|\cdot\|_{\mathbf{H}^{1,2}} \leq C\|\cdot\|_\alpha \tag{3.12}$$

for some positive constant C (see, e.g., Chapter 11 in [169]). We can thus obtain that the first type terms are locally Lipschitz in $\mathbf{X}_\alpha$. The other two types of terms can be treated similarly. The terms involving $\mathbf{f}$ and inequality (3.11) can be estimated by using (3.2), (3.3) and relationship (3.12) above. Using the Cauchy and Hölder inequalities, we obtain (3.10) and (3.11).

In order to prove Theorem 3.1, we will need some results from linear systems of second order hyperbolic equations. Consider the system of three equations for $\mathbf{w}(x,t) = (w_1, w_2, w_3)^T$:

$$\begin{cases} \mathbf{w}_{tt} + \mu \mathbf{w}_t = \Delta \mathbf{w} + B(x)\mathbf{w} & \text{for } (x,t) \in \Omega \times R, \\ \mathbf{w}(x,t) = 0 & \text{for } (x,t) \in \partial\Omega \times R. \end{cases} \tag{3.13}$$

Here, μ is any real constant and $B(x)$ is a 3×3 matrix with each entry b_{ij} bounded in L^∞. Let Ξ be the operator on $\mathbf{Y}$ defined by:

(3.14)
$$\Xi\{(y_1, .., y_6)^T\} = (y_2, \Delta y_1, y_4, \Delta y_3, y_6, \Delta y_5)^T \qquad \text{for } \mathbf{y} = (y_1, .., y_6)^T \in \mathbf{Y},$$

with domain $D(\Xi) = [(H^{2,2}(\Omega) \bigcap H_0^{1,2}(\Omega)) \times H_0^{1,2}(\Omega)]^3$ dense in $\mathbf{Y}$. Define $\tilde{B}(x)$ to be the 6×6 matrix:

$$\tilde{B}(x) = \begin{bmatrix} 0 & 0 & 0 & 0 & 0 & 0 \\ b_{11} & -\mu & b_{12} & 0 & b_{13} & 0 \\ 0 & 0 & 0 & 0 & 0 & 0 \\ b_{21} & 0 & b_{22} & -\mu & b_{23} & 0 \\ 0 & 0 & 0 & 0 & 0 & 0 \\ b_{31} & 0 & b_{32} & 0 & b_{33} & -\mu \end{bmatrix}.$$

Here, the entries of $\tilde{B}(x)$ are described above in terms of those of $B(x)$. The multiplication operator $\tilde{B}$ defined by $\mathbf{y} \to \tilde{B}\mathbf{y}$ is a bounded linear operator on $\mathbf{Y}$. If we set $w_1 = y_1, (w_1)_t = y_2, w_2 = y_3, (w_2)_t = y_4, w_3 = y_5, (w_3)_t = y_6$, the system (3.13) can be written in the form of a differential equation in $\mathbf{Y}$ as follows:

$$\mathbf{y}_t = \Xi\mathbf{y} + \tilde{B}\mathbf{y}, \qquad \text{for } \quad t \in R. \tag{3.15}$$

To analyze (3.15), we deduce the following properties:

Lemma 3.2. *The operator* $\Xi + \tilde{B}$ *with domain* $D(\Xi + \tilde{B}) = [(H^{2,2} \bigcap H_0^{1,2}) \times H_0^{1,2}]^3$ *is an infinitesimal generator of a* C_0 *group of bounded linear operators* $T(t)$ *on* $\mathbf{Y}$ *satisfying* $\|T(t)\| \leq Me^{k|t|}, -\infty < t < +\infty$, *for some positive constants* $M \geq 1$ *and* $k > 0$.

Proof. Let $\lambda > 0$, and $\mathbf{f} = (f_1, \dots, f_6)^T \in \mathbf{Y}$ be arbitrary. Consider the equation:

$$\lambda \mathbf{y} - \Xi \mathbf{y} = \mathbf{f}, \qquad \text{or} \tag{3.16}$$

$$\begin{cases} \lambda y_i - y_{i+1} = f_i, \\ \lambda y_{i+1} - \Delta y_i = f_{i+1} \end{cases} \tag{3.17}$$

for $i = 1, 3, 5$. These lead to $\lambda^2 y_i - \Delta y_i = \lambda f_i + f_{i+1}$, which has a unique solution $y_i \in H^{2,2} \bigcap H_0^{1,2}$ for each $i = 1, 3, 5$. Define $y_{i+1} = \lambda y_i - f_i \in H_0^{1,2}$. We obtain a unique $\mathbf{y} \in D(\Xi)$ satisfying (3.16), which can also be expressed in terms of the resolvent as $\mathbf{y} = R(\lambda, \Xi)\mathbf{f}$. Multiplying the second equation in (3.17) by y_{i+1} we obtain for $i = 1, 3, 5$:

$$\lambda \|y_{i+1}\|_{L^2}^2 + \int_\Omega \nabla y_{i+1} \cdot \nabla y_i \, dx = \int_\Omega f_{i+1} y_{i+1} \, dx.$$

Substituting $y_{i+1} = \lambda y_i - f_i$ in the second integral, we find

$$\begin{aligned} \lambda(\|y_{i+1}\|_{L^2}^2 + \|\nabla y_i\|_{L^2}^2) &= \textstyle\int_\Omega (\nabla f_i \nabla y_i + f_{i+1} y_{i+1}) \, dx \\ &\leq \|\nabla f_i\|_{L^2} \|\nabla y_i\|_{L^2} + \|f_{i+1}\|_{L^2} \|y_{i+1}\|_{L^2} \\ &\leq (\|\nabla f_i\|_{L^2}^2 + \|f_{i+1}\|_{L^2}^2)^{1/2} (\|\nabla y_i\|_{L^2}^2 + \|y_{i+1}\|_{L^2}^2)^{1/2}. \end{aligned}$$

For $\lambda < 0$, we deduce analogously that:

$$\lambda(\|y_{i+1}\|_{L^2}^2 + \|\nabla y_i\|_{L^2}^2) \geq -(\|\nabla f_i\|_{L^2}^2 + \|f_{i+1}\|_{L^2}^2)^{1/2} (\|\nabla y_i\|_{L^2}^2 + \|y_{i+1}\|_{L^2}^2)^{1/2}.$$

Defining $|\mathbf{y}| := (\|\nabla y_1\|_{L^2}^2 + \|y_2\|_{L^2}^2 + \|\nabla y_3\|_{L^2}^2 + \|y_4\|_{L^2}^2 + \|\nabla y_5\|_{L^2}^2 + \|y_6\|_{L^2}^2)^{1/2}$, from above we obtain the following:

$$|\mathbf{y}| \;=\; |R(\lambda, \Xi)\mathbf{f}| \;\leq\; |\lambda|^{-1} |\mathbf{f}| \quad \text{for all } \lambda \neq 0.$$

This leads to:

$$|R(\lambda,\Xi)^n\mathbf{f}| \le |\lambda|^{-n}|\mathbf{f}| \quad \text{for all } \lambda \neq 0, \ n=1,2,\ldots \tag{3.18}$$

Using (3.18) and the Poincaré inequality, we obtain for $\lambda \neq 0$,

$$\|R(\lambda,\Xi)^n\mathbf{f}\|_{\mathbf{Y}} \le M|R(\lambda,\Xi)^n\mathbf{f}| \le M|\lambda|^{-n}|\mathbf{f}| \le M|\lambda|^{-n}\|\mathbf{f}\|_{\mathbf{Y}}, \tag{3.19}$$

$n = 1,2,3\ldots$, for some constant $M > 0$. From Theorem 6.3, Chapter 1 in [184] (i.e. Theorem A4-4 in Chapter 6 below), (3.19) implies that Ξ is the infinitesimal generator of a C_0 group of bounded linear operators $\hat{T}(t)$ on $\mathbf{Y}$ satisfying $\|\hat{T}(t)\| \le M, -\infty < t < +\infty$. Here, the norm is the usual operator norm for bounded linear operators on $\mathbf{Y}$. Further, by the perturbation Theorem 1.1 of Chapter 3 in [184] (i.e. Theorem A4-5 in Chapter 6 below), we can assert that the operator $\Xi + \tilde{B}$ is an infinitesimal generator of a semigroup of bounded linear operators $T(t)$ on $\mathbf{Y}$ satisfying $\|T(t)\| \le M\exp\{M\|\tilde{B}\|t\}$, for $t > 0$. By Theorem 5.3 of Chapter 1 in [184] (i.e. Theorem A4-3 in Chapter 6), this implies that the operator norm of its resolvent satisfies

$$\|R(\lambda,\Xi+\tilde{B})^n\| \le M(\lambda - M\|\tilde{B}\|)^{-n} \quad \text{for } \lambda > M\|\tilde{B}\|, \ n=1,2,\ldots \tag{3.20}$$

For $\lambda > 0, \mathbf{f} \in \mathbf{Y}$, one readily find $R(\lambda,-\Xi)\mathbf{f} = R(-\lambda,\Xi)(-\mathbf{f})$, and

$$R(\lambda,-\Xi)^n\mathbf{f} = R(-\lambda,\Xi)^n((-1)^n\mathbf{f}), \text{ for } n=1,2\ldots$$

Thus from (3.19), we obtain:

$$\|R(\lambda,-\Xi)^n\| \le M\lambda^{-n}, \quad \text{for } \lambda > 0, \ n=1,2,\ldots \tag{3.21}$$

Theorem A4-3 in Chapter 6 thus implies that $-\Xi$ generates a semigroup of bounded linear operators $\hat{S}(t)$ satisfying $\|\hat{S}(t)\| \le M$ for $t > 0$. By means of the aforementioned perturbation Theorem A4-5 and of the converse part of Theorem A4-3 in Chapter 6 again, we obtain:

$$\|R(\lambda,-\Xi-\tilde{B})^n\| \le M(\lambda - M\|\tilde{B}\|)^{-n} \quad \text{for } \lambda > M\|\tilde{B}\|, \ n=1,2,\ldots \tag{3.22}$$

For $\lambda < -M\|\tilde{B}\|$, $\mathbf{f} \in \mathbf{Y}$, consider the relation:

$$R(\lambda,\Xi+\tilde{B})^n(\mathbf{f}) = R(-\lambda,-\Xi-\tilde{B})^n((-1)^n\mathbf{f}).$$

Thus we obtain from (3.22) that for $\lambda < -M\|\tilde{B}\|$ (or $-\lambda > M\|\tilde{B}\|$),

$$\begin{aligned}\|R(\lambda, \Xi + \tilde{B})^n(\mathbf{f})\|_{\mathbf{Y}} &= \|R(-\lambda, -\Xi - \tilde{B})^n((-1)^n\mathbf{f})\|_{\mathbf{Y}} \\ &\leq M(|\lambda| - M\|\tilde{B}\|)^{-n}\|\mathbf{f}\|, \qquad \text{for } n = 1, 2, \dots. \end{aligned} \tag{3.23}$$

From (3.20) and (3.23), we deduce from Theorem A4-4 in Chapter 6, that the operator $\Xi + \tilde{B}$ is an infinitesimal generator of a C_0 group of bounded linear operators $T(t)$ as described in the statement of Lemma 3.2 with $k = M\|\tilde{B}\|$.

We will need some estimates concerning solutions of nonlinear equations related to (3.13). For a given continuous function $\mathbf{z} : (-\infty, \tau] \to \mathbf{X}_\alpha$, $\boldsymbol{\theta} \in \mathbf{H}_0^{1,2}$, and $\tilde{\boldsymbol{\theta}} \in \mathbf{L}^2$, consider the problem with three components $\mathbf{w} = (w_1, w_2, w_3)^T$:

$$\begin{cases} \mathbf{w}_{tt} + \mu\mathbf{w}_t = \Delta\mathbf{w} + B(x)\mathbf{w} + \mathbf{G}(\mathbf{z}, \mathbf{w}) & \text{for } t \leq \tau, x \in \Omega, \\ \mathbf{w}(x, t) = 0 & \text{for } t \leq \tau, x \in \partial\Omega, \\ \mathbf{w}(x, \tau) = \boldsymbol{\theta}(x), \mathbf{w}_t(x, \tau) = \tilde{\boldsymbol{\theta}}(x) & \text{for } x \in \Omega. \end{cases} \tag{3.24}$$

Let $\boldsymbol{\eta} = (\eta_1, \eta_2, \eta_3, \eta_4, \eta_5, \eta_6)^T := (\theta_1, \tilde{\theta}_1, \theta_2, \tilde{\theta}_2, \theta_3, \tilde{\theta}_3)^T$, where $\boldsymbol{\theta} = (\theta_1, \theta_2, \theta_3)^T$, $\tilde{\boldsymbol{\theta}} = (\tilde{\theta}_1, \tilde{\theta}_2, \tilde{\theta}_3)^T$. Denote the solution of problem (3.24) by:

$$\boldsymbol{\phi}(t; \tau, \boldsymbol{\eta}, \mathbf{z}(\cdot, t)) = (w_1, \partial w_1/\partial t, w_2, \partial w_2/\partial t, w_3, \partial w_3/\partial t)^T, \text{ for } t \leq \tau.$$

Lemma 3.3. *For $t \leq \tau$, the solution $\boldsymbol{\phi}(t; \tau, \boldsymbol{\eta}, \mathbf{z}(\cdot))$ of problem (3.24) satisfies:*

$$\begin{aligned}\|\boldsymbol{\phi}(t; \tau, \boldsymbol{\eta}, \mathbf{z}(\cdot)) - \boldsymbol{\phi}(t; \tau, \hat{\boldsymbol{\eta}}, \hat{\mathbf{z}}(\cdot))\|_{\mathbf{Y}} &\leq Me^{(k+ML)(\tau-t)}\|\boldsymbol{\eta} - \hat{\boldsymbol{\eta}}\|_{\mathbf{Y}} \\ &\quad + ML\int_t^\tau e^{(k+ML)(r-t)}\|\mathbf{z}(r) - \hat{\mathbf{z}}(r)\|_\alpha dr. \end{aligned} \tag{3.25}$$

Here L, M and k are the constants described in Lemmas 3.1 and 3.2 above.

Proof. For simplicity, we denote $\boldsymbol{\phi}(t; \tau, \boldsymbol{\eta}, \mathbf{z})$ and $\boldsymbol{\phi}(t; \tau, \hat{\boldsymbol{\eta}}, \hat{\mathbf{z}})$ respectively by $\boldsymbol{\phi}(t)$ and $\hat{\boldsymbol{\phi}}(t)$. We will consider $\boldsymbol{\phi}(t)$ as a mild solutions of

$$\begin{aligned} &\boldsymbol{\phi}_t = \Xi\boldsymbol{\phi} + \tilde{B}\boldsymbol{\phi} + \tilde{\mathbf{G}}(\mathbf{z}, \boldsymbol{\phi}) \qquad t \leq \tau, \\ &\boldsymbol{\phi}(\tau) = \boldsymbol{\eta}, \end{aligned}$$

where $\tilde{\mathbf{G}} = (0, G_1, 0, G_2, 0, G_3)^T$, and $\mathbf{G} = (G_1, G_2, G_3)^T$. We use Lemma 3.2 to obtain an integral representation of $\boldsymbol{\phi}(t)$ in terms of the group $T(t)$, and estimate by means of Lemma 3.1. We also use the fact that the odd components of $\tilde{\mathbf{G}}$ are zero. We thus obtain for $t \leq \tau$:

$$\|\phi(t) - \hat{\phi}(t)\|_{\mathbf{Y}} \leq Me^{k(\tau-t)}\|\boldsymbol{\eta} - \hat{\boldsymbol{\eta}}\|_{\mathbf{Y}}$$

$$+ML \int_t^\tau e^{-k(t-s)}[\|\mathbf{z}(s) - \hat{\mathbf{z}}(s)\|_\alpha + \|\phi(s) - \hat{\phi}(s)\|_{\mathbf{Y}}]\, ds.$$

From the Gronwall's inequality (cf. [13]), we obtain:

$$e^{kt}\|\phi(t) - \hat{\phi}(t)\|_{\mathbf{Y}} \leq Me^{k\tau}\|\boldsymbol{\eta} - \hat{\boldsymbol{\eta}}\|_{\mathbf{Y}} + ML \int_t^\tau e^{ks}\|\mathbf{z}(s) - \hat{\mathbf{z}}(s)\|_\alpha ds$$

$$+ \int_t^\tau MLe^{ML(s-t)}[\ Me^{k\tau}\|\boldsymbol{\eta} - \hat{\boldsymbol{\eta}}\|_{\mathbf{Y}} + ML \int_s^\tau e^{kr}\|\mathbf{z}(r) - \hat{\mathbf{z}}(r)\|_\alpha dr\ ]ds.$$

Interchanging the order of integration and canceling some terms, we obtain (3.25).

Proof of Theorem 3.1. Recall the definition of the operator $A = A(\epsilon) = -\epsilon^{-1}\Delta$ on $\mathbf{X}$ and the related inequalities (3.7) and (3.8). We will construct the function σ^*, which represents the invariant manifold, as a fixed point of an appropriate mapping on a class of functions. For this purpose, we will use the estimates in Lemma 3.3 above to deduce that the mapping is a contraction.

Let K_1 and K_2 be arbitrary positive constants. Let Λ be the class of continuous functions $\sigma : \mathbf{Y} \to \mathbf{X}_\alpha$ satisfying:

$$\|\sigma(\mathbf{y})\|_\alpha \leq K_1, \quad \|\sigma(\mathbf{y}) - \sigma(\hat{\mathbf{y}})\|_\alpha \leq K_2\|\mathbf{y} - \hat{\mathbf{y}}\|_{\mathbf{Y}}, \tag{3.26}$$

for all $\mathbf{y}, \hat{\mathbf{y}} \in \mathbf{Y}$. For $\boldsymbol{\eta} = (\eta_1, \eta_2, \eta_3, \eta_4, \eta_5, \eta_6)^T \in \mathbf{Y}$, let $\boldsymbol{\omega}(t) = \boldsymbol{\omega}(t; 0, \boldsymbol{\eta}, \sigma) = (w_1, \partial w_1/\partial t, w_2, \partial w_2/\partial t, w_3, \partial w_3/\partial t)^T$ be the solution of:

$$\begin{cases} \mathbf{w}_{tt} + \mu\mathbf{w}_t = \Delta\mathbf{w} + B(x)\mathbf{w} + \mathbf{G}(\sigma(\boldsymbol{\omega}(t)), \mathbf{w}) & \text{for } t \leq 0,\ x \in \Omega, \\ \mathbf{w}(x,t) = 0 & \text{for } t \leq 0, x \in \partial\Omega, \\ \mathbf{w}(x,0) = (\eta_1, \eta_3, \eta_5)^T, \mathbf{w}_t(x,0) = (\eta_2, \eta_4, \eta_6)^T & \text{for } x \in \Omega, \end{cases} \tag{3.27}$$

where $\mathbf{w} = (w_1, w_2, w_3)^T$. As in Lemma 3.3, the growth of $\boldsymbol{\omega}(t)$ for each $t \leq 0$ can be estimated, and we can assert that the solution exists for all $t \leq 0$ (cf. [184]). For a given $\sigma \in \Lambda$, define $\mathbf{Q}(\sigma) : \mathbf{Y} \to \mathbf{X}_\alpha$ by:

$$\mathbf{Q}(\sigma)(\boldsymbol{\eta}) := \int_{-\infty}^0 e^{A(\epsilon)s} P\mathbf{F}(\sigma(\boldsymbol{\omega}(s)), \mathbf{w}(s))ds, \tag{3.28}$$

where $\boldsymbol{\omega}(s) = \boldsymbol{\omega}(s; 0, \boldsymbol{\eta}, \sigma)$ is defined above for $s \leq 0$. Observe that:

$$\mathbf{F}(\sigma(\boldsymbol{\omega}(s)), \mathbf{w}(s)) = \mathbf{f}(\sigma(\boldsymbol{\omega}(s)) + \mathbf{u}^0, \mathbf{w}(s) + \mathbf{v}^0) - (\sigma(\boldsymbol{\omega}(s)) + \mathbf{u}^0) \cdot \nabla(\sigma(\boldsymbol{\omega}(s)) + \mathbf{u}^0).$$

Let $\hat{K}_1 = K_1 + \|\mathbf{u}^0\|_\alpha$. By hypothesis (3.4) and inequalities (3.12) above, we find

$$\|P\mathbf{F}(\sigma(\boldsymbol{\omega}(s)), \mathbf{w}(s))\|_{\mathbf{X}} \leq \|\mathbf{F}(\sigma(\boldsymbol{\omega}), \mathbf{w})\|_{\mathbf{L}^2} \leq N(\hat{K}_1) + 4\hat{K}_1^2 C^2 \tag{3.29}$$

for $\|\sigma(\boldsymbol{\omega})\|_\alpha \leq K_1$. Inequalities (3.7) and (3.29) imply that formula (3.28) is well-defined. Moreover, for $\epsilon > 0$ sufficiently small, we have:

$$\|\mathbf{Q}(\sigma)(\boldsymbol{\eta})\|_\alpha \leq \int_{-\infty}^{0} \hat{M}(-s)^\alpha e^{\beta(\epsilon)s}[N(\hat{K}_1) + 4\hat{K}_1^2 C^2]\, ds \leq K_1.$$

We next prove that the function $\mathbf{Q}(\sigma)$ also satisfy the second property in (3.26), and will thus belong to the class Λ of functions. Let $\boldsymbol{\vartheta}(t) := \boldsymbol{\omega}(t; 0, \boldsymbol{\eta}, \sigma(\boldsymbol{\vartheta}))$, $\hat{\boldsymbol{\vartheta}}(t) := \boldsymbol{\omega}(t; 0, \hat{\boldsymbol{\eta}}, \hat{\sigma}(\hat{\boldsymbol{\vartheta}}))$ be solutions of problem (3.27) as described above. By Lemma 3.3, we find that for $t \leq 0$, we have

$$\begin{aligned}
&\|\boldsymbol{\vartheta}(t) - \hat{\boldsymbol{\vartheta}}(t)\|_{\mathbf{Y}} \\
&\quad \leq Me^{-(k+ML)t}\|\boldsymbol{\eta} - \hat{\boldsymbol{\eta}}\|_{\mathbf{Y}} + ML\textstyle\int_t^0 e^{(k+ML)(s-t)}\|\sigma(\boldsymbol{\vartheta}(s)) - \hat{\sigma}(\hat{\boldsymbol{\vartheta}}(s))\|_\alpha ds \\
&\quad \leq Me^{-(k+ML)t}\|\boldsymbol{\eta} - \hat{\boldsymbol{\eta}}\|_{\mathbf{Y}} \\
&\qquad + ML\textstyle\int_t^0 e^{(k+ML)(s-t)}\{K_2\|\boldsymbol{\vartheta}(s) - \hat{\boldsymbol{\vartheta}}(s)\|_{\mathbf{Y}} + \|\sigma - \hat{\sigma}\|_\Lambda\}ds \\
&\quad \leq Me^{-(k+ML)t}\|\boldsymbol{\eta} - \hat{\boldsymbol{\eta}}\|_{\mathbf{Y}} + ML(k+ML)^{-1}[e^{-(k+ML)t} - 1]\|\sigma - \hat{\sigma}\|_\Lambda \\
&\qquad + ML\textstyle\int_t^0 e^{(k+ML)(s-t)}K_2\|\boldsymbol{\vartheta}(s) - \hat{\boldsymbol{\vartheta}}(s)\|_{\mathbf{Y}} ds,
\end{aligned}$$

where $\|\sigma - \hat{\sigma}\|_\Lambda := \sup\{\|\sigma(\mathbf{y}) - \hat{\sigma}(\mathbf{y})\|_\alpha : \mathbf{y} \in \mathbf{Y}\}$. For convenience, let $\lambda = (k + ML)$; we obtain from above, by means of Gronwall's inequality:

$$\begin{aligned}
\|\boldsymbol{\vartheta}(t) - \hat{\boldsymbol{\vartheta}}(t)\|_{\mathbf{Y}} &\leq M\|\boldsymbol{\eta} - \hat{\boldsymbol{\eta}}\|_{\mathbf{Y}} e^{-(\lambda+MLK_2)t} \\
&\quad + e^{-(\lambda+MLK_2)t} ML(MLK_2 + \lambda)^{-1}\|\sigma - \hat{\sigma}\|_\Lambda \\
&\quad - ML(MLK_2 + \lambda)^{-1}\|\sigma - \hat{\sigma}\|_\Lambda.
\end{aligned} \tag{3.30}$$

For $\boldsymbol{\vartheta} = (\vartheta_1, \vartheta_2, \vartheta_3, \vartheta_4, \vartheta_5, \vartheta_6)^T$ and $\hat{\boldsymbol{\vartheta}} = (\hat{\vartheta}_1, \hat{\vartheta}_2, \hat{\vartheta}_3, \hat{\vartheta}_4, \hat{\vartheta}_5, \hat{\vartheta}_6)^T$ as above, we define for convenience the odd components $\boldsymbol{\vartheta}_0(t) := (\vartheta_1, \vartheta_3, \vartheta_5)^T$ and $\hat{\boldsymbol{\vartheta}}_0(t) := (\hat{\vartheta}_1, \hat{\vartheta}_3, \hat{\vartheta}_5)^T$. From (3.8), (3.10), (3.28) and (3.30) we deduce the following:

(3.31)
$$\|\mathbf{Q}(\sigma)(\boldsymbol{\eta}) - \mathbf{Q}(\hat{\sigma})(\hat{\boldsymbol{\eta}})\|_\alpha$$

$$\leq \int_{-\infty}^0 \hat{M}(-s)^{-\alpha} e^{\beta s} \|P\mathbf{F}(\sigma(\boldsymbol{\vartheta}(s)), \boldsymbol{\vartheta}_0(s)) - P\mathbf{F}(\hat{\sigma}(\hat{\boldsymbol{\vartheta}}(s)), \hat{\boldsymbol{\vartheta}}_0(s))\|_{\mathbf{X}} ds$$

$$\leq \int_{-\infty}^0 \hat{M}(-s)^{-\alpha} e^{\beta s} C(K_1)\{\|\sigma - \hat{\sigma}\|_\Lambda + K_2 \|\boldsymbol{\vartheta}(s) - \hat{\boldsymbol{\vartheta}}(s)\|_{\mathbf{Y}}$$

$$+ \|\boldsymbol{\vartheta}_0(s) - \hat{\boldsymbol{\vartheta}}_0(s)\|_{\mathbf{H}^{1,2}}\} ds$$

$$\leq \hat{M}C(K_1) \int_0^\infty u^{-\alpha} e^{-\beta u} du \|\sigma - \hat{\sigma}\|_\Lambda$$

$$+ \hat{M}C(K_1)[K_2 + 1] \int_{-\infty}^0 (-s)^{-\alpha} e^{\beta s} [\, M e^{-(\lambda + MLK_2)s} \|\boldsymbol{\eta} - \hat{\boldsymbol{\eta}}\|_{\mathbf{Y}}$$

$$+ ML(MLK_2 + \lambda)^{-1} \|\sigma - \hat{\sigma}\|_\Lambda \{e^{-(\lambda + MLK_2)s} - 1\}\,]\, ds$$

$$= \hat{M}C(K_1)[K_2 + 1]M\, \|\boldsymbol{\eta} - \hat{\boldsymbol{\eta}}\|_{\mathbf{Y}}\, \theta_2 + \hat{M}C(K_1)\|\sigma - \hat{\sigma}\|_\Lambda\, [\, \theta_2(1 + K_2)ML \cdot$$

$$(MLK_2 + \lambda)^{-1} + \theta_1\{1 - (1 + K_2)ML(MLK_2 + \lambda)^{-1}\}\,]$$

$$\leq \hat{M}C(K_1)[K_2 + 1]\, M \|\boldsymbol{\eta} - \hat{\boldsymbol{\eta}}\|_{\mathbf{Y}}\, \theta_2 + \hat{M}C(K_1)\|\sigma - \hat{\sigma}\|_\Lambda\, \theta_2,$$

where

$$\theta_1 = \int_0^\infty u^{-\alpha} e^{-\beta u} du, \quad \theta_2 = \int_0^\infty u^{-\alpha} e^{-\beta u} e^{(\lambda + MLK_2)u} du. \tag{3.32}$$

The last inequality in (3.31) is justified by $0 < \theta_1 < \theta_2$. Moreover, in view of the properties of β stated for (3.7), $\theta_2 \to 0^+$ as $\epsilon \to 0^+$. Hence, for sufficiently small $\epsilon > 0$, we have

$$\hat{M}C(K_1)[K_2 + 1]M\theta_2 < K_2, \quad \text{and} \quad \hat{M}C(K_1)\theta_2 < 1. \tag{3.33}$$

The mapping $\mathbf{Q}(\sigma)$ is thus a contraction in the class of functions Λ with norm $\|\cdot\|_\Lambda$; and there exists a fixed point σ^0 where $\mathbf{Q}(\sigma^0) = \sigma^0$.

We now prove that if $\tilde{\boldsymbol{\omega}}(t)$ is a solution of (3.27) for $-\infty < t < \infty$ with σ replaced by σ^0, then $(\sigma^0(\tilde{\boldsymbol{\omega}}(t)), \tilde{\mathbf{w}}(t))$ satisfies (3.9) for $-\infty < t < \infty$, with $\tilde{\mathbf{w}}(t) := \tilde{\boldsymbol{\omega}}_0(t)$, which consists of the odd components of $\tilde{\boldsymbol{\omega}}(t)$. From (3.28) and the autonomous property of (3.27), we have

$$\sigma^0(\tilde{\boldsymbol{\omega}}(t)) = \mathbf{Q}(\sigma^0)(\tilde{\boldsymbol{\omega}}(t))$$

$$= \int_{-\infty}^0 e^{As} P\mathbf{F}(\sigma^0(\tilde{\boldsymbol{\omega}}(s + t)), \tilde{\mathbf{w}}(s + t)) ds$$

$$= \int_{-\infty}^t e^{A(\tau - t)} P\mathbf{F}(\sigma^0(\tilde{\boldsymbol{\omega}}(\tau)), \tilde{\mathbf{w}}(\tau)) d\tau.$$

This shows that $\tilde{\mathbf{u}} = \sigma^0(\tilde{\boldsymbol{\omega}}(t))$ is a bounded solution in $\mathbf{X}_\alpha$ of:

$$\tilde{\mathbf{u}}_t = -A\tilde{\mathbf{u}} + P\mathbf{F}(\sigma^0(\tilde{\boldsymbol{\omega}}(t), \tilde{\mathbf{w}}(t)), \quad -\infty < t < \infty. \tag{3.34}$$

(Note that the nonhomogeneous linear system (3.34) cannot have more than one bounded solution for $-\infty < t < \infty$, since 0 is not in the spectrum of $\epsilon^{-1}\Delta$). Finally for $(z_1, z_2, z_3)^T \in \mathbf{H}_0^{1,2}$ and $(z_4, z_5, z_6)^T \in \mathbf{L}^2$, define

$$\sigma^*((z_1, z_2, z_3, z_4, z_5, z_6)^T) = \sigma^0(\mathbf{y}), \tag{3.35}$$

where $\mathbf{y} = (y_1, y_2, y_3, y_4, y_5, y_6)^T := (z_1, z_4, z_2, z_5, z_3, z_6)^T$. Then we have the existence of an invariant manifold as described by $\mathbf{S}$ in the statement of Theorem 3.1. Moreover,
the function σ^* has the boundedness and Lipschitz property as described. This completes the proof of Theorem 3.1.

Remark 3.3. The local existence and continuation of solution of (3.9) in semigroup setting with $(\tilde{\mathbf{u}}, \tilde{\mathbf{v}}, \tilde{\mathbf{v}}_t)$ in $\mathbf{X}_\alpha \times \mathbf{H}_0^{1,2} \times \mathbf{L}^2$ under the conditions of Lemma 3.1 for arbitrary initial conditions can be proved by fixed point methods similar to that in the proof of Theorem 3.1 or in [84] and [184]. Since the emphasis in this section is not on this issue, the details will be omitted. A solution $(\mathbf{u}(\cdot, t), \mathbf{v}(\cdot, t))$ of (3.1) in this section will always be interpreted as $\mathbf{u} = \tilde{\mathbf{u}} + \mathbf{u}^0, \mathbf{v} = \tilde{\mathbf{v}} + \mathbf{v}^0$, where $(\tilde{\mathbf{u}}, \tilde{\mathbf{v}}, \tilde{\mathbf{v}}_t)$ is a semigroup solution of (3.9) in $\mathbf{X}_\alpha \times \mathbf{H}_0^{1,2} \times \mathbf{L}^2$.

Remark 3.4. Under the hypotheses of Theorem 3.1, the size of small ϵ for the existence of the invariant manifold is determined by the two inequalities in (3.33).

Part B: Dependence on Initial Conditions, Asymptotic Stability of the Manifold, and Applications.

We next proceed to estimate the dependence of the solutions on initial conditions, and study the asymptotic stability of the manifold. We first investigate the dependence of solutions on initial conditions and the relationship between solutions of the original problem (3.1) and solutions on the manifold found in Part A of this section. In Theorem 3.2, we will consider the asymptotic stability of the manifold. Let $(\mathbf{u}(\cdot, t), \mathbf{v}(\cdot, t))$ be a solution of (3.1), and $\tilde{\mathbf{u}} = \mathbf{u} - \mathbf{u}^0, \tilde{\mathbf{v}} = \mathbf{v} - \mathbf{v}^0 = (\tilde{v}_1, \tilde{v}_2, \tilde{v}_3)^T$ be the corresponding of solution of (3.9). Define:

$$\gamma(\cdot, t) := (\tilde{v}_1, \partial\tilde{v}_1/\partial t, \tilde{v}_2, \partial\tilde{v}_2/\partial t, \tilde{v}_3, \partial\tilde{v}_3/\partial t)^T. \tag{3.36}$$

Recall that for any $\mathbf{y} = (y_1, y_2, y_3, .., y_6)^T \in \mathbf{Y}$, we denote $\mathbf{y}_0 = (y_1, y_3, y_5)^T \in \mathbf{H}^{1,2}$ to be the odd components of $\mathbf{y}$. For $0 \leq s \leq t$, let $\boldsymbol{\psi}(s; t) := (\psi_1, \psi_2, \psi_3, \ldots$

$\psi_6)^T \in \mathbf{Y}$ such that $\partial\psi_i/\partial s = \psi_{i+1}$, for $i = 1, 3, 5$, and $\boldsymbol{\psi}_0(s;t) = \boldsymbol{\psi}_0(s;t,(\tilde{\mathbf{v}}(\cdot,t),$ $\tilde{\mathbf{v}}_t(\cdot,t)), \sigma^0(\boldsymbol{\psi}(s;t))$ described above satisfy:

(3.37)
$$\begin{cases} \partial^2\boldsymbol{\psi}_0/\partial s^2 + \mu\partial\boldsymbol{\psi}_0/\partial s = \Delta\boldsymbol{\psi}_0 + B(x)\boldsymbol{\psi}_0 + \mathbf{G}(\sigma^0(\boldsymbol{\psi}),\boldsymbol{\psi}_0) & \text{for } s \le t,\ x \in \Omega, \\ \boldsymbol{\psi}_0(s;t)|_{x\in\partial\Omega} = 0 & \text{for } s \le t, \\ \boldsymbol{\psi}_0(t;t) = \tilde{\mathbf{v}}(\cdot,t),\ (\partial\boldsymbol{\psi}_0/\partial s)(t;t) = \tilde{\mathbf{v}}_t(\cdot,t) & \text{for } x \in \Omega. \end{cases}$$

The function $\boldsymbol{\gamma}(s)$ describes the components in $\mathbf{Y}$ for a solution of (3.9), which is converted from (3.1). The function $\boldsymbol{\psi}(s;t)$ is the $\mathbf{Y}$ components of a solution on the manifold. For $s = t$, we have $\boldsymbol{\psi}(t;t) = \boldsymbol{\gamma}(t)$. We now compare the solutions $\boldsymbol{\psi}(s;t)$ with $\boldsymbol{\gamma}(s)$ for $s \le$ t. They will be related to the function:

$$\boldsymbol{\xi}(t) := \tilde{\mathbf{u}}(\cdot,t) - \sigma^0(\boldsymbol{\gamma}(\cdot,t)). \tag{3.38}$$

Lemma 3.4. *For $0 \le s \le t$, the functions $\boldsymbol{\psi}(s;t)$ and $\boldsymbol{\gamma}(s)$ described above satisfy:*

$$\|\boldsymbol{\psi}(s;t) - \boldsymbol{\gamma}(s)\|_{\mathbf{Y}} \le ML\int_s^t \exp\{[k + ML(K_2+1)](r-s)\}\|\boldsymbol{\xi}(r)\|_\alpha dr. \tag{3.39}$$

Moreover, for $s \le 0 \le t, \boldsymbol{\psi}$ satisfies:

$$\|\boldsymbol{\psi}(s;0) - \boldsymbol{\psi}(s;t)\|_{\mathbf{Y}} \le M^2L\int_0^t \exp\{[k + ML(K_2+1)](r-s)\}\|\boldsymbol{\xi}(r)\|_\alpha dr. \tag{3.40}$$

Recall that L, M, k and K_2 are the constants described in Lemmas 3.1, 3.2 and Theorem 3.1.

Proof. For $0 \le s \le t$, we obtain from Lemma 3.3:

$$\begin{aligned} \|\boldsymbol{\psi}(s;t) - \boldsymbol{\gamma}(s)\|_{\mathbf{Y}} &\le ML\textstyle\int_s^t e^{(k+ML)(r-s)}\|\sigma^0(\boldsymbol{\psi}(r;t)) - \tilde{\mathbf{u}}(\cdot,r)\|_\alpha\, dr \\ &\le ML\textstyle\int_s^t e^{(k+ML)(r-s)}\{\|\sigma^0(\boldsymbol{\psi}(r;t)) - \sigma^0(\boldsymbol{\gamma}(r))\|_\alpha + \|\sigma^0(\boldsymbol{\gamma}(r)) - \tilde{\mathbf{u}}(\cdot,r)\|_\alpha\}\, dr. \end{aligned}$$

Letting $\boldsymbol{\xi}(r) = \tilde{\mathbf{u}}(\cdot,r) - \sigma^0(\boldsymbol{\gamma}(r))$, we obtain the following inequality for $0 \le s \le t$:

$$e^{(k+ML)s}\|\boldsymbol{\psi}(s;t) - \boldsymbol{\gamma}(s)\|_{\mathbf{Y}} \le ML\int_s^t e^{(k+ML)r}\{\|\boldsymbol{\xi}(r)\|_\alpha + K_2\|\boldsymbol{\psi}(r;t) - \boldsymbol{\gamma}(r)\|_{\mathbf{Y}}\}\, dr.$$

By the Gronwall's inequality, we find

$$\begin{aligned} e^{(k+ML)s}\|\boldsymbol{\psi}(s;t) - \boldsymbol{\gamma}(s)\|_{\mathbf{Y}} &\le ML\textstyle\int_s^t e^{(k+ML)r}\|\boldsymbol{\xi}(r)\|_\alpha\, dr \\ &\quad + MLK_2\textstyle\int_s^t exp\{MLK_2(r-s)\}[ML\int_r^t e^{(k+ML)\tau}\|\boldsymbol{\xi}(\tau)\|_\alpha d\tau]\, dr \\ &= \textstyle\int_s^t exp\{MLK_2(r-s)\}MLe^{(k+ML)r}\|\boldsymbol{\xi}(r)\|_\alpha\, dr. \end{aligned}$$

This yields inequality (3.39) for $0 \leq s \leq t$.

For $s \leq 0 \leq t$, we use Lemma 3.3 again to obtain:

$$\|\psi(s;0) - \psi(s;t)\|_{\mathbf{Y}} \leq Me^{-(k+ML)s}\|\psi(0;0) - \psi(0;t)\|_{\mathbf{Y}}$$

$$+ML\int_s^0 e^{(k+ML)(r-s)}\|\sigma^0(\psi(r;0)) - \sigma^0(\psi(r;t))\|_\alpha \, dr.$$

This gives

$$e^{(k+ML)s}\|\psi(s;0) - \psi(s;t)\|_{\mathbf{Y}} \leq M\|\gamma(0) - \psi(0;t)\|_{\mathbf{Y}}$$

$$+ML\int_s^0 e^{(k+ML)r}K_2\|\psi(r;0) - \psi(r;t)\|_{\mathbf{Y}} \, dr.$$

Gronwall's inequality gives for $s \leq 0 \leq t$,

$$e^{(k+ML)s}\|\psi(s;0) - \psi(s;t)\|_{\mathbf{Y}} \leq M\|\gamma(0) - \psi(0;t)\|_{\mathbf{Y}}$$

$$+MLK_2\int_s^0 exp\{MLK_2(r-s)\}M\|\gamma(0) - \psi(0;t)\|_{\mathbf{Y}} \, dr$$

$$= M\|\gamma(0) - \psi(0;t)\|_{\mathbf{Y}} \, exp\{-MLK_2 s\}$$

$$\leq M\, exp\{-MLK_2 s\}ML\int_0^t exp\{[k + ML(K_2+1)]r\}\|\boldsymbol{\xi}(r)\|_\alpha \, dr.$$

(Note that the last line above is due to (3.39), with $s = 0$.) This leads to inequality (3.40), and completes the proof of this Lemma.

We now prove that solutions of (3.1), which start close to the invariant manifold described in Theorem 3.1 at $t = 0$, will tend to the manifold as $t \to +\infty$. For convenience, we define

$$q(\epsilon) := \left(C(2K_1)\hat{M}\{1 + \frac{M^2L(1+K_2)}{\beta(\epsilon) - k - ML(1+K_2)}\}\Gamma(1-\alpha)\right)^{1/(1-\alpha)}, \tag{3.41}$$

where Γ is the Gamma function, and K_1, K_2 are the constants defined in the proof of Theorem 3.1. Also, recall the definition of $\beta(\epsilon)$ and $\hat{M}$ in (3.7). We will use Lemma 3.4 to prove the following asymptotic stability theorem.

Theorem 3.2 (Asymptotic Stability of the Invariant Manifold). *Under hypotheses (3.2)–(3.4), suppose that $\epsilon > 0$ is small enough for Theorem 3.1 to hold. Let $(\mathbf{u}(\cdot,t), \mathbf{v}(\cdot,t))$ be a solution of (3.1) such that $(\mathbf{u}(t), \mathbf{v}(t), \mathbf{v}_t(t)) \in \mathbf{X}_\alpha \times \tilde{\mathbf{Y}}$ for $t \geq 0$ and, possibly taking a smaller ϵ, assume that:*

$$q(\epsilon) < \beta(\epsilon). \tag{3.42}$$

Then the invariant manifold through

$$(0, \mathbf{u}^0 + \sigma^*(\mathbf{v}(\cdot,0) - \mathbf{v}^0, \mathbf{v}_t(\cdot,0)), (\mathbf{v}(\cdot,0), \mathbf{v}_t(\cdot,0)))$$

is asymptotically stable in the sense that

$$\|\mathbf{u}(\cdot,t) - [\mathbf{u}^0 + \sigma^*(\mathbf{v}(\cdot,t) - \mathbf{v}^0, \mathbf{v}_t(\cdot,t))]\|_\alpha \leq \tag{3.43}$$
$$\hat{M}\hat{R}\|\mathbf{u}(\cdot,0) - [\mathbf{u}^0 + \sigma^*(\mathbf{v}(\cdot,0) - \mathbf{v}^0, \mathbf{v}_t(\cdot,0))]\|_\alpha e^{(q-\beta)t}$$

for $t \geq 0$, *provided that the norm of the initial data* $\|\mathbf{u}(\cdot,0) - [\mathbf{u}^0 + \sigma^*(\mathbf{v}(\cdot,0) - \mathbf{v}^0, \mathbf{v}_t(\cdot,0))]\|_\alpha$ *is small enough. Here* $\hat{R}$ *is a constant depending on* α. *(Note that (3.42) is always satisfied for* ϵ *sufficiently small, since* $\beta(\epsilon) \to +\infty$ *as* $\epsilon \to 0^+$; *also, recall* $3/4 < \alpha < 1$.*)*

Proof. Using (3.38) and the integral representation of $\tilde{\mathbf{u}}(\cdot,t)$, we deduce for $t \geq 0$ that:

$$\begin{aligned}\boldsymbol{\xi}(t) &= e^{-At}\boldsymbol{\xi}(0) + \textstyle\int_0^t e^{-A(t-s)}[P\mathbf{F}(\tilde{\mathbf{u}}(\cdot,s), \tilde{\mathbf{v}}(\cdot,s)) - P\mathbf{F}(\sigma^0(\boldsymbol{\psi}(s,t)), \boldsymbol{\psi}_0(s,t))]ds \\ &\quad + \textstyle\int_0^t e^{-A(t-s)}P\mathbf{F}(\sigma^0(\boldsymbol{\psi}(s,t)), \boldsymbol{\psi}_0(s,t))ds - \sigma^0(\boldsymbol{\gamma}(\cdot,t)) + e^{-At}\sigma^0(\boldsymbol{\gamma}(\cdot,0)).\end{aligned} \tag{3.44}$$

Using the autonomous property

$$\sigma^0(\boldsymbol{\gamma}(\cdot,t)) = \mathbf{Q}(\sigma^0(\boldsymbol{\gamma}(\cdot,t)) = \int_{-\infty}^t e^{-A(t-s)}P\mathbf{F}(\sigma^0(\boldsymbol{\psi}(s,t)), \boldsymbol{\psi}_0(s,t))ds$$

and

$$e^{-At}\sigma^0(\boldsymbol{\gamma}(\cdot,0)) = e^{-At}\int_{-\infty}^0 e^{As}P\mathbf{F}(\sigma^0(\boldsymbol{\psi}(s,0)), \boldsymbol{\psi}_0(s,0))ds,$$

we can rewrite the last three terms of (3.44) to obtain:

$$\begin{aligned}\boldsymbol{\xi}(t) &= e^{-At}\boldsymbol{\xi}(0) + \textstyle\int_0^t e^{-A(t-s)}[P\mathbf{F}(\tilde{\mathbf{u}}(\cdot,s), \tilde{\mathbf{v}}(\cdot,s)) - P\mathbf{F}(\sigma^0(\boldsymbol{\psi}(s,t)), \boldsymbol{\psi}_0(s,t))]ds \\ &\quad + \textstyle\int_{-\infty}^0 e^{-A(t-s)}[P\mathbf{F}(\sigma^0(\boldsymbol{\psi}(s,0)), \boldsymbol{\psi}_0(s,0)) - P\mathbf{F}(\sigma^0(\boldsymbol{\psi}(s,t)), \boldsymbol{\psi}_0(s,t))]ds.\end{aligned} \tag{3.45}$$

As long as $\|\tilde{\mathbf{u}}(\cdot,s)\|_\alpha \leq 2K_1$, we can use (3.8), (3.10), (3.39) and the fact that

$\sigma^0 \in \Lambda$ to estimate:

(3.46)

$$\int_0^t \|e^{-A(t-s)}[P\mathbf{F}(\tilde{\mathbf{u}}(\cdot,s),\tilde{\mathbf{v}}(\cdot,s)) - P\mathbf{F}(\sigma^0(\boldsymbol{\psi}(s;t),\boldsymbol{\psi}_0(s;t))]\|_\alpha\, ds$$

$$\le \hat{M} \int_0^t (t-s)^{-\alpha} e^{-\beta(t-s)} C(2K_1)[\|\tilde{\mathbf{u}}(\cdot,s) - \sigma^0(\boldsymbol{\psi}(s;t))\|_\alpha +$$

$$ML \int_s^t exp\{[k+ML(K_2+1)](r-s)\}\|\boldsymbol{\xi}(r)\|_\alpha dr]\, ds$$

$$\le \hat{M} \int_0^t (t-s)^{-\alpha} e^{-\beta(t-s)} C(2K_1)[\|\boldsymbol{\xi}(s)\|_\alpha +$$

$$ML(K_2+1) \int_s^t exp\{[k+ML(K_2+1)](r-s)\}\|\boldsymbol{\xi}(r)\|_\alpha dr]\, ds$$

$$= \hat{M} \int_0^t C(2K_1)(t-s)^{-\alpha} e^{-\beta(t-s)} \|\boldsymbol{\xi}(s)\|_\alpha ds$$

$$+ \hat{M}C(2K_1)ML(K_2+1) \int_0^t exp\{[k+ML(K_2+1)](r-t)\}\|\boldsymbol{\xi}(r)\|_\alpha \cdot$$

$$\int_{t-r}^t z^{-\alpha} exp\{-(\beta - k - ML(K_2+1))z\} dz\, dr.$$

Moreover, from inequality (3.40), we can estimate:

(3.47)

$$\int_{-\infty}^0 \|e^{-A(t-s)}[P\mathbf{F}(\sigma^0(\boldsymbol{\psi}(s;t)),\boldsymbol{\psi}_0(s;t)) - P\mathbf{F}(\sigma^0(\boldsymbol{\psi}(s;0)),\boldsymbol{\psi}_0(s;0))]\|_\alpha\, ds$$

$$\le \hat{M} \int_{-\infty}^0 (t-s)^{-\alpha} e^{-\beta(t-s)} C(2K_1)[\|\sigma^0(\boldsymbol{\psi}(s;t) - \sigma^0(\boldsymbol{\psi}(s;0))\|_\alpha$$

$$+\|\boldsymbol{\psi}(s;t) - \boldsymbol{\psi}(s;0)\|_{\mathbf{Y}}]ds$$

$$\le \hat{M} \int_{-\infty}^0 (t-s)^{-\alpha} e^{-\beta(t-s)} C(2K_1) M^2 L(K_2+1) \cdot$$

$$\int_0^t exp\{[k+ML(K_2+1)](r-s)\}\|\boldsymbol{\xi}(r)\|_\alpha dr\, ds$$

$$= \hat{M}C(2K_1)M^2L(K_2+1) \int_0^t exp\{[k+ML(K_2+1)](s-t)\}\|\boldsymbol{\xi}(s)\|_\alpha \cdot$$

$$\int_t^\infty z^{-\alpha} exp\{-(\beta - k - ML(K_2+1))z\} dz\, ds.$$

Combining (3.45) to (3.47), we obtain:

$$\|\boldsymbol{\xi}(t)\|_\alpha \le \hat{M}e^{-\beta t}\|\boldsymbol{\xi}(0)\|_\alpha + \hat{M}C(2K_1)\int_0^t \|\boldsymbol{\xi}(s)\|_\alpha I(s,t)ds,$$

where

$$I(s,t) = (t-s)^{-\alpha}e^{-\beta(t-s)} + M^2L(K_2+1)exp\{[k+ML(K_2+1)](s-t)\}\cdot$$
$$\textstyle\int_{t-s}^{\infty} z^{-\alpha}exp\{-[\beta-k-ML(K_2+1)]z\}\,dz$$
$$\le (t-s)^{-\alpha}e^{-\beta(t-s)} + M^2L(K_2+1)exp\{[k+ML(K_2+1)](s-t)\}\cdot$$
$$(t-s)^{-\alpha}\textstyle\int_{t-s}^{\infty} exp\{-[\beta-k-ML(K_2+1)]z\}\,dz$$
$$= (t-s)^{-\alpha}e^{-\beta(t-s)}\{1+\tfrac{M^2L(K_2+1)}{\beta-k-ML(K_2+1)}\}.$$

We obtain

$$e^{\beta t}\|\boldsymbol{\xi}(t)\|_\alpha \le \hat{M}\|\boldsymbol{\xi}(0)\|_\alpha + b\int_0^t (t-s)^{-\alpha}e^{\beta s}\|\boldsymbol{\xi}(s)\|_\alpha\,ds, \tag{3.48}$$

where $b = \hat{M}C(2K_1)\{1+M^2L(K_2+1)[\beta-k-ML(K_2+1)]^{-1}\}$. Inequality (3.48) holds as long as $\|\tilde{\mathbf{u}}(\cdot,s)\|_\alpha \le 2K_1$ for $0 \le s \le t$. Hence, inequality (3.43) follows from (3.48) and the generalized Gronwall inequality (cf. [84]), as long as $\|\tilde{\mathbf{u}}\|_\alpha \le 2K_1$. Here $\hat{R}$ depends on K_1, α and b. Recall that $\|\sigma^*\|_\alpha \le K_1$ by Theorem 3.1. Consequently, (3.43) holds for all $t \ge 0$, provided that $\|\mathbf{u}(\cdot,0) - [\mathbf{u}^0 + \sigma^*(\mathbf{v}(\cdot,0) - \mathbf{v}^0, \mathbf{v}_t(\cdot,0))]\|_\alpha \le K_1(\hat{M}\hat{R})^{-1}$.

Remark 3.5. For the reader's convenience, the generalized Gronwall's inequality can be found in Theorem 7.1.1 of [84]. Let $w(t) := e^{\beta t}\|\boldsymbol{\xi}(t)\|_\alpha$, which satisfies inequality (3.48). Then the generalized Gronwall's inequality implies that $w(t)$ satisfies:

$$w(t) \le \hat{M}\|\boldsymbol{\xi}(0)\|_\alpha C e^{\theta t} \qquad \text{for } t \ge 0,$$

where C is a constant which depends on α and b, and $\theta = (b\Gamma(1-\alpha))^{1/(1-\alpha)}$. Here Γ is the Gamma function.

We finally consider some special cases when the theories of the last two theorems can be readily applied. We will restrict the damping constant to be positive, i.e. $\mu > 0$, in the remaining part of this article. Such sign condition is appropriate for application of (3.1) to most physical problems.

Theorem 3.3 (Asymptotic Stability for Equilibrium on the Invariant Manifold). *Assume all the hypotheses of Theorem 3.2 and* $\mathbf{h} = \mathbf{q} = 0, \mathbf{u}^0 = \mathbf{v}^0 = 0$, $\mu > 0$. *Suppose that the origin is on the invariant manifold and is asymptotically stable with respect to flows on the manifold as* $t \to +\infty$, *then it is asymptotically stable with respect to solutions of the full system (3.1).*

Remark 3.6. Roughly speaking, if the origin attracts neighboring solutions on the invariant manifold, then it will attract all neighboring solutions of the full system. The theorem will be proved by means of Lyapunov's method.

Proof. The evolution on the manifold is described by the system:

$$\begin{cases} \mathbf{z}_{tt} + \mu \mathbf{z}_t = \Delta \mathbf{z} + B(x)\mathbf{z} + \mathbf{G}(\sigma^*(\mathbf{z}, \mathbf{z}_t), \mathbf{z}) & \text{for } x \in \Omega, -\infty < t < \infty, \\ \mathbf{z}(x,t) = 0 & \text{for } x \in \partial\Omega, -\infty < t < \infty. \end{cases} \tag{3.49}$$

Let $\chi_0 = (\chi_0^1, \chi_0^2, \chi_0^3, ..., \chi_0^6)^T \in \mathbf{Y}$, and $\tilde{\chi}(t; t_0, \chi_0) := (\mathbf{z}(t; t_0, (\chi_0^1, \chi_0^3, \chi_0^5, \chi_0^2, \chi_0^4, \chi_0^6)^T, \mathbf{z}_t(t; t_0, (\chi_0^1, \chi_0^3, \chi_0^5, \chi_0^2, \chi_0^4, \chi_0^6)^T))$ has its first three components as the solution of (3.49) with initial conditions $(\mathbf{z}(t_0), \mathbf{z}_t(t_0)) = (\chi_0^1, \chi_0^3, \chi_0^5, \chi_0^2, \chi_0^4, \chi_0^6)^T$. Let $\chi(t; t_0, \chi_0) := (z_1(t), \partial z_1(t)/\partial t, z_2(t), \partial z_2(t)/\partial t, z_3(t), \partial z_3(t)/\partial t)^T$ for convenience. There exists a small $r_0 > 0$ such that for $\|\chi_0\|_\mathbf{Y} \leq r_0$, we have $\|\chi(t; 0, \chi_0)\|_\mathbf{Y} \leq \theta(t) \to 0$ as $t \to +\infty$. We may assume that $\theta(t)$ has continuous negative derivative, thus it has continuous inverse $T(\epsilon), 0 < \epsilon < \theta(0)$ so that $T(\theta(t)) = t$, with $T(\epsilon) \to +\infty$ as $\epsilon \to 0^+$. From (3.3), Lemma 3.2 and the representation by semigroup, we can show that the mapping $\chi_0 \to \chi(t; 0, \chi_0), t \geq 0$ has Lipschitz constant Je^{Dt} for some constants J and D, if $\|\chi_0\|_\mathbf{Y} \leq r_0$. As in Theorem 19.3 in Yoshizawa [239], define $g(\epsilon) = exp\{-1(1 + D)T(\epsilon)\}$, $g(0) = 0$; and for $k = 1, 2, 3, \ldots$

$$Q_k(z) = max\{0, z - k^{-1}\},$$

$$V_k(\chi_0) = g(1/(k+1)) sup_{t \geq 0}\{e^t Q_k(\|\chi(t; 0, \chi_0)\|_\mathbf{Y}\} \quad \text{for } \|\chi_0\|_\mathbf{Y} \leq r_0.$$

(Note that the supremum may be taken only on $0 \leq t \leq T_k := T(1/(k+1))$.) One can verify that $0 \leq V_k(\chi_0) \leq g(1/(k+1))e^{T_k}\theta(0)$, and

$$V_k(\chi(h; 0, \chi_0)) = e^{-h} g(1/(k+1)) sup_{t \geq h}\{e^t Q_k(\|\chi(t; 0, \chi_0)\|_\mathbf{Y})\} \leq e^{-h} V_k(\chi_0)$$

for small $h \geq 0$. Defining $V(\chi_0) = \Sigma_{k=1}^{\infty} 2^{-k} V_k(\chi_0)$, for $\|\chi_0\|_\mathbf{Y} \leq r_0$, we can show as in [239] that

$$|V(\chi_1) - V(\chi_2)| \leq J\|\chi_1 - \chi_2\|_\mathbf{Y}, \tag{3.50}$$

$$\dot{V}(\chi_0) := \limsup_{h \to 0^+} h^{-1}[V(\chi(h; 0, \chi_0)) - V(\chi_0)] \leq -V(\chi_0), \tag{3.51}$$

$$a(\|\chi_0\|_\mathbf{Y}) \leq V(\chi_0) \leq J\|\chi_0\|_\mathbf{Y}, \tag{3.52}$$

where $a(s)$ is a continuous and strictly increasing function for $0 \leq s \leq r_0$, with $a(0) = 0$. Moreover (3.51) and the comparison Theorem 4.1 in [239] imply that

$$V(\chi(T; 0, \chi_0)) \leq e^{-T} V(\chi_0).$$

For $\|\boldsymbol{\chi}_0\|_{\mathbf{Y}} \le r_0, \|\mathbf{u}_0 - \sigma^0(\boldsymbol{\chi}_0)\|_\alpha$ small, define

$$W(\mathbf{u}_0, \boldsymbol{\chi}_0) := V(\boldsymbol{\chi}_0) + \hat{P}\|\boldsymbol{\xi}_0\|_\alpha, \quad \text{where } \boldsymbol{\xi}_0 = \mathbf{u}_0 - \sigma^0(\boldsymbol{\chi}_0),$$

and $\hat{P}$ is a large positive constant to be chosen below. Let $T > 0$ be a large number such that:

$$e^{-T} \le 1/2 \qquad \text{and} \qquad \hat{M}\hat{R}e^{(q-\beta)T} \le 1/4,$$

where $\hat{M}, \hat{R}, q$ and β are described in Theorem 3.2 above. Let $(\mathbf{u}(\cdot,t), \mathbf{v}(\cdot,t))$ be a solution of (3.1) satisfying $\mathbf{u}(\cdot,0) = \mathbf{u}_0, (\mathbf{v}(\cdot,0), \mathbf{v}_t(\cdot,0)) = (\chi_0^1, \chi_0^3, \chi_0^5, \chi_0^2, \chi_0^4, \chi_0^6)^T$. Denote $\boldsymbol{\omega}(\cdot,t) := (v_1(\cdot,t), \partial v_1(\cdot,t)/\partial t, v_2, \partial v_2/\partial t, v_3, \partial v_3/\partial t)^T, \boldsymbol{\xi}(t) = \mathbf{u}(\cdot,t) - \sigma^0(\boldsymbol{\omega}(\cdot,t))$, and consider the following:

$$\begin{aligned}
&W(\mathbf{u}(\cdot,T), \boldsymbol{\omega}(\cdot,T)) = V(\boldsymbol{\omega}(\cdot,T)) + \hat{P}\|\boldsymbol{\xi}(T)\|_\alpha \\
&\le e^{-T}V(\boldsymbol{\chi}(0;T,\boldsymbol{\omega}(\cdot,T)) - e^{-T}V(\boldsymbol{\omega}(\cdot,0)) + e^{-T}V(\boldsymbol{\omega}(\cdot .0)) \\
&\quad + \hat{P}\hat{M}\hat{R}e^{(q-\beta)T}\|\boldsymbol{\xi}(0)\|_\alpha \\
&\le e^{-T}J\|\boldsymbol{\chi}(0;T,\boldsymbol{\omega}(\cdot,T)) - \boldsymbol{\omega}(\cdot,0)\|_{\mathbf{Y}} + 2^{-1}V(\boldsymbol{\omega}(\cdot,0)) + (1/4)\hat{P}\|\boldsymbol{\xi}(0)\|_\alpha \\
&\le e^{-T}JML\hat{M}\hat{R}[k + ML(K_2+1) + q - \beta]^{-1}(e^{[k+ML(K_2+1)+q-\beta]T} - 1)\|\boldsymbol{\xi}(0)\|_\alpha \\
&\quad + 2^{-1}V(\boldsymbol{\omega}(\cdot,0)) + (1/4)\hat{P}\|\boldsymbol{\xi}(0)\|_\alpha.
\end{aligned}$$

The last inequality is due to (3.39) and (3.43). Thus by choosing

$$\hat{P} \ge 4e^{-T}JML\hat{M}\hat{R}[k + ML(K_2+1) + q - \beta]^{-1}(e^{[k+ML(K_2+1)+q-\beta]T} - 1),$$

we obtain

$$W(\mathbf{u}(\cdot,T), \boldsymbol{\omega}(\cdot,T)) \le (1/2)W(\mathbf{u}(\cdot,0), \boldsymbol{\omega}(\cdot,0)).$$

Estimating by (3.43) the norm $\|\boldsymbol{\xi}(t)\|_\alpha$ for $t \in [0,T]$, and possibly reducing $\|\boldsymbol{\xi}(0)\|_\alpha$, we obtain

$$W(\mathbf{u}(\cdot,t), \boldsymbol{\omega}(\cdot,t)) \le Q2^{-t/T}, \qquad \text{for all } t \ge 0,$$

where Q is some positive constant. This proves the asymptotic stability of the origin for the full system (3.1).

Remark 3.7. Consider system (3.1) under hypotheses (3.2) to (3.3), without the boundedness assumption (3.4) on $\mathbf{f}$. Suppose that $B(x)$ is symmetric, $\mu > 0$ and $\mathbf{g}$ satisfies the additional condition:

$$\|\mathbf{g}(\mathbf{u},\mathbf{v})\|_{\mathbf{L}^2} \le \hat{K}\|\mathbf{u}\|_\alpha$$

for all $(\mathbf{u}, \mathbf{v}) \in \mathbf{X}_\alpha \times \mathbf{H}^{1,2}$ where $\hat{K}$ is a positive constant. Let Q_1 and Q_2 be arbitrary positive constants. We can show by integral representation using semigroup as in [128], that there exists $\hat{Q}_2$ (which depends on Q_1 and Q_2) such that $\|(\mathbf{v}(\cdot,t), \mathbf{v}_t(\cdot,t))\|_{\tilde{\mathbf{Y}}} \leq \hat{Q}_2$ for all $t \geq 0$, as long as both $\|(\mathbf{v}(\cdot,0), \mathbf{v}_t(\cdot,0))\|_{\tilde{\mathbf{Y}}} \leq Q_2$ and $\|\mathbf{u}(\cdot,t)\|_\alpha \leq 2Q_1$ for all $t \geq 0$. Define $\mathbf{f}_{\hat{Q}_2}(\mathbf{u}, \mathbf{v}) = \mathbf{f}(\mathbf{u}, \mathbf{v}\psi(\mathbf{v}/\hat{Q}_2))$ for $\mathbf{u} \in \mathbf{X}_\alpha, \mathbf{v} \in \mathbf{H}_0^{1,2}$, where $\psi : \mathbf{H}_0^{1,2} \to [0,1]$ is a Lipschitzian function satisfying $\psi(\tilde{\mathbf{v}}) = 1$ if $\|\tilde{\mathbf{v}}\|_{\mathbf{H}^{1,2}} \leq 1$ and $\psi(\tilde{\mathbf{v}}) = 0$ if $\|\tilde{\mathbf{v}}\|_{\mathbf{H}^{1,2}} \geq 2$. Let $(3.1 - \hat{Q}_2)$ denote the system (3.1) with only $\mathbf{f}(\mathbf{u}, \mathbf{v})$ replaced with $\mathbf{f}_{\hat{Q}_2}(\mathbf{u}, \mathbf{v})$. System $(3.1 - \hat{Q}_2)$ satisfies hypotheses (3.2)-(3.4). For sufficiently small $\epsilon > 0$, we can thus apply Theorem 3.1 to obtain an invariant manifold for $(3.1 - \hat{Q}_2)$ represented by $\sigma^* \in \Lambda$, satisfying $\|\sigma^*(\mathbf{v} - \mathbf{v}^0, \mathbf{v}_t)\|_\alpha \leq Q_1$ for all $(\mathbf{v}, \mathbf{v}_t) \in \tilde{\mathbf{Y}}$. If $\|(\mathbf{v}(\cdot,0), \mathbf{v}_t(\cdot,0))\|_{\tilde{\mathbf{Y}}} \leq Q_2$ and $(\mathbf{u}(\cdot,0), \mathbf{v}(\cdot,0), \mathbf{v}_t(\cdot,0))$ is close enough to the manifold represented by σ^*, Theorem 3.2 and the above consideration imply that $\|\mathbf{u}(\cdot,t)\|_\alpha \leq 2Q_1$ and $\|(\mathbf{v}(\cdot,t), \mathbf{v}_t(\cdot,t))\|_{\tilde{\mathbf{Y}}} \leq \hat{Q}_2$ for $t \geq 0$. From the bound on $\mathbf{v}$ we conclude that such $(\mathbf{u}(\cdot,t), \mathbf{v}(\cdot,t))$ is actually a solution of the original problem (3.1) for $t > 0$. Such solution of (3.1) thus tends to the part of the manifold represented by σ^* for $(3.1 - \hat{Q}_2)$ with $\|(\mathbf{v}, \mathbf{v}_t)\|_{\tilde{\mathbf{Y}}} \leq \hat{Q}_2$ and $\|\sigma^*\|_\alpha \leq Q_1$, as $t \to +\infty$.

5.4 Existence and Global Bounds for Fluid Equations of Plasma Display Technology

This section considers the fluid model for the discharge of plasma particle species in display technology. The fluid equations are coupled with Poisson's equation, which describes the effect of the charged particles on the electric field. The diffusion and mobility coefficients for the positive ion particles depend on the electric field, while those for the electrons depend on the electron mean energy. The reaction rates are proportional to the products of the densities of the reacting particles involved in the particular ionization, conversion or recombination reactions. The ionization and discharge reactions are described by an initial-boundary value problem for a system of coupled parabolic-elliptic partial differential equations. The model is adopted from the various investigations on discharges in plasma display panels by Raul and Kushner [191], Hagelaar [77], Hagelaar, Klein et al. [78], Veronis and Inan [223]. The system is first analyzed by upper-lower solution method. By means of the a-priori bounds obtained for an arbitrary time, the existence of solution for the initial-boundary value problem is proved in an appropriate Hölder space. The rigorous analytical development in this section follows the presentation in Leung and Chen [134].

Part A: Model Introduction, Upper and Lower Estimates.

Many types of plasma particles occupy the cells between the front and rear glass panels. The voltage in the cell is influenced by the electrodes attached to the front and rear glass panels, while the electric field inside the cell is also affected by the charged particles. The space and time variation of the densities of the plasma particles, electrons and electron energy are described by the fluid equations. The fluid equations are coupled with the Poisson's equation which describes the effect of the charged particles on the electric field, which is the gradient of the voltage in the cell. We will take only four key positively charged particles into consideration. They are the two xenon ions Xe^+, Xe_2^+ and the two neon ions Ne^+, Ne_2^+. Let $n_1(x,t), ..., n_4(x,t)$ respectively represent the densities of the particles $Xe^+, Xe_2^+, Ne^+, Ne_2^+$. Let $n_5(x,t), n_6(x,t)$ respectively represent the densities of electron e and electron energy; and V(x,t) represents the voltage in the cell. We obtain the following seven coupled parabolic-elliptic equations for the 4 positively charged particles, electrons, energy and voltage.

(4.1)

$$\begin{cases} \partial n_1/\partial t - \nabla\cdot(D_1(|\nabla V|^2)\nabla n_1) + (\mu_1(|\nabla V|^2)/\epsilon)[\sum_{i=1}^4 q_i n_i - q_5 n_5]n_1 \\ \qquad -\mu_1(|\nabla V|^2)[(\nabla V)\cdot\nabla n_1] = c_{11}n_5 - c_{12}n_1, \\ \partial n_2/\partial t - \nabla\cdot(D_2(|\nabla V|^2)\nabla n_2) + (\mu_2(|\nabla V|^2)/\epsilon)[\sum_{i=1}^4 q_i n_i - q_5 n_5]n_2 \\ \qquad -\mu_2(|\nabla V|^2)[(\nabla V)\cdot\nabla n_2] = c_{21}n_1 - c_{22}n_5 n_2, \\ \partial n_3/\partial t - \nabla\cdot(D_3(|\nabla V|^2)\nabla n_3) + (\mu_3(|\nabla V)|^2/\epsilon)[\sum_{i=1}^4 q_i n_i - q_5 n_5]n_3 \\ \qquad -\mu_3(|\nabla V|^2)[(\nabla V)\cdot\nabla n_3] = c_{31}n_5 - c_{32}n_3, \\ \partial n_4/\partial t - \nabla\cdot(D_4(|\nabla V|^2)\nabla n_4) + (\mu_4(|\nabla V|^2)/\epsilon)[\sum_{i=1}^4 q_i n_i - q_5 n_5]n_4 \\ \qquad -\mu_4(|\nabla V|^2)[(\nabla V)\cdot\nabla n_4] = c_{41}n_3 - c_{42}n_5 n_4, \\ \partial n_5/\partial t - \nabla\cdot(D_5(n_6/n_5)\nabla n_5) - (\mu_5(n_6/n_5)/\epsilon)[\sum_{i=1}^4 q_i n_i - q_5 n_5]n_5 \\ \qquad + \mu_5(n_6/n_5)[(\nabla V)\cdot\nabla n_5] = c_{11}n_5 + c_{31}n_5 - c_{22}n_2 n_5 - c_{42}n_4 n_5, \\ \partial n_6/\partial t - (5/3)\nabla\cdot(D_5(n_6/n_5)\nabla n_6) \\ \qquad -(5/3)(\mu_5(n_6/n_5)/\epsilon)[\sum_{i=1}^4 q_i n_i - q_5 n_5]n_6 + (5/3)\mu_5[(\nabla V)\cdot\nabla n_6] \\ \qquad = e\mu_5(\nabla V\cdot\nabla V)n_5 - eD_5[\nabla V\cdot\nabla n_5] - n_5[\sum_{i=1}^5 d_i n_i], \\ \Delta V = \epsilon^{-1}(\sum_{i=1}^4 q_i n_i - q_5 n_5), \end{cases}$$

for $(x,t) \in \Omega \times [0,\infty)$. The quantities D_i and μ_i for $i = 1, ..., 5$ are the diffusion and mobility coefficients for the corresponding i-th species. For $i = 1, .., 4$, they will be assumed to be functions of grad $V(x,t) = \nabla V$, and are thus coupled nonlinearly with the seventh equation. The product of $\epsilon^{-1}\mu_i(|\nabla V|^2)$ with the quadratic terms on the left of the equations expresses part of the influence of the electric field (i.e. grad $V(x,t)$) on the evolution of the particles. Note that this quadratic expression is exactly $\mu_i$div(grad V)n_i as described in the seventh equation. The parameter ϵ represents dielectric permitivitty, e represents the elementary charge, and q_i are positive constants for $i = 1, ..., 5$. For $i = 5$, we will assume the diffusion and mobility coefficients to depend on the so called electron mean energy η_6/η_5. The right-hand side of the six parabolic equations describes the reaction among the ions. The reaction rates are proportional to the products of the densities of the reacting particles involved in the particular ionization, conversion or recombination reactions. In the first equation for Xe^+ density, the reactions are due to ionization from Xe to Xe^+, *i.e.* $e + Xe \to 2e + Xe^+$, and ion conversions from Xe^+ to Xe_2^+ or other particles e.g. $Xe^+ + 2Xe \to Xe_2^+ + Xe$ and $Xe^+ + 2Ne \to NeXe^+ + Ne$. Note that we assume the densities of Xe and Ne to be constants. In the second equation for Xe_2^+ density, the reactions are due to ion conversion to Xe_2^+ (this includes e.g. $Xe^+ + 2Xe \to Xe_2^+ + Xe$ or $Xe^+ + Xe + Ne \to Xe_2^+ + Ne$), and recombination $e + Xe_2^+ \to Xe^{**} + Xe$. In the third equation for Ne^+ density, the reactions are due to ionization $e + Ne \to 2e + Ne^+$ and ion conversion e.g. $Ne^+ + 2Ne \to Ne_2^+ + Ne$, and $Ne^+ + Ne + Xe \to NeXe^+ + Ne$. In the fourth equation for Ne_2^+ density, the reactions are due to ion conversion $Ne^+ + 2Ne \to Ne_2^+ + Ne$ and recombination $e + Ne_2^+ \to Ne^{**} + Ne$. In the fifth equation for electron density, the reactions are due to ionizations $e + Xe \to 2e + Xe^+$, $e + Ne \to 2e + Ne^+$, recombinations $e + Xe_2^+ \to Xe^{**} + Xe$ and $e + Ne_2^+ \to Ne^{**} + Ne$. In the sixth equation for energy, the first two source terms on the right-hand side are due to heating by electric field, and the last term is due to energy loss in collisions. The terms involving diffusion and mobility on the left-hand side are due to appropriate approximation of drift-diffusion, contributing to the electron energy flux.(See e.g. [77], [78] and [191] for more detailed explanations).

To fix ideas, we assume Ω is a bounded domain in R^m, with boundary $\partial\Omega$ in the Hölder space $C^{2+\alpha}$, $0 < \alpha < 1$. As described above, we assume in (4.1), for $i = 1, .., 4$, that $D_i = D_i(|\nabla V)|^2)$, $\mu_i = \mu_i(|\nabla V|^2)$ and $D_5 = D_5(\eta_6/\eta_5)$, $\mu_5 = \mu_5(\eta_6/\eta_5)$. We will assume that, for $i = 1, ..., 5$, $D_i(z)$ and $\mu_i(z)$ are respectively positive $C^{1+\alpha}$ and C^α functions in compact subsets of all real numbers; and further, $\mu_i(z)$ are bounded for all real z, $i = 1, ..., 5$. All the parameters c_{ij} in (4.1) are assumed to be constants. We will consider the system (4.1) with prescribed initial-boundary conditions:

$$n_i(x,0) = g_i(x,0) \geq 0 \qquad \text{for } x \in \Omega,\ i = 1,...,6,$$

$$(4.2) \qquad n_i(x,t) = g_i(x,t) \geq 0 \qquad \text{for } (x,t) \in \partial\Omega \times [0,\infty),\ i = 1,...,6,$$

$$V(x,t) = \phi(x,t) \qquad \text{for } (x,t) \in \partial\Omega \times [0,\infty).$$

In Theorem 4.1, we show that if appropriate coupled upper-lower solutions for problem (4.1) and (4.2) exist in a certain time interval, they will serve as a-priori upper and lower bounds for the solutions of the initial-boundary value problem. We will give examples to show that under appropriate ranges of the reaction rates c_{ij}, Theorem 4.1 can establish bounds for the densities of the charged particles. Such bounds can determine whether the reaction rates will lead the densities of the particles to generate light through discharge. Under more stringent conditions on the coefficients D_i and μ_i, the a-priori bounds will also be used in Main Theorem 4.4 to prove the existence of the solutions of (4.1), (4.2) in appropriate Hölder space on the time interval involved. These methods can be used to study more elaborate systems involving more particles. They are however too lengthy for this present analysis. Better understanding of these problems is beneficial for further study of the control of light distribution on the display panel.

We now define convenient coupled upper-lower solutions for problem (4.1) and (4.2). They will be used as upper and lower bounds for solutions of the initial-boundary value problem.

Definition 4.1. A pair of vector functions $(\hat{n}_1(x,t),...,\hat{n}_5(x,t))$ and $(\tilde{n}_1(x,t),..., \tilde{n}_5(x,t))$ with each component in $C^{21}(\bar{\Omega}\times[0,T])$, where $\tilde{n}_i \leq \hat{n}_i, i = 1,...,5$, with

$$0 \leq \tilde{n}_i(x,t) \leq g_i(x,t) \leq \hat{n}_i(x,t), \quad i = 1,...,5, \text{ and}$$

$$(4.3) \qquad 0 < \tilde{n}_5(x,t) \qquad \text{for (x,t)} \in (\Omega \times \{0\}) \cup (\partial\Omega \times [0,T]),$$

are called coupled upper-lower solutions for problem (4.1), (4.2), if $(\hat{n}_1,...,\hat{n}_5)$ satisfies in $\Omega \times (0,T)$:

(4.4)

$$\begin{cases} \partial\hat{n}_1/\partial t - \nabla\cdot(D_1(|\nabla W|^2)\nabla\hat{n}_1) + (\mu_1(|\nabla W|^2)/\epsilon)[\,q_1\hat{n}_1 \\ \quad + \sum_{i=1,i\neq 1}^4 q_i z_i - q_5 z_5]\hat{n}_1 - \mu_1(|\nabla W|^2)[(\nabla W)\cdot\nabla\hat{n}_1] \;\geq\; c_{11}z_5 - c_{12}\hat{n}_1, \\ \partial\hat{n}_2/\partial t - \nabla\cdot(D_2(|\nabla W|^2)(\nabla\hat{n}_2) + (\mu_2(|\nabla W|^2)/\epsilon)[\,q_2\hat{n}_2 \\ \quad + \sum_{i=1,i\neq 2}^4 q_i z_i - q_5 z_5]\hat{n}_2 - \mu_2(|\nabla W|^2)[(\nabla W)\cdot\nabla\hat{n}_2] \;\geq\; c_{21}z_1 - c_{22}z_5\hat{n}_2, \\ \partial\hat{n}_3/\partial t - \nabla\cdot(D_3(|\nabla W|^2)\nabla\hat{n}_3) + (\mu_3(|\nabla W|^2)/\epsilon)[\,q_3\hat{n}_3 \\ \quad + \sum_{i=1,i\neq 3}^4 q_i z_i - q_5 z_5]\hat{n}_3 - \mu_3(|\nabla W|^2)[(\nabla W)\cdot\nabla\hat{n}_3] \;\geq\; c_{31}z_5 - c_{32}\hat{n}_3, \\ \partial\hat{n}_4/\partial t - \nabla\cdot(D_4(|\nabla W|^2)\nabla\hat{n}_4) + (\mu_4(|\nabla W|^2)/\epsilon)[\,q_4\hat{n}_4 \\ \quad + \sum_{i=1,i\neq 4}^4 q_i z_i - q_5 z_5]\hat{n}_4 - \mu_4(|\nabla W|^2)[(\nabla W)\cdot\nabla\hat{n}_4] \;\geq\; c_{41}z_3 - c_{42}z_5\hat{n}_4, \\ \partial\hat{n}_5/\partial t - \nabla\cdot(D_5(Z)\nabla\hat{n}_5) - (\mu_5(Z)/\epsilon)[\sum_{i=1}^4 q_i z_i - q_5\hat{n}_5]\hat{n}_5 \\ \quad + \mu_5(Z)[(\nabla W)\cdot\nabla\hat{n}_5] \;\geq\; c_{11}\hat{n}_5 + c_{31}\hat{n}_5 - c_{22}z_2\hat{n}_5 - c_{42}z_4\hat{n}_5, \end{cases}$$

while $(\tilde{n}_1, ..., \tilde{n}_5)$ satisfy (4.4) with corresponding $\hat{n}_i$ replaced by $\tilde{n}_i$ and all the inequalities reversed, at all $(z_1, ..., z_5)$ with $\tilde{n}_i \leq z_i \leq \hat{n}_i, i = 1, ..., 5$. Here, for $i = 1, ..., 4, D_i = D_i(|\nabla W|^2), \mu_i = \mu_i(|\nabla W|^2)$ and W is any C^{21} function in $\bar{\Omega} \times [0, T]$. In the 5-th inequality, D_5 and μ_5 are allowed to be evaluated at any $C^{21}(\bar{\Omega} \times [0, T])$ function Z.

Theorem 4.1 (Searching for A-priori Bounds). *Consider the initial-boundary value problem (4.1), (4.2). Let $(\hat{n}_1, ..., \hat{n}_5)$ and $(\tilde{n}_1, ..., \tilde{n}_5)$ be a pair of coupled upper-lower solutions as described in Definition 4.1, with $\hat{n}_i > \tilde{n}_i$ in $\bar{\Omega} \times [0, T], i = 1, ..., 5$; and $(n_1, ..., n_6, V)$ be a solution of the initial-boundary value problem with components in $C^{21}(\bar{\Omega} \times [0, T])$. Then the following inequalities are satisfied:*

$$\tilde{n}_i(x, t) \leq n_i(x, t) \leq \hat{n}_i(x, t), \qquad i = 1, ..., 5 \tag{4.5}$$

for all $(x, t) \in \bar{\Omega} \times [0, T]$.

Proof. For convenience, let $f_i = f_i(x, t, w_1, w_2, ..., w_5), i = 1, ..., 5$ be defined by (4.6)

$$f_1 = c_{11}w_5 - c_{12}w_1 - \epsilon^{-1}\mu_1(|\nabla V(x,t)|^2)[\textstyle\sum_{i=1}^4 q_i w_i - q_5 w_5]w_1,$$

$$f_2 = c_{21}w_1 - c_{22}w_5 w_2 - \epsilon^{-1}\mu_2(|\nabla V(x,t)|^2)[\textstyle\sum_{i=1}^4 q_i w_i - q_5 w_5]w_2,$$

$$f_3 = c_{31}w_5 - c_{32}w_3 - \epsilon^{-1}\mu_3(|\nabla V(x,t)|^2)[\textstyle\sum_{i=1}^4 q_i w_i - q_5 w_5]w_3,$$

$$f_4 = c_{41}w_3 - c_{42}w_5 w_4 - \epsilon^{-1}\mu_4(|\nabla V(x,t)|^2)[\textstyle\sum_{i=1}^4 q_i w_i - q_5 w_5]w_4,$$

$$f_5 = c_{11}w_5 - c_{31}w_5 - c_{22}w_2 w_5 - c_{42}w_4 w_5$$

$$+\epsilon^{-1}\mu_5(n_6(x,t)/n_5(x,t))[\textstyle\sum_{i=1}^4 q_i w_i - q_5 w_5]w_5.$$

Since the functions μ_i are assumed bounded for $i = 1, ..., 5$, the functions f_i satisfies:

$$(4.7) \qquad |f_i(x, t, w_1^\alpha, ..., w_5^\alpha) - f_i(x, t, w_1^\beta, ..., w_5^\beta)| \leq K(\sum_{j=1}^{5}[w_j^\alpha - w_j^\beta]^2)^{1/2}$$

for some $K > 0$, for all $(x,t) \in \bar{\Omega} \times [0,T]$ and $(w_1^\alpha, ..., w_5^\alpha), (w_1^\beta, ..., w_5^\beta)$ with both $w = w_i^\alpha$ and w_i^β satisfying

$$min\{\tilde{n}_i(x,t)|(x,t) \in \bar{\Omega} \times [0,T]\} \leq w \leq max\{\hat{n}_i(x,t)|(x,t) \in \bar{\Omega} \times [0,T]\},$$

$i = 1, ..., 5$. Let $k = min_{1 \leq i \leq 5}\{\hat{n}_i(x,t) - \tilde{n}_i(x,t)|(x,t) \in \bar{\Omega} \times [0,T]\}$. For $0 < \sigma < (k/2)[1 + 3K5^{1/2}T]^{-1}, i = 1, ..., 5$, define

$$n_i^{\pm\sigma}(x,t) = n_i(x,t) \pm [1 + 3K5^{1/2}t]\sigma \qquad \text{for } (x,t) \in \bar{\Omega} \times [0,T].$$

By hypothesis, at $t = 0$ for $i = 1, ..., 5$, we have

$$(4.8) \qquad \tilde{n}_i(x,t) < n_i^{+\sigma}(x,t) \text{ and } n_i^{-\sigma}(x,t) < \hat{n}_i(x,t) \quad \text{for } (x,t) \in \bar{\Omega} \times [0,T].$$

Suppose one of these inequalities fails at some point in $\bar{\Omega} \times (0, \tau_1)$, where $\tau_1 = min\{T, 1/(3K5^{1/2})\}$; and (x_1, t_1) is a point in $\bar{\Omega} \times (0, \tau_1)$, with minimal t_1 where (4.8) fails. At $(x_1, t_1), \tilde{n}_i = n_i^{+\sigma}$ or $n_i^{-\sigma} = \hat{n}_i$ for some i. Assume that the latter is the case; a similar proof hold for the former case. (Also, assume that $i = 1$; same arguments hold for other cases of i). First, let $\hat{n}_1(x_1, t_1) = n_1^{-\sigma}(x_1, t_1)$ with $x_1 \in \Omega$. In the differential inequality (4.4) for upper solution, we evaluate W at the function $V(x,t)$ and Z at the function $n_6(x,t)/n_5(x,t)$, which are obtained from the components of the solution itself. At (x_1, t_1), we deduce:

$$
\begin{aligned}
&(\partial/\partial t)(\hat{n}_1 - n_1^{-\sigma})|_{(x_1,t_1)} = \partial\hat{n}_1/\partial t - \partial n_1/\partial t + 3\sigma K 5^{1/2} \\
&\geq \nabla\cdot(D_1(|\nabla V|^2)\nabla(\hat{n}_1 - n_1)) + \mu_1(|\nabla V|^2)[(\nabla V)\cdot\nabla(\hat{n}_1 - n_1)] \\
&\quad - \mu_1(|\nabla V|^2)\epsilon^{-1}(q_1\hat{n}_1 + \textstyle\sum_{i=2}^4 q_i z_i - q_5 z_5)\hat{n}_1 \\
&\quad + \mu_1(|\nabla V|^2)\epsilon^{-1}(\textstyle\sum_{i=1}^4 q_i n_i - q_5 n_5) n_1 \\
&\quad + c_{11} z_5 - c_{12}\hat{n}_1 - c_{11} n_5 + c_{12} n_1 + 3\sigma K 5^{1/2} \\
&= f_1(x_1, t_1, n_1^{-\sigma}, z_2, ..., z_5) - f_1(x_1, t_1, n_1, ..., n_5) + 3\sigma K 5^{1/2},
\end{aligned} \tag{4.9}
$$

where $z_i = max\{n_i^{-\sigma}(x_1,t_1), \tilde{n}_i(x_1,t_1)\}$ or $min\{n_i^{+\sigma}(x_1,t_1), \hat{n}(x_1,t_1)\}$ for $i = 2, ..., 5$. (Here, we apply the inequality (4.4).) Since $|n_1^{-\sigma} - n_1|$ and $|z_i - n_i|, i = 2, ..., 5$ are $\leq (1 + 3K5^{1/2}t_1)\sigma \leq 2\sigma$, we obtain

$$(\partial/\partial t)(\hat{n}_1 - n_1^{-\sigma})|_{(x_1,t_1)} \geq \sigma K 5^{1/2} > 0,$$

contradicting the definition of (x_1, t_1). Hence, we must have $x_1 \in \partial\Omega$. However, by (4.3), we must have $n^{-\sigma}(x,t) < \hat{n}_i(x,t)$ for $(x,t) \in \partial\Omega \times [0,T]$. Thus (4.8) must hold for $(x,t) \in \bar{\Omega}\times(0,\tau_1)$. Passing to the limit as $\sigma \to 0^+$, we obtain (4.5) for $i = 1, ..., 5$, $(x,t) \in \bar{\Omega}\times[0,\tau_1]$. If $\tau_1 < T$, we repeat the above arguments in the shifted time interval $[\tau_1, \tau_2]$, where $\tau_2 = min\{T, \tau_1 + 1/(3K5^{1/2})\}$. Eventually, we obtain (4.5) in $\bar{\Omega}\times[0,T]$ for $i = 1, ..., 5$.

Corollary 4.2. *Assume that $\tilde{\mu}_i$ and $\hat{\mu}_i$ are constants such that*

$$0 < \tilde{\mu}_i \leq \mu_i(\xi) \leq \hat{\mu}_i \quad \text{for all } \xi \in R,\ i = 1, ..., 5.$$

Suppose there exist positive constants δ and $C_i, i = 1, ..., 5$ with $\delta < C_5$, $\sum_{i=1}^4 q_i C_i < q_5 C_5$ satisfying:

$$
\begin{cases}
-(\hat{\mu}_1/\epsilon)(q_1C_1 - q_5C_5)C_1 + c_{11}C_5 - c_{12}C_1 < 0, \\
-(\hat{\mu}_2/\epsilon)(q_2C_2 - q_5C_5)C_2 + c_{21}C_1 - c_{22}\delta C_2 < 0, \\
-(\hat{\mu}_3/\epsilon)(q_3C_3 - q_5C_5)C_3 + c_{31}C_5 - c_{32}C_3 < 0, \\
-(\hat{\mu}_4/\epsilon)(q_4C_4 - q_5C_5)C_4 + c_{41}C_3 - c_{42}\delta C_4 < 0, \\
+(\tilde{\mu}_5/\epsilon)(q_1C_1 + q_2C_2 + q_3C_3 + q_4C_4 - q_5C_5)C_5 + c_{11}C_5 + c_{31}C_5 < 0,
\end{cases} \tag{4.10}
$$

and

$$-(\hat{\mu}_5/\epsilon)q_5\delta + c_{11} + c_{31} - c_{22}C_2 - c_{42}C_4 > 0. \tag{4.11}$$

Moreover, suppose that the initial-boundary functions in (4.2) satisfy:

$$\begin{aligned} &0 \leq g_i(x,t) \leq C_i, && i = 1, ..., 4, \\ &\delta \leq g_5(x,t) \leq C_5 && \text{for } (x,t) \in (\Omega \times \{0\}) \cup (\partial\Omega \times [0,T]); \end{aligned} \tag{4.12}$$

and $(n_1, ..., n_6, V)$ *is a solution of the initial-boundary value problem (4.1), (4.2) with each component in* $C^{21}(\bar{\Omega} \times [0,T])$. *Then the following inequalities:*

$$\begin{aligned} &0 \leq n_i(x,t) \leq C_i && \text{for } i = 1, ..., 4, \text{ and} \\ &\delta \leq n_5(x,t) \leq C_5 \end{aligned} \tag{4.13}$$

are satisfied for all $(x,t) \in \bar{\Omega} \times [0,T]$. (Note that $V(x,t)$ can be readily estimated by using the bounds for $n_i(x,t), i = 1, ..., 5$ in (4.13) and the last equation in (4.1).)

Proof. Choose $\tilde{n}_i(x,t) := 0, \hat{n}_i(x,t) := C_i$, for $i = 1, ..., 4$, and $\tilde{n}_5(x,t) := \delta, \hat{n}_5(x,t) := C_5$ for $(x,t) \in \bar{\Omega} \times [0,T]$. Then the inequalities in (4.13) follow readily from Theorem 4.1.

Remark 4.1. Under further assumptions, we will also deduce bounds for $n_6(x,t)$ in $\bar{\Omega} \times [0,T]$ in Part B below.

Example 4.1. Suppose that:

$$\begin{aligned} &q_1 = q_2 = q_3 = q_4 = q_5 = 1.6 \times 10^{-19} \text{ coulombs}, \\ &c_{11} = c_{31} = 6 \times 10^{-9} \text{ sec}^{-1}, \ c_{12} = c_{32} = 2.5 \times 10^{-8} \text{ sec}^{-1}, \\ &c_{21} = c_{41} = 10^{-16} \text{ sec}^{-1}, \ c_{22} = c_{42} = 5 \times 10^{-14} \text{ cm}^3/\text{sec}, \\ &\tilde{\mu}_i/\epsilon = \hat{\mu}_i/\epsilon = 1 \text{ cm}^3/\text{coulomb-sec}, \text{ for } i = 1,2,3,4, \\ &\tilde{\mu}_5/\epsilon = 0.8 \times 10^7 \text{ cm}^3/\text{coulomb-sec}, \ \hat{\mu}_5/\epsilon = 10^7 \text{ cm}^3/\text{coulomb-sec}. \end{aligned}$$

Then, the inequalities (4.10) and (4.11) are satisfied by choosing:

$$C_1 = C_2 = C_3 = C_4 = 10^5 \text{ cm}^{-3}; \ C_5 = 4.1 \times 10^5 \text{ cm}^{-3}; \ \delta = 1 \text{ cm}^{-3}.$$

Consequently, Corollary 4.2 can be applied to obtain bounds for solutions of (4.1), (4.2). (Here, some of the rates are similar to those in e.g. [78].)

Example 4.2 (Relation of Bounds with Glow Discharge). From experimental observations, glow discharge occurs when the ions reach concentration level of the order 10^{14}. This example shows that when the boundary concentrations of electrons are above a certain sufficient level for δ, one will need large constants of the order 10^{14} for upper solutions for the five concentration components. Suppose that:

$$\begin{aligned}
&q_1 = q_2 = q_3 = q_4 = q_5 = 1.6 \times 10^{-19},\\
&c_{11} = c_{31} = 6 \times 10^{-3}, c_{12} = c_{32} = 2.5 \times 10^{-2},\\
&c_{21} = c_{41} = 10^{-7}, c_{22} = c_{42} = 5 \times 10^{-17}\\
&\tilde{\mu}_i/\epsilon = \hat{\mu}_i/\epsilon = 1, \text{ for } i = 1,2,3,4, \ \ \tilde{\mu}_5/\epsilon = 0.8 \times 10^4, \ \ \hat{\mu}_5/\epsilon = 10^4.
\end{aligned}$$

Then, the inequalities (4.10) and (4.11) are satisfied by choosing:

$$C_1 = C_2 = C_3 = C_4 = 10^{14}; \ C_5 = 4.1 \times 10^{14}; \ \delta = 10^{12}.$$

Consequently, Corollary 4.2 can be applied to obtain bounds for solutions of (4.1), (4.2). (Here, the units for corresponding quantities are the same as those in the last example.)

Remark 4.2. In [134], some of the parameters c_{ij} are allowed to depend on $|\nabla V|^2$; and the examples there are slightly more complicated.

Part B: Existence of Classical Solutions for the Initial-Boundary Value Problem.

We now consider the existence of solution for the initial-boundary value problem (4.1), (4.2) for an arbitrary time interval. We will need additional assumptions concerning the dependence of diffusivity and mobility of the species on the electric field and mean electron energy. In the problem (4.1), (4.2), we assume that for $i = 1, ..., 5$:

$$D_i = D_i(z, \lambda) = \alpha_i + \lambda f_i(z), \qquad \mu_i = \mu_i(z, \lambda) = \beta_i + \lambda \tilde{f}_i(z) \tag{4.14}$$

where α_i and β_i are positive constants, and λ is a non-negative real parameter. For $i = 1,2,3,4$, the functions $f_i(z)$ and their first two derivatives are continuous and bounded on bounded subsets in R; the function $f_5(z)$ and its first two derivatives are bounded for all real z. The functions $\tilde{f}_i(z), i = 1, ..., 5$ and its first derivative are bounded for all real z. We will assume that the initial-boundary functions $g_i(x,t)$ defined on $(\partial\Omega \times [0,T]) \cup (\Omega \times \{0\}), i = 1, ..., 6$, can be extended to be functions $\hat{g}_i$ in the Hölder space $C^{2+\alpha,(2+\alpha)/2}(\bar{\Omega} \times [0,T]), 0 < \alpha < 1$. The boundary function $\phi(x,t)$ will be assumed to be in the class $C^{2+\alpha,(2+\alpha)/2}(\partial\bar{\Omega}_T)$.

We will search for solutions of problem (4.1), (4.2) so that the functions $n_i(x,t) - \hat{g}_i(x,t)$, $i = 1, ..., 6$ are inside sets of the form:

(4.15)
$$S(Q) := \{u \in C^{1+\alpha,(1+\alpha)/2}(\bar{\Omega} \times [0,T]) : u_i = 0 \text{ on } (\partial\Omega \times [0,T]) \cup (\Omega \times \{0\}),$$
$$\|u_i\|^{1+\alpha} < Q+1, i = 1, ..., 6\}$$

for some large constant Q. Deforming D_i, $i = 1, ..., 5$ in (4.14) by $0 \leq s \leq 1$:

$$D_i^s = D_i^s(z, \lambda) = \alpha_i + \lambda s f_i(z), \tag{4.16}$$

and accordingly transforming problem (4.1), (4.2), we will consider a family of problems for $0 \leq s \leq 1$:

(4.17)
$$\begin{cases} \partial n_1^s/\partial t - \nabla \cdot (D_1^s(|\nabla V^s|^2)\nabla n_1^s) + s(\mu_1(|\nabla V^s|^2)/\epsilon)[\sum_{i=1}^4 q_i n_i^s - q_5 n_5^s]n_1^s \\ \qquad - s\mu_1(|\nabla V^s|^2)[(\nabla V^s) \cdot \nabla n_1^s] = (c_{11}n_5^s - c_{12}n_1^s)s, \\ \partial n_2^s/\partial t - \nabla \cdot (D_2^s(|\nabla V^s|^2)\nabla n_2^s) + s^2(\mu_2(|\nabla V^s|^2)/\epsilon)[\sum_{i=1}^4 q_i n_i^s - q_5 n_5^s]n_2^s \\ \qquad - s\mu_2(|\nabla V^s|^2)[(\nabla V^s) \cdot \nabla n_2^s] = c_{21}n_1^s s^2 - c_{22}n_5^s n_2^s s, \\ \partial n_3^s/\partial t - \nabla \cdot (D_3^s(|\nabla V^s|^2)\nabla n_3^s) + s(\mu_3(|\nabla V^s|^2)/\epsilon)[\sum_{i=1}^4 q_i n_i^s - q_5 n_5^s]n_3^s \\ \qquad - s\mu_3(|\nabla V^s|^2)[(\nabla V^s) \cdot \nabla n_3^s] = (c_{31}n_5^s - c_{32}n_3^s)s, \\ \partial n_4^s/\partial t - \nabla \cdot (D_4^s(|\nabla V^s|^2)\nabla n_4^s) + s^2(\mu_4(|\nabla V^s|^2)/\epsilon)[\sum_{i=1}^4 q_i n_i^s - q_5 n_5^s]n_4^s \\ \qquad - s\mu_4(|\nabla V^s|^2)[(\nabla V^s) \cdot \nabla n_4^s] = c_{41}n_3^s s^2 - c_{42}n_5^s n_4^s s, \\ \partial n_5^s/\partial t - \nabla \cdot (D_5^s(n_6^s/n_5^s)\nabla n_5^s) - s(\mu_5(n_6^s/n_5^s)/\epsilon)[\sum_{i=1}^4 q_i n_i^s - q_5 n_5^s]n_5^s \\ \qquad + s\mu_5(n_6^s/n_5^s)[(\nabla V^s) \cdot \nabla n_5^s] = (c_{11}n_5^s + c_{31}n_5^s - c_{22}n_2^s n_5^s - c_{42}n_4^s n_5^s)s, \\ \partial n_6^s/\partial t - (5/3)\nabla \cdot (D_5^s(n_6^s/n_5^s)\nabla n_6^s) \\ \qquad -s(5/3)(\mu_5(n_6^s/n_5^s)/\epsilon)[\sum_{i=1}^4 q_i n_i^s - q_5 n_5^s]n_6^s + s(5/3)\mu_5[(\nabla V^s) \cdot \nabla n_6^s] \\ \qquad = (e\mu_5(\nabla V^s \cdot \nabla V^s)n_5^s - eD_5^s[\nabla V^s \cdot \nabla n_5^s] - n_5^s[\sum_{i=1}^5 d_i n_i^s])s, \\ \Delta V^s = \epsilon^{-1}(\sum_{i=1}^4 q_i n_i^s - q_5 n_5^s), \end{cases}$$

with prescribed initial-boundary conditions:

$$(4.18)\qquad \begin{aligned} &n_i^s(x,0) = g_i^s(x,0) \geq 0 && \text{for } x \in \Omega,\ i = 1,...,6,\\ &n_i^s(x,t) = g_i^s(x,t) \geq 0 && \text{for } (x,t) \in \partial\Omega \times [0,\infty),\ i = 1,...,6,\\ &V^s(x,t) = \phi(x,t) && \text{for } (x,t) \in \partial\Omega \times [0,\infty). \end{aligned}$$

We will assume that the initial-boundary functions $g_i^s(x,t)$ defined on $(\partial\Omega \times [0,T]) \cup (\Omega \times \{0\}), i =, 1, ..., 6$, can be extended to be functions $\hat{g}_i^s$ in the Hölder space $C^{2+\alpha,(2+\alpha)/2}(\bar{\Omega} \times [0,T])$. We will let $g_i^0(x,t) = 0$, $g_i^1(x,t) = g_i(x,t)$ for $(x,t) \in (\Omega \times \{0\}) \cup (\partial\Omega \times [0,T])$, $i = 1, ..., 6$. Further, for $0 \leq s \leq 1, (x,t) \in (\Omega \times \{0\}) \cup (\partial\Omega \times [0,T])$, the functions will be assumed to satisfy:

$$(4.19)\qquad \begin{aligned} &0 \leq g_i^s(x,t) \leq C_i, \quad i = 1, ..., 4,\\ &s(1-\sigma)\delta \leq g_5^s(x,t) \leq C_5 \end{aligned}$$

for a small positive constant σ. For convenience, we will denote:

$$\Gamma_T := (\partial\Omega \times [0,T]) \cup (\bar{\Omega} \times \{0\}), \qquad \bar{\Omega}_T := \bar{\Omega} \times [0,T].$$

To simplify writing, we will not display the dependence of D_i^s, μ_i on the parameter λ in (4.17) and some following formulas. Let $\bar{\lambda} > 0$, and for this part we assume:

$$0 < \tilde{\mu}_i \leq \mu_i(z,\lambda) \leq \hat{\mu}_i$$

for all $0 \leq \lambda \leq \bar{\lambda}, z \in R$, $i = 1, ..., 5$.

In order to deform problem (4.1), (4.2) to a simple form, we first deduce bounds for the spatial derivatives of the solutions for the family of problems (4.17), (4.18) by the methods presented in Ladyzhenskaya et al. [113].

Lemma 4.1. *Suppose that there exist positive constants $\delta, C_i, i = 1, ..., 5$, with $\delta < C_5, \sum_{i=1}^4 q_iC_i < q_5C_5$ satisfying (4.10), (4.11); and the initial-boundary functions g_i^s satisfy (4.19), for $0 \leq s \leq 1$. Let $(n_1^s, ..., n_6^s, V^s)$, be a solution of problem (4.17), (4.18) with components in $C^{21}(\bar{\Omega}\times[0,T])$, and $n_i^s - \hat{g}_i^s \in S(Q)$ for $i = 1, ..., 6$. Then for each $Q > 0$ there exists $\hat{\lambda} \in (0, \bar{\lambda}]$ sufficiently small, so that for $0 \leq \lambda < \hat{\lambda}$, the solution must satisfy for $0 \leq s \leq 1$, $i = 1, ..., 5, j = 1, ..., m$:*

$$(4.20)\qquad max_{(x,t)\in\Gamma_T} |\partial n_i^s/\partial x_j| \leq \tilde{M}_1,$$

where $\tilde{M}_1$ is a constant independent of Q, determined by δ, and $C_i, i = 1, ..., 5$.

Proof. In (4.10), the second and fourth inequalities can be modified to:

$$(4.21)\qquad \begin{cases} -(\hat{\mu}_2/\epsilon)(q_2C_2 - q_5C_5)C_2 + c_{21}C_1 - c_{22}(1-\sigma)\delta C_2 < 0,\\ -(\hat{\mu}_4/\epsilon)(q_4C_4 - q_5C_5)C_4 + c_{41}C_3 - c_{42}(1-\sigma)\delta C_4 < 0 \end{cases}$$

provided σ is a sufficiently small positive number. For each $0 \le s \le 1$, define the coupled upper-lower solution pair for problem (4.17), (4.18) to be $(C_1, ..., C_5)$ and $(0, 0, 0, 0, s[1-\sigma]\delta)$ as described in Definition 4.1, modified by appropriate scaling with s or s^2 as in (4.17). By (4.10), (4.11) and (4.21), one readily verifies that the pair indeed satisfies the appropriate inequalities described in (4.4) for an upper-lower solution pair. Consequently, using condition (4.19) and Theorem 4.1 we assert that:

$$0 \le n_i^s(x,t) \le C_i,\ i = 1,2,3,4, \qquad s(1-\sigma)\delta \le n_5^s(x,t) \le C_5$$

for $(x,t) \le \bar{\Omega}_T$. These bounds and the last equation in (4.17) then give a bound for the first spatial derivatives of V^s in $\bar{\Omega}_T$ by means of W^{2p} estimates and embedding theorem. In the first 5 equations in (4.17), the first spatial derivatives of n_i^s are multiplied by $(\partial D_i^s/\partial z)(|\nabla V^s|^2)V_{x_j}^s V_{x_j x_k}^s$ or $(\partial D_5^s/\partial z)(n_6^s/n_5^s)(n_6^s/n_5^s)_{x_k}$, and also by the first spatial derivatives of V^s. The last equation in (4.17) gives estimate of $V_{x_j x_k}^s$ in terms of the Hölder norm of $n_i^s, i = 1,..,5$, which are, by hypotheses of this lemma, limited by Q. Consequently, using the assumptions on D_i^s, for $0 \le \lambda \le \hat{\lambda}$ sufficiently small, the factors multiplying with the first spatial derivatives of n_i^s in the first five equations in (4.17) are bounded by a constant independent on Q, determined by $C_i, i = 1, ..., 5$ and δ. Moreover, the assumption on f_i implies that for sufficiently small positive λ, we have $D_i^s \ge \alpha_i/2$ whenever they are evaluated in the equations.

Using the growth estimate of the terms in the i-th equation with respect to ∇n_i^s and the lower estimate of the ellipticity, we can apply Lemma 3.1, Chapter VI, on p. 535 of [113], to obtain the estimate (4.20) in this Lemma. The method of S.N. Bernstein is used to estimate the gradient on the boundary in proving the Lemma 3.1 referred in [113].

Lemma 4.2 (Bounds for Spatial Derivatives). *Assume all the hypotheses of Lemma 4.1, and $(n_1^s, ..., n_6^s, V^s)$ is a solution of (4.17), (4.18) with properties as described there. Then for each $Q > 0$ there exists $\hat{\lambda} > 0$ sufficiently small, so that for $0 \le \lambda < \hat{\lambda}$, the solution must satisfy for $0 \le s \le 1$, i=1,...,5, j=1,...,m :*

$$max_{(x,t)\in\bar{\Omega}_T} |\partial n_i^s/\partial x_j| \le \tilde{M}_2, \tag{4.22}$$

where $\tilde{M}_2$ is a constant independent of Q, determined by $\tilde{M}_1, \delta$, and $C_i, i = 1, ..., 5$.

Proof. As in Lemma 4.1, the first spatial derivatives of V^s are bounded by constants dependent on δ and $C_i, i = 1, ..., 5$. Moreover, we have $n_5^s \ge s(1-\sigma)\delta$ in $\bar{\Omega}_T$; and for each $1 \ge s > 0$, $|n_6^s/n_5^s|$ is bounded by a constant dependent on Q and s. In each of the first five equations in (4.17), we interpret ∇V^s and n_6^s/n_5^s to be given functions on (x,t) substituted into $D_k^s, k = 1, ..., 4$ and D_5^s respectively. We will apply Theorem 4.1 in p. 443 of [113]. In relation to the notations of that Theorem, we identify u with n_k^s, p with $(n_k^s)_{x_i}$, and $a_i(x,t,u,p)$ with $D_k^s(|\nabla V^s|^2)(x,t)(n_k^s)_{x_i}$ or $D_5^s(n_6^s/n_5^s)(x,t)(n_5^s)_{x_i}$. For instance, the derivative $\partial D_k^s(|\nabla V^s|^2)(x,t)/\partial x_i$ involves terms of the form $(\partial D_k^s(|\nabla V^s|^2)/\partial z)V_{x_j}^s V_{x_j x_i}^s$, while as explained in the previous lemma the quantities $V_{x_j x_i}^s$ are bounded by a constant determined by Q. Consequently, by assumption (4.16), we can readily verify that for $0 < \lambda < \hat{\lambda}$ sufficiently small, the expressions $|D_k^s(|\nabla V^s|^2)(x,t)|$, $|\partial D_k^s(|\nabla V^s|^2)(x,t)/\partial x_i|$, $|D_5^s(n_6^s/n_5^s)(x,t)|$ and $|\partial D_5^s(n_6^s/n_5^s)(x,t)/\partial x_i|$ are all bounded by a constant independent of Q. On the other hand, the growth of the other terms in the each of the equations with respect to n_k^s and $(n_k^s)_{x_i}$ are bounded by $K_1 + K_2|\nabla n_k^s|$ for some constants K_1, K_2, since n_i^s are bounded for $i = 1, ..., 5$. Thus for each equation in (4.17), $k = 1, ..., 5$, we can apply Theorem 4.1 on p. 443 in [113], to assert that $max_{(x,t)\in\bar{\Omega}_T}|\partial n_i^s/\partial x_j|$ can be estimated in terms of δ, $C_i, i = 1, ..., 5$ and $max_{\Gamma_T}|\partial n_k^s/\partial x_j|$. Finally, by Lemma 4.1, we obtain inequality (4.22).

Lemma 4.3 (Bounds for Electron Energy). *Assume all the hypotheses of Lemma 4.1, and $(n_1^s, ..., n_6^s, V^s)$ is a solution of (4.17), (4.18) with properties as described there. Then for each $Q > 0$ there exists $\hat{\lambda} > 0$ sufficiently small, so that for $0 \le \lambda < \hat{\lambda}$, the solution must satisfy for $0 \le s \le 1$, j=1,...,m:*

$$max_{(x,t)\in\bar{\Omega}_T}|n_6^s| \le K_1, \quad max_{(x,t)\in\bar{\Omega}_T}|\partial n_6^s/\partial x_j| \le K_2, \tag{4.23}$$

where K_1, K_2 are constants independent of Q, determined by $\tilde{M}_1, \tilde{M}_2, \delta$, and $C_i, i = 1, ..., 5$.

Proof. Consider the 6-th equation in (4.17). The function $z = n_6^s$ can be viewed as a solution of the linear nonhomogeneous parabolic problem:

$$\begin{cases} \partial z/\partial t - (5/3)\nabla \cdot (D_5^s(n_6^s/n_5^s)\nabla z) \\ \quad -s(5/3)(\mu_5(n_6^s/n_5^s)/\epsilon)[\sum_{i=1}^4 q_i n_i^s - q_5 n_5^s]z + s(5/3)\mu_5[(\nabla V^s)\cdot\nabla z] \\ = (e\mu_5(\nabla V^s \cdot \nabla V^s)n_5^s - eD_5^s[\nabla V^s \cdot \nabla n_5^s] - n_5^s[\sum_{i=1}^5 d_i n_i^s])s \\ \qquad\qquad\qquad\qquad \text{for } (x,t) \in \Omega \times (0,T); \\ z(x,t) = g_6^s(x,t) \qquad\qquad \text{for } (x,t) \in \Gamma_T. \end{cases} \tag{4.24}$$

By Lemmas 4.1, 4.2 and formula (4.14), the coefficients and nonhomogeneous terms are all bounded independent of Q, for $0 \le \lambda \le \hat{\lambda}$ sufficiently small. Let B_1, B_2, Y_0, W_0 be constants such that for $(x,t) \in \bar{\Omega}_T$,

$$(4.25)\quad \begin{cases} |s(5/3)(\mu_5(n_6^s/n_5^s)/\epsilon)[\sum_{i=1}^4 q_i n_i^s - q_5 n_5^s]| \le B_1, \\ |(e\mu_5(\nabla V^s \cdot \nabla V^s)n_5^s - eD_5^s[\nabla V^s \cdot \nabla n_5^s] - n_5^s[\sum_{i=1}^5 d_i n_i^s])s| \le B_2; \\ Y_0 \le |g_6^s(x,t)| \le W_0 \qquad \text{for } (x,t) \in \Gamma_T. \end{cases}$$

Define $Y(x,t) := Y(t)$, $W(x,t) := W(t)$, for $(x,t) \in \bar{\Omega}_T$, to be functions respectively satisfying:

$$(4.26)\quad \begin{cases} dY/dt = B_1 Y - B_2, \ Y(0) = Y_0; \\ dW/dt = B_1 W + B_2, \ W(0) = W_0. \end{cases}$$

The functions Y, W are respectively lower and upper solutions of problem (4.24) as described in [125] and [183]. Thus we find:

$$(4.27)\quad Y(t) \le n_6^s(x,t) \le W(t) \quad \text{for} \quad (x,t) \in \bar{\Omega}_T.$$

We obtain the first inequality in (4.23). The second inequality follows by applying Lemma 3.1 on p. 535 and Theorem 4.1 on p. 443 in [113], as in the proof of Lemma 4.1 and Lemma 4.2 above.

By means of the lemmas above, we can now deduce estimates for the time derivatives of the functions η_i^s.

Theorem 4.3 (Bounds for Time Derivatives). *Suppose that there exist positive constants $\delta, C_i, i = 1, ..., 5$ satisfying (4.10), (4.11) as described in Lemma 4.1; and the initial-boundary functions g_i^s satisfy (4.19), for $0 \le s \le 1$. Let $(n_1^s, ..., n_6^s, V^s)$, be a solution of problem (4.17), (4.18) with components in $C^{21}(\bar{\Omega} \times [0,T])$, and $n_i^s - \hat{g}_i^s \in S(Q)$ for $i = 1, ..., 6$. Then for each $Q > 0$ there exists $\hat{\lambda} > 0$ sufficiently small, so that for $0 \le \lambda < \hat{\lambda}$, the solution must satisfy for $0 \le s \le 1$, i=1,... 6 :*

$$(4.28)\quad max_{(x,t)\in\bar{\Omega}_T} |\partial n_i^s/\partial t| \le \tilde{M}_3,$$

where $\tilde{M}_3$ is a constant independent of Q, determined by δ, and $C_i, i = 1, ..., 5$.

Proof. For $(x,t) \in \bar{\Omega} \times [0,T)$, $i = 1, \ldots 6$, $0 \le s \le 1$ and $h > 0$ small, let:

$$(4.29)\quad n_i^{s,h}(x,t) = h^{-1}[n_i^s(x,t+h) - n_i^s(x,t)].$$

For convenience, we denote:

$$\tilde{q}_i = q_i, i = 1, ..., 4; \ \tilde{q}_5 = -q_5.$$

Using the first equation of (4.17) at $(x, t+h)$ and (x,t), and writing e.g.

$$[D_1^s(|\nabla V^s|^2)(x,t+h) - D_1^s(|\nabla V^s|^2)(x,t)]h^{-1}$$

$$= [\int_0^1 (D_1^s)'(\tau|\nabla V^s|^2(x,t+h) + (1-\tau)|\nabla V^s|^2(x,t))d\tau] \tag{4.30}$$

$$\times [\sum_{i=1}^m (V_{x_i}^s(x,t+h) + V_{x_i}^s(x,t))(V_{x_i}^s(x,t+h) - V_{x_i}^s(x,t))h^{-1}],$$

we deduce that $n_1^{s,h}(x,t)$ satisfy:

(4.31)

$$\partial n_1^{s,h}/\partial t - \nabla \cdot \{D_1^s(|\nabla V|^2(x,t))\nabla n_1^{s,h}$$

$$+ \nabla n_1^s(x,t+h)[\int_0^1 (D_1^s)'(\tau|\nabla V^s|^2(x,t+h) + (1-\tau)|\nabla (V^s)|^2(x,t))d\tau]$$

$$\times [\sum_{i=1}^m (V_{x_i}^s(x,t+h) + V_{x_i}^s(x,t)][\int_\Omega \epsilon^{-1} G_{x_i}(x,\xi) \sum_{k=1}^5 \tilde{q}_k n_k^{s,h}(\xi,t)d\xi]\}$$

$$+ s\epsilon^{-1} \int_0^1 \mu_1'(\tau|\nabla V^s|^2(x,t+h) + (1-\tau)|\nabla V^s|^2(x,t))d\tau$$

$$\times [\sum_{i=1}^m (V_{x_i}^s(x,t+h) + V_{x_i}^s(x,t)) \int_\Omega \epsilon^{-1} G_{x_i}(x,\xi) \sum_{k=1}^5 \tilde{q}_k n_k^{s,h}(\xi,t)d\xi]$$

$$\times [\sum_{k=1}^5 \tilde{q}_k n_k^s(x,t+h)]n_1^s(x,t+h)$$

$$+ s\epsilon^{-1}\mu_1(|\nabla V^s|^2(x,t))[\sum_{k=1}^5 \tilde{q}_k n_k^{s,h}(x,t)]n_1^s(x,t+h)$$

$$+ s\epsilon^{-1}\mu_1(|\nabla V^s|^2(x,t))[\sum_{k=1}^5 \tilde{q}_k n_k^s(x,t)]n_1^{s,h}(x,t)$$

$$- s \int_0^1 \mu_1'(\tau|\nabla V^s|^2(x,t+h) + (1-\tau)|\nabla V^s|^2(x,t))d\tau$$

$$\times [\sum_{i=1}^m (V_{x_i}^s(x,t+h) + V_{x_i}^s(x,t)][\int_\Omega \epsilon^{-1} G_{x_i}(x,\xi) \sum_{k=1}^5 \tilde{q}_k n_k^{s,h}(\xi,t)d\xi]$$

$$\times [\nabla V^s(x,t+h) \cdot \nabla n_1^s(x,t+h)]$$

$$- s\mu_1(|\nabla V^s|^2(x,t))[(\int_\Omega \sum_{k=1}^5 \tilde{q}_k n_k^{s,h}(\xi,t)\nabla G(x,\xi)d\xi) \cdot \nabla n_1^s(x,t+h)$$

$$+ \nabla V^s(x,t) \cdot \nabla n_1^{s,h}(x,t)]$$

$$= (c_{11}n_5^{s,h}(x,t) - c_{12}n_1^{s,h}(x,t))s.$$

The integral terms over Ω above arise from expressing $V_{x_i}^s$ by means of Newtonian potential and the derivative of the Green's function G using the last equation in (4.17), (see e.g. p. 54 in Gilbarg and Trudinger [71]).

From the first equation in (4.17), we find that $n_1^s(x,t)$ is a solution of a linear nonhomogeneous parabolic equation with the coefficients and nonhomogeneous terms dependent on $n_k^s(x,t)$. Thus using the fact that $n_k^s - \hat{g}_k^s$ are in $S(Q)$, we can obtain by Schauder's estimate that $|\Delta n_1^s|$ is bounded. Note that $|V_{x_i x_j}^s|$ can also be shown to have a bound dependent on Q by using Schauder's estimate with the last equation in (4.17). Consequently, from (4.31) and similar equations for $n_i^{s,h}, i = 2, ..., 6$, we can verify that $z_i(x,t) = n_i^{s,h}(x,t)$ satisfy in $\Omega \times [0,T)$:

(4.32)
$$\frac{\partial z_i(x,t)}{\partial t} - \sigma_i^s(x,t)\Delta z_i + [R_1^{si}(x,t) + \lambda R_2^{si}(x,t)] \cdot \nabla z_i + \sum_{j=1}^6 R_{3j}^{si}(x,t) z_j(x,t)$$

$$+ \sum_{j=1}^5 \int_\Omega \sum_{k=1}^m [R_{4k}^{si}(x,t) + \lambda R_{5k}^{si}(x,t)] G_{x_k}(x,\xi)\tilde{q}_j z_j(\xi,t) d\xi$$

$$+\lambda \sum_{j=1}^5 \int_\Omega \sum_{r=1}^m \sum_{k=1}^m H_{rk}^{si}(x,t) G_{x_r x_k}(x,\xi)\tilde{q}_j [z_j(\xi,t) - z_j(x,t)] d\xi$$

$$+\lambda \sum_{j=1}^5 \tilde{q}_j z_j(x,t) \int_{\partial\Omega} \sum_{r=1}^m \sum_{k=1}^m H_{rk}^{si}(x,t) G_{x_r}(x,\xi)\nu_k(\xi) d\xi = 0,$$

where $\sigma_i^s(x,t)$, $R_1^{si}(x,t), R_{3j}^{si}(x,t), R_{4k}^{si}(x,t), H_{rk}^{si}(x,t)$ are bounded continuous functions in $\bar{\Omega}_T$. $R_2^{si}(x,t), R_{5k}^{si}(x,t)$ are also bounded continuous functions in $\bar{\Omega}_T$, however with bounds determined by Q; they are multiplied by the factor λ which arises from differentiating the functions D_1^s and μ_1. Here, ν_k denotes the k-th component of the outward unit normal. The last two expressions in (4.32) arise from applying the operator ∇ in the second term of (4.31) on the integral over Ω. Here, we have also used Lemma 4.1 to Lemma 4.3, and the assumptions on D_i and μ_i in (4.14) in the assertion concerning the bounds of the coefficients. We also have the coefficient in (4.32) satisfying $\sigma_i^s(x,t) > 0$ in $\bar{\Omega}_T$ for $0 \le \lambda < \hat{\lambda}$ sufficiently small.

In order to simplify writing for equation (4.32), we denote:

$$\tilde{F}_i^s(x,t,z_1(x,t),...,z_6(x,t);z_1(\cdot,t),...,z_5(\cdot,t),\lambda) = \sum_{j=1}^6 R_{3j}^{si} z_j(x,t)$$

(4.33)

$$+ \sum_{j=1}^5 \int_\Omega \sum_{k=1}^m [R_{4k}^{si}(x,t) + \lambda R_{5k}^{si}(x,t)] G_{x_k}(x,\xi)\tilde{q}_j z_j(\xi,t) d\xi,$$

$$F_i^s(x,t,z_1(x,t),...,z_6(x,t);z_1(\cdot,t),...,z_5(\cdot,t),\lambda) = \tilde{F}_i^s(x,t,...,\lambda)$$

(4.34)
$$+\lambda \sum_{j=1}^5 \int_\Omega \sum_{r=1}^m \sum_{k=1}^m H_{rk}^{si}(x,t) G_{x_r x_k}(x,\xi)\tilde{q}_j [z_j(\xi,t) - z_j(x,t)] d\xi$$

$$+\lambda \sum_{j=1}^5 \tilde{q}_j z_j(x,t) \int_{\partial\Omega} \sum_{r=1}^m \sum_{k=1}^m H_{rk}^{si} G_{x_r}(x,\xi)\nu_k(\xi) d\xi$$

for $i = 1, ..., 6$. The function $\tilde{F}_i^s$ satisfies:

$$
\begin{aligned}
&|\tilde{F}_i^s(x, t, \bar{z}_1(x,t), ..., \bar{z}_6(x,t); \bar{z}_1(\cdot,t), \ldots \bar{z}_5(\cdot,t), \lambda) \\
&- \tilde{F}_i^s(x, t, \tilde{z}_1(x,t), ..., \tilde{z}_6(x,t); \tilde{z}_1(\cdot,t), \ldots \tilde{z}_5(\cdot,t), \lambda)| \qquad (4.35) \\
&\leq L_1 \cdot max_{\xi\in\Omega}(\textstyle\sum_{j=1}^6 |\bar{z}_j(\xi,t) - \tilde{z}_j(\xi,t)|^2)^{1/2}
\end{aligned}
$$

for some constant $L_1 > 0$, for all $(x,t) \in \bar{\Omega}_T, 0 \leq \lambda < \hat{\lambda}$. Let

$$
L_2 = \hat{\lambda} 6^{-1/2} \sum_{j=1}^{5} \sum_{r=1}^{m} \sum_{k=1}^{m} max_{\Omega_T} | \int_{\partial\Omega} H_{rk}^{si} G_{x_r}(x,\xi)\nu_k(\xi) d\xi | \tilde{q}_j.
$$

The proof of Theorem 4.3 is complete by using the following lemma to obtain bounds for the difference quotients in (4.29). The parameter λ is fixed in $[0, \hat{\lambda})$.

Lemma 4.4. *Let $z_j(x,t), v_j(x,t), w_j(x,t), j = 1, \ldots 6$ be functions in $C^{21}(\bar{\Omega} \times [0,T])$, and has $w_j(x,t) - v_j(x,t) > 0$ for all $(x,t) \in \bar{\Omega}_T$. Suppose that the functions z_j satisfy for $(x,t) \in \Omega \times (0,T]$:*

$$
\begin{aligned}
&\partial z_j(x,t)/\partial t - \sigma_j^s(x,t)\Delta z_j + [R_1^{sj}(x,t) + \lambda R_2^{sj}(x,t)] \cdot \nabla z_j \\
&+ F_j^s(x, t, z_1(x,t), ..., z_6(x,t); z_1(\cdot,t), ..., z_5(\cdot,t), \lambda) = 0, \qquad (4.36)
\end{aligned}
$$

and

$$
v_j(x,t) < z_j(x,t) < w_j(x,t) \qquad \text{for } (x,t) \in (\Omega \times \{0\}) \cup (\partial\Omega \times [0,T]). \qquad (4.37)
$$

Moreover, the functions $v_j(x,t)$ and $w_j(x,t), j = 1, ..., 6$ satisfy:

$$
\begin{aligned}
&\partial v_j(x,t)/\partial t - \sigma_j^s(x,t)\Delta v_j + [R_1^{sj}(x,t) + \lambda R_2^{sj}(x,t)] \cdot \nabla v_j \\
&+ F_j^s(x, t, \tilde{v}_1(x,t), ..., \tilde{v}_6(x,t); \tilde{v}_1(\cdot,t), ..., \tilde{v}_5(\cdot,t), \lambda) < 0 \qquad (4.38)
\end{aligned}
$$

for $(x,t) \in \Omega \times (0,T]$, where for each $k = 1, ..., 6$,

$$
\begin{aligned}
&\tilde{v}_k(\cdot,t) := z_k(\cdot,t) + \sigma[1 + 3(L_1 + L_2)6^{1/2}t], \quad \textit{provided} \\
&v_r(x,\tau) \leq \tilde{v}_r(x,\tau) \leq w_r(x,\tau) + \sigma[1 + 3(L_1 + L_2)6^{1/2}T] \qquad (4.39)
\end{aligned}
$$

for all $\tau \in [0,t)$, $x \in \bar{\Omega}$, $1 \leq r \leq 6$. (Here, σ is any constant satisfying $0 < \sigma < min_{1\leq i\leq 6}\{w_i(x,t) - v_i(x,t) | (x,t) \in \bar{\Omega} \times [0,T]\}(1/2)[1 + 3(L_1 + L_2)6^{1/2}T]^{-1})$; and

$$
\begin{aligned}
&\partial w_j(x,t)/\partial t - \sigma_j^s(x,t)\Delta w_j + [R_1^{sj}(x,t) + \lambda R_2^{sj}(x,t)] \cdot \nabla w_j \\
&+ F_j^s(x, t, \tilde{w}_1(x,t), ..., \tilde{w}_6(x,t); \tilde{w}_1(\cdot,t), ..., \tilde{w}_5(\cdot,t), \lambda) > 0 \qquad (4.40)
\end{aligned}
$$

for $(x,t) \in \Omega \times (0,T]$, *where for each* $k = 1, ..., 6$,

$$\tilde{w}_k(\cdot,t) := z_k(\cdot,t) - \sigma[1 + 3(L_1 + L_2)6^{1/2}t], \quad \textit{provided} \tag{4.41}$$
$$v_r(x,\tau) - \sigma[1 + 3(L_1 + L_2)6^{1/2}T] \le \tilde{w}_r(x,\tau) \le w_r(x,\tau)$$

for all $\tau \in [0,t)$, $x \in \bar{\Omega}$, $1 \le r \le 6$. *Then the following is satisfied:*

$$v_j(x,t) \le z_j(x,t) \le w_j(x,t) \qquad \textit{for all } (x,t) \in \bar{\Omega}_T,\ j = 1, ..., 6. \tag{4.42}$$

Proof of Lemma 4.4. For $(x,t) \in \bar{\Omega}_T$, define

$$z_i^{\pm\sigma}(x,t) = z_i(x,t) \pm [1 + 3(L_1 + L_2)6^{1/2}t]\sigma.$$

Note that

(4.43)
$$F_i^s(x,t;z_1(x,t),...,z_6(x,t);z_1(\cdot,t),...,z_5(\cdot,t))$$
$$-F_i^s(x,t,z_1^{-\sigma}(x,t),...,z_6^{-\sigma}(x,t);z_1^{-\sigma}(\cdot,t),...,z_5^{-\sigma}(\cdot,t))$$
$$= \tilde{F}_i^s(x,t;z_1(x,t),...,z_6(x,t);z_1(\cdot,t),...,z_5(\cdot,t))$$
$$- \tilde{F}_i^s(x,t,z_1^{-\sigma}(x,t),...,z_6^{-\sigma}(x,t);z_1^{-\sigma}(\cdot,t),...,z_5^{-\sigma}(\cdot,t))$$
$$+\sigma[1 + 3(L_1 + L_2)6^{1/2}t]\lambda \textstyle\sum_{j=1}^5 \tilde{q}_j \int_{\partial\Omega} \sum_{r=1}^m \sum_{k=1}^m H_{rk}^{si} G_{x_r}(x,\xi)\nu_k(\xi)d\xi.$$

By hypotheses, at $t = 0$ for $i = 1, ..., 6$, we have

$$v_i(x,t) < z_i^{+\sigma}(x,t) \ \text{ and } \ z_i^{-\sigma}(x,t) < w_i(x,t) \ \text{ for } (x,t) \in \bar{\Omega}_T. \tag{4.44}$$

Suppose one of these inequalities fails at some point in $\bar{\Omega} \times (0,\tau_1)$ where $\tau_1 = min\{T, (3(L_1 + L_2)6^{1/2})^{-1}\}$; and (x_1,t_1) is a point in $\bar{\Omega} \times (0,\tau_1)$ with minimal t_1 where (4.44) fails. At (x_1,t_1), $v_i = z_i^{+\sigma}$ or $z_i^{-\sigma} = w_i$ for some i. Assume that the latter is the case; a similar proof holds for the former case.(Also assume that $i = 1$, while the same arguments hold for other cases of i). First, let $w_1(x_1,t_1) = z_1^{-\sigma}(x_1,t_1)$ with $x_1 \in \Omega$. At (x_1,t_1), we deduce

(4.45)
$$(\partial/\partial t)(w_1 - z_1^{-\sigma})|_{(x_1,t_1)} = \partial w_1/\partial t - \partial z_1/\partial t + 3\sigma(L_1 + L_2)6^{1/2}$$
$$\ge \sigma_1^s \Delta(w_1 - z_1) + [R_1^{s1} + \lambda R_2^{s1}] \cdot \nabla(w_1 - z_1)$$
$$+ F_1^s(x_1,t_1,z_1^{-\sigma}(x_1,t_1),...,z_6^{-\sigma}(x_1,t_1);z_1^{-\sigma}(\cdot,t_1),...,z_5^{-\sigma}(\cdot,t_1))$$
$$- F_1^s(x_1,t_1,z_1(x_1,t_1),...,z_6(x_1,t_1);z_1(\cdot,t_1),...,z_5(\cdot,t_1)) + 3\sigma(L_1 + L_2)6^{1/2}.$$

We have

$$\begin{aligned}
&|F_1^s(x_1,t_1,z_1^{-\sigma}(x_1,t_1),...,z_6^{-\sigma}(x_1,t_1);z_1^{-\sigma}(\cdot,t_1),...,z_5^{-\sigma}(\cdot,t_1)) \\
&\quad - F_1^s(x_1,t_1,z_1(x_1,t_1),...,z_6(x_1,t_1);z_1(\cdot,t_1),...,z_5(\cdot,t_1))| \qquad (4.46)\\
&\le (L_1+L_2)6^{1/2}\sigma[1+3(L_1+L_2)6^{1/2}t_1] \le (L_1+L_2)6^{1/2}\sigma 2.
\end{aligned}$$

Thus from (4.45) and (4.46), we have

$$(\partial/\partial t)(w_1 - z_1^{-\sigma})|_{(x_1,t_1)} \ge \sigma(L_1+L_2)6^{1/2} > 0,$$

contradicting the definition of (x_1,t_1). The remaining part of the proof of Lemma 4.4 is the same as the last part of the proof of Theorem 4.1, and will thus be omitted. The dependence on λ is suppressed in the writing. The lemma is actually true for each $\lambda \in [0,\hat{\lambda}), 0 \le s \le 1$.

The proof of Theorem 4.3 will be complete by choosing appropriate v_i and w_i for the solution $z_i = n_i^{s,h}$ of equation (4.32), and then apply Lemma 4.4 to obtain the bound (4.28). For $i = 1,..,6$, $(x,t) \in \bar{\Omega}_T$, define

$$v_i(x,t) := -Ke^{Pt}, \quad w_i(x,t) := Ke^{Pt},$$

where the constants K and P will now be chosen. The assumptions on g_i^s implies that (4.37) is satisfied by choosing K sufficiently large. In order to verify (4.38) and (4.40), we first note that one term in (4.34) satisfies:

$$\begin{aligned}
&\lambda \textstyle\sum_{j=1}^5 \int_\Omega \sum_{r=1}^m \sum_{k=1}^m H_{rk}^{si}(x,t) G_{x_r x_k}(x,\xi)\tilde{q}_j[z_j(\xi,t) - z_j(x,t)]d\xi \\
&\le \lambda \textstyle\sum_{j=1}^5 \int_\Omega \sum_{r=1}^m \sum_{k=1}^m H_{rk}^{si}(x,t) G_{x_r x_k}(x,\xi)\tilde{q}_j \bar{K}|\xi - x|^{\alpha/2} d\xi \le \hat{R}
\end{aligned}$$

for some constant $\hat{R}$, where $\bar{K}$ is some Hölder constant for the functions $z_j = n_j^{s,h}$. We are able to find such a $\bar{K}$ for all $n_j^{s,h}$ under consideration by using the fact that $(n_j^s - \hat{g}_j^s) \in S(Q)$, and thus

$$|n_j^{s,h}(\xi,t) - n_j^{s,h}(x,t)| = (\partial/\partial t)(n_j^s(\xi,t^*) - n_j^s(x,t^*)| \le \bar{K}|\xi - x|^{\alpha/2}.$$

Here, t^* depends on the pair (x,ξ), and it satisfies $t^* \in (t,t+h)$. The constant $\bar{K}$ depends on Q. We have used the fact that n_j^s are solutions of linear equations related to (4.17), (cf. (4.47) below), so that the bound on the Hölder norms of the coefficients give rise to a bound to the $C^{2+\alpha,(2+\alpha)/2}$ norms of the solutions by means of Schauder's estimates. We may then reduce the size of $\hat{\lambda}$ so that $\hat{R}$ is independent of Q for all $\lambda \in [0,\hat{\lambda})$. Estimating the other terms of $F_j^s(x,t,\tilde{v}_1(x,t),...,\tilde{v}_6;\tilde{v}_1(\cdot,t),...,\tilde{v}_5(x,t),\lambda)$, we obtain by means of (4.33), (4.34) and the second inequality in (4.39) that

$$\begin{aligned}
&|F_j^s(x,t,\tilde{v}_1(x,t),...,\tilde{v}_6;\tilde{v}_1(\cdot,t),...,\tilde{v}_5(\cdot,t),\lambda)| \\
&\le \bar{B}[Ke^{Pt} + \sigma(1+3(L_1+L_2)6^{1/2}T)] + \hat{R},
\end{aligned}$$

where $\bar{B}$ is a constant independent of Q, as long as $\lambda \in [0, \hat{\lambda})$. Here, $\hat{\lambda}$ is determined by Q, and we have used the boundedness properties of $R^{si}_{jk}(x,t)$ as described for (4.32). Thus (4.38) is satisfied if

$$-PKe^{Pt} + \bar{B}[Ke^{Pt} + \sigma(1 + 3(L_1 + L_2)6^{1/2}T)] + \hat{R} < 0,$$

for $0 \le t \le T$, which is true by choosing sufficiently large P. Inequality (4.40) can be verified in the same way as (4.38). This completes the proof of Theorem 4.3.

In order to investigate the solutions for (4.17) and (4.18), we will search for fixed points of the following related mappings: $T_s : (\tilde{n}_1^s, ..., \tilde{n}_6^s) \longrightarrow (\hat{n}_1^s, ..., \hat{n}_6^s)$, with $(\tilde{n}_1^s - \hat{g}_1^s, ..., \tilde{n}_6^s - \hat{g}_6^s) \in S(Q)$, where $n_i^s = \hat{n}_i^s, i = 1, ..., 6$, $0 \le s \le 1$ is defined to be the solution in $\Omega \times (0, T)$ of the linear system:

(4.47)
$$\begin{cases}
\partial n_1^s/\partial t - \nabla \cdot (D_1^s(|\nabla \tilde{n}_7^s|^2)\nabla n_1^s) + s(\mu_1(|\nabla \tilde{n}_7^s|^2)/\epsilon)[\sum_{i=1}^4 q_i\tilde{n}_i^s - q_5\tilde{n}_5^s]n_1^s \\
\qquad - s\mu_1(|\nabla \tilde{n}_7^s|^2)[(\nabla \tilde{n}^s)_7 \cdot \nabla n_1^s] + sc_{12}n_1^s = c_{11}\tilde{n}_5^s s, \\
\partial n_2^s/\partial t - \nabla \cdot (D_2^s(|\nabla \tilde{n}_7^s|^2)\nabla n_2^s) + s^2(\mu_2(|\nabla \tilde{n}_7^s|^2)/\epsilon)[\sum_{i=1}^4 q_i\tilde{n}_i^s - q_5\tilde{n}_5^s]n_2^s \\
\qquad - s\mu_2(|\nabla \tilde{n}_7^s|^2)[(\nabla \tilde{n}_7^s) \cdot \nabla n_2^s] + sc_{22}\tilde{n}_5^s n_2^s = c_{21}\tilde{n}_1^s s^2, \\
\partial n_3^s/\partial t - \nabla \cdot (D_3^s(|\nabla \tilde{n}_7^s|^2)\nabla n_3^s) + s(\mu_3(|\nabla \tilde{n}_7^s|^2)/\epsilon)[\sum_{i=1}^4 q_i\tilde{n}_i^s - q_5\tilde{n}_5^s]n_3^s \\
\qquad - s\mu_3(|\nabla \tilde{n}_7^s|^2)[(\nabla \tilde{n}_7^s) \cdot \nabla n_3^s] + sc_{32}n_3^s = c_{31}\tilde{n}_5^s s, \\
\partial n_4^s/\partial t - \nabla \cdot (D_4^s(|\nabla \tilde{n}_7^s|^2)\nabla n_4^s) + s^2(\mu_4(|\nabla \tilde{n}_7^s|^2)/\epsilon)[\sum_{i=1}^4 q_i\tilde{n}_i^s - q_5\tilde{n}_5^s]n_4^s \\
\qquad - s\mu_4(|\nabla \tilde{n}_7^s|^2)[(\nabla \tilde{n}_7^s) \cdot \nabla n_4^s] + sc_{42}\tilde{n}_5^s n_4^s = c_{41}\tilde{n}_3^s s^2, \\
\partial n_5^s/\partial t - \nabla \cdot (D_5^s(\tilde{n}_6^s/\tilde{n}_5^s)\nabla n_5^s) - s(\mu_5(\tilde{n}_6^s/\tilde{n}_5^s)/\epsilon)[\sum_{i=1}^4 q_i\tilde{n}_i^s - q_5\tilde{n}_5^s]n_5^s \\
\qquad + s\mu_5(\tilde{n}_6^s/\tilde{n}_5^s)[(\nabla \tilde{n}_7^s) \cdot \nabla n_5^s] - s(c_{11} + c_{31} - c_{22}\tilde{n}_2^s - c_{42}\tilde{n}_4^s)n_5^s = 0, \\
\partial n_6^s/\partial t - (5/3)\nabla \cdot (D_5^s(\tilde{n}_6^s/\tilde{n}_5^s)\nabla n_6^s) \\
\qquad - s(5/3)(\mu_5(\tilde{n}_6^s/\tilde{n}_5^s)/\epsilon)[\sum_{i=1}^4 q_i\tilde{n}_i^s - q_5\tilde{n}_5^s]n_6^s + s(5/3)\mu_5[(\nabla \tilde{n}_7^s) \cdot \nabla n_6^s] \\
\quad = (e\mu_5(\nabla \tilde{n}_7^s \cdot \nabla \tilde{n}_7^s)\tilde{n}_5^s - eD_5^s[\nabla \tilde{n}_7^s \cdot \nabla \tilde{n}_5^s] - \tilde{n}_5^s[\sum_{i=1}^5 d_i\tilde{n}_i^s])s, \\
\Delta \tilde{n}_7^s = \epsilon^{-1}(\sum_{i=1}^4 q_i\tilde{n}_i^s - q_5\tilde{n}_5^s); \\
n_i^s(x,t) = g_i^s(x,t), \quad \text{for } (x,t) \in (\partial\Omega \times [0,T]) \cup (\Omega \times \{0\}, i = 1, \ldots 6, \\
\tilde{n}_7^s(x,t) = \phi(x,t) \quad \text{for } (x,t) \in \partial\Omega \times [0,T].
\end{cases}$$

Defining $\hat{z}_i^s = \hat{n}_i^s - \hat{g}_i^s$ Then $z_i^s = \hat{z}_i^s$ is the solution in $\Omega \times (0,T)$ of the linear system:

(4.48)

$$
\begin{cases}
\partial z_1^s/\partial t - \nabla\cdot(D_1^s(|\nabla\tilde{n}_7^s|^2)\nabla z_1^s) + s(\mu_1(|\nabla\tilde{n}_7^s|^2)/\epsilon)[\sum_{i=1}^4 q_i\tilde{n}_i^s - q_5\tilde{n}_5^s]z_1^s \\
\qquad - s\mu_1(|\nabla\tilde{n}_7^s|^2)[(\nabla\tilde{n}_7^s)\cdot\nabla z_1^s] + sc_{12}z_1^s \\
\quad = -\partial\hat{g}_1^s/\partial t + G_1(s,\tilde{n}_1^s,...,\tilde{n}_5^s,[\nabla\tilde{n}_7^s],[\hat{g}_1^s]), \\
\partial z_2^s/\partial t - \nabla\cdot(D_2^s(|\nabla\tilde{n}_7^s|^2)\nabla z_2^s) + s^2(\mu_2(|\nabla\tilde{n}_7^s|^2)/\epsilon)[\sum_{i=1}^4 q_i\tilde{n}_i^s - q_5\tilde{n}_5^s]z_2^s \\
\qquad - s\mu_2(|\nabla\tilde{n}_7^s|^2)[(\nabla\tilde{n}_7^s)\cdot\nabla z_2^s] + sc_{22}\tilde{n}_5^s z_2^s \\
\quad = -\partial\hat{g}_2^s/\partial t + G_2(s,\tilde{n}_1^s,...,\tilde{n}_5^s,[\nabla\tilde{n}_7^s],[\hat{g}_2^s]), \\
\partial z_3^s/\partial t - \nabla\cdot(D_3^s(|\nabla\tilde{n}_7^s|^2)\nabla z_3^s) + s(\mu_3(|\nabla\tilde{n}_7^s|^2)/\epsilon)[\sum_{i=1}^4 q_i\tilde{n}_i^s - q_5\tilde{n}_5^s]z_3^s \\
\qquad - s\mu_3(|\nabla\tilde{n}_7^s|^2)[(\nabla\tilde{n}_7^s)\cdot\nabla z_3^s] + sc_{32}z_3^s \\
\quad = -\partial\hat{g}_3^s/\partial t + G_3(s,\tilde{n}_1^s,...,\tilde{n}_5^s,[\nabla\tilde{n}_7^s],[\hat{g}_3^s]), \\
\partial z_4^s/\partial t - \nabla\cdot(D_4^s(|\nabla\tilde{n}_7^s|^2)\nabla z_4^s) \\
\qquad + s^2(\mu_4(|\nabla\tilde{n}_7^s|^2)/\epsilon)[\sum_{i=1}^4 q_i\tilde{n}_i^s - q_5\tilde{n}_5^s]z_4^s \\
\qquad - s\mu_4(|\nabla\tilde{n}_7^s|^2)[(\nabla\tilde{n}_7^s)\cdot\nabla z_4^s] + sc_{42}\tilde{n}_5^s z_4^s \\
\quad = -\partial\hat{g}_4^s/\partial t + G_4(s,\tilde{n}_1^s,...,\tilde{n}_5^s,[\nabla\tilde{n}_7^s],[\hat{g}_4^s]), \\
\partial z_5^s/\partial t - \nabla\cdot(D_5^s(\tilde{n}_6^s/\tilde{n}_5^s)\nabla z_5^s) - s(\mu_5(\tilde{n}_6^s/\tilde{n}_5^s)/\epsilon)[\sum_{i=1}^4 q_i\tilde{n}_i^s - q_5\tilde{n}_5^s]z_5^s \\
\qquad + s\mu_5(\tilde{n}_6^s/\tilde{n}_5^s)[(\nabla\tilde{n}_7^s)\cdot\nabla z_5^s] - s(c_{11}+c_{31}-c_{22}\tilde{n}_2^s - c_{42}\tilde{n}_4^s)z_5^s \\
\quad = -\partial\hat{g}_5^s/\partial t + G_5(s,\tilde{n}_1^s,...,\tilde{n}_4^s,[\tilde{n}_5^s],[\tilde{n}_6^s],\nabla\tilde{n}_7^s,[\hat{g}_5^s]), \\
\partial z_6^s/\partial t - (5/3)\nabla\cdot(D_5^s(\tilde{n}_6^s/\tilde{n}_5^s)\nabla z_6^s) \\
\qquad - s(5/3)(\mu_5(\tilde{n}_6^s/\tilde{n}_5^s)/\epsilon)[\sum_{i=1}^4 q_i\tilde{n}_i^s - q_5\tilde{n}_5^s]z_6^s \\
\qquad + s(5/3)\mu_5[(\nabla\tilde{n}_7^s)\cdot\nabla z_6^s] \\
\quad = -\partial\hat{g}_6^s/\partial t + G_6(s,\tilde{n}_1^s,...,\tilde{n}_4^s,[\tilde{n}_5^s],[\tilde{n}_6^s],\nabla\tilde{n}_7^s,[\hat{g}_6^s]), \\
\Delta\tilde{n}_7^s = \epsilon^{-1}(\sum_{i=1}^4 q_i\tilde{n}_i^s - q_5\tilde{n}_5^s); \\
z_i^s(x,t) = 0, \quad \text{for } (x,t)\in(\partial\Omega\times[0,T])\cup(\Omega\times\{0\}, i=1,\ldots 6, \\
\tilde{n}_7^s(x,t) = \phi(x,t) \quad \text{for } (x,t)\in\partial\Omega\times[0,T],
\end{cases}
$$

where

(4.49)

$$G_1(s, \tilde{n}_1^s, ..., \tilde{n}_5^s, [\nabla \tilde{n}_7^s], [\hat{g}_1^s]) = c_{11}\tilde{n}_5^s s + \nabla \cdot (D_1^s(|\nabla \tilde{n}_7^s|^2)\nabla \hat{g}_1^s)$$

$$- s(\mu_1(|\nabla \tilde{n}_7^s|^2)/\epsilon)[\textstyle\sum_{i=1}^4 q_i \tilde{n}_i^s - q_5 \tilde{n}_5^s]\hat{g}_1^s + s\mu_1(|\nabla \tilde{n}_7^s|^2)[(\nabla \tilde{n}_7^s) \cdot \nabla \hat{g}_1^s] - sc_{12}\hat{g}_1^s,$$

$$G_2(s, \tilde{n}_1^s, ..., \tilde{n}_5^s, [\nabla \tilde{n}_7^s], [\hat{g}_2^s]) = c_{21}\tilde{n}_1^s s^2 + \nabla \cdot (D_2^s(|\nabla \tilde{n}_7^s|^2)\nabla \hat{g}_2^s)$$

$$- s^2(\mu_2(|\nabla \tilde{n}_7^s|^2)/\epsilon)[\textstyle\sum_{i=1}^4 q_i \tilde{n}_i^s - q_5 \tilde{n}_5^s]\hat{g}_2^s$$

$$+ s\mu_2(|\nabla \tilde{n}_7^s|^2)[(\nabla \tilde{n}_7^s) \cdot \nabla \hat{g}_2^s] - sc_{22}\tilde{n}_5^s \hat{g}_2^s,$$

$$G_3(s, \tilde{n}_1^s, ..., \tilde{n}_5^s, [\nabla \tilde{n}_7^s], [\hat{g}_3^s]) = c_{31}\tilde{n}_5^s s + \nabla \cdot (D_3^s(|\nabla \tilde{n}_7^s|^2)\nabla \hat{g}_3^s)$$

$$- s(\mu_3(|\nabla \tilde{n}_7^s|^2)/\epsilon)[\textstyle\sum_{i=1}^4 q_i \tilde{n}_i^s - q_5 \tilde{n}_5^s]\hat{g}_3^s + s\mu_3(|\nabla \tilde{n}_7^s|^2)[(\nabla \tilde{n}_7^s) \cdot \nabla \hat{g}_3^s] - sc_{32}\hat{g}_3^s,$$

$$G_4(s, \tilde{n}_1^s, ..., \tilde{n}_5^s, [\nabla \tilde{n}_7^s], [\hat{g}_4^s]) = c_{41}\tilde{n}_3^s s^2 + \nabla \cdot (D_4^s(|\nabla \tilde{n}_7^s|^2)\nabla \hat{g}_4^s)$$

$$- s^2(\mu_4(|\nabla \tilde{n}_7^s|^2)/\epsilon)[\textstyle\sum_{i=1}^4 q_i \tilde{n}_i^s - q_5 \tilde{n}_5^s]\hat{g}_4^s$$

$$+ s\mu_4(|\nabla \tilde{n}_7^s|^2)[(\nabla \tilde{n}_7^s) \cdot \nabla \hat{g}_4^s] - sc_{42}\tilde{n}_5^s \hat{g}_4^s,$$

$$G_5(s, \tilde{n}_1^s, ..., \tilde{n}_4^s, [\tilde{n}_5^s], [\tilde{n}_6^s], \nabla \tilde{n}_7^s, [\hat{g}_5^s]) = \nabla \cdot (D_5^s(\tilde{n}^6/\tilde{n}_5^s)\nabla \hat{g}_5^s)$$

$$+ s(\mu_5(\tilde{n}_6^s/\tilde{n}_5^s)/\epsilon)[\textstyle\sum_{i=1}^4 q_i \tilde{n}_i^s - q_5 \tilde{n}_5^s]\hat{g}_5^s - s\mu_5(\tilde{n}_6^s/\tilde{n}_5^s)[(\nabla \tilde{n}_7^s) \cdot \nabla \hat{g}_5^s]$$

$$+ s(c_{11} + c_{31} - c_{22}\tilde{n}_2^s - c_{42}\tilde{n}_4^s)\hat{g}_5^s,$$

$$G_6(s, \tilde{n}_1^s, ..., \tilde{n}_4^s, [\tilde{n}_5^s], [\tilde{n}_6^s], \nabla \tilde{n}_7^s, [\hat{g}_6^s]) = (e\mu_5(\nabla \tilde{n}_7^s \cdot \nabla \tilde{n}_7^s)\tilde{n}_5^s - eD_5^s[\nabla \tilde{n}_7^s \cdot \nabla \tilde{n}_5^s]$$

$$- \tilde{n}_5^s[\textstyle\sum_{i=1}^5 d_i \tilde{n}_i^s])s + (5/3)\nabla \cdot (D_5^s(\tilde{n}_6^s/\tilde{n}_5^s)\nabla \hat{g}_6^s)$$

$$+ s(5/3)(\mu_5(\tilde{n}_6^s/\tilde{n}_5^s)/\epsilon)[\textstyle\sum_{i=1}^4 q_i \tilde{n}_i^s - q_5 \tilde{n}_5^s]\hat{g}_6^s - s(5/3)\mu_5[(\nabla \tilde{n}_7^s) \cdot \nabla \hat{g}_6^s].$$

In order to obtain existence of a solution in the appropriate class, we will need to clarify the assumptions on the initial and boundary functions $g_i(x,t)$ in (4.2) for $(x,t) \in \Gamma_T$. We will assume that $g_i(x,t)$ is the restriction of a family of functions $g_i^s(x,t) := g_i(x,t,s), 0 \le s \le 1$ with:

$$g_i(x,t,1) = g_i(x,t); \;\; g_i(x,t,0) = 0, \;\text{ for } (x,t) \in \Gamma_T.$$

Moreover, the functions $g_i(x,t,s)$ has extentions $\hat{g}_i(x,t,s)$ to be defined as a member of $C^{2+\alpha,(2+\alpha)/2}(\bar{\Omega} \times [0,T])$ for each $s \in [0,1]$; and as a function of

$s \in [0,1]$ into the Banach space $C^{2+\alpha,(2+\alpha)/2}(\bar{\Omega} \times [0,T])$, the function $\hat{g}_i$ is continuous in s. We will also assume the following usual compatibility condition at $x \in \partial\Omega$, $t = 0$.

(4.50)

$$\frac{\partial \hat{g}_i}{\partial t}(x,0,s)|_{x\in\partial\Omega} = G_i(s, \tilde{n}_1^s(x,0), .., \tilde{n}_5^s(x,0), [\nabla \tilde{n}_7^s(x,0)], [\hat{g}_i(x,0,s)])|_{x\in\partial\Omega}$$

$$\text{for } i = 1,..,4,$$

$$\frac{\partial \hat{g}_5}{\partial t}(x,0,s)|_{x\in\partial\Omega}$$

$$= G_5(s, \tilde{n}_1^s(x,0), .., [\tilde{n}_5^s(x,0)], [\tilde{n}_6^s(x,0)], \nabla \tilde{n}_7^s(x,0), [\hat{g}_5(x,0,s)])|_{x\in\partial\Omega},$$

$$\frac{\partial \hat{g}_6}{\partial t}(x,0,s)|_{x\in\partial\Omega}$$

$$= G_6(s, \tilde{n}_1^s(x,0), .., [\tilde{n}_5^s(x,0)], [\tilde{n}_6^s(x,0)], \nabla \tilde{n}_7^s(x,0), [\hat{g}_6(x,0,s)])|_{x\in\partial\Omega}.$$

Remark 4.3. Note that $\tilde{n}_1^s(x,0), ..., \tilde{n}_6^s(x,0), \nabla \tilde{n}_7^s(x,0)$ above actually only depends on $\hat{g}_i(x,0,s)$ because $z_i^s(x,0) = 0$ for $x \in \Omega$. They do not depend on the preimage of T_s as in appearance. Moreover, the conditions are only imposed for $x \in \partial\Omega$ at $t = 0$, and can be readily satisfied in numerous occasions.

Main Theorem 4.4 (Existence of Classical Solution). *Let $T > 0$ be arbitrary, and assume there exist positive constants $\delta, C_i, i = 1, ..., 5$ satisfying (4.10), (4.11) as described in Corollary 4.2. Let $\sigma > 0$ be small enough so that (4.21) remains true, and suppose that the initial-boundary functions $g_i^s, i = 1, ..., 5$ satisfy (4.19), with extention properties as described above satisfying compatibility conditions (4.50) for $i = 1, ..., 6$. Then for $\lambda \in (0, \hat{\lambda})$, $\hat{\lambda}$ sufficiently small, the initial-boundary value problem (4.1), (4.2) has a solution with $n_i \in C^{2+\alpha,(2+\alpha)/2}(\bar{\Omega}_T), i = 1, ..., 6$. Moreover, we have*

$$0 \le n_i(x,t) \le C_i, \;\; i = 1, ..., 4, \;\; (1-\sigma)\delta \le n_5(x,t) \le C_5 \;\text{ for } (x,t) \in \bar{\Omega}_T.$$

(Recall that the coefficient and boundary functions $D_i(z,\lambda), \mu_i(z,\lambda), \phi(x,t)$ respectively are assumed to satisfy properties as described at the first paragraph of Part B. The dependence of D_i and μ_i on λ in (4.1) is not explicitly displayed.)

Proof. Let B be the Banach space:

$$B := \{u \in C^{1+\alpha,(1+\alpha)/2}(\bar{\Omega}_T) : u_i = 0 \;\text{ on } \Gamma_T, \; i = 1, ..., 6\}$$

with norm $\|u\| = \sum_{i=1}^{6} \|u_i\|_{\Omega_T}^{1+\alpha}$. For $\xi = (\xi_1, ..., \xi_6) \in B$, define the mapping: $\hat{T}_s : B \longrightarrow B, 0 \le s \le 1$, by $\hat{T}_s(\xi) = \hat{z}^s := (\hat{z}_1^s, ..., \hat{z}_6^s)$ where $z_i = \hat{z}_i^s, i = 1, ..., 6$ is the solution of (4.48) with

$$\tilde{n}_i^s = \xi_i + \hat{g}_i^s \text{ in } \bar{\Omega}_T, \ i = 1, ..., 6. \tag{4.51}$$

Thus the function

$$\hat{n}_i^s := \hat{z}_i^s + \hat{g}_i^s, \text{ in } \bar{\Omega}_T, \ i = 1, ..., 6, \tag{4.52}$$

is a solution of (4.47). In case $\xi = \hat{T}_s(\xi) = \hat{z}_s$, then (4.51), (4.52) give $\tilde{n}_i^s = \hat{n}_i^s$, for $i = 1, .., 6$. We thus have $(n_1^s, ..., n_6^s) = (\hat{n}_1^s, ..., \hat{n}_6^s)$ as a solution of (4.17), (4.18). Under the assumptions of δ and C_i of this Theorem, the proof of Corollary 4.2 gives:

$$0 \le \hat{n}_i^s(x,t) \le C_i, \ i = 1,2,3,4, \quad s(1-\sigma)\delta \le \hat{n}_5^s(x,t) \le C_5 \tag{4.53}$$

for $\sigma > 0$ sufficiently small, $(x,t) \in \bar{\Omega}_T$. Moreover, if the solution is in S(Q), for some $Q > 0$, we can apply Lemma 4.1 to obtain a bound independent of Q for $max_{\Gamma_T}|\partial \hat{n}_i^s/\partial x_j|$ for $i = 1, ..., 5$, for $0 \le \lambda < \hat{\lambda}$, $\hat{\lambda}$ sufficiently small. Subsequently, we can apply Lemma 4.2, Lemma 4.3 and Theorem 4.3 to obtain bounds, independent of Q, (determined only by δ and $C_i, i = 1, .., 5$) for $max_{\bar{\Omega}_T}|\hat{n}_6^s|$, $max_{\bar{\Omega}_T}|\partial \hat{n}_i^s/\partial x_j|, i = 1, ..., 6$ for each j, and $max_{\bar{\Omega}_T}|\partial \hat{n}_i^s/\partial t|, i = 1, ..., 6$. By Theorem 4.3, on p. 448 in [113], these bounds determine a bound for the norm of the function $(\hat{n}_1^s, ..., \hat{n}_6^s)$ in B. Since they are independent of Q, we conclude that the solution $(\hat{n}_1^s, ..., \hat{n}_6^s)$ cannot be on the boundary of the set $S(\tilde{Q})$ in B for $\tilde{Q} > 0$ sufficiently large, for all $s \in [0,1]$. Consider the map $H : [0,1] \times B \longrightarrow B$ defined by:

$$H(s, \xi) := \xi - \hat{T}_s(\xi).$$

The mapping $\hat{T}_s(\xi)$ is compact, and the equation $H(0, \xi) = 0$ has a unique solution $\xi = 0$. Thus, by the homotopic invariance principle due to Leray-Schauder, we conclude that the equation $H(1, \xi) = 0$ must have a solution in the interior of $S(\tilde{Q})$. (See e.g. [113], [3] or [125].) Since the solution $(n_1^1, ..., n_6^1)$ is a solution of the linear problem (4.47) for $s = 1$, Schauder's theory asserts that $n_i = n_i^1$ is in $C^{2+\alpha,(2+\alpha)/2}(\bar{\Omega}_T)$.

Notes.

Theorem 1.1 and Corollary 1.2 are obtained from the beginning of Hou and Leung [90]. Lemmas 1.2 to 1.5 and Theorem 1.3 are gathered from Wu and Zou [234], [235] and Boumenir and Nguyen [14]. Theorem 1.4 and Corollary 1.5 are modifications of part of Leung, Hou and Li [136]. Theorems 2.1 to 2.9 can be found in Leung [127]. Theorems 3.1 to 3.3 are proved in Leung [130]. Theorem 4.1 to Theorem 4.4 are obtained from Leung and Chen [134].

Chapter 6

Appendices

6.1 Existence of Solution between Upper and Lower Solutions for Elliptic and Parabolic Systems, Bifurcation Theorems

In the first part of this section, we describe two theorems concerning the existence of solution between upper and lower solutions for elliptic and parabolic systems used in various chapters in this book. Let Ω be a bounded domain in R^N with boundary $\partial\Omega \in C^{2+\alpha}, 0 < \alpha < 1$, and

$$L \equiv \sum_{j,k=1}^{N} a_{jk}(x)\partial^2/\partial x_j \partial x_k + \sum_{j=1}^{N} b_j(x)\partial/\partial x_j + c(x)$$

is an uniformly elliptic operator in $\bar{\Omega}$, where a_{jk}, b_j, c are all in $C^\alpha(\bar{\Omega})$, with $c(x) \leq 0$ in $\bar{\Omega}$. We consider the boundary value problem

$$Lu = f(x,u) \quad \text{in } \Omega, \quad u = \phi(x) \text{ on } \partial\Omega, \tag{1.1}$$

with $u = (u_1, ..., u_m), f = (f_1, ..., f_m), \phi = (\phi_1, ..., \phi_m)$ and the operator L is applied componentwise. For each i we assume $f_i \in C^1(\bar{\Omega} \times R^m)$ and $\phi_i \in C^{2+\alpha}(\partial\Omega)$. The following theorem was found by Tsai in [221], and is presented as Theorem 1.4-2 in [125].

Theorem A1-1. *Let L, f and ϕ be as described above. Suppose that there exist $\alpha(x) = (\alpha_1(x), ..., \alpha_m(x))$ and $\beta(x) = (\beta_1(x), ..., \beta_m(x)), \alpha_i(x), \beta_i(x)$ in $C^2(\bar{\Omega}), i = 1, ..., m.$ such that for each i, $\alpha_i(x) < \beta_i(x)$ in $\bar{\Omega}$,*

$$\begin{cases} 0 \leq L\alpha_i(x) - f_i(x, u_1, ..., u_{i-1}, \alpha_i(x), u_{i+1}, ..., u_m), \\ \\ 0 \geq L\beta_i(x) - f_i(x, u_1, ..., u_{i-1}, \beta_i(x), u_{i+1}, ..., u_m), \end{cases} \tag{1.2}$$

for all $x \in \bar{\Omega}, \alpha_j(x) \le u_j \le \beta_j(x), j \ne i$; and

$$\alpha_i(x) \le \phi_i(x) \le \beta_i(x) \quad on\ \partial\Omega. \tag{1.3}$$

Then the boundary value problem (1.1) has a solution u with $u_i \in C^{2+\alpha}(\bar{\Omega})$ satisfying

$$\alpha_i(x) \le u_i(x) \le \beta_i(x) \text{ in } \bar{\Omega}, \quad \text{for each } i = 1, ..., m.$$

The following theorem considers quasimontone systems. It shows the existence of steady-state when the components of the initial condition of the corresponding parabolic system are upper and lower solution of the elliptic problem. As $t \to \infty$, the solution of the parabolic problem converges monotonically to a max-min pair of steady state. The result can be found in Pao [182] and Sattinger [200], p. 998-999. We define

$$L_i \equiv \sum_{j,k=1}^{N} a^i_{j,k}(x)\partial^2/\partial x_j \partial x_k + \sum_{j=1}^{N} b^i_j(x)\partial/\partial x_j \quad \text{for } i = 1, 2,$$

where the operators L_i are uniformly elliptic in $\bar{\Omega}$, and its coefficients are in $C^\alpha(\bar{\Omega}), 0 < \alpha < 1$. We consider the following two related systems

$$\begin{cases} u_t - L_1 u = f_1(x, u, v) & \\ & \text{in } \Omega \times [0, \infty), \\ v_t - L_2 v = f_2(x, u, v) & \\ u = v = 0 & \text{on } \partial\Omega \times [0, \infty), \\ u(x, 0) = u_0(x), \ v(x, 0) = v_0(x) & \text{in } \Omega, \end{cases} \tag{1.4}$$

(here, we assume $u_0(x) = v_0(x) = 0$ for $x \in \partial\Omega$);

$$\begin{cases} -L_1 u = f_1(x, u, v) & \\ & \text{in } \Omega, \\ -L_2 v = f_2(x, u, v) & \\ u = v = 0 & \text{on } \partial\Omega. \end{cases} \tag{1.5}$$

Theorem A1-2. *Let $(\hat{u}(x), \hat{v}(x))$ and $(\tilde{u}(x), \tilde{v}(x))$ be ordered upper and lower solutions of (1.5) with each component in $C^2(\bar{\Omega})$ and vanishing on the boundary $\partial\Omega$, i.e. $\tilde{u} \le \hat{u}$ and $\tilde{v} \le \hat{v}$ in $\bar{\Omega}$, and let (f_1, f_2) be a quasimonotone nonincreasing C^1 function between $(\tilde{u}, \tilde{v})$ and $(\hat{u}, \hat{v})$. That is, assume*

$$\begin{cases} -L_1\hat{u} \ge f_1(x, \hat{u}, \tilde{v}), \ -L_2\hat{v} \ge f_2(x, \tilde{u}.\hat{v}) & \\ & \text{for } x \in \Omega, \\ -L_1\tilde{u} \le f_1(x, \tilde{u}, \hat{v}), \ -L_2\tilde{v} \le f_2(x, \hat{u}.\tilde{v}) & \\ \hat{u} = \tilde{u} = \hat{v} = \tilde{v} = 0 & \text{for } x \in \partial\Omega. \end{cases}$$

Then the solution $(\bar{U}(x,t), \underline{V}(x,t))$ of (1.4) with initial condition $(u(x,0), v(x,0)) = (u_0(x), v_0(x)) = (\hat{u}(x), \tilde{v}(x))$ converges monotonically from above and below respectively as $t \to +\infty$ in each component to a solution $(\bar{U}_s(x), \underline{V}_s(x))$ of (1.5); and the solution $(\underline{U}(x,t), \bar{V}(x,t))$ of (1.4) with initial condition $(u(x,0), v(x,0)) = (\tilde{u}(x), \hat{v}(x))$ converges monotonically as $t \to +\infty$ from below and above respectively to a solution $(\underline{U}_s(x), \bar{V}_s(x))$ of (1.5). Moreover, $\bar{U}_s \geq \underline{U}_s, \underline{V}_s \leq \bar{V}_s$ in $\bar{\Omega}$, and if (u^*, v^*) is any solution of (1.5) between $(\tilde{u}, \tilde{v})$ and $(\hat{u}, \hat{v})$, then it satisfies

$$\underline{U}_s \leq u^* \leq \bar{U}_s, \ \underline{V}_s \leq v^* \leq \bar{V}_s \ \text{ in } \bar{\Omega}.$$

Many more theorems analogous to the two above can be found in e.g. Leung [125] and Pao [183].

In the remaining part of this section we present a few bifurcation theorems which are used in many chapters in the book in order to obtain positive solutions in addition to the trivial solution. When the hypotheses of the implicit function theorem fail at a certain point, one might have more than one solution in its neighborhood. These theorems discuss sufficient conditions for such phenomenon to occur.

Let X, Y be Banach spaces and let $f \in C(O, Y)$ where O is an open subset of X. (Here, $C(O, Y)$ denotes the set of continuous function from O into Y). We say f is (Fréchet) differentiable at $a \in O$ if there exists a bounded linear transformation $df_a : X \to Y$ (denoted by $df_a \in L(X, Y)$) such that

$$||f(a+\xi) - f(a) - df_a\xi|| = o(||\xi||) \qquad \text{as } ||\xi|| \to 0.$$

(Here $||\cdot||$ denotes both the norms in Y or X, whichever is appropriate). The linear transformation df_a is called the (Fréchet) derivative of f at a. We say $f \in C^1(O, Y)$ if the map $a \to df_a$ is continuous form O into $L(X, Y)$.

We denote the second derivative $d^2 f_a$ of $f \in C(O, Y)$ at a point $a \in O$ to be the continuous bilinear form $d^2 f_a : X \times X \to Y$ which satisfies

$$||f(a+\xi) - f(a) - df_a\xi - (1/2)d^2 f_a(\xi, \xi)|| = o(||\xi||^2) \ \text{ as } ||\xi|| \to 0.$$

Here $d^2 f_a$ can also be interpreted as a bounded linear transformation from X into $L(X, Y)$, i.e. $d^2 f_a \in L(X, L(X, Y))$. We say $f \in C^2(O, Y)$ if the map $a \to d^2 f_a$ is continuous from $O \to L(X, L(X, Y))$.

Let B_1, B_2, B_3 be Banach spaces and U be open in $B_1 \times B_2$. Suppose $f : U \to B_3$ and $(u_1, u_2) \in U$. We say f is differentiable with respect to the first variable at (u_1, u_2) if the function $g(x) = f(x, u_2)$ is differentiable at $x = u_1$. We write $dg_{u_1} = D_1 f(u_1, u_2)$. Similarly we define $D_2 f(u_1, u_2)$ as the derivative with respect to the second variable.

The following local bifurcation theorem is due to Crandall and Rabinowitz in [34].

Theorem A1-3. *Let X and Y be Banach spaces and $f \in C^2(U, Y)$, where $U = S \times V$ is an open subset of $R \times X$ containing $(\lambda_0, 0)$. Let $L_0 = D_2 f(\lambda_0, 0), L_1 = D_1 D_2 f(\lambda_0, 0)$; and $N(L_0), R(L_0)$ denotes the null space and range of L_0 respectively. Suppsoe that:*

(i) $f(\lambda, 0) \equiv 0$ for all $\lambda \in S$,
(ii) $N(L_0)$ is one-dimensional, spanned by u_0,
(iii) $dim[Y/R(L_0)] = 1$,
(iv) $L_1 u_0 \notin R(L_0)$.

Then there is $\delta > 0$ and a C^1-curve $(\lambda, \phi) : (-\delta, \delta) \to R \times Z$ (here, Z is a closed subspace of X with the property that any $x \in X$ is uniquely representable as $x = \alpha u_0 + z$ for some $\alpha \in R, z \in Z$) such that:

(a) $\lambda(0) = \lambda_0, \phi(0) = 0$, and
(b) $f(\lambda(s), s(u_0 + \phi(s)) = 0$ for $|s| < \delta$.

Moreover, there is a neighborhood of $(\lambda_0, 0)$ such that any solution of $f(\lambda, u) = 0$ is either on this curve or is of the form $(\lambda, 0)$.

Note that using Taylor's Theorem, we may write

$$f(\lambda, u) = L_0 u + (\lambda - \lambda_1) L_1 u + r(\lambda, u),$$

where $L_0 = D_2 f(\lambda_0, 0), L_0 = D_1 D_2 f(\lambda_0, 0)$ and $r \in C^2$ satisfies

$$r(\lambda, 0) = 0, D_2 r(\lambda_0, 0) = D_1 D_2 r(\lambda_0, 0) = 0.$$

The next two theorems lead to bifurcation results which are more general and nonlocal. Theorem A1-4 is due to Krasnosel'skii and Theorem A1-5 is the general bifurcation theorem of Rabinowitz [190]. More detailed exposition can be found in Smoller [209] and López-Gómez [161].

Theorem A1-4. *Let B be a Banach space and $f \in C(U, B)$, where U is an open subset of $R \times B$. Assume that f is expressible as*

$$f(\lambda, u) = u - \lambda L u + h(\lambda, u), \quad \textit{where} \tag{1.6}$$

(a) $L : B \to B$ is a compact, linear operator,
(b) the equation $Lv = \rho v$ has $\rho = 1/\lambda_0$ as an eigenvalue of odd multiplicity,
(c) $h : U \to B$ is a compact,
(d) $h(\lambda, u) = o(||u||)$ as $u \to 0$, uniformly on bounded $\lambda-$intervals.

Then $(\lambda_0, 0)$ is a bifurcation point of $f(\lambda, u) = 0$.

To be specific, let $\Gamma : (\lambda, u(\lambda))$ be a curve of solutions of $f(\lambda, u) = 0$ with (λ_0, u_0) as an interior point on Γ. We say (λ_0, u_0) is a bifurcation point with respect to Γ if every neighborhood of (λ_0, u_0) in $R \times B$ contains solutions of $f(\lambda, u) = 0$ not on Γ.

Theorem A1-5. *Suppose that all the hypotheses (a) to (d) of Theorem A1-4 are satisfied. Let G denotes the closure of the set of solutions of $f(\lambda, u) = 0$ with $u \neq 0$. Then G contains a component S which meets $(\lambda_0, 0)$, and either*

(i) S is noncompact in U (Hence, if $U = R \times B$, the compactness of L and h together with formula (1.6) imply that S is unbounded), or
(ii) S meets $u = 0$ in a point $(\bar{\lambda}, 0)$, where $\bar{\lambda} \neq \lambda_0$ and $1/\bar{\lambda}$ is an eigenvalue of L.

6.2 The Fixed Point Index, Degree Theory and Spectral Radius of Positive Operators

In the first part of this section, we introduce some basic theory involving fixed point index of a compact map f with respect to a cone. A nonempty subset A of a metric space X is called a retract of X if there exists a continuous map $r : X \to A$ (called a retraction), such that $r|A = id_A$. It is easily seen that every retract is a closed subspace of X. By a theorem of Dugundji [54], every nonempty closed convex subset of a Banach space E is a retract of E. There is a theory of fixed point index of compact map over an open subset U of a retract X of a Banach space, with respect to X. The index can be expressed in terms of the well known Leray-Schauder degree as given below in (2.1).

Theorem A2-1. *Let X be a retract of a Banach space E. For every open subset U of X and every compact map $f : \bar{U} \to X$ which has no fixed points on ∂U, there exists an integer $i_X(f, U)$ satisfying the following conditions:*
(i) (Normalization) For every constant map f mapping $\bar{U}$ into U, $i_X(f, U) = 1$;
(ii)(Additivity) For every pair of disjoint open subsets U_1, U_2 of U such that f has no fixed points on $U \backslash (U_1 \cup U_2)$,

$$i_X(f, U) = i_X(f, U_1) + i_X(f, U_2),$$

where $i_X(f, U_k) := i_X(f|\bar{U}_k, U_k), k = 1, 2$;
(iii)(Homotopy invariance) For every compact interval $\Lambda \subset R$, and every compact map $h : \Lambda \times U \to X$ such that $h(\lambda, x) \neq x$ for $(\lambda, x) \in \Lambda \times \partial U$, the integer $i_X(h(\lambda, \cdot), U)$ is well-defined and independent of $\lambda \in \Lambda$;
(iv)(Permanence) If Y is a retract of X and $f(\bar{U}) \subset Y$, then

$$i_X(f, U) = i_Y(f, U \cap Y),$$

where $i_Y(f, U \cap Y) = i_Y(f|\overline{U \cap Y}, U \cap Y)$. The family $\{i_X(f, U)|X$ retract of E, U open in X, $f : \bar{U} \to X$ compact without fixed points on $\partial U\}$ is uniquely determined by the properties (i)-(iv), and $i_X(f, U)$ is called the fixed point index of f (over U with respect to X).

Remark A2-1. In the Theorem above, the topological notions (open, closed, boundary, etc) refer to relative topology of X as a subspace of E.

Outline of Proof. Let $\{i_X(f,U)\}$ be any family satisfying conditions (i)-(iv). Then by choosing $X = E$, conditions (i)-(iii) are precisely the properties characterizing the Leray-Schauder degree. Consequaently, we have

$$i_E(f,U) = deg(id - f, U, 0),$$

where $deg(id - f, U, 0)$ denotes the Leray-Schauder degree with respect to zero of the compact map $(id - f)$, defined on the closure of the open set $U \subset E$.

Next, suppose X is an arbitrary retract of E, and denote by $r : E \to X$ an arbitray restraction. it is readily seen that due to the permanence property, we have

$$i_X(f,U) = i_E(f \circ r, r^{-1}(U)) = deg(id - f \circ r, r^{-1}(U), 0).$$

Thus every fixed point index of f over U with respect to X is equal to $deg(id - f \circ r, r^{-1}(U), 0)$. We might thus define

$$i_X(f,U) = deg(id - f \circ r, r^{-1}(U), 0), \tag{2.1}$$

where $r : E \to X$ is an arbitrary retraction. It can be readily verified by the excision property of the Leray-Schauder degree (see e.g. Schwartz [205] and Amann [3]) that this definition is indepedent of the choice of the retraction r, and (2.1) is well-defined. By means of (2.1) and the basic properties of Leray-Schauder degree, we can also readily verify the properties (i)-(iv).

From the above Theorem, one can readily deduce the following properties of the fixed point index.

Corollary A2-2. *The fixed point index defined in the last Theorem has the following further properties:*

(iv)(Excision) For every open subset $V \subset U$ such that f has no fixed point in $U \backslash V$,

$$i_X(f,U) = i_X(f,V);$$

(v) (Solution property) If $i_X(f,U) \neq 0$, then f has at least one fixed point in U.

Suppose U is an open subset of X, and is also open in E. Assume $x_0 \in U$ is an isolated fixed point of some compact map $f : \bar{U} \to X$. Then there exists a positive number ρ_0 such that $x_0 + \rho\bar{B} \subset U$ for all $\rho \in [0, \rho_0]$, where B denotes the open unit ball of E. Moreover ,we may assume x_0 is the only fixed point of f in $x_0 + \rho_0 B$. Consequently, by the excision property,

$$i(f, x_0) := i_X(f, x_0 + \rho B),$$

is the local index of f at x_0, is well-defined and independent of $\rho \in [0, \rho_0]$. Moreover, we can show by means of the permanence property that the local index $i(f, x_0)$ coincides with the local index as defined by

$$lim_{\rho \to 0}\, deg(id - f, x_0 + \rho B, 0),$$

which is the standard definition of the local index in the Leray-Schauder degree theory.

Using the theory presented above and the classical Leray-Schauder theory, we obtain the following theorem concerning diffferentiable compact maps.

Theorem A2-3. *Let X be a retract of some Banach space E, U be an open subset of X, and $f : \bar{U} \to X$ be a compact map. Suppose that $x_0 \in U$ is a fixed point of f, and that there exists a positive number ρ such that $x_0 + \rho B \subset U$, where B denotes the open unit ball of E. Further, assume that f is differentiable at x_0, such that 1 is not an eigenvalue of the derivative $f'(x_0)$. Then x_0 is an isolated fixed point of f and*

$$i(f, x_0) = (-1)^m,$$

where m is the sum of the multiplicities of all the eigenvalues of $f'(x_0)$ which are greater than one.

Let E be a ordered Banach space, with ordering denoted by $\geq$, and $P := \{x \in E | x \geq 0\}$. For any $\rho > 0$, let P_ρ be the positive part of the ball ρB, that is $P_\rho := \rho B \bigcap P = \rho B^+$. Observe that P_ρ is an open neighborhood of 0 in P. Due to the convexity of B and P, the closure $\bar{P}_\rho$ of P_ρ in P coincides with $\rho \bar{B} \bigcap P$. Hence the boundary S_ρ^+ in P is the same as $\rho S \bigcap P$, where S is the unit sphere in E. Since the positive cone is a closed convex subet of E, we know by Dungundji's theorem that it is a retract of E. Consequently, by Theorem A2-1 above, for any open subset U of P and any compact map $f : \bar{U} \to P$, the fixed point index $i_P(f, U)$ is well-defined, provided that f has no fixed points on ∂U. The following is a very simple useful consequence of the Theroem A2-1.

Theorem A2-4. *Let $f : \bar{P}_\rho \to P$ be a compact map such that $f(x) \neq \lambda x$ for every $x \in S_\rho^+$ and every $\lambda \geq 1$. Then $i_P(f, P_\rho) = 1$.*

Proof. Define a compact map $h : [0,1] \times \bar{P}_\rho \to P$ by $h(\tau, x) := \tau f(x),$. Then, by the homotopy invariance and the normalization properties, we have $i_P(f, P_\rho) = i_P(0, P_\rho) = 1$.

More details of the theories presented in Theorem A2-1 to Theorem A2-4 can be found in Amann [3]. The following theorem is concerned with standard properties of positive operators and the general Krein-Rutman Theorem. Recall that an eigenvalue λ of a linear operator T is called simple if

$$dim(\cup_{k=1}^{\infty} ker(\lambda - T)^k) = 1. \qquad (2.2)$$

Theorem A2-5. *Let E be a Banach space ordered by a cone $P = \{x \in E | x \geq 0\}$, where P is closed with nonempty interior. Suppose $T : E \to E$ is a strongly positive compact bounded linear operator. Then the following are true:*
(i) The spectral radius $r(T) := lim_{k\to\infty}||T^k||^{1/k}$ is positive;
(ii) $r(T)$ is a simple eigenvalue of T having a positive eigenvector (i.e. in $P \backslash \{0\}$) and there is no other eigenvalue with a positive eigenvector;
(iii) For every $y \in P\backslash\{0\}$, the equation

$$\lambda x - Tx = y$$

has exactly one positive solution if $\lambda > r(T)$, and no positive solution for $\lambda \leq r(T)$. The equation $r(T)x - Tx = -y$ has no positive solution. (Here, T is strongly positive means $T(P\backslash\{0\})$ is contained in the interior of P.)

The theorem is proved by applying results in Kranosel'skii [108] or Krein-Rutman [110], as explained in Amann [3]. Let Ω be a bounded domain in R^N with $C^{2+\mu}, 0 < \mu < 1$, boundary. In the situation when $T : C^{\mu}(\bar{\Omega}) \to C^{2+\mu}(\bar{\Omega})$ is the solution operator denoting for every $f \in C^{\mu}(\bar{\Omega})$, by Tf the unique solution of the problem

$$-\Delta u = f \quad \text{in } \Omega, \quad u = 0 \quad \text{on } \partial\Omega,$$

the theorem above does not apply immediately to the extension to $C(\bar{\Omega})$. For an application of the previous theorem, we let e be the unique solution of the problem

$$-\Delta e = 1 \quad \text{in } \Omega, \quad e = 0 \quad \text{on } \partial\Omega.$$

Define the subspace $C_e(\bar{\Omega})$ to consist of $u \in C(\bar{\Omega})$ such that there exists a positive constant so that $-\alpha e \leq u \leq \alpha e$. This space is given the natural ordering and topology defined by means of the norm

$$||u||_e := inf\{\alpha > 0 | -\alpha e \leq u \leq \alpha e\}.$$

It can be shown that the solution operator T has a unique extension, again denoted by T, to a compact, strongly positive linear operator from $C(\bar{\Omega})$ into $C_e(\bar{\Omega})$. Moreover, the restriction of T to $C_e(\bar{\Omega})$ is a compact, strongly positive bounded linear operator on $C_e(\bar{\Omega})$. Theorem A2-5. above can then be applied to this solution operator.

The following theorem provides a convenient method to deduce the spectral radius of some positive linear operators. It can be found in [151].

Theorem A2-6. *Let E be a Banach space ordered by a cone P with properties as described in the above theorem, and T is a strongly positive compact bounded linear operator on E. Suppose there exists a positive element $u > 0$ (i.e. $u \in P\backslash\{0\}$) with one of the following properties:*
(i) $Tu > u$, then the spectral radius of T satisfies $r(T) > 1$;

(ii) $Tu < u$, *then the spectral radius of* T *satisfies* $r(T) < 1$;
(iii) $Tu = u$, *then the spectral radius of* T *satisfies* $r(T) = 1$.

Proof. (iii) $Tu = u$ implies that 1 is an eigenvalue of T with positive eigenvector u. The conclusion that $r(T) = 1$ follows from the theorem above.
(i) Suppose $Tu > u$ for some $u > 0$. Assume that $r(T) = 1$, then for some $y > 0$ the equation $r(T)x - Tx = -y$ has a positive solution $x = u$. This contradicts the above theorem; thus $r(T) \neq 1$. Suppose $1 > r(T)$, then $y := (-u) - T(-u) > 0$, and the equation $1x - Tx = y$ has exactly one positive solution by the above theorem. Moreover, the equation $1x - Tx = y$ has a unique solution because all eigenvalues are less than $r(T)$, which is < 1. Consequently the solution $x = (-u)$ is positive, and we have $u < 0$. This contradiction implies that we cannot have $1 > r(T)$ also. Thus $r(T) > 1$.
(ii) The proof is the same as in part (i), by reversing the inequality $>$ to $<$, obtaining $r(T) < 1$.

6.3 Theorems Involving Maximum Principle, Comparison and Principal Eigenvalues for Positive Operators

In this section we assume that Ω is a bounded domain in R^N with boundary of class C^2. Recall the definition of eigenvalues denoted by $\hat{\rho}_1(L)$ and $\rho_1(-L)$ for an elliptic operator L in Chapter 1. The first theorem in this section involves maximum pricinple for $W^{2,p}(\Omega)$ solution of elliptic scalar equation. The details can be found in Theorem 2.2 in Delgado, López-Gómez and Suarez [45].

Theorem A3-1. *Consider a uniformly elliptic operator of the form*

$$L = \sum_{i,j=1}^{N} a_{ij}(x)\partial_i\partial_j + \sum_{j=1}^{N} b_j(x)\partial_j + c(x), \tag{3.1}$$

with

$$a_{ij} \in C(\bar{\Omega}),\ b_j,\ c \in L^\infty(\Omega),\ i, j \in \{1, ..., N\}. \tag{3.2}$$

Suppose that

$$\hat{\rho}_1(L) < 0.$$

Let $p > N$ *and* $u \in W^{2,p}(\Omega)$ *satisfies* $Lu \le 0$ *a.e. in* Ω *and* $u \ge 0$ *in* $\partial\Omega$; *then, unless* $u(x) \equiv 0$ *in* $\bar{\Omega}$, *it must satisfy* $u(x) > 0$ *for* $x \in \Omega$, *and for those* $x \in \partial\Omega$ *with* $u(x) = 0$ *its outward normal derivative satisfies* $(\partial u/\partial \nu)(x) < 0$.

The next three theorems concern properties of $W^{2,p}(\Omega)$ solution of Volterra-Lotka scalar equation and comparison with upper solutions. Let L be the uniformly elliptic operator of the form (3.1) with coefficients satisfying (3.2) in a domain as described in Theorem A3-1. Consider the equation

$$\begin{cases} -Lw = \gamma w - f(x)w^2 & \text{in } \Omega, \\ w = 0 & \text{on } \partial\Omega, \end{cases} \tag{3.3}$$

where $\gamma \in R$ and $f \in C(\bar{\Omega})$ satisfies $f(x) > 0$ in $\bar{\Omega}$. By means of the maximum principle above, we readily prove the following theorem by iteration and upper-lower solution method.

Theorem A3-2. *If $p > N$, then the problem (3.3) has a positive solution in $W^{2,p}(\Omega) \cap W_0^{1,p}(\Omega)$ if and only if $\gamma > \rho_1(-L)$. Moreover, it is unique if it exists. Let the solution be denoted by $\theta_{[-L,\gamma,f]}$, then*

$$lim_{\gamma \to \rho_1(-L)^+} \theta_{[-L,\gamma,f]} = 0$$

uniformly in $\bar{\Omega}$.

More details of the above theorem can be found in Theorem 3.1 in [45]. The next Theorem can be found in Lemma 3.2 of [45].

Theorem A3-3. *Consider the problem (3.3) above. Suppose $\gamma > \rho_1(-L)$ and $\bar{w} \in W^{2,p}(\Omega)$, $p > N$, is a positive strict upper solution of the above problem, i.e. $L\bar{w} + \gamma\bar{w} - f(x)\bar{w}^2 \le 0$ a.e. in Ω, $\bar{w} \ge 0$ on $\partial\Omega$, $\bar{w} \not\equiv \theta_{[-L,\gamma,f(x)]}$ in $\bar{\Omega}$; then $\bar{w} - \theta_{[-L,\gamma,f(x)]} > 0$ for $x \in \Omega$, and $(\partial/\partial\nu)(\bar{w} - \theta_{[-L,\gamma,f(x)]})(x) < 0$ for those $x \in \partial\Omega$, where $\bar{w}(x) = 0$.*

Proof. We have

$$[L - f(x)(\bar{w} + \theta_{[-L,\gamma,f]}) + \gamma](\bar{w} - \theta_{[-L,\gamma,f]}) \le 0 \text{ a.e. in } \Omega,$$
$$\bar{w} - \theta_{[-L,\gamma,f]} \ge 0 \qquad \text{on } \partial\Omega.$$

Moerover, we have

$$\hat{\rho}_1(L - f(x)(\bar{w} + \theta_{[-L,\gamma,f]}) + \gamma) < \hat{\rho}_1(L - f\theta_{[-L,\gamma,f]} + \gamma) = 0.$$

Thus the conclusion follows from Theroem A3-1 above with L and u respectively replaced by $[L - f(x)(\bar{w} + \theta_{[-L,\gamma,f]}) + \gamma]$ and $(\bar{w} - \theta_{[-L,\gamma,f]})$.

Theorem A3-4. *Let L and Ω be as described in Theorem A3-1 and $\theta_{[-L,a,b(x)]}$ be defined as Theorem A3-2, where we assume $b \in C(\bar{\Omega})$ and $b(x) > 0$ for each $x \in \bar{\Omega}$. Then the following holds*

$$lim_{a \to \infty} \frac{\theta_{[-L,a,b(x)]}}{a} = \frac{1}{b(x)}$$

uniformly in compact subsets of Ω.

The above theorem concerning the limit of the solution of the diffusive Volterra-Lotka equation as the growth rate tends to infinty can be found in Theorem 3.4 in [45]. The next theorem involves comparison of eigenvalues, and can be found in e.g. Theorem 2.3 in [45].

Theorem A3-5. *Let the operator L be as described in Theorem A3-1. Let $V_1, V_2 \in L^\infty(\Omega)$ such that $V_1 \leq V_2$ and $V_1 < V_2$ on a set of positive measure. Then*

$$\rho_1(-L+V_1) < \rho_1(-L+V_2).$$

The next two theorems concern the solution of scalar equations on R^N or R^N_+. The following is a variant of Liouville type theorem asserting that a bounded solution has to be exactly zero, and can be found in Lemma 7.5 in Delgardo, López-Gómez and Suarez [45].

Theorem A3-6. *Assume that $D = R^N$ or $D = R^N_+$, where*

$$R^N_+ = \{x \in R^N : x_N \geq 0\}.$$

If $V \in L^\infty(D) \cap C^\alpha(D), 0 < \alpha < 1, V \geq 0, V \not\equiv 0$, then $\theta = 0$ is the only bounded solution of

$$(-\Delta + V)\theta = 0 \text{ in } D.$$

Theorem A3-7. *Let $u(x)$ be a non-negative C^2 solution of*

$$\Delta u + u^\alpha = 0 \quad in\ R^N, N > 2,$$

with $1 < \alpha < \frac{N+2}{N-2}$. Then $u(x) \equiv 0$.

The above theorem can be found in Theorem 1.1 in Gidas and Spruck [70]. The following two theorems involve cooperative quasimonotone systems.

Let $L_k, k = 1, 2$, be uniformly elliptic operators as described for L in (3.1) and (3.2). Let $\alpha(x), \beta(x), \gamma(x)$ and $\rho(x)$ be functions in $C(\bar{\Omega})$ with the two functions $\beta(x)$ and $\gamma(x)$ positive almost everywhere in Ω. Consider the linear eigenvalue problem for λ in the following cooperative system; that is those λ such that

$$\text{(3.4)} \qquad \begin{cases} L_1 u + \alpha(x)u + \beta(x)v = \lambda u & \\ & \text{in } \Omega, \\ L_2 v + \gamma(x)u + \rho(x)v = \lambda v & \\ u = v = 0 & \text{on } \partial\Omega, \end{cases}$$

has some solution $(u, v) \in W^{2,p}_0(\Omega) \times W^{2,p}_0(\Omega), (u, v) \neq (0, 0), p > N$.

Theorem A3-8. *(i) There exists a largest eigenvalue of problem (3.4), denoted by λ^* and is called the principal eigenvalue of the problem. This eigenvalue is simple and has a unique positive eigenfunction, up to a multiplicative constant, called principal eigenfunction. Moreover, each component of the principal eigenfunction is strictly positive in Ω; and λ^* is the only eigenvalue of (3.4) with such positive eigenfunction. Furthermore, any other eigenvalue λ of (3.4) must satisfy:*

$$Re\ \lambda < \lambda^*.$$

(ii) The principal eigenvalue λ^* of (3.4) is negative if and only if there exists $(\bar{u}, \bar{v}) \in W^{2,p}(\Omega) \times W^{2,p}(\Omega)$ such that

$$\begin{cases} L_1\bar{u} + \alpha(x)\bar{u} + \beta(x)\bar{v} \leq 0 & \\ & \text{in } \Omega, \\ L_2\bar{v} + \gamma(x)\bar{u} + \rho(x)\bar{v} \leq 0 & \\ \bar{u} \geq 0, \ \bar{v} \geq 0 & \text{on } \partial\Omega, \end{cases}$$

where at least one of the four inequalities above is not identically zero almost everywhere.

The above theorem can be found in Theorem 8.3 and 8.4 in [45]. The next theorem is a generalization of the sweeping principle to cooperative quasimontone systems. It can be found in e.g. McKenna and Walter [168] or Lemma 9.3 of [45]

Theorem A3-9. *Let $z = (u, v) \in W_0^{2,p}(\Omega) \times W_0^{2,p}(\Omega), p > N$, be a solution of the problem*

$$\begin{cases} -L_1 u = f(x, u, v) & \\ & \textit{in } \Omega, \\ -L_2 v = g(x, u, v) & \\ u = v = 0 & \textit{on } \partial\Omega, \end{cases} \tag{3.5}$$

where f and g are two continuous functions in x and of class C^1 in (u, v), f is increasing in v, and g increasing in u. For each $t \in (t_0, t_1]$, let $\bar{z}^t := (\bar{u}^t, \bar{v}^t) \in W_0^{2,p}(\Omega) \times W_0^{2,p}(\Omega)$ be a strict upper solution of problem (3.5). Assume that $\bar{z}^t$ is continuous and strictly increasing in $t \in [t_0, t_1]$. Moreover, suppose that each component of $\bar{z}^{t_1} - z$ is strictly positive for all $x \in \Omega$, and the outward normal derivative $\partial \bar{z}^t / \partial \nu$ at $\partial\Omega$ is continuous in t. Then we have the following inequality componentwise:

$$z \leq \bar{z}_{t_0} \quad \text{in } \bar{\Omega}.$$

We now present a few more general theorems concerning eigenvalues for positive operators. Let E be a Banach space with a cone K. An operator A acting in the space E is called completely continuous if it is continuous and if it maps every bounded set into a compact set. It is called positive if it maps the cone K into itself.

Theorem A3-10. *Let A be a completely continuous linear positive operator acting on the real Banach space E with cone K. Let the relation*

$$A^p u \geq \alpha u \quad (\alpha > 0)$$

be satisfied for some non-zero element u such that $-u \notin K, u = v - w$, $(v, w \in K)$ where p is some natural number. Then the operator A has at least one eigenvector x_0 in K:

$$Ax_0 = \lambda_0 x_0,$$

where the positive eigenvalue λ_0 satisfies the inequality

$$\lambda_0 \geq (\alpha)^{1/p}.$$

A cone K in the real Banach space E is called reproducing if every element $x \in E$ can be represented in the form

$$x = u - v \qquad \text{for some } u, v \in K,$$

where the elements u and v are not defined uniquely in the representation above. Let u_0 be some fixed non-zero element of K. We call a linear positive operator A u_0−bounded below if for every non-zero $x \in K$ a natural number n and a positive number α can be found such that

$$\alpha u_0 \leq A^n x.$$

Analogously, we call the positive operator A u_0−bounded above if for every non-zero $x \in K$ an m and a β can be found such that

$$A^m x \leq \beta u_0.$$

Finally, if for every non-zero $x \in K$

$$\alpha u_0 \leq A^n x \leq \beta u_0$$

for some n, then we call the operator A u_0−positive.

Theorem A3-11. *Let $\phi_0 \in K$ and ϕ_0 be an eigenvector of a u_0−positive linear operator A:*

$$A\phi_0 = \lambda_0 \phi_0.$$

Suppose that K is a reproducing cone. Then λ_0 is a simple eigenvalue of the operator A. (Recall the definition of simple eigenvalue in (2.2) of Section 6.2.)

Theorem A3-12. *Let the conditions of Theorem A3-11 be satisfied. Then the positive eigenvalue corresponding to a positive eigenvector $\phi \in K$ is greater than the absolute magnitudes of the remaining eigenvalues.*

Theorems A3-10, A3-11 and A3-12 correspond respectively to Theorem 2.5, Theorems 2.10 and 2.13 in Chapter 2 of Krasnolsel'skii [108].

6.4 Theorems Involving Derivative Maps, Semigroups and Stability

We first present a theorem concerning mappings between L^p spaces and then a theorem relating the Fréchet and Gauteau derivatives of mappings between Banach spaces.

Let $g_i(u_1, ..., u_n, x), i = 1, ..., n$ be continuous with respect to $(u_1, ..., u_n)$ for almost all $x \in \bar{\Omega}$, and measurable with respect to x in $\bar{\Omega}$ for every fixed $(u_1, ..., u_n)$, $u_i \in (-\infty, +\infty)$. Let $\tilde{u}_i(x), i = 1, ..., n$, be n real functions in $L^p(\Omega)$, $p > 0$. Consider the operator $h = (h_1, ..., h_n)$ defined for $\tilde{u} = (\tilde{u}_1, ..., \tilde{u}_n)$ by:

$$h_i[\tilde{u}] = g_i(\tilde{u}_1(x), ..., \tilde{u}_n(x), x), \quad \text{for } i = 1, ..., n, \tag{4.1}$$

to give n measurable scalar functions in Ω.

Theorem A4-1. *The operator $h = (h_1, ..., h_n)$ in (4.1) maps $[L^p(\Omega)]^n$ into $[L^{p_1}(\Omega)]^n$ if and only if the the functions $g_1, ..., g_n$ described above satisfy for each $i = 1, ..., n$:*

$$|g_i(u_1, ..., u_n, x)| \le a_i(x) + b \sum_{k=1}^{n} |u_k|^r,$$

where $a_i(x) \in L^{p_1}(\Omega), p_1 > 0, b > 0, r = p/p_1$.

The above theorem is given in Theorem 19.2 of Vainberg [222].

Let X and Y be Banach spaces and F is an operator where $F : X \to Y$. The operator is said to have a Gateaux deriviative at the point $x_0 \in X$ if there exists a bounded linear operator denoted by $DF(x_0, h)$ such that $h \to DF(x_0, h) \in Y$ for all $h \in X$ and

$$lim_{t \to 0} \frac{F(x_0 + th) - F(x_0)}{t} = DF(x_0, h).$$

The following theorem relates the Gateaux and Fréchet derivatives, and is presented in Theorem 3.3 in Vainberg [222].

Theorem A4-2. *If the operator $F : X \to Y$ has Gateaux derivatives for all points in a neighborhood of $x_0 \in X$ and the Gateaux derivative $DF(x, h)$ is continuous at $x = x_0$, then F has a Fréchet derivative at x_0, denoted by $F'(x_0)$, and*

$$DF(x_0, h) = F'(x_0)h$$

for all $h \in X$.

Semigroup theory is used throughout this book to study parabolic and hyperbolic systems. An excellent exposition on the semigroup theory of linear operators and application to partial differential equations can be found in Pazy [184]. Here, we review a few basic definitions.

Definition A4.1(a). Let X be a Banach space. A one parameter family $T(t), 0 \leq t < \infty$, of bounded linear operators from X into X is a semigroup of bounded linear operators on X if
(i) $T(0) = I$, (I is the identity operator on X),
(ii) $T(t + s) = T(t)T(s)$ for every $t, s \geq 0$.

The linear operator A defined by

$$A(x) = lim_{t \to 0^+} \frac{T(t)x - x}{t} = \frac{d^+T(t)x}{dt}|_{t=0} \quad \text{for } x \in D(A), \ \text{ where}$$

$$D(A) = \{x \in X : lim_{t \to 0^+} \frac{T(t)x - x}{t} \text{ exists}\} \ \text{ is the domain of } A,$$

is called the infinitesimal generator of the semigroup $T(t)$.

Defintion A4.2. A semigroup $T(t), 0 \leq t < \infty$ of bounded linear operators on X is a strongly continuous semigroup of bounded linear operators if

$$lim_{t \to 0^+} T(t)x = x \quad \text{for every } \ x \in X.$$

A strongly continuous semigroup of bounded linear operators on X will be called a semigroup of class C_0 or simply C_0 semigroup.

If A is a linear, not necessarily bounded, operator on X, the resovlent set of A is the set of all complex numbers λ for which $\lambda - A$ is invertible, i.e. $(\lambda - A)^{-1}$ is a bounded linear operator in X. The family $R(\lambda : A) = (\lambda - A)^{-1}$ of bounded linear operators, for all λ in the resolvent set of A, is called the resolvent of A.

The following four well-known theorems in semigroup theory are used in various parts of this book.

Theorem A4-3. (Hille-Yosida) *A linear operator A is the infinitesimal generator of a C_0 semigroup $T(t)$ on the Banach X satisfying $||T(t)|| \leq Me^{\omega t}$, if and only if*
(i) A is closed and its domain $D(A)$ is dense in X,

(ii) the resolvent set of A contains the interval (ω, ∞), and the resolvent $(\lambda I - A)^{-1}$ satisfies

$$||(\lambda I - A)^{-n}|| \leq \frac{M}{(\lambda - \omega)^n} \qquad \text{for } \lambda > \omega, \ \ n = 1, 2, \ldots .$$

Definition A4.1(b). Let X be a Banach space. A one parameter family $T(t), -\infty < t < \infty$, of bounded linear operators from X into X is a C_0 group of bounded linear operators on X if
(i) $T(0) = I$, (I is the identity operator on X),
(ii) $T(t + s) = T(t)T(s)$ for $-\infty < t, s < \infty$,
(iii) $lim_{t \to 0} T(t)x = x$ for $x \in X$.

The infinitesimal generator A of the group $T(t)$ is defined by

$$A(x) = \lim_{t \to 0} \frac{T(t)x - x}{t}$$

whenever the limit exists; the domain of A is the set of all elements $x \in X$ for which the above limit exists.

Theorem A4-4. *A linear operator A is the infinitesimal generator of a C_0 group $T(t)$ on the Banach X satisfying $||T(t)|| \leq Me^{\omega|t|}$, if and only if*
(i) A is closed and its domain $D(A)$ is dense in X,
(ii) every real $\lambda, |\lambda| > \omega$, is in the resolvent set $\rho(A)$ of A and for such λ

$$||R(\lambda : A)^n|| \leq \frac{M}{(|\lambda| - \omega)^n} \qquad \text{for } \lambda > \omega, \ \ n = 1, 2, \ldots .$$

Theorem A4-5. *Let X be a Banach space and let A be the infinitesimal generator of a C_0 semigroup $T(t)$ on X, satisfying $||T(t)|| \leq Me^{\omega t}$. If B is a bounded linear operator on X then $A+B$ is the infinitesimal generator of a C_0 semigroup $S(t)$ on X, satisfying $||S(t)|| \leq Me^{(\omega + M||B||)t}$.*

Theorem A4-6. *For each $i = 1, ..., k$, let A_i be an infinitesimal generator with domain $D(A_i)$ of a C_0 semigroup $S_i(t)$ on Banach space X satisfying $||S_i(t)|| \leq M_i e^{\omega_i t}$. Suppose that $\cap_{i=1}^k D(A_i)$ is dense in X and*

$$||(S_1(t)S_2(t) \cdots S_k(t))^n|| \leq Me^{\omega nt}, \quad n = 1, 2, \ldots .$$

for some constant $M \geq 1$ and $\omega \geq 0$. If for some λ with $Re\lambda > \omega$ the range of $\lambda - (A_1 + A_2 + \cdots + A_k)$ is dense in X, then the closure of $A_1 + A_2 + \cdots + A_k$ is an infinitesmal generator of a C_0 semigroup, denoted by $S(t)$, on X satisfying $||S(t)|| \leq Me^{\omega t}$. Moreover, we have

$$S(t)x = lim_{n \to \infty}(S_1(t/n)S_2(t/n) \cdots S_k(t/n))^n x \qquad \text{for } x \in X,$$

and the limit is uniform on bounded t intervals.

Theorems A4-3 and A4-4 are respectively Theorem 5.3 and Theorem 6.3 in Chapter 1 of Pazy [184]. Theorems A4-5 and A4-6 are respectively Theorem 1.1 and Corollary 5.5 in Chapter 3 in Pazy [184]. Theorem A4-6 involves the so called Trotter product formula. The following theorem is known as the Trotter-Neveu-Kato semigroup convergence theorem and can be found in Th. 7.2 of Chapter 1 in Goldstein [74].

Theorem A4-7. *Let $A_n, n = 0,1,2,\ldots,$ generate a C_0 semigroup T_n on the Banach space X, satisfying the condition:*

$$||T_n(t)|| \leq Me^{\omega t}, \quad for\ n = 0,1,2,..., \ t \geq 0, \tag{4.2}$$

where M and ω are independent of n and t. Let D be a subspace of X, such that A_0 is the closure of its restriction to D. Assume that for each $f \in D$, we have

$$f \in lim\,inf_{n\to\infty}D(A_n), \ and\ lim_{n\to\infty}A_n f = A_0 f.$$

Then $lim_{n\to\infty}T_n(t)g = T_0(t)g$ for each $g \in X$, uniformly for t in compact subsets of $[0,\infty)$. Moreover the resolvents satisfy $lim_{n\to\infty}R(\lambda : A_n)f = R(\lambda : A_0)f$ for each $f \in X$, uniformly for λ in compact subsets of (ω,∞).

For more sophisticated use of the semigroup theory, we need the following definitions.

Definition A4-3. Let $T(t)$ be a C_0 semigroup on a Banach space X. The semigroup $T(t)$ is called differentiable for $t > t_0$ if for every $x \in X, t \to T(t)x$ is differentiable for $t > t_0$. $T(t)$ is called differentiable if it is differentiable for $t > 0$.

Definition A4-4. Let $II = \{z : \psi_1 < arg z < \psi_2, \psi_1 < 0 < \psi_2\}$ and for $z \in II$, let $T(z)$ be a bounded linear operator. The family $T(z), z \in II$ is analytic semigroup in II if

(i) $z \to T(z)$ is analytic in II,
(ii) $T(0) = I$ and $lim_{z\to 0, z\in II}T(z)x = x$ for every $x \in X$,
(iii) $T(z_1 + z_2) = T(z_1)T(z_2)$ for $z_1, z_2 \in II$.

The semigroup $T(t)$ will be called analytic if it is analytic in some sector II containing the non-negative real axis.

We next present some useful theory concerning semigroup of linear operators generated by a system of elliptic linear operator on the m-vector space of continuous functions. Let Ω be a bounded domain in $R^N, N \geq 1$; if $N > 1$ we assume that Ω is uniformly of class $C^{1+\alpha}$ for some $\alpha > 0$ (See for example Ladyzhenskaya, Solonnikov and Ural'ceva [113], section 4 in Chapter IV). We define the operators

$$L_i \equiv \sum_{j,l=1}^{N} a^i_{j,l}(x)\partial^2/\partial x_j \partial x_l + \sum_{j=1}^{N} b^i_j(x)\partial/\partial x_j \quad \text{for } i = 1,...,m,$$

where the boundary $\partial\Omega$ is uniformly of class $C^{2+\alpha}$ if $N > 1$; $a^i_{j,l}$ and b^i_l are all in the class $C^{\alpha}(\bar{\Omega})$, $0 < \alpha < 1$, and each L_i is uniformly elliptic for $x \in \bar{\Omega}$. We first consider the linear problem:

$$(4.3) \quad \begin{cases} \frac{\partial u_i}{\partial t} = L_i u_i & \text{in } \Omega \times [0,\infty), \ i = 1,...,m, \\ u_i(x,t) = 0 & \text{for } (x,t) \in \partial\Omega \times [0,\infty), \ i = 1,...,m. \end{cases}$$

We have the following theorem from Theorem 2.4 of Mora [176].

Theorem A4-8. *The solution of the linear problem (4.3) determines an analytic semigroup on the Banach space* $[\hat{C}(\bar{\Omega})]^m := \{u = (u_1,...,u_m) | u_i \in C(\bar{\Omega}), u_i|_{\partial\Omega} = 0, i = 1,...,m.\}$, *with generator A; and its domain is given by* $D(A) = \{u \in [\hat{C}(\bar{\Omega})]^m | \ (L_1 u_1,...,L_m u_m) \in [\hat{C}(\bar{\Omega})]^m\}$.

We next consider the nonlinear problem:

$$(4.4) \quad \begin{cases} \frac{\partial u_i}{\partial t} = L_i u_i + f_i(x,u_1,...,u_n), & \text{in } \Omega \times [0,\infty), \ i = 1,...,m, \\ u_i(x,t) = 0 & \text{for } (x,t) \in \partial\Omega \times [0,\infty), \ i = 1,...,m. \end{cases}$$

Here, the functions $f_i(x,u_i,...,u_m)$ has partial derivatives $D^{\beta}_{u_j} f_i$ for $|\beta| \leq 1$ and $D^{\beta}_{u_j}(x,u_1,...,u_m) \in C(\bar{\Omega})$ for all $(u_1,...,u_m) \in R^m$. Moreover, The functions $D^{\beta}_{u_j}(x,u_1,...,u_m)$ are locally Hölder continuous with respect to $(u_1,...,u_m) \in R^m$, with Hölder constants independent of $x \in \Omega$. The following stability result is given in Theorem 4.2 in Mora [176].

Theorem A4-9. *Under the conditions on Ω, L_i and $f_i, i = 1,2,...,m$ described above, the nonlinear problem (4.4) determines a semiflow of class C^1 on the Banach space $[\hat{C}(\bar{\Omega})]^m$ described in the last Theorem. Suppose w_0 is a stationary (or steady) state of this semiflow, and let $\hat{L} := A + DF(w_0)$ be the linearized operator for (4.4) at w_0 on $[\hat{C}(\bar{\Omega})]^m$, where A is defined in the last Theorem. Let $\sigma^* := \sup Re\, \sigma(\hat{L})$ where $\sigma(\hat{L})$ denotes the spectrum of $\hat{L}$. If $\sigma^* > 0$, then the stationary state is unstable in the semiflow on $[\hat{C}(\bar{\Omega})]^m$; if $\sigma^* < 0$, then the stationary state is asymptotically stable in the same flow.*

Here $F = (f_1,...,f_m)$, and $DF(w_0) : [\hat{C}(\bar{\Omega})]^m \to [\hat{C}(\bar{\Omega})]^m$ can be given by $\bar{u} \to D_u F(x,w_0)\bar{u}$, where $D_u F(x,w_0)$ is the Jacobian matrix for $F = (f_1,...,f_m)$ with respect to u evaluated at $u = w_0$. From the boundedness of Ω, we find that the spectrum of $\hat{L}$ consists only of eigenvalues. For the definition of stability and asymptotic stability, refer to the clarification below immediately before Remark 4.1, with the Banach space X^{α} replaced by $[\hat{C}(\bar{\Omega})]^m$. The system in (4.3) above is diagonal; actually the theory in Mora [176] is given in much more general system which is not diagonal.

In order to estimate the location of the spectrum for the linearized equation, the following theorem related to the semicontinuity of the spectrum of closed operator is useful. It can be found in Chapter 4, Theorem 3.6 in p. 209 of Kato [102].

Theorem A4-10. *Let T be a closed linear operator on a Banach space B, and S be a bounded linear operator on B, such that S commutes with T. Then the distance between the spectrum $\sigma(T)$ and $\sigma(T+S)$ does not exceed the spectral radius of S, and a fortiori, $||S||$.*

Stabilities of steady states of the parabolic systems (or semiflows) are also considered in function spaces other than the continuous functions. We now define some spaces induced by certain linear operators A.

Definition A4-5. A linear operator A in a Banach space X is called a sectorial operator if it is a closed densely defined operator such that, for some ϕ in $(0, \pi/2)$ and some $M \geq 1$ and a real a, the sector

$$S_{a,\phi} = \{\lambda | \ \phi \leq |arg(\lambda - a)| \leq \pi, \lambda \neq a\}$$

is in the resolvent set of A and

$$||(\lambda - A)^{-1}|| \leq M/|\lambda - a| \ \text{ for all } \lambda \in S_{a,\phi}.$$

Definition A4-6. Let A be a sectorial operator in a Banach space X and $Re\, \sigma(A) > 0$; then for any $\alpha > 0$

$$A^{-\alpha} = \frac{1}{\Gamma(\alpha)} \int_0^\infty t^{\alpha-1} e^{-At} dt.$$

Here, $\{e^{-At}\}_{t \geq 0}$ denotes the analytic semigroup generated by $-A$, where

$$e^{-At} = \frac{1}{2\pi i} \int_C (\lambda + A)^{-1} e^{\lambda t} d\lambda,$$

where C is the contour in $\rho(-A)$ with $arg\, \lambda = \pm\theta$ as $|\lambda| \to \infty$ for some θ in $(\pi/2, \pi)$.

It can be shown that if A is a sectorial operator in X with $Re\, \sigma(A) > 0$, then for any $\alpha > 0$, $A^{-\alpha}$ is a bounded linear operator on X which is one-one and satisfies $A^{-\alpha}A^{-\beta} = A^{-(\alpha+\beta)}$ whenever $\alpha > 0, \beta > 0$.

Definition A4-7. Let A be a sectorial operator in a Banach space X and $Re\, \sigma(A) > 0$, then for $\alpha > 0$, A^α is defined as the inverse of $A^{-\alpha}$, with domain $D(A^\alpha)$ as the range of $A^{-\alpha}$; A^0 is defined as the identity operator on X.

Definition A4-8. Let A be a sectorial operator in a Banach space X. For each $\alpha \geq 0$ define X^α to be the domain of A_1^α with the graph norm

$$||u||_\alpha = ||A_1^\alpha u||, \ u \in X^\alpha,$$

where $A_1 = A + aI$ with a chosen so that $Re\,\sigma(A_1) > 0$. Different chioces of a give equivalent norms on X^α, so the dependence on the choice of a is supressed in writing.

Let A be a sectorial linear operator in a Banach space X, and $f : \tilde{U} \to X$ where $\tilde{U}$ is a cylindrical neighborhood in $R \times X^\alpha$ (for some $0 \leq \alpha < 1$) of $(\tau, \infty) \times \{u_0\}$. We say u_0 is an equilibrium point if $u(t) \equiv u_0$ is a solution of

$$\frac{du}{dt} + Au = f(t, u), \quad t > t_0, \tag{4.5}$$

i.e. if $u_0 \in D(A)$ and $Au_0 = f(t, u_0)$ for all $t > t_0$.

A solution $\bar{u}(\cdot)$ on $[t_0, \infty)$ is stable (in X^α) if, for any $\epsilon > 0$, there exists $\delta > 0$ such that any solution u with $||u(t_0) - \bar{u}(t_0)||_\alpha < \delta$ exists on $[t_0, \infty)$ and satisfies $||u(t) - \bar{u}(t)||_\alpha < \epsilon$ for all $t \geq t_0$; that is, if $\hat{u}_0 \to u(t; t_0, \hat{u}_0)$ is continuous (in X^α) at $\hat{u}_0 = \bar{u}(t_0)$, uniformly in $t \geq t_0$. The solution $\bar{u}$ is uniformly stable if $u_1 \to u(t; t_1, u_1)$ is continuous as $u_1 \to \bar{u}(t_1)$, uniformly for $t \geq t_1$ and $t_1 \geq t_0$.

The solution $\bar{u}(\cdot)$ is uniformly asymptotically stable if it is uniformly stable and $u(t; t_1, u_1) - \bar{u}(t) \to 0$ as $t - t_1 \to \infty$, uniformly in $t_1 \geq t_0$ and $||u_1 - \bar{u}(t_1)||_\alpha < \delta$, for some constant $\delta > 0$.

Remark 4.1. It is shown in Henry [84] or Chapter 7 in Pazy [184], that the operator $-L := diag.(-k\Delta, ..., -k\Delta), k > 0$, is a sectorial operator on $X = L^p(\Omega) \times \cdots \times L^p(\Omega)$, $1 < p < \infty$, and $L := diag.(k\Delta, ..., k\Delta)$ generates an analytic semigroup on X. Here, Ω is a bounded domain in R^N with smooth boundary. Consequently, we can apply the theory for (4.5) with A replaced by $-L$ for parabolic systems. The following two theorems concerning stability of solutions in X^α are used throughout this book. Note that by choosing p large and α close to 1, we can obtain information concerning the spatial derivatives of the solutions.

Theorem A4-11. *Let A, f be as described above for (4.5), and further $f(t,u)$ is locally Hölder continuous in $t, t > \tau$, locally Lipschitzian in u, on $\tilde{U}$. Let u_0 be an equilibrium point, and*

$$f(t, u_0 + z) = f(t, u_0) + Bz + g(t, z)$$

where B is a bounded linear map from $X^\alpha, 0 \leq \alpha < 1$, to X and $||g(t,z)|| = o(||z||_\alpha)$ as $||z||_\alpha \to 0$, uniformly in $t > \tau$.

If the spectrum of $A - B$ *lies in* $\{Re\,\lambda > \beta\}$ *for some* $\beta > 0$, *or equivalently if the linearization*

$$\frac{dz}{dt} + Az = Bz$$

has the zero solution uniformly asymptotically stable, then the problem (4.5) has the solution u_0 *uniformly asymptotically stable in* X^α. *More precisely, there exist* $\rho > 0, M \geq 1$ *such that if* $t_0 > \tau$ *and* $||u_1 - u_0||_\alpha \leq \rho/(2M)$ *then there exists a unique solution of (4.5) with initial condition* $u(t_0) = u_1$ *defined on* $t_0 \leq t < \infty$ *and satisfying for* $t \geq t_0$

$$||u(t; t_0, u_1) - u_0||_\alpha \leq 2Me^{-\beta(t-t_0)}||u_1 - u_0||_\alpha.$$

Theorem A4-12. *Let* A, f *satisfy properties described in Theorem A4-11. Assume* $Au_0 = f(t, u_0)$ *for* $t \geq t_0 > \tau$,

$$f(t, u_0 + z) = f(t, u_0) + Bz + g(t, z), \quad g(t, 0) = 0,$$

$$||g(t, z_1) - g(t, z_2)|| \leq k(\rho)||z_1 - z_2||_\alpha \text{ for } ||z_1|| \leq \rho, \; ||z_2||_\rho \leq \rho$$

with $k(\rho) \to 0$ *as* $\rho \to 0^+$, *and* B *is a bounded linear map from* X^α *to* X.

If the spectrum of $L = A - B$ *has the property that* $\sigma(L) \cap \{Re\,\lambda < 0\}$ *is a nonempty set. Then the equilibrium solution* u_0 *is unstable. More precisely, there exist* $\epsilon_0 > 0$ *and* $\{u_n, n \geq 1\}$ *with* $||u_n - u_0||_\alpha \to 0$ *as* $n \to \infty$, *such that for all* n,

$$sup_{t \geq t_0}||u(t; t_0, u_n) - u_0||_\alpha \geq \epsilon_0 > 0.$$

Here the supremum is taken over the maximal interval of existence of $u(\cdot; t_0, u_n)$.

Theorems A4-11 and A4-12 above can be found respectively as Theorems 5.1.1 and 5.1.3 in Henry [84].

The following theorem which can be found in Billoti and LaSalle [10], is used for the study of persistence in Chapter 4. Let X be a complete metric space and $T(t)$ be a C_0 semigroup on X. The semigroup $T(t)$ is called point dissipative in X if there is a bounded nonempty set B in X such that for any $x \in X$, there is a $t_0 = t_0(x, B)$ such that $T(t)x \in B$ for $t \geq t_0$.

Theorem A4-13. *If*

(i) $T(t)$ *is point dissipative in* X,
(ii) there is a $t_0 \geq 0$ *such that* $T(t)$ *is compact for* $t > t_0$,

then there is a nonempty global attractor A *in* X. *(Here, global attractor is defined in Part A of Section 4.1 in the familiar way.)*

6.5 $W_p^{2,1}$ Estimates, Weak Solutions for Parabolic Equations with Mixed Boundary Data, Theorems Related to Optimal Control, Cross-Diffusion and Traveling Wave

For the convenience of the reader, we present in this section a few theorems used in Chapters 3 to 5. The following theorem provides estimates for an arbitrary $W_p^{2,1}$ solution of a parabolic problem on a given time interval. The theorem is found in Ladyzhenskaya, Solonnikov and Ural'ceva [113]. It is used to deduce the existence of time periodic solution of a parabolic problem in Chapter 3. Let Ω be a bounded domain in R^N with C^2 boundary $\partial\Omega$.

Theorem A5-1. *Let T_1, T_2 be any two numbers satisfying $0 < T_1 < T_2$, $f \in L^p(\Omega \times (0, T_2])$, and $c \in C^\alpha(\bar{\Omega} \times [0, T_2])$. Suppose that $u \in W_p^{2,1}(\Omega \times (0, T_2])$ satisfies*

$$\begin{cases} u_t - \Delta u + cu = f & \text{in } \Omega \times (0, T_2), \\ \frac{\partial u}{\partial \nu} = 0 & \text{on } \partial\Omega \times (0, T_2), \end{cases}$$

then we have

$$||u||_{W_p^{2,1}(\Omega \times (T_1, T_2))} \leq C_1 ||f||_{L^p(\Omega \times (0, T_2])} + C_2 ||u||_{L^p(\Omega \times (0, T_2])}.$$

Here $p > 1$ is any positive number and $0 < \alpha < 1$.

The following two theorems are used for the theory of optimal control of parabolic systems with mixed boundary data.

Consider the operator:

$$Ay := -\sum_{i,j=1}^{N} \frac{\partial}{\partial x_i}(a_{ij}(x,t)\frac{\partial y}{\partial x_j}),$$

where $a_{ij}(x,t) \in C(\bar{\Omega} \times [0,T])$, and $-A$ is uniformly elliptic operator on the bounded domain Ω with smooth boundary $\partial\Omega$ in R^N. We consider the problem

$$(5.1) \quad \begin{cases} \frac{\partial y_h}{\partial t} + Ay_h = f & \text{in } \Omega \times (0,T),\ f \in L^2(\Omega \times (0,T)), \\ \frac{\partial y_h}{\partial \nu_A} + hy_h = 0 & \text{on } \partial\Omega \times (0,T),\ h \in L^2(\partial\Omega \times (0,T)), \\ y_h(x,0) = y_0(x), & y_0 \in L^2(\Omega) \end{cases}$$

where h is prescribed in an admissible set $S := \{h | 0 < \beta \leq \xi_0(x,t) \leq h(x,t) \leq$

$\xi_1(x,t)$ a.e. on $\partial\Omega \times (0,T)$, $\xi_0, \xi_1 \in L^\infty(\partial\Omega \times (0,T))\}$; and

$\frac{\partial u}{\partial \nu_A} = \sum_{i,j} a_{ij} \frac{\partial u}{\partial x_j} cos(\nu, x_j)$ on $\partial\Omega$,

$cos(\nu, x_j)$ is the j-th direction cosine of the outward normal ν on $\partial\Omega$.

Problem (5.1) is solved in the following sense: we set

$$V := \{\psi | \psi \in H^1(\Omega)\},$$

$$a_h(t;\phi,\psi) := \sum_{i,j=1}^{N} \int_\Omega a_{ij}(x,t) \frac{\partial \phi}{\partial x_j} \frac{\partial \psi}{\partial x_i} dx + \int_{\partial\Omega} h\phi\psi ds.$$

Theorem A5-2. *There exists one and only one function y_h such that*

$$(5.2) \qquad \begin{cases} y_h \in L^2(0,T;H^1(\Omega)), \\ y_h|_{\partial\Omega\times(0,T)} \in L^2(\partial\Omega \times (0,T)), \\ \frac{d}{dt}(y_h,\psi) + a_h(t;y_h(t),\psi) = (f(t),\psi) \text{ for all } \psi \in V, \\ y_h(0) = y_0. \end{cases}$$

From the first three lines of (5.2), we obtain the first two lines of (5.1). Then

$$\frac{\partial}{\partial t} y_h \in L^2(0,T;H^{-1}(\Omega)),$$

and thus we can deduce that $u \in C([0,1], L^2(\Omega))$, and consequently the fourth line concerning initial condition in (5.2) makes sense. Therefore it is reasonable to define y_h as the solution of (5.1) by means of the above theorem.

The above theorem can be found in Section 15.4 of Chapter 3 in Lions [158]. The next theorem concerns compact embedding involving $u, du/dt$ is given in Proposition 4.2 of Chapter 4 in Lions [157].

Let V and H be real Hilbert spaces. Assume that $V \subset H$ and the injection of V into H is continuous. If V' denotes the dual space of V, H may be identified with a subspace of V', and we may write

$$V \subset H \subset V'.$$

Let $H_1(-\infty,+\infty;V,V')$ denotes the normed linear space consisting of elements of $u \in L^2(-\infty,+\infty;V)$ such that $|\tau|\hat{u}(\tau) \in L^2(-\infty,+\infty;V')$, with the norm

$$||u||_{H_1(-\infty,+\infty;V,V')} = (||u||^2_{L^2(-\infty,+\infty;V)} + |||\tau|\hat{u}(\tau)||^2_{L^2(-\infty,+\infty;V')})^{1/2},$$

where $\hat{u}(t)$ is the Fourier transform of $u(t)$. Let $H_1(0,T;V,V')$ denotes the space formed by restriction of elements in $H_1(-\infty,+\infty;V,V')$ to $(0,T)$, with norm

$$||u||_{H_1(0,T;V,V')} = inf.||v||_{H_1(-\infty,+\infty;V,V')} \text{ such that } v = u \ a.e. \text{ in } (0,T).$$

Theorem A5-3. *Suppose that the injection of V into H is compact. Then the injection of $H_1(0,T;V,V')$ into $L^2(0,T;H)$ is also compact.*

The following two theorems are used for the study of positive steady-states in elliptic systems with cross-diffusion. Theorem A5-4 can be found in Zeidler [241].

Theorem A5-4. *Let M and N be two metric spaces and a mapping $G \in C(M,N)$ satisfies the following two conditions:*
(i) $G^{-1}(K)$ is compact in M for any compact subset K of N;
(ii) G is locally invertible on M.
Suppose also that M is arcwise connected and N is simply connected, then G is a homeomorphism from M onto N.

Theorem A5-5. *Let Ω be a bounded smooth domain in $R^N, N \geq 1$. Suppose that $w \in C^2(\Omega) \cap C^1(\bar{\Omega})$ is a positive solution to $\Delta + c(x)w = 0$ in Ω subject to homogeneous Neumann boundary condition with $c \in C(\bar{\Omega})$. Then there exists a positive constant $C_* = C_*(N, \Omega, ||c||_\infty)$ such that*

$$max_{\bar{\Omega}}\, w \leq C_*\, min_{\bar{\Omega}}\, w.$$

The Harnack type inequality in the theorem above is obtained by Lin, Ni and Takagi [155], Lemma 4.3.

We now consider a parabolic equation with degenerate diffusion. We first give the definition of solution and upper or lower solution. Then we state a comparison result in Theorem A5-6.

For $\Omega = (-L, L)$, $Q_T = \Omega \times (0,T]$, $Q = \Omega \times R^+$, we consider the problem:

$$\begin{cases} u_t = \eta(u)_{xx} + f(u) & (x,t) \in Q, \\ u(\pm L, t) = 0 & t \in (0,\infty), \\ u(x,0) = u_0(x) & x \in \Omega, \end{cases} \tag{5.3}$$

where η, f and u_0 satisfy the following assumptions:

A1. $\eta : R \to R$ is locally Lipschitz continuous, nondecreasing and $\eta(0) = 0$,
A2. $f : R \to R$ is locally Lipschitz continuous,
A3. $u_0 \in L^\infty(\Omega)$.

Definition A5-1. A solution u of (5.3) on $[0,T]$ is a function u with the following properties:
(i) $u \in C([0,T]:L^1(\Omega)) \cap L^\infty(Q_T)$,
(ii) $\int_\Omega u(t)\phi(t)\,dx - \int\int_{Q_t}(u\phi_t + \eta(u)\phi_{xx})\,dxdt = \int_\Omega u_0\phi(0)\,dx + \int\int_{Q_t} f(u)\phi\,dxdt$
for all $\phi \in C^2(\bar{Q}_T)$ such that $\phi \geq 0$, $\phi = 0$ at $x = \pm L$ and $0 \leq t \leq T$. A solution on $[0,\infty)$ means a solution on each $[0,T]$, and a lower solution (upper solution) is defined by (i), and (ii) with equality replaced by $\leq$ ($\geq$).

Theorem A5-6. *Let u be lower solution and $\hat{u}$ be an upper solution of problem (5.3) in $[0,T]$ with initial conditions u_0 and $\hat{u}_0$ respectively. Suppose $u_0 \leq \hat{u}_0$ in Ω, then $u \leq \hat{u}$ in Q_T.*

The theory here is valid if Ω is a bounded domain in R^N and $\eta(u)_{xx}$ is replaced by $\Delta\eta(u)$ or $E\eta(u)$ where E is a suitable elliptic operator. The above theorem can be found in Aronson, Crandall and Peletier [5].

The following theorem considers the existence of solution with prescribed limits at $\pm\infty$ for second order ODE monotone systems in R. They are used for finding traveling wave solutions for parabolic systems with prescribed limits at $\pm\infty$.

Let $w_i^+ < w_i^-, i = 1,...,n$, and $G := (w_1^+, w_1^-) \times \cdots \times (w_n^+, w_n^-)$. Let $F(u) = (F_1(u),...,F_n(u))$; and for each i, assume F_i is continuously differentiable in $\bar{G}$ with

$$\frac{\partial F_i}{\partial u_k} > 0 \ \text{ in } G, \ \text{ for } k = 1,...,n, \ k \neq i.$$

Moreover, for $w_+ := (w_1^+,...,w_n^+), w_- := (w_1^-,...,w_n^-)$, we suppose

$$F(w_+) = F(w_-) = 0.$$

We consider finding montonically decreasing solution of the problem

$$Dw'' + cw + F(w) = 0, \tag{5.4}$$

where D is a given diagonal $n \times n$ matrix with positive diagonal elements, with the conditions at the infinities

$$lim_{x\to\pm\infty} w(x) = w_\pm, \tag{5.5}$$

and c is a constant. Here c is the wave speed of traveling wave solution of the form

$$u(x,t) = w(x - ct)$$

for the monotone system

$$\frac{\partial u}{\partial t} = D\frac{\partial^2 u}{\partial x^2} + F(u).$$

We introduce a functional ω^*; with the aid of this functional we may determine the minimum speed of a wave. Let K be the class of vector-valued functions $\rho(x) \in C^2(-\infty, \infty)$, decreasing montonically and satisfying the conditions at $\pm\infty$:

$$lim_{x\to\pm\infty}\, \rho(x) = w_{\pm}.$$

We set

$$\psi^*(\rho) = sup_{x,k} \frac{d_k \rho_k''(x) + F_k(\rho(x))}{-\rho_k'(x)}, \tag{5.6}$$

$$\omega^* = inf_{\rho \in K}\, \psi^*(\rho). \tag{5.7}$$

Here d_k and F_k are the diagonal elements of the matrix D and the elements of the vector F appearing in system (5.4).

Theorem A5-7. *Assume that there exists a vector $p \geq 0$, $p \neq 0$ such that*

$$F(w_+ + sp) \geq 0 \;\; for \;\; 0 < s \leq s_0, \tag{5.8}$$

where s_0 is a positive number.(Here, all vector inequalities are interpreted componentwise). Furthermore, suppose that in the interval $[w_+, w_-]$ (i.e. for $w_+ \leq w \leq w_-$) the vector-valued function $F(w)$ vanishes only at the points w_+ and w_- in $\bar{G}$. Then for all $c \geq \omega^*$ there exists a montonically decreasing solution of system (5.4) satisfying the conditions (5.5). When $c < \omega^*$, such solutions do not exist.

The above result is shown in Theorem 4.2 of Chapter 3 in Volpert, Volpert and Volpert [224].

Bibliography

[1] S. Agmon, A. Douglis, and L. Nirenberg, Estimates near the boundary for solutions of elliptic partial differential equations satisfying general boundary conditions I, Comm. Pure Appl. Math. 12 (1959), 623-727.

[2] N. Alikakos, An application of the invariance principle to reaction-diffusion equations, J. Diff. Eqs. 33 (1979), 201-225.

[3] H. Amann, Fixed point equations and nonlinear eigenvalue problems in ordered Banach spaces, SIAM Rev. 18 (1976), 620-709.

[4] P. Angelstam, Predation on ground-nesting birds' nests in relation to predator densities and habitat edge, Oikos, 47 (1986), 365-373. S. Anita, Analysis and Control of Age-Dependent Population Dynamics, Mathematical Modelling: Theory and Applications, Vol. 11, Kluwer Academic Publishers, Boston, 2000.

[5] D. Aronson, M. Crandall and L. Peletier, Stabilization of solutions of a degenerate nonlinear diffusion problem, Nonlinear Anal. T.M.A. 6 (1982), 1001-1022.

[6] D. Aronson and H. Weinberger, Nonlinear diffusion in population genetics, combustion, and nerve propagation, Lecture Notes in Mathematics, 446, 5-49, Springer, New York (1975).

[7] H. Berestycki and P. Lions, Some applications of the method of sub- and supersolutions, Bifurcation and Nonlinear Eigenvalue Problems, p. 16-41, Lecture Notes in Mathematics 782, Springer-Verlag, Berlin 1980.

[8] N. Bhatia and G. Szego, Stability Theory of Dynamical Systems, Springer, Berlin, 1970.

[9] R. Bierregaard, Jr., T. Lovejoy, T. Kapos, A. dos Santos and R. Hutchings, The biological dynamics of tropical rain forest fragments, Biosci. 42 (1992), 859-866.

[10] J. Billoti and J. P. LaSalle. Periodic dissipative processes, Bull. Amer. Math. Soc., 6 (1971), 1082-1089.

[11] J. Blat and K. Brown, Bifurcation of steady-state solutions in predator-prey and competition systems, Proc. Roy. Soc. Edinburgh, Sect. A 97 (1984), 21-34.

[12] J. Blat and K. Brown, A reaction-diffusion system modeling the spread of bacterial infections, Math. Methods Appl. Sci. 8 (1986), 234-246.

[13] F. Brauer and J. Nohel, Qualitative Theory of Ordinary Differential Equations, W.A. Benjamin, N.Y., 1969.

[14] A. Boumenir and V. Nguyen, Perron theorem in monotone iteration method for traveling waves in delayed reaction-diffusion equations, J. Diff. Eqs. 244 (2008), 1551-1570.

[15] K. Brown, Spatially inhomogeneous steady state solutions for systems of equations describing interacting populations, J. Math. Anal. Appl. 95 (1983), 251-264.

[16] G. Butler, H. Freedman and P. Waltman, Uniformly persistent dynamical systems, Proc. Amer. Math. Soc. 96 (1986), 425-430.

[17] A. Canada, P. Drabek and A. Fonda, Handbook of Differential Equations: Ordinary Differential Equations, Vol. 2, Elsevier Science and Technology, New York, 2004.

[18] R. Cantrell and C. Cosner, On the steady state problem for the Volterra-Lotka competition model with diffusion, Houston J. Math., 13 (1987), 337-352.

[19] R. Cantrell and C. Cosner, Practical persistence in ecological models via comparison methods, Proc. R. Soc. Edinburgh, 126A (1996), 247-272.

[20] R. Cantrell and C. Cosner, Spatial Ecology via Reaction-Diffusion Equations, Wiley Series in Mathematical and Computational Biology, Wiley, West Sussex, 2003.

[21] R. Cantrell, C. Cosner and V. Hutson, Permanence in ecological systems with spatial heterogeneity, Proc. R. Soc. Edinburgh, 123A (1993), 533-559.

[22] R. Cantrell, C. Cosner and V. Hutson, Permanence in some diffusive Lotka-Volterra models for three interacting species, Dynamic Sys. and Appl. 2 (1993), 505-530.

[23] V. Capasso and L. Maddalena, A non-linear diffusion system modelling the spread of oro-faecal diseases, in "Nonlinear Phenomena in Mathematical Sciences" (V. Laksmikantham, Ed.), Academic Press, New York, 1981.

[24] V. Capasso and L. Maddalena, Convergence to equilibrium states for a reaction-diffusion system modelling the spread of a class of bacterial and viral diseases, J. Math. Biology 13 (1981), 173-184.

[25] A. Castro and A. Lazer, Results on periodic solutions of parabolic equations suggested by elliptic theory, Boll. Un. Mat. Ital. B(I) 6 (1982), 1089-1104.

[26] S. Chen and S.T. Yau, Ed., Geometry and Nonlinear Partial Differential Equations, AMS/IP Studies in Advanced Mathematics Series, 2002.

[27] M. Chipot, M. Fila and P. Quittner, Stationary solutions, blowup and convergence to stationary solutions for semilinear parabolic equations with nonlinear boundary conditions, Acta Math. Univ. Comenianae, LX (1991), 35-105.

[28] S. Chow and J. Hale, Methods of Bifurcation Theory, Springer-Verlag, N.Y., 1982.

[29] G. Christensen, S. Soliman and R. Nieva, Optimal Control of Distributed Nuclear Reactor, Plenum, New York, 1990.

[30] C. Clark, Mathematical Bioeconomics: The Optimal Management of Renewable Resources, Wiley, New York, 1976.

[31] C. Conley and R. Gardner, An application of generalized Morse index to traveling wave solutions of a competitive reaction diffusion model, Indiana Univ. Math. J. 33 (1984), 319-345.

[32] C. Cosner and A. Lazer, Stable coexistence states in the Volterra-Lotka competition model with diffusion, SIAM J. Appl., 44 (1984), 1112-1132.

[33] M. Crandall and P. Rabinowitz, Bifurcation from simple eigenvalues, J. Funct. Anal. 8 (1971), 321-340.

[34] M. Crandall and P. Rabinowitz, Bifurcation, perturbation of simple eigenvalues and linearized stability, Arch. Rat. Mech. Anal. 52 (1973), 161-181.

[35] S. Cui, Analysis of a mathematical model for the growth of tumors under the action of external inhibitors, J. Math. Biol. 44 (2002), 395-426.

[36] S. Cui and A. Friedman, Analysis of a mathematical model of the growth of necrotic tumors, J. Math. Anal. Appl. 255 (2001), 636-677.

[37] E. N. Dancer, On the indices of fixed points of mappings in cones and applications, J. Math. Anal. Appl. 91 (1983), 131-151.

[38] E. N. Dancer, Counterexamples to some conjectures on the number of solutions of nonlinear equations, Math. Ann. 272 (1985), 421-440.

[39] E. N. Dancer, Multiple fixed points of positive maps, J. Reine Ang. Math. 371 (1986), 46-66.

[40] E. N. Dancer, On the existence and uniqueness of positive solutions for competing species models with diffusion, Trans. Amer. Math. Soc. 36 (1991), 829-859.

[41] E. Dancer and Y. Du, Competing species equations with diffusion, large interactions and jumping nonlinearities, J. Diff. Eqs. 114 (1994), 434-475.

[42] E. Dancer and Z. Guo, Uniqueness and Stability for solutions of competing species equations with large interactions, Comm. Appl. Nonlin. Anal. 1 (1994), 19-45.

[43] P. Darlington, Zoogeography: The Geographical Distribution of Animals, Wiley, New York, 1957.

[44] C. DeLisi, Antigen Antibody Interactions, Lecture Notes in Biomath., Vol. 8, Springer-Verlag, New York, 1976.

[45] M. Delgado, J. López-Gómez and A. Suarez, On the symbyotiic Lotka-Volterra model with diffusion and transport effects, J. Diff. Eqs. 160 (2000), 175-262.

[46] P. DeMottoni, A. Schiaffino and A. Tesei, Attractive properties of non-negative solutions of a class of nonlinear degenerate parabolic problems, Ann. Mat. Pura. Appl. 36 (1984), 35-48.

[47] J. Diaz and J. Hernandez, On the existence of a free boundary for a class of reaction-diffusion systems, SIAM J. Math. Anal. 15 (1984), 670-685.

[48] O. Diekmann, R. Durrett, K. Hadeler, R. Maini and H. Smith, Mathematics Inspired by Biology, Springer-Verlag, Berlin, 2000.

[49] K. Diemling, Nonlinear Functional Analysis, Springer-Verlag, New York, 1985.

[50] X. Ding and T. P. Liu, Ed., Nonlinear Evolutionary Partial Differential Equations, AMS/IP Studies in Advanced Mathematics Series, 1997.

[51] D. Doak, Source-sink models and the problem of habitat degradation: General models and applications to Yellowstone Grizzly, Conservation Biol. 9 (1995), 1370-1379.

[52] J. Duderstadt and L. Hamilton, Nuclear Reactor Analysis, John Wiley, New York, 1976.

[53] J. Dugundji, An extension of Tietze's theorem, Pacific J. Math. 1 (1951), 353-367.

[54] J. Dugundji, Topology, Allyn and Bacon, Boston, 1966.

[55] S. Dunbar, Traveling wave solutions of diffusive Lotka-Volterra equations, J. Math. Biol. 17 (1983), 11-32.

[56] S. Dunbar, Traveling wave solutions of diffusive Lotka-Volterra equations: A heteroclinic connection in R^4, Trans. Am. Math. Soc., 286 (1984), 557-594.

[57] L. Evans, Partial Differential Equations, Graduate Studies in Math. 19, Amer. Math. Soc., R. I., 1998.

[58] P. Fife, Solutions of parabolic boundary problems existing for all time, Arch. Rational Mech. Anal. 16 (1964), 155-186.

[59] P. Fife, Mathematical Aspects of Reacting and Diffusing Systems, Lecture Notes in Biomathematics 28, Springer, Berlin, 1979.

[60] M. Fila and P. Quitter, The blow up rate for heat equation with nonlinear boundary condition, Math. Meth. Appl. Sci. 14 (1991), 197-205.

[61] J. Filo, Diffusivity verus absorption through the boundary, J. Diff. Eqs. 99 (1992), 281-305.

[62] W. Fleming, A selection-migration model in population genetics, J. Math. Biol. 2 (1975).

[63] A. Friedman, Partial Differential Equations of Parabolic Type, Prentice-Hall, 1964.

[64] A. Friedman and F. Reitich, Analysis of a mathematical model for the growth of tumors, J. Math. Biol. 38 (1999), 262-284.

[65] L. Friesen, P. Eagles and R. MacKay, Effects of residential development on forest-dwelling neotropical migrant songbirds, Conservation Biol. 9 (1995), 1408-1411.

[66] Y. Furusho, Existence of positive entire solutions for weakly coupled semilinear elliptic systems, Proc. Roy. Soc. Edinburgh, 120A (1992), 79-91.

[67] G. Galdi, An Introduction to Mathematical Theory of Navier-Stokes Equations, Springer Tracts in Natural Philosophy, Vols 38, 39, Springer, NY, 1994.

[68] R. Gardner, Existence and stability of traveling wave solutions of competition models: A degree theoretic approach, J. Diff. Eqs. 44 (1982), 56-79.

[69] R. Gardner, Review on traveling wave solutions of parabolic systems by A.I. Volpert, V.A. Volpert and V.A. Volpert, Bull. Amer. Math. Soc. 32 (1995), 446-452.

[70] B. Gidas and J. Spruck, A priori bounds for positive solutions of nonlinear elliptic equations, Comm. Partial Diff. Eqs. 6 (1981), 883-901.

[71] D. Gilbarg and N. Trudinger, Elliptic Partial Differential Equations of Second Order, Second Ed., Springer-Verlag, New York, 1983.

[72] S. Glasstone and A. Sesonske, Nuclear Reactor Engineering, Van Nostrand, New York, 1981.

[73] B. Goh, Sector stability of complex ecosystem model, Math. Biosci. 40 (1978), 157-166.

[74] J. Goldstein, Semigroups of Linear Operators and Applications, Oxford Mathematical Monographs, Oxford University Press, New York, 1985.

[75] A. Granas, Points fixes pour less applications compact, University of Montreal Press, Montreal, 1980.

[76] M. Gurtin, Some mathematical models for population dynamics that lead to segregation, Quart. J. Appl. Math. 32 (1974), 1-9.

[77] G. Hagelaar, Modeling of microdischarges for display technology, Proefschrift, Technische Universiteit Eindhoven, The Netherlands, 2000.

[78] G. Hagelaar, M. Klein, R. Snijkers and G. Kroesen, Energy loss mechanisms in the microdischarges in plasma display panels, J. Appl. Phys 89 (2001), 2033-2039.

[79] J. Hale, Ordinary Differential Equations, Wiley Interscience, NY, 1969.

[80] J. Hale, Large diffusivity and asymptotic behaviour in parabolic systems, J. Math. Anal. Appl. 118 (1986), 455-466.

[81] J. Hale and P. Waltman, Persistence in infinite-dimensional systems, SIAM J. Math. Anal. 20 (1989), 388-395.

[82] F. He, A. Leung and S. Stojanovic, Periodic optimal control for parabolic Volterra-Lotka type equations, Math. Methods in Appl. Sci. 18 (1995), 127-146.

[83] F. He, A. Leung and S. Stojanovic, Periodic optimal control for competing parabolic Volterra-Lotka type systems, J. of Computational and Appl. Math. 52 (1994), 199-217.

[84] D. Henry, Geometric Theory of Semilinear Parabolic Equations, Lecture Notes in Math. 840, Springer-Verlag, N.Y., 1981.

[85] P. Hess, Periodic-Parabolic Boundary Value Problems and Positivity, Pitman R.N.M., Longman, Harlow, 1991.

[86] M. Hirsch, Stability and convergence in strongly monotone dynamical systems, J. Reine Ang. Math. 383 (1988), 1-58.

[87] J. Hofbauer and K. Sigmund, Dynamical Systems and the Theory of Evolution, Cambridge Univ. Press, Cambridge, 1988.

[88] J. Hofbauer and K. Sigmund, Evolutionary Games and Population Dynamics, Cambridge Univ. Press, Cambridge, 1998.

[89] F. Hoppensteadt, Asymptotic stability in singular perturbation problems, J. Diff. Eqs. 4 (1968), 350-358.

[90] X. Hou and A. Leung, Traveling wave solutions for a competitive reaction-diffusion system and their asymptotics, Nonlin. Anal: Real World Appl. 9 (2008), 2196-2213.

[91] B. Hu and H. Yin, The profile near blowup time for solutions of heat equation with nonlinear boundary conditions, IMA preprint series no. 1115, 1993.

[92] V. Hutson, A theorem on average Liapunov functions, Monatsh. Math. 98 (1984), 267-275.

[93] V. Hutson. J. Pym and M. Cloud, Applications of Functional Analysis and Operator Theory, Vol. 200, 2nd Edition, Mathematics in Science and Engineering, Elsevier Science, N.Y., 2005.

[94] V. Hutson and K. Schmitt, Permanence and the dynamics of biological systems, Math. Biosci. 111 (1992), 1-71.

[95] T. Jackson, Vascular tumor growth and treatment: Consequences of polyclonality, competition and dynamic vascular support, J. Math. Biol. 44 (2002), 201-226.

[96] J. Jorné, The diffusive Lotka-Volterra oscillating system, J. Theor. Biol. 65 (1977), 133-139.

[97] Y. Kan-on, Stability of singularly perturbed solutions to nonlinear diffusion system arising in population dynamics, Hiroshima Math. J. 23 (1993), 509-536.

[98] Y. Kan-on, Existence and instability of Neumann layer solutions for a 3-component Lotka-Volterra model with diffusion, J. Math. Anal. Appl. 243 (2000), 357-372.

[99] Y. Kan-on and M. Mimura, Singular perturbation approach to a 3-component reaction-diffusion system arising in population dynamics, SIAM J. Math. Anal. 29 (1998), 1519-1536.

[100] Y. Kan-on and E. Yanagida, Existence of non-constant stable equilibria in competition diffusion equations, Hiroshima Math. J. 23 (1993), 193-221.

[101] W. Kastenberg, Stability criterion for space-dependent nuclear reactor-systems with variable temperature feedback, Nuc. Sci. Eng. 37 (1969), 19-29.

[102] T. Kato, Perturbation Theory for Linear Operators, Springer-Verlag, New York, 1966.

[103] E. Kerner, Further considerations on the statistical mechanics of biological association, Bull. Math. Biophys. 21 (1959), 217-255.

[104] D. Kirschner, S. Lenhart and S. Serbin, Optimal control of the chemotherapy of HIV, J. Math. Biology 35 (1997), 775-792.

[105] K. Kishimoto and H. Weinberger, The spatial homogeneity of stable equilibria of some reaction-diffusion systems on convex domains, J. Diff. Eqs. 58 (1985), 15-21.

[106] A. Kolmogorov, A. Petrovskii and N. Piskunov, A study of the equation of diffusion with increase in the quantity of matter, and its application to a biological problem, Bjul. Moskovskovo Gov. Univ. 17 (1937), 1-72.

[107] P. Korman and A. Leung, On the existence and uniqueness and positive steady-states in the Volterra-Lotka ecological models with diffusion, Appl. Anal. 26 (1987), 145-160.

[108] M. Krasnosel'skii, Positive Solutions of Operator Equations, P. Noordhoff, Groningen, The Netherlands, 1964.

[109] M. Krasnoselski and P. Zabrieko, Geometrical Methods of Nonlinear Analysis, Springer-Verlag, Berlin, 1984.

[110] M. Krein and M. Rutman, Linear operators leaving invariant in a cone in a Banach space, Amer. Math. Soc. Transl., 10 (1962), 199-325.

[111] J. Kuby, Immunology, third ed., Freeman, New York, 1997.

[112] G. Ladde, V. Lakshmikanthm and A. Vatsala, Monotone Iteration Techniques for Nonlinear Differential Equations, Pitman, 1985.

[113] O. Ladyzhenskaya, V. Solonnikov and N. Ural'ceva, Linear and Quasilinear Equations of Parabolic Type, Amer. Math. Soc. Transl. 23, 1968.

[114] I. Lasiecka and R. Triggiani, Control Problems for Systems described by Partial Differential Equations and Applications, Lecture Notes in Control and Information Sciences, No. 97, Springer-Verlag, New York, 1988.

[115] A. Lazer, Some remarks on periodic solutions of parabolic equations, In Dynamical Systems II, eds A. Bednarek and L Cesari, 227-246, Academic Press, New York, 1982.

[116] S. Lenhart and M. Bhat, Application of distributed parameter control model in wildlife damage management, Math. Models and Methods in Appl. Sciences 2 (1992), 423-439.

[117] S. Lenhart, M. Liang and V. Protopopescu, Optimal control of boundary habitat hostility for interacting species, Math. Meth. Appl. Sci. 22 (1999), 1061-1077.

[118] S. Lenhart and V. Protopopescu, Optimal control for parabolic systems with competitive interactions, Math. Meth. Appl. Sci. 17 (1994), 509-524.

[119] S. Lenhart and D. Wilson, Optimal control of a heat transfer problem with convective boundary condition, J. Optim. Theory Appl. 79 (1993), 581-587.

[120] S. Lenhart and J. Workman, Optimal Control Applied to Biological Models, Chapman and Hall/CRC Mathematical and Computational Biology, 2007.

[121] A. Leung, Equilibria and stabilities for competing-species reaction-diffusion equations with Dirichlet boundary data, J. Math. Anal. 73 (1980), 204-218.

[122] A. Leung, Stabilities for equilibria of competing-species reaction-diffusion equations with homogeneous Dirichlet condition, Funk. Ekv. (Ser. Interna.) 24 (1981), 201-210.

[123] A. Leung, Monotone schemes for semilinear elliptic systems related to ecology, Math. Meth. Appl. Sci. 4 (1982), 272-285.

[124] A. Leung, Nonlinear density-dependent diffusion for competing species interactions: Large-time asymptotic behaviour, Proc. Edinburgh Math. Soc. 27 (1984), 131-144.

[125] A. Leung, Systems of Nonlinear Partial Differential Equations, Applications to Biology and Engineering, Kluwer Academic Publishers, Dordrecht/Boston, Springer-Verlag, N.Y., 1989.

[126] A. Leung, Optimal harvesting-coefficient control of steady-state prey-predator diffusive Volterra-Lotka systems, Appl. Math. Optimization 31 (1995), 219-241.

[127] A. Leung, Bifurcating positive stable steady-states for a system of wave equations, Diff. Int. Eqs. 16 (2003), 453-471.

[128] A. Leung, Asymptotically stable invariant manifold for coupled nonlinear parabolic-hyperbolic partial differential equations, J. Diff. Eqs. 187 (2003), 184-200.

[129] A. Leung, Positive solutions for large elliptic systems of interacting species groups by cone index methods, J. Math. Anal. Appl. 291 (2004), 302-321.

[130] A. Leung, Stable invariant manifolds for coupled Navier-Stokes and second-order wave systems, Asymptotic Analysis 43 (2005), 339-357.

[131] A. Leung and G. S. Chen, Positive solutions of temperature-dependent two group neutron flux equations: Equilibrium and stabilities, SIAM J. Math. Anal. 15 (1984), 131-144.

[132] A. Leung and G. S. Chen, Elliptic and parabolic systems for neutron fission and diffusion, J. Math. Anal. Appl. 120 (1986), 655-669.

[133] A. Leung and G. S. Chen, Optimal control of multigroup neutron fission systems, Appl. Math. Optim. 40 (1999), 39-60.

[134] A. Leung and G. S. Chen, Existence and global bounds for a fluid model of plasma display technology, J. Math. Anal. Appl. 310 (2005), 436-458.

[135] A. Leung and G. Fan, Existence of positive solutions for elliptic systems-degenerate and nondegenerate ecological models, J. Math. Anal. Appl. 151 (1990), 512-531.

[136] A. Leung, X. Hou and Y. Li, Exclusive traveling waves for competitive reaction-diffusion systems and their stabilities, J. Math. Anal. Appl. 338 (2008), 902-924.

[137] A. Leung and L. Ortega, Bifurcating solutions and stabilities for multi-group neutron fission systems with temperature feedback, J. Math. Anal. Appl. 194 (1995), 489-510.

[138] A. Leung and L. Ortega, Bifurcating solutions and stabilities for multi-group neutron fission systems with temperature feedback, J. Math. Anal. Appl. 194 (1995), 489-510.

[139] A. Leung and L. Ortega, Positive steady-states for large systems of reaction-diffusion equations: Synthesizing from smaller subsystems, Canadian Appl. Math. Quar. 4 (1996), 175-195.

[140] A. Leung and S. Stojanovic, Optimal control for elliptic Volterra-Lotka type equations, J. Math. Anal. Appl. 17 (1993), 603-619.

[141] A. Leung and B. Villa, Reaction-diffusion systems for multigroup neutron fission with temperature feedback: Positive steady-state and stability, Diff. Int. Eqs. 10 (1997), 739-756.

[142] A. Leung and B. Villa, Bifurcation of reaction-diffusion systems, application to epidemics of many species, J. Math. Anal. Appl. 244 (2000), 542-563.

[143] A. Leung and Q. Zhang, Reaction diffusion equations with non-linear boundary conditions, blowup and steady states, Math. Meth. Appl. Sci. 21 (1998), 1593-1617.

[144] S. Levin, Models of population dispersal, differential equations and applications, Ecology, Epidemics and Population Problems, edited by S. Busenberg and K. Cooke, Academic Press, N.Y., 1981.

[145] H. Levine and L. Payne, Nonexistence theorems for the heat equations with nonlinear boundary conditions for porous medium equation backward in time, J. Diff. Eqs. 16 (1974), 319-334.

[146] J. Lewins, Nuclear Reactor Kinetics and Control, Plenum, New York, 1978.

[147] B. Li, H. Weinberger and M. Lewis, Spreading speeds as slowest wave speeds for cooperative systems, Math. Biosci. 196 (2005),82-98.

[148] L. Li, Coexistence theorems of steady states for predator-prey interacting systems, Trans. Amer. Math. Soc., 305 (1988), 143-166.

[149] L. Li and A. Ghoreishi, On positive solutions of general nonlinear elliptic symbiotic interacting systems, Appl. Anal. 40 (1991), 281-295.

[150] L. Li and Y. Liu, Spectral and nonlinear effects in certain elliptic systems of three variables, SIAM J. Math Anal. 24 (1993), 480-498.

[151] L. Li and R. Logan, Positive solutions to general elliptic competition models, Diff. Int. Eqs., 4 (1991), 817-834.

[152] L. Li and A. Ramm, A singular perturbation result and its applications to mathematical ecology, Proc. Amer. Math. Soc. 111 (1991), 1043-1050.

[153] X. Li and J. Yong, Optimal Control Theory for Infinite Dimensional Systems, Birkhäuser, Boston, 1995.

[154] C. Lin, Z. Lin and W. Jiang, Optimal control of boiling water reactor loading-following operation, Nuclear Sci. Eng. 102 (1989), 134-139.

[155] C. Lin, W. Ni and I. Takagi, Large amplitude stationary solutions to a chemotaxi system, J. Diff. Eqs. 72 (1988), 1-27.

[156] D. Linthicum and N. Farid, Anti-idiotypes, Receptors, and Molecular Mimicry, Springer-Verlag, New York, 1988.

[157] J. L. Lions, Equations Differentielles Operationnelles, et problemes aux Limites, Springer-Verlag, Berlin 1961.

[158] J. L. Lions, Optimal Control of Systems Governed by Partial Differential Equations, Springer-Verlag, Berlin, 1972.

[159] J. L. Lions and E. Magenes, Nonhomogeneous Boundary Value Problems and Applications I, Springer-Verlag, New York, 1972.

[160] Y. Liu, Positive solutions to general elliptic systems, Nonlinear Anal. 25 (1995), 229-246.

[161] J. López-Gómez, Spectral Theory and Nonlinear Functional Analysis, Chapman and Hall/CRC Research Notes in Mathematics 426, London, 2001.

[162] J. López-Gómez and R. Pardo San Gil, Coexistence regions in Lotka-Volterra models with diffusion, Nonl. Anal. TMA 19 (1992), 11-28.

[163] J. López-Gómez and R. Pardo San Gil, Coexistence in simple food chain with diffusion, J. Math. Biology 30 (1992), 655-668.

[164] Y. Lou and W. Ni, Diffusion, self-diffusion and cross-diffusion, J. Diff. Eqs., 131 (1996), 79-131.

[165] Y. Lou and W. Ni, Diffusion vs cross-diffusion: An elliptic approach, J. Diff. Eqs. 154 (1999), 157-190.

[166] R. Lui, Biological growth and spread modeled by systems of recursion,I. mathematical theory, Math. Biosci. 93 (1989), 269-295.

[167] H. Matano and M. Mimura, Pattern formation in competition-diffusion systems in non-convex domains, Publ, RIMS. Kyoto Univ. 19 (1983), 1049-1079.

[168] P. McKenna and W. Walter, On the Dirichlet problem for elliptic systems, Appl. Anal. 21 (1986), 207-224.

[169] R. McOwen, Partial Differential Equations, Methods and Applications, Prentice-Hall, Upper Saddle River, NJ, 1996.

[170] M. Mimura, Stationary pattern of some density-dependent diffusion system with competitive dynamics, Hiroshima Math. J. 11 (1981), 621-635.

[171] M. Mimura, S. Ei and Q. Fang, Effect of domain-type on the coexistence problems in competition-diffusion system, J. Math. Biol. 29 (1991), 219-237.

[172] M. Mimura and K. Kawasaki, Spatial segregation in competitive interaction-diffusion equations, J. Math. Biol. 9 (1980), 49-64.

[173] M. Mimura, Y. Nishiura, A. Tesei and T. Tsujikawa, Coexistence problem for two competing species models with density-dependent diffusion, Hiroshima Math. J. 14 (1984), 425-449.

[174] K. Mischaikow and V. Hutson, Travelling waves for mutualist species, SIAM J. Math. Anal. 24 (1993), 987-1008.

[175] K. Mischaikow, J. Reineck and F. James, Travelling waves in predator-prey systems, SIAM J. Math. Anal. 24 (1993), 1179-1214.

[176] X. Mora, Semilinear parabolic problems define semiflows on C^k spaces, Trans. Am. Math. Soc. 278 (1983), 21-55.

[177] J. Murray, Mathematical Biology II: Spatial Models and Biomedical Applications, Springer-Verlag, New York, 2003.

[178] R. Nussbaum, The fixed point index for locally condensing maps, Ann. Mat. Pura Appl. 87 (1971), 217-258.

[179] A. Okubo and S. Levin, Diffusion and Ecological Problems: Modern Perspectives, 2nd ed., Springer-Verlag, Berlin, 2001.

[180] L. S. Ortega, On the Leung-Chen feedback model for nuclear fission, Nonlinear Anal. TMA 18 (1992), 353-360.

[181] R. Paine, Food web complexity and species diversity, Amer. Natur. 100 (1966), 65-75.

[182] C. V. Pao, Coexistence and stability of a competition-diffusion system in population dynamics, J. Math. Anal. Appl., 83 (1981), 54-76.

[183] C. V. Pao, Nonlinear Parabolic and Elliptic Equations, Plenum Press, New York, 1992.

[184] A. Pazy, Semigroup of Linear Operators and Applications to Partial Differential Equations, Springer Verlag, N.Y., 1983.

[185] L. Peletier, The porous medium equations, in "Applications of Nonlinear Analysis in Physical Science" (H. Amann, N. Bazley, and Kirchgassner Eds), 229-241, London, 1981.

[186] M. Pozio and A. Tesei, Degenerate parabolic problems in population dynamics, Japan J. Appl. Math. 2 (1985), 351-380.

[187] M. Pozio and A. Tesei, Support properties of solutions for a class of degenerate parabolic equations, Comm. Partial Differential Equations 12 (1987), 47-75.

[188] M. Protter and H. Weinberger, Maximum Principles in Differential Equations, Prentice Hall, Englewood Cliff, N.Y. 1967.

[189] P. Quittner, On global existence and stationary solutions for two classes of semilinear parabolic problems, Comment Math. Univ. Carolin 34 (1993), 105-124.

[190] P. Rabinowitz, Some global results for nonlinear eigenvalue problems, J. Funct. Anal., 1 (1971), 487-513.

[191] S. Raul and M. Kushner, Dynamics of coplanar-electrode and plasma display panel cell, I. Basic operation, J. Appl. Phys. 85 (1999), 3460-3469.

[192] R. Redheffer and Z. Zhou, Global asymptotic stability for a class of many-variable Volterra prey-predator systems, J. Nonlinear Anal. 5 (1981), 1309-1329.

[193] I. Roitt, J. Brostoff and D. Male, Immunology, 2nd ed., Mosby Gower Med. Publishing, 1989.

[194] W. Ruan and W. Feng, On fixed point index and multiple steady-state solutions of reaction-diffusion systems, Differential Integral Eqs. 8 (1995), 371-391.

[195] W. Ruan and C. Pao, Positive steady-state solutions of competing reaction diffusion system, J. Diff. Eqs. 117 (1995), 411-427.

[196] K. Ryu and I. Ahn, Coexistence theorem for steady states for nonlinear self-cross diffusion systems with competitive dynamics, J. Math. Anal. Appl. 283 (2003), 46-65.

[197] K. Ryu and I. Ahn, Positive solutions for ratio-dependent predator-prey interaction systems, J. Diff. Eqs. 218 (2005), 117-135.

[198] K. Ryu and I. Ahn, Positive solutions of certain elliptic systems with self-diffusions: Non-degenerate vs. degenerate diffusions, Nonl. Anal. 63 (2005), 247-259.

[199] M. Salvatori and E. Vitillaro, Decay for the solutions of nonlinear abstract damped evolution equations with applications to partial and ordinary differential systems, Diff. Int. Eqs. 11 (1998), 223-262.

[200] D. Sattinger, Monotone methods in nonlinear elliptic and parabolic boundary value problems, Indiana Univ. Math. J. 21 (1972), 979-1000.

[201] D. Sattinger, Topics in Stability and Bifurcation Theory, Lecture Notes in Mathematics, Springer Verlag, N.Y., 1973.

[202] A. Selvadurai, Partial Differential Equations in Mechanics 1: Fundamentals, Laplace's Equation, Diffusion, Wave Equation, Springer-Verlag, N.Y., 2000.

[203] H. Schaefer, Topological Vector Spaces, Macmillan, New York, 1966.

[204] H. Schatzman, Stationary solutions and asymptotic behavior of quasilinear degenerate parabolic equation, Indiana Univ. Math. J. 33 (1984), 1-30.

[205] J. Schwartz, Nonlinear Functional Analysis, Gordan and Breach, New York, 1968.

[206] N. Shigesada, K. Kawasaki and E. Teramoto, Spatial segregation of interacting species, J. Theoret. Biol. 79 (1979), 83-99.

[207] D. Simberloff, Experimental zoogeorgraphy of islands: Effects of island size, Ecology 57 (1976), 629-648.

[208] H. Smith, Monotone Dynamical Systems: An Introduction to the Theory of Coooperative Systems, Am. Math. Soc., Providence, R.I., 1995.

[209] J. Smoller, Shock Waves and Reaction-Diffusion Equations, Springer-Verlag, New York, 1983.

[210] R. Sperb, Maximum Principles and Their Applications, Academic Press, New York, 1981.

[211] B. Stewart, Generation of analytic semigroups by strongly elliptic operators, Trans. Amer. Math. Soc. 199 (1974), 141-161.

[212] B. Stewart, Generation of analytic semigroups by strongly elliptic operators under general boundary conditions, Trans. Amer. Math. Soc. 259 (1980), 299-310.

[213] S. Stojanovic, Optimal damping control and nonlinear parabolic systems, Numer. Funct. Anal. Optim. 10 (1989), 573-591.

[214] W. Strauss, Nonlinear Wave Equations, CBMS No. 71, Amer. Math. Soc., 1989.

[215] M. Tang and P. Fife, Propagating fronts for competing species with diffusion, Arch. Rational Mech. Anal. 73 (1978), 69-77.

[216] R. Temam, Navier-Stokes Equations and Nonlinear Functional Analysis, SIAM, Philadelphia, 1983.

[217] R. Temam, Infinite Dimensional Dynamical Systems to Mechanics and Physics, Appl. Math. Sciences 68, Springer-Verlag, N.Y., 1997.

[218] B. Terney and D. Wade, Optimal control applications in nuclear reactor design and operation, Adv. Nuclear Sci. Tech. 10 (1997).

[219] W. Thompsom, An Introduction to Plasma Physics, Addison-Wesley, London, 1964.

[220] A. Tihononov, Systems of differential equations containing small parameters in the derivatives, Mat. Sbornik 27 (1952), 147-156 (in Russian).

[221] L.Y. Tsai, Existence of solutions of nonlinear elliptic systems, Bulletin Inst. Math. Academia Sinica, 8 (1980), 111-127.

[222] M. Vainberg, Variational Methods for the Study of Nonlinear Operators, Holden-Day Series in Mathematical Physics, San Francisco, 1964.

[223] G. Veronis and U. Inan, Simulation studies of coplanar electrode and other plasma display panel cell designs, J. Appl. Phys. 91 (2002), 9502-9512.

[224] I. Volpert, V. Volpert and V. Volpert, Traveling Wave Solutions of Parabolic Systems, Translations of Math. Mono., Vol. 140, Am. Math. Soc., Providence, R.I., 1994.

[225] W. Walter, On the existence and nonexistence in the large of solutions of parabolic differential equations with nonlinear boundary conditions, SIAM J. Math. Anal. 6 (1975), 85-90.

[226] M. Wang, Stationary patterns of strongly coupled prey-predator models, J. Math. Anal. Appl., 292 (2004), 484-505.

[227] M. Wang and Y. Wu, Global exist and blow up problems for quasi-linear parabolic equations with nonlinear boundary conditions, SIAM J. Math. Anal. 24 (1993), 1515-1521.

[228] J. Ward and J. King, Mathematical modelling of avascular-tumor growth, IMA J. Math. Appl. Med. Bio. 15 (1998), 1-42.

[229] W. Wasow, Asymptotic Expansions for Ordinary Differential Equations, 2nd edn, R.E. Krieger Publ., Huntington, NY, 1976.

[230] G. Webb, A bifurcation problem for a nonlinear hyperbolic partial differential equation, SIAM J. Math. Anal. 10 (1979), 922-932.

[231] S. Williams and P. Chow, Nonlinear reaction-diffusion models, J. Math. Anal. Appl. 62 (1978), 157-169.

[232] R. Wong, Asymptotic and Computational Analysis, Lecture Notes in Pure and Applied Mathematics, Marcel Dekker Inc., N.Y. 1990.

[233] J. Wu, Theory and Applications of Partial Functional Differential Equations, Applied Mathematical Sciences, Springer-Verlag, N.Y., 1996.

[234] J. Wu and X. Zou, Traveling wave fronts of reaction-diffusion systems with delay, J. Dynamics and Diff. Eq. 13 (2001), 651-687.

[235] J. Wu and X. Zou, Erratum to "traveling wave fronts of reaction-diffusion systems with delays", J. Dynamics and Diff. Eq. 2 (2008), 531-533.

[236] Y. Wu, Existence of stationary solutions with transition layers for a class of cross-diffusion systems, Proc. Roy. Soc. Edinburgh, Sect. A, 132 (2002), 1493-1511.

[237] S. Xu, (Editor), Neurobiochemistry, Shanghai Medical University Press, 1989.

[238] Q. Ye and M. Wang, Travelling wave front solutions of Noyes-Field system for Belousov-Zhabotinskii reaction, Nonlin. Anal. TMA, 11 (1987), 1289-1302.

[239] T. Yoshizawa, Stability Theory of Liapunov's Second Method, University of Tokyo Press, 1966.

[240] J. Young, The Life of Mammals, Oxford University Press, New York, 1957.

[241] E. Zeidler, Nonlinear Functional Analysis and its Applications, Springer-Verlag, Berlin/New York, 1985.

[242] E. Zeidler, Applied Functional Analysis: Applications to Mathematical Physics, Applied Mathematical Sciences, Springer-Verlag, N.Y. 1995.

[243] X. Zou and J. Wu, Existence of traveling wavefronts in delayed reaction-diffusion system via monotone iteration method, Proc. Amer. Math. Soc. 125 (1997), 2589-2598.

Index